Island Biogeography

Island Biogeography

Geo-environmental dynamics, ecology, evolution, human impact, and conservation

Robert J. Whittaker

Professor of Biogeography, School of Geography and Environment, University of Oxford, Oxford, UK

José María Fernández-Palacios

Ecology Professor, Island Ecology and Biogeography Group, Departamento de Botánica, Ecología y Fisiología Vegetal e Instituto Universitario de Enfermedades Tropicales y Salud Pública de Canarias (IUETSPC), Universidad de La Laguna, Canary Islands, Spain

Thomas J. Matthews

Senior Research Fellow, GEES (School of Geography, Earth and Environmental Sciences), University of Birmingham, UK

OXFORD
UNIVERSITY PRESS

Great Clarendon Street, Oxford, OX2 6DP,
United Kingdom

Oxford University Press is a department of the University of Oxford.
It furthers the University's objective of excellence in research, scholarship,
and education by publishing worldwide. Oxford is a registered trade mark of
Oxford University Press in the UK and in certain other countries

Published in the United States of America by Oxford University Press
198 Madison Avenue, New York, NY 10016, United States of America

British Library Cataloguing in Publication Data
Data available

Library of Congress Control Number: 2023930773

ISBN 978–0–19–886856–9
ISBN 978–0–19–886857–6 (pbk.)

DOI: 10.1093/oso/9780198868569.001.0001

Printed and bound by
CPI Group (UK) Ltd, Croydon, CR0 4YY

Links to third party websites are provided by Oxford in good faith and
for information only. Oxford disclaims any responsibility for the materials
contained in any third party website referenced in this work.

Preface and Acknowledgements

Writing this volume has provided an excuse to immerse ourselves in a rich and rapidly expanding literature, in which the pace of discoveries and of new syntheses has proven to be simultaneously a challenge and a stimulus. We hope to convey within the book the importance of islands as model systems in biogeography, ecology, and evolution, as repositories of biodiversity, much of it threatened, and as homes for diverse human societies.

Although we have drawn on case studies from insular systems from around the world, our strength of interest in Macaronesia and oceanic islands in general has inevitably flavoured this selection. Undoubtedly, we have neglected many insular systems and scholarly contributions that fully merit inclusion, for which we can only express regret. However, our hope has been to cover the waterfront of ideas, theories, and concepts, selecting case studies and syntheses that assist in this goal. By returning multiple times to a few archipelagos in different sections of the book, we hope to have built a fuller picture of how different strands of evidence and theory intersect.

Our goal has been to make the content, some of which is drawn from technically challenging primary sources, as accessible as possible to a diverse academic readership. This inevitably has required some simplification of complex literature and often a light-touch approach to the caveats that pertain to all scientific studies. One specific challenge we faced in compiling the material was to decide on the dating system to be used. In general, we have adopted Ma to refer to events a given number of millions of years ago, ka for dates thousands of years before present, and sometimes BC and AD where it seemed helpful. For events across the last few thousand years, the primary source literature we drew from employs a variety of dating systems. For simplicity, we have typically translated dates expressed in systems such as calibrated radiocarbon years before present (with a baseline date of AD 1950) somewhat crudely into ka. Readers for whom the precise dating is important should refer to the original sources for more specific information on the dates of events discussed.

This book owes its origins and basic framework to two previous editions of *Island Biogeography: Ecology, Evolution, and Conservation*, the first published in 1998 and the second in 2007. Encouraged by our editor at Oxford University Press, Ian Sherman, we first began to plan for a major refresh during the Island Biology meeting in Terceira Island, Azores, in July 2016. Our colleague Kostas Triantis helped in these early planning meetings but changing work commitments meant that he was unable to continue as part of the authorship team. We are deeply grateful to him for his contribution to the development of this book. In practice, the overall approach taken herein will be recognizable to those familiar with the second edition of *Island Biogeography*, but more than two thirds of the references and figures are new and the text has been very largely rewritten. In recognition of the extent of new content and the enhanced attention given, first, to the relevance of geo-environmental dynamics to insular biodynamics, and second, to the history and impact of human societies on islands, we have furnished this volume with additional chapters and a revised title.

There are many individuals and some institutions we need to thank. We have been fortunate to work with many remarkable scientists who have helped shape our ideas and understanding of island biogeography. Although too many to name more than a few, we would particularly like to thank for sharing ideas and/or for generously providing revised figures for our use: Jason Ali, Paulo Borges, Michael Borregaard, Juliano Cabral, Richard Field, Artur Gil, Rosemary Gillespie, François Guilhaumon, Larry Heaney, Rubén Heleno, David Hembrey, Thomas Ibanez, Walter Jetz, Holger Kreft, Frederic Lens, Thomas Leppard, Mark Lomolino, Antonio Machado, Lisa Matisoo-Smith, David Mouillot, Lea de Nascimento, Sandra Nogué, Jens Olesen, Rudi Otto, Jonathan Price, Nigel Purse, François Rigal, James Russell, Manuel Steinbauer, Kostas Triantis, Luis Valente, Catherine Wagner, Joe Wayman, Patrick Weigelt, and Kathy Willis. We also thank Ailsa Allen, Alison Pool, and Stephanie Ferguson for figure preparation, Angela Whittaker for technical assistance with the text, and Paul Nash for copyediting. At OUP we thank Ian Sherman for encouragement throughout, and Adam Breivik for helping land the project.

We thank the many journals, publishers, and individuals who kindly granted permission to reproduce versions of diagrams and photographs in the present work. While we have made efforts to contact all concerned, if we have inadvertently omitted to include specific statements of copyright that should have been included, we would be pleased to correct such omissions in any future reprint.

RJW acknowledges the support of St Edmund Hall, and of the School of Geography and the Environment, University of Oxford, and also the Centre for Macroecology, Evolution, and Climate, University of Copenhagen, where he was welcomed as a part-time staff member for the early stages of this project. JMFP acknowledges La Laguna University and Jersey International Centre of Advanced Studies for their constant support. TJM acknowledges the support of the School of Geography, Earth and Environmental Sciences, University of Birmingham, as well as the Azorean Biodiversity Group, University of the Azores, and the Biodiversity & Conservation Biogeography lab, University of the Ryukyus, both of which supported his island work at various stages of this project.

For Angela, for Neli, and for Emilie, Leo, and Alice.

R. J. Whittaker, J. M. Fernández-Palacios,
and T. J. Matthews
Oxford, UK; La Laguna, Spain; Birmingham, UK
20 November 2022

Contents

Part II Island Ecology 91

9 Evolutionary diversification across islands and archipelagos   **225**

PART I

Setting the Scene: Islands as Natural Laboratories

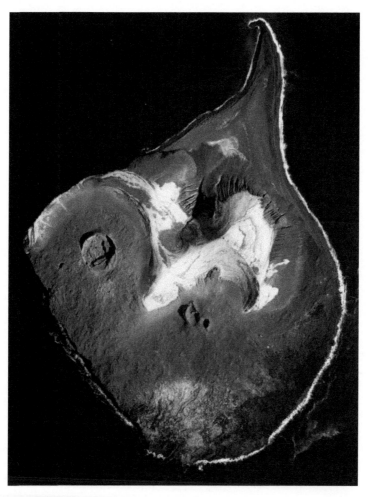

Frontispiece: Aerial view of Surtsey Island (Iceland) taken 12 August 2004 (from Baldursson and Ingadottir 2007, © Loftmyndir) The green area towards the bottom of the image reveals the impact of the establishment of a seabird colony on the build-up of biomass and the development of ecosystems on the island

CHAPTER 1

The natural laboratory paradigm

> The natural history of this archipelago is very remarkable: it seems to be a little world within itself.
>
> **(Darwin 1839, *Journal of Researches*, p. 454)**

> It is not too much to say that when we have mastered the difficulties presented by the peculiarities of island life we shall find it comparatively easy to deal with the more complex and less clearly defined problems of continental distribution.
>
> **(Wallace 1902, p. 242)**

1.1 Scope

Island biogeography is the study of the geographical distribution of life's diversity and how it has developed through time, understood through the lens of insular systems. The quotations above, from Charles Darwin's *Journal of Researches* and from Alfred Russel Wallace's seminal work *Island Life*, together encapsulate an over-arching concept central to this book and to island biogeography as a discipline. It is that islands, being discrete, internally quantifiable, numerous, and varied entities, provide us with a suite of natural laboratories, from which the discerning natural scientist can make a selection to simplify the natural world, enabling theories of general importance to be developed and tested. This notion of islands as model systems has been applied to a great range of insular habitats, not limited to islands in the sea, although real islands provide our main playground through the following chapters. The concepts explored are diverse (Table 1.1) and their investigation calls upon a broad array of research tools within the life and environmental sciences. By reviewing these topics and their interrelationships within the bounds of a single book, we hope to persuade the reader of the profound and continuing importance of the study of islands to our understanding of the natural world.

We have divided our discussion into four parts. The first part, 'Setting the Scene: Islands as Natural Laboratories', reviews the physical characteristics and dynamics of islands as natural laboratories, concentrating principally on islands *sensu stricto*, providing an essential framework for understanding the biogeographical data derived from them. We also describe the basic biogeographical characteristics, diversity levels, and degrees of affinity among clusters of islands and between islands and mainlands. The second part, 'Island Ecology', is concerned with pattern and process on ecological timescales and is focused on the number and composition of species on islands and how they vary between islands and through time. The chapters making up this part introduce the many theoretical, conceptual, and methodological contributions the study of islands has fed into mainstream ecology and biodiversity studies. They also explore how accounting for geo-environmental dynamism can improve the realism of macroecological models. The third part, 'Island Evolution', explores evolutionary pattern and process, at all levels from the instantaneous loss of genetic variation that may mark the moment of island colonization, through to the longthier and often dramatic changes associated with the great radiations of island lineages on remote oceanic archipelagos such as Hawaii. As

Island Biogeography. Robert J. Whittaker, José María Fernández-Palacios, and Thomas J. Matthews, Oxford University Press.

Table 1.1 Selected scientific concepts of different disciplines in which island research has played a seminal role (modified from Fernández-Palacios 2010). We do not offer definitions of most of these terms in this introductory chapter but do so as they are introduced throughout the book.

Discipline	Concepts
Earth Sciences (Vulcanology)	atoll formation, caldera, guyot, hotspot, island arcs, island ontogeny, mega-landslides, subsidence, tsunamis
Biogeography	back-colonization, colonization, disharmony, dispersal filters, equilibrium theory, immigration, kipuka, long-distance dispersal, neoendemism, palaeoendemism, progression rule, relictualism, rescue effect, species impoverishment, species turnover, stepping-stones, sweepstake dispersal, target-area effect, turnover
Ecology	community assembly rules, density compensation, double mutualism, ecological cascades, ecological release, incidence functions, niche shifts, primary succession, r versus K selection, species–area relationships
Evolution	adaptive radiation, character displacement, competitive displacement, evolution by natural selection, founder effects, genetic drift, geographic speciation, hybridization and introgression, inbreeding depression, island evolutionary syndromes (e.g. island body size rule), non-adaptive radiation, taxon cycle
Conservation Biology	ecological naivety, ecosystem collapse, extinction debt, habitat fragmentation, invasion, invasional meltdown, meso-predator release, metapopulation, nestedness, reserve design, single large or several small (SLOSS) debate, small population paradigm, species relaxation

will become evident, islands are among the greatest theatres of evolution, provide influential exemplars of evolutionary analysis, and again illustrate the importance of geo-environmental dynamics in understanding biological dynamics. The fourth and final part, 'Human Impact and Conservation', incorporates two contrasting literatures, concerned respectively with the threats to biodiversity derived from the increased insularization of continental ecosystems (Chapter 12), and the transformation arising from the loss of insularization of remote islands and how some of the acute threats to island biodiversity may be mitigated (Chapters 13–15).

1.2 Insularity: 'the state or condition of being an island'

From cowpats to South America, it is difficult to see what is not or at some time has not been an island.

(Mabberley 1979, p. 261)

The condition of insularity can be broadly defined in relation to a focal individual, population, species, or assemblage of species, as any favourable place entirely surrounded by a hostile environment preventing or constraining the movement or persistence of the organism(s) of interest. We may describe many biological systems as insular, from individual thistle plants (islands of sorts for the arthropods that visit them) in an abandoned field, through to remote volcanic archipelagos like the Galápagos and Hawaii (Table 1.2, Fig. 1.1). The former are very ephemeral, even in an ecological context, but individual volcanic islands are also quite short-lived and dynamic platforms when judged in an evolutionary time frame.

Consider the following example. The Juan Fernández archipelago consists of two main islands: Robinson Crusoe Island (Masatierra), an island of 48 km^2 and 915 m height, lying some 670 km from mainland Chile; and Alejandro Selkirk Island (Masafuera), an island of 50 km^2 and 1300 m height, arising from the ocean floor a further 180 km out into the Pacific. We might be interested in analysing how biological properties of these islands, such as species richness and endemism, relate to potentially important controlling variables such as island area, elevation, and isolation. However, when Robinson Crusoe Island first arose from a hotspot on the Nazca plate around 4 Ma, it probably formed an island of some 1000 km^2 and perhaps 3000 m in height (Stuessy et al. 1998, 2022). Since then, it has been worn down and diminished in area by subaerial erosion, wave action, subsidence, and catastrophic flank failures, losing both habitats and species in the process. The more remote Alejandro Selkirk Island is much younger, originating 1–2 Ma, but has also been subject to the same processes

Table 1.2 Classification of island types distinguishing: (1) important types of 'real' island (i.e. land surrounded by open water), from (2) types of habitat islands, for which the contrast between the 'island' and the surrounding matrix is generally less stark, but still sufficient to represent a barrier or strong filter to population movements, and (3) lakes, the most effectively isolated of which are the aquatic equivalent of real islands rather than habitat islands.

Type of island	Examples
Land surrounded by water	
Island continent*	Australia
Oceanic islands	Hawaii, Line Islands, Austral-Cook Islands, Galápagos, Azores, Canaries, St Helena
Continental fragments	Madagascar, New Caledonia, granitic Seychelles
Continental shelf islands	British Isles, Newfoundland, Singapore, and islands of the Java Sea
Islands in lakes or rivers	Isle Royale (Lake Superior), Barro Colorado Island (Lake Gatún), Gurupá Island (River Amazon)
Habitat islands	
Patches of a distinct terrestrial habitat isolated by a hostile matrix	Great Basin (USA) mountaintops ('sky islands') surrounded by desert
	Woodland fragments surrounded by agricultural land
	Thistle heads in a field
Marine habitat islands	Fringing reefs around an isolated oceanic island
	Coral reefs separated from other reefs by stretches of seawater
	Seamounts, i.e. not yet emerged mountains below sea level, or former islands that have become submerged (e.g. guyots)
Lakes	e.g. lakes Baikal, Titicaca, Tanganyika

*Australia, given its huge size, is essentially continental in character and we do not consider it as an island system herein.

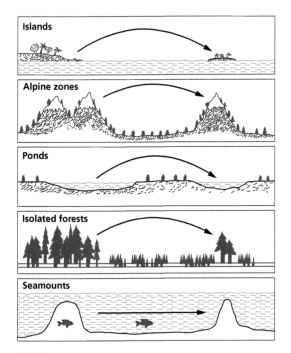

Figure 1.1 There are many different types of insular environment in addition to the principal class we discuss here: restricted areas of land surrounded by water. This figure illustrates just a few of them. Modified from an original in Wilson and Bossert (1971).

of attrition. Both islands have also been subject to repeated sea-level fluctuations, both isostatic (local uplift/downwarping) and eustatic (global water volume changes) in origin, generating phases of increase and decrease in both area and elevation, superimposed on the intrinsic ontogeny (Stuessy et al. 2022). Modelling efforts that ignore these environmental histories could potentially produce quite misleading accounts of factors controlling biotic dynamics on such islands.

Islands (*sensu stricto*) come in many shapes and sizes, and their arrangement in space, their geological origins and dynamics, their environments, and biotic characteristics vary greatly (Chapters 2–4). While this makes them marvellous natural laboratories, it also means that attempting to extract generalizations about islands analytically, without first delimiting the characteristics of the islands included, holds the danger of confounding factors influencing the outcome. We therefore need to pay some attention to classifying and describing island systems. And, despite our inclusion of other types of island within the book, we focus first and foremost on the dictionary definition of an island as a 'piece of land surrounded by water'. Of these classically defined islands, some authors regard land areas that are < 10 ha (too small to sustain a supply of fresh water) as merely beaches or sand bars rather than proper islands. At the other end of the scale, Australia, although surrounded by water, and around 8000 islands of its own, is classed as the smallest of the continents at 7.7 M km². Adopting a definition of proper islands as being land areas within the seas and oceans that are bigger than sand bars and no larger than New Guinea (discounting the largely ice-covered Greenland), then proper islands constitute approximately 5.3% of the Earth's land area and there are rather a lot of them (Box 1.1; Mielke 1989; Weigelt et al. 2013).

Following A. R. Wallace's lead, islands in the sea are generally subdivided biogeographically into three principal types: oceanic islands, continental fragments, and continental shelf islands, as discussed in detail in Chapter 2. In brief, **oceanic islands** are those that have formed over oceanic plates and have never been connected to

---

Box 1.1 The size distribution of the world's islands

Herein, we consider New Guinea (785,750 km²) to be the largest island. Although Greenland is larger (2,166,086 km²) and increases the global land area of islands to around 6.67% (Fernández-Palacios et al. 2021a), it is 80% ice covered and largely barren. We follow the convention of considering Australia (7,686,850 km²) to be the smallest continent and so both Greenland and Australia are excluded for present purposes.

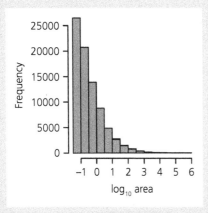

Box Figure 1.1 Distribution in km² of size of island for 80,604 marine islands smaller than Greenland, extracted from a global database of islands. The dataset includes 65,730 islands < 1 km² and 19,392 > 1 km², just 17 of which exceed 100,000 km². Islets of < 10⁻¹·⁵ km² were excluded from the plot.
Redrawn from Weigelt et al. (2013).

On this basis, marine islands contribute around 5% of global land surface area, but it is hard to determine how many there are. Weigelt et al. (2013) provide an analysis based on a global compilation, which demonstrates the abundance of small islands (Box Figure 1.1). Even so,

the number of smaller islands and islets in the graphic is undoubtedly an underestimate, as indicated by the following estimates from Wikipedia.org. In Indonesia alone there are 17,507 islands (of which 8844 have been named and 6000 are inhabited) and there are another 7017 in the Philippines. In Finland there are an estimated 190,000 lakes and within them 180,000 islands. Similarly, Canada possesses at least 32,000 lakes larger than 3 km². Other large clusters of islands include the South China Sea (30,000), the Stockholm archipelago (24,000), Japan (14,125), the Åland archipelago (6500), the Aegean archipelago (6000), the Chinese Zhoushan islands (4700), the Australian Great Barrier (3800), the Maldives (1322) and the Dalmatian coast in Croatia (1185), just to include those clusters with more than a thousand islands and islets.

Furthermore, there are approximately 30,000 *known* seamounts in the world's oceans and an *estimated* 100,000 of > 1 km height (Wessel et al. 2010). Although the tally will depend on definitions, there are clearly vast numbers of insular habitats to pick from. It follows that there is a huge potential range in island properties of biogeographical relevance (e.g. area, age, elevation, isolation, or climate/latitude).

• Area: New Guinea (785,750 km²) down to rocks less than 100 m² (ten orders of magnitude).
• Island age: Madagascar (*c.* 150 Ma) to Surtsey (formed in 1963).
• Island elevation: Hawaii (> 4000 m) to flat atolls just centimetres above sea level.
• Isolation: Marquesas (*c.* 4800 km) to e.g. Anglesey (hundreds of metres).
• Climate/latitude: from 84°N (Oodak, Greenland, the world's northernmost emerged land) to 81°S (Berkner Island, Antarctica).

continental landmasses. **Continental fragments** are those islands that by their location would pass for oceanic islands but which, at least in terms of their foundation, are ancient fragments of continental rock stranded out in the oceans by plate tectonic processes. **Continental (shelf) islands** refer to islands on the continental shelf (i.e. surrounded by sea depths up to around 140 m, beyond which the steeper slopes of the continental margin commence).

The current geological period, the Quaternary, which commenced 2.588 Ma, has been marked by multiple ice ages, although the trend of global cooling began rather earlier than this (Lomolino et al. 2017). The Quaternary is divided into the Pleistocene (which featured lengthy glacial periods interspersed with shorter interglacials) and the current interglacial, the Holocene, which began 11.7 ka. Many continental shelf islands were joined to the mainland during glacial episodes, as these were periods of significantly lower sea levels. The most recent period of connection for these so called **'land-bridge'** islands was achieved through a slow fall to around –120 or possibly –134 m (compared with the present) during the Last Glacial Maximum *c.* 29–21 ka. The subsequent deglaciation saw sea levels rise, with some irregularity, over a span of many thousands of years, with the greater part of the adjustment spanning 16.5–8.2 ka, with a slower phase of adjustment lasting until *c.* 2.5 ka (Lambeck et al. 2014). The timing of (re-)separation, as continental shelf hilltops became islands once again, was therefore highly variable, being dependent upon the regional patterns of isostatic adjustment (Section 3.9) and topography (reflected in contemporary bathymetry). Hence, many islands regained their insularity long after the onset of interglacial conditions *c.* 11.7 ka.

In addition to islands *sensu stricto*, we may also distinguish a wide variety of **habitat islands**, an umbrella term for all other circumstances in which discrete patches of habitat are surrounded by strongly contrasting environments (Table 1.2). Classic habitat islands include woodland 'islands' surrounded by cereal fields, or treeless alpine mountaintops surrounded by mid-elevation bands of woodland (Matthews 2021). For the marine realm, discrete habitat types separated by strongly

contrasting aquatic environments (e.g. shallow benthic environments on seamounts isolated by deep water) can be considered marine habitat islands (Dawson 2016). As one type of island can occur within another type (e.g. habitat islands or lakes within an oceanic island), there are, in practice, innumerable specific forms and contexts of insularity.

All types of island are grist to the island biogeographical mill, from thistle-head habitat islands (Brown and Kodric-Brown 1977) to Hawaii (e.g. Wagner and Funk 1995). However, habitat islands often have blurred edges and exist within complex landscape matrices, which themselves may change dramatically in ecological properties over just a few years. Also, matrix landscape elements may be hostile to some but not all species of the habitat islands. Hence the degree of isolation of habitat islands is typically less clear-cut than is the case for real islands, which arguably thus provide more tractable and productive model systems. Key properties, such as their area, elevation, isolation, geological age, developmental stage, and species membership can be quantified with some degree of objectivity, even if some of these properties can vary hugely over the lengthy life span of an island (Stuessy et al. 1998; Fernández-Palacios et al. 2011, 2016a).

1.3 Geo-environmental dynamics, ecology, evolution, human impact, and conservation: key themes in island biogeography

Our goal in this section is first to provide a short route map to the rest of Part I—'Setting the Scene'—and second, to identify some prominent themes developed within the book. To this end, we introduce some simple graphical models that we have selected for their pedagogic value in communicating some key ideas and how they fit together.

Our focus in the second and third chapters of Part I is mostly on the origins and geo-environmental dynamics of classically defined islands. It is these island systems that contribute disproportionately to global biodiversity, and which have been the inspiration and test bed for most island biogeographical,

Table 1.3 What makes islands interesting? Based on Fleming and Racey (2009), Fernández-Palacios et al. (2021a), and material included in the present volume.

Topic	Examples of key features
Natural laboratories	Vast number of islands, analytical options to control variables and replicate, potential to measure inputs/outputs in small isolates
Physical features	Distinctive geology, geomorphology, and geodynamics, limited size, number of habitats, isolation, disturbance regimes
Biological features	Impoverishment, disharmony, dispersal filtering, niche shifts (e.g. loss of dispersal capacity, body-size changes, generalism), model systems of meta-community assembly and evolution
Biodiversity features	Disproportionate contribution to global diversity, e.g. about one sixth of plants occur on islands, half of all *Anolis* lizards; 3 of 17 families of bats endemic to islands; overall *c.* 20% of species
Conservation concerns	> 60% of historical extinctions across plants, invertebrates and birds are from islands, with islands contributing *c.* 91–94% of bird extinctions; half of the world's threatened species are island endemics
Human societies	10% of human population and one quarter of the sovereign states of the world, contributing cultural and linguistic diversity; contested narratives of sustainability of indigenous cultures and impact of colonialism and globalization

ecological, and evolutionary theory (Tables 1.1, 1.3). We therefore explore in these chapters the varied contexts in which such islands form, develop, and eventually founder, highlighting the distinctive features of their environments, in particular their geological and climatic characteristics, and their sensitivity to climate change and sea-level oscillations.

In Chapter 4 we provide a fuller introduction to the basic biogeographical properties of island systems, noting that islands tend to decline in species richness and become increasingly peculiar in taxonomic composition ('disharmonic') and rich in endemics (species unique to islands) with increasing remoteness. We can understand these patterns as the consequence of filter effects and of the gradual loss of taxa with limited powers of dispersal. Tracing the relationship between island biotas and their source regions has always been a core biogeographical task and has seen historical biogeographers disputing the relative importance of long-distance dispersal and vicariance explanations that invoke barrier formation to explain close relationships between populations of disjunct distribution. Both processes have patently had their part to play and the changing degrees of isolation of islands over time remains of central importance to understanding the biogeography of particular islands (Section 4.7; and e.g. Stace 1989; Heads 2011, 2017; Katinas et al. 2013; Machado 2022).

To complete the task of setting the scene, we conclude Chapter 4 with a review of the global significance of island biodiversity, which can scarcely be overstated. Large islands, especially large tropical islands such as Cuba, Madagascar, New Guinea, Borneo, and Sulawesi are particularly important hotspots of biodiversity, but remote oceanic archipelagos such as Galápagos and Hawaii are also famed for their high degree of endemism and for their spectacular illustrations of insular radiation. Scientists continue to discover new species on islands, some of which, unfortunately, are species that have gone extinct since human arrival, and are known only from fossils or subfossils. These ongoing discoveries sometimes challenge pre-existing ideas and interpretations of island biogeography. They also underline the accelerated rate of attrition of island biotas through human action, which qualifies many islands today as 'threatspots' (below) as well as centres of endemism.

Having set the scene in the first four chapters, we proceed to examine the ecological and evolutionary insights provided by islands in depth within Part II ('Island Ecology') and Part III ('Island Evolution'). Of course, ecological communities operate within the constraints of evolutionary forcing and in turn evolution cannot work without ecology to drive it, so the separation of *ecology* and *evolution* into separate parts of the book is to divide a continuum

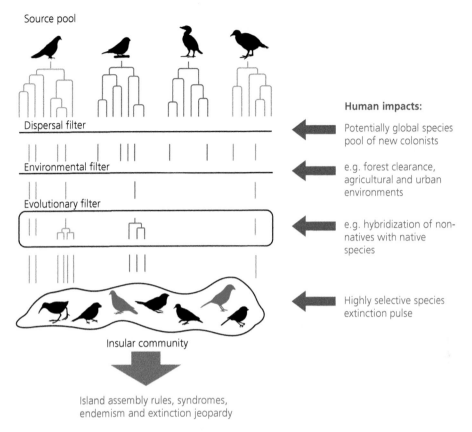

Figure 1.2 What are the winning insular combinations and how do they arise? Island communities are structured by dispersal, environmental, and evolutionary filters to generate communities of distinctive character (phylogenetic and functional traits, syndromes, and endemism), which historically have been subject to anthropogenic 'filtering' processes that have typically modified the mix significantly on islands globally through extinction of native species (the birds shown in red) and introduction of non-natives combined with land-use and other changes.
Loosely based on original figures by Weigelt et al. (2015) and Triantis et al. (2022).

of intertwined processes. Figure 1.2 explicitly spans both sets of processes, setting out that the insular communities we study are the result of successive filters that shape how the biota is composed. First, the species that reach a remote island will be biased towards particularly dispersive species. Second, they must find suitable habitats, environmental conditions, and biotic resources within the island to be able to establish a breeding population. Third, evolutionary change, through a combination of random and selective processes, generates changes in the colonist species to produce novel insular forms, often involving diversification of a subset of those colonists into a suite of new species. The outcome is a particular suite of diversity patterns expressed

per island, per archipelago, and in the relationships among the islands of an archipelago. These emergent features include particular macroecological patterns, (e.g. characteristic variation in island species–area relationships), island assembly rules (focused on the first two filters in the figure), and evolutionary syndromes (repeated shifts in species trait combinations).

A key challenge within island biogeography is to distinguish meaningful repeated diversity patterns across islands from those that might arise by chance, and having done so, to attribute the patterns we observe to relevant ecological and evolutionary processes. To assist in this task, it is helpful, if not essential, to pay some attention to the scale parameters of the systems we are studying and within which the biological processes are playing out. Scale refers to both the space and time parameters of the system (i.e. the range in the properties of island area, island isolation, and time). Figure 1.3 provides three graphical sketches that together provide just such a framework, alongside Table 1.4, which picks out the characteristics of archipelagos for which particular theoretical frameworks and theories appear to be most relevant.

Following Haila's (1990) reasoning, 'islands' of very limited isolation, such as an isolated tree in a garden, may be utilized by a particular individual bird from time to time and so the presence of the individual can be understood as a consequence of its behaviour, constituting the first scale of insularity in Fig. 1.3a. As islands become more isolated and larger, we may study in turn the dynamics of part of a metapopulation, of a distinct island population, and ultimately the evolution of a novel species or suite of species on an oceanic island. Heaney's model picks up the third and fourth of these scales, showing how anagenesis (dominant between curves 1 and 2) is expected to give way increasingly to cladogenesis as an engine of increasing species richness (curves 3 to 5, with inset phylogenetic trees) on increasingly remote islands (Fig. 1.3b).

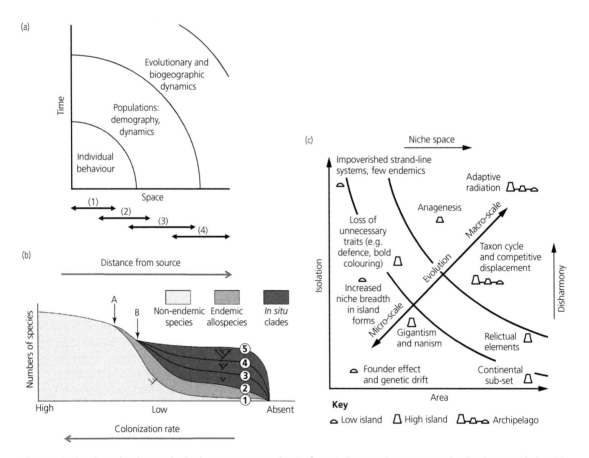

Figure 1.3 The relationships between key biodynamic processes and scale of area, isolation, and time as encapsulated in three simplified models. (a) Haila's (1990) distinction between four scales of insularity: (1) individual scale; (2) population scale 1: dynamics; (3) population scale 2: differentiation; (4) evolutionary scale. (b) Heaney's (2000) model (slightly modified) of the impact of reduced immigration rate with distance from source on richness and diversification within island systems, in which point A represents the point mean gene flow drops below one individual per generation, and B the point cladogenesis becomes significant. (c) Whittaker's (1998) representation of how particular evolutionary phenomena tend to be more prominent in different portions of the span of island isolation and island area across ocean basins.

Table 1.4 Some prominent island biogeographical theories, the geographical configurations of islands for which they hold greatest relevance, and chapters in which they are discussed.

Type of archipelago	Prominent theories/concepts
Small, very near	Metapopulation dynamics (Chapter 12)
Small, near	Equilibrium model of island biogeography (Chapter 5)
Medium, mid-distance	Assembly rules; stepping-stones and sweepstake dispersal (Chapters 4 and 6)
Large, distant	Island syndromes, taxon cycle (Chapters 9–11)
Large, very distant	Island syndromes, cladogenesis (adaptive and non-adaptive); general dynamic model of oceanic island biogeography (Chapters 7–11)

Figure 1.3c complements and extends Heaney's model, providing a perhaps over-ambitious attempt to place island evolutionary ideas and models into a simple island area–isolation context. The aim was to highlight the geographical circumstances in which specific evolutionary phenomena are most prominent. This is not to imply that, for instance, founder events and drift are insignificant in large and isolated oceanic islands, as they do form elements of the macroevolutionary models. Rather, it is that they emerge as the main evolutionary storyline principally for small, near-mainland islands.

Working up the isolation axis within Fig. 1.3c, small islands of limited topographic range and degree of dissection are liable to be poorly buffered from pronounced environmental fluctuations, such as are associated with El Niño events or cyclonic storms. Those species or varieties that become endemic on such islands are likely to have relatively broad niches, at least in so far as tolerating disruptions of normal food supplies is concerned, as illustrated by the Samoan fruit bat, *Pteropus samoensis* (see Pierson et al. 1996). Large, near-continent islands are likely to have very low levels of endemism. Mainland-island founder effects and drift are also unlikely to be prominent in their biota. Islands fitting this category, such as (mainland) Britain, typically possess a subset of the adjacent continental biota, most species having colonized prior to the sea level rise in the early Holocene that returned the area to its island condition

Large islands, with a somewhat greater degree of isolation and considerable antiquity, are the sites where relictual elements are most frequently claimed in the biota (see evaluation in Section 4.7; Kim et al. 1996; Kondraskov et al. 2015; Gibbs 2016).

The most rapid and extensive evolutionary change occurs on remote, high islands. Where these islands are found singly, or in very widely spread archipelagos, speciation is frequent, but often without the greatest radiation of lineages (Chapter 9), fitting the model of anagenesis. Non-adaptive radiation features principally in the literature on land snails, for high and remote islands: where the fine-scale space usage of snails and their propensity to become isolated *within* topographically complex islands allows for genetic drift of allopatric populations. The adaptive radiation of lineages is best seen on large, diverse, remote archipelagos (or at least archipelagos towards the outer limits of reach of a particular taxon) in which inter-island movements within the archipelago allow a mix of allopatric and sympatric speciation processes as a lineage expands into an array of habitats.

Most cases proposed for the taxon cycle (Section 9.2) are from archipelagos that are strung out in stepping-stone fashion from a continental landmass (the Antilles) or equivalent (the smaller series of islands off the large island of New Guinea). In these contexts, the degree of disharmony is less than for systems such as Hawaii and the Galápagos, and, crucially, there are likely to be repeat colonization events by organisms quite closely related, taxonomically or ecologically, to the original colonists.

Of course, Fig. 1.3c is a simplification, and the placing of islands into this framework is likely to vary depending on the taxon under consideration, because of the different spatial scale at which different organisms (snails, beetles, birds, etc.) interact with their environment, both in terms of space occupied by individuals, and dispersal abilities. Indeed, the scalars for time, area, and isolation for each panel of Fig. 1.3 should vary between taxa, such that, for instance, the radiation zones for different taxa may coincide with different archipelagos. Such radiation is, however, best accomplished on higher islands, and archipelagos of such islands, than it is on archipelagos of low islands or single high islands. Hence, the emphasis within Fig. 1.3c is on the additive effects of increased area and isolation from the mainland, as signified by the shift in emphasis from micro-evolution (lightest grey) towards more significant 'macro' evolutionary changes (darkest grey).

Anthropogenic changes to biodiversity are evident—and of growing concern—across the planet. Islands provide numerous important lessons concerning the impact of particular processes, such as habitat loss, or of the impact of introduced species, and indeed of the synergistic impacts of multiple drivers. In the final section of the book, Part IV ('Human Impact and Conservation'), we tackle two distinct aspects of the biodiversity crisis. First, in Chapter 12, we consider the application of island theory to fragmented landscapes. This chapter leans heavily on the theoretical and empirical work reviewed in Chapters 5 and 6, and those readers particularly interested in the application of island theory within conservation may prefer to consider these three chapters in sequence, skipping the intervening chapters. The final three chapters of Part IV cover, in turn, the human transformation of island ecosystems (Chapter 13), anthropogenic extinctions on islands (Chapter 14), and some of the further issues and potential solutions involved in meeting the conservation challenge (Chapter 15). As highlighted in Fig. 1.4, a remarkably high proportion of global species losses in the historic period have been island species; indeed, they supply the *majority* of extinctions in all seven major taxa shown. Not only have humans 'de-assembled' insular systems by selectively eliminating species with particular sets of traits (e.g. ground-nesting and large-bodied birds),

but we have also introduced non-random sets of non-native species that have profoundly altered the compositional profile of island communities—in effect representing an additional *anthropogenic filter* on insular assemblages (Fig. 1.2). Island extinctions typically began not in the historic period, but concurrent with the first arrival of modern humans on islands and the threats posed to island species are continuing to mount. It is important that we identify some of the ways in which we can mitigate these losses by targeted and effective conservation action—a task we set ourselves in the final chapter of the book.

1.4 A very brief history of island biogeography

The small size of the island, together with its vast distance from either the eastern or western continent, did not admit of a great variety of animals.
(G. Forster, 1777, Book I, Chapter VIII, p. 156)

Islands only produce a greater or less number of species, as their circumference is more or less extensive.
(J. R. Forster, 1778, Chapter V, p. 169)

We close this introductory chapter with a very brief history of island biogeography, which serves to illustrate that many of the ideas discussed in this book have a lengthy history of development. More detailed accounts of the development of specific topics can be found elsewhere (e.g. for the species–area relationship, see: Tjørve et al. 2021) and for a fuller picture of the place of islands within the development of biogeography as a whole, see Lomolino et al. (2017), and the essays by John

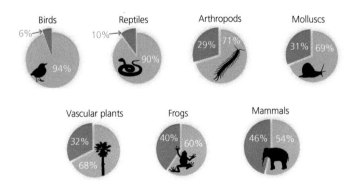

Figure 1.4 The proportion of terrestrial species that have become extinct (or which are extinct in the wild) that are island species (green) in relation to continental losses (blue), according to IUCN data collated in 2017, and where the total is the known post-description extinct species from the historic period.
Fernández-Palacios et al. 2021a, their figure 4; CC BY 4.0.

C. Briggs and Christopher J. Humphries and by Robert J. Whittaker in *Foundations of Biogeography* (Lomolino et al. 2004), as well as the selection of island-themed papers reproduced in that volume. We regret that space does not permit a fuller account or the recognition of the many other scientists past and present who have contributed to the field.

As each island or archipelago constitutes a distinct geological and biological 'natural experiment', it is easy to understand why islands have attracted interest ever since the first modern European transoceanic expeditions (commencing in 1492 AD with Columbus' first voyage). Indeed, from a very early point of the age of exploration, ships' naturalists were charged with collecting and describing the plants and animals they encountered as they crisscrossed the world's oceans. This period constitutes the pre-paradigm phase of the natural sciences, where theoretical frameworks were largely lacking and the emphasis was on discovery and documentation (—and for some—increasingly futile attempts to square these discoveries with biblical narratives such as the Ark). From the observations made on these voyages, the first insights into island biogeography were generated. Sailing with Captain Cook on his second voyage, the Prussian-born naturalist Johann Reinhold Forster and his son Georg Forster described two fundamental patterns in accounts published in 1777 and 1778, namely the species–area relationship and the decrease in richness with isolation (see the above epigraph quotations). Some decades later, Leopold von Buch, a German geologist, who visited Madeira and the Canaries in 1815, reported what may have been the first inference of geographic speciation on islands (see Box 1.2).

These and other early natural science contributions (including Lyell's *Principles of Geology* and the work of Alexander von Humboldt and Augustin de Candolle) influenced three British naturalists—Joseph Dalton Hooker, Charles Darwin, and Alfred Russel Wallace—each of whom went on to make key contributions to the natural sciences that were

---

Box 1.2   Leopold von Buch and the idea of geographic speciation

As a result of his visit to the Galápagos Islands in 1835 during the voyage of the *Beagle*, the Galápagos' (Darwin's) finches (*Geospizinae*) and tortoises (*Chelonoidis nigra* complex) played a pivotal role in Charles Darwin's subsequent development of the theory of evolution by natural selection (Darwin, 1859). Less well known is that the Canarian endemic daisies (*Argyranthemum* spp.) and sea lavenders (*Limonium* spp.) played a similar role, some 20 years before Darwin's trip (1815), in prompting Leopold von Buch's thoughts about geographic speciation.

Leopold von Buch (1774–1853) was a German geologist who began his career defending Neptunism (the idea that all rocks were initially deposited as sediment under water), but influenced by visiting the Canaries, he abandoned this view and argued for an important role for volcanos in shaping the Earth. He also introduced the Spanish word *caldera* as a scientific term for craters produced by volcanic collapse. He visited Madeira and the Canaries in 1815 with the Norwegian botanist Christen Smith, subsequently writing a book that contains a chapter about the Canarian flora (von Buch 1825). In it (pp. 133–135) he wrote what we suspect may be the first account of populations diverging in isolation, having dispersed to occupy locations differing in landscape, soil, and resources; and subsequently coexisting with similar derived species after secondary contact has been established. He recognized that in large and abrupt islands, mountain ridges may create isolating barriers comparable to the sea itself, causing isolated populations to form new species, not existing elsewhere in the island, which will be more differentiated from the original continental ancestor the longer they have been isolated. He pointed out that although the members of the groups discussed are diverse in form, the similarity among all those species, each occupying a different, non-overlapping location, is so evident, that it allows the conclusion that all these species stem from a common continental ancestor. Here von Buch is effectively making a first reference to a monophyletic radiation event.

We can only speculate on what inspiration Darwin might have drawn from a visit to the Canaries. It is known that he was looking forward to visiting Tenerife during his voyage. Unfortunately, owing to strict quarantine rules introduced by the local authorities as a precaution against cholera, the stay on Tenerife was cancelled, and the *Beagle* continued its voyage without Darwin setting foot in the Canaries.

in some measure inspired by their encounters with island flora and fauna. Hooker's work on islands began as surgeon-botanist on the Antarctic expedition led by James Clark Ross in 1839–1843. His observations of disjunct distributions of plants with seemingly limited dispersal powers occurring across remote islands in the Southern Ocean, led him to support ideas of extensionism, postulating former land connections that have since sunk, in what was a precursor to vicariance biogeography. These ideas were reinforced by observations of apparent relictual (palaeoendemic) elements in oceanic floras, as reported in his seminal lecture on island floras delivered in Nottingham, England, in 1866 (Hooker 1866).

Charles Darwin's journey on the *Beagle*, during which he famously visited the Galápagos Islands in 1835, brought many contributions to science in addition to its pivotal role in his thinking on evolution, not least a theory for the origin of fringing atolls and inferences on the importance of long-distance dispersal in the colonization of remote islands (Darwin 1839, 1842, 1859). Darwin's grasp of the value of islands as natural laboratories put them to the fore within biology. Hooker was a strong supporter of Darwinism and a long-standing correspondent of Darwin's, notwithstanding their difference of view on the matter of extensionism versus long-distance dispersal. This debate can be seen as a forerunner to the ongoing vicariance–dispersalism debate. A key difference is that extensionism has been largely discredited through the revolution in understanding within Earth Sciences that has its origins in Alfred Wegener's theory of continental drift. Wegener's theory was first published in 1912, but these ideas came to maturity only following the development and general acceptance of plate tectonic theory in the 1960s and 1970s (Lomolino et al. 2017).

The third key British naturalist of the 19th century was, of course, Alfred Russel Wallace, co-discoverer (in parallel to Darwin), of the theory of evolution by natural selection and recognized by many as the founding father of biogeography. Although Wallace had previously made extensive observations and collections (which were lost in a shipwreck) in the Amazon, it was his work in the South-East Asian islands that led him to his key insights on evolution. Wallace's writings include *The Malay Archipelago* (1869), chronicling his experiences and observations collecting in the region, *Tropical Nature and Other Essays* (Wallace 1878), and his classic *Island Life* (first published in 1880 and updated for a third time in 1902). Wallace created a structured theory of biogeography, in which islands held a central place. In addition, his work on regionalization, while not the first global scheme (that was published by Philip Sclater in 1858), was particularly influential. While travelling along the Sunda Islands, Wallace had been struck by the dramatic turnover in mammal fauna between Bali and Lombok. His subsequent division of the region became known as Wallace's line, and in due course the insular area between the South-East Asian and Sahul regions became known as Wallacea in his honour (Section 4.6). Wallace's synthesis of biogeographical information was and remains hugely influential. He recognized the distinction between the three main types of island (oceanic, continental fragment, and continental shelf) and he also wrote extensively about the influence of past glaciations on the biogeography of islands.

The 19th- and early 20th-century literature on islands was not of course confined to works in English or by British scientists (Box 1.2). An example of an early synthesis of emergent island biogeographical phenomena written in German is provided by Kny (1867), who discussed geographical patterns of island species' richness with isolation and latitude, and commented on patterns of endemism, and insular plant syndromes. The eruption of Krakatau in 1883, although causing a humanitarian disaster, presented an ideal opportunity for monitoring the colonization process of a sterilized island, demonstrating the capacity of plant and animal species for effective over-water dispersal. Under the auspices of the Dutch colonial administration, French, German, and Dutch scientists (such as Melchior Treub, Alfred Ernst, and Willem Docters van Leeuwen) began the work of documenting this process. Their seminal studies led to Krakatau becoming a classic long-running natural experiment both for primary succession and for island recolonization and turnover (see reviews by Whittaker et al. 1989; Thornton 1996). Their data were later exploited by MacArthur and Wilson (1967) as a key test-system for their equilibrium model of island biogeography. Although Krakatau provides merely a single case study, it serves to illustrate a continuity of insular research stretching back to the 19th century. Indeed, for numerous islands and archipelagos around the world, contemporary island biogeographical research is generally built upon decades and indeed often centuries of collecting records and observations (see e.g. Machado's 2022 account of the long history of work on *Laparocerus* weevils in Macaronesia).

In parallel with such empirical observations, the first half of the 20th century was productive in the development of ecological theory, including the first mathematical treatments and debates about the form of island species–area relationships, in which key proponents included Olof Arrhenius, Henry Gleason, and Frank Preston (Tjørve et al. 2021). Somewhat later, key contributions were made by Ernst Mayr, founder of the modern synthesis in evolution, whose work embraced islands and included analyses of the genetic consequences of founder events, inferences of double invasions, and the allopatric speciation dynamics of island birds (e.g. Mayr 1942, 1963). Other notable island scientists of the period include Philip Darlington, a zoologist interested in the flightlessness of island beetles (Darlington 1957), Charles Elton, an ecologist who provided key syntheses of the consequences of nonnative species introductions in islands (Elton 1958), and David Lack, who undertook the first detailed studies of the evolution of Galápagos finches (Lack 1947a).

As this short review makes clear, island biogeography has deep roots, but the field gained identity relatively recently, through the work of two North American biologists, Robert H. MacArthur (who was born in Toronto, Canada) and Edward O. Wilson (born in Birmingham, Alabama), who advocated and led a shift from what they characterized as a historical, descriptive approach to island studies towards a more quantitative, predictive science. Wilson's (1959, 1961) seminal work on the taxon cycle in Melanesian ants sought to infer the evolutionary patterns and processes of diversification of island ants from their distribution across islands. His model represents a bridge from the historical, narrative approach towards an emphasis on repeated patterns and turnover. In their famous

Theory of Island Biogeography, MacArthur and Wilson (1963, 1967) went a great deal further, developing a mathematical model that stripped down the biology to a minimum, in the form of their seminal equilibrium model, before demonstrating how the biology could be added back in more complex amendments to their model. Their work has achieved paradigmatic status and is continuing to guide island biogeographical research today.

Both their 1967 monograph and island biogeography as a field is, of course, far richer than a simple model of immigration versus extinction and the influence of the key insular properties of area and isolation. However, the significance of this model and approach will be evident from the attention given to it in Part II of the present volume. On the whole, however, evolutionary research on islands continued along largely separate tracks through the second half of the 20th century. A particularly influential contribution was made by the American botanist Sherwin Carlquist, as summarized in his two books, *Island Life* and *Island Biology* (Carlquist 1965, 1974). Based on an outstanding knowledge of natural history, especially for Polynesia, Carlquist was a champion of the derivate nature of island woodiness and of the trend towards sexual dimorphism in island plants, as well as of the tendency towards loss of dispersibility in island plants and animals (Chapters 10 and 11; and see Traveset et al. 2016). Another significant contribution during this important period in the development of the field was made by Jared Diamond (1974, 1975), who developed an approach to the study of bird distributions in Melanesia that sought to identify assembly rules (i.e. patterns in the composition of island avifaunas), which he controversially linked strongly to the role of interspecific competition (reviewed in Chapter 6). The critical response to this work was based around the use of null models to question whether the patterns described were truly non-random in the first place (e.g. Connor and Simberloff 1979). The use of null or neutral models in island biogeography prompted by the work of MacArthur and Wilson and the debate over assembly rules has since grown apace, bringing important insights to the field (Warren et al. 2015)

Building upon these foundations, island biogeographical research has entered a remarkably profitable phase over the last four decades. In part this has been built on continuing traditions of long-term study, such as the decades of work by Peter and Rosemary Grant (e.g. 2008) on the microevolutionary processes shaping the diversification of the Galápagos (also known as Darwin's) finches. It has also benefitted from a number of overlapping and intersecting technical developments, advances in genetics, in computer power, in the digitization, collation, and sharing of data, in the development of statistical and modelling tools, in remote sensing, geographical information systems, and the capacity to determine bathymetry accurately and precisely, in plate tectonic theory, in dating of rocks, and of fossil and subfossil material, and in the understanding of palaeoenvironmental change. Along the way, we have increasingly seen an integration between equilibrium and non-equilibrium models and concepts, and between historical and dynamic narratives, recognizing that island geo-environmental dynamics need to be modelled and integrated into models of island biodynamics. These advances have drawn in an increasingly large and international body of scientists who work partly or mostly on islands. Even if our concerns for the future of island biodiversity are acute, it is an exciting time to be working on islands. The remainder of this volume hints at, but cannot do full justice to, the diversity of authors and approaches and of the remarkable body of recent work that has been published on the themes of the geo-environmental dynamics, ecology, evolution, human impact, and conservation of island systems and which we synthesize under the overarching embrace of *island biogeography*.

1.5 Summary

In this introductory chapter, we outline the scope of this volume, as the study of islands as model biogeographical systems. That is, our central concern is with establishing the patterns of distribution of island life, at levels varying from traits and sub-species up to emergent properties of the biotas of entire islands or archipelagos, and with using these patterns to develop and test explanatory and predictive models linked to causal mechanisms. We highlight the numerous contributions made by the study of islands in the life sciences and the disproportionately large contribution islands make to global biodiversity and to the tally of anthropogenic extinctions and to the list of threatened species.

We extend the concept of the island from the classic 'small area of land surrounded by water' to recognize the great diversity of terrestrial and aquatic systems that possess (sometimes temporarily) the key property of insularity, which is a significant degree of isolation from other areas of equivalent habitat(s). We establish that there are vast numbers of real islands and most are small, making them tractable systems for biogeographical quantification and analysis. We highlight that they are typically short-lived compared with continents and that understanding their geo-environmental dynamics can be crucial to understanding their biological dynamics.

In briefly outlining the structure of the book, we pick out the importance of recognizing the scale of space and time of the island system we are studying, arguing that different processes and patterns are likely to emerge to different and predictable degrees as a function of the range in area and isolation, and the duration of time over which the patterns assemble. We also introduce the concept of different filters on island community membership, namely: the dispersal, environmental, evolutionary, and, finally, anthropogenic controls that structure the assemblage and subsequent dynamics of island systems.

We close the chapter with a short account of the development of the field, commencing with the role of natural scientists in the age of European expansion and exploration, picking out the contributions of figures such as Johann Reinhold Foster, Georg Forster, Leopold von Buch, Joseph Hooker, Charles Darwin, and Alfred Russel Wallace. We go on to recognize the importance of natural history and descriptive traditions, the development of mathematical approaches to phenomena such as species–area relationships, and the turn towards quantitative and predictive models initiated by Robert H. MacArthur and Edward O. Wilson. Finally, we reflect briefly on trends towards integration of multiple disciplines and approaches in recent decades, constituting a modern renaissance in the study of island life.

CHAPTER 2

Island types, origins, and dynamics

2.1 Classifying marine islands

In his seminal book *Island Life* Alfred Russel Wallace (1902) classified islands by geological origin and biological properties into three main categories. Although derived long before the development of plate tectonic theory and at a time where scientists were only just beginning to understand the importance of the glaciation events, his classification remains biogeographically useful for its simplicity (Box 2.1; Table 2.1).

- **Continental islands** (or recent continental islands *sensu* Wallace) are emergent fragments of the continental shelf, typically similar geologically and biologically to the continents, from which they are separated by comparatively narrow, shallow waters. Pronounced sea-level change driven by glacial/interglacial cycles has repeatedly joined many such islands to—and separated them from—the adjacent continental landmass during the last 2.588 Ma (the Quaternary). These (past) land-bridge islands typically attained their current insular status during the > 100 m eustatic sea-level rise just a few thousand years ago, following the transition to the present interglacial (the Holocene).

- **Continental fragments** or micro-continents (ancient continental islands *sensu* Wallace) were once part of larger continental landmasses, but over tens of millions of years, tectonic drift has isolated these fragments. Now the waters between them and the continents are typically wide and deep. Long isolation has allowed both the persistence of some ancient lineages and the development of new species *in situ*. They have

---

Box 2.1 The biogeographical properties of continental shelf, continental fragment islands, and oceanic islands

Here, we highlight the biogeographical relevance of A. R. Wallace's tripartite classification of islands by generalizing key features, while breaking down oceanic islands into two main subsets: high volcanic islands and atolls (Box Table 2.1).

- **Continental (shelf) islands** are geologically varied, containing both ancient and recent stratified rocks. They are rarely remote from a continent and typically contain some land mammals and amphibians as well as representatives of the other classes and orders in considerable variety (Wallace 1902), partly legacies of their mostly being **past land-bridge islands** (i.e. areas connected to mainland during the Last Glacial Maximum and

often later—during the earlier part of the Holocene transgression).

- **Continental fragment islands** are in some respects intermediate between the other two classes, but being typically ancient and long-isolated, they are in other respects biologically peculiar and tend to feature a higher proportion of relictual forms.

- **Oceanic islands** are built over the oceanic plate, are of volcanic or coralline formation, are remote and have never been connected to mainland areas, from which they are separated by deep sea. They generally lack indigenous land mammals and amphibians but typically have a fair number of birds and insects and usually some reptiles.

Island Biogeography. Robert J. Whittaker, José María Fernández-Palacios, and Thomas J. Matthews, Oxford University Press.
© Robert J. Whittaker, José María Fernández-Palacios, and Thomas J. Matthews (2023). DOI: 10.1093/oso/9780198868569.003.0002

Box 2.1 *Continued*

Box Table 2.1 General features of islands of different origin of relevance to their biogeography.

Features	Continental islands	Continental fragments	Oceanic islands	Atolls
Origin	Interglacial sea-level rise	New mid-ocean rift creation	Between or within plate submarine volcanic activity	Volcanic erosion/subsidence in warm oceans*
End	Glacial sea-level drop	Collision with a continent	Erosive break up and subsidence	Subduction/arrival at Darwin's line**
Degree of isolation from mainland	Small (zero during low sea-level stands)	Variable; small for Cuba, large for New Zealand	Mostly large (exceptions include Canaries)	Large
Size	From very small to very large	Some small, mostly large	Small (except Iceland)	Very small
Longevity	Short (20–30 ka) as separate entities	Long (50–150 Ma)	Variable (hours–20 Ma)	Variable, depending on sea-level changes
Water gap depth	Small ($\leq$ 130 m)	Large (> 1000 m)	Large (> 1000 m)	Large (> 1000 m)
Basal parent rocks	Granites	Granites	Basalts	Calcareous
Erodibility	Low	Low	High	High
Archipelagic clustering	Low	Low (except Balearics and Seychelles)	High (with exceptions)	High
Original biota	Present	Present	Absent	Absent
Relictualism	Lacking	High	Moderate	Absent
Biogeographical emphasis	Relaxation	Vicariance	Dispersal and radiation	Marine subsidies
Endemism	Variable: increases towards equator	Very high	High	Lacking, except in raised atolls

*Atolls can be drowned and re-emerge re-iteratively with fluctuations in sea level.
**Darwin's line is the latitude at which coral growth fails to counteract subsidence due to low water temperatures.

The biogeographical distinction between islands of differing geological origins depends to some large degree on the ability of potential inhabitants of an island to disperse to it, whether over land or sea. As a simplification, we may conceptualize an oceanic island as one for which evolution is faster than immigration ('Darwinian islands'), and a continental island as one where immigration is faster than evolution (Williamson 1981). By this reasoning, a particular island may be thought of as essentially continental for some highly dispersive groups of organisms (e.g. ferns and some types of birds), yet essentially oceanic in character for organisms with poor powers of dispersal through or across seas (e.g. conifers, terrestrial mammals, and freshwater fish). Moreover, there are some groups of islands of geologically mixed origins and types (e.g. the Seychelles) or which appear to have been partially connected at times of lowered sea levels, such that they have biological characteristics part way between oceanic and continental, land-bridge islands (e.g. the avifauna of the Philippines (Diamond and Gilpin 1983)).

Table 2.1 Examples of islands corresponding to the three types identified by Alfred Russel Wallace. Some archipelagos defy easy classification, e.g. Japan and the Philippines provide examples with a mixed (continental–oceanic) origin.

Ocean	Continental islands	Continental fragments	Oceanic islands
Arctic	Svalbard Novaya Zemlya Baffin		Iceland Jan Mayen
North Atlantic	Britain Ireland Newfoundland		Azores Madeira Canaries
Mediterranean	Elba Rhodos Djerba	Balearic Archipelago Corsica–Sardinia Crete	Santorini Aeolian
Caribbean	Trinidad Tobago	Cuba Jamaica Hispaniola	Martinique Guadeloupe Montserrat
South Atlantic	Falklands Tierra del Fuego	South Georgia	Ascension St Helena Tristan da Cunha
Indian	Zanzibar Sri Lanka Sumatra	Madagascar the granitic Seychelles Kerguelen	Réunion Mauritius Saint Paul
North Pacific	Vancouver Haida Gwaii St Lawrence		Kuril Hawaii Northern Marianas
Central and South Pacific	Borneo New Guinea Tasmania	New Zealand New Caledonia	Galápagos Society Islands Marquesas

also invariably received colonists by over-water dispersal, post-isolation. These continental fragments lose their insular status only when, after tens of millions of years of drift, they finally collide with a different continent, forming a new peninsula, with which they may exchange species. A classic example is India (once part of Gondwana), which collided with Eurasia c. 50–40 Ma (Lomolino et al. 2017).

- **Oceanic islands** originate as seamounts, created by submarine volcanic activity, mostly with basaltic foundations. They have never been connected to the continents and so are colonized via over water dispersal. In time, *in situ* speciation contributes to the tally of species by adding new endemic forms. Hence, they are sometimes termed Darwinian Islands (e.g. Gillespie and Roderick 2002). Oceanic islands form in connection with plate boundaries and, in certain circumstances, within a tectonic plate. Of the millions of seamounts, only a small fraction reach sea level to form oceanic islands. These islands typically are short-lived and even substantial ones may only exist for a few million years before subsiding and eroding back into the ocean. Where sea temperature permits, they may persist through coralline ring formation, as atolls. Occasionally, non-volcanic islands, such as Cyprus, emerge due to oceanic floor uplift and may present as oceanic in biogeographical character.

Plate tectonics theory has revolutionized our understanding of the Earth's surface and, along with it, our understanding of the origins, dynamics, and configuration of islands. The following is a simplified typology alongside a short introduction to the major reconfiguration of the Earth's surface over the last c. 200 Myr and why this is so important

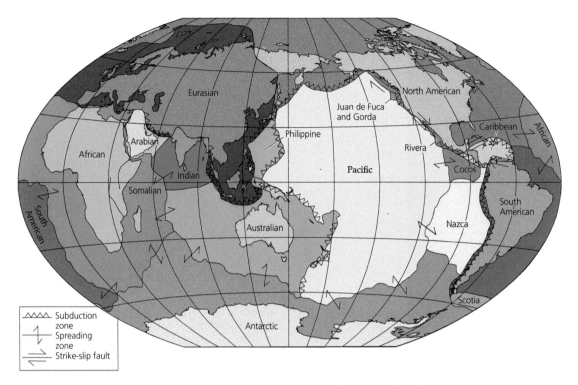

Figure 2.1 Map of the seven major and some of the more important microplates, with indication of the plate boundary interactions involved. From Lomolino et al. (2017), their figure 8.17, which in turn was derived from http://pubs.usg.gov/publications/text/slabs.html.

to biogeographical debate. We then introduce a recently proposed framework that seeks to provide a more process-driven island classification alongside lineage-specific analysis of the biotas to be found on these islands.

There are seven major tectonic plates > 20 M km^2 in area and numerous smaller fragments (Fig. 2.1). With the exception of the Pacific plate and its subordinates (Nazca, Cocos, Juan de Fuca, and Philippines), the plates typically comprise an oceanic part and a continental part (Gillespie and Clague 2009). The silicon/aluminium-rich granitic parts of the plates are of relatively low density and support the continents themselves (consisting of a highly varied surface geology), extending to about 200 m below sea level. The zone from 0 to c. −140 m forms the continental shelf and supports islands such as the British Isles and the Frisian Islands. Such islands tend to be geologically similar to the nearby continent, involving a mix of rock types and modes of formation, such that any combination of

sedimentary, metamorphic, and igneous rocks may be found.

In places, continental plate can be found at much greater depths than 200 m below sea level and can then be termed **sunken continental shelf**. Islands on these sections of shelf (ancient continental islands *sensu* Wallace: Table 2.1) are thus formed, or built upon basements of, continental rocks. A recently confirmed example is Zealandia, a 4.9 M m^2 area in the south-west Pacific comprising sunken continental crust, 94% submerged and with a modal depth of −1100 m (substantially higher than the surrounding oceanic crust) and supporting the islands of New Zealand, New Caledonia, and Lord Howe, among others (Mortimer et al. 2017) (Fig. 2.2).

There is generally a steeply sloping transition zone from shallow continental shelf down to around 2000 m, where oceanic, **basaltic** crust is found and the true oceanic islands occur. Around half of the Earth's surface takes the form of abyssal plains, at some 3000 to 6000 m below sea level,

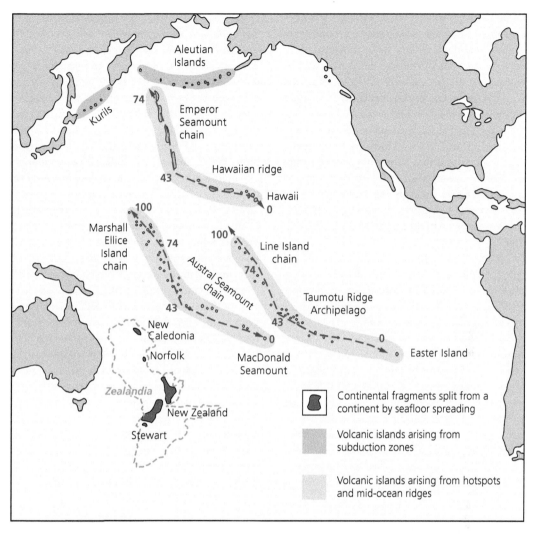

Figure 2.2 Map of the Pacific Ocean, depicting the major modes of origin of islands. The dashed lines and dates indicate the ages of the islands/seamounts at either end of three island chains, with a third date showing the point at which plate movement orientation altered. New Zealand (part of the mostly sunken Zealandia continent) has been physically isolated from Australia and Antarctica by up to 2000 km for 60 million years, as a result of Late Cretaceous to Late Palaeocene seafloor spreading.
Compiled from figures in Mielke 1989; Lomolino et al. 2017; Mortimer et al. 2017.

comprising basaltic magmas formed at mid-ocean ridges and coated in fine grain sediments as the plate is pulled sideways by seafloor spreading. The average depth of the oceans has been estimated as 3800 m, with the extensive abyssal plains being broken up by mid-ocean ridges, trenches, and volcanic **seamounts**. There may be 25 million seamounts of > 100 height and 100,000 of > 1 km height, although only around 10% of even those larger features have

been identified and mapped (Wessel et al. 2010). Oceanic islands are merely the tallest of these features: volcanic in origin, but in cases composed of sedimentary material (principally limestones), formed as the volcanic core has sunk below sea level.

True oceanic plate islands are those that have never been attached to a greater landmass. They may grow through further volcanism, or subside, erode, and disappear below the waves. In a geological sense, they tend to be transitory. Some may last only a few days, others some millennia. Relatively few last tens of millions of years. And, as a consequence of the sea-level fluctuations of the Quaternary, many have also repeatedly emerged and submerged, merged, and separated from one another (Norder et al. 2019). Thus, in addition to the islands that sustain terrestrial biota today, there are also many seamounts that represent past and future islands (Fernández-Palacios et al. 2011). Flat-topped seamounts, formed through the submergence of limestone-topped volcanos, are called **guyots**, after the Swiss geologist Arnold Guyot (Jenkyns and Wilson 1999).

Islands may originate by existing areas of land becoming separated from other landmasses to which they were formerly connected, by erosion, or by changes in relative sea level from a variety of causes. A great example of both is provided by the British Isles. At the time of the Last Glacial Maximum (LGM, *c.* 21 ka), global sea level was depressed *c.* 135 m and Britain was (and had long been) part of the continent. Gradual warming led to eustatic sea-level adjustment (i.e. driven by change in the volume of water in the oceans) over several thousand years. Although the warming was briefly reversed during the Younger Dryas cold snap, interglacial conditions were established by the onset of the Holocene epoch (*c.* 11.7 ka). The Dover Straits

were breached, with Great Britain separating from the continent rather later, 8–7.5 ka, as the final, eastern connection via Doggerland failed (Fig. 2.3; Sturt et al. 2013). Hence, for thousands of years, temperate species of plants and animals were able to colonize Great Britain across land, before the terrestrial link was severed. Going further back, prior to around 450 ka Britain remained connected to mainland Europe even during the interglacial high-stands, through a chalk ridge spanning the Dover Straits between southern England and France (Gupta et al. 2017). This ridge was breached and the Strait opened through at least two major episodes of erosion, involving an initial phase of lake formation behind (and to the east of) the rock ridge, lake overspill, and episodes of catastrophic flooding associated with the final breaching of the Strait.

Many islands originate from volcanism associated with plate movements (Box 2.2). The form this volcanism takes and the pattern of island genesis involved depend on the nature of the contact zone between plates; that is, whether plates are moving apart, moving towards each other, or moving past each other laterally. One particularly important context of island genesis is where oceanic plate collides with the oceanic crust of an adjacent plate, subducting the older (colder and denser) oceanic crust beneath the younger (warmer and lighter), generating an abyssal trench and a volcanic arc (e.g. Marianas, Aleutian, Kurils, Lesser Antilles, or South

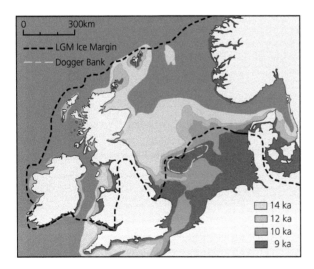

Figure 2.3 Reconstruction of successive shorelines during ice retreat following the LGM (*c.* 21 ka), showing that Britain remained connected to mainland Europe following the onset of the Holocene (11.7 ka) via an area to the east known as Doggerland. The narrowest gap between Britain and France is marked by deeper overspill channels.
Updated slightly from Bailey et al. 2020, courtesy of Geoff Bailey.

Box 2.2 Plate tectonics and the long-term dynamics of the world's islands

Oceanic crust tends to be denser and thinner than continental crust and when oceanic crust collides with continental crust, the former tends to be subducted under the latter. Hence, the world's oceanic lithosphere is continually, if slowly, recycled, such that very little is left that exceeds 150 million years in age (Box Figure 2.2a) and most islands are very much younger than this. In illustration, Weigelt et al. (2013) have reported a mean age (based on oldest rocks) of just under 7 Ma for their database of a few hundred oceanic islands.

split from Gondwana during the early Cretaceous (c. 130 Ma), becoming a vast island and moving northwards to collide with Eurasia c. 50–40 Ma, in the process forming the Himalayan range (Box Figure 2.2b; Lomolino et al. 2017).

By the Late Cretaceous (c. 80 Ma) a circum-equatorial seaway, the Tethys Sea, had formed and landmasses identifiable as South America and Africa were already discrete entities (Lomolino et al. 2017). It was also during the Cretaceous that first Madagascar, and later the Seychelles and Kerguelen micro-fragments, began their trip to their present

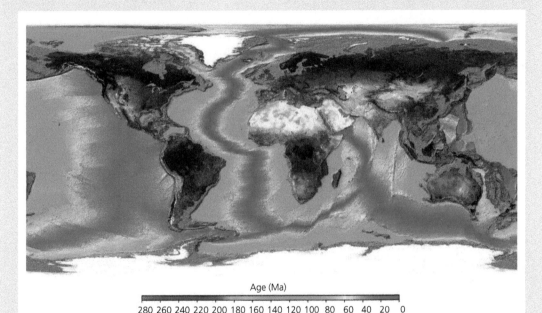

Box Figure 2.2a The age of the oceanic lithosphere.
From Lomolino et al. (2017), after an original in Müller et al. (2008).

If we briefly trace the distribution of the major landmasses from the start of the Jurassic (about 200 Ma), the northern and southern supercontinent, Laurasia and Gondwana, had long since formed a single massive continental landmass, Pangea, which was beginning to break apart. Over the Jurassic and the early Cretaceous, both Laurasia and Gondwana gradually split, while North and later the South Atlantic began to open. The Indian subcontinent

isolated geographic locations in the south-west, north-west, and south of the Indian Ocean, respectively. Australia and New Zealand rifted from Antarctica about 100 Ma, and subsequently, New Zealand broke first from Australia and then from Antarctica about 80 Ma (Lomolino et al. 2017).

While other parts of Gondwana drifted towards the lower latitudes, Antarctica slowly moved towards its present position over the South Pole. In the process it became

Box 2.2 *Continued*

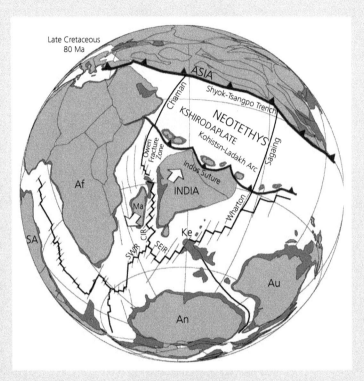

Box Figure 2.2b A Late Cretaceous (*c*. 80 Ma) snapshot of the position of the Indian plate as Gondwana broke apart and the Indian plate collided with the Kohistan–Ladakh island arc, prior to joining Asia. The position of south-western and south-eastern Indian ridges are shown, along with Kerguelen, Madagascar, Africa, South America, Antarctic, Australia and Asia. Madagascar split from Africa during the Early Cretaceous and subsequently parted from India around 90–85 Ma.
Slightly modified from figure 13 of Charterjee et al. (2013), with permission from Elsevier and thanks to the authors.

isolated by the development of a clockwise circum-Antarctic marine current, creating the conditions during the Miocene (23–5.3 Ma) for the build-up of a great polar icecap. This contributed to falling sea level and to the disappearance of the epicontinental seaways and to a slow drift downwards in global temperatures. Following the joining of North America and South America and the loss of the circum-equatorial warm sea current, this cooling resulted in the major glaciation episodes of the Quaternary (from 2.588 Ma). Hence, plate tectonics has not only driven major changes in the absolute and relative positions of islands and continents through time, but these changes have, in turn, influenced climatic regimes and major oceanic and atmospheric circulation patterns, controlling insular biotic sources and colonization rhythms.

Even if the islands we are interested in are just a few million years old, a longer-term view may be essential to biogeographical interpretation. For example, the oldest Canary and Hawaiian islands emerged *c*. 23 Ma and 5.1 Ma, respectively, but the archipelagos may have been around since *c*. 68 and 70–85 Ma, respectively (e.g. Wagner and Funk 1995; Fernández-Palacios et al. 2011; Hunt and Jarvis 2017). Indeed, recently Grehan (2017) has argued for even older origins of elements of the Macaronesian biota, pointing to evidence of pre-cursor islands in the region. Molecular dating indicates that some Hawaiian lineages pre-date the formation of all current high islands in the archipelago, although few crown lineages appear to have established before *c*. 10 Ma, reflecting a lack of continuity of availability of (high) islands over time. Similarly,

---

Box 2.2 *Continued*

García-Verdugo et al. (2019a) report a lack of crown ages pre-dating emergence of the current oldest Canarian island. This highlights the importance of distinguishing between the geological age (of rocks) and biological age (time since emergence or sterilization) of islands. Hence, reconstructing the biogeographical history of islands and archipelagos requires detailed geological information on the current island(s) but also for the surrounding ocean basin, as legacies may be evident today that pre-date even the oldest current island of an archipelago (compare Heads 2011; Hunt and Jarvis 2017; Ali and Hedges 2021).

Sandwich Islands) pointing towards the subducting plate. When the oceanic crust subducts below the continental crust of an adjacent plate (e.g. Nazca–South America) an arc of volcanos is also formed, but on the continent (e.g. the Andes). Another important context is where a string of islands is generated by plate movement across hotspots in the mantle (e.g. the Hawaiian Islands).

Relatively few islands are ancient biogeographically. Some have experienced a complex history of both lateral movements and alternating emergence and submergence, with profound importance to the resulting biogeography, not just of particular islands, but also of the wider region (Box 2.1; Keast and Miller 1996; Katinas et al. 2013; Musgrave 2013; Ali 2017; Heads 2017).

Oceanic islands may be subdivided into two major categories: plate boundary and intraplate island types, each of which may also be subdivided, based on their relationship to the plate boundaries (Table 2.2). This approach has utility in describing recognized archipelagos but has the limitation that it can result in grouping islands that may turn out to have rather different geodynamics (below).

2.2 Plate boundary islands

Islands at divergent plate boundaries

Divergent plate boundaries produce islands (i) along mid-ocean ridges (divergent boundaries) and (ii) along the axes of back-arc or marginal basins behind island arcs that are themselves associated with convergent plate boundaries. Although divergent plate boundaries are constructive areas involving more magma output than occurs in any other situation, seamounts rarely breach the ocean surface in connection with these boundaries. This is because of decreasing magma supply as the seamounts drift away from the plate boundary, and the great depth of the ocean floor. In some cases, mid-ocean ridge islands may also be associated with hotspots in the mantle, providing the seemingly exceptional conditions necessary for their conversion from seamounts to large islands. Iceland, which emerged *c.* 15 Ma, is the largest such example (103,106 km^2), being a product of the mid-Atlantic ridge and a hotspot that may have been active since *c.* 55 Ma. Another context in which mid-ocean ridge islands can be found

---

Table 2.2 P. D. Nunn's (1994) classification of oceanic islands: highlighting distinct configurations of islands of volcanic origin formed over oceanic plate.

Level 1	Level 2	Examples
Plate boundary islands	Islands at divergent plate boundaries Islands at convergent plate boundaries Islands along transform plate boundaries	Iceland, St Paul (Indian Ocean) Antilles, South Sandwich (Atlantic) Cikobia, Clipperton (Pacific)
Intraplate islands	Linear groups of islands Clustered groups of islands Isolated islands	Hawaii, Marquesas, Tuamotu Canaries, Galápagos, Cabo Verde (Cape Verde) St Helena, Christmas Island (Indian Ocean), Rapa Nui (Easter Island)

is in association with triple junctions in the plate system, a notable example being the Azores, where the North American, Eurasian, and African plates meet. Two islands, Corvo and Flores, sit over the North American plate, west of the mid-Atlantic ridge, the other seven are associated with the Azores plateau and are aligned over or parallel with the boundary between the Eurasian and African plate (Fig. 2.4; Hildenbrand et al. 2014). The second form of divergent plate boundary island, exemplified by the Tongan island of Niuafo'ou, is that sometimes formed in back-arc basins, which develop as a result of plate convergence, but which produce areas of seafloor spreading (Nunn 1994).

Islands at convergent plate boundaries

Where two oceanic plates converge, a trench in the ocean floor forms at the point of subduction and an arc of volcanic islands develops parallel with the trench axis on the surface of the upper of the two plates (Fig. 2.5). This mechanism accounts for some of the classic island arcs in the Pacific and Caribbean. The magma involved in island arc volcanism derives from the melting of the subducted oceanic crust, complete with its sedimentary load. Where basaltic crust and water-bearing sediment are subducted, this commonly leads to explosive andesitic volcanism, as typified within the Sunda Island arc, including the Indonesian islands. Basaltic volcanism in island arcs is less common, possibly reflecting a relative paucity of sedimentary load in the subducted crust. The Sandwich arc in the South Atlantic involves both basaltic and andesitic volcanism.

Islands along transverse plate boundaries

This is a fairly rare context for island formation as, by definition, less divergence or convergence of plates is involved. However, strike–slip movement, compression, or both, can occur between adjacent parts of plates. An example of an island believed to have been produced by these forces is the Fijian island of Cikobia, in the south-west Pacific.

2.3 Islands in intraplate locations

This class of islands contains some of the most biogeographically notable islands of all, such as the Hawaiian, Galápagos, and Canary islands.

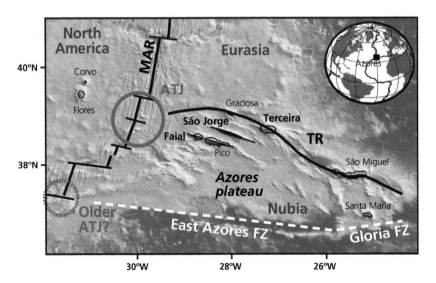

Figure 2.4 The Azorean archipelago is located near the triple junction between the North American, Eurasian, and African (Nubian) plates. The figure reveals that the archipelago comprises a western group and eastern group, either side of the mid-Atlantic ridge and also shows the association of the eastern islands with the Azorean plateau (ATJ: Azores Triple Junction; TR: Terceira Rift; FZ: Fracture Zone). Hildenbrand et al. (2014).

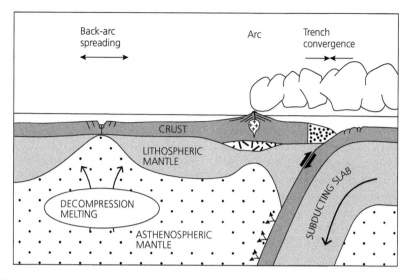

Figure 2.5 Simplified representation of an intra-oceanic subduction zone, in which the densely stippled area represents the outer forearc (which is subject to either sediment accretion or subduction erosion).
Modified from figure 2 in Larter and Leat (2003).

Linear island chains

The Hawaiian chain is the classic example of an archipelago with a significant age sequence arranged linearly across a plate (Table 2.3; Fig. 2.2). Among the larger islands, the youngest is Hawaii, dating to *c.* 0.5 Ma and the oldest, Kauai, dates to 5.1 Ma. Edifices older than Kauai stretch away to the north-west, but through combinations of erosion, subsidence, and coral formation, they are now reduced to forming coral atolls, or seamounts. The chain extends around 6000 km, from the Loihi Seamount, which is believed to be closest to the hotspot centre, through the Hawaiian Islands them-selves and along the Emperor Seamount Chain, to the Meiji Seamount (*c.* 74–85 Ma) near the Kam-chatka Peninsula (Keast and Miller 1996; Regelous et al. 2003). The antiquity of the chain may have allowed for colonization of the present archipelago from islands that have since sunk. However, there appears to have been a recent period without high islands: molecular data indicate that few lineages have crown ages > 10 Ma and the pre-Kauai legacy in the biota is rather limited (Wagner and Funk 1995; Keast and Miller 1996).

The building of linear island groups such as the Hawaiian chain is explained by J. T. Wilson's (1963)

Table 2.3 Approximate ages of selected islands and of the Meiji Seamount. This table illustrates the wide variation in the ages of oceanic islands, the age sequence of members moving north-west along the Hawaii–Emperor chain, and the apparent lack of a similarly simple sequence moving south in the Austral–Cook cluster.

Island group	Age (Ma)
Hawaii–Emperor chain	
Mauna Kea, Hawaii (the Big Island)	0.38
Kauai	5.1
Laysan	19.9
Midway	27.7
Meiji Seamount	74–85
Austral–Cook cluster	
Aitutaki	0.7
Rarotonga	1.1–2.3
Mitiaro	12.3
Rimatara	4.8–28.6
Rurutu	0.6–12.3
Isolated intraplate islands	
Ascension	1.5
St Helena	12–14
Christmas Island (Indian Ocean)	37.5

Sources: Nunn (1994), Wagner and Funk (1995), Regelous et al. (2003), Gillespie and Clague (2009).

hotspot hypothesis, which postulates that 'station-ary' thermal plumes in the Earth's upper mantle lie beneath the active volcanos of the island chain (Wagner and Funk 1995). A volcano builds over the hotspot, then drifts away from it as a result of plate movement and eventually becomes separated from the magma source, and is subject to erosion by waves and subaerial processes, and to subsidence under its own weight. It has been calculated that the Hawaiian Islands have been sinking at a rate of approximately 2.6–2.7 mm a year over the last 475,000 years (cited in Whelan and Kelletat 2003). As each island moves away, a new island begins to form more directly over the hotspot. This process is thought to have been operating over some 75–80 million years in the case of the Hawaiian hotspot. The alteration in orientation of the chain (Fig. 2.2) has been attributed to past changes in the direction of plate movement and occurred about 43 Ma. The Society and Marquesas island groups, also in the Pacific, provide further examples of hotspot chains.

The classic hotspot model may not easily account for the dynamics of all linear intraplate archipelagos. For example, the tectonic-control model, attributed to Jackson et al. (1972), postulates that, instead of lying along a single lineation, the islands lie along shorter lines of crustal weakness (termed *en echelon* lines), which are sub-parallel to each other. More recently, small-scale sub-lithospheric convection has also been proposed to account for the long duration of volcanism, synchronous over significant distances within some volcanic chains. Such mechanisms may explain clusters of islands and chains such as the Line Islands in the Central Pacific, which fail to demonstrate a simple age–distance relationship along the length of the chain (Ballmer et al. 2009).

Clustered groups of islands

Intraplate archipelagos that form clusters rather than linear chains of islands have been the subject of much debate. Some clustered groups, such as the Galápagos, are nonetheless attributable to a mantle hotspot and do show a broad east–west age gradient connected to the drift of the Nazca plate over the hotspot (Ali and Aitchison 2014). The Canaries and Madeira, in the North-eastern Atlantic, also provide examples of intraplate island and seamount clusters. Their fit with the classic hotspot model has been questioned as they again do not appear

to show all the classic features: remaining active far longer than expected, lacking a simple age sequence, and failing to show evidence of rapid subsidence (being built on older, more rigid crust). However, the larger picture is that an older set of seamounts is associated with each archipelago, stretching north towards the Iberian Peninsula, providing a broadly coherent age sequence stretching back for 60–70 Myr (Fig. 2.6; Fernández-Palacios et al. 2011; Troll et al. 2015) and indeed there are seamounts of twice this age in the region (Bogaard 2013). While the geological models emerging are often more complex than the simple hotspot scenario sketched above (e.g. Bogaard 2013; Miller et al. 2015), from a biogeographical point of view we can reasonably consider both the Canarian and Madeira volcanic provinces to constitute linear arrays of islands and former islands, the dynamics of which have followed a hotspot-style progression or ontogeny in relation to a fixed mantle plume (Troll et al. 2015; Carracedo and Troll 2021).

Isolated islands

As knowledge of the bathymetry of the seafloor has improved, a number of islands believed to be isolated have been shown rather to be part of a seamount/island chain or cluster, an example being the islands of the Trindade–Martin Vaz Insular Complex in the Southern Atlantic Ocean. These islands are joined by nine others (currently seamounts) during Pleistocene sea-level minima, so that the currently exposed islands are at these times merely the eastern end of a lengthy chain of islands that extends towards the Atlantic coast of Brazil (e.g. Pinheiro et al. 2017). In general, truly isolated intraplate islands form only at or close to mid-ocean ridges. They result from a single volcano breaking off the ridge with part of the sub-ridge magma chamber beneath it (Nunn 1994). Examples include Ascension, Gough, and St Helena in the Atlantic, Christmas Island in the Indian Ocean, and Guadalupe in the Pacific. Isolated intraplate islands or island clusters may also, in certain cases, be the product of small continental fragments being separated from the main continental mass. The granitic Seychelles provide a good example: there is good evidence that they sit atop a continental

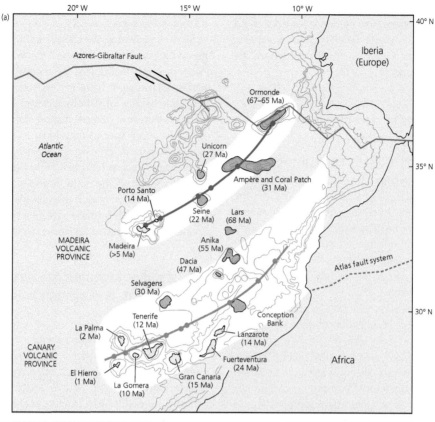

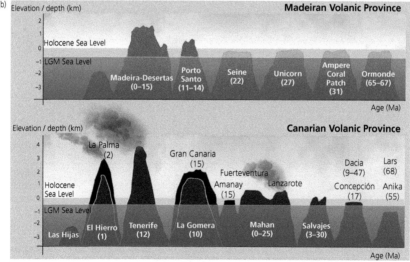

Figure 2.6 Development of the Canarian and Madeiran volcanic provinces. (a) The islands (sand-coloured) and seamounts (dark blue) of the Madeira and Canary volcanic provinces, with ages of origin (emergence in the case of the contemporary islands) indicated, and the inferred plume tracks, which are believed to curve slightly due to the rotation of the African plate. Contours indicate 500 m depth intervals. From Lomolino et al. (2017, their figure 8.29) based on Troll et al. (2015). (b) Schematic (not to scale or position) of the profiles and ages of the islands, modified from Fernández-Palacios et al. (2011, their figure 2). NB Island age estimates are not fully resolved and in some cases vary between sources

basement affiliated to the Madagascar–India part of Gondwana, but there is no evidence for underlying continental crust under a majority of the Seychelles Islands which are not located over the topographic plateau (Hammond et al. 2013).

2.4 Next steps in classifying marine islands and their biota

Jason Ali (2017) recently suggested that it is more helpful to recognize four major categories of island: continental, island arc, composite terrane, and mantle-plume hotspot, within which several minor types can be distinguished, sitting alongside a fifth basket of other minor types (Table 2.4). We include this classification for completeness, although it has yet to be widely applied within island biogeography.

As a separate step, Ali (2017) also suggested an accompanying classification of island life-forms based on how a particular organism's ancestors originally colonized a particular island: recent land-bridge, recent ice-sheet, overwater-dispersed, and deep-time vicariant. Deep-time vicariant distributions refer to ancient distributional separations through barrier formation (vicariance events)

brought about primarily, in this context, by plate tectonic processes (Box 2.2).

By distinguishing the classification of an island's geodynamics from analyses of particular lineages, Ali was emphasizing that the incumbents of a particular island may have arrived on the island by differing means, at different times, thereby experiencing differing geo-environmental histories. For example, some species may have crossed dry land during a period of lowered sea level during the Pleistocene (recent land-bridge), while others colonized following the same island's separation by sea-level rise (overwater dispersed), or, if considering certain high-latitude systems, they may have crossed an ice-connection (recent ice-sheet).

2.5 Island ontogenies: the birth, development, and disappearance of islands

As the Earth's tectonic plates shift, carrying islands and continents with them, in the simplest terms they do one of the following: rub along one another (conservative margin), move apart (constructive margin), or collide (destructive margin), such that the denser crust (typically oceanic) gets subducted

Table 2.4 Jason Ali's (2017) typology of island origins, with an example of each type.

Continental		Ocean plate hotspot	
Shelf	*Great Britain*	Within-plate hotspot	*Hawaii*
Orogenic margin	*New Guinea*	Hotspot atoll	*Cocos-Keeling*
Isolated	*Madagascar*	Ocean ridge hotspot	*Iceland*
Micro-continental terrane	*Timor*	Oceanic LIP remnants	*Kerguelen*
Continental-arc	*Krakatau*	Oceanic LIP atoll	*Ontong Java*
Rifted arc-raft	*Crete*	**Composite terrane**	
Continental fore-arc	*Ryukyu*		
Isolated raft atoll	*Paracels*	Thinned continental-continental rafts	*South Island, New Zealand*
Shelf volcano	*Bioko*	Thinned continental raft and ophiolite	*Cuba*
		Arc and oceanic LIP	*Solomons*
Intra-oceanic arc		**Left-over**	
Arc	*Montserrat*	Outer-trench high	*Christmas*
Arc-apron	*Guadaloupe*	Transpressional transform	*Macquarie*
Fore-arc basement	*Bonin Ridge*	Leaky transform	*Clipperton*
Accretionary prism	*Barbados*	Volcano atop continental ribbon	*Norfolk*
Remnant arc	*Guam*	Enigmatic	*Offshore Gulf of Guinea*
Remnant arc atoll	*Aves*		

Footnotes: LIP, large igneous province; terrane, a fragment of crust that has been transported laterally; a composite terrane comprises two or more geotectonic elements.

(above). This is one key reason islands are typically short-lived. However, many islands disappear as platforms for terrestrial life long before the plate they are on gets subducted, through processes such as terrestrial and marine erosion, subsidence, and sea-level rise (e.g. Fig. 2.5, 2.6).

Intraplate hotspots are like a conveyor belt of island generation and destruction. Each island goes through a life cycle, or **ontogeny**, of birth, maturation, decline, and disappearance, in which the environmental and ecological properties develop in harness along broadly predictable lines (Fig. 2.7). Having built from the ocean floor over the hotspot, achieving great heights, the island is carried away from the hotspot, gradually becoming inactive. In tandem, it is subject to surface and marine erosion, initially increasing the topographic complexity, but over time, and in combination with island subsidence, reducing the elevational range (Fig. 2.7). This idea of a predictable island ontogeny is integral to the general dynamic model of oceanic island biogeography, which seeks to represent the way in which the core biogeographical processes of migration, speciation, and extinction respond to the sequenced birth and death of islands within remote archipelagos (Whittaker et al. 2008; Chapter 9). At the outset, we should recognize that islands typically depart from this generalized ontogeny; for example, sometimes experiencing periods of uplift as well as subsidence (e.g. Ramalho et al. 2017) and

in cases with more prolonged and complex histories of secondary volcanism (below). In the following subsections we provide details of the long-term dynamics of three systems, which are wholly or partly attributable to hotspot dynamics and which provide classic case-studies systems in island evolution. The idea of predictable, distinctive ontogenies can be extended to other forms of island, such as island arcs, or seamounts (e.g. Borregaard et al. 2016, 2017; Pinheiro et al. 2017), but as will be demonstrated in the Caribbean section, some islands, archipelagos, and regions experience very particular, individualistic ontogenies.

The Canaries and Palaeo-Macaronesia

The oldest parts of the present-day Canary archipelago are the Northern, Betancuria, and Jandía basaltic shields of the island of Fuerteventura, which are estimated to have emerged above sea level between 20 and 24 Ma (Fig. 2.6a) (Anguita et al. 2002; Troll et al. 2015; Hunt and Jarvis 2017; Carracedo and Troll 2021). This is pretty ancient for an oceanic archipelago. However, the Canary Islands volcanic province has a far earlier start point, dating at least to the emergence of what is now the Lars (Essaouira) Seamount *c.* 68 Ma, since when it is possible that there has always been at least one high island available within the Canary

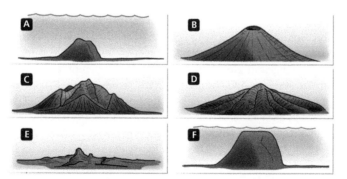

Figure 2.7 Stages in the idealized ontogeny of a volcanic island (based on sketches in Ollier 1988 and Nunn 1994): (a) after a lengthy period building as a seamount; (b) an island may break surface, building an intact volcano, with radial grooves (e.g. Tristan da Cunha); (c) planeze stage, in which some of the radial drainage channels have undertaken headwall capture, leaving triangular-shaped remnants relatively untouched by fluvial erosion (examples can be seen on St Helena); (d) residual volcano, where planezes have been removed by erosion, but the original form of the volcano is still evident (examples can be seen in the Eastern Canaries (Fuerteventura) or Cabo Verde (Sal, Boat Vista, Maio)); (e) volcanic skeleton, where only the necks and dykes remain as prominent features (e.g. Bora-Bora, French Polynesia); (f) subsidence and erosion reduce the island once again to a seamount.

Volcanic Province. At this time, the Mediterranean region was dotted with islands, and another island, the Ormonde Seamount, from the Madeiran Province, was to be found to the north, providing a stepping-stone between Lars and potential source regions around the Tethys Sea (Fernández-Palacios et al. 2011).

As recent phylogenetic evidence points to colonization routes for particular lineages that may be indirect (e.g. Iberia to Azores to Madeira to Canaries), we should be prepared to broaden our perspective from the contemporary archipelago, not only to the now submerged parts of the Canary Volcanic Province, with its potential 68-Myr history, but also to encompass a regional-scale analysis of past island availability and continental configurations. Biogeographically, based on some shared floristic elements, Macaronesia refers to the Azores, Madeira, Selvagens (Salvage or Savage Islands), Canaries, and Cabo Verde Islands. Hence the term Palaeo-Macaronesia is given to reconstructions of the islands of the region extending back through the Cenozoic era (Fernández-Palacios et al. 2011), or potentially even longer (Grehan 2017). In practice, given the ages of the islands and seamounts established to date, the possibilities of a continuity of insular lineages from Palaeo-Macaronesia to contemporary Macaronesia depend largely on the older islands of the Madeiran and Canarian provinces (Fig. 2.6b). Molecular phylogenetic analyses to date do not appear to provide much support for such legacies, reflecting the likelihood of turnover linked to continued colonization from outside the insular species pools (but see Grehan 2017). Biological innovation, combined with changes in global and regional climate and current systems, and the re-emergence of seamounts to form stepping-stone islands during Pleistocene sea-level minima have doubtless further impacted rates of colonization and turnover.

Returning to the development of the contemporary Canary Islands (Fig. 2.6), c. 14 Ma, the basaltic massif of Femés (Los Ajaches) emerged, forming the first shield of what became the island of Lanzarote, which was thus available for colonization simultaneously from the older island and mainland source regions. Lanzarote continued to build via the Famara volcanic shield, active 10–5.7 Ma

(Carracedo and Troll 2021). The general pattern of westwards expansion of the archipelago into the Atlantic continued, with the emergence of Gran Canaria c. 14.5 Ma, Tenerife c. 12 Ma, and La Gomera c. 10 Ma (Hunt and Jarvis 2017; Carracedo and Troll 2021). These islands, once much larger and higher than today, have suffered many catastrophic landslides during their geological history (below). Tenerife is formed from three basaltic shield volcanos, the Miocene Central and Teno massifs, and the Pliocene Anaga massif. Activity in the centre of Tenerife resumed after long quiescence in the form of Las Cañadas complex from 3.5 Ma, building (or re-building) anew the territory between Anaga and Teno in a series of large-scale eruptions and flank collapse events.

As this account illustrates, the processes of island-building and eventual decline are not smooth. For example: (i) large-scale flank collapses, both submarine and subaerial are associated with island emergence and subsequent decline (Canals et al. 2000; Hunt and Jarvis 2017) and (ii) c. 3.7 Ma, the Roque Nublo eruption sequence denuded large parts of Gran Canaria, although complete sterilization is considered unlikely (Anderson et al. 2009). Following any such high-magnitude event, new surfaces become available for colonization. In the case of Gran Canaria, the recolonization process will have started both from surviving refugia and—for the mountain and summit ecosystems—especially from the nearby island massifs of Anaga (Tenerife) and Jandía (Fuerteventura), respectively 60 and 90 km away. Similarly, Tenerife has doubtless both donated and received colonists into disturbed territory at different points in time (e.g. Machado 2022), while some 10% of the endemic plants of the Canaries are restricted to one or more of the palaeoislands of Tenerife, likely including a mix of early diverging and recently diverging lineages (Trusty et al. 2005).

The Quaternary (2.588 Ma to present) has been a period of great geo-environmental dynamism in the Canaries. Volcanic mountain-building has constructed the two westernmost islands, La Palma (1.7 Ma) and El Hierro (1.2 Ma). Both have been prone to catastrophic mega-landslides, the most recent of which, El Golfo, redistributed the northwest half of El Hierro across some 1500 km^2 of the ocean floor around 15 ka (Canals et al. 2000). It is only within the last 1.5 Ma that the great acidic volcanic cycle of Las Cañadas (re-)unified the old basaltic massifs of Teno, Adeje, and Anaga into today's Tenerife (Fig. 7.6 inset). The new land surface enabled enhanced biotic exchange between the old massifs. Several landslides (Icod, Las Cañadas, La Orotava, Güímar) followed the construction of this edifice, having catastrophic effects both within Tenerife and on nearby islands. The Teide stratovolcano formed c. 175 ka, with its highest tip (3718 m) being built only around 1200 years ago. Such events leave their imprint on the genetic architecture of lineages, through repeated local vicariance (geographical splitting of populations) and recolonization (e.g. Moya et al. 2004; Machado 2022).

Finally, some 50 ka, the small islets north of Lanzarote (La Graciosa, Montaña Clara, and Alegranza) and Fuerteventura (Lobos) were built, giving the Canary archipelago approximately its present shape. Apart from La Gomera, which has been dormant since c. 2–3 Ma, each island shows evidence of volcanic activity within the Holocene. Since the start of the 16th century, there have been 14 confirmed volcanic eruptions, distributed across La Palma (7, the most recent commencing in September 2021), Tenerife (4), Lanzarote (2) and El Hierro (1, on the submarine flank) (Longpré and Felpeto 2021).

In addition to the building, collapse, and erosion of the islands, eustatic sea-level changes during the Pleistocene have alternately increased and decreased the emerged area of the archipelago, such that the present area of 7841 km^2 represents just 63% of the 12,462 km^2 achieved during the LGM sea-level minimum, the lowest point of the Pleistocene (Rijsdijk et al. 2014; Norder et al. 2019). At that point, the islands were about 130 m higher in elevation and Lanzarote and Fuerteventura were joined, along with nearby islets, to form the largest island (Mahan, c. 5000 km^2). The shortest distance from the archipelago to the African mainland was reduced from 95 km to 60 km. The present-day submarine bank of Amanay, north of Jandía, formed an island of some 315 km^2, and a 'stepping-stone' corridor of islands provided enhanced connectivity between the Canaries, Madeira, and the Iberian Peninsula (Fig. 2.8; Rijsdijk et al. 2014).

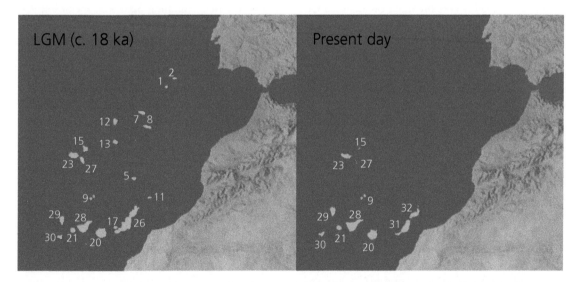

Figure 2.8 The Canary and Madeira island systems as they are today and as they were c. 21–18 ka (LGM), when global sea levels were around 134 m lower than today. Numbers represent islands and seamounts, while missing numbers represent seamounts that are too deep to emerge at the LGM lowstand. 1 Gettysburg, 2 Ormonde, 5 Dacia, 7 Ampere, 8 Coral Patch, 9 Selvagens, 11 Conception, 12 Unicorn, 13 Seine, 15 Porto Santo, 17 Amanay, 20 Gran Canaria, 21 La Gomera, 23 Madeira, 26 Mahan (= 32 + 31), 27 Desertas, 28 Tenerife, 29 La Palma, 30 El Hierro, 31 Fuerteventura, 32 Lanzarote.
Redrawn from Fernández-Palacios et al. (2011).

Hawaii and the Emperor Seamount Chain

The Hawaiian chain is characterized by the growth of shield volcanos that go through a well-established life cycle (Price and Clague 2002):

1 Volcanos form over the hotspot and are subsequently removed from it by the movement of the Pacific plate, producing a linear array of volcanic summits increasing in age to the north-west.
2 After achieving their maximum heights some 0.5 Ma after their emergence, volcanos rapidly subside as they move away from the hotspot.
3 Erosive processes reduce volcanic peaks to sea level over several million years.
4 Small atolls may remain while conditions are suitable for coral growth, or they are reduced to guyots if they drown and sink below the sea surface, carried down by the spreading and subsiding plate (Jenkyns and Wilson 1999).

During island growth, the lava deposited below sea level forms a steeper slope than that deposited subaerially. Sonar surveys of the seafloor have therefore enabled the reconstruction of past island configurations by reference to submerged breaks-in-slope (Moore et al. 1994, Clague 1996). The maximum elevation achieved can also be reconstructed assuming a 7° angle for subaerial lava deposits. Knowing the age, the original elevation of rocky outcrops, and the subsidence rate, it is possible to estimate a rough rate of erosion per island.

Using this information, Price and Clague (2002) have reconstructed the history of the Chain from 32 Ma, the point of emergence of the first Hawaiian island (the earlier islands of the Emperor Seamount Chain were either submerged or just atolls by this time). Between *c.* 32 and 18 Ma, a few volcanos (Kure, Midway, Lisianski, Laysan), briefly exceeded 1000 m elevation, although at any given moment most islands have been small. The most important island that pre-dated the present high islands was Gardner, which formed around 16 Ma and was likely comparable in size (10,000 km^2) and height (> 4000 m) to today's Big Island (Hawaii). After its formation, a series of mid-sized volcanos formed (French Frigate Shoals, La Perouse Pinnacle), culminating with Necker, *c.* 11 Ma.

After the formation of Necker, there was a period when only smaller islands formed, leading to an archipelago diminished in height and area and thus impoverished in habitats. The emergence of Kauai (5.1 Ma) constituted the beginning of the 'second peak period', which continues today, during which land area has been at a maximum (within the last 32 Myr) and there have always been multiple volcanic island summits > 1000 m and some > 2000 m. Around 1.2 Ma, the now separated islands of Maui, Lanai, Molokai, and Kahoolawe, as well as the Penguin Bank Seamount, constituted the Maui-Nui complex ('Big Maui' in Polynesian), an island larger than the current island of Hawaii. By 0.6 Ma subsidence had divided it into separate islands during sea-level high-stands, although they reunite during glacial era sea-level minima, when the area of the archipelago increases from 17,000 to 21,800 km^2 (Price and Elliott-Fisk 2004; Norder et al. 2019).

The Caribbean Islands

The Caribbean meta-archipelago, comprising the Bahamas, Greater, Lesser, and Leeward Antilles, forms a crescent of islands between North and South America, spanning 3200 km, including 900 islands of a total area of about 230,000 km^2 (Fig. 2.9a). Many of these islands are geologically quite old and have complex histories involving cycles of emergence and submergence and changes in relative position and degree in isolation, but the precise sequence of events and degree of overall connectivity is much debated (Ali and Hedges 2021; Iturralde-Vinent 2006; Ali 2012; Lomolino et al. 2017; Roncal et al. 2020). The following account is thus tentative. Islands began to emerge along the eastern edge of the Caribbean Plate in the Early Cretaceous (140–120 Ma), shifting eastwards over time through plate movement (Fig. 2.9b). By around 50 Ma (during the Eocene), the Greater Antilles were approaching their present positions, although they appear to have been permanently subaerial only since the collision of the Caribbean Plate with the Bahamas platform a few million years later.

It has been proposed that around 34 Ma, a remnant arc of volcanos in the eastern Caribbean known as the Aves Ridge, was uplifted to form—with the Greater Antilles—a more or less continuous

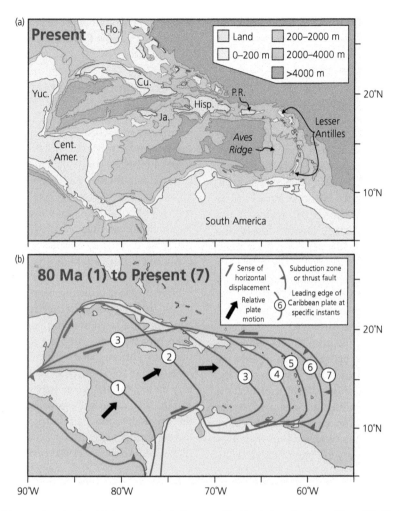

Figure 2.9 The Caribbean region, showing: (a) the present-day bathymetry, with abbreviations showing positions of Florida, Yucatan, Central America, South America, Cuba, Jamaica, Hispaniola, Puerto Rico; and (b) schematic of the development of the region from 1 to 7 indicating, respectively, 80 Ma, 60 Ma, 44 Ma, 30 Ma, 14 Ma, 6 Ma, 5 Ma, and present.
From Ali (2012), his figure 1.

land-bridge, lasting some 2 million years, linking South America to Puerto Rico and thence to Cuba, Hispaniola, and eastern Jamaica (e.g. Iturralde-Vinent 2006). This GAARlandia hypothesis thus allows for some biotic exchange between North and South American regions in advance of the completion of the Panamanian isthmus *c.* 3 Ma (discussion in Lomolino et al. 2017) and may account for colonization of the Caribbean by particular clades—such as *Deinopis* spiders—the timing of which is congruent with the proposed dates of emergence (Chamberland et al. 2018). However, based on

review of recent palaeogeographical data, Ali and Hedges (2022) highlight that the southern and central part of the Aves Ridge was almost certainly submerged at the very point in time that the GAARlandia hypothesis requires its presence, hence, refuting the hypothesis. They argue that the strongly filtered nature of the vertebrate fauna of the Greater Antilles is in fact consistent with over-water dispersal rather than land-bridge connections to South America. Although the geo environmental history of the region is far from settled, this brief account serves to highlight that geological age and current configuration of islands may be misleading of long-term biogeographical history.

This is well illustrated by Jamaica, which comprises two blocks, western and eastern, of separate origin and developmental history. While Jamaican terranes appear to have been connected to Central America in the Early Eocene, any legacies of earlier faunas were lost through submergence. The re-emergence of persistent land forming part of present-day Jamaica occurred *c.* 35 Ma, in the Late Eocene, in the form of the Blue Mountains block, although a large part of the island surfaced as late as 12–10 Ma, in the Late Miocene (Iturralde-Vinent 2006; Carstensen et al. 2012). Hence, despite the undoubted geological evidence of horizontal terrane movements, vertical changes of land and sea have arguably been of more general significance to Caribbean biogeography than horizontal land movements. A classic early insight into the role of past sea-level change was provided by Williams' (1972) work on anoline lizards. Anoles are small green or brown lizards that are more or less the only diurnal arboreal lizards found in most of the region. Williams' work showed that much of the distributional pattern within this group is related to the submarine banks on which today's islands stand. Each bank has many islands on it, but as recently as 7–10 ka, many of these islands were joined together, as they would have been for lengthy periods during the Pleistocene.

2.6 Coralline islands: reefs, atolls, and guyots

In tropical seas, coral reefs built upon sunken volcanos form an important and distinctive category of tropical islands—atolls. For many hotspot islands, in particular, these sedimentary toppings represent a persistent end stage to their life cycle, although such low-lying islands may also come and go as a function of changing sea level, their own capacity for upward growth (which depends mainly on the ocean sea surface temperature) and further erosion or subsistence.

Darwin (1842) distinguished three main reef types: fringing reefs, barrier reefs, and atolls. He explained atolls by invoking a developmental series from one type into the next as a result of subsidence of volcanic islands: fringing reefs are coral reefs around the shore of an island, barrier reefs feature an expanse of water between reef and island, and the final stage is the formation of an atoll, where the original island has disappeared, leaving only the coral ring surrounding a shallow water lagoon (Fig. 2.10). As Ridley (1994) noted, Darwin thought

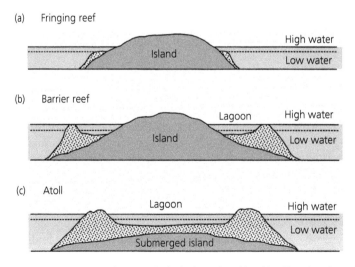

Figure 2.10 The developmental sequence of coral reefs as a result of subsidence, from (a) to (b) to (c), as hypothesized by Darwin (1842). Redrawn from Mielke 1989, Fig. 7.10.

out the outline of his theory on the west coast of South America before he had seen a true coral reef! Although the theory may not be globally applicable and requires modification in the light of contemporary understanding of the significance of past sea-level change in shaping the developmental history of contemporary atolls, his basic model holds some relevance to oceanic atolls and can account for most of the massive coral reefs (Steers and Stoddart 1977; but see Dickinson 2009; Toomey et al. 2016).

Coral reefs are built by small coelenterate animals (corals) that secrete a calcareous skeleton. Within the tissues and calcareous skeleton, numerous algae and small plants are lodged. The algae are symbionts critical to reef formation, providing the food and oxygen supplement necessary to account for the energetics of coral colonies, while obtaining both growth sites and nutrients from the coral (Mielke 1989). Reef-building corals require light and generally grow in waters less than 100 m deep (exceptionally they can be found as deep as 300 m). They are also restricted to tropical and subtropical regions as they require water temperatures between 23°C and 29°C.

Coral growth has been found to vary between 0.5 and 2.8 cm/year, with the greatest growth rates occurring in water of < 45 m depth (Mielke 1989). In suitable areas, they grow sufficiently fast to have maintained reefs in shallow water as either the underlying sea floor subsides or the sea level has risen (or both). Drilling evidence from islands with barrier reefs demonstrates this ability, as predicted by Darwin's (1842) theory. For example, the thicknesses of coral that have accumulated on the relatively young islands of Moorea (1.5–1.6 Ma), Raiatea (2.4–2.6 Ma), and Kosrae (4.0 Ma) vary between 160 and 340 m, whereas the rather older islands of Mangareva (5.2–7.2 Ma), Ponape (8.0 Ma), and Truk/Chuuk (12.0–14.0 Ma) have accumulated depths of 600–1100 m of coral (Menard 1986). Massive reef accretion also occurs in the Western Atlantic on continental shelves that have not experienced long-term subsidence (Mielke 1989). This lends support to another theory: reefs established on wave-cut platforms during glacial sea-level minima and were subsequently able to keep pace with rising interglacial sea levels. This

illustrates that several processes—subsidence, eustatic sea-level changes, temperature changes, and wave action—must each be considered with respect to the influence that they have had on the growth of coral reefs (cf. Nunn 1997, 2000).

Over time, sea levels have fluctuated markedly, particularly during the glacial–interglacial cycles of the Quaternary (Chapter 3). These fluctuations, in combination with island subsidence, wave action, and subaerial erosion, have resulted in many islands declining below sea level. In nautical parlance, they are known as banks if they are less than 200 m from the surface (Menard 1986). Flat-topped sunken islands are termed guyots, and although they were once thought to be predominantly erosional features, most are in fact flat-topped through accretion of carbonate sediments (Jenkyns and Wilson 1999). Once seamounts have sunk below about 200 m, they are well below the range of eustatic sea-level changes, even in the coldest phases of the Pleistocene, and such sinking seamounts will rarely re-emerge as islands. However, islands can become elevated through isostatic rebound or tectonic uplift. Christmas Island (maximum elevation 360 m), in the Indian Ocean, is an example of a submerged coral-capped island that has been uplifted to form a plateau of 150–250 m above sea level. The general aspect of the island is of coastal limestone cliffs 5–50 m high, which carries with it the implication that the area currently available for seaborne propagules to come ashore is considerably less than the island perimeter, a feature shared with many volcanic islands.

Finally, we should note that the relative elevation of particular islands may be subject to influence by the behaviour of others in the vicinity. The loading of the ocean floor by a volcano appears to cause flexure of the lithosphere, producing both near-volcano moating and compensatory up-arching some distance from the load. Indeed, one repeated pattern in the South Pacific is the association within island groups of young volcanos (< 2 million years) with raised reef, or makatea islands, as exemplified respectively by Pitcairn and Henderson Islands (Benton and Spencer 1995). There are many other examples of islands (e.g. Jamaica) that contain substantial amounts of limestone as a product of uplift.

2.7 Summary

Understanding the biogeographical dynamics of islands requires a grasp of the geodynamics of the island and archipelago of interest. This chapter focuses on the origins and geodynamics of islands within the oceans, which are traditionally subdivided by biogeographers into oceanic islands, continental fragments, and continental shelf islands. Islands are rarely ancient and in many cases are significantly less ancient biologically than geologically. Advances in Earth Sciences permit an increasingly refined classification of island dynamics, including better understanding of the history of isolation of the atypically old, continental fragment islands.

Oceanic islands are concentrated in a suite of distinctive inter- and intraplate settings and have commonly experienced a dynamic history involving varying degrees of lateral and vertical displacement, the latter confounded by (but often in excess of) eustatic sea-level changes. Simplifying, for islands following classic hotspot dynamics, we may identify a general ontogeny from steep, high active volcanos in youth, through less active, dissected terrain during intermediate ages, to subsiding, eroded, low-relief old age. In the tropics and subtropics, upward growth of reef-forming corals at times of relative or actual subsidence has led to the formation of numerous islands of only a few metres elevation, representing an extended old-age stage for many old, subsided volcanic islands. Alongside the hotspot volcanos and atolls, there are several major forms of plate boundary situations in which islands are created and they each generate their own dynamics of island and archipelago formation. We illustrate something of the complexity that can be involved by summarizing the developmental history of the Canarian and Hawaiian archipelagos and the Caribbean region.

CHAPTER 3

Island environments

3.1 Varied platforms

The benefits of viewing islands as model systems may sometimes tempt us in the direction of over-simplifying island environments. In this chapter we attempt to tread a path between generalizing environmental characteristics of islands and describing some of the variation in form that can be encountered. The greatest degree of insular distinctiveness is typically to be found in off-shelf archipelagos, the origins of which were described in Chapter 2. Focusing on island functional attributes rather than origins, Nunn et al. (2016) have helpfully classified 1779 Pacific islands into eight main types based on their prevailing lithology and elevation (Fig. 3.1). These types were continental islands, reefs, plus high and low islands characterized, by, in turn, each of volcanic, limestone, and composite lithologies. Such classifications may be helpful from the perspective of ecological analysis and environmental management as they group together functionally similar systems with similar ecological properties, climate variation, natural hazard profiles, etc. Across the vastness of the Pacific Basin, just 18 of the islands were classed as continental in type and low-lying composite islands were equally rare, whereas a third were reefs and another third were high volcanic islands. Atolls and high volcanic islands are prominent in terms of occupation by human societies and of their place in island biogeographical studies (e.g. Keast and Miller 1996; Dickinson 2009). Hence the attention paid to their geological dynamics in Chapter 2. However, we must also consider continental shelf islands, which are also extremely numerous and biogeographically informative (e.g. Itescu et al. 2019; Hammoud et al. 2021).

3.2 Topographic characteristics

The British Isles exemplify continental islands as essentially small offcuts of continent, possessing a mix of lithologies and with varied but often subdued topographic properties. Oceanic islands, being of volcanic origin, are more distinctive. In youth, they tend to develop great height and steep slopes. The precise form depends on the nature of the volcanism and its duration, but typically hotspot islands become increasingly inactive with age and, in their landform processes, dominated by erosion, leading to the development of highly dissected forms and eventually, far lower relief (Fig. 2.7). Height of island can be important in relation to changes in sea level, receipt of rainfall, presence of surface water, and other climatic characteristics (Craig 2003; Weigelt et al. 2013). In addition, large, high volcanic islands are subject to periodic catastrophic landslips, which can radically reshape substantial areas of the island (Hürlimann et al. 2004; below).

Topographic characteristics for a selection of islands in the New Zealand area were summarized by Mielke (1989), who noted that, while the largest islands have the highest peaks, smaller islands display variable area–elevation relationships. Coral or limestone islands and atolls tend to be very low-lying and flat. Those which have been uplifted to more than a few metres above sea level are termed makatea islands. Examples include Makatea itself (Tuamotu Archipelago), Atiu (Cook Islands), and most of the inhabited islands of Tonga (Nunn 1994; Nunn et al. 2016). They are characterized by rocky coralline substrates, but some are partly volcanic. Many feature commercially exploitable deposits

Island Biogeography. Robert J. Whittaker, José María Fernández-Palacios, and Thomas J. Matthews, Oxford University Press.
© Robert J. Whittaker, José María Fernández-Palacios, and Thomas J. Matthews (2023). DOI: 10.1093/oso/9780198868569.003.0003

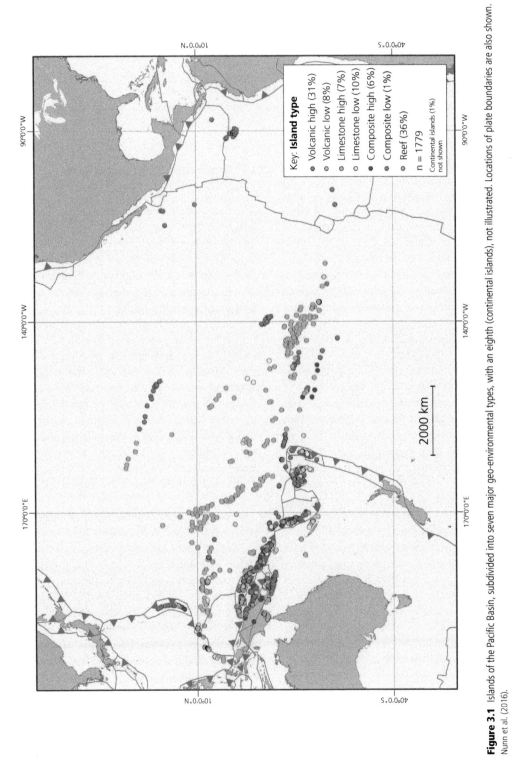

Figure 3.1 Islands of the Pacific Basin, subdivided into seven major geo-environmental types, with an eighth (continental islands), not illustrated. Locations of plate boundaries are also shown. Nunn et al. (2016).

of phosphate-rich seabird droppings (guano) (e.g. Nauru; see Chapter 15).

3.3 Edaphic characteristics

The edaphic (soil) properties of volcanic islands depend upon their climatic and geological setting, the nature of the volcanism involved, and the geochemistry and physical form of the initial substrates. In illustration, as a consequence of the sterilizing eruptions of 1883, the Krakatau islands were mantled in thick deposits of fine-textured pyroclastic material that provided an unconsolidated, fertile material for plant growth, essentially requiring just the fixing of atmospheric nitrogen to enable the rapid vegetation succession observed (Schlesinger et al. 1998). However, solid lava flows can be observed in islands such as Tenerife (Canaries) and Hawaii that take decades to develop a vegetation cover, with windblown material gathering in cracks and depressions often key to the initiation of pedogenesis.

Ibáñez and Effland (2011) demonstrate that just as island biodiversity scales with island area (Chapter 5), so too does the diversity of soil types, which tend to assemble in predictable order with increasing range of elevation, habitat diversity, and age range, while retaining a strong overall signal of the tectonic setting. Long-term change in soil properties with increasing substrate age has been most clearly shown for the Hawaiian Islands, where biological processes and weathering produce the most fertile soils in intermediate-aged sites, with fertility declining through further weathering and leaching on ancient (> 2 Ma) lava flows (Box 3.1; Chadwick et al. 1999).

Island soils and communities may also receive significant external inputs of nutrients, constituting external subsidies, either marine or terrestrial in origin. One classic example of a marine subsidy is provided by the volcanic island of Surtsey, which formed in 1963 off the south coast of Iceland. In these climatically challenging conditions, ecosystem development has been extremely slow, except in the area of a seagull colony, which first established in 1986, gradually growing to around 300 pairs by 2003. The gull colony had a big impact in terms of nutrient input, especially contributing

to soil nitrogen, likely also contributing to seed transport, and in sum greatly advancing the development of soil fertility, plant and animal biomass, and diversity in the vicinity of the colony (Frontispiece to Part I; see also Section 6.8; Magnússon et al. 2009, 2020). Grey seals using the island to breed, also contribute to the marine subsidy of terrestrial ecosystems on Surtsey. The role of seabirds in transferring nutrients on to small, otherwise extremely nutrient-limited atolls can be hugely important and may well play a key role in ecosystem function of small islands (Box 6.2; Graham et al. 2018; Obrist et al. 2020). Exemplifying terrestrial subsidies, wind transport of aerosols and dust provides another important input to many island ecosystems. In Box 3.1, we mention how the Canary Islands (96–445 km from Africa) regularly receive inputs of windblown dust, some of it rich in organic matter and originating from extensive, dried North African lakebeds (Kis and Schweitzer 2010). Even in Hawaii, dust from central Asia (> 6000 km distant) provides an important source of phosphorus to soils while marine aerosols also contribute a source of cations (Chadwick et al. 1999).

> ### Box 3.1 Space for time substitution: the example of island soil chronosequences
>
> In oceanic archipelagos, the active volcanism of the islands continually recreates similar starting conditions over intervals of hundreds through to millions of years. As we cannot monitor change over such durations, biologists often use the chronosequence approach of substituting space for time: taking a set of sites (or whole islands) of similar starting point but different age and analysing the data as though looking at a temporal sequence.
>
> In many environments, phosphorus and/or nitrogen are key limiting nutrients. In young volcanic soils, apatite within the parent material provides the initial store of phosphorus, which then becomes progressively unavailable as the soil ages by: (i) its incorporation into biomass; (ii) soil occlusion as insoluble Fe and Al phosphates; and (iii) leaching from the system (Walker and Syers 1976). New volcanic soils lack nitrogen, which has to be accumulated through fixation, or by accumulation through organic matter incursion (aerial supply of dead insects,

etc.). Hence, it is expected that in soils derived from volcanic ejecta, nitrogen will initially be limiting, but over time useable nitrogen supply increases while phosphorus declines so as to become limiting.

These ideas are supported by a soil chronosequence study from *Metrosideros polymorpha* forests in the Hawaiian Islands, comprising soil ages extending from Kilauea volcano, Big Island (c. 40 yrs) up to sites on Kauai (c. 4 Ma) (Vitousek, 2004). The maximum availability of phosphorus peaked between 0.1 and 1 Myr. A more recent study in the Canaries (Gallardo et al. 2018), of a chronosequence spanning from Teneguía volcano, La Palma (40 yrs) to Tamadaba massif, Gran Canaria (11 Ma), within Canarian pine (*Pinus canariensis*) forest, produced less straightforward results, possibly because of the impact of phosphorus-rich dust deposition from the Sahara. The origin of this dust is through post-glacial drying of former lake beds in North Africa. These areas have provided a consistent wind-borne supply of phosphorus to the Canary Islands over the last several thousand years. This has counteracted the expected decrease in phosphorus expected according to the Walker and Syers (1976) model.

3.4 Climatic characteristics

Island climates have, inevitably, a strong oceanic influence, and quite often are considered anomalous for their latitude, as a consequence of their location in the path of major ocean- or atmospheric-current systems (e.g. the Galápagos; Darwin 1839). Weigelt et al. (2013) provide systematic comparison of insular and mainland climates demonstrating that islands are typically low lying, slightly cooler than continents, with less variation in annual temperature, less seasonality in precipitation, and greater exposure to past temperature change (Fig. 3.2a). However, these differences reflect the inclusion of a large number of high-latitude islands, which accounts for the lower average temperature and to some degree the apparent exposure to past climate change. In practice, insular endemics on high-elevation islands in the tropics and subtropics will often have been able to migrate vertically to mitigate impacts of global climate change. Broadly speaking, in terms of contemporary climate, islands are representative of the range of global climate, albeit particular categories of island have differing distributional concentrations in an ordination

of climate indices (Fig. 3.2b–e). Notwithstanding their generally dampened temperature variation, many islands, especially in the tropics and subtropics, experience quite large interannual variations in precipitation. For example, variability in rainfall linked to the El Niño–Southern Oscillation (ENSO), can be of real ecological importance to islands such as the Galápagos, while the Greater Antilles and Philippines exemplify archipelagos regularly lashed by extreme weather events in the form of hurricanes/typhoons (below) (Stoddart and Walsh 1992).

Low islands tend to have relatively dry climates. High islands tend to generate heavy rainfall (with accompanying river systems), although they may also have extensive dry areas in the rain shadows, providing a considerable environmental range in a relatively small space. Even an island of only moderate height, such as Christmas Island (Indian Ocean), with a peak of 360 m and general plateau elevation of only 150–250 m, benefits from orographic rainfall, allowing rainforest to be sustained through a pronounced dry season (Renvoize 1979). Islands may also be expected to receive rainfall of a generally different chemical content than experienced over continental interiors (Waterloo et al. 1997).

The range of climatic conditions within an island is determined largely by the elevation of the highest mountain peaks in combination with its geographical position. Tenerife is an excellent example of a high volcanic island with a steep central ridge system and a pronounced rain shadow. The island experiences a Mediterranean-type climate of warm, dry summers and mild, wet winters, despite its latitudinal proximity to the Sahara. The Canaries lie on the subsiding eastern side of the semi-permanent Azores anticyclone at about 28°N. The subsidence produces a warm, dry atmosphere aloft, separated at 1500–1800 m from a lower layer of moist, southward-streaming air (the north-east trade winds), producing a mid-elevation temperature **(trade-wind) inversion** (Gillespie and Clague 2009). The mountain backbone forces the lower layer of air to rise until trapped by the descending dry air mass. It cools, and this produces a 300–500-m-thick band of orographic clouds, the *mar de nubes*, between about 600 and 1500 m, on the windward slope of

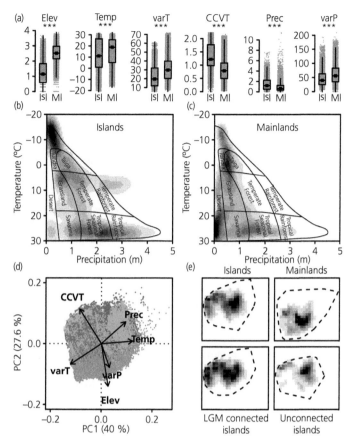

Figure 3.2 Climate properties based on 17,883 islands > 1 km² and 42,984 equal area mainland grid cells. (a) Boxplots (median and interquartile ranges), for islands (Isl) and mainlands (Ml), from left to right, for elevation (log 10 m), annual mean temperature (°C), annual temperature range (in °C), past temperature change velocity (CCVT, rate of movement in log 10 m/year needed to track temperature change since the LGM), annual precipitation (m), and variation in monthly precipitation. (b, c) Density of (b) island- and (c) mainland-grid cells in relation to temperature, precipitation, and biomes. (d) Distributions of mainland cells (grey), continental-shelf islands (magenta), and oceanic islands (cyan) in a two-dimensional ordination (Principal Components Analysis) of the variables shown in the boxplots. (e) The density of different categories of island and of mainland cells in the PCA space.
Figure 3 from Weigelt et al. (2013).

the island. This 'sea of clouds' builds up more or less daily, providing the forests with the moisture surplus (through fog drip and reduced evapotranspiration) needed to overcome the dry Canarian summers. The southern sector, in the rain shadow, experiences less precipitation and differential rates of air mass cooling, such that the vegetation is generally more xerophytic and both lower and upper limits of forest growth are higher on the southern side (Fernández-Palacios and de Nicolás 1995).

Contraction, or 'telescoping', of elevational zones is typical of smaller islands, especially in the tropics,

potentially increasing the number of major habitat types and therefore species that may be supported. For example, on Krakatau, Indonesia, the plentiful atmospheric moisture and the cooling influence of the sea result in the near permanent presence of cloud in the upper parts, further lowering temperatures and allowing the development of a montane mossy forest, rich in epiphytes, at around 600 m above sea level, a much lower elevation than the continental norm of 1200–2000 m (Whittaker et al. 1989). In his review of upper limits of forests on tropical and warm-temperate oceanic islands,

Leuschner (1996) cited a range of 1000–2000 m for the lowering of the forest line compared with similar latitudes within continental areas. Factors that are involved include steepened lapse rates (and associated trends in atmospheric humidity levels) and droughts on trade-wind-exposed island peaks above temperature inversions, such as the case with the upper parts of Tenerife. However, factors other than climate can be important in relation to elevational zonation patterns. For instance, some remote islands are also lacking in high-elevation tree cover because of the youth of their high-elevation habitats and because they lack well-adapted, high-elevation tree species. Island summits tend to provide rather ephemeral habitats, geographically far distant from source pools, and which soon disappear through erosion and subsidence, often failing to accrue a full complement of species before doing so (Fernández-Palacios et al. 2014). In addition, human settlers may act to deforest island environments, sometimes thereby reducing tree lines, although the opposite is also the case, as the introduction of non-native species may lead to an increase in the tree line elevation.

3.5 Water resources

Water availability shapes the ecology and human use of islands (Whitehead and Jones 1969; Ecker 1976; Menard 1986; Dickinson 2009). Most oceanic islands, whether they are high volcanos or atolls, contain large reservoirs of fresh water. Fresh lava flows are highly permeable, but over time the permeability and porosity of the rock decreases due to weathering and subsurface depositional processes. The residence time of groundwater in the fractured aquifers of large volcanic islands spans decades to centuries. Hydrologically, we may divide these systems into two zones: the **vadose zone** and the **saturated** or **basal water zone**. The vadose zone contains groundwater compartments, often linked into chains, interspersed with dry zones. The basal zone is characterized by many closely placed groundwater compartments and by a high percentage of saturated secondary fractures. Both fresh water and seawater occur in this zone, with tides influencing the water level up to 4–5 km from the coast (Ecker 1976). In the absence of rain, the seawater within islands would be at sea level, but rainwater percolates through to float on the denser saline water permeating the base of the island, forming a **Ghyben–Herzberg lens** (Menard 1986).

As indicated above, annual rainfall varies greatly within the island of Tenerife, from > 800 mm in the highest parts of the Anaga peninsula (a sum that might be doubled if fog drip were considered), to < 100 mm in the extreme south of the island. These gradients have profound effects on the ecology and human use of the island. The steep north side, which is wetter, has traditionally been intensively cultivated, whereas much of the flatter east and west coastal zones is essentially useless for cultivation without irrigation, but in recent decades has been favoured for a rampant sun-seeking tourist industry. Thus, even on this large, high island, groundwater (which increasingly has been tapped from aquifers deep in the volcano) is the most important source of water for the human inhabitants, and water constitutes a key limiting resource for development (Fernández-Palacios et al. 2004). Even low-lying atolls maintain a freshwater lens but those of less than around 10 ha can lack a permanent lens. Such habitats are liable to be hostile to plants other than halophytes specialized in strandline habitats, thus limiting the variety of plant species that can survive on them (Whitehead and Jones 1969).

3.6 Isolation: tracks in the ocean

The geographical isolation of any island determines its relationship with the rest of the biosphere. In most island biogeographical analyses, isolation is measured simply as distance to mainland but in practice the biological isolation of an island is determined by: (i) the disposition of other islands and archipelagos, (ii) atmospheric and ocean surface currents (Fig. 3.3), (iii) distance to habitat-specific source pools, and (iv) position in relation to migration routes (e.g. Weigelt and Kreft 2013). Many species of migratory birds follow distinct corridors involving regular stopovers on particular islands, as notably the case for many Mediterranean islands but also true of more isolated archipelagos. One passerine, the bobolink *Dolichonyx oryzivorus* regularly passes through the Galápagos on migration,

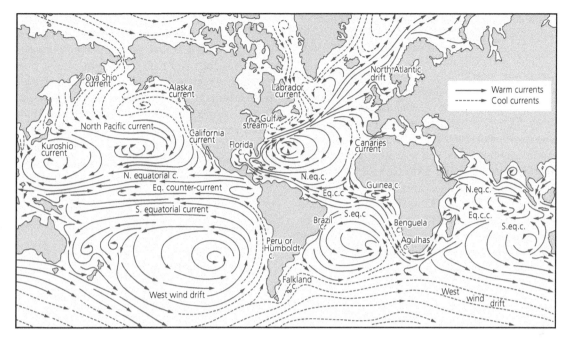

Figure 3.3 Surface drifts and ocean currents in January.
Redrawn from Nunn (1994), Fig. 4.

where it may be responsible for the introduction of *Plasmodium* bird parasites (Levin et al. 2013).

The major ocean- and atmospheric-current systems (Fig. 3.3) vary both seasonally and inter-annually, but in some areas are strongly directional and persistent. Indeed, current systems sometimes provide a parsimonious explanation of biogeographical patterns (Cook and Crisp 2005). The 62 km² island of Mona is equidistant between Hispaniola and Puerto Rico. Its 46 species of butterflies feature nine subspecies in common with Puerto Rico and none with the larger Hispaniola, whereas the ratio of source island areas is 9:1 in favour of Hispaniola. The explanation appears to lie in a remarkably constant bias in wind direction from Puerto Rico towards Hispaniola (Spencer-Smith et al. 1988). However, over thousands and millions of years, significant changes in ocean- and atmospheric-circulation patterns can occur, driven by changing climate systems and land/sea configuration (below; Vitousek et al. 1995; Gillespie and Clague 2009). Such past dynamics may be crucial to understanding of contemporary biogeography but

are typically too poorly known to be formally incorporated into statistical analyses. For instance, it is suspected that during Pleistocene glacial maxima, the polar front pushed the Hadley cell (responsible for the trade winds) equator-wards, creating the conditions for Westerlies, which usually blow at 40° latitude, to blow at 30°, opening west–east dispersal windows at these latitudes and perhaps contributing to the instances of reverse colonization ('boomerang events') between Macaronesia and Africa (e.g. Sun et al. 2016).

3.7 Insular disturbance regimes

As is evident from the often catastrophic impacts of certain irregular climatic phenomena on oceanic island landscapes, the degree to which particular environments can accommodate the impacts of particular uncommon or extreme events without fundamental alteration, while variable, is generally low. For those oceanic island landscapes in a state of dynamic equilibrium, the effects of such events may be to cause a threshold of landscape development to be crossed … The effects of irregular climatic phenomena are so variable, so site-specific, that it is pointless to attempt a broad generalization.

(Nunn 1994, pp. 159–160)

Islands are particularly dynamic platforms, so much so that it is challenging to determine the relative importance of all the differing drivers (Table 3.1). Over long spans of time—thousands of years or more—any island will experience substantial environmental change, often gradual but profound in impact. But disturbance events (more or less discrete events that cause a pulse of mortality and open up space for other individuals/species to occupy) right across the magnitude/frequency spectrum also hold significance for understanding ecological and evolutionary dynamics (Fig. 3.4; Pickett and White 1985). Often, a single event, such

as a hurricane (tropical cyclone), will impact on both mainland and island systems; however, because of the geographical and geological idiosyncrasies of small islands and their location in oceans, their disturbance regimes when scaled to island size are frequently atypical of continental landmasses (Lugo 1988). Examples ordered by increasing magnitude would include: (i) pronounced droughts, (ii) hurricanes, (iii) individual volcanic eruptions, and (iv) partial sterilization in large-scale eruptions, and/or flank collapses (mega-landslides) of oceanic islands.

Extreme weather events

Between 1871 and 1964, an average of 4.6 hurricanes per year were recorded in the Caribbean, with a

Table 3.1 Summary of major geo-environmental factors affecting islands, with crude indication of their relative importance (modified from Fernández-Palacios et al. 2016a). Glacial cycles encompass climate, current system, and sea-level change.

Island type	Plate tectonics (100–1 Myr)	Island ontogeny (10–1 Myr)	Glacial cycles (100–10 kyr)	Volcanic activity (10–0.01 kyr)
Continental-shelf	–	–	++	–
Continental fragments	++	–	+	–
Oceanic (volcanic)	+	++	+	++
Oceanic (atolls)	–	+	++	–

–, relatively unaffected; +, moderately affected; ++ considerably affected

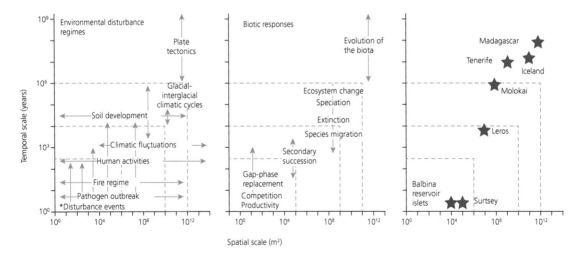

Figure 3.4 Environmental disturbance regimes and biotic responses, viewed in the context of four space–time domains (shown here bounded by dashed lines), named micro-, meso-, macro-, and mega-scales by the scheme's authors. The third panel simply exemplifies the size and age of a small selection of islands.

*Disturbance events such as e.g. storm damage, earthquakes.

Partially based on a figure in Delcourt and Delcourt (1991).

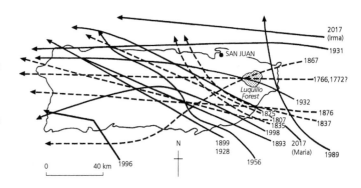

Figure 3.5 Depiction of the many hurricane tracks that have crossed directly across Puerto Rico since around AD 1700. Hurricane Hugo (1989) caused great damage to the El Yunque National Forest (Luquillo forest), where the last wild population of the Puerto Rican parrot was seriously impacted. Most recently, the combination of Hurricane Irma and the even more devastating Maria, both in 2017, had a massive ecological and economic impact on the island. Updated from Scatena and Larsen (1991) based on various sources.

mean return period for the island of Puerto Rico of 21 years (Walker et al. 1991; Fig. 3.5). With wind speeds > 120 km/h and paths tens of kilometres wide, hurricanes can have a profound impact, fundamental to an understanding of the structure of natural ecosystems in the region. A well-studied example is Hurricane Hugo, which in 1989 caused complete defoliation of a large part of the El Yunque National Forest (Luquillo forest)—the only substantial remaining area of rain forest on Puerto Rico. Despite a rapid 'greening-up' of the forest, the signature of such an event will be evident in the unfolding vegetation mosaic for many decades (Walker et al. 1996). The forest was devastated once again by Hurricane Maria in September 2017. Moreover, the Caribbean by no means corners the market in tropical storms. Hurricanes (also known as typhoons) develop in all tropical oceanic areas where sea surface temperatures exceed 27–28°C, although they are generally absent within 5° north and south of the equator, where persistent high pressure limits their development (Nunn 1994). Wind defoliation and large blow-downs are important and frequent disturbances for forested tropical islands throughout the hurricane belts, 10–20° north and south of the equator. They may have particularly destructive impacts on high islands, an example being the devastation of large areas of forest on the Samoan island of Upolu by tropical cyclones Ofa, Val, and Lyn in 1990, 1991, and 1993, respectively (Elmqvist et al. 1994). Although storm damage to lower islands can also be severe, storms can occasionally be important to island growth by throwing up rubble ramparts (Stoddart and Walsh 1992). Thus, some very small islands, such as the so called *motu* (sand islands), may be in a kind

of dynamic equilibrium with both extreme and normal climatic phenomena (Nunn 1994).

A comprehensive analysis of disturbance regimes requires quantification of both magnitude and frequency of events (Stoddart and Walsh 1992). Lugo (1988) provided a synthesis of Caribbean disturbance phenomena, concluding that whereas hurricanes have the highest frequency of recurrence among his major 'stressors', the susceptibility of Caribbean islands to hurricanes is intermediate, partly because they do not change the base energy signature over extensive periods. Also, with the largest Caribbean islands, hurricane damage rarely impacts catastrophically on the ecology of the entire island area. Given their recurrence interval, these island ecosystems should be evolutionarily conditioned by such storm events (Walker et al. 1996). Recent years have, however, seen some particularly devastating Caribbean storm systems, notably including Hurricanes Irma and Maria in August and September 2017, respectively. Such storms can have massive impacts on human infrastructure, health, and livelihoods, with the average hurricane strike having an estimated regional economic impact (from data for Central America and Caribbean regions) of 0.83 percentage points (Strobl 2012).

Not all extreme weather phenomena take the form of storms. ENSO phenomena are large-scale interannual events that are the product of variations in air pressure and associated rainfall patterns between the Indonesian and South Pacific regions, coupled with ocean-current and temperature variation. In the Asia–Pacific region, ENSO events are associated with heightened amplitude of interannual climatic variation—that is, more intense droughts and wet periods. As an illustration, the island of Aitutaki (Cook Islands) received 3258 mm rainfall in December 1937, which is in excess of the average annual total (Nunn 1994). Such events have detectable impacts even on ecosystems that experience them regularly, driving pulses of mortality and/or of increase in population size (Gibbs and Grant 1987; Grant and Grant 1996a). Analyses of historical data have shown that the frequency of climatic phenomena such as hurricanes and ENSO events varies over decadal timescales and there are concerns that 21st-century climate change will see increased incidence of such extreme weather phenomena in particular regions.

Disturbance from volcanism and mega-landslides

Volcanic eruptions are the main constructive forces of oceanic islands: they bring them into existence, and subsequently make them larger and higher, whereas the processes of erosion (by rain, wind, and sea), subsidence, and collapse provide the opposing forces that reduce them to sea level or below (e.g. Figures 2.8, 2.9). The same volcanic processes that enlarge islands simultaneously damage or destroy island ecosystems. The impacts can be broadly distributed and diffuse, as when volcanic ash is deposited in thin layers over a wide area, or can be intense and devastatingly destructive (e.g. Partomihardjo et al. 1992; Hilton et al. 2003). In some cases, destructive volcanic ash flows can be deposited over the entire island, more or less completely eliminating the existing biota. This is known as **island sterilization**, and implies that colonization has to begin from nothing. It can be very difficult to demonstrate complete sterilization, even for historical events such as the 1883 Krakatau event (Whittaker et al. 1989; Thornton 1996). The challenge is even greater for events as distant as the Roque Nublo eruptive period on Gran Canaria (c. 5.3–3.7 Ma). Although the volcanism was protracted and at times violent, involving large eruptions and collapse events, wholesale sterilization now appears improbable (Anderson et al. 2009). Santorini in the Aegean Sea, the collapse of which in 1628 BC may have destroyed the Minoan culture on Crete, or,

more recently, Martinique and Montserrat in the Caribbean, provide further examples of the destructiveness of island volcanos. Indeed, the eruptions of the Soufrière Hills volcano Montserrat in 1995–1998, after a lengthy period of dormancy, provide a classic illustration of the extensive cross-cutting impact that volcanic action can have in transforming the physiography, ecology, and human use of islands (Hilton et al. 2003; Le Friant et al. 2004). More than half the island was rendered uninhabitable, and two-thirds of the human population left the island. The volcano remained active after the 1995–1998 eruptions, with a further partial dome collapse occurring in 2010 (Burns et al. 2017).

Volcanic islands are built in a variety of distinct tectonic settings and remain active over varying lengths of time, often spanning many millions of years (Chapter 2). A corollary of this is that a number of major types of volcanic eruption can be identified. Indeed, any particular island may exhibit variation in the nature of volcanism over time. Often, radically different forms of volcanism occur in rapid order at times of peak activity. In increasing degrees of explosiveness (Decker and Decker 1991) the major forms are as follows:

- **Icelandic eruptions** are fluid outpourings from lengthy fissures, and they build flat plateaux of lava, such as typify much of Iceland itself.
- **Hawaiian eruptions** are similar, but occur more as summit eruptions than as rift eruptions, thereby building **shield volcanos**.
- **Strombolian eruptions** take their name from a small island off Sicily that produces small explosions of bursting gas that throw clots of incandescent lava into the air.
- **Vulcanian eruptions**, named after the nearby island of Vulcano, involve the output of dark ash clouds preceding the extrusion of viscous lava flows, thus building a **stratovolcano** or **composite cone**.
- **Peléan eruptions** produce pyroclastic flows termed **nuées ardentes**, high-speed avalanches of hot ash mobilized by expanding gases and travelling at speeds in excess of 100 km/h.
- **Plinian eruptions** are extremely explosive, involving the sustained projection of volcanic ash into a high cloud. They can be so violent,

and involve so much movement of magma from beneath the volcano, that the summit area collapses, forming a great circular basin, termed a **caldera**.

Like all generalizations, this classification is over-simplistic; for example, flank collapse events appear to be the principal cause of many caldera collapse events (Hunt et al. 2018). However, it serves to illustrate that there are great differences in the nature of volcanism, both between islands and within a single island over time.

Even fairly small volcanic eruptions can have very important consequences for island ecosystems. For instance, eruptions on Tristan de Cunha in 1961 covered only a few hectares in ejecta, but toxic gases affected one-quarter of the island area. The evacuation of the human population left behind uncontrolled domestic and feral animals, which transformed the effects of the eruption and produced a lasting ecological impact (Stoddart and Walsh 1992). The eruptive action within Hawaii over the past few centuries has been both more consistent and less ecologically destructive than that of Krakatau, which in 1883 not only sterilized itself but also, through caldera collapse, created a tsunami killing an estimated 36,000 inhabitants of the coastlines of Java and Sumatra. This death toll was tragically surpassed by an order of magnitude on 26 December 2004, by a tsunami that swept across island and mainland coastal settlements in an arc from north Sumatra, in this case generated not by caldera collapse but by submarine faulting.

Over their life span, oceanic islands experience alternating phases of construction and downwasting, and to varying degrees subsidence and uplift (Ramalho et al. 2017; Stuessy et al. 2022). The progressive accumulation of volcanic material upon young oceanic islands tends to build up relatively steep slopes within a short geological time span, and at a faster rate than 'normal' subaerial erosion processes can bring into dynamic equilibrium (cf. Le Friant et al. 2004). Exceeding critical slope values produces gravitational instabilities that contribute to the collapse of the slopes through debris avalanches that transfer hundreds of cubic kilometres of debris into the sea (Carracedo and Tilling 2003; Whelan and Kelletat 2003)

Hypothesized in 1962 by Telesforo Bravo, to explain the origin of the Las Cañadas caldera in Tenerife, our knowledge of catastrophic landslides has recently developed apace (Fig. 3.6), aided by three-dimensional bathymetric analysis of the ocean floor. These analyses have led to the discovery of the blocks and debris avalanches resulting from the flank collapses, often extending over areas of hundreds of square kilometres around the islands (Table 3.2). Extensive deposits in the sea around oceanic islands can also be produced by pyroclastic flows, or indeed by a combination of landslips and volcanic ejecta. Flank collapses are now well documented in numerous volcanic archipelagos, including Hawaii, French Polynesia, Mauritius, the Lesser Antilles, Azores, Cabo Verde, and the Canaries. Many of these events have been catastrophic, high-magnitude events, often producing tsunami deposits on the flanks of other islands in their path (Costa et al. 2015; Hunt et al. 2018; Paris et al. 2018). Such collapses may occur unexpectedly and suddenly and may lead to the disappearance of a significant part of an island, as much as a quarter of it, in just minutes (Carracedo and Tilling 2003), with debris reaching velocities over 100 km/h and tsunami waves striking far inland on neighbouring islands.

Major volcanic slope failures are fairly common globally, occurring at least four times a century and they are a particular feature of volcanic islands in excess of 1000 or 1500 m height (Hürlimann et al. 2004). As many as 20 mega-landslides, with volumes varying between 30 and 5000 km^3, have been recognized for both the Canaries (Canals et al. 2000) and for Hawaii (Fig. 3.6; Moore et al. 1994). These events occur reiteratively in actively forming islands. For example, El Hierro, the youngest of the Canaries, with an age only slightly greater than 1 million years, has already suffered four such catastrophic events, the last of them (El Golfo landslide *c.* 15 ka), carried more than half the island into the sea (Carracedo et al. 1999). Similarly, the La Orotava valley slide on Tenerife is estimated to have been up to 1000 m in thickness, and created an amphitheatre up to 10 km wide and 14 km long and an offshore prolongation of *c.* 100 km in length (Hürlimann et al. 2004). The frequency of recurrence of large landslides on the Canaries has been about

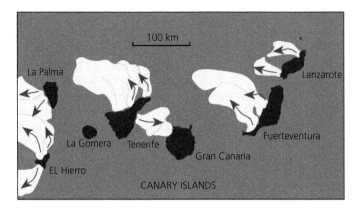

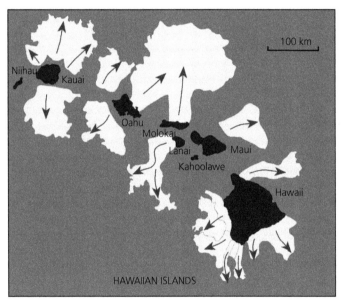

Figure 3.6 Mega-landslides or flank collapses are common over geological time within hotspot archipelagos. The figure shows the approximate extent of mega-landslide deposits in two oceanic island archipelagos.
Redrawn from Carracedo and Tilling (2003) their Figure 2.18.

one per 100,000 years across the whole archipelago, or once every 300,000 years for each single island (Masson et al. 2002).

Such landslides can modify the subsidence regime of the affected islands, as well as promoting new constructive volcanic activity (Ancochea et al. 1994). When not eroded or buried by the emission of new volcanic material, the scars created by such collapses are easily recognized. They include calderas (Taburiente in La Palma or Las Cañadas in Tenerife), valleys with knife-cut cliffs—called *laderas* in Spanish—delimiting the collapse-formed valleys (Laderas de Güímar or Ladera de Tigaiga, in Tenerife) or concave cliff-arches such as those in the Anaga (Tenerife), Tamadaba (Gran Canaria), or Jandía (Fuerteventura) massifs.

The sudden collapse of volcanic island slopes and the consequent debris avalanches have the potential to generate tsunamis of such extreme magnitude (mega-tsunamis), that they not only affect nearby islands (Paris et al. 2018) but also impact far-distant landmasses. For example, it is estimated that the El Golfo collapse (El Hierro) produced a mass movement with a volume of *c*. 400 km³, generating a

Table 3.2 Examples of large debris-avalanche landslides occurring around oceanic islands (slightly modified from Canals et al. 2000 and Whelan and Kelletat 2003).

Island	Landslide name	Vertical distance moved (m)	Seafloor area covered (km^2)	Volume (km^3)	Event age (ka)
Réunion	Ralé-Poussée	1700	200	30	4.2
Réunion	Eastern Plateau	3000	–	500	15–60
					100–130
Fogo (C. Verde)		1200	–	100	> 10
El Hierro	El Golfo	5000	1500	400	15–20
					& 100–130
El Hierro	El Julan	4600	1800	130	> 160
El Hierro	Las Playas II	4500	950	< 50	145–176
La Palma	Cumbre Nueva	6000	780	200	125–536
Tenerife	Güímar	4000	1600	120	780–840
Tenerife	La Orotava (Icod)		2100	< 500	540–690
Tenerife	Las Cañadas	> 4000	1700	150	150–170
Kauai	South Kauai	> 4000	6800	–	> 13
Kauai	North Kauai	> 4000	14000	–	–
Oahu	Nuuanu	5000	23000	5000	–
Hawaii	Pololu	> 4000	3500	–	370
Hawaii	Alika 1–3	4800	4000	2000	247

mega-tsunami that lifted boulders weighing more than 1000 t to elevations of 11 m on the Bahamas Archipelago, more than 3000 km westwards (Whelan and Kelletat 2003). Ward and Day (2001) have highlighted the great instability of the Cumbre Vieja massif, the active southern summit of La Palma, where seven confirmed volcanic eruptions have occurred since the start of the 16th century, the most recent, Volcán de Tajogaite, in 2021. It has been predicted that Cumbre Vieja will collapse to the west in the coming millennia, generating a mega-tsunami likely to devastate the east coast of North America on the far side of the Atlantic. In addition to subaerial flank collapses, a further hidden threat is submarine collapses, which may also occur during island building (Hunt and Jarvis 2017).

3.8 Quaternary climate change on islands

In the case of these islands we see the importance of taking account of past conditions of sea and land and past changes of climate, in order to explain the relations of the peculiar or endemic species of their fauna and flora.

(Wallace 1902, p. 291)

Short-term variations in climate tend to have lower amplitude than long-term variation; that is, the variance within decades is less than the variance within centuries, which is again less than the variance within millennia (Williamson 1981). Given the comparative youth of the vast majority of islands, the most relevant large-scale climatic shifts are those of the glacial–interglacial cycles of the Quaternary period (comprising the Pleistocene and Holocene), which commenced 2.588 Ma (Lomolino et al. 2017). For much of the last glacial period, for instance, southern parts of the British Isles resembled arctic tundra, while the northern and western regions were ice covered. Subpolar islands such as the Aleutians (north Pacific) and Marion Islands (southwest Indian Ocean) also supported extensive icecaps at the Last Glacial Maximum (LGM), and there is evidence that the Pleistocene cold phases caused extinction of plant species on remote high-latitude islands, such as the sub-Antarctic Kerguelen (Moore 1979). The classical model of four major Pleistocene glaciations has long been replaced by an appreciation that there have been multiple changes between glacial and interglacial conditions over the Quaternary (Lomolino et al. 2017). Although high-latitude islands have been the most dramatically affected by these climatic oscillations, even low-latitude oceanic islands were affected by these events, both through climate and linked sea-level changes (Douville et al. 2010; Fernández-Palacios et al. 2016a; Caujapé-Castells et al. 2022).

Biogeographers and palaeoecologists have long been interested in the impacts of these events and in particular whether they have driven variations in colonization, speciation, and extinction rates on islands (Caujapé-Castells et al. 2022). Within large landmasses one important way in which species respond to high-amplitude climate change is by range displacement. The possibilities for this mode of response within isolated oceanic islands are of course limited in terms of horizontal shifts, with the corollary that persistent island endemic forms must have found habitable environments within the limited confines of their island(s) through elevational shifts, while less broad-niched or adaptable forms have gone extinct. In general, the impact of Quaternary climate change on the key biogeographical rates must depend on features such as the position of the island in relation to global climate and current systems through time, the elevational (and latitudinal) range and topographic complexity of the island, and the extent to which the change in climate was accompanied by expansion or contraction of the island's area (McGlone 2002; Fernández-Palacios et al. 2016a; Weigelt et al. 2016; Norder et al. 2019). High-latitude islands were particularly impacted by Ice Age cooling, as evidenced by a general paucity of endemics (Sadler, 1999), but the notion that mid-latitude, subtropical, or tropical oceans buffered the effects of glacial-era climate excursions is challenged by evidence from such islands as Tahiti, Rapa Nui (Easter Island), Galápagos Islands, and the Canaries demonstrating significant ecological change, in cases including the apparent disappearance of whole vegetation types (Margalef et al. 2013; Prebble et al. 2016; Castilla-Beltrán et al. 2021).

The Galápagos Islands are desert-like in their lower regions, with moist forests in the highlands, but palaeoenvironmental data from lacustrine sediments demonstrate that the highlands were dry during the last glacial period. The moist conditions returned to the highlands c. 10,000 ka, although the palaeo-pollen data indicate a lag of some 500–1000 years before vegetation similar to that of the present day occupied the higher elevations (Colinvaux 1972). This delay may reflect the slow progress of primary succession after expansion from relict populations in limited refugia in more moist valleys, or the necessity of many plants having to disperse over great expanses of ocean (the group is approximately 1000 km west of mainland Ecuador) to reach the archipelago. More detailed analyses from the same site provide evidence of further climatic shifts during the Holocene, especially in precipitation regime, suggesting changing regional climatic conditions impacted on these remote islands (Conroy et al. 2008).

A 9.6-k-year pollen record from La Gomera (Canary Islands) demonstrates a mid-Holocene climatic shift, c. 5.5 ka, towards drier conditions, coinciding with a general regional aridification trend across Northern and Western Africa (Nogué et al. 2013). In the Pacific, analyses of lake sediment cores from subtropical Rapa Nui also show the local effects of global climate change. Conditions between 34 and 28 ka were cold and humid, becoming colder still from 28 ka until the end of the LGM conditions around 17 ka. The Holocene has being generally warm and moist, but with some drier phases, demonstrating a record of ongoing climatic change (Sáez et al. 2009; Rull et al. 2016).

As more records are documented, expanding the temporal and spatial coverage of island systems and applying ever more sophisticated palaeoenvironmental techniques, it is becoming increasing apparent that island systems have experienced significant shifts in climate throughout not only the Pleistocene but also the post-glacial or Holocene period. While the global shift from LGM to interglacial climate systems provided a broadly synchronous and general shift in island climates, it is also evident that there have been many regionally lagged or asynchronous shifts in climatic regimes of sufficient magnitude to alter ecosystem dynamics on islands around the world (Table 3.1).

3.9 Changes in sea level

The level of the ocean surface determines the form and size of the island estate. Indeed, within the Quaternary, changing sea level has been the most important reason for the formation and loss of islands. Changes in sea level may be eustatic, those caused by changing volume of water in the world's oceans (Fig. 3.7a), or isostatic, where the change is caused

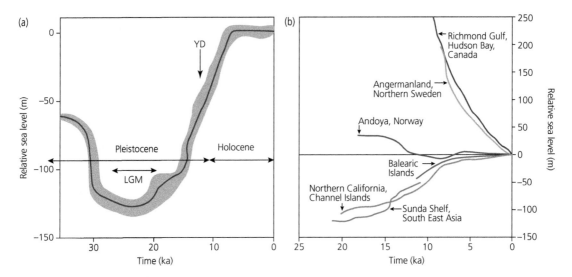

Figure 3.7 Change in relative sea level for sites around the world over the last several thousand years. (a) A generalized relative global sea-level curve (YD: Younger Dryas (cold) interval; LGM: Last Glacial Maximum, *c.* 21 ka). (b) Reconstructions from several regions showing the variability generated by the interaction of isostatic and eustatic change.
Panel a, simplified from Lambeck et al. (2014); panel b, based on data provided by Lambeck and Chappell (2001), Reeder-Myers et al. (2015), and Vacchi et al. (2018).

by a relative adjustment of the elevation of the land surface. Reconstruction of the changing config-uration of islands and archipelagos is challenging because eustatic and isostatic changes are inter-twined and the latter are particularly complex to reconstruct (Ali and Aitchison 2014; Ali and Hedges 2021, 2022).

Isostatic change can be brought about by the removal of mass from the land causing uplift, as when an icecap melts, or by tectonic adjustments. Subsidence of the lithosphere can be caused by increased mass (e.g. increased ice, water, or rock loading) or by the movement of an island away from mid-ocean ridges and other areas that can sup-port anomalous mass (Nunn 1994; Toomey et al. 2016). Adjustments to the elevation of an island may also come about by erosion, deposition, or renewed volcanism.

The formation and melting of major ice sheets not only produce significant global change eustat-ically, but also these processes generate connected regional isostatic change through loading and unloading, typically with lagged response times. The upshot of all this is that there is huge varia-tion in how relative sea level has changed across the world (especially within high latitude regions)

in response to glacial–interglacial cycles (Fig. 3.7b). That isostatic effects associated with these cycles have not only been confined to high-latitude con-tinents and their margins, but also extend to low latitudes and ocean basins, necessitates a cautious approach to the construction of regional eustatic curves (Nunn 1994). Stratigraphic data from iso-lated oceanic islands have been of particular value in such analyses.

Sea levels during the Cenozoic (the last 66 Myr) have generally been significantly higher than cur-rent levels, with levels in the Eocene (50 Ma) around 70–100 m higher, but with fluctuations linked at least partly to waxing and waning of ice sheets. Low sea-level excursions to 50–125 m below present occurred during the late Miocene *c.* 11.6–5.33 Ma, followed by a period of higher levels (*c.* +30 m) during the Pliocene (5.33–2.588 Ma), tailing down to the more pronounced fluc-tuations and low sea levels driven by Quaternary glaciation (Lambeck et al. 2014; Rohling et al. 2014; Toomey et al. 2016). Within the Quaternary, the pattern has been one of glacial episodes in the north-ern latitudes corresponding with lowered sea lev-els, and interglacials associated with levels similar to those of the present day, yet superimposed on a falling trend, such that the LGM provided the lowest point.

As already outlined, relative sea-level excursions can vary in magnitude and timing across the Earth's surface, because of the complex interaction of isostatic and eustatic processes. One further contributing factor is that the oceanic geoid surface (i.e. the sea surface itself) is actually rather irregular, with an amplitude of about 180 m relative to the Earth's centre. Relatively minor shifts in the configuration of the geoid surface, which might be produced by underlying tectonic movements, would be sufficient to cause large amounts of noise in the glacio-eustatic picture (Nunn 1994; Benton and Spencer 1995). However, there is some measure of agreement that stand levels have not exceeded present levels by more than a few metres within the past 340,000 years. Eustatic minima varied in magnitude from glacial to glacial, with an estimated low of 120–134 m below present at the peak of the LGM (Lambeck et al. 2014), although with considerable regional variation (Nunn 1994; Lambeck and Chappell 2001).

The order of sea-level depression is sufficient to connect many (but not all) present-day continental shelf islands—such as mainland Britain—to continents, thus allowing biotic exchange between land areas that are now disconnected. Equally important, many islands existed across the oceans which are now below sea level. Simply drawing lines on maps in accord with a particular bathymetric level provides only a crude basis for reconstructing past island–mainland or archipelagic configurations (Ali and Aitchison 2014; Ali and Hedges 2021, 2022) but the implications of these fluctuations are hugely important biogeographically. Across the globe, the LGM low-stand connected many continental-shelf islands to much larger mainland areas, restocking their biotas. Even true oceanic islands were greatly reduced in isolation at those times by the temporary emergence of many present-day seamounts as islands and by some nearby islands becoming joined together (e.g. Fernández-Palacios et al. 2011; Norder et al. 2019).

The eustatic rise from the LGM minimum of *c.* −120–134 m commenced around 15 ka, with near-current levels attained some 7000 years later (Fig. 3.7a; Lambeck et al. 2014). The timing of separation of islands was highly individualistic, being dependent on the local topography and also, in places being influenced by isostatic adjustments, which are ongoing in the most heavily ice-loaded areas (Lambeck and Chappell 2001). Gradual global isostatic adjustments may also explain a mid-Holocene high-stand of around 2 m above current sea levels, which peaked *c.* 4–2 ka (Dickinson 2009). The fall back from this high-point exposed atoll reefs that had been growing upwards atop previously exposed limestone plateaus, inheriting their ring shape from the saucer profile generated through solution weathering during exposure within glacial sea-level low-stands. Many were awash during the Holocene high-stand. From a human perspective, they only became suitable for settlement following the post-high-stand fall in sea level and most have indeed been occupied only within the last 1500 years. This serves to emphasize that for such small, low-lying islands, comparatively small adjustments in sea level, of the order predicted as a consequence of 21st-century climate change, can have huge implications for island societies and ecosystems (Courchamp et al. 2014; Storlazzi et al. 2018; Section 15.2).

3.10 Anthropogenic environmental change and disturbance

As many island biogeographical studies follow the rationale of islands as model systems, authors often distinguish the 'native' and the 'natural', from the non-native and the altered, focusing analysis on the system as it was or might have been. Yet, it is an undeniable fact that as humans have spread across the islands of the world, we have become an agent of profound environmental change. Moreover, the human influence on island ecosystems is typically measurable and separable from the backdrop of natural environmental dynamics from an early point following human colonization of islands (Nogué et al. 2021; and see Fig. 13.4), although the attribution of ecological change is rarely straightforward (e.g. de Nascimento et al. 2009; Rull et al. 2016; Burjachs et al. 2017). It is a general truth that our impact on island ecosystems has steadily increased alongside our increased population size, improved transport networks, and—recently—the burgeoning island tourism industry.

Humans today impact island ecosystems through the usual array of activities: use of fire for the clearance of native ecosystems; introduction of non-native plants, animals, fungi, and diseases; introduction of pastoral and arable agricultural practices, leading to alteration to rates of erosion and deposition; alteration of soil properties, drainage, and water use; through application of herbicides, pesticides, and other chemical products and by-products, alteration of nutrient levels and of the chemical environment generally; sequestering of significant amounts of net primary productivity; construction of urban environments and transport infrastructure; building of ports, harbours, and airports; light and noise pollution; etc. (Gillespie and Clague 2009). In the 21st century, islands are in many ways becoming increasingly vulnerable to anthropogenic change. While mitigation efforts have been stepped up by many island governments to combat environmental degradation, the impacts of non-native species, and other *in situ* challenges, climate change and predicted sea-level increase provide an existential threat, especially to low-lying islands and island societies (Dickinson 2009; Courchamp et al. 2014). We consider these themes more fully in Chapters 13–15.

3.11 Summary

Island environments are distinctive from the vast bulk of continental landmasses, particularly when examining distinct classes of oceanic islands such as reefs, atolls, and high volcanic islands. Many volcanic islands have remarkable topographic amplitude for such comparatively small land areas. Island soils are also distinctive, owing to their volcanic (or, for atolls, coralline) origins, with evidence that productivity can become limiting through diminishing availability of phosphorus on ageing volcanic substrates. Inward transfer of nutrients through atmospheric transport and seabird activity (on smaller islands) can override such intrinsic features.

Island climates tend to be comparatively mild, with dampened temperature variation but many tropical and subtropical islands experience extreme weather events, such as generated by ENSO and

hurricane activity. The height of many volcanic islands is of crucial importance in generating a high diversity of eco-climatic zones, while the combination of high elevation and coastal proximity leads to compression of elevational zonation. Water availability is often limited on islands, owing to a lack of large catchments and natural lakes, but volcanic islands also typically contain large reservoirs of fresh water within their mass.

The position of an island in relation not only to other islands and mainland, but also ocean and atmospheric circulation systems, is a crucial influence not only on weather and climate, but also the effective biological isolation of that island. Variability of circulation systems over time may thus be important to understanding the degree of potential connectivity ecologically.

The image of tranquillity often associated with island environments has to be replaced with recognition that, for many islands, the reality features intermittent disturbances, often of high magnitude and severity of impact. Some of these disturbances change the base energy signature or physical features of an island, whereas others are impactful but relatively short-lived. Among the most important to people and ecosystem function are high-energy tropical storms, epitomized by Caribbean hurricanes, which hit the region annually, devastating local economies and defoliating large swathes of woodland and forest. Other weather vagaries, such as ENSO events, are less dramatic but also have detectable ecological impact on island ecosystems in affected areas. However, it is volcanic activity and linked mega-geomorphological processes which have the biggest impact on island systems, at their worst leading to island sterilization, large-scale loss of human life, and permanent reconfiguration of affected islands. Biogeographically speaking, we should regard these high-magnitude but intermediate-/low-frequency events as being integral to the evolutionary theatre of islands and archipelagos. They both destroy and renew.

Although buffered to a degree by their oceanic setting, the palaeoecological record indicates significant climatic changes within the Late Pleistocene and Holocene on a range of islands, including examples from the tropics and subtropics. Allied to global climatic change is a record of pronounced fluctuations in sea level, including significantly higher and lower levels than present within the last 10 Myr, and with a biogeographical key period around the LGM with a low-stand around 120–134 m below present sea level. In general, post-glacial warming drove a lagged sea-level rise over several thousand years, drowning lowlands that at the time of the LGM linked islands to each other and/or to continents. In addition, islands that emerged at the LGM to become stepping-stones were drowned, once again returning them to their interglacial seamount status. These changes in connectivity have huge implications for the movement of species among and between islands and continents.

Finally, we turn to humans as agents of environmental change on islands, noting the extensive range of ways in which we have altered the physical, biological, and chemical environments of islands and the existential threat posed, in particular to the societies and ecosystems of low-lying islands and atolls by projected climate change and sea-level rise in the coming decades.

CHAPTER 4

The biogeography of island life: biodiversity hotspots in context

The scarcity of kinds—the richness in endemic forms in particular classes or sections of classes,—the absence of whole groups, as of batrachians, and of terrestrial mammals notwithstanding the presence of aerial bats,—the singular proportions of certain orders of plants,—herbaceous forms having developed into trees, &c.,—seem to me to accord better with the view of occasional means of transport having been largely efficient in the long course of time, than with the view of all our oceanic islands having been formerly connected by contiguous land with the nearest continent.

(Darwin 1859, p. 384)

4.1 Are islands rich or poor?

To appreciate the special significance of island biotas we need to consider in what ways they are peculiar and how they contribute to global **biodiversity**. Biodiversity—a contraction of biological diversity—refers to the variability of life from all sources, including within species, between species, and of ecosystems. The most commonly used biodiversity unit is the species, although at times we are concerned with higher (e.g. genus or family) or lower (e.g. subspecies or haplotypes) points in the taxonomic hierarchy, or with slightly different concepts such as functional or phylogenetic diversity. As a first-order generalization, islands are species poor for their size but rich in species found nowhere else (i.e. **endemic** to that island or archipelago). They are thus 'hotspots' of biodiversity (see Box 4.1). Given that the number of species on the planet remains unknown to within an order of magnitude (Lomolino et al. 2017), we cannot precisely determine the absolute contribution of islands, but we can state emphatically that *taken collectively*, islands contribute disproportionately for their area to global species totals (below).

In addition to the relative species poverty and high proportions of endemism displayed with increasing isolation, islands are also compositionally distinct or 'disharmonic' relative to continental areas. These properties have long fascinated biogeographers (e.g. Hooker 1866), providing key insights for the development of evolutionary theory (Darwin 1859; Wallace 1902). Hence, the first steps in island biogeography were (and remain) to establish how island biotas are related both to continental biotas and those of other islands, and to establish how different the unique (i.e. endemic) forms found on islands are from the colonizing forms. As a first step in this endeavour, we explore below the application of biogeographical regionalization tools to island systems.

The general explanation for reduced richness, increased disharmony, and higher endemism is that crossing larger gaps of ocean is challenging and thus greater distance filters out less dispersive taxa, although it is also likely that some potential colonists fail on arrival because of a lack of suitable habitat, mutualistic partners, appropriate food, or resources. By way of contrast, some islands (e.g. Madagascar and the British Isles) were once connected to larger landmasses, from which they have been severed by plate movements and/or sea-level adjustments (Chapters 2 and 3). As we outline

Island Biogeography. Robert J. Whittaker, José María Fernández-Palacios, and Thomas J. Matthews, Oxford University Press.
© Robert J. Whittaker, José María Fernández-Palacios, and Thomas J. Matthews (2023). DOI: 10.1093/oso/9780198868569.003.0004

below, determining the respective roles of dispersal and barrier formation provides a fundamental task within island biogeography.

Islands are not only biodiverse but they are also repositories for many of the world's threatened species (Box 4.1), having experienced disproportionately high levels of anthropogenic species extinction in historic and prehistoric times. Hence, some of the biogeographical structure that we seek to interpret has been distorted or lost. Lack of full knowledge of pre-human contact island biodiversity patterns can be problematic, in that we may misinterpret fragmentary data and reach erroneous conclusions. Therefore, before we embark on a consideration of ecological and evolutionary patterns and processes, we provide at the end of this chapter some brief illustrations of island losses as a cautionary note. A fuller account of anthropogenic impacts and management responses for threatened species follows in Chapters 13–15.

Box 4.1 Island hotspots: geological, biodiversity, and conservation

Islands may qualify as hotspots in at least three distinct ways. First, some islands are the product of **geological hotspots**, localized but long-lived areas of volcanism overlying plumes in the mantle (Chapter 2). Second, many islands are **biodiversity hotspots** in the sense of possessing concentrations of locally endemic species. For example, BirdLife International's Endemic Bird Area (EBA) scheme delimits areas supporting two or more species of birds with breeding ranges of < 50,000 km^2. Of the 218 EBAs designated in the 1996 version of the scheme, 113 are island EBAs (21 continental islands, 21 large oceanic islands, 55 small oceanic islands) (Long et al. 1996). Many islands are hotspots for multiple taxa in this sense and, collectively they hold a disproportionately high degree of global diversity.

The third distinct usage stems from an approach to conservation prioritization put forward by the British environmentalist Norman Myers in the late 1980s and since promoted by the international environmental NGO Conservation International (Myers et al. 2000). Myers' approach was to combine a measure of the concentration of biodiversity with a simple index of threat; that is, it is really a **biodiversity hotspot–threatspot** scheme. The CI hotspot scheme as published in 2000 identified 25 such hotspots based on two criteria: (1) the area possessed > 0.5% (1500) of the world's plant species as endemics, and (2) the area had lost > 70% of its primary vegetation. The CI hotspot approach attracted unprecedented levels of funding following the 2000 paper (a decade later, Mittermeier et al. 2011 estimated that it may have focused US$1 billion on the regions covered), and became extremely influential in strategic conservation planning.

The CI scheme was subsequently extended to include 36 hotspot areas (Box Figure 4.1; Noss et al. 2015), with some effort made to increase the coverage of the world's islands, particularly in the Pacific. The addition of the East Melanesian islands hotspot was explicitly based on estimates of increased habitat loss (Mittermeier et al. 2011). It has been calculated that the 36 areas (representing 16% of the global land surface area) contain > 50% of the world's plant species as endemics, along with 77% of terrestrial vertebrates, of which 43% are endemic (Mittermeier et al. 2011).

Of the 36 hotspots, nine are composed exclusively of islands. These are: (1) the Caribbean islands, (2) Madagascar and adjacent islands (the Comoros, Mascarenes, and Seychelles), (3) East Melanesia, (4) Japan, (5) New Caledonia, (6) New Zealand, (7) Philippines, (8) Polynesia-Micronesia, and (9) Wallacea. Three more have a substantial proportion of their diversity within islands: (10) the Mediterranean Basin (including the Atlantic islands of Macaronesia), (11) the Western Ghats and Sri Lanka, and (12) Sundaland. Almost all tropical and subtropical islands of high biodiversity value have been included (with the exception of New Guinea), although it is evident that, were many of the individual archipelagos considered as separate biogeographical entities, they would not have reached the qualifying diversity-plus-threat criteria.

The world's islands not only represent disproportionate amounts of diversity, but they also account for a high proportion of recorded global extinctions over the last few hundred years, and a high proportion of globally threatened species (Chapters 13–15). A stress on islands is thus an appropriate part of any global conservation assessment based on the currency of species and indeed is common to schemes promoted by the major international conservation NGOs (e.g. the WWF, Conservation International, BirdLife International) (see review in Ladle and Whittaker 2011).

Box 4.1 *Continued*

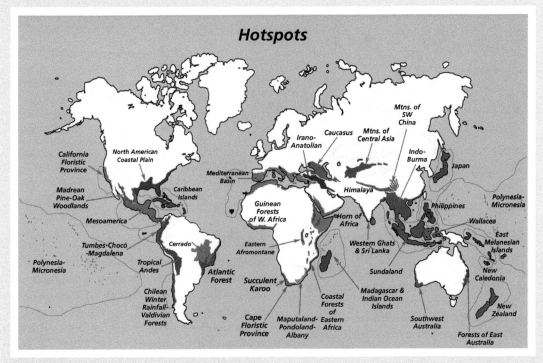

Box Figure 4.1 The Conservation International Biodiversity hotspots scheme as per 2016 (revised to include 36 areas: see figure 1 of Harrison and Noss (2017).
© Conservation International.

4.2 Species poverty

In general, the species richness of an island is a positive function of the size of the island, such that the island species–area relationship (ISAR) for an archipelago can be represented by a linear model providing area and species richness are each logged prior to analysis (Chapter 5; Triantis et al. 2012; Matthews et al. 2016a). For these log–log (power) models, the slope of the relationships represents the rate of increase of richness with area, while the intercept represents the richness on an island of unit area. As archipelago isolation increases, the ISAR tends to steepen, meaning that the difference in richness between smaller islands and large ones is greater. Hence, the slope of the ISAR tends to increase, and the intercept decreases, from habitat islands

to inland or continental shelf islands, to oceanic islands (Triantis et al. 2012; Matthews et al. 2016a). Habitat islands, in turn, tend to be less species rich for their area than continuous patches of habitat, especially at small patch size.

These generalizations will be examined more fully in Chapter 5, and here we provide just two examples. First, plant data from California show what is considered a general pattern: whether constructing nested or non-nested sequences of data points from the mainland, they lie above the steeper regression line for the island data (Fig. 4.1). Second, St Helena (15°56′ S, 5°42′ W), a remote island dating to c. 12–14 Ma, about 122 km^2 in area, 1800 km from the nearest continent (Africa) and 1300 km from the nearest island (Ascension), has a known

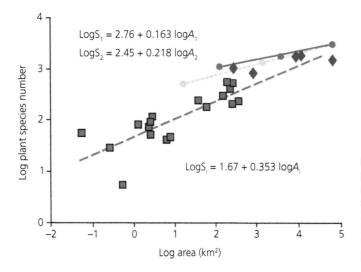

Figure 4.1 Species–area curves for Californian plants, showing the steeper slope for islands (log S_1 red squares), compared with two alternative nested sets from within the mainland (log S_1 mid blue circles, San Francisco area; log S_2, light blue circles, Marin County). The diamonds represent other areas from the mainland, which lie in neither nested set. Redrawn from Rosenzweig 1995, Fig. 2.9.

indigenous flora of only around 43 flowering plant species, 37 of them endemic (6 of which are extinct), plus 30 ferns, of which 13 are endemic (Gillespie and Clague 2009). That this poverty of species may be linked to its natural state of isolation is supported by the observation that some 250 introduced (i.e. non-native) plant species have become established on the island since its discovery in AD 1502.

The successful spread of species transported to islands such as St Helena in recent centuries (Seebens et al. 2017), might appear to indicate that such islands are under-saturated in their natural state and could support higher diversity (Sax et al. 2002; Seebens et al. 2017). However, while overall diversity often increases, some **turnover** is generally involved, with endemic plant and especially animal species becoming extinct as part of this process (indeed, losses can be devastating). As discussed in Parts II and IV, the assessment of such changes in diversity is rarely straightforward as there are often multiple drivers operating in tandem, such as forest clearance, agricultural activities, and urbanization. Typically, at least some native and endemic species became extinct through human actions before scientific surveys were undertaken, so that the known native biota is an incomplete record. Analysis of fossil evidence has shown this to have happened on numerous other oceanic islands, including Easter Island, Mauritius, and the Canaries (Nogué et al. 2017).

4.3 Disharmony

Islands tend to have a different balance of species compared to equivalent patches of mainland: they are thus said to be **disharmonic** (Williamson 1981). One aspect of this disharmony is that island climates are often noticeably different from the nearest continental region, even if at the same latitude, so that the successful colonists may reflect more distant source pool areas. Thus, the Canaries, located west of the Sahara, feature a substantial Mediterranean element to their flora and vegetation, Kerguelen in the Indian Ocean is bleak and Antarctic-like for its latitude, and the Galápagos, although equatorial, are closer to desert (i.e. subtropical) islands, being influenced by upwellings of cool subsurface waters. Hence, there is a climatic 'filter' constraining species assembly (Fig. 3.2). A second aspect is that islands sample disproportionately from the dispersive portion of the mainland pool (i.e. there is an isolation or dispersal filter (Fig. 4.2)). Third, there may be biotic filters operating, as for example the absence of necessary mutualists preventing other species from colonizing, with species poverty or disharmony in one taxon leading to the absence of representatives of another. In addition, *in situ* diversification within particular colonist lineages may contribute further to insular disharmony (Weigelt et al. 2014).

To demonstrate disharmony (difference in compositional balance) as distinct from simply impoverishment (fewer species), it is necessary to

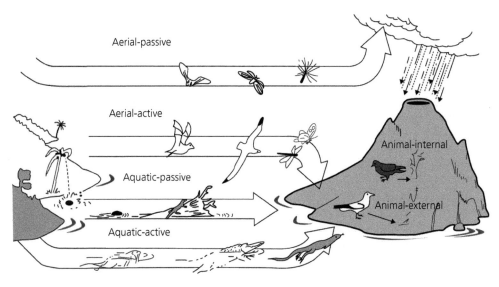

Figure 4.2 The various modes of dispersal by which species may naturally disperse to and colonize a remote island through or across a seawater gap (redrawn from Aguilera et al. 1994, based on an original in Rodríguez de la Fuente 1980). Alternative nomenclature for the top three categories as they apply to plants: seed dispersal by wind = anemochory; in the gut of an animal = endozoochory; attached to the fur or feathers = exozoochory; flotation in the sea = thalassochory.

demonstrate non-random assembly in relation to a given mainland source pool (e.g. Heleno and Vargas 2015; Chapter 6). A global analysis of vascular plant species across 178 oceanic islands by König et al. (2020) exemplifies this approach. They examined the representation of certain plant families on the islands using a null model based on identified source pool regions. The Urticaceae (nettles) and Convolvulaceae (bindweed) were found to be generally over-represented, while the Asteraceae (daisies) and Orchidaceae (orchids) were found to be generally under-represented, although not on all islands. Disharmony was found to be greater on large, high-elevation islands, which could indicate that *in situ* speciation, which is more likely to occur in larger, more heterogeneous islands, is an important driver of island compositional distinctiveness (König et al. 2020).

It is hard to estimate the absolute cross-ocean dispersal distances that can be attained given thousands of years of chance dispersal, but in general, terrestrial mammals (excluding bats), amphibians, and freshwater fishes have very limited capacity to survive a lengthy journey across ocean, with occasional opportunities being provided by large vegetation rafts for some mammals or amphibians.

Hence remote oceanic islands such as Hawaii, the Azores, and the islands of Tristan da Cunha possessed no amphibians or non-volant land mammals when humans first reached them.

Below the level of major taxa, further filter effects can be seen, for example, in the general paucity of large-bodied spiders or beetles that have colonized remote islands such as the Azores (Whittaker et al. 2014). The partial sampling of mainland pools is also exemplified by the flora of distant Hawaii, which, relative to other tropical floras, has few orchids, only a single genus of palms (*Pritchardia*), and is altogether lacking in gymnosperms and primitive flowering plant families. An estimated 263–270 colonists have given rise to 1030 flowering plant species (90% endemic) and 162 fern species (70% endemic), of which the ten-richest lineages of flowering plants, within the families Campanulaceae, Lamiaceae, Rutaceae, Asteraceae (two separate lineages), Caryophyllaceae, Gesneriaceae, Arecaceae, Rubiaceae, and Myrsinaceae, have diversified to supply 41% of the flora (Price and Wagner 2004; Keeley and Funk 2011). The insects of Hawaii also provide numerous examples of disharmony (Hembry et al. 2021). For example, cosmet moths (Cosmopterigidae), a family providing perhaps 1% of global Lepidopteran species, represent 30–40% of Hawaiian Lepidoptera, in large part due to one endemic genus, *Hyposmocoma*, which has diversified to form over 600 species (Hembry et al. 2021).

Interestingly, recent studies have found a stronger role for climate, environment, and biotic filters than for dispersal filters in driving island community assembly (Carvajal-Endara et al. 2017; König et al. 2020). For example, an analysis of the Galápagos flora incorporating data on species dispersal strategies and climatic suitability found the archipelago's flora represented a strongly clustered subset of the mainland species pool. However, this clustering was more a result of the matching of species' climate niches between the mainland and the archipelago, rather than variation in species dispersal ability (Carvajal-Endara et al. 2017). This is surprising given the relative isolation of the Galápagos (almost 1000 km from the nearest point on the South American mainland), but likely reflects that the climatic conditions on the Galápagos are more arid than much of the closest mainland.

4.4 Dispersal

All forms of life have some capacity to move around at some point or points in their life cycle, a necessary attribute for replenishing populations and for colonizing suitable but unoccupied areas. There are various different categories of movement that may bring animal species to islands, including migratory movements related to the seasons, nomadic movements within a range, winter movements across ice bridges, and vagrancy (accidental extralimital occurrences resulting from unpredictable movements of individuals, often linked to extreme weather). Each of the foregoing mechanisms may, occasionally, result in a species establishing on an island and forming a breeding population. However, among birds, vagrancy is relatively common behaviour, but rarely results in a population establishing. For example, over 300 species from the approximately 627 species recorded for Britain are considered vagrants (McInerny et al. 2018). Moreover, in a study based on 66 oceanic islands/archipelagos, Lees and Gilroy (2014) show that propensity to vagrancy among birds is rather a poor predictor of insular colonization success,

but that large range size and migratory behaviour appear to increase chances of colonization.

Species may also have occupied islands without crossing large water gaps, by diffusing across contiguous land connections that have since been severed (a form of vicariance; Section 4.7). However, the majority of the species that colonize remote islands are neither migratory, nor nomadic, and do not get there across land-bridges. Rather, colonization comes about through the inherent tendency to disperse that is connected with establishing new territorial occupancy at specific points in the life cycle, typically for plants as seeds or spores, and for animals as young adults (sometimes as larvae, or eggs).

Some biogeographers are sceptical of the evidence for really long-distance dispersal events. They particularly object if it appears that several taxa have developed very similar disjunct distributions, requiring repeated long-distance dispersal between the same remote points by different species: especially if several of them lack evident long-distance dispersal traits. They look to evidence from plate tectonics and sea-level change to join the dots and permit long-term persistence combined with movements that constitute either cross-land connections (extreme vicariance), or far more modest dispersal across narrow water gaps (Heads 2011, 2017; Grehan 2017). At the other extreme, it has been suggested with respect to microbes that they have such huge dispersal and persistence capabilities that . . . 'everything is everywhere, but the environment selects . . .' (the Baas Becking hypothesis; Lomolino et al. 2017).

Dispersal can be active, involving energy expenditure, as undertaken by flying birds or insects. Or it can be passive, meaning that little or no energy is expended by the dispersing individual, rather the work is done by external forces such as wave or wind power (Fig. 4.2). A lot of species disperse by means of a third party; for example, seeds that are packaged to provide a reward to a frugivorous bird or bat, or ectoparasites that attach themselves to the feathers of birds.

Many plants and animals have evident adaptations to dispersal, such that categorization of, for example, fruit and seed traits, provides an effective lens by which to analyse the long-distance dispersal potential of a suite of species (Heleno and Vargas 2015). In seed plants, there are four primary sets of traits (syndromes) that can promote long-distance dispersal. These involve traits that: (i) attract frugivores that consume and disperse the seeds in a viable state (endozoochory); (ii) allow the seeds to attach (e.g. via hooks) to animal bodies (epizoochory); (iii) enable dispersal by wind (anemochory); and (iv) provide buoyancy and protect the seeds against saltwater, thus enabling dispersal via water (thalassochory) (e.g. Heleno and Vargas 2015). However, relating 'dispersal' traits to likelihood of colonization is problematic because: (i) dispersal kernels for individual species typically exhibit a strongly skewed distribution, with a very long tail (estimating the mean/modal dispersal distance is fairly easy but predicting the maximum distance is incredibly difficult); (ii) such analysis may fail to predict the actual means of island colonization (below); and (iii) effective isolation of islands varies due to long-term environmental change in ways that are challenging to estimate for longer timescales (Chapter 2; Katinas et al. 2013; Ali and Hedges 2022).

While many organisms have evident adaptations for a particular mode of dispersal, some plant species exhibit an evolutionary two-way bet, **diplochory**, meaning that they have two evident modes of dispersal (Vargas et al. 2015). In such cases it is generally obvious which mode is the more likely agent of long-distance transportation. Classic examples include species of the Indo-Pacific strand-line flora that have colonized the Krakatau Islands by seeds/fruit borne in the sea, yet also have adaptations for seed spread into island interiors by birds or bats (Whittaker and Jones 1994b).

Perhaps the biggest challenge with dispersal is that it is so difficult to observe the rare events that result in potentially viable propagules reaching a remote island. Examples include records of individual birds noted by citizen scientist birdwatchers on many islands, although they rarely result in population establishment (above; Lees and Gilroy 2014). More often, dispersal has to be inferred from phylogenetic evidence linking an island species to a related ancestral or sister species (e.g. Keeley and Funk 2011; Le Roux et al. 2014), or simply from the absence of alternative explanations for the native

flora or fauna of truly remote islands for which no possibility of a past land connection exists.

The most remote islands in terms of distance from the mainland are to be found in French Polynesia, the largest of which, within the Society Islands group (current oldest high island c. 4.6 Ma) is the comparatively young (c. 1.3 Ma) island of Tahiti, which is some 5900 km from the nearest mainland (Weigelt et al. 2013). Phylogenetic evidence for 49 plant and animal taxa from the Society Islands indicates their closest relationships to be with taxa in other tropical Pacific archipelagos, particularly in the Cook, Austral, and Marquesas groups, thus involving apparent dispersal distances of up to 2000 km (Fig. 4.3; Hembry and Balukjian 2016). Being so remote, the colonists have been restricted to groups with exceptional dispersal powers: bryophytes and ferns are particularly important components of the flora, native conifers are absent, and flowering plant colonists have been sufficiently uncommon that those arriving have led to some significant radiations of endemic species (as per Box 4.2). There have also been some spectacular radiations of arthropods and land snails, with most colonist lineages being of small-bodied species.

For the Atlantic Ocean, we have already mentioned St Helena (above), 1350 km from the nearest landmass and rising from the deep ocean floor, as illustrative of implicit evidence of long-distance over-water dispersal, given that it has always been a really isolated island throughout its c. 14.6-Myr life span (Katinas et al. 2013). Another powerful example is provided by the Azores, which are also around 1350 km from the nearest mainland (the Iberian Peninsula) and nearly 900 km from the nearest of the other Macaronesian archipelagos. The oldest island, Santa Maria, first emerged around 6 Ma, but subsequently submerged and then re-emerged, so that its biological age is estimated to be < 4 Ma (Hildenbrand et al. 2014; Ramalho et al. 2017). Given which and notwithstanding Quaternary sea-level minima and the emergence of a few seamounts within the Macaronesian region (Fernández-Palacios et al. 2011), the Azores have necessarily accumulated their biota by long-distance colonization events. Phylogenetic evidence indicates that the islands have received colonists from Iberia, as well as from other parts of

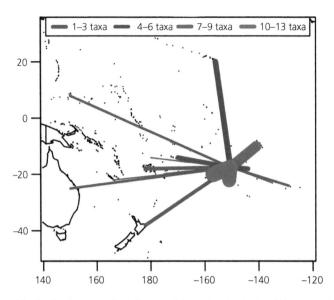

Figure 4.3 The distributions of closely related taxa to endemic lineages of the Society Islands. Line thicknesses and colours indicate how many plant or animal taxa show particular sister-group relationships; direction of colonization not indicated. The connected regions are: Cook Islands, Marquesas Islands, Tuamotu Islands, Austral Islands, Gambier Islands, Pitcairn Islands, Hawaii, Samoa, Tonga, Fiji, Wallis and Futuna, Caroline Islands (Micronesia), New Caledonia, Australia, and New Zealand.
Hembry and Balukjian (2016; their figure 6a).

Box 4.2 Palaeoendemism and neoendemism, with examples of explosive radiation

Some island endemics belong to groups that formerly had more extensive, continental distributions. Since colonizing their island(s) in the remote past (over either land or sea), they have failed to persist in their original form in their mainland range; that is, they are **palaeoendemics**, or 'relictual' forms that have not evolved very far from their ancestral template (Cronk 1992). By contrast, species that have evolved their distinctiveness *in situ* are termed **neoendemics**. As with all such distinctions, this is an oversimplification, as it implies a lack of change in either island or mainland forms, respectively, which may not strictly be the case (see e.g. Carlquist 1995). Moreover, endemic-rich islands of any antiquity commonly include a mixture of palaeoendemics and neoendemics (see evaluation in Veron et al. 2019). We continue the theme of palaeoendemism in Box 4.3. Here we exemplify neoendemism in the form of some numerically spectacular cases of evolutionary radiation.

Much of the contribution of insular environments to global biodiversity comes from cases of so-called 'explosive radiation' and thus also of neoendemism. Outstanding examples of such radiation include Hawaiian *Drosophila* fruit flies, African Great Lakes cichlid fish, Madagascan lemurs, Caribbean *Anolis* lizards, New Zealand land snails, and New Guinean microhylid frogs. Palms have undergone explosive radiation both in Madagascar (*Dypsis*) and in the Caribbean (*Coccothrinax* and *Copernicia*), and examples for dicotyledons include several Hawaiian (*Cyanea, Cyrtandra, Haplostachys*), New Caledonian (*Phyllanthus, Psychotria, Pittosporum*), and Macaronesian (*Aeonium, Sonchus, Echium*) clades (Box Table 4.2). Although not illustrated in the table, other largely insular environments, such as many mountains, also possess restricted range endemic species and many recently uplifted ranges feature impressive radiations of species, a classic example being Andean *Lupinus*, within which a monophyletic group of 81 species have radiated in the 'sky-islands' of the high Andes within the Pleistocene (Hughes and Eastwood 2006).

Box 4.2 *Continued*

Box Table 4.2 Selected examples of explosive radiation events that have happened in insular (including both continental fragment and oceanic islands) and lacustrine environments, with the number of species indicated in brackets (various sources). For further examples, see Fernández-Palacios et al. (2021a).

Island/archipelago or lake	Animal taxa	Plant taxa
African Great lakes	Cichlidae fishes (*c.* 1500)	
Caribbean islands	*Anolis* lizards (*c.* 400)	*Coccothrinax* palms (48)
		Copernicia palms (19)
Galápagos	Darwin's (*Geospiza*) finches (15)	*Scalesia* (Asteraceae) (15)
Hawaii	*Drosophila* flies (*c.* 1000)	*Brighamia, Cyanea, Clermontia, Delissea,*
	Hyposmocoma moths (up to 600)	*Lobelia, Trematolobelia*
	Mecyclothorax beetles (at least 239)	(Campanulaceae) (125)
	Nesophrosyne leafhoppers (up to 200)	*Cyrtandra* (Gesneriaceae) (58)
	Trigonidiine crickets (173)	*Haplostachys, Phyllostegia, Stenogyne*
	Orthotylus bugs (95)	(Lamiaceae) (57)
	Honeycreeper-finches (*c.* 60)	*Melicope, Platydesma* (Rutaceae) (52)
	Rhyncogonus weevils (47)	
Macaronesia	*Laparocerus* beetles (236)	*Aeonium* clade (Crassulaceae) (55)
	Napaeus snails (74)	*Sonchus* clade (Asteraceae) (35)
	Dysdera spiders (72)	*Echium* (Boraginaceae) (29)
Madagascar	Lemurs (99)	*Dombeya* (Malvaceae) (170)
	Vangidae birds (21)	*Dypsis* palms (162)
		Psychotria (Rubiaceae) (100)
		Helichrysum (Asteraceae) (100)
		Pandanus (Pandanaceae) (78)
Mascarene	*Phelsuma* geckos (40)	*Psiadia* (Asteraceae) (26)
		Pandanus (Pandanaceae) (22)
New Caledonia	Diplodactylid geckos (58)	*Phyllanthus* (Phyllanthaceae) (111)
	Lygosomine skinks (51)	*Psychotria* (Rubiaceae) (85)
		Pittosporum (Pittosporaceae) (50)
New Guinea	Microhylid frogs (215)	*Cyathea* (Cyatheaceae) (20)
		Dolianthus (Rubiaceae) (13)
New Zealand	Land snails (*c.* 1000)	*Olearia* (Asteraceae) (42)
	Cicadas (40)	*Chionochloa* (Poaceae) (26)
	Diplodactylid geckos (37)	

Macaronesia, and even from the New World, while also donating colonists to Madeira, Canaries, and to Iberia (Patiño et al. 2015; Price et al. 2018).

One of the most remarkable cases allowing inference of long-distance dispersal of higher plants is provided by a dated phylogeny showing that the supposedly endemic *Acacia heterophylla* of Réunion Island, in the Indian Ocean, is closely related to and almost certainly the same species as the supposedly endemic *Acacia koa* of Hawaii, some 18,000 km away

in the Pacific (Le Roux et al. 2014). It seems most likely that the ancestral species to *A. koa* reached Hawaii from Australia by bird dispersal. A species of petrel is the most plausible vector for the subsequent colonization of Réunion from Hawaii. More direct efforts to catch dispersal in action have been made, an interesting example being the use of falcons to catch migratory birds in the Canary Islands during their passage from Europe to Sub-Saharan Africa. Around 1% of these migrants carried seeds in their guts, mostly being species not found within the archipelago, thereby demonstrating the long-distance transport in action, but also illustrating that dispersal and establishment are two very distinct processes (Viana et al. 2016).

Alongside those cases where dispersal mechanisms appear straightforward, such as many strand-line plants, the seeds or fruits of which float and retain viability, there are as many puzzling cases of species for which neither close ancestors, nor the island species itself has any obvious adaptation for long-distance dispersal (Fig. 4.4; Heleno and Vargas 2015). Possession of a particular dispersal syndrome may even be misleading as to the actual means by which species reach islands. An example of how **non-standard dispersal** can work is provided by Padilla et al. (2012), who demonstrate that viable seeds of dozens of plant species eaten by the Eastern Canary Islands lizard (*Gallotia atlantica*) can be found in the pellets of predatory birds (shrikes *Lanius meridionalis* and kestrels *Falco tinnunculus*), capable of moving between islands, effecting 'secondary dispersal' of the plants. Vargas et al. (2012) identified 36% of Galápagos native plant species as having no obvious specialization for long-distance dispersal, slightly in excess of the proportion of zoochores (32%). A further compelling example of dispersal through a seemingly unlikely mechanism,

is the finding that small snails can survive passage through the gut of small passerine birds, providing a plausible means of their colonizing oceanic islands (Wada et al. 2012), although this is not to rule out other possible colonization mechanisms.

Finally, the role of rafting in island colonization has long been recognized. For example, vegetation rafts transported down major river drainages (e.g. the Congo and Ogooué) have been proposed as the most likely way in which amphibians and reptiles reached the oceanic islands of the Gulf of Guinea, off the western coast of Central Africa (Bell et al. 2015). Lomolino et al. (2017) provide a summary of another classic example of a chance rafting event: a huge raft of vegetation cast adrift from the island of Guadeloupe as a result of Hurricane Luis in September 1995. Around a month later the raft beached on the island of Anguilla, some 320 km away, delivering a cargo of at least 15 green iguanas (*Iguana iguana*). More recently, phylogenomic analyses, coupled with oceanographic simulation modelling, has shown that kelp rafting events linking islands (e.g. New Zealand, the Kerguelen Islands) and mainland in the Southern Hemisphere are much more common than previously assumed. Rafts have been shown to transport numerous invertebrate passengers over journeys of thousands of kilometres (Fraser et al. 2022).

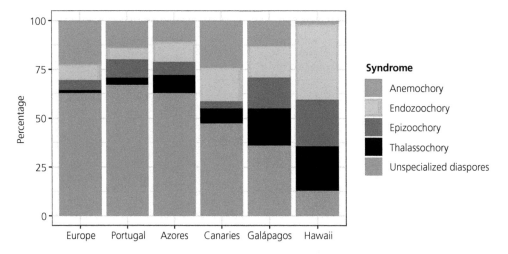

Figure 4.4 Relative proportions (%) of dispersal syndromes for different floras. Isolation of archipelagos from the nearest continent (Azores 1356 km, Canaries 96 km, Galápagos 972 km, Hawaii 3650 km) fails to explain the frequency of unspecialized diaspores. Hawaiian estimates were based on different criteria from the other floras.
Based on data from Heleno and Vargas (2015).

4.5 Filter effects and sweepstake dispersal

In cases, it is possible to identify a route of faunal interchange involving significant barriers that are only occasionally crossed by rare, chance events involving unpredictable winners—this is **sweepstake dispersal** *sensu* George Gaylord Simpson (1940). The aforementioned cases of relatively poor dispersers colonizing islands via vegetation rafts exemplify such chance dispersal events.

While there is certainly a large element of stochasticity to which species cross barriers, the thinning out of a species pool with distance from a dominant source area is often broadly explicable as a function of dispersal traits and propensity. This can be seen in the decline of families and subfamilies of land and freshwater birds found in New Guinea with increasing distance into the Pacific. For example, according to Williamson (1981), 14 taxa don't make it beyond New Guinea (pelicans, snakebirds, storks, larks, pipits, logrunners, shrikes, orioles, mudnesters, butcherbirds, birds of paradise, bowerbirds, Australian nuthatches, Australian tree-creepers); two don't make it beyond New Britain and Bismarck islands (cassowaries, quails, and pheasants), 10 reach their limits at the Solomon Islands (owls, frogmouths, crested swifts, bee-eaters, rollers, hornbills, pittas, drongos, sunbirds, flower-peckers), and so on. In other regions a two-way filter appears to operate (Carlquist 1974), best exemplified for linearly configured archipelagos connecting two different larger source regions, such as documented across the islands of Wallacea, between South-East Asia and the Sahul region (New Guinea and Australia), as we discuss in more detail shortly.

4.6 Biogeographical regionalism, modules, and nodes

Filter effects also occur in continental areas, especially where two long-isolated regions have come back into contact; for example, following the formation of the Central American isthmus towards the end of the Pliocene (Lomolino et al. 2017). Transition zones and filter effects reflect the biogeographical imprint of often distant events in the Earth's history, intimately connected with the plate tectonic processes that have seen the break-up of supercontinents, followed by the collision of the fragments many millions of years later. Early students of biogeography, notably Philip Sclater and Alfred Russel Wallace, recognized the discontinuities and, on the basis of the distribution patterns of particular taxa, were able to divide the world into a number of major zoogeographic or phytogeographic regions. The basis of all regionalization schemes is that the membership of a fauna or flora is used to determine the degree of relationship among different areas and thus to determine where to draw the natural boundaries between natural regions or subregions (Figs. 4.5, 4.6, 4.7). Comparisons can be based largely on families, genera, or species and some recent schemes have incorporated phylogenetic information representing the evolutionary relationships between taxa in order to delineate regions (Holt et al. 2013). Almost irrespective of the approach, maps broadly similar to the foundational regional schemes of Wallace can be recovered, but they all differ in details, especially so in transition regions.

The boundary between the Oriental and Australian zoogeographic regions is marked by what has long been known as **Wallace's line** (Fig. 4.5), a discontinuity in the distribution of mammals dividing the Sunda Islands between Bali and Lombok, despite their being less than 40 km apart. Alternative versions of the line have been drawn, with further demarcation lines proposed by scientists such as Webber, Pelseneer, Lydekker, and others (Ali and Heaney 2021). For instance, Weber's line (or Pelseneer's line) distinguishes a barrier for Australian mammals akin to Wallace's line for Oriental mammals. It has also recently been proposed, based on data for mammals and squamates, that Wallace's line should be extended into the Indian Ocean to place Christmas Island on the Australasian side of the divide (Ali et al. 2020).

Although biotic turnover is pronounced across these regions, the precise placing of lines can be contentious. In part this is because the area between the Asian and Australian continental shelves actually contains relatively few Asian or Australian mammals, and in part because other groups of more dispersive animals, such as birds, butterflies, and reptiles, show a filter effect rather than an abrupt

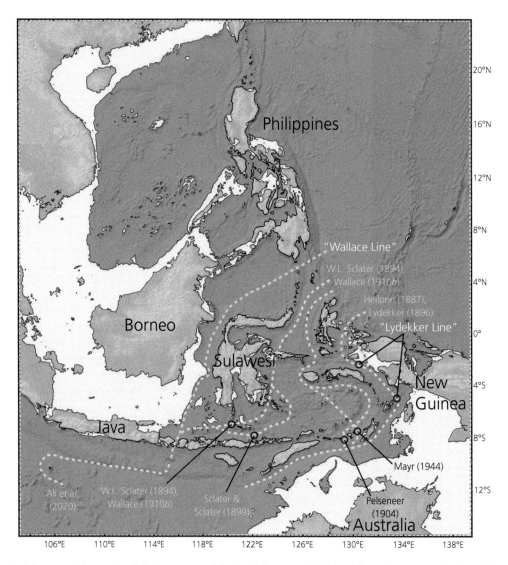

Figure 4.5 South-East Asia and Australasia showing some of the boundaries proposed by A. R. Wallace and others, dividing the two faunal regions (the dates of description are given; for the source references see Ali and Heaney (2021). The area between Wallace's line and Lydekker's line demarcates Wallacea. The continental shelves down to *c.* 120 m below sea level are shown in white. The most recent addition (denoted Ali et al. 2020) is the extension of Wallace's line to the north of Christmas Island.
The figure is a composite of Figs. 5 and 6 of Ali and Heaney (2021), with courtesy of Jason Ali.

line. Another problem for the drawers of lines is that the insects of New Guinea are mainly of Asian origin. Hence, the treatment of this complex region varies a little between taxa and schemes. Nonetheless, biogeographers generally view the Oriental and Australia regions as being limited to the respective continental shelves, with the intervening transition area being designated **Wallacea** (after A. R.

Wallace) by Elmer Merrill in 1924 (Fig. 4.5; and see Keast and Miller 1996; Michaux 2010; Lomolino et al. 2017; Ali and Heaney 2021). The complex geological history of the region results from the collision of the Australian and Eurasian plate and their interactions with the Philippine Sea plate, a small plate of oceanic crust. The biogeographical development of the region is equally complex (Michaux

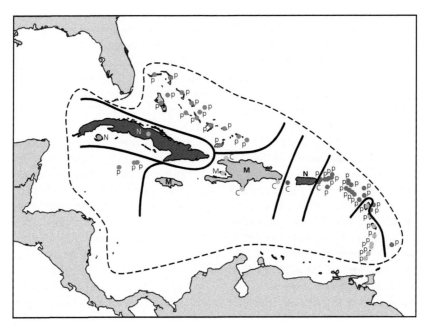

Figure 4.6 Biogeographical modules and island roles of the West Indies. Each colour corresponds to one of six detected biogeographical modules: (1) Cuba and nearby Isla de la Juventud (purple module), (2) Bahamas and the Cayman Islands (pink module), (3) Jamaica and Hispaniola and its satellite islands (yellow module), (4) Puerto Rico and nearby Mona island (blue module), (5) the northern and outer younger low-lying Lesser Antillean Islands (red), and (6) the southern inner Lesser Antillean arc consisting of Guadeloupe in the north and all main volcanic islands south to Grenada (green module). The biogeographical modules fit well with both island geography and the geological history of the region. Island roles are indicated with letters: P, peripheral; C, connector; M, module hub; N, network hub. The biogeographical modules and island roles are based on a network approach and the distribution of terrestrial breeding birds across the West Indies, as explained further in Dalsgaard et al. (2014).

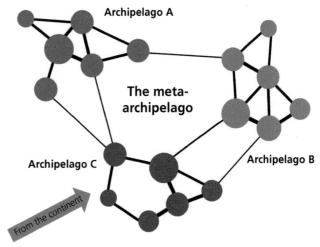

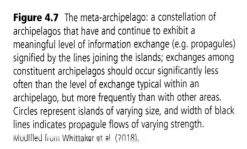

Figure 4.7 The meta-archipelago: a constellation of archipelagos that have and continue to exhibit a meaningful level of information exchange (e.g. propagules) signified by the lines joining the islands; exchanges among constituent archipelagos should occur significantly less often than the level of exchange typical within an archipelago, but more frequently than with other areas. Circles represent islands of varying size, and width of black lines indicates propagule flows of varying strength. Modified from Whittaker et al (2018).

2010). Hence, several efforts have been made to determine the precise boundaries of Oriental and Australian regions and thus of Wallacea (reviewed in Ali and Heaney 2021). In some treatments (e.g. the CI biodiversity hotspot scheme (Box 4.1; Myers et al. 2000)), the Philippines are entirely excluded from a 'reduced' Wallacea, although Ali and Heaney (2021) argue that this is without justification.

Notwithstanding the challenges involved, finer-scale analyses within complex island regions can produce valuable insights into the patterns of biogeographical relationships. One such approach focuses on the links between islands provided by shared species to identify networks and modules. Cartensen et al. (2012) and Dalsgaard et al. (2014) illustrate this network approach in analyses of bird data for Wallacea and for the West Indies (Fig. 4.6). They identify four categories of island: (i) network hubs possess both many local species and many shared across the region; (ii) module hubs have many local species but few of regional distribution; (iii) connector islands possess a few local species along with many shared across the region; and finally, (iv) peripheral islands have few local species and few shared regionally. Their analyses identified four modules within Wallacea and six within the West Indies. More remote large islands tended to possess high richness of endemics and thus to feature local linkage, whereas stronger regional topological linkages, reflecting richness of non-endemics, was characteristic of smaller islands distant from mainland sources but situated near the boundaries between modules. These analyses thus help determine, within complex island regions, how best to delimit archipelago membership from a biogeographical perspective. Where the strengths of biogeographical flows and linkages between two clusters of islands is above background levels, but less than within a cluster, we can think of the islands as forming two archipelagos within a meta-archipelago (Fig. 4.7; Whittaker et al. 2018).

The placing of more distant oceanic islands into biogeographical regionalization frameworks can be problematic, as they have often gained their biota from multiple sources and directions (Fig. 4.3). Molecular evidence for Hawaiian plant lineages demonstrates links with sources all around the Pacific, from the Arctic south to New Zealand, and even with Africa. Based on molecular phylogenies of 37 groups, accounting for over half the native flora, some 64% of species derive from sources to the south-west (Asia, Australasia, Malesia, or the South Pacific), 35% from the east to north-east (27% from North America), and there is just the one taxon (*Hesperomannia*) of African origin (Keeley and Funk 2011). Most of the species rich groups are monophyletic (radiations from a single colonist) and North American colonists have contributed disproportionately to the diversification of the flora in relation to the number of colonist events. In practice, phytogeographers generally assign the flora to its own floristic region (Davis et al. 1995).

It is likely that ocean currents, major wind systems, and bird migration routes have influenced colonization biases and resulting patterns of affinities in most island biotas. To provide an example from human biogeography, Madagascar, may have first been colonized by humans as late as *c.* AD 830 (see Section 13.2 for a wider range of estimates), by a small group of people from Indonesia rather than from East Africa, which is much closer, perhaps due in part to favourable ocean currents (Cox et al. 2012). The flora of Gough Island provides another example (Moore 1979). The predominant wind circulation system around the South Pole provides a plausible explanation for the affinities with likely source areas South America and New Zealand (Fig. 4.8).

Changes in landmass positions through plate tectonics, and changes in relative sea level, and in oceanic and atmospheric circulation linked to climate change, can mean that the contributions made by particular regions to the biota of a focal island group (and vice versa) have varied over time (below; Pokorny et al. 2015). The Canary Islands appear to exemplify this (below). What we learn from such biogeographical analyses is that while Waldo Tobler's so-called first law of geography ('everything is related to everything else, but near things are more related than distant things') typically applies fairly well within biogeography, distance from a larger source pool can sometimes be an imperfect or incomplete indicator of biogeographical affinity.

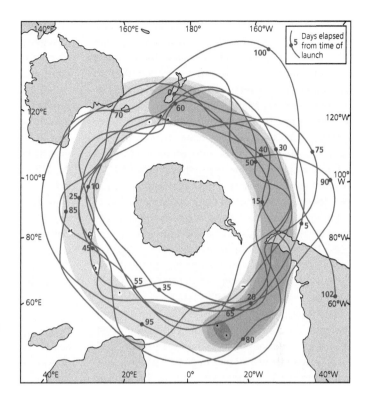

Figure 4.8 Floristic affinities of Gough Island (Tristan da Cunha) in relation to predominant wind direction as revealed by the flight of a balloon. Areas sharing 5–10, 10–15, 15–20, 20–25 species, respectively, indicated by increasing density of green shading.
After Moore (1979): his figures 1 and 2.

4.7 Disjunct distributions and the tussle between vicariance and dispersalism

Biogeographers are fascinated with disjunct distributions; that is, species or groups of species that share common ancestry and which occur today in widely separated locations. Large disjunctions across oceans can be explained either by **vicariance** (the breaking of a past land connection severing a once continuous distribution) or **long-distance dispersal** across the ocean gap: two hypotheses that have long been in opposition (sometimes fiercely debated) within biogeography. Prior to the acceptance of continental drift and specifically plate tectonics theory, the challenge for biogeographers was to resolve this conundrum without moving the pieces around. Those 19th-century scientists who considered it implausible that chance dispersal across huge distances could account for shared disjunct patterns, promoted the idea of extensionism, sunken intercontinental land-bridges, to explain them. The development of plate tectonic theory has since provided for a far more nuanced

and sometimes complex set of biogeographical scenarios: in effect the pieces can now be moved around, created and destroyed, joined and parted, and allowed to submerge and re-emerge (whether through uplift and subsidence or repeated sea-level change) (see review Lomolino et al. 2017).

Some ancient continental fragment islands, including Madagascar, New Zealand, and New Caledonia, are characterized by high endemism and some interesting ancient lineages, which may be described as relictual or palaeoendemics. Yet, just as lineages may have older origins than the platforms on which we find them, others (even quite old ones) may have reached their islands following their initial isolation. For instance, several Madagascan mammal taxa such as the lemurs, the Tenrecidae, and the Viverridae, although relatively primitive, are nonetheless estimated to have diverged from their ancestral lineages during the Palaeogene, roughly between 56 and 26 Ma: long after the c. 130 Ma separation from Gondwana and following the global re-setting of the Cretaceous–Palaeogene mass extinction event

(Samonds et al. 2013; Everson et al. 2016; Ali and Hedges 2022). Samonds et al. (2013) sums up that '… Madagascar's extant tetrapod fauna owes more to colonization during the Cenozoic than to earlier arrivals'. Throughout this time, Madagascar has remained an island. Even so, the degree of isolation has not been unchanged. There have probably been periods of enhanced opportunities for exchange of colonists with other landmasses due to changing climate, current systems, and sea level (at times exposing a few seamounts within the Mozambique channel) over this long period of isolation. Although efforts have recently been made to revive the hypothesis of land-bridge connections to explain some colonization events (e.g. Masters et al. 2020), Ali and Hedges (2022) contend that the latest geological evidence refutes the idea of Cenozoic-era terrestrial walkways.

New Zealand and New Caledonia, both sit upon Zealandia, a 5 M km^2 area of continental crust, 94% of which is now submerged (Mortimer et al. 2017). Zealandia was at one stage part of the southeast margin of the Gondwana supercontinent, from which it began to separate during the Late Cretaceous. That these islands are built upon ancient continental crust and contain many unique, and in cases old lineages, explains why there has been such interest in vicariance hypotheses for at least some of their biota. According to Gibbs (2016), candidates for Gondwanan origin and persistence include the kiwi (Apterygidae), tuatara (Sphenodontia), huia (Callaeatidae), and the kauri (Araucariaceae). Yet many features of New Zealand's biogeography are comparable with those of true oceanic island archipelagos such as Hawaii, with compelling evidence that much of its biota has colonized long after the break-up of Gondwana and in cases much more recently than that, following a period of partial inundation of New Zealand itself (Goldberg et al. 2008; McCulloch et al. 2016).

An illustration of the wider significance of islands in trans-ocean biotic transfer is provided by Katinas et al. (2013) who present a scenario for the early evolution of the Asteraceae (daisy family) in the isolated continent of South America around 50 Ma, followed by stepwise dispersal along an island chain formed by the Walvis Ridge and Rio Grande Rise, across an already wide South Atlantic Ocean. This is only a scenario and while they put forward reasons to regard it as credible, like many biogeographical case studies it should be seen as a hypothesis in want of further testing.

Greater certainty can be applied to cases such as the Azores, a small group of young islands, the biological age of the oldest of which (Santa Maria) is probably no more than 4 Myr, with the other eight islands being of Pleistocene age (Hildenbrand et al. 2014; Ramalho et al. 2017). These islands have never been within 1000 km of a continental landmass. Molecular phylogenies support the bathymetry in pointing to long-distance dispersal, including in both directions across the Atlantic, as the explanation for the biota of this archipelago (Patiño et al. 2015), notwithstanding that some taxa have been exchanged between the Azores and other contemporary Macaronesian Islands.

More recent examples of barrier formation and breaking of barriers involve eustatic changes in sea level, especially the reductions associated with major periods of glaciation, when huge ice sheets have formed, locking up vast amounts of water. These changes can turn islands into extensions of continents. For instance, during Pleistocene sea-level minima, Tasmania, Australia, and New Guinea formed one landmass, called Sahul. Something similar happened with the huge banks that emerged in the Indian Ocean during the last glaciation, including the granitic Seychelles (41,615 km^2), Saya de Malha (26,405 km^2), and Nazaret (22,099 km^2) platforms, lying between Madagascar and the Indian subcontinent. This in part explains the strong Indian flavour of the Madagascan and Mascarene biotas (Warren et al. 2010). Many seamounts also emerged at these times, providing additional 'stepping-stone' islands that may have facilitated the colonization of remote islands.

As this very brief review illustrates, the route to explaining disjunctions and resolving the dispersalist–vicariance controversy lies in first accepting a number of propositions: (i) some extraordinary long-distance dispersal and colonization events do indeed happen; (ii) some clades do survive very long periods on islands having colonized without long-distance dispersal; (iii) some movements of terrestrial species occur across modest water gaps, permitting long-term lineage persistence in remote island regions through a process akin to handing on a baton in a relay race (cf. Heads 2017); and (iv) the relative position and degree of connectivity of particular islands and other landmasses varies significantly over time such that which of the foregoing applies may be clade and time specific for the same island system. In short, both vicariance and dispersal (including over remarkable distances) operate, and both contribute to biogeographical patterns (see also: Keast and Miller 1996; Galley and Linder 2006; Katinas et al. 2013; Chamberland et al. 2018). The challenge is to agree on the methods to determine which of these propositions applies to which cases (see review in Lomolino et al. 2017).

4.8 Macaronesia—the biogeographical affinities of the Happy Islands

In this section, we pull several of the themes we have been discussing together, in a more detailed look at one particular insular region: Macaronesia. The term Macaronesia refers to a biogeographical subregion including all the north-west Atlantic archipelagos off the South European and North African coasts, thus comprising the Azores, Madeira, Selvagens, Canaries, and the Cabo Verde Islands (Fig. 4.9). Some authors extend the envelope to include a narrow coastal strip of the African continent, and affinities with south-west Iberia have also sometimes been recognized.

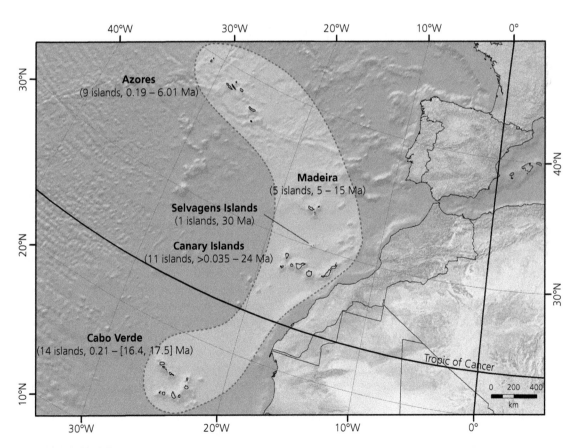

Figure 4.9 The biogeographical area known as Macaronesia consists of the island groups as shown (slightly modified after Florencio et al. 2021) and a narrow coastal strip of North-West Africa. The affinities with the Iberian Peninsula are also recognized by some as warranting a link to the peninsula.

Geologically, the Azores are distinctive as their origins are associated with a triple junction. The islands have formed (i) both west and east of the mid-Atlantic ridge and (ii) in association with the junction of the African and Eurasian plates (Chapter 2). The origins of all the other Macaronesian islands lie within the African plate. There are over 100 seamounts in the Canary Island Seamount Province and the oldest date back to *c.* 140 Ma, formed at a comparatively early stage of the rifting of the Atlantic (van den Bogaard 2013). However, this doesn't mean that high islands have been continually available for insular lineages to persist over such time periods: many seamounts never break surface and those that do may take remarkably long periods to do so, for example, Fuerteventura may have begun building *c.* 70 Ma as a seamount but emerged as an island *c.* 24 Ma (van den Bogaard 2013; Hunt and Jarvis 2017; Carracedo and Troll 2021). Lars (Essaouira) Seamount (68 Ma, Canarian province) and Ormonde Seamount (65–67 Ma, Madeiran province) were probably islands by the start of the Palaeogene, and thus available for colonization following the global biological re-set of the Cretaceous–Palaeogene (K–Pg) mass extinction event. It is likely that there have been some emergent high islands in the region continually since then, as new islands have emerged and gradually older ones have submerged (see detailed account in Section 2.5; Fernández-Palacios et al. 2011). However, to date, García-Verdugo et al. (2019a) report a lack of crown ages among Canarian lineages older than the contemporary oldest Canarian island. The Azores is the youngest of these archipelagos, with a continuous terrestrial history of less than 4 Myr (Chapter 2). Eruptions in recent decades attest to the active nature of these volcanic systems (e.g. Capelinhos, Faial in the Azores in 1957; Teneguía, just offshore from El Hierro in the Canaries in 2011–12; Caldeira do Pico, Fogo, in the Cabo Verde archipelago in 2014; and Tajogaite, La Palma in 2021).

The Macaronesian archipelagos span the cool-oceanic climate of the Azores (Corvo at 40° N) to the oceanic tropical monsoon-drift climate of the Cabo Verde Islands (Brava at 15° N), with the largely Mediterranean climates of Madeira and the Canaries in between. Only 96 km separates Fuerteventura (Canaries) and Stafford Point (Western Sahara), Boa Vista (Cabo Verde) is 570 km from the nearest mainland at Dakar (Senegal), Porto Santo (Madeira) is 630 km from the nearest mainland at Cape Sim (Morocco), and São Miguel (Azores) is some 1370 km from mainland landfall in Portugal. Furthermore, the Azorean islands of Corvo and Flores, although occurring within the North American plate, are virtually equidistant from Cape Race (Newfoundland) and Lisbon.

Although the nucleus of the Canaries and Madeira are recognized to have zoogeographical affinities, on the whole Macaronesia is a phytogeographical concept, having been considered a distinct phytogeographical region within the Holarctic Kingdom for more than a century. In practice, the degree of shared membership across the region is limited, as there are transitions from strong Eurosiberian–Atlantic affinities for the Azores, to a Mediterranean flavour for Madeira and the Canaries and, finally, to a Saharan–Sudanian character for the Cabo Verde Islands (Torre et al. 2019).

Canarian endemic plants share affinities with several distinct regions: 18% are Macaronesian, 35% Mediterranean, 25% North-West Africa, and 22% have affinities with more distant regions, including East and South Africa and also the New World (Sun et al. 2016). As typical of remote archipelagos, many endemics have not diversified or have done so only a little between islands, while some have repeatedly speciated within islands. Several species rich genera are monophyletic within this archipelago or the different Macaronesian archipelagos (*Aeonium* clade, *Argyranthemum*, *Cheirolophus*, *Echium*, *Pericallis*, *Sonchus* clade, etc.) but others are polyphyletic and comprise a mix of non-endemic and endemic lineages, an example being the spurges, genus *Euphorbia*, which features nine Canarian endemics. These species are considered by Sun et al. (2016) to have arisen from several colonization events (i.e. they are polyphyletic), with only one lineage diversifying within the Canary Islands (section *Aphyllis* subsection *Macaronesicae*). The ancestral distribution for this group appears to have been the Cabo Verde Islands or Tenerife. Within the Canaries, Tenerife and La Gomera acted as source areas for the other islands, with one species, *Euphorbia regis-jubae*, colonizing North Africa from Gran Canaria (Sun et al. 2016).

The Canary Islands generally decrease in age with greater distance from Africa (Fig. 2.6). The simplest biogeographical model would have the nearest/oldest island being stocked from the nearest continental source pool, and in turn supplying a high proportion of colonists to younger, more distant islands. Although this pattern can be found, as the above data indicate, the assembly of island biotas is not always so unidirectional or simple. We can further illustrate this point by reference to floristic elements shared between Macaronesia and very distant parts of Africa.

The 19th-century plant geographers A. Engler and H. Christ noted the existence of an African-Macaronesian element, which became known as the 'Rand flora' (meaning rim or edge flora), referring to unrelated floristic elements that shared similar disjunct distributions around the continental margins of Africa and the adjacent islands of Macaronesia (Fig. 4.10; Pokorny et al. 2015). Hypotheses for how

disjunctions arise variously invoke earth system changes driven by plate tectonic processes, climate change and their interaction, and the biogeographical processes of vicariance, long-distance dispersal, and various intermediate scenarios of corridors and stepping stones. Pokorny et al. (2015) estimate the timing of disjunction of 17 lineages from 12 families using a combination of nuclear and chloroplast DNA to test these alternative scenarios. They concluded that rather than being explicable by one generalized track, or vicariance scenario, the Rand flora demonstrates biogeographical pseudocongruence. That is, the distributions of related species shared between Macaronesia and Eastern Africa arose at different points over the last 20 Myr, corresponding to contrasting climate and biome distributions. There appears to be a general connection of the pattern to the long-term northward drift of the continent and associated aridification episodes, involving episodes of range expansions and contractions. In short, it seems necessary to invoke both vicariance and varying degrees of dispersal (not least

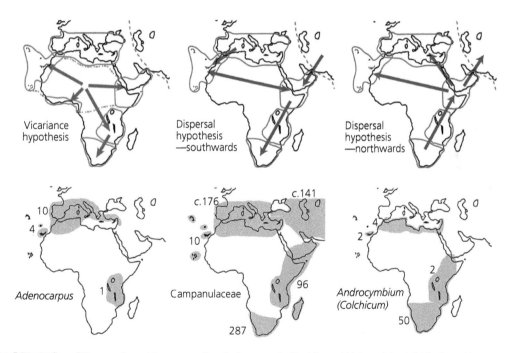

Figure 4.10 The flora of Macaronesia contains a group of species known as the Rand flora, which have disjunct distributions with eastern and southern Africa. Top row: three main hypotheses put forward to explain their origin(s). Bottom row: three plant groups that may provide examples of each. The vicariance hypothesis invokes climate change to break up a formerly continuous flora; dispersal southwards invokes immigration from the Mediterranean; and northwards dispersal a pattern of spreading from southern and eastern Africa. The numbers indicate the species richness of each region.
From Sanmartín et al. (2010), their figure 1.

between Africa and Macaronesia), to explain these floristic connections.

Recent developments in our understanding of the environmental history of the islands (Chapter 2) and the phylogeography of Macaronesian lineages have illuminated these debates, providing the following general picture. Although there are older seamounts in the region, the contemporary islands (and precursors) have emerged sequentially over the last *c.* 68 Myr. There is limited evidence of such lengthy biogeographical legacies or of crown ages exceeding the ages of the oldest current islands. Opportunities for island-hopping colonization have been ever present, but all the Macaronesian Islands are true oceanic islands, which have been colonized across open ocean, in many instances over great distances. Their degree of biogeographical affinity decreases largely with distance (greater within archipelago than between archipelago), but there are elements in their biotas linking Macaronesia to remarkably distant land areas. We consider these observations and assumptions in more detail in Box 4.3.

Box 4.3 Macaronesian palaeoendemism: a role for both dispersal and vicariance?

Palaeoendemism of the Macaronesian laurel forest: Macaronesia features numerous recent (neoendemic) radiations in plants and animals (Box 4.2; Caujapé-Castells et al. 2017, 2022; Machado 2022). However, the region's laurel forest trees have been considered to provide several examples of palaeoendemism.

The laurel forests of the Azores, Madeira, and the Canaries are examples of the evergreen broadleaved subtropical biome that can be found between 25° and 35° latitude in both hemispheres (i.e. Texas, Florida, Atlantic islands, Chile, Argentina, South Africa, South-East Asia, Japan, South Australia, and New Zealand) growing, at least in Macaronesia, under the influence of orographic cloud layers. These Macaronesian laurel forests are considered to be the only surviving remnants of a once rather richer forest flora, which was, for example, distributed in central Europe (fossils found in Austria and Hungary) and southern Europe (fossils found in France, Italy, and Spain) during the Oligocene–Miocene. Today we know that the laurel forest has been present on the Atlantic islands at least from the late Pliocene and quite possibly much longer (Fernández-Palacios et al. 2019).

The failure of these forests to persist in Europe has been attributed to the dramatic geological and climatic events that took place in Europe following the transition of the Tethys Ocean seaway into the largely enclosed Mediterranean basin during the Neogene. Around the end of the Miocene, between *c.* 5.6 and 5.3 Ma, the connection between the Mediterranean and Atlantic temporarily closed, leading to the Mediterranean shrinking massively, with associated drought and salinization: an episode known as the Messinian Salinity Crisis. The crisis ended with the Zanclean flood, when a breach occurred to form the Straits of Gibraltar (Lomolino et al. 2017). This spectacular event was followed by the glacial cycles of the last 2–3 million years and the desertification of the Saharan region. With the closure of the Panama Strait, and the associated shifts in global ocean currents, the present-day Mediterranean climate developed around 3 Ma and with it emerged the Mediterranean sclerophyllous forest, which was better adapted to the new prevailing climatic conditions of hot, dry summers and cool, wet winters.

The explanation for the survival of this 'relict' forest flora on the Atlantic islands, albeit in rather species-poor form, has been attributed to three factors:

- The islands may have been buffered from the climatic extremes of mainland southern and central Europe and the Mediterranean by their mid-Atlantic location.
- The topographic amplitude of the islands has enabled vertical migrations of several hundred metres in response to such changes in climate as the islands have experienced.
- The existence in the higher islands of a stable orographic cloud layer has counteracted the aridity of the generally Mediterranean climate, especially so in the Northern Hemisphere summer.

As a result, the laurel forests of the Atlantic islands possess a number of plant species that are considered to be palaeoendemics. These include the following Madeiran/Canarian species known from Neogene (Tertiary) deposits in Europe (5.3 Ma in southern France): *Persea indica*, *Laurus novocanariensis*, *Ilex canariensis*, and *Picconia excelsa*. Quite how old and 'relictual' these forests are is, however, the subject of some debate. Stem age estimates for a suite of 18 laurel forest species (mostly trees or shrubs) placed 14 of the branching events within the Plio-Pleistocene (< 6 Ma), with just four in the Upper Miocene (< 11.5 Ma) (Kondraskov et al. 2015). However, Grehan (2017) has promoted an

Box 4.3 *Continued*

alternative view arguing for far longer-term *in situ* persistence of several Macaronesian lineages, including—but not limited to—endemic laurel forest tree genera (*Ocotea*, *Persea*, *Apollonias*, *Clethra*, *Ilex*, *Morella*, etc.). He calls upon recognition of geological evidence for seamounts in

the region dating back to > 100 Ma, as evidence in support of the likelihood of long-term survival on a succession of long-lost islands, and as part of more extensive palaeo-distribution areas spanning the Old and New worlds across a far narrower Atlantic (Box Figure 4.3). By this view (and

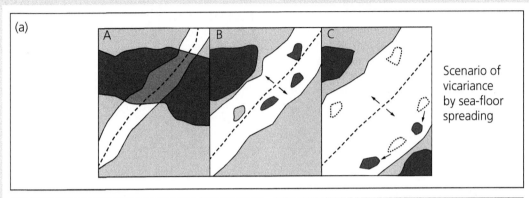

Scenario of vicariance by sea-floor spreading

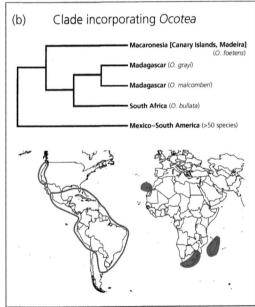

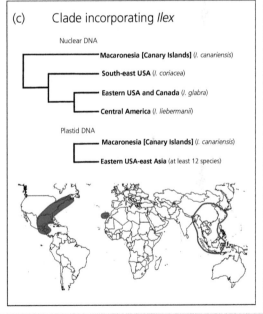

Box Figure 4.3 Biogeography of Macaronesian palaeoendemics: (a) hypothesized vicariance origin of relictual or palaeoendemic species within Macaronesia by the fragmentation of a once widespread ancestral distribution (A) by seafloor spreading, resulting in stranding on islands (B) that in time serve to populate later emerging islands before themselves submerging (C): where light grey is land, white is ocean and dark grey is the range of the clade. (b, c) The phylogenies and distribution maps show two clades, each incorporating a Macaronesian laurel forest tree species; (a) *Ocotea foetens*, showing on the map the New World clade (outline) and the Old World clade (shaded); (b) *Ilex canariensis*, showing the map for the *I. canariensis* nuclear DNA clade (shaded) and an East Asian plastid DNA sister group (outline)

Based on figures 5, 13, and 23 of Grehan (2017).

taking a radically different approach to phylogenetic calibration), the ancestral distributions were attained by spread
overland and dispersal across only limited spans of water,
with the species subsequently persisting for tens of millions
of years by moving modest distances from island to island as
plate tectonics reconfigured the relationships between the
regions.

As exemplification of the vicariance models that Grehan prefers to chance long-distance dispersal hypotheses
as explanations for widely disjunct distributions, we have
picked out two species, *Ilex canariensis* and *Ocotea foetens*.
The phylogenies and accompanying maps are shown in the
figure. According to Grehan's (2017) reconstruction, the
distribution of *Ocotea* is most likely a result of the break-
up of a widespread New World, African and Madagascan

distribution, with initial divergence of *Ocotea* occurring
between Macaronesia and mainland Africa. This view contrasts with that of Kondraskov et al. (2015), who conclude
that Macaronesian *Ocotea foetens* has a mean stem age
of 1.83 Ma, suggesting a Pleistocene origin, and also a
Pleistocene loss of the population from the Mediterranean.
Although Kondraskov et al. (2015) find several such cases
of Pleistocene (rather than earlier) origins within Macaronesian laurel forest trees, they do note that some taxa may
indeed be relictual. Biogeographical interpretations of the
relationships of these older elements of the Macaronesian
biota cannot yet be considered settled, but it is clear that
Macaronesian biota include a mix of both palaeoendemic
and neoendemic elements.

Successful, structured long-distance dispersal becomes easier to contemplate when we consider the following factors (de Nicolás et al. 1989):

- First, as a result of continuing volcanic activity, new territory has been provided for colonization throughout the history of the various archipelagos.
- Second, the general persistence of the north-easterly trade winds and the cool Canarian marine current since the closure of the Panama Strait (*c.* 3.5–5 Ma), has favoured the constant arrival of propagules from the adjacent continents.
- Third, during the repeated and lengthy low sea-level stands in the Pleistocene, a number of sub-marine banks have emerged as islands, serving as stepping stones at intervals of 200 km or less, facilitating the flow of colonists among these archipelagos (Fig. 2.8; Carine et al. 2004).

The banks to the north of Madeira (Seine, Ampere, Gettysburg, Ormonde) provided stepping-stone connections between Madeira and the Iberian Peninsula, and the Dacia and Concepción banks similarly linked Madeira with Africa (Fig. 2.8). These connections may have contributed to the biotic richness of Madeira and the Canaries and to the sharing of some 11 genera and 37 plant

species (and 8 subspecies) between the Canaries, Selvages, and Madeira (del Arco and Rodríguez-Delgado 2018). In contrast, the Azores and Cabo Verde Islands were less well served by former stepping stones, again consistent with the pattern of affinities.

Molecular phylogenetic analyses can assist in indicating the most likely colonization pathways for particular taxa both between the continent and Macaronesia and within the region. For example, it appears that some plants of Canarian origin (e.g. species of *Aeonium*, *Echium*, and *Sonchus*) dispersed across the *c.* 1300 km from the Canaries to the Cabo Verde Islands, where endemic forms developed (Santos-Guerra 1999). Furthermore, at least six plant taxa appear to have colonized Madeira from the Canaries: *Bystropogon*, *Aeonium*, *Convolvulus*, *Crambe*, *Pericallis*, and the woody *Sonchus* alliance (Trusty et al. 2005). Some lineages (e.g. *Aichryson*) subsequently colonized the Azores, most likely via the southernmost Azorean island, Santa Maria, some 850 km from Madeira. In contrast, *Argyranthemum* and *Crambe*, two Asteraceae genera that have radiated profusely in Macaronesia, are believed to have colonized the Canaries from Madeira (Santos-Guerra 1999).

Among birds, it has been suggested that the chaffinches colonized the Azores from the Iberian

Peninsula, and then colonized the western Canaries via Madeira (Recuerda et al. 2021), whereas the African blue tit (*Cyanistes teneriffae*) appears to have independently colonized the Canaries (different parts of the archipelago each time) from the African mainland three times, leading to substantial differentiation among populations of the species in the archipelago (Stervander et al. 2015). The robin (*Erithacus rubecula*) followed a slightly different pattern again, colonizing in three waves: (i) the first *c.* 2.95 Ma, giving rise to the recently recognized Gran Canarian robin *Erithacus marionae*; (ii) the second *c.* 2.17 Ma resulting in the Tenerife robin, *E. superbus* (also recently elevated to species rank); (iii) and the third occupying the remaining Canary Islands *c.* 0.75 Ma (Sangster et al. 2022). According to Rodrigues et al. (2013), this final lineage subsequently went on to colonize the Azores, likely by a single founder event followed by a fast expansion across most of the archipelago.

Discoveries made in the Anti-Atlas valleys of typical Macaronesian floristic elements, including the Dragon tree (*Dracaena draco*) and the laurel (*Laurus* cf. *novocanariensis*), highlight the significance of North-West Africa in relation to Macaronesia. Another very interesting piece of the jigsaw is the discovery of North African endemic species of genera considered to be Canarian in origin. Specific examples include several species of *Aeonium*: *A. arboreum*, *A. korneliuslemsii* (both in Morocco), *A. leucoblepharum*, and *A. stuessyi* (both in Ethiopia). A further example is a species of *Sonchus* (*S. pinnatifidus*), found in the easternmost Canaries, from where it has colonized the African coast, indicating a return trip for lineages that developed on the Canaries having originally colonized from Africa some 4 Ma (Santos-Guerra 2001).

Carine et al. (2004) report that, in the majority of plant genera investigated so far, Macaronesian endemic taxa form a single monophyletic group, in most cases with western Mediterranean sister groups, but also with affinities revealed with North Africa and the Iberian Peninsula, and indeed with more widely separated regions, such as the New World (Madeiran *Sedum*), East Africa (*Solanum*), and southern Africa (*Phyllis*). They also note cases of multiple congeneric colonization events, and further cases of back-colonization ('boomerang

events'—see Section 9.8) from Macaronesia to continental areas (e.g. in *Tolpis* and *Aeonium*).

To sum up:

- There have been multiple and indeed conflicting colonization pathways followed by different taxa.
- The Macaronesian flora and fauna comprise lineages diverging from their closest continental relatives at different times, some ancient and some relatively recent.
- There may well have been distinct waves of colonization related to the greatly changing geological and climatic history of the region.
- Although the islands are truly oceanic, such that long-distance dispersal is necessary to the exchange of lineages with the mainlands, this is not to rule out a role for past variation in the strength of barriers between these islands and mainland areas, and in this sense both dispersal and vicariance reasoning has relevance to the region's biogeography.
- Because of the complexity of the colonization scenarios within Macaronesia, assigning these islands to a particular phyto- or zoo-geographic region, or mainland source area, is inevitably going to be problematic; nevertheless, there is clearly a strong Mediterranean flavour to the biota.

4.9 Island endemism

Island endemism simply refers to species that have their natural distributions confined exclusively to islands. They can usefully be subdivided for purposes of analysis into single-island endemics, multiple-island endemics (i.e. archipelagic endemics), or even multi-archipelago endemics. They may also be subdivided into neoendemics (evolved *in situ*) or palaeoendemics (relictual species), as elaborated in Box 4.3.

There is no doubt as to the disproportionate contribution to global biodiversity of island endemics, especially so for remote oceanic and ancient continental islands. But our estimates of richness and endemism on islands have changed over time and remain incomplete. Much island biogeographical analysis rests on the assumption that our distributional and taxonomic data are fit for purpose. This does not mean that we must have a 100% complete dataset of species per island, but that the data we do have provide a relatively complete, largely unbiased, and reasonably comparable estimate across the set of islands being analysed. Unfortunately, this is also hard to determine, especially for taxa that are harder to sample or identify, although there are statistical approaches that can be used to aid us in the task (e.g. Gray and Cavers 2014). Unsurprisingly, retrospective analyses of past states of knowledge have shown sensitivity of analyses to survey completeness, and cases where the richness through time plots fail to show evidence of the desired approach to asymptote. Even with perfect specimen sampling, changes occur over time as a result of taxonomic revisions which may, for example, result in an island form being recognized as merely a subspecies or variety rather than a species, or alternatively of subspecies being proposed for elevation to the status of a species (e.g. Lifjeld et al. 2016). The following sections provide statistics that are indicative of the global importance of islands for some reasonably well-studied taxa.

Plants

Approximately 308,300 vascular plant species have been described, of which 295,380 are flowering plants (Christenhusz and Byng 2020). Taking New Guinea to be the largest island for these purposes, islands constitute about 5% of the land surface area of the world. The top ten of the islands and archipelagos listed in Table 4.1 provide at least 36,700 endemic species, amounting to a minimum of 12% of the world's vascular plants. New island species continue to be discovered and described. Empirical (Fig. 5.15) and modelled data show the significance of concentrations of plant diversity on tropical islands, especially on large complex islands in South-East Asia and the Caribbean (Weigelt et al. 2013). Around one in six of the Earth's plant species grows on islands, and one in three of all known threatened plants are island forms (Fernández-Palacios et al. 2021a).

The percentage of endemics varies greatly among islands. Atolls and other very small ocean islands, have smaller floras, and typically lower proportions of endemics. This is due in part to the reduced

Table 4.1 Estimated native higher plant species richness and endemism of selected islands or archipelagos. These data, compiled from various sources and updated (in cases significantly) from Whittaker and Fernández-Palacios (2007), should be considered as indicative rather than definitive: divergent values can be found in different sources, indicative of the need for further inventory and taxonomic effort in many island systems.

Island or archipelago	Number of species	Number of endemic species	% Endemic
Borneo	15,000	6000	40
New Guinea	13,630	9300	68
Madagascar	11,260–14,000	9180–12,180	87
Japan	7100	1900	27
Cuba	6514	3229	50
Jamaica	3308	906	27
New Caledonia	3400	2500	75
New Zealand	2300	1955	85
Seychelles	1640	250	15
Fiji	1628	812	50
Balearic	1569	173	11
Canaries	1300	511	39
Hawaii	1006	905	90
Mascarenes	959	691	72
Madeira	793	118	15
Galápagos	522	236	45
Cook Islands	284	3	1
Cabo Verde	224	66	29
Azores	197	67	34
St Helena	73	50	69

variety of habitats and to the wide mixing of the typically sea-dispersed strandline species that dominate their floras. Ancient continental fragment islands, such as Madagascar, New Zealand, and New Caledonia, feature especially high rates of endemism (Table 4.1). But, large numbers and high proportions of endemics are also associated with the larger, higher oceanic islands in tropical and warm-temperate regions. Where they form part of an ancient, persistent archipelago, even if the existing islands are relatively young, they can possess some lineages that have an insular evolutionary history pre-dating the contemporary islands (Keeley and Funk 2011; Borregaard et al. 2016).

Hawaii, to provide a classic example of a large oceanic archipelago, has about 1180 native vascular plants, and of the 1006 angiosperms (flowering plants) about 90% are considered endemic (Wagner and Funk 1995). Some lineages have radiated spectacularly. The silversword alliance comprises

33 endemic species, of monophyletic origin (i.e. a single common ancestor), but placed in three genera: *Argyroxiphium*, *Dubautia*, and *Wilkesia*. They display broad variation in functional traits and occupy a great diversity of habitats (Blonder et al. 2016). The genus *Cyrtandra* (Gesneriaceae) is represented by 58 species, again monophyletic and again displaying high morphological diversity (Knope et al. 2012).

The Macaronesian flora provides the best developed Atlantic equivalent to Hawaii, with some exquisite examples of largely monophyletic radiations. For instance, there are more than 60 monophyletic endemic species of Crassulaceae, distributed across the genera *Aeonium*, *Greenovia*, *Aichryson*, and *Monanthes*. There are also more than 20 species representing monophyletic taxa within four genera of Asteraceae, namely *Pericallis*, *Cheirolophus*, *Argyranthemum*, and *Sonchus*. And there are almost equally impressive radiations within the genera *Echium* (Boraginaceae), *Lotus* (Fabaceae), *Sideritis* (Lamiaceae), and *Limonium* (Plumbaginaceae) (del Arco and Rodríguez-Delgado 2018; Price et al. 2018).

Land snails

Although estimates as high as 80,000 species have been published, at a conservative estimate around 24,000 species of land snail have been described, of which around 11,000 (46%) occur on islands (excluding New Guinea) (Proios et al. 2021). The selection of 12 islands/archipelagos listed in Table 4.2 account for approximately 28% of the world's land snails (Proios 2021). Studies in the Pacific suggest that, once again, it is the larger, higher oceanic islands that are richest in species and endemism, whereas low-lying atolls are neither species rich nor endemic rich. On most islands with high snail diversity, the snails appear to be concentrated in the interiors, especially mountainous regions with 'primary' forest cover.

Insects

We provide a few case studies to illustrate insect diversity on islands. Around 5818 insect species are considered native to Hawaii, 91% endemic (see Hembry et al. 2021). In the family Drosophilidae

Table 4.2 Land snail species richness and endemism for selected islands/archipelagos for which the data are complete enough for the proportion of endemics to be estimated (from Proios 2021).

Island or Archipelago (no. of islands)	Number of species	Number of endemic species	% Endemic
Hawaiian Islands (14)	749	744	99
Japan (4)	582	471	81
Madagascar (1)	1147	1124	98
New Caledonia (1)	181	175	97
Madeira (4)	178	176	99
Canary Islands (12)	289	262	91
Mascarene Islands (3)	144	127	88
Austral Islands (5)	159	137	86
Greater Antilles (4)	1929	1832	95
Socotra (4)	88	88	100
Sri Lanka (1)	237	195	82
Greater Sunda (35)	1057	893	84

(fruit flies), which have persisted on the Hawaiian Islands for approximately 25 million years, around 690 species are currently named and described for Hawaii, within two major lineages (*Drosophila* and *Scaptomyza*) (Hembry et al. 2021). The Hawaiian Heteroptera ('typical bugs') fauna is disharmonic (Section 4.3) in lacking many entire species rich continental families. It comprises 382 species, 380 of which (99%) are believed to be endemic. Within the Coleoptera (beetles), there are four major radiations, each involving more than 100 species: *Blackburnia* and *Mecyclothorax* Carabidae, *Proterhinus* weevils, and *Plagithmysus* longhorned beetles (Hembry et al. 2021). Other major radiations of Hawaiian insects include the tree crickets, with three endemic genera and some 68 species, and the moth genus *Hyposmocoma*, with up to 600 estimated species.

To provide an Atlantic example, the Canarian insect fauna is increasingly well studied, and comprises some 5950 species, of which 40% are endemic (Gillespie and Clague 2009). Among them, perhaps the most explosive example of radiation is offered by the weevil genus *Laparocerus*, comprising some 261 species and subspecies in insular Macaronesia, plus three that result from a back-colonization to Morocco—around 95% of the species are single-island endemics (Machado et al. 2017; Machado 2022). A recent comprehensive review of the insects of the Seychelles catalogued approximately 3000 species (J. Gerlach, personal communication), with a variable level of endemism: 24% in Dermaptera (earwigs), 49% in Lepidoptera, 51% in Diptera (flies), 87% in Dictyoptera (termites, cockroaches, and mantises), and 100% in Phasmatodea (stick insects), although the latter only number six species (Gerlach 2008).

By contrast with some of the examples just given of highly localized speciation within islands, the tropical Pacific butterfly fauna consists of 285 species, of which 157 (55%) also occur on continental landmasses. One hundred are endemic to a single island/archipelago and 28 are regional endemics; that is, they are found on more than one archipelago but not on the mainlands (Adler and Dudley 1994). Interestingly, butterfly speciation in the Pacific archipelagos is primarily a result of limited inter-archipelago speciation. Intra-archipelago speciation appears important only in the Bismarcks and Solomons, which contain, respectively, 36 and 35 species endemic at the archipelago level. Of the other 24 archipelagos in the survey, only New Caledonia makes double figures, with 11 endemic species, and the others contribute just 18 species between them. The Bismarcks and Solomons are the two largest archipelagos in land area and the closest to continental source areas. Butterflies are generally specialized herbivores, their larvae feeding only on a narrow range of plants, or even a single species. On the smaller and more remote archipelagos, the related host plants simply may not be available for the evolution of new plant associations, thereby impeding the formation of new butterfly species (Adler and Dudley 1994).

Lizards

In their survey of the 27 oceanic archipelagos and isolated islands in the tropical Pacific Ocean for which detailed herpetological data existed, Adler et al. (1995) recorded 100 species of skinks in 23 genera, of which 66 species are endemic to a single island/archipelago and a further 13 are regional island endemics. Of the 23 genera, 9 are endemic to the islands of the tropical Pacific. A few species are widespread; for example, *Emoia cyanura* occurs on 24 of 27 archipelagos in the survey. The Bismarcks, Solomons, and New Caledonia appear particularly rich in skinks, but most archipelagos contain

fewer than 10 species and it seems that no lizards of any sort managed to colonize Hawaii without human assistance. Another important group of island lizards are the Caribbean anolines, which are typically small, arboreal insectivores. Around 379 species of *Anolis* have been described thus far, around half of which are insular Caribbean species and half mainland species. Notwithstanding the impressive diversity of island *Anolis*, speciation rates of *Anolis* in island and mainland settings appear comparable (Poe et al. 2018). On the whole, while lizards have diversified on medium-sized islands, especially in the Caribbean, large radiations are comparatively unusual. However, island lizards do show a variety of interesting trends in ecological habits (e.g. several act as pollinators) and in body size (Sections 11.12 and 10.3, respectively).

Birds

The examples of adaptive radiation best known to every student of biology from high-school days are probably the Galápagos finches and the Hawaiian honeycreepers (Chapter 9). Therefore, it will come as no surprise that islands are important for bird biodiversity. The exact numbers depend on the avian taxonomy used; the BirdTree taxonomy of Jetz et al. (2012) containing a total of 9993 global bird species, while the Birdlife International taxonomy (also used by the International Union for Conservation of Nature (IUCN)) contains 10,994 extant bird species. Based on the latter, a recent study, which defined an island endemic as any species that only occurs on islands that were not connected to the mainland during the last glacial period, reported 1715 true island endemics (i.e. 16%) (Sayol et al. 2020). Of these, 32% are currently considered to be threatened (i.e. listed by the IUCN as Critically Endangered, Endangered, or Vulnerable), considerably in excess of the 10% of non-island endemic species worldwide (Matthews et al. 2022).

Earlier studies by Adler (1992, 1994), although doubtless in need of update, indicate some general patterns across the Indian Ocean and tropical Pacific based on then known distributions of extant and recently extinct species. Using data for 14 archipelagos in the Indian Ocean, he found that richness of each of regional endemics species shared with continents, and total species number increased with the number of large islands and total land area (Adler 1994). In addition, less remote and low-lying islands tended to have more continental species (of the low islands, only Aldabra had its own endemics), and higher islands tended to have more local endemics and proportionately fewer continental species.

Similar results were obtained for the birds of the tropical Pacific, but the Indian Ocean avifauna contained fewer species, had lower proportions of endemics, and had several families that were much more poorly represented, despite being diverse in the mainland source areas (Adler 1992). The generally small size of the Indian Ocean islands may be significant in these differences. The total land area is only 7767 km^2 (excluding Madagascar), compared with 165,975 km^2 for the Pacific study, and there are also fewer island archipelagos in the Indian Ocean and they tend to be less isolated. The poor representation of certain families, notably the ducks and kingfishers, might be due to a shortage of suitable habitats, such as freshwater streams and lakes. Intra-archipelago speciation has been rare in the Indian Ocean avifauna, possibly occurring once or twice in the Comoros and a few times in the Mascarenes. Adler (1994) suggested that there may be too few large islands to have promoted intra-archipelago speciation in the Indian Ocean. In contrast, it has been common within several Pacific archipelagos, accounting for most of the endemic avifauna of Hawaii, being important in the Bismarcks and Solomons, and occurring at least once in New Caledonia, Fiji, the Society Islands, Marquesas, Cooks, Tuamotus, and Carolines.

Mammals

With the exception of bats, native mammals are not a feature of the most isolated oceanic islands, but less isolated islands include many that either once hosted or still do host interesting assemblages of native and/or endemic mammals. Indeed, a global analysis of mammalian richness on all marine islands smaller than Greenland identified 782 single-island endemic species, the majority of which (86%) were non-volant (Barreto et al. 2021). Using generalized linear models, it was found that

island area and isolation (both past and current isolation) were positively related to the number of single-island endemics, as were a range of climatic variables, such as mean temperature (positive effect) and precipitation (negative effect). In addition, islands that experienced greater climate change velocity (i.e. the change in temperature since the LGM, 21 ka) were also found to support a larger number of single-island endemics (Barreto et al. 2021).

Perhaps the most impressive examples of mammalian island endemism are Madagascar and the Philippines. In terms of the former, 88% of the mammalian fauna has been identified as being endemic (Barreto et al. 2021). In terms of the latter, the Philippines is an archipelago of 7000 mostly true oceanic islands, featuring 170 species of mammals in 84 genera, of which 24 (29%) genera and 111 (64%) species are endemic (Heaney et al. 2005). Particularly impressive radiations have occurred in fruit bats and murid rodents, with patterns of endemism in the non-volant mammals clearly linked to the configuration of the islands as they were during the major low sea-level stands of the Pleistocene. Heaney et al. (2016) reported that between 2000 and 2012, intensive survey efforts doubled the number of native non-flying mammals known from the large island of Luzon, from 28 to 56. Nearly all the newly discovered species were found to be restricted in distribution to a single mountain or mountain range, highlighting a remarkable degree of within-island diversification and indicating that more species remain to be discovered in these islands.

As an example of a less isolated group, the islands in the Sea of Cortés (off the western coast of the USA and Mexico), the terrestrial mammal fauna, including known extinctions, comprises 45 species (across 14 genera), with the majority of these being rodents, and of which 18 are considered endemics (40%) (Lawlor et al. 2002). Twelve species of bats have also been recorded, although no endemic forms are reported, likely due to the vagility of bats and the close proximity of the islands to the mainland (Lawlor et al. 2002). The evolution of 18 non-volant endemic mammal species on these islands highlights, however, how speciation can occur at relatively high rates even on less isolated islands when the taxon under focus has relatively low mobility (Chapter 9).

In the tropical Pacific, Carvajal and Adler (2005) note that a number of species of marsupials and rodents occur on the Solomons and Bismarcks, neither archipelago being particularly remote from New Guinea. But for the more remote islands, beyond the reach of non-volant terrestrial mammals, mammal representation is restricted to bats, of which flying foxes are arguably the most important group, although other types of bat occur on islands. Indeed, some 925 species (60% of all bat species) occur on islands and about 25% of all bat species are island endemics, 8% being endemic to particular islands (Fleming and Racey 2009).

There are believed to be 161–174 species of flying fox (Chiroptera: Pteropodidae) and they are distributed throughout the Old World Tropics (but not the New World). In number of species, range, and population sizes, flying foxes have been the most successful group of native mammals to colonize the islands of the Pacific. The Pacific land region consists of about 25,000 islands, most of them very small. In addition to occupying small islands, most flying foxes have restricted distributions. Thirty-eight of the 55 island species occupy land areas of less than 50,000 km², 22 less than 10,000 km², while 35 occur only on a single island or group of small islands (Wilson and Graham 1992). Island bats are particularly important for their ecological roles as seed dispersers and pollinators: unfortunately, many island endemic bats are considered to be threatened, amounting to half of all threatened bat species (Fleming and Racey 2009).

Comparisons between taxa at the regional scale

Within the figures cited above, those from G. H. Adler and his colleagues, although now a little dated, provide a common methodological approach and enable comparisons of patterns of endemism of different taxa at the regional scale (Table 4.3). The studies are based on 26–30 tropical Pacific Ocean islands or island archipelagos, plus in one case 14 Indian Ocean islands/archipelagos. The data include all known extant species plus others known only from subfossils.

Both birds and butterflies are capable of active dispersal over long distances and have colonized virtually every archipelago and major island within the tropical Pacific. Birds have a higher frequency of endemism than butterflies. Adler and Dudley (1994) consider that birds have superior dispersal ability, and that this might be anticipated to have maintained higher rates of gene flow than in butterflies, contrary to the higher degree of endemism. One plausible explanation is that the lower rate of endemism in butterflies might be a consequence of the constraints of the co-evolutionary ties with particular host plants required by butterfly larvae. This may limit their potential for rapid evolutionary change on islands (see above). On the other hand, the capacity of lepidopterans to reach moderately remote oceanic islands should not be too lightly dismissed, as evidenced by Holloway's (1996) long-term light-trap study on Norfolk Island, which recorded 38 species of non-resident Macrolepidoptera over a 12-year period. Norfolk Island is 676 km from New Caledonia, 772 km from New

Table 4.3 Degree of species endemism among tropical Pacific and Indian Ocean island faunas. (From Adler 1992, 1994; Adler and Dudley 1994; Adler et al. 1995; Carvajal and Adler 2005.)

Group	Total number	Continental	Regional endemics	Local endemics
Pacific Ocean butterflies	285	157 (55%)	28 (10%)	100 (35%)
Pacific Ocean skinks	100	21 (21%)	13 (13%)	66 (66%)
Pacific Ocean birds	592	145 (25%)	59 (10%)	388 (65%)
Pacific Ocean mammals	106	42 (40%)	7 (6%)	57 (54%)
Indian Ocean birds	139	60 (43%)	10 (7%)	69 (50%)

Continental, species also occurring on continental landmasses; **regional endemics**, species occurring on more than one archipelago within the region but not on continents or elsewhere; **local endemics**, species restricted to a single archipelago or island. NB: Carvajal and Adler (2005) mistakenly give 292 as the figure for local endemic Pacific Ocean birds. The correct figure is as given above (G. H. Adler, personal communication, 2005).

Zealand, and 1368 km from the source of most of the migrants, Australia. Most of the arrivals appear to be correlated with favourable synoptic situations, such as the passage of frontal systems over the region.

Lizards are also widely distributed in the tropical Pacific. The proportions of skinks in the three categories of endemic are remarkably similar to the equivalent figures for Pacific birds (Table 4.3). In general, in each of skinks, geckos, birds, and butterflies, the proportion of species endemic to an archipelago is best explained by reference to the number of large (high) islands and to total land area. However, scrutiny of the data on an archipelago-by-archipelago basis reveals greater differences. For instance, 100% of New Caledonia's skinks are endemic to the Pacific Ocean islands, and as many as 93% are endemic to the New Caledonian islands themselves. The equivalent figures for birds (including subfossils) are 47% and 33%, respectively (Adler et al. 1995). These data contrast with those for Hawaii, on which most birds are endemic, but where there are no endemic lizards. Adler et al. (1995) suggest that these differences are explicable in relation to the differing dispersal abilities of lizards and birds. Birds, being better dispersers, reached Hawaii relatively early and have radiated spectacularly, whereas the three species of skink may only have colonized fairly recently and most likely with human assistance. New Caledonia, in contrast, may not be sufficiently isolated to have allowed such a degree of avifaunal endemism to have developed.

The proportions of Pacific island mammals in each of the endemism categories falls closer to the figures for birds and skinks than those for butterflies. However, the five marsupial species make it no further than the Bismarcks, and the 18 species of native rodents no further than the Bismarck or Solomon islands where 14 of the 18 species are endemics (Carvajal and Adler 2005). The remaining native mammal species indicated in Table 4.3 are all bats. Variation in richness in the mammal data thus appeared explicable as a function of a combination of intra-archipelago speciation in archipelagos that include several large islands, and inter-archipelago speciation, particularly among more isolated archipelagos.

To summarize, while different taxa have radiated to different degrees on particular islands or archipelagos, it appears to be the case that the greatest degree of endemism is found towards the extremity of the dispersal range of each taxon (the **'radiation zone'** *sensu* MacArthur and Wilson 1967). Islands that are large, high, and remote typically have the highest proportion of endemics. In total, islands account for significant slices of the global biodiversity cake. We will return to the theme of the generation of island endemism in Part III.

4.10 Cryptic and extinct island endemics: a cautionary note

The foregoing account made only passing note of the 'state of health' of island endemics, and some of the assessments reviewed have included many highly endangered species and others already believed to be extinct. Before moving on from the geography of endemism to an examination of the theories that may account for the patterns of island ecology and evolution, it is important to consider both the problem of rare cryptic forms and the extent of the losses already suffered. Otherwise, there may be a danger of misinterpreting evolutionary patterns that are either incompletely described or are merely the more resistant (or lucky) fragments of formerly rather different tapestries (Steadman 1997, 2006; Duncan et al. 2013; Illera et al. 2016; Sayol et al. 2020). Our goal here is merely to introduce the topic of insular extinction, which we return to in full in Part IV.

Given the vast number of islands, especially in the tropics, and the general tendency for diversity to concentrate in less apparent taxa (beetles, and the like), it is clear that many undiscovered species remain for collectors and taxonomists (the so-called Linnean shortfall) among these relatively cryptic taxa. Only slightly more surprising is that we continue to discover new species of vertebrates and that many of the newly discovered species turn out to have remarkably confined montane ranges (Heaney et al. 2016). In the Canaries, which might be expected to be very well known scientifically, two quite sizeable endemic lizards new to science but of cryptic habits were discovered during the 1990s, *Gallotia intermedia* on Tenerife (Hernández et al. 2000) and *Gallotia gomerana* on La Gomera (Valido et al. 2000). Following the first discovery of *G. intermedia* in the Teno peninsula of Tenerife, a further population was found in another location adjacent to a teeming tourist resort.

Even in birds, probably the best-studied vertebrate group, 'new' species have been described very recently on islands. For example, a six-week expedition to three poorly studied island groups to the north-east of Sulawesi, Indonesia, discovered five new species and five new subspecies of birds, including the Peleng fantail (*Rhipidura habibiei*) and the Taliabu leaf-warbler (*Phylloscopus emilsalimi*) (Rheindt et al. 2020). Unfortunately, no sooner had these taxa been discovered than some of them were flagged as being of conservation concern given that many of the islands they are endemic to have suffered substantial habitat loss. Modern-day island discoveries not only relate to species completely new to science, but also to the re-discovery of species not seen for decades, and even centuries (so-called 'Lazarus species'). Perhaps the most famous recent example of this is in the context of islands is the black-browed babbler (*Malacocincla perspicillata*). The species was known from only a single specimen collected in the 1840s, until finally re-discovered in the rainforests of Borneo in 2020 (Akbar et al., 2021).

The current rate of description of new terrestrial animal species for the Canaries is one species every 10 days (Machado 2022). This represents a combination of new collection and the description and analysis of specimens leading to the splitting of taxa. For example, the beetle *Nesotes fusculus*, known from Gran Canaria, Tenerife, and La Gomera and thought to be a single species, was found to be paraphyletic, the *fusculus* phenotype having evolved independently in the dry coastal lowlands of the three islands (Rees et al. 2001). There are a number of forms of evidence that may be used to determine that two geographically distinct (allopatric) populations are 'good species' as opposed to varieties or subspecies. Given that the capacity to interbreed cannot be determined in the field, traditional morphological data have increasingly been supplemented or largely replaced by genetic evidence. In the case of the endemic blue chaffinches (*Fringilla teydea*) found in two of the Canary Islands, Lifjeld et al. (2016) base their case to distinguish *Fringilla*

teydea (Tenerife form) from *Fringilla polatzeki* (Gran Canaria form) on a combination of genetic evidence, body morphometrics, song and plumage characteristics, and sperm lengths. Illera et al. (2016) argue that other Canarian bird species showing variation between islands should similarly be subdivided.

Turning to already extinct island species, it has recently been established by analysis of cave sediments, etc., that around 66% of the endemic and 13% of the native avifauna of the Canaries have become extinct since human colonization around 2500 years ago, with ground nesting, flightless, and weak-flying species faring particularly badly (Illera et al. 2016). In Hawaii, the original avifauna of at least 111 species was reduced by 51% following Polynesian colonization around 1500 years ago, and a further 20% have become extinct since then, leaving just 29% extant (Boyer et al. 2014). Such dramatic losses greatly reduce functional diversity and have cascading impacts on ecological processes such as pollination and seed dispersal on islands (Boyer and Jetz 2014; Heinen et al. 2018).

As the potential archives of prehistoric bird bones on many prehistorically occupied islands, especially in the Pacific, have yet to be systematically analysed, there are likely to be many more extinct species yet to be catalogued. Indeed, empirically calibrated models indicate around 1000 species of non-passerine land birds alone became extinct from remote Pacific islands following human colonization (Duncan et al. 2013). Overall, since AD 1500, 149 of the 164 bird extinctions (91%) listed by the IUCN have been island endemic species (Matthews et al. 2022). Extending this to include what we know of extinctions pre-AD 1500, 477 out of 595 known avian extinctions (80%) are of probable island endemics (Sayol et al. 2020; Matthews et al. 2022).

Anthropogenic island extinctions have by no means been restricted to birds (Fernández-Palacios et al. 2021a). For example, the islands of the Caribbean held around 140 terrestrial mammal species at the onset of the Holocene, of which 80% of non-volant species and 25% of endemic bats have since become extinct (Upham 2017). The first wave of extinctions, particularly impacting larger-bodies species, occurred following Amerindian settlement from around 6 ka, with around 29 species becoming extinct following European arrival at the end of the 15th century AD.

It is unsurprising, but significant biogeographically, that in addition to species becoming globally extinct during the Holocene, many species have gone extinct from particular islands, while surviving on others (e.g. Illera et al. 2016; Steadman and Franklin 2020), and the population sizes of many surviving species will also have been reduced. As such, it is not just island species richness that will have been affected by human impact, but also archipelago composition patterns, patterns of relative abundance across species, and species distribution patterns: all topics discussed at length in subsequent chapters.

Some of these losses will have occurred as a result of climatic change and the associated sea-level changes and habitat alterations at the end of the Pleistocene, but it is increasingly evident that human colonization inevitably has generated a significant, often massive, extinction spike, especially among island vertebrates, driven by human activities, including hunting, habitat alteration, and the introduction of exotics. Interpretation of the present-day biogeography of many island regions thus requires knowledge not only of past faunas but also of the extent to which humans have been involved in their alteration. As Steadman (2006, p. 467) argues in relation to birds in Oceania, 'in Remote Oceania, where human residency does not exceed 3500 years, major islands have lost from 20 to 100% of their land bird populations; the Polynesian "heartland" of Tonga, Cook, Society, and Marquesas Islands, has modern avifaunas so depleted from their condition at human contact as to challenge biologically cogent analyses'. Unfortunately, mitigation of this deficiency via the fossil record is challenging because it is so patchy. For example, in the Society Islands, when historical and prehistorical records are included, the island of Huahine has the same number of land bird species (19) as Tahiti, despite the former being more than an order of magnitude smaller than the latter (Steadman 2006). This disparity is driven by the presence of the Fa'ahia fossil excavation on Huahine, where prehistoric fossil specimens have increased the island's richness by 11 species (i.e. 11 species are only known from prehistoric fossils). In contrast, Steadman (2006) reported

that there was no prehistoric evidence of birds available from Tahiti at his time of writing.

The difficulty of this task is further illustrated by a study from Henderson Island. In his investigation of 42,213 bird bones, 31% of which were identifiable, Wragg (1995) found that 12% of the fossil bird species recorded were accounted for by just 0.05% of the total number of identifiable bones, indicating them to be of uncertain status, quite possibly vagrants. Consequently, biogeographical studies that rely simply on a list of fossil birds might, exceptionally, assign resident status to temporary inhabitants. A different form of bias probably occurs in other studies because of the large mesh sizes (6 or 13 mm) commonly used in sieving soil samples. Larger-boned species such as seabirds, pigeons, and rails are usually found, but small passerines and hummingbirds are much less likely to be recovered (Milberg and Tyrberg 1993). Our knowledge of anthropogenic island extinctions in the prehistoric period for invertebrates and plants is undoubtedly even patchier (e.g. Nogué et al. 2017).

Currently we have only a fragmentary picture of past losses, biased towards vertebrates and especially birds, which tend to have a better-studied and more informative insular fossil record. Much is still unknown. What, for instance, has been the impact of the losses from New Zealand of the giant moas on the dynamics of forests they formerly browsed? How might the disappearance of hundreds of populations and a few entire species of Pacific seabirds have influenced marine food webs, in which seabirds are top consumers (Steadman 1997; Graham et al. 2018)? What of the island fruit bats and land birds (including many now extinct) and their roles as plant pollinators, dispersers, and (some birds) seed predators? The loss of these critically important (often keystone) vertebrates must have had important repercussions for other taxa, many of which may be ongoing (Olesen et al. 2002; Boyer and Jetz 2014; Heinen et al. 2018; Carpenter et al. 2020).

Steadman (1997) has pointed out that the Kingdom of Tonga did not qualify in a recent attempt to identify the Endemic Bird Areas of the world because of its depauperate modern avifauna, yet bones from just one of the islands of Tonga, 'Eua, indicate that at least 27 species (increased to 35 or 36 in Steadman 2006) of land birds lived there before humans arrived, around 3 ka. Forest frugivores/granivores have declined from 12 to 4, insectivores from 6 to 3, nectarivores from 4 to 1, omnivores from 3 to none, and predators from 2 to 1 (Steadman 1997). As is consistent with predation by humans, rats, dogs, and pigs, the losses have been more complete for ground-dwelling species. Hence, it is highly likely that the means of pollination and seed dispersal, not just of the plants of 'Eua, but of islands across the globe, have over the past few thousand years been diminished greatly as a consequence of such losses (e.g. Heinen et al. 2018).

Finally, not only have humans eliminated many native and endemic species from islands in historic and prehistoric periods, but we have also introduced huge numbers of species that had not reached and colonized them naturally (Seebens et al. 2017). This includes everything from species native to adjacent islands within the same archipelago to commensals, weeds, and pests, deliberate and accidental introductions from far distinct points on the Earth. It is not always easy to distinguish the native from the anthropogenically introduced (Nogué et al. 2017). We have thus distorted and overridden the effects of geographical isolation and created new anthropogenic patterns of island biogeography (Helmus et al. 2014; Sayol et al. 2020; Gleditsch et al. 2023). These patterns can be quite different from those predicted under classic island biogeography theory. For example, introduced species have been shown to exhibit positive species–isolation relationships (Moser et al. 2018; Gleditsch et al. 2023)—possibly due to more isolated islands (i) being unsaturated in terms of the number of species, and (ii) containing a less resistant and more naive biota, and thus providing greater opportunities for successful invasions—the opposite of the negative relationships commonly predicted for native species (e.g. MacArthur and Wilson 1967). That the attrition of native and endemic species from oceanic islands and the gain of non-native species may have greatly altered distributional patterns, compositional structure, and ecological processes, to degrees largely unknown (Nogué et al. 2017), serves as a caution on our attempts in the following chapters to generalize biogeographical and evolutionary patterns across islands.

4.11 Summary

This chapter set out to establish the biogeographical and biodiversity significance of island biotas, particularly remote island biotas, and to provide an indication of the ways in which island assemblages are distinctive. In global terms, for a variety of taxa, islands make a contribution to biodiversity out of proportion to their land area, and in this sense collectively they can be thought of as 'hotspots'. Some of course, are considerably hotter in this regard than others. Their high biodiversity value is widely acknowledged, as is the threat to island biodiversity, and islands thus typically feature prominently in global and regional conservation prioritization schemes.

Accompanying their high contribution to global diversity islands are, however, typically species poor for their area in comparison with equivalent-sized areas of mainland. The more isolated the island, and the less the topographic relief, the greater the impoverishment. Remote island biotas are typically disharmonic (filtered) assemblages, differing in relation to source areas, for example, by climatic filtering and by subsampling preferentially more dispersive taxa. To add nuance, an island may be a remote isolate (oceanic in character) for a taxon with poor ocean-going powers, but be linked by fairly frequent population flows (continental in character) to a source area for a highly dispersive taxon.

Traditional zoogeographic and phytogeographic analyses have enabled biogeographers to identify unidirectional dispersal filters (so-called sweepstake dispersal routes), filter effects between different biogeographical regions, and cases where islands have been subject to the influence of multiple source areas. We appraise the issues involved in placing oceanic islands into traditional biogeographical regionalism schemes, showing that remote islands typically receive colonists from multiple sources and via multiple routes.

Historical biogeography has for some time been embroiled in debates between those advocating a prominent role for long-distance transoceanic dispersal and those largely dismissive of the biogeographical significance of such processes, favouring plate tectonic and/or sea-level transgression scenarios of barrier creation, involving the splitting of ranges (vicariance events). Work reviewed in this and the previous chapter on the environmental history of islands and on molecular phylogenies, argues for both sets of processes playing a role, while emphasizing the prime importance of long-distance dispersal in supplying colonists to remote islands.

In some cases, island forms may have changed less than the mainland lineages from which they sprung, or have persisted while the mainland population has failed, in which case they may be termed palaeoendemics. This contrasts with the more common neoendemism, where novelty has arisen predominantly in the island context. Molecular phylogenetic analyses are in the process of transforming our understanding of island lineages and their relationships to continental sister clades, as illustrated herein by extensive reference to the complex colonization and evolutionary history of the Atlantic islands of Macaronesia.

In this chapter we have avoided discussion of the mechanisms of evolutionary change on islands, although noting some correlates of high proportions of endemics in island biotas. With increasing isolation, island size, and topographic variety, the number and proportion of endemic species increases. Endemism within a taxon appears to be at its greatest in regions near the edge of the effective dispersal range.

Some lineages have done particularly well on oceanic islands, land snails and fruit bats among them. Particular genera, such as the Hawaiian *Drosophila* (fruit flies) and *Cyrtandra* (a plant example) have radiated spectacularly. Terrestrial mammals have made a good showing only on less remote islands, often those insufficiently remote to have escaped the attention of *Homo sapiens* at an early stage, and many have thus failed to survive to historical times. New discoveries continue to be made of living (extant) and fossil (extinct) island forms and these serve as a cautionary note: such data must be appraised carefully when assessing biogeographical theories.

PART II

Island Ecology

Frontispiece: Peak of Teide, from near Teno Alto, Tenerife, March 2016. The Teno massif is one of the oldest parts of Tenerife and the steep and dissected terrain of the foreground is typical of volcanic islands in the intermediate stage of their ontogeny. The variation in climate from the coast to the summit (c. 3715 m asl) provides a great range in habitats. Photograph by Robert J. Whittaker.

CHAPTER 5

Island macroecology

5.1 From pattern to process

Having established the biogeographical significance of islands in Chapter 4, we turn in Parts II and III to the theories that have been developed from their study. We take our lead from Robert H. MacArthur and Edward O. Wilson's (1967) seminal monograph *The Theory of Island Biogeography*, by working from a consideration of ecological patterns and processes, through island evolutionary processes, to the emergent evolutionary outcomes. Their theory persists as an active research framework—arguably reinvigorated by the application of new analytical and data science tools—despite multiple claims of its refutation (e.g. Gilbert 1980; Bush and Whittaker 1991) and calls for replacement paradigms to be devised (Lomolino 2000a). As a research programme, its remarkable longevity reflects: (i) its focus on the key biogeographical processes of (im)migration, local extinction, and speciation; (ii) that these processes must relate in some degree to island area and isolation; and (iii) its simplicity (Losos and Ricklefs 2010; Warren et al. 2015). By treating species initially as equivalent, the core MacArthur–Wilson model provides a largely neutral model for understanding patterns in diversity across islands, onto which increasing biological and geo-environmental realism can then be built.

We therefore begin with an account of the MacArthur–Wilson theory, following which we explore current understanding of the patterns in diversity it seeks to explain. As this exploration exploits the statistical analysis of emergent properties of island diversity datasets, we have labelled the chapter island **macroecology** (*sensu* Brown 1995) We will consider multiple diversity components (e.g. chorotypes; from single-island endemics to non-native species) across different categories of island (from habitat to oceanic) and the explanatory power of island area, isolation, elevation, habitat diversity, and climate. As the general rationale is to understand the natural ecological processes governing island diversity, introduced (non-native) species are typically excluded from analysis, although, as we shall also see, their study can provide valuable insights and tests of concepts.

Much of the chapter is concerned with **island species–area relationships (ISARs)**, but we also consider less-well researched macroecological properties, such as the species abundance distribution (SAD) and species–occupancy relationship, as they are important to building a more complete theory of insular diversity. While it is easy to stress the noisiness of macroecological patterns, the extent to which repeated emergent patterns can be identified is perhaps more remarkable: it is worth continuing the search for the dynamic models and biological laws that must underpin them.

In the final section we focus on the evidence for turnover through time as island systems approach dynamic equilibrium—a key premise of the MacArthur–Wilson theory. We argue that this condition is approximated to varying degrees, allowing us to identify a range of insular system descriptions along two axes, first from low to high environmental variability and second from low to high biodynamism. In the present chapter, we take a deliberately simplified approach, following MacArthur and Wilson (1967), by focusing on ecological dynamics in the absence of ideas of predictable geo-environmental dynamism. We will add detail in Chapter 6 by looking at patterns in island community membership, including

Island Biogeography. Robert J. Whittaker, José María Fernández-Palacios, and Thomas J. Matthews, Oxford University Press.
© Robert J. Whittaker, José María Fernández-Palacios, and Thomas J. Matthews (2023). DOI: 10.1093/oso/9780198868569.003.0005

through the analysis of functional and phylogenetic diversity, before going on to consider how geo-environmental dynamics may influence the operation of island biodynamics over extended time frames in Chapter 7.

5.2 MacArthur and Wilson's equilibrium theory of island biogeography

One of the most influential contributions to ecological theory of the past century, the equilibrium theory of island biogeography was first published as a zoogeographic model in the journal *Evolution* and subsequently developed as a general theory covering all taxa and all forms of island in their classic monograph (MacArthur and Wilson 1963, 1967). It represented a bold attempt to reformulate (island) biogeography around fundamental processes and principles, relating distributional pattern to population ecological processes, and at its heart lies a very simple, but powerful concept. The theory states that: (i) the number of species of a given taxon found on an island will be the product of

opposing forces leading respectively to the gain and loss of species, (ii) resulting in the development of a **dynamic equilibrium**, involving a continued turnover of the species present on each island through time, with (iii) the rates of the key processes varying in specific predictable ways as a function of island area and of island isolation.

Several ingredients were combined in the derivation of the dynamic equilibrium model (Williamson 1981; Losos and Ricklefs 2010; Warren et al. 2015). First, the existence and apparent regularity of form of species–area relationships (Fig. 5.1), suggestive of a powerful general mechanism by which larger areas provide more resources and can thus support more species. Second, the distance effect, exemplified by the role that increasing island isolation appeared to have in driving reduced richness of Pacific island birds. Third, the notion of the turnover of species on islands, a force that Wilson (1959, 1961) had previously invoked on evolutionary timescales in the taxon cycle (Section 9.2), and which other biologists had described as occurring on ecological timescales; for instance, during the recolonization

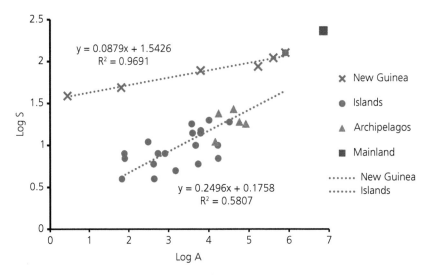

Figure 5.1 Species–area relationships (SARs) for ponerine (including cerapachyine) ants in Melanesia. The uppermost regression line bisects six data points that represent a species accumulation curve (SAC) for New Guinea, the final point being the whole of New Guinea. The lower regression line is the island species–area relationship (ISAR) for 20 islands (including New Guinea), excluding the five datapoints representing entire archipelagos. The SAC, being constructed from contiguous plots in a nested sampling system, has a low slope but a tight fit and thus a high R^2 value, whereas the ISAR, being for isolates and non-nested in construction, is steeper with characteristically larger scatter, although the fit is significant. As both area (km^2) and species number have been logged, the equations represent the (log) power model. Note that SACs and ISARs are distinct types of SAR and as explained later in the text their parameters are thus not strictly comparable.
Data points digitally captured (with thanks to Kostas Proios and Kostas Triantis) from Wilson (1961) and replotted with newly fitted regression models.

of the Krakatau Islands (below). Noting the general prevalence of rare species as established by work on SADs (Section 5.9, below), they made the inference that those species present within an island at very low densities would have little chance of persisting and should frequently go locally extinct. And, the further potential colonists had to travel from the mainland, the lower their chance of colonizing (or re-colonizing). That is, turnover (Box 5.1) should be a consistent and continual feature of island systems and the rate of turnover at equilibrium should vary predictably with island area and with isolation.

They integrated all these features in the famous graphical model (Fig. 5.2a), in which immigration rate declines exponentially and extinction rate rises exponentially as an initially empty island fills up towards its equilibrium richness (Fig. 5.2b). They argued that the immigration rate curve will flatten with increasing distance from source pools, and the extinction rate will rise more slowly with increasing island area, as larger islands have more resources and can accommodate more viable species populations. Hence, the model generates a family of curves predicting unique combinations of equilibrium richness and turnover for each combination of area and isolation.

Colonization curve. The change through time of numbers of species found together on an island.

Extinction. The total disappearance of a species from an island (does not preclude recolonization).

Extinction rate. Number of species on an island that become extinct per unit time.

Turnover rate. The number of species eliminated and replaced per unit time.

MacArthur and Wilson's definitions are fine in theory but are difficult to apply (Section 5.10). Propagule arrival is incredibly hard to measure and typically in empirical studies the immigration data used are closer in form to the definition of colonists given above. Survey data also often lack the detail to allow researchers to distinguish between a species that has immigrated and one that is observed in transit or in insufficient numbers or condition to establish a breeding population. Similarly, the total disappearance of a relatively inconspicuous plant or animal from a large, topographically complex island may also be hard to prove. MacArthur and Wilson were, of course, aware of these problems and suggested workarounds to enable tests of the theory.

Box 5.1 Definitions of terms involved in analyses of species turnover

MacArthur and Wilson's (1967, pp. 185–191) definitions are as follows:

Immigration. The process of arrival of a propagule on an island not occupied by the species. The fact of an immigration implies nothing concerning the subsequent duration of the propagule or its descendants.

Immigration rate. Number of new species arriving on an island per unit time.

Propagule. The minimal number of individuals of a species capable of successfully colonizing a habitable island. A single mated female, an adult female and a male, or a whole social group may be propagules, provided they are the minimal unit required.

Colonization. The relatively lengthy persistence of an immigrant species on an island, especially where breeding and population increase are accomplished

---

Assuming that turnover is a purely stochastic process, this graphical model provides a simple, biologically near-neutral model that appeared to explain the phenomena outlined above and which generated novel predictions (e.g. concerning rates of turnover). This is the core **equilibrium model of island biogeography (EMIB)**. However, MacArthur and Wilson's (1967) *theory* contained a considerable amount of additional development, including: (i) how 'stepping-stone' islands between a mainland and a focal island would boost immigration rates and modify equilibrium points and turnover rates, (ii) efforts to distinguish the traits of successful colonists, and (iii) the contribution of *in situ* evolution (cladogenesis). The highly simplified general model at the core of their theory was thus shown to be a building block on to which other ideas could be bolted and through which new hypotheses could be tested.

The EMIB is 'near' rather than perfectly neutral because the hollow curves in Fig. 5.2a recognize that species are not entirely equivalent. When the number of species on an island is small relative to

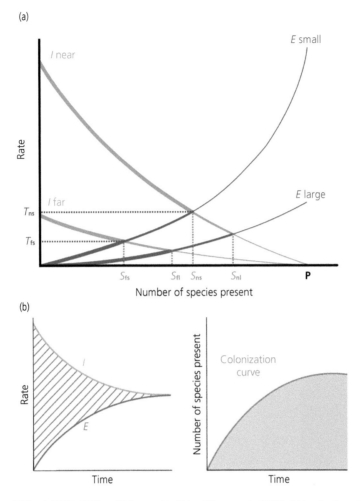

Figure 5.2 MacArthur and Wilson's (1963, 1967) equilibrium model of island biogeography (EMIB). (a) Immigration rates (I) are expected to vary as a function of distance from the source pool (P), and extinction rate (E) as a function of island area. Considering a newly created or sterilized island of given area and isolation, it will begin with zero species richness and we may then follow the colonization process to the point the curves intersect. Each combination of island area and isolation provides a unique combination of S and T (turnover) at equilibrium. To prevent clutter, only two values for T are shown. (b) Colonization of an empty island (zero richness) should lead to declining immigration as the proportion of the mainland species pool not on the island declines over time, so leaving fewer and less dispersive species yet to reach the island, while the extinction rate rises over time, as the island fills, until balancing immigration. Once attained, equilibrium should be maintained indefinitely.
Modified from MacArthur and Wilson 1967, their figures 8 and 20.

the mainland source pool, P, a high proportion of propagules arriving on an island will be of species not resident on the island but this proportion must decline as richness increases. Early colonists will typically be drawn from the most dispersive members of the source pool. As the island assemblage becomes richer in species, the potential colonists will be drawn from a remnant pool of decreasingly dispersive species and so the rate of immigration

(as per Box 5.1) declines exponentially. The hollow form of the extinction curve recognizes that some species are more likely to die out than others, and that in accord with our knowledge of SADs, the more species there are, the rarer on average each will be, and so the more likely each is to die out. At equilibrium, immigration and extinction rates should be approximately in balance and thus $I = E = T$ (Fig. 5.2a).

MacArthur and Wilson (1967) went further in developing a quantitative biological profile of the superior colonist and how island environments should 'select' for different species types at early and late stages of colonization. They borrowed from the standard notation of population models (r, the intrinsic rate of population increase; K, the carrying capacity), to describe early colonists as **r-selected**—highly mobile, opportunistic colonizers capable of rapid growth, maturity, and population increase, and later colonists as **K-selected**—slower dispersing and growing, but with greater ability to sustain their populations in resource-limited systems approaching the environmental carrying capacity. These biological differences are important to understanding ideas of island assembly, which we consider in detail in Chapter 6.

As isolation increases, the chance of propagules reaching an island must diminish, simultaneously limiting gene flow and leaving unfilled niche space. MacArthur and Wilson (1967) therefore argued (i) that contributions to island diversity from *in situ* diversification (cladogenesis) should increase with distance from mainland sources, but (ii) because of the relatively slow pace of phylogenesis, evolutionary adjustment on remote islands would not fully compensate for the reduction in equilibrium richness generated through the lower pace of immigration shown in Fig. 5.2. Hence, their model was consistent with two established patterns: the reduction in species richness with increased isolation and the propensity for clades to diversify in what they termed the **radiation zone**—the islands towards the outer limits of their dispersion into a given ocean. They also argued that rather than the species richness of such remote islands remaining fixed, evolutionary species packing would lead over time to a gradual elevation of the ISAR (Box 5.2).

Box 5.2 The MacArthur and Wilson model, time, and speciation

Perhaps because of its omission from the famous graphical model (Fig. 5.2a), it has often (incorrectly) been remarked that the equilibrium theory takes no account of speciation. The following excerpt, the first formal statement of the model at the core of the theory, shows that *in situ* cladogenesis was included alongside immigration, as a driver of increasing diversity.

'We start with the statement that $\Delta s = M + G - D$ where s is the number of species on an island. M is the number of species successfully immigrating to the island per year. G is the number of new species being added per year by local speciation (not including immigrant species that mainly diverge to species level without multiplying), and D is the number of species dying out per year' (MacArthur and Wilson 1963, p. 378).

In the more familiar notation of the 1967 monograph, the equation for equilibrium is:

$$S_{t+1} = S_t + I + V - E,$$

where S is the number of species at time t, S_{t+1} is the number at time t + 1, I denotes additions through immigration, V denotes additions through speciation (where applicable), and E denotes losses by extinction.

MacArthur and Wilson (1967) reserved to later chapters of their monograph consideration of the gain in species richness through cladogenesis within a remote archipelago. They argued that local adaptation would permit increase in the equilibrium point for each island over evolutionary time. This would serve to elevate the island species–area relationship (ISAR) of a remote archipelago so that were it sampled at several distinct time points, perhaps tens of thousands of years apart, the relationship would remain consistent in slope but increase in intercept (Box Figure 5.2a).

Wilson (1969) sketched how this might look from the perspective of the richness of a single island through time, modifying the colonization curve in Fig. 5.2b to recognize increasing complexity (Box Figure 5.2b). Here he was inspired by insights arising from a classic experimental test of the EMIB (Simberloff and Wilson 1969, 1970), but also from consideration of evolutionary change. The figure shows an initial short-lived non-interactive equilibrium (T1), followed by a slight reduction in diversity as competitive exclusion takes hold (T2). As new candidate species arrive over time, increased niche fitting can occur, leading to an assortative expansion in species richness (T3). This process would continue in evolutionary time as adaptive speciation occurs, raising species richness (T4) as per Box Figure 5.2a.

Box 5.2 *Continued*

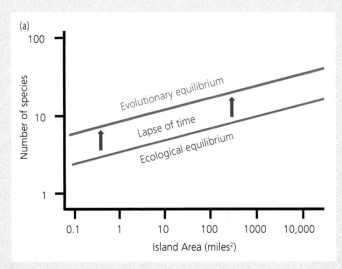

Box Figure 5.2a The potential evolutionary elevation of an ISAR (island species–area relationship) over time for a remote archipelago, based on figure 59 of MacArthur and Wilson (1967). The original sketch was an attempt to estimate the eventual upper limit (red) on the species density of the Polynesian ant fauna for a group of islands to which ants have been introduced by human commerce, forming an initial 'quasi-equilibrium' (blue): diversity levels were predicted to creep steadily upwards (arrows) over thousands of generations.

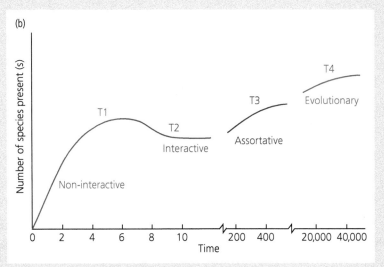

Box Figure 5.2b Wilson's (1969) figure 5 (redrawn) showing changing species richness over time for a hypothetical island, recognizing that the equilibrium value for an island (and by extension across an archipelago) may change over time (T1–T4). The units of time were said to be arbitrary but might correspond with years or generations.

These ideas about temporal change in equilibrium over time focus on still quite simplified biological processes. More recently, efforts have been made to consider how these biological dynamics may intersect with long-term geo-environmental dynamics of island systems (Chapter 7; Whittaker et al. 2008; Fernández-Palacios et al. 2016a),

To sum up, MacArthur and Wilson (1967) provided a highly simplified stochastic macroecological model (the EMIB) representing core processes responsible for regulating island diversity. But they also anticipated many of the developments and criticisms of such an approach. Within their monograph they systematically built upon the EMIB by adding biological detail—from short-term population dynamics up to emergent evolutionary outcomes. Fundamental to their theory is (i) the notion that islands typically tend towards dynamic equilibrium, (ii) even if that equilibrium may be subject to adjustment over time, it should remain dynamic (i.e. turnover should continue), and (iii) the rate of species richness increase with island area should persist largely unchanged. Although often criticized and found wanting in early empirical tests, their theory initiated a highly productive research programme that continues to hold relevance (Losos and Ricklefs 2010; Warren et al. 2015; Valente et al. 2020), as is evidenced below.

5.3 Species richness and area: the basics

Long before MacArthur and Wilson's collaboration, the father-and-son naturalists Johann Reinhold Forster and Georg Forster (who both sailed with James Cook on his second Pacific voyage in 1772–1775) made the following prescient observations about island diversity: 'Islands only produce a greater or lesser number of species, as their circumference is more or less extensive' (J. R. Forster 1778) and 'The small size of the island, together with its vast distance from either the eastern or western continent, did not admit of a great variety of animals' (G. Forster 1777). Note the reference here both to island *size* and *isolation*.

By the early years of the 20th century, building on further observations during the 19th century, the first mathematical depictions of the **species–area relationship** (SAR) were being debated (Matthews et al. 2021a). Olof Arrhenius (1920) led the way in proposing that SARs can be described by a power law, a non-asymptotic convex-upward curve commonly expressed using the formula $S = cA^z$ (where S is species number, A is area, and c and z are the parameters). Log transformation of the power model provides the formulation

$\log S = \log c + z \log A$. Objections were raised to the power model by Henry Gleason, who argued in papers published in 1922 and 1925 for what became known as the logarithmic model, $S = \log c + d \log A$ (where c and d are parameters). Whereas the power model provides a simple linear fit (useful for secondary analyses) when both area and species number are logged, the logarithmic model is linear in form when using a log–linear plot (i.e. area is logged but species number is not), as shown in Fig. 5.3. Many other formulas and approaches have since been proposed and the precise form taken by SARs has been subject to continuing theoretical and empirical attention (below; Preston 1962, Triantis et al. 2012, Matthews et al. 2021a). For a comprehensive history of the SAR see Tjørve et al. (2021).

Notwithstanding frequent reference to its status as one of ecology's few general laws, the SAR is not a singular phenomenon but refers to a suite of distinct types of relationship between area and diversity (Box 5.3). In much of the literature, little attention has been paid to the possibility of a systematic difference in form and/or parameters as a function of the different properties of SACs and ISARs (but see e.g. Preston 1962, Matthews et al. 2016b).

Herein, we are principally concerned with ISARs as defined in Box 5.3. A large literature quantifying

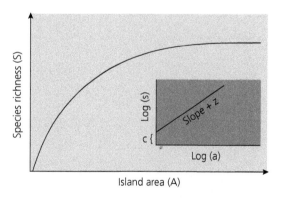

Figure 5.3 A hypothetical power model ISAR (island species–area relationship), revealing the curved form of the arithmetic version of the model, initially rising steeply but gradually flattening towards (yet not quite reaching) an asymptote (main panel), contrasting with the linear fit of the log version of the model (inset). NB Those interested in applications of the power model may wish to note that for a given data set the parameters of the log and arithmetic versions of the power model may not be identical.

Box 5.3 Types of species–area relationship: important variants on a theme

The term **species–area relationship** (SAR) refers to several distinct ways of measuring how the number (richness) of species varies with the area sampled, among which the configuration of areas and the approach to tallying both properties vary (Box Figure 5.3). Whether the sampled areas are contiguous or isolated from other areas of the same habitat type is one key property. As, or more important, is whether the number of species is tallied separately for each sampling unit, or a cumulative total is recorded, so that for each increment in area, only those additional species not previously encountered are added to the tally of species richness (Preston 1962). The latter approach commonly produces a convex upward curve, rising steeply and gradually flattening.

Scheiner (2003) identifies four sampling layouts commonly used, of which the first three (Box Figure 5.3a–c) are typically used to construct **species accumulation curves (SACs)**, while the final layout (d) refers typically to **island species–area relationships (ISARs)**. As many SARs and especially many SACs are curved when plotted, they are often termed **species–area curves**.

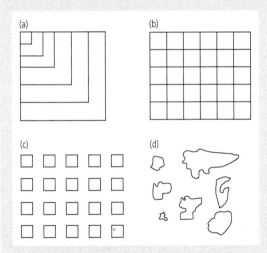

Box Figure 5.3 Sampling layouts commonly used in SAR research: (a) strictly nested quadrats, (b) a contiguous grid of quadrats, (c) a regular but non-continuous grid, and (d) a set of areas of varying size, often islands.
Source: Scheiner (2003).

Species accumulation curves: SACs

The species accumulation (or collectors) curve is a graph of the *cumulative* number of species (*y* axis) with increasing sampling effort (*x* axis), through, for instance: (i) adding

contiguous sampling plots (Box Figure 5.3, panels a or b), (ii) additional sample plots of a particular type from within a general locality (panel c), or (iii) additional sampling effort within the same sampling area (not illustrated). SACs showing the accumulation of richness with increased spatial extent of sample are forms of SAR (also called nested SARs or sample-area SARs), but those that involve increased sampling effort/duration within the same area are not. Researchers use such SACs to estimate and compare the species richness of different localities and to determine the appropriate sample size for community ecological programmes (Magurran 2004).

Island species–area relationships: ISARs

We use the term ISAR to refer to a particular type of SAR: analyses of how many species were found in each island as a function of island area. Thus, the only permitted sampling design is the non-nested type (Box Figure 5.3, panel d). To make this explicit we add the term island (isolate), to create the acronym ISAR. However, it is worth noting that ISARs can be built using data points that are not isolated in the true sense of the word, such as countries or biomes (e.g. Gerstner et al. 2014). It is primarily with ISARs (of island data!) that we are concerned in this book.

Application

Wilson's (1961) seminal graph (our Fig. 5.1) combines examples of both principal relationship types, with the nested sampling SAC from New Guinea providing a flatter, elevated (intra-provincial) curve in comparison to the ISAR from islands around New Guinea. The difference in form stems in part from the impact of isolation on richness of each area but also in part from the different mode of construction of the curves (Preston 1962; Rosenzweig 1995). We aren't aware of a systematic comparison of the two sampling structures from contiguous habitats, but Matthews et al. (2016b) have compared fitted SACs and ISARs for 97 habitat island datasets (as per Box Figure 5.3d). They found that the fitted slopes of the power model were steeper for the SAC than the ISAR for 77% of datasets. The differences in slope were related to the degree to which the datasets formed nested subsets (Section 6.4): the SAC is steeper when the system tends to anti-nestedness (linked to greater compositional dissimilarity between islands), and as the degree of nestedness increases, the differences in slope are reduced until the ISAR becomes steeper than the SAC. These findings provide a caution on comparisons of species–area models that include a mix of SAC and ISAR data structures: this appears to be the equivalent of comparing apples and

---

Box 5.3 *Continued*

---

oranges. Unfortunately, much of the literature fails to distinguish clearly between the different data structures and forms of SAR, which has arguably confounded many attempts at synthesis. It should also be noted that some authors who do distinguish different types of SAR nonetheless use different labels than we have adopted here. For a full review of the SAR literature in theory and practice, see Matthews et al. (2021a).

NB We use the acronyms SAC and ISAR—and later SAD (for species abundance distribution)—for convenience, but the reader should beware, the 'A' stands for different terms, accumulation, area, and abundance, respectively.

---

the ISAR for hundreds of archipelago/taxon combinations allows us to conclude that Arrhenius' power model is the best overall model, providing a satisfactory (often the best fit) to a large proportion of ISAR datasets (Triantis et al. 2012; Matthews et al. 2016a, 2021a). Island biogeography is not unique in this regard as power laws are known to characterize many phenomena across a range of fields, including physics, computer science, economics, and linguistics. Other models sometimes provide better fits to the ISAR, and there are datasets for which area fails to provide a satisfactory fit at all. This is far from surprising, given that so many other factors are known to influence diversity: island isolation, elevation, habitat diversity, climate, and age being some of the more important ones. As our capacity to gather and analyse such data has improved, many studies have made use of multivariate statistical tools that provide more complete models weighing up these and other variables alongside area (e.g. Kalmar and Currie 2006; Matthews et al. 2019a). We return to these more complex island diversity models below but first we explore variation in ISAR form in greater depth.

5.4 What shape is the island species–area relationship?

The exchange between Arrhenius and Gleason in favour, respectively, of the power and logarithmic models, marked the beginning of an enduring enquiry into the form of SARs (both SACs and ISARs). Both these models imply a simplicity of form, permitting the curve to be transformed into a straight line fit (respectively by log–log and log–linear transformation of area and species), thereby enabling evaluation by linear regression. Lomolino (2000b, 2002) has questioned this approach, arguing that there may be good ecological reasons to posit more complex scale-dependent relationships.

Considering a far fuller range in island area, Lomolino (2000b) proposed that untransformed ISARs should exhibit a sigmoidal form. He envisaged a phase across low values of area, where species numbers scarcely increase, followed by a rapid increase with area and a subsequent flattening as the number of species approaches the richness of the species pool (Fig. 5.4). Tjørve and Tjørve (2021) also argue that the ISAR should be sigmoidal given that, they argue, individual species incidence functions (the probability of occurrence of a species as a function of area; Section 6.3) are sigmoidal, and the ISAR is an additive combination of these curves. We will briefly review the background and evidence of such complexities, beginning with a focus on the left-hand end of the curve.

The small-island effect

The small-island effect, a phenomenon first graphically described by Niering (1956) before being integrated more formally into island theory by MacArthur and Wilson (1967) and Whitehead and Jones (1969), describes datasets in which there is no significant gain in species number across the smallest islands in a dataset, or, according to some studies, where the relationship between area and richness is different on small relative to larger islands (discussed in Dengler 2010). The small island effect has been observed for both true island (Lomolino and Weiser 2001; Morrison 2014, Matthews et al. 2020; Schrader et al. 2020) and habitat island

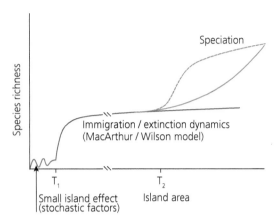

Figure 5.4 Mark Lomolino's proposal for a more complex two or three-phase island species–area relationship. In red and continuing in blue, a hypothetical sigmoidal model: as area increases from the origin there is stochastic fluctuation but no trend in species number (red) until beyond a first threshold area T_1, following which species number follows a convex form towards an asymptote (in blue), as a function of the immigration/extinction dynamics of the EMIB. Lomolino argued that for islands larger than area T_2, species number may once again begin to rise, through the additional contributions of *in situ* speciation. Lomolino (2000b) originally proposed a convex upward form for the contribution of speciation beyond T_2 (dashed orange line), with a tendency towards an asymptote, but Lomolino et al. (2017, their figure 13.20) show a concave upward rising curve (solid orange line).

datasets (Matthews et al. 2014b; Wang et al. 2018) but is far from universal. For those cases where a small-island effect has been found, the breakpoint area tends to be smaller where immigration is theoretically stronger; that is (i) for taxonomic groups that are more dispersive (Lomolino and Weiser 2001; Chisholm et al. 2016), and (ii) on less isolated archipelagos (Chisholm et al. 2016; Schrader et al. 2020).

Explanations for the small-island effect invoke (i) stochastic effects (e.g. disturbance) that generate variation in species number and drown out the gain expected from increased space on smaller islands (Fig. 5.4), or (ii) habitat/niche limitations in tiny islands of few, simple habitats (Whitehead and Jones 1969). Here, the argument is that islands are only able to accrue multiple, complex habitats above a critical threshold area, at which point richness can begin to scale with area. Evaluating such mechanisms has proven difficult. More generally, several studies have reported evidence supporting

the role of habitat diversity in driving the small-island effect (e.g. Triantis et al. 2006; Chen et al. 2020; Matthews et al. 2020). Taken together, these studies demonstrate that area is only ever a partial explanation for species richness variation on islands.

Chisholm et al. (2016) have put forward a biphasic model for the small-island effect, which contends that niche diversity increases only slowly with area, while immigration occurs at a low rate on small islands and then rapidly increases. Hence, niche effects dominate on smaller islands and immigration becomes more important on larger islands. Within the model there is a switch from a niche-structured regime for the 'flat' phase operating over the small islands, followed by an accelerating upward curve (in semi-log space) as colonization–extinction dynamics kick in for larger islands. The authors report good fits using 100 datasets (Chisholm et al. 2016). The idea that immigration rate varies with island *area* is, of course, at variance with the EMIB—see discussion of the 'target-area effect', Section 5.10.

Attempts to evaluate the small-island effect have thrown up several issues that have been hard to resolve (Dengler 2010; Matthews et al. 2014b; Wang et al. 2016; Gao et al. 2019; Chen et al. 2021; Matthews and Rigal 2021). First, there are several models, of varying complexity, that have been proposed to test for thresholds in the ISAR and there is no consensus yet on which is the best approach. Second, the smallest sampled islands sometimes contain no species, and it is unclear whether (or how) those zeros should be included in analysis. Third, detection of the effect may depend on arbitrary decisions taken regarding data transformation. Assessing this third issue, Matthews et al. (2014b) used breakpoint regression to test for thresholds in a meta-analysis of 76 habitat island datasets (Fig. 5.5). They repeated their analyses for: (i) no data transformation, (ii) log transformation of area only (semi-log), and (iii) log transformation of area and species richness (log–log) (Fig. 5.5). Threshold models were the best fits in 28%, 37%, and 32% of cases using untransformed, semi-log, and log–log transformations of the data, respectively. Thresholds were more detectable for datasets with larger numbers of habitat islands, likely reflecting the increased

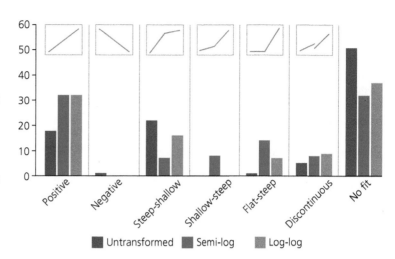

Figure 5.5 An analysis testing for threshold effects in 76 habitat island datasets using regression and breakpoint regression to determine the best model from a choice of simple linear (positive and negative) and four alternative threshold models. The analyses were undertaken with untransformed species and area data and repeated using semi-log (species number vs log-area) and log–log data. The inset graphs show the form of the models tested for ($x =$ area; $y =$ species richness) and the histograms show the percentage of systems best fitted by each model (or by no model). Data from Matthews et al. (2014b).

resolution provided. Outcomes consistent with the small-island effect (i.e. flat–steep in Fig. 5.5) were detected in 11 of 76 datasets using semi-log transformation, five datasets using log–log transformation, and one case using untransformed datasets. The breaks in rate of species gain, when occurring, were detected at areas of < 0.5 km² (50 ha); that is, among quite small habitat islands. Overall, threshold detection was shown to be highly sensitive to the initial approach to data transformation. This is illustrated by one dataset which returned a small-island effect (flat–steep) model using semi-log data and a steep–flat model for untransformed data (see Fig. 12.10b). In effect, we see how rescaling the data via log- or semi-log transformation focuses on different portions of the range of variation, thereby determining the shape we perceive the ISAR to be. A subsequent study of 90 habitat island datasets by Wang et al. (2018), using a flexible combination of semi-log and log–log transformation and different analytical protocols, was slightly more supportive of a small-island effect, which they detected in 40% of cases.

Revealing the full shape of the ISAR?

Expanding our focus from thresholds mostly found at small island sizes, two recent meta-analyses have analysed overall ISAR shape for a compilation of 601 true island datasets and 297 habitat island datasets (Triantis et al. 2012; Matthews et al. 2016a).

For each dataset the best model was determined from 20 alternatives, representing three shapes: (a) linear, (b) convex (e.g. the power and logarithmic models), and (c) sigmoid. The presence (a feature of 11 of the models) or absence (9 models) of an asymptote was also tested for. Among true island datasets, satisfactory models were obtained for 465 datasets, often with multiple models being deemed adequate. Assessing the best models by reference to their generality and efficiency, Triantis et al. (2012) report that: (i) most datasets can be best described as convex (upward) in form (as per main panel in Fig. 5.3) rather than sigmoidal (best in only 6% of cases), with linear models performing least well, while (ii) 87% of datasets showed no evidence of an asymptote.

So, how general is the sigmoidal ISAR proposed in Fig. 5.4? In defence of the proposition, as most datasets analysed are for specific archipelagos, it is possible that there is simply insufficient range in island area and isolation to reveal the underlying sigmoidal form posited in Fig. 5.4 (e.g. Tjørve and Tjørve 2021). Although it was the 'best' model in just 6% of cases in the Triantis et al. (2012) study, sigmoidal models nonetheless provided significant fits for many more datasets, especially where there was a larger range of island area. However, when scrutinized in detail, few curve forms in either of the meta-analyses bore much resemblance to the idealized sigmoidal form (RJW/TJM personal observation). At this point, for generality, simpler (convex) models win. Nonetheless, we should not regard debate on the form of ISARs as at an end. Interestingly, an analysis of ISARs for cichlid fishes in African lakes by Wagner et al. (2014) provides some support for the hypothesis that *in situ* diversification can at least in some circumstances, lead to an increased pace of species gain once beyond a critical area threshold (compare Figs. 5.4, 5.6; see also Losos and Schluter 2000).

To sum up, ISAR shape varies. A particular dataset can often be successfully fit by different models, sometimes by *both* a simple model and a more complex or multiphase model, whether showing the flat–steep form inherent to the small-island effect, or the steep–flat form that approximates the classic convex model of Fig. 5.3. There may be a biological case for more complicated models (Chisholm et al. 2016; Matthews et al. 2021a), but ISAR data are often noisy, other confounding variables are at play, and more complex models are correspondingly hard to demonstrate (if indeed they are valid). We therefore return next to consideration of what we can learn from the application of one of the simplest ISAR models, the power model.

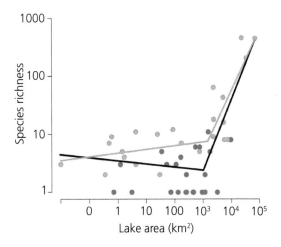

Figure 5.6 The biphasic island species–area relationship (ISARs) for cichlid fishes in African lakes, showing lakes without speciation (red dots), lakes with speciation (blue dots), and breakpoint regression models showing an increase in slope for lakes larger than around 1030 km² (black line, all lakes; blue line, lakes with speciation only). Redrawn from Wagner et al. (2014; their figure 3b).

5.5 The power model, *c* and *z*, and scale dependency of ISARs

In its log form ($\log S = \log C + z\log A$) the power model has the useful property of linearizing the curve (Fig. 5.3), which enables estimation of c and z to be determined by simple ordinary least squares (linear) regression. In this equation, z is the slope of the log–log relationship and $\log C$ the intercept. Thus, a lower z implies reduced sensitivity to island area, while $\log C$ indicates the richness of an island of unit area, which we expect to vary with taxon, climate, and biogeographical region. Values of both parameters may be compared across studies providing only that area is expressed in a standard scale of measurement (e.g. km²) and using the same logarithmic base, as otherwise $\log C$ values are incomparable.

In their seminal monograph, MacArthur and Wilson (1967) reported that z ranged between 0.20 and 0.35 for islands (ISARs), but that for non-isolated sample areas on continents (or within large islands), a range of 0.12–0.17 was obtained. Thus, slopes appeared to be steeper for ISARs, or, in the simplest terms, any reduction in island area lowers the diversity more than a similar reduction of sample area from a contiguous mainland habitat. It also follows that the intercepts will be lower for islands than for contiguous habitat. However, this analysis is confounded by the data structure differences between ISARs and SACs (above) and so it is important to establish how the ISAR parameters vary with increasing degree of isolation.

We now have both more datasets and more advanced statistical tools to do just that. In the meta-analyses discussed above, median values for slope were found to increase across the categories of island as expected based on their increasing isolation: habitat island systems $z = 0.22$, inland (lake) islands $z = 0.28$, continental-shelf $z = 0.28$, and oceanic islands $z = 0.35$, although there is a lot of variation around these values (Fig. 5.7a; Matthews et al. 2016a). Alongside variation in ISAR slope there is a corresponding degree of variation in the intercept, with the differences in average values of z and of $\log C$ between island categories conforming with theoretical expectation (Fig. 5.8). Nonetheless, there is also an enormous degree of scatter and a great

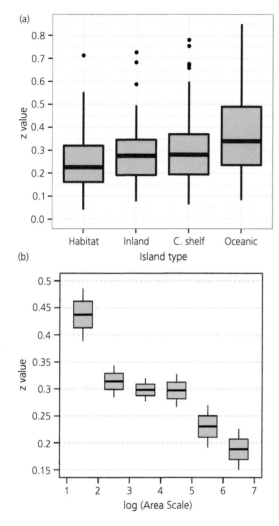

Figure 5.7 Variation in island species–area relationship (ISAR) slope (*z*) using the (log–log) power model fitted to datasets representing a wide variety of taxa. (a) Variation in *z* for habitat (132 datasets), inland lake (58), continental-shelf (277), and oceanic (125) islands. Box plots display the median, first and third quartiles, and outliers (dots). (b) Variation in *z* with increasing range in island area per dataset, for 465 true island datasets (of all types). Box plots display the mean, standard error of the mean, and 95% confidence intervals (whiskers).
Based on Figure 3a from Matthews et al. (2016a) and Figure 4 from Triantis et al. (2012).

range in key island properties encompassed within each dataset may influence the resulting model fits. This is shown by the scale-dependency in *z* values revealed in Fig. 5.8b, in which ISAR slope declines as a function of the range in island area encompassed within each dataset (Fig. 5.7b). In turn, range in island area may be related to other properties of the insular system (Triantis et al. 2012). Second, the combination of slope and intercept values observed can be partly explained by the overall species richness of the archipelago, as shown in Fig. 5.9. Note that for a given archipelagic richness (size of sphere): (i) there appears to be a trade-off between increasing slope and decreasing intercept values, with (ii) a general increase being seen in the values of both with increasing archipelagic richness, and that (iii) this appears to occur across taxa (see Matthews et al. 2019a, 2021b). These observations suggest that we need to pay more attention to other confounding properties of islands (not just area and isolation), and we need also to consider archipelagos as units of analysis. We do both in the following sections.

5.6 Towards some generalizations about ISARs

In comparing archipelagos, we would expect that as a rule *z* will increase with increasing degree of isolation, both of the islands from each other and of the archipelago as a whole from the nearest neighboring archipelago or continent.

(MacArthur and Wilson 1967, p. 16)

We have established that there is a general, if very messy, tendency for ISAR slope to increase and intercepts to decrease with the isolation of the system, from habitat islands to continental shelf, to remote oceanic islands (Fig. 5.8, 5.10a). In the case of archipelagos that are scarcely isolated, such as many habitat island systems, species are prevented from becoming extinct, or are rescued from having done so by very frequent immigration events and thus the insular penalty of small size is limited. In the case of very remote islands, rescue effects are vanishingly rare and so the penalty of small size is great, while larger islands gain the benefit of *in situ* cladogenesis. Remote systems thus *in*

deal of overlap between archipelago types evident in the plot. The theory works, but it appears to be weak.

Let's look at two features within the data that hint at why. First, as we have already mentioned, the

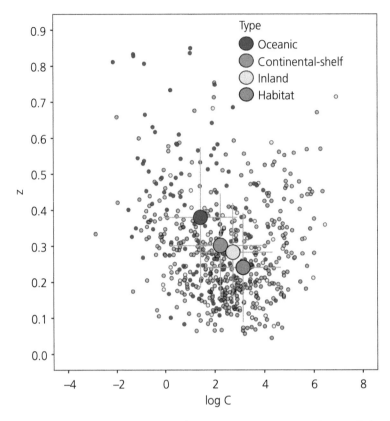

Figure 5.8 Variation in slope (*z*) and intercept (log *C*) for 588 ISARs fitted with the log–log power model, showing both individual dataset values (small circles) and the group means (large circles) and standard deviations (grey bars) for four categories of island (overplotting obscures many points).
Modified from Matthews et al. (2021b; their Figure 3.1).

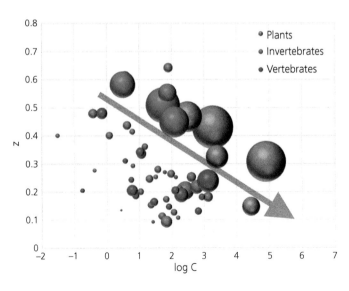

Figure 5.9 Variation in slope (*z*) and intercept (log *C*) for 57 ISARs fitted with the log–log power model, where the size of the sphere is scaled to reflect the overall species richness (gamma diversity) of each archipelago. The pattern suggests a trade-off (arrow), for a given gamma, between decreasing slope with increasing intercept. For reference, the largest gamma values are, plants 1994 species, invertebrates 1354 species, and vertebrates 552 species, and the smallest sphere represents a gamma of 5 species.
The data shown are drawn from those reported by Matthews et al. (2021b), here limited to archipelagos that are either entirely volcanic oceanic islands or a mix of oceanic and continental islands.

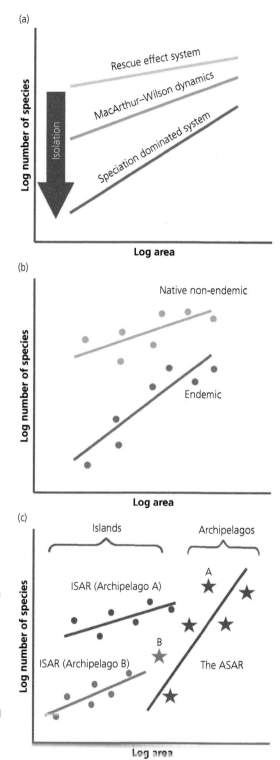

Figure 5.10 Generalizations about island species–area relationships (ISARs). (a) As island isolation increases, the ISAR slope (z) increases, with intermediate isolation generating the highest rates of species turnover. (b) Within remote archipelagos, the endemics subsets have steeper slopes and lower intercepts than do non-endemic native species. Among habitat islands, the same distinction can be made between habitat specialist and generalist species, respectively (not illustrated). (c) The slope of the archipelago species–area relationship (ASAR) should generally exceed the slopes of the constituent archipelago ISARs. Points A and B on the ASAR represent the archipelago diversity for archipelagos A and B, respectively. NB. Panel (c) was inspired by Rosenzweig's (1995) three-tiered SAR model but unlike his model does not include the within-province SAR because it is an EAC (a nested species accumulation curve) rather than an ISAR Whittaker et al. (2017).

general have the steepest ISARs and lowest inter-cepts, with islands of intermediate isolation closest to following the dynamic turnover inherent in the MacArthur–Wilson EMIB (Fig. 5.10a). One of the likely reasons this pattern is not clearer, and why we see the trade-off shown in Fig. 5.9, is that islands are typically clustered together in archipelagos, and as the archipelago becomes more isolated from main-land source regions, other islands (such as those within the archipelago) increasingly act as the main source for one another, as in fact was anticipated by MacArthur and Wilson (1967, pp. 29–31). This effect is detectable in comparative analyses of ISAR parameters for oceanic islands, in which Matthews et al. (2021b) report that lower intra-archipelago isolation tends to move the archipelago along the

line of the arrow in Fig. 5.9, i.e. producing higher intercepts in combination with lower slopes.

ISARs may also be decomposed into different **chorotypes** (distributional groups), such as endemic (to islands or to the archipelago), native but not endemic, and non-native (anthropogenic introduc-tions). Intriguingly, the ISARs of non-native species are often strongly aligned with those of the native fauna or flora, showing a similar slope, especially for plants (Sax et al. 2002; Baiser and Li 2018); that is, richness scales in the same way with area. This is shown for the beetles and spiders of the Azores in Fig. 5.11 (top panels) (Whittaker et al. 2014). These patterns suggest a flexibility of island carrying capacity in response to the combination of changing immigration rates and profound habitat

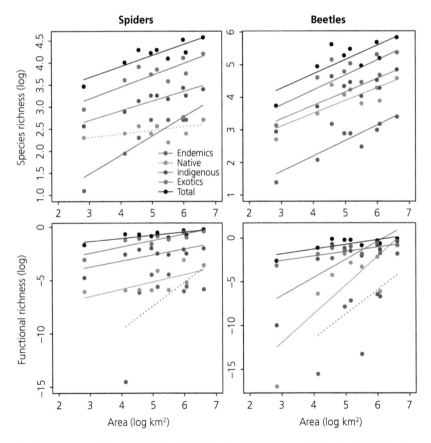

Figure 5.11 The island species–area relationships (ISARs) and island functional diversity–area relationships (IFDARs) for Azorean spiders and beetles, by chorotype (*n* = 9 islands, except for IFDARs for endemics, where *n* = 8 as Corvo Island was excluded). Indigenous species are the sum of native but non-endemic species (natives) plus Azorean endemics (endemics); Total refers to all indigenous plus non-native species. Solid lines significant, dashed not (*P* < 0.05). The IFDAR values used the FRic index, which is based on an ordination analysis of functional traits (for further details see Whittaker et al. 2014). The IFDAR results are discussed in Chapter 6 but are included here for ease of comparison.

modification, but that island area nonetheless has strong influence on the emergent diversity patterns (even if the mechanisms involved may vary; Blackburn et al. 2021). This influence carries over to shape non-native species diversity as non-native species become incorporated in island interaction networks (cf. Olesen et al. 2002; Traveset et al. 2015).

Another general pattern shown by the Azorean spiders (but not by the beetles) is for ISARs of endemics to be steeper, with lower intercepts than those for native but non-endemic species (Figs. 5.10b, 5.11). This effect is most clearly seen for single-island endemics, for which there is no possibility of inter-island rescue effects (Triantis et al. 2008). Similarly, as demonstrated in a comparative analysis of some 20 habitat island datasets, the subset of specialist woodland-dependent birds tend to have steeper ISARs, with lower intercepts, than do the more generalist species that are able to make use of resources in the environment matrix surrounding the woodlands (Matthews et al. 2014a).

Archipelagos can also be considered as units of analysis in island biogeography and have been included alongside islands in some analyses (e.g. Figure 5.1). However, the ISAR is an imperfect predictor of overall archipelago richness and it is therefore preferable to analyse islands separately from archipelagos. Let's expand on this point. If we imagine that we are seeking to predict the archipelago richness given only the number of species per island, we can constrain the possible archipelagic richness as lying between that of the richest island (the minimum archipelago value if the data form a perfectly nested compositional series) and the summed richness values of all islands (if each island had no species overlap with any other islands). Hence, the residual variation between the ISAR prediction for archipelago richness and the actual archipelago richness provides a crude index of nestedness (see Section 6.4). Santos et al. (2010) undertook an analysis of 97 datasets, showing reasonable concordance between ISAR predictions and archipelagic richness for many data sets but also some emergent patterns in the distribution of residuals. For example, among oceanic archipelagos, the tendency for the ISAR to underpredict the archipelago richness (tending to non nestedness) appeared to be higher among more isolated archipelagos. However, both over- and under-prediction could be found for different taxa within the same archipelago, perhaps indicative of their contrasting scales of dispersability and mobility.

Continuing the theme of archipelagos as units of analysis, Triantis et al. (2015) have collated data for 14 oceanic island archipelagos distributed across the world, estimating the overall archipelago species–area relationships (ASARs). Their method controls for variation in such factors as isolation and number of islands. For both native and endemic species there is a remarkable uniformity of ASAR slope between the different taxa, estimated at around 0.57 and 0.72, respectively (Fig. 5.12). These oceanic ASARs are thus steeper than constituent ISARs, consistent with each archipelago acting like an independent province, within which a significant proportion of species are generated *in situ* (as predicted by Rosenzweig 1995). This is captured in Fig. 5.10c, which shows the ISARs for two contrasting archipelagos, one larger and richer than the other, each with their corresponding archipelagic richness, and in turn forming part of the (steeper) multi-archipelago ASAR (Whittaker et al. 2017).

In sum, Fig. 5.10 provides generalizations about ISAR form based on application of the (log) power model, how form varies with isolation of an archipelago, and with the distributional extent or habitat breadth of the constituent species subsets. However, area is never a complete explanation for island (or archipelago) diversity, and in cases fails to provide a significant fit in the face of variation in such factors as isolation, elevation, island age, climate, habitat diversity, and past disturbance events (e.g. Leihy et al. 2018). We now go on to consider some of these factors and how they combine into more complete models of island diversity.

5.7 Island species richness and distance

The depauperate nature of isolated archipelagos, the absence of land mammals, and the filtering out of gradually less dispersive groups of taxa have long been recognized (Wallace 1902). MacArthur and Wilson (1963, 1967) argued that this should lead to steeper ISARs for more isolated islands, providing supporting evidence of this in data for Polynesian avifaunas (but see Box 5.4). Subsequent work has

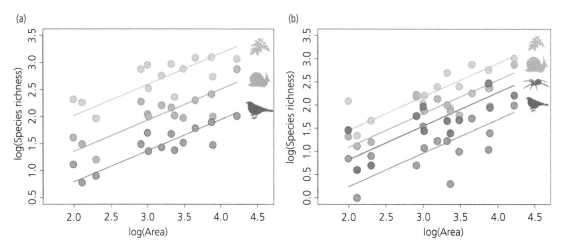

Figure 5.12 **Archipelago species–area relationships** (ASARs) using data for 14 oceanic archipelagos for (a) native species for plants, land snails and land birds and (b) archipelagic endemics, for the same taxa plus spiders (in this case $n = 12$ archipelagos). The lines are derived from mixed effect models, controlling for additional predictor variables.
Modified from Triantis et al. (2015).

generally confirmed the expected negative relationship between species richness and geographical isolation when analysing large datasets and multiple causal variables (e.g. Kreft et al. 2008). Similarly, the expected steeper slope of ISARs with increased isolation has been validated as a general pattern (Fig. 5.7 and 5.8). Yet, we also know these patterns to be noisy, partly due to variation in other island properties that may be confounded with variation in isolation (see Fig. 5.13; Power 1972; Kalmar and Currie 2006; Matthews et al. 2019a). Even the position of islands in relation to dominant wind and ocean currents and migratory pathways (for birds) may influence immigration rates (Spencer-Smith et al. 1988; Cook and Crisp 2005), while for habitat islands, immigration rates may vary depending on the characteristics of the landscape matrix in which they are embedded (Watson et al. 2005; Matthews 2021).

Box 5.4 Anthropogenic disruption of ISARs

The actions of humans following their colonization of remote islands has generally depleted their native faunas and faunas (Section 4.10; Chapter 14). What impact may this have had on ISAR form? The case of birds on island groups in Oceania provides a pertinent example.

Most major islands in Oceania for which good fossil data are available, have lost between 20 and 100% of their pre-human contact avifauna (Steadman 2006). Among higher islands, a greater proportional loss has been experienced on smaller islands, lacking refugia from extinction drivers. Steadman points out that this will have artificially inflated ISAR z values. Indeed, he suggests that the relatively small inter-island distances in many archipelagos would have produced quite flat ISARs. This is exemplified by a study using data on prehistoric (fossil) and current birds from seven islands in the Kingdom of Tonga by Franklin and Steadman (2008). They found that (i) there was no significant positive relationship between island area and richness, and (ii) the minimum island area required to support all species in the island group may have been as small as 12 km^2.

Humans have also introduced many non-native species to remote islands, constituting a massive increase in immigration rates compared to pre-human occupancy (Chapters 13–15). The inclusion of introduced species in island species lists represents a potential source of noise, given that introduced species have been found to exhibit positive species–isolation relationships (Moser et al. 2018; Gleditsch et al. 2023).

With increasing distance from continents, the clustering of islands within archipelagos becomes increasingly significant in respect of propagule exchange. As immigration rates from the mainland

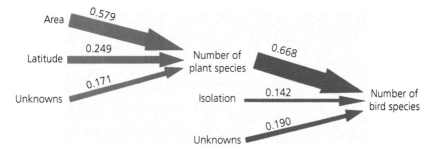

Figure 5.13 Path diagram showing relationships among variables in explanation of species numbers of plants and birds on islands off the coast of California as determined by stepwise multiple regression (modified from Power 1972). The coefficient associated with each path is the proportion of the variation in the variable at the end of the path explained by the independent variable at the beginning of the path, while holding constant the variation accounted for by other contributing variables. Although area was an important determinant of plant species number, it was a relatively poor direct predictor of bird species numbers. The role of climate in mediating plant species richness was indicated by the role of latitude. Power (1972) suggested that near-coastal islands with richer and structurally more complex floras tended to support a larger avifauna.

decline the proportional contribution of intra-archipelago movements must increase. Hence, those remote archipelagos that are tightly clustered should have reduced ISAR slopes compared to well-spaced archipelagos and within-archipelago isolation metrics should feature more when included in statistical models (e.g. Weigelt and Kreft 2013). Earlier we showed the pattern of ISAR slope and intercept for a large set of oceanic archipelagos (Fig. 5.9) and how it was linked to overall archipelago diversity. Structural equation models to explain these interrelationships were developed by Matthews et al. (2021b) and are shown in Fig. 5.14. When considered in a multivariate path model, distance from mainland was not selected in the final model. But, there was a detectable effect of increasing within-archipelago isolation, leading to lower intercepts and steeper slopes. Hence, isolation matters, it is just manifesting in this analysis at the within-archipelago rather than island–mainland scale.

Simulation analyses have also highlighted the importance of archipelago configuration for island diversity patterns, and by extension the ISAR. For example, using a spatially structured neutral model, Gascuel et al. (2016) showed that, given a set distance from the mainland, islands within a group will have higher species richness than a single island of the same size (see also Aguilée et al. 2021; Chapter 7) due to intra-island rescue effects in the former. Islands in the centre of the group, being more connected, will also have higher richness than those on the

edges. Interestingly, their simulations showed that islands in a group displayed a hump-shaped relationship between richness and distance from the mainland, indicating that within-archipelago processes can disrupt the simple theoretically expected negative relationship between isolation and island richness (Gascuel et al. 2016).

Given these findings and notwithstanding the central role of distance from mainland source pool in the MacArthur–Wilson model, it will not surprise the reader to learn that distance from mainland source pools is often found to explain little independent variation in island richness: especially when considering very remote archipelagos. The decline in immigration rate with increasing distance from source pools is doubtless a general feature. The distance effect is real. It is just that frequently it is confounded by other island and archipelago properties.

It should be noted that most studies analysing the role of isolation on island species richness have used the distance from the island/archipelago to the nearest mainland as the measure of isolation. This was the primary isolation measure used by MacArthur and Wilson (1967) and is generally the easiest measure to calculate. However, alternative isolation metrics have since been proposed and evaluated on large numbers of datasets (Weigelt and Kreft 2013; Itescu et al. 2019; Carter et al. 2020). They include inter alia distance to mainland, distance from nearest large island, landscape connectivity metrics taken from circuit theory, least cost paths,

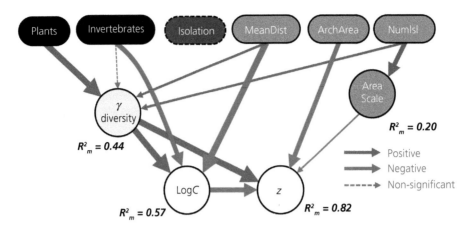

Figure 5.14 Path diagrams for a set of 39 oceanic archipelago ISARs showing the relative contribution of taxon (black boxes: plants, invertebrates, as against vertebrates as the base taxon), distance from mainland (mid-grey box: isolation), and archipelago configuration (light-grey boxes), specifically, mean intra-archipelago distance (MeanDist), archipelago area and number of islands, with intermediate roles for archipelago diversity (γ or gamma) and the range in area per archipelago (Area Scale). The arrow widths indicate the relative contribution of each variable, while the marginal R^2 values indicate the variance explained in the endogenous variables. Note that variation in slope (z) is better explained than variation in intercept (Log$C$). The data are a very slightly reduced subset (excluding mixed archipelagos) of those shown in Fig. 5.9. Modified from Matthews et al. (2021b).

distance to the nearest climatically similar land-mass, amount of land in a buffer around islands, and metrics that account for wind and ocean currents. While Weigelt and Kreft (2013) found that the proportion of land area within a buffer had the highest explanatory power of the metrics they considered, perhaps the most important general conclusion across these various studies is that there are different dimensions to island isolation, reflecting relationships within and between archipelagos and with more distant mainlands.

It is also important to consider temporal dynamics (i.e. temporal isolation) when thinking about island isolation, given that many islands were previously connected to the mainlands or other islands during glacial periods (Fernández-Palacios et al. 2016a; Weigelt et al. 2016; Itescu et al. 2019; Flantua et al. 2020). Much the same applies to habitat islands, where matrix permeability may be highly changeable over time (Flantua et al. 2020; Matthews 2021).

Finally, it is worth pointing out that almost all island biogeography studies of isolation, and all of the metrics described above, consider what landscape ecologists call 'structural connectivity' (i.e. the spatial relationships between islands and mainlands), and there has been far less work on

'functional connectivity'; that is, how well islands are connected from the perspectives of different organisms (Flantua et al. 2020). The latter would generally involve the study of the actual movement of individuals between islands and between islands and the mainland, and it is thus not surprising that it has been less studied. It is recommended that researchers incorporating island isolation as a predictor of island richness should in future consider including different isolation indices, and particularly those that represent different facets of isolation, such as distance to mainland area, availability of stepping-stones, and past land-connections (Itescu et al. 2019; Carter et al. 2020).

5.8 Towards more complete models of island diversity variation

Area and habitat diversity

Area effects have sometimes been split into: (i) a direct area or **area per se effect**, and (ii) a **habitat diversity effect**. (i) Area per se refers to the idea that very small amounts of space of a particular habitat can support only a limited number of individuals and thus very few viable species populations. As the area increases, more space and resources are available and so more individuals can accumulate, sufficient for breeding populations of more species. The range in area over which richness will continue to rise for this reason will vary depending on the ecological characteristics of the system. It may be very small, less than 0.1 ha for patches of grassland plants in Britain, or very large (many tens of km^2) for top consumers. (ii) The habitat diversity effect is that as area accumulates, so typically does the variability of habitat and indeed entirely distinct habitat types may be added, such that species specializing in, for example, freshwater, boggy, or upland habitats are added as the size and complexity of the island increases.

Generally, both effects operate in tandem and they are analytically hard to separate, although sometimes thresholds in ISAR models may be indicative of additions of new distinct habitats (above). A simple approach to the contribution of habitat diversity was introduced by Triantis et al. (2003), consisting of replacing area with the term K (choros, 'dimensional space'), which is obtained by multiplying island area with the number of different habitats present. Species richness is then expressed as a power function of the choros; that is, $S = cK^z$, in place of the more familiar $S = cA^z$. They compared the two models for 22 datasets of varied plant and animal taxa, finding significantly improved fits in all but two cases. Subsequent applications suggest that the choros model will often generate improved fits for both habitat and real island datasets and can be applied to more complex scenarios (e.g. Proença and Pereira, 2013; Carey et al. 2020).

The effects of habitat diversity on islands are likely to vary as a function of species' ecological specialization; habitat specialists being more limited by the presence–absence of specific habitat types. Sfenthourakis et al. (2021) have developed a metric—ecorichness—to assess this. It is calculated as the sum of the number of habitats that each species exploits, across all species present on a particular island, standardized by species richness. In their analyses of terrestrial isopods from the Aegean islands, they found a hump-shaped relationship between ecorichness and island area, habitat heterogeneity, and the K parameter of the choros model. Their explanation is that it was the product of changing contributions of habitat specialists and generalists to island richness along the area gradient, with generalists representing high proportions of species on small- and intermediate-sized islands, while the contribution of specialists increases with area (see also Matthews et al. 2014a).

We have thus far covered area, contrasting scales of isolation, and habitat diversity but there are many variants on the variables that have been assessed, ranging from alternative aspects of island environmental diversity, through climate and productivity, disturbance regime, island age, and human impact (Box 5.4). Some of these properties are integral to competing (often non-mutually exclusive) hypotheses for island diversity variation (Box 5.5). Comparative studies of different taxonomic groups from the same islands frequently return contrasting statistical models, defying easy synthesis. For example, multiple regression analyses of Lesser Antillean data selected habitat diversity and maximum island elevation for herptiles, isolation for bats, and isolation, area, and habitat diversity for birds (Morand 2000), reflecting the different forms

---

Box 5.5 Competing explanations for island species richness patterns

This is a selection of the theoretical effects and ideas put forward in explanation of island species richness patterns: while some are broad frameworks potentially widely applicable, others are rather specific effects that may not be expected to be widely detectable. This list is far from comprehensive but serves to illustrate the generative value of islands as model systems.

- **MacArthur and Wilson's equilibrium model (EMIB)** postulates the number of species on an island as a dynamic equilibrium between immigration and extinction, involving continuing turnover of species, rates being dependent on island isolation and area.
- **Random placement**. If individuals are distributed at random, larger samples will contain more species. An

Box 5.5 *Continued*

island can be regarded as a sample of such a random community, without reference to particular patterns of turnover. Larger islands are theorized to sample more individuals and thus contain more species. Connor and McCoy (1979) term this **passive sampling** and advocate its use as a null hypothesis against all alternatives (see also Chase et al. 2019; Almeida-Gomes et al. 2022).

- **Neutral theory**. Neutral theory applies the concept of equivalence to individuals rather than, as in EMIB, to the species, postulating equal probability of immigration for any individual, so that species probabilities become a function of the prevalent species abundance distribution (SAD). It provides an alternative level of null model for island biogeography (Hubbell 2001, 2010), offering the promise of linking SADs to emergent patterns of island diversity (e.g. Rosindell and Harmon 2013).

- **Habitat diversity**. The number of species may scale with the number/diversity of habitats. The choros model (Triantis et al. 2003) incorporates area per se effects alongside habitat diversity (cf. species–energy theory, below).

- **Incidence functions**. Some species can occur only on large islands because they need large territories; others only on small islands where they can escape from competition (Chapter 6; and Diamond 1974).

- **Species–energy theory**. According to this theory, the capacity for richness is considered a function of the resource base of the island, where the latter may be estimated, for example, using total primary productivity multiplied by area (Wright 1983). This mechanism may account for variation in species richness but is essentially neutral on the issue of turnover.

- **Small-island effect**. Whitehead and Jones (1969) argue that very small atolls lacking fresh water are limited to strandline species and tend not to show a significant increase in the number of species with area until a critical threshold area is passed. This describes a particular form of threshold effect; others are possible (see text).

- **Small-island habitat effect**. In contrast to the previous idea, it has been suggested that in some systems, small islands may be different in character because of their smallness, so that they actually possess habitats not possessed by larger islands and thus sample an extra little 'pool' of species. Alternatively, they may just have a greater diversity of habitats than anticipated from their area (e.g. via telescoping of elevational zones).

- The **subsidized island hypothesis** was proposed by Anderson and Wait (2001) as an explanation for non-standard relationships in datasets involving small islands that may benefit from input of marine nutrients (e.g. by deposition of guano). Notwithstanding the evidence of dramatic effects of such transfers on ecosystem function in particular systems (e.g. Kurle et al. 2021), an analysis of plant richness across a dataset of 790 islands worldwide, failed to find much support for this effect (Menegotto et al. 2020). However, Obrist et al. (2020), in a study of 91 islands in Canada, found that bird density was higher, and richness lower, on islands with higher marine subsidies and that there was a negative interaction between island size and the amount of marine subsidy. It is thus possible that this hypothesis is more relevant to certain taxonomic groups than others.

- The **disturbance hypothesis** postulates that small islands or 'habitat islands' suffer greater disturbance, and disturbance removes species or makes sites less suitable for a portion of the species pool (e.g. McGuinness 1984, and for a contrary example Wardle et al. 1997). Disturbance might also open up sites to invasion by new members.

- The **trophic theory of island biogeography** argues that the MacArthur–Wilson model has been applied with insufficient attention to trophic interdependencies, and that trophic complexity increases with area, providing potential for refined diversity models that encapsulate such multi-trophic interactions within ecosystems (Holt 2010; Roslin et al. 2014). The potential importance of food chain/resource dependency across trophic levels in this context was raised in response to EMIB by Lack (1969, 1970) and see also Bush and Whittaker (1991).

- The **general dynamic model of oceanic island biogeography** argues that volcanic oceanic islands experience a humped trend in carrying capacity between emergence and eventual submergence, leading to predictable trends in diversity over time (Chapter 7; Whittaker et al. 2008).

- **Pleistocene lag effects**. Reduced Pleistocene sea levels, expanded connectivity, and altered climates have long been suggested to have impacted contemporary diversity patterns (e.g. Preston 1962; Fernández-Palacios et al. 2016a) and have been considered analytically in a number of studies (e.g. Weigelt et al. 2016; Norder et al. 2019).

and scales of dependency of different groups on resources. We will now briefly examine some of the more important additional properties, specifically climate/energy, and island age.

Climate and energy

The term **species–energy theory** appears to have been coined by Wright (1983), who modified the ISAR approach by replacing area in the species–area regression with the product of energy and island area (cf. the choros model, above). He analysed data for angiosperms on 24 islands worldwide, ranging from Jamaica (12,000 km^2) to the island continent of Australia (7,705,000 km^2), using actual evapotranspiration (a measure of water balance) to estimate energy availability. He also analysed land bird species richness for an overlapping set of 28 islands, using net primary productivity as the energy variable. The models outperformed standard ISARs, explaining 70% and 80% of the variance, respectively. As the islands spanned a climatically heterogeneous set, from the equator to the Arctic, it is perhaps unremarkable that incorporating estimates of energy regime should improve model fits.

Subsequent work validates a prominent role of climate and/or energy flow at the global scale and at lesser scales where there is a substantial climatic gradient involved. For example, Wylie and Currie (1993) have applied the species–energy approach to mammals (excluding bats) on land-bridge islands from across the world, again reporting an improved fit. Kalmar and Currie (2006) analysed land-bird species richness for 346 marine islands, ranging from 10 ha to 800,000 km^2, and representative of global variation in climate, topography, and isolation. They developed a multivariate model based on area, elevation, various isolation metrics, mean annual temperature, and annual precipitation, capable of describing as much as 85–90% of the variation in bird species numbers. Starting with the individual variables, they found constraining (triangular) relationships between richness and area, temperature, precipitation, and (inversely) distance from the nearest continent, indicating that these variables may each set upper limits to richness, but that richness frequently falls below those limits because of the effects of the other limiting variables. An example of a regional-scale study also supporting a prominent role for climate is provided by Leihy et al. (2018), who analysed species richness of plants and insects on Southern Ocean Islands spanning 37°S to 62°S. Richness of both taxa declined with isolation and increased with mean land surface temperature, while area did not enter the models.

It has long been evident that species richness is influenced by climate on a global scale and indeed the geographical ('latitudinal') gradient in plant species richness is detectable across the world's islands, with higher richness in tropical regions and lower values on high-latitude islands (Fig. 5.15; Weigelt et al. 2013). Larger and less isolated tropical islands in the Caribbean and South-East Asia have particularly high plant species richness values. Oceanic islands are generally less rich for their size than continental islands. Weigelt et al. (2013) developed a multivariate model based on a sample of 475 islands globally, using 10 physical (including island area, elevation, and isolation) and bioclimatic variables, which was able to largely replicate the observed richness patterns (pseudo-R^2 = 0.94) and generate reasonable predictions globally.

Trophic differences

The trophic theory of island biogeography (Box 5.5) predicts that the number of trophic levels will increase with island area and linked to this that ISAR slope should increase with trophic rank (Roslin et al. 2014; Holt et al. 2021). The latter is predicted to be more prevalent for specialists, but generalists have also been found to follow this theorized pattern in some cases (reviewed in Holt et al. 2021). In their meta-analysis of ISAR form, Triantis et al. (2012) reported mean z values of 0.355, 0.323, and 0.287 for plants, invertebrates, and vertebrates, respectively, which is inconsistent with this expectation. However, the Triantis et al. analysis is not a direct test of trophic theory as it fails to limit the comparison to the same archipelago or to tight trophic groups. When doing so, mixed results can be obtained (see data in Matthews et al. 2019a). However, the effects are most likely to be detected when the study system contains very small islands or islets that are too small to sustain viable populations of species of higher trophic level. The predictions of

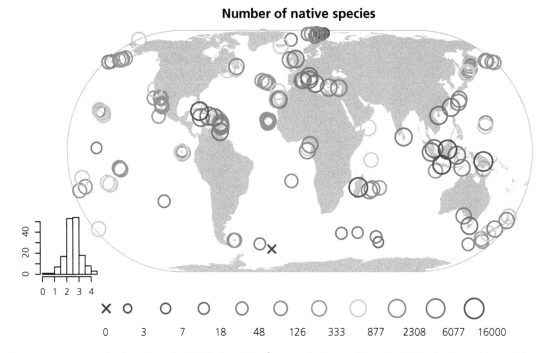

Figure 5.15 Native vascular plant richness for 177 islands > 100 km², extracted and mapped from the GIFT database by Patrick Weigelt (pers. comm. 4 June 2021). Symbol size and colour indicate richness. The histogram shows the distribution of log10 species richness (i.e. each bar gives the number of islands for a given range of richness values). Large islands of lowest richness values, with a few remote island exceptions, are concentrated at high latitudes, and the highest values are in the tropics.
For further information on the data see: Weigelt and Kreft (2013) and Weigelt et al. (2020).

trophic theory are supported in analyses for 20 near-shore Finnish continental islands by Roslin et al. (2014), in which food chain length and ISAR slope increased from plants to herbivores, to primary parasitoids, and to secondary parasitoids. By comparison the consistency of slope for plants, land snails, and birds when using whole oceanic archipelagos as the unit of analysis is striking (Fig. 5.12). See further discussion of trophic theory in Section 6.6, and Holt et al. (2021).

System age and geo-environmental dynamics: a first look

Island age can be an important determinant of island species diversity (Box 5.2). In the earliest stage of emergence of an island, as ecosystems build, it may take centuries for the processes of ecological assembly and succession to develop to the point of an initial equilibrium. Over longer timescales, there are not only evolutionary processes to consider but also the implications of environmental change intrinsic to and external to the geodynamics of a focal archipelago. Islands may expand and contract, disappear and reappear, merge with other islands or with the mainland, and then separate again. In island macroecological analyses, island age might therefore be estimated as the time since isolation of a 'land-bridge' island, disconnected some thousands of years ago from a larger landmass by rising sea levels, or by reference to the geological age of an emergent volcanic island, or sometimes to the biological age of a re-set system (Chapters 2 and 3).

Younger islands and those disturbed by volcanic eruptions may have anomalously low species richness compared with older or undisturbed islands, as they have yet to reach their potential equilibrium. Conversely, islands that have shrunk in size or increased in isolation (as has happened to innumerable land-bridge islands), may for a time at least, hold more species than expected (termed 'supersaturation'). If such effects are truly long-lasting, then it may be possible to detect the impact of long-vanished island configurations in statistical analyses (e.g. Hammoud et al. 2021). In their glacial-sensitive model of island biogeography, Fernández-Palacios et al. (2016a) suggest testing for such effects even in oceanic island systems. Subsequent analyses by Norder et al. (2019) tested alternative mixed effects models for land snail and plant data for 53 oceanic islands from 12 archipelagos. They found that palaeo-configurations of islands had limited explanatory power in models of non-endemic native species. But richness of single-island endemics appeared to reflect lag effects, being higher than expected for their area when islands had benefitted from expanded area in the Late Pleistocene, and lower than expected when the islands had been connected to other islands, thereby permitting endemics to be shared. Similar results have been reported for plants on islands worldwide (Weigelt et al. 2016).

The role of island age in influencing island diversity patterns over the entire timespan of an island or archipelago is most fully conceptualized in the **general dynamic model of oceanic island biogeography**, which intersects MacArthur–Wilson dynamics with a simplified volcanic island ontogeny, generating a number of novel predictions as to the behaviour of island diversity over time (Whittaker et al. 2007, 2008; Borregaard et al. 2016, 2017). We reserve discussion of this model to Chapter 7.

Scale effects

The theme of scale dependency has run through much of our consideration of ISARs, and, for example, is integral to the generalizations in Fig. 5.10. ISAR form has been shown to vary as a function of the range in island areas considered, both in terms of shape and the parameters of the power model. The importance of isolation in island diversity models also demonstrates scale dependency. As the isolation of an archipelago increases, the frequency of exchange with the mainland must decline to the point that first, the differences in the distance between each island and the mainland

becomes immaterial while the configuration within the archipelago achieves ever greater importance to the rates of exchange between the islands within the archipelago. Indeed, we should recognize that even this argument is over-simplified, as there are multiple archipelagos in the oceans, and we know from studies of relatedness of their biotas that successful colonization is often an inter-archipelago event, leading to the notion of the meta-archipelago (Fig. 4.7; Whittaker et al. 2018), an idea akin to the notion of stepping-stone islands (MacArthur and Wilson 1967).

Although many models we have discussed are statistical rather than mechanistic, they are based on the inclusion of candidate variables of established importance. In sum, they demonstrate that island area, habitat diversity, local and island–mainland isolation, and climate, when combined within flexible multivariate models, provide highly effective models of island diversity for a range of taxa. The combination of variables that emerge varies because each group of islands has its own unique spatial configuration and range of environmental conditions. The *range* of variation is important not just in terms of the orders of magnitude of each variable encompassed, but also the *effective range* in relation to the biology of the organisms being considered.

The effective mobility of different taxa (e.g. birds versus terrestrial mammals) and of different ecological guilds (e.g. sea-dispersed vs bird-dispersed plants) can differ greatly and mobilizing this insight may help us understand diversity patterns across archipelagos. For a moderately dispersive taxon, it might be assumed that very small distances have no discernible impact on colonization and thus on species richness. At the other extreme, beyond a certain limit, in the order of hundreds of kilometres, the group might be absent, and thus further degrees of isolation again have no impact. It is thus unsurprising that comparison of regression models, in which richness has been analysed against an array of independent variables, shows a diversity of answers as to which factors enter the models in which order, and with what relationship to richness. Even variables of known ecological importance may not have significance over all scales of variation and/or their effects are drowned out by other variables.

5.9 Rarity within island biotas: SADs and range size/occupancy

We have focused our attention on the richness of islands and archipelagos. These are the emergent patterns from dynamics operating from the scale of local patches and communities upwards (although top-down effects also operate). Rosenzweig and Ziv (1999) described the resulting linked diversity patterns as the echo pattern of diversity. One way to analyse these phenomena is to construct and compare nested species accumulation curves (e.g. top data series in Fig. 5.1), although it can require significant sampling effort in order to generate perhaps a single such curve, which may not be representative.

A second approach is to develop comparative sampling programmes using standardized plot-based sampling within and across a set of islands: an approach exemplified by Ibanez et al. (2018), who analysed data for tropical woody plant communities from 41 islands representing 19 archipelagos. They report a great deal of within-island variability in species density per plot, reflecting environmental heterogeneity and stochasticity at this scale. But they also found that the maximum species density per island increased as a function of island area, with a secondary contribution attributed to variation in dry season precipitation. Moreover, when re-running the analysis at archipelago level

(Fig. 5.16), maximum species density was a positive function of area, and diminished with increasing isolation and range in annual temperature. This is a satisfying result as it illustrates the echo of the patterns found for entire island and archipelago faunas/floras and discussed above: a role for area and climate at both island and archipelago scale, and distance from mainland entering models for archipelagos only. Analyses of species density hold as yet largely untapped promise for connecting the plot, patch or community level to the island level.

Another approach to the problem is to look at the profile of rarity/abundance of the constituent species and it is this we turn to next. In the great majority of plant and animal communities there are only a few species represented by many individuals and many more species of few individuals. This is one of the most robust macroecological patterns and despite our turning to it deep within the chapter, it formed a key part of MacArthur and Wilson's (1967) thinking in developing their theory.

There are many proposed theories concerning **species abundance distributions (SADs)** (reviewed in McGill et al. 2007; Matthews and Whittaker 2014). Of these, two are worth summarizing here. First, Fisher et al. (1943) suggested that the largest class of species is of those that are individually rarest. This gives rise to the **logarithmic**

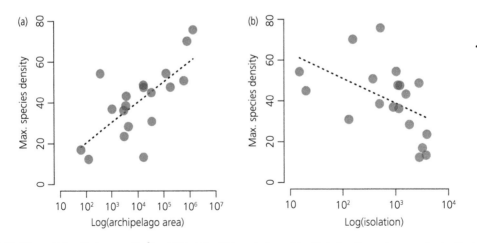

Figure 5.16 Effects of (a) archipelago area (km²) and (b) isolation (distance to the mainland in km) on the maximum species density of woody plant species observed per archipelago (*n* = 19), derived from 113 plots of > 0.1 ha distributed across 41 islands. The relationships are each of the expected form. Both are significant, if noisy trends (especially in the case of isolation).
Modified from figure 3 of Ibanez et al. (2018).

series of abundance. Second, Preston (1948, 1962) argued that species more typically fit a **lognormal abundance distribution**; that is, that the most numerous species in a community are those of intermediate abundance. Although both are merely statistical distributions, the logseries and lognormal distributions are frequently predicted by ecological and statistical theories (McGill et al. 2007; Matthews and Whittaker 2014). Other, more mechanistic, SAD models have been proposed based on processes such as niche partitioning (e.g. MacArthur 1957), or on neutral dynamics (Hubbell 2010).

Preston argued that insufficient sampling was responsible for the apparent fit of the logseries model to many datasets. This can be illustrated diagrammatically for a hypothetical community by plotting histograms of the abundances of individual species against number of species using a log-scale of number of species (Fig. 5.17a). If the sample is small, then the sparsest species of the lognormal distribution will not be sampled, and the abundance distribution will be that shown to the right of the **veil line** (Preston 1948). On increasing the sampling effort, more of the rarer species of the system will be sampled, pushing the veil line to the left. By analogy, a small area 'veils'—or excludes—the existence of all the species whose total abundance falls below a critical minimum.

To put this into an insular context, first imagine that we sample the birds of a patch of mature woodland, finding that some of the specialist woodland species have only a handful of breeding pairs. Over time, temporary adversity, such as a harsh winter, will increase mortality and cause some of these rarer species to disappear from the patch. Subsequent breeding success in woodland 2 km distant soon permits the return of the rare species to the sample patch. Now let us imagine that the woodland patch is 20 km from the nearest source. When the rare species populations fail in this context it may take years or decades for the lost species to return (e.g. Paine 1985). Hence, small, isolated patches of habitat should always be missing some of the rare species, an argument that suggests linkage between the slope of the ISAR and the form of the SAD, as first proposed by Preston (1962), who demonstrated that a slope (z) of 0.263 should follow from a lognormal distribution of abundance: a value well in the range of typical z values from islands and habitat islands (MacArthur and Wilson 1967; and see our Fig. 5.7).

In practice, as elsewhere, island community SADs are sometimes best fit by a logseries model, sometimes by a lognormal, and sometimes by another competing model. In illustration, Matthews et al. (2014c) report that bimodal SADs sometimes provide improved fits for arthropod data from remnant forest patches in the Azores. One interpretation is that 'tourist' species, mostly non-natives from surrounding modified habitats, have spilled over into the fragments in low abundance generating a superimposed second peak in the distribution (Fig. 5.17b; see also Matthews 2021).

Recent work has found that many of the same drivers of island species richness affect variation in the form of island SADs. For example, in an analysis of Azorean arthropod data, from 16 forest fragments distributed across the archipelago, Matthews et al. (2017) tested for the effect of patch isolation on the shape of the SAD (this time using unimodal SAD models). Contrary to our conjecture above, they found that increasing patch isolation leads to a shift from higher to lower gambin alpha values (Fig. 5.17c), indicating communities with a higher proportion of rare species and SADs that increasingly resemble logseries rather than lognormal distributions. Again, this perhaps reflects incursions of non-native arthropods and species more associated with anthropogenic land-uses into the native forest patches. These findings remind us that many islands are comparatively large, contain complex landscapes of variegated habitats, and for smaller organisms, such as arthropods, samples likely reflect these local controlling conditions rather than being representative of the overall pattern of abundance across the entire island. Moving from isolation to island area, Ibanez et al. (2020), in a study of the SADs of tree communities in 1-ha plots distributed across 20 islands in the Indo-Pacific, found that island area was a driver of SAD form, primarily indirectly through its effect on species richness. Larger islands were richer and had more logseries-type local community SADs. However, the most important driver of SAD form in these islands was found to be cyclone disturbance, which was found to result in more lognormal-like

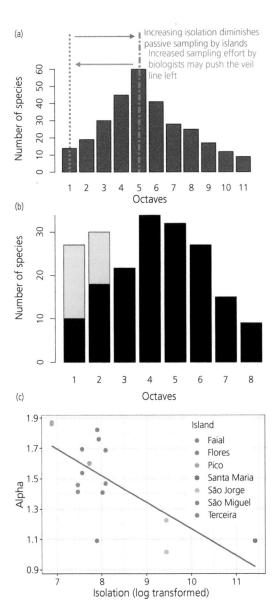

Figure 5.17 Species abundance distributions (SADs) on islands. (a) A hypothetical SAD that roughly approximates a lognormal distribution. The abundance data are grouped into octaves (e.g. octave 1 contains the number of species with 1 individual, octave 2 the number of species with 2–3 individuals, octave 3 the number of species with 4–7 individuals, etc.). The simulated dataset comprised 300 species. The portion of this hypothetical abundance distribution that is sampled may be a function of either active sampling effort (by the biologist) or passive sampling effort (by an island). Only the portion to the right of the veil line will be sampled. (b) A hypothetical bimodal SAD. The data are simulated but can be imagined to represent a sample of arthropods from a forest habitat island surrounded by a matrix of pasture. Abundances are grouped into the same octaves as in the first panel. The blue bars represent tourist species from the matrix habitat that are present in the patch but in low abundance. The presence of these species has the effect of increasing the number of very rare species, resulting in a second peak in the distribution. Adapted from Matthews (2021). (c) SADs of arthropod communities sampled in 16 isolated forest remnants within the Azores show a shift from higher to lower gambin alpha values with increasing fragment isolation. Each data point represents a single fragment's SAD denoted by the standardized gambin alpha. Fragment isolation ranged from 0.97 km to 90.78 km; the line of best fit is given.
For explanation of gambin alpha and further details, see Matthews et al. (2017).

SADs. This disturbance effect, which had a larger impact on smaller islands, worked by increasing the number of individuals in the plot, but decreasing the number of species, likely via changing the island environmental conditions (Ibanez et al. 2020). Thus, it seems that disturbance can be a particularly important factor in driving both the richness (Box 5.5) and relative abundances of species on smaller islands.

For situations where effort-controlled abundance data are also available for species across islands, Chase et al. (2019) recently proposed a framework to identify the mechanisms driving the ISAR, including distinguishing between compositional heterogeneity effects (which include habitat diversity but also heterogeneity due to dispersal limitation) and area per se. As a first step the framework analyses the relationship between rarefied richness and island area to test the simplest explanation for the ISAR: that larger islands are simply sampling more individuals from the species pool and thus have more species than smaller islands ('passive sampling'; Box 5.5). If this is rejected, examination of within-island beta-diversity patterns (Chapter 6) can then be used to help identify the relative roles of area per se and heterogeneity effects in shaping the ISAR (Chase et al. 2019). A test of a modified version of this framework using data from 505 islands across 34 island groups around the globe found that the passive sampling hypothesis could be rejected, and that higher habitat heterogeneity on larger islands was the likely driver of ISAR form in many cases (Gooriah et al. 2021).

For larger islands, it may be more relevant to analyse variation in prevalence of species at coarser scales than provided by local counts of individuals, and instead to scrutinize the occupancy of grid cells (i.e. occupancy frequency distributions) across whole islands (e.g. Borges et al. 2018a, b). Doing so reveals a similarly skewed distribution of range size/occupancy for whole faunas and floras as is found for SADs. This is shown for the Canaries in Fig. 5.18. We have to caution that (i) these data are only for the endemic species, (ii) these islands have been heavily impacted by humans for over 2000 years, and in particular, the avifauna has been depleted by anthropogenic extinctions, and (iii) this is just one archipelago. Yet, the general form of these

distributions is consistent with many datasets from continental regions: there are few really widespread species and a large number of very limited distribution (McGeoch and Gaston 2002). The Canarian distributions differ principally in being 'scaled down'; that is, the entire distribution is left shifted towards smaller range sizes.

A more nuanced analysis of multiple aspects of rarity within the native Canarian flora is provided by Fernández-Palacios et al. (2021b). They first divided the flora into (i) those species that have colonized the archipelago naturally but failed to undergo cladogenesis (some of these species have nonetheless become endemic species), which they term the non-diversified group; (ii) those that belong to lineages that have diversified *in situ* to produce between 2 and 7 endemic species, the moderately diversified group; and (iii) those in groups of > 7 endemic species, the highly diversified group (Fig. 5.19).

Whether in terms of grid cell occupancy, climate niche, or local abundance, most species are rare and there is a strong tendency for rarity to increase as the degree of lineage diversification increases (Fig. 5.19b–d). Species within highly diversified lineages (the evolutionary winners) were found to provide important elements of local cover only in two major ecosystem types: (i) the highest elevation scrub and (ii) rocky, steep terrain of low overall cover across the elevational gradient. Everywhere else, within the four other core major ecosystem types, the vegetation tended to be dominated by species of the non-diversified group. When combining the different elements into a composite index of rarity (panel e), the non-diversified lineages once again stand out as more successful in ecological terms. It is unknown how general these 'evolutionary winners: ecological losers' patterns are, but it is intriguing to note that Joseph Dalton Hooker made some similar general observations about the flora of Madeira in an article published in 1867 (Fernández-Palacios et al. 2021b).

More work is needed to develop our understanding of how: (i) the profiles of commonness/rarity of island systems change from the scale of local SADs, up to whole island or whole archipelago patterns of range size and occupancy, and (ii) how these patterns are linked to the emergent patterns of richness

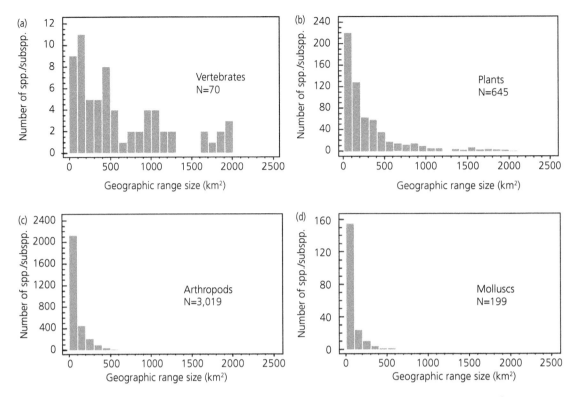

Figure 5.18 Range size distributions of endemic species and subspecies for the Canary archipelago, based on presence in 2 km² grid cells: (a) vertebrates, (b) plants, (c) arthropods, and (d) molluscs.
Modified from Martín (2009); his figure 4.

of different chorotypes across archipelagos (e.g. Fattorini et al. 2016; Theuerkauf et al. 2017; Craven et al. 2021; Fernández-Palacios et al. 2021b). What portion of the variance in island species richness is explained by bottom-up controls from very local processes involving energy capture and flow from plants through invertebrate herbivores to larger vertebrate taxa? And to what degree is the diversity of the patch a product of top-down ecological constraints resulting from the area, isolation, and environmental diversity of the island and indeed the archipelago? Our instinct is that both directions of influence operate, but to what relative degree remains unquantified.

Neutral theory provides an alternative, potentially powerful approach to exploring the linkages between the population level, as encapsulated in the form of the SAD, and the level of species colonization and turnover on islands (Box 5.5; Hubbell 2010). Moreover, simulations based on

neutral theory can be applied to different insular situations (e.g. land-bridge versus volcanic oceanic islands), generating novel predictions for subsequent empirical evaluation (e.g. Rosindell and Harmon 2013). These and other simulation models linking population level processes to emergent diversity patterns provide promising new lines enabling exploration of ideas that can otherwise be hard to test empirically (Cabral et al. 2019a, b).

5.10 Species turnover, equilibrium, and non-equilibrium

We have thus far confined our exploration of island macroecology largely to emergent *spatial* patterns and comparisons. Yet, the MacArthur–Wilson (1967) theory is, at its core, a dynamic model. And so we turn next to the central question: to what extent do the patterns of island diversity

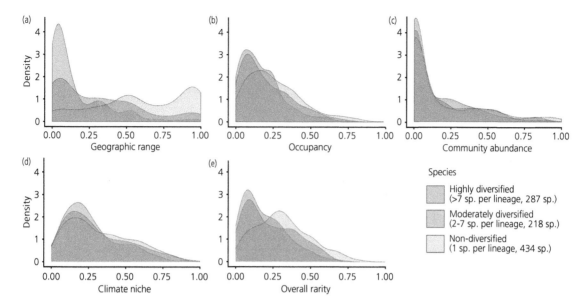

Figure 5.19 The distributions of species belonging to non-diversified, moderately and highly diversified lineages within the native Canarian vascular flora (939 species) according to different aspects of rarity: (a) geographical range size, (b) occupancy of 500 × 500 m grid cells, (c) peak abundance recorded in local communities, (d) climate niche breadth, and (e) a weighted mean of a–d. The measure of geographical range reflects the overall geographical distance spanned by the extremes of the range of each species. Many non-diversified species are found in multiple islands, but the diversified groups include a comparatively high proportion of species found on a single island, especially so in the highly diversified lineages. From figure 2 of Fernández-Palacios et al. (2021b). Creative Commons CC BY.

reflect dynamic equilibria of the form proposed in equilibrium theory?

The MacArthur–Wilson theory holds that islands approach a dynamic equilibrium by means of a gradually declining immigration rate, matched by a gradually rising extinction rate. Both trends are monotonic (without reversals of direction). The compositional turnover occurring over time is assumed to be largely stochastic and the theory holds that once attained, islands will remain at a dynamic equilibrium, unless disturbed, whereupon a dynamic readjustment will occur, restoring the ecological equilibrium.

It may be useful to start by asking what are the alternatives to this dynamic equilibrium? The core EMIB assumes that environmental dynamics are relatively unimportant most of the time, and that once an island has formed, the ecological processes they invoke have the power to maintain an equilibrium richness marked by continued species turnover: the bottom left corner of Fig. 5.20. However, we can conceive of situations where the processes

of immigration, speciation, and ecological adjustment are clearly detectable yet comparatively slow in relation to fluctuations in island environments. Such islands will show progress towards equilibrium but rarely hit it: this describes the dynamic, non-equilibrium condition at the top left of the figure. Extending the time frame to evolutionary timescales, MacArthur and Wilson (1967) argued that the equilibrium point for remote islands may drift upwards over time (Box 5.2), thereby recognizing island systems with biogeographical processes of low dynamism and hence lengthy lag times in adjustment. Such systems may be essentially static in species membership over the duration of ecological studies yet are nonetheless slowly adjusting. As shown in the top right of the figure, they are low dynamism, non-equilibrium systems. We may also hypothesize a fourth combination, one in which an equilibrium is established on an island reflecting habitat controls and system carrying capacity but in which internal ecological processes produce communities resistant both to newly arriving

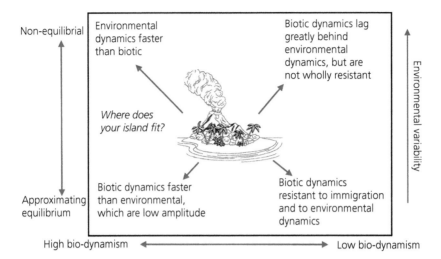

Figure 5.20 A representation of the conceptual extremes of island species turnover. The bottom left corner corresponds to MacArthur and Wilson's (1963, 1967) theory; the bottom right corner equates to Lack's (1969) ideas on island turnover of birds; the top right corner to Brown's (1971) work on non-equilibrium mountain tops; and the top left corner to Bush and Whittaker's (1991, 1993) interpretations of Krakatau plant and butterfly data. Considering a single taxon, different positions in this diagram may correspond to different islands or archipelagos, and different taxa in the same island group may also correspond to different positions (Table 5.1).
Modified from an original in Whittaker (1998).

Table 5.1 Exemplification of classification of studies of island richness and turnover as per Fig. 5.20, showing that different taxa from within a single island group (Krakatau) might be informally attributed to different models, and that in different contexts, data from the same taxa (terrestrial vertebrates, birds, invertebrates, plants) have been interpreted as supporting different models.

High biodynamism, non-equilibrial	Low biodynamism, non-equilibrial
1. Krakatau plants (Whittaker et al. 1989; Bush and Whittaker 1991)	7. Krakatau reptiles—several species introduced by people, only two species have been lost, both related to habitat losses (data in Rawlinson et al. 1992)
2. Krakatau butterflies (Bush and Whittaker 1991)	8. Great Basin mountain tops, North America, small mammals (Brown 1971)
3. Bahamas, plants on small islands (Morrison 2002b)	9. Lesser Antilles—small land birds, examined over an evolutionary time frame (Ricklefs and Bermingham 2001)*

High biodynamism, equilibrial	Low biodynamism, equilibrial
4. Krakatau birds are a reasonable fit, but show successional structure in assembly and turnover (Bush and Whittaker 1991; Thornton et al. 1993)	10. Krakatau terrestrial mammals—no recorded extinctions to date (Thornton 1996)
5. Mangrove islets of the Florida Keys, arthropods (Simberloff and Wilson 1970)	11. Bahamas, ants on small islands (Morrison 2002a)
6. British Isles, birds on small islands (Manne et al. 1998; McCollin 2017)	13. At archipelago scale, land birds of a sample of oceanic archipelagos (Valente et al. 2020)*

* Evolutionary dynamics involve speciation processes that take thousands of years to play out combined with extremely rare immigration and extinction events, hence systems that conform with dynamic models on long time frames exhibit stability of membership in the time frame of ecological monitoring programmes and are labelled here 'low biodynamism'.

species and to fluctuations in environment: a low dynamism or static equilibrium (Fig. 5.20, Table 5.1).

Even for island systems that approximate the dynamic equilibrium, the criticism has been levelled that the EMIB is 'true but trivial' (Williamson 1989a), in that most of the turnover is typically of species that never fully established, and could be considered ephemeral or transient (e.g. Schoener and

Spiller 1987; Williamson 1989b; Whittaker et al. 2000). How fair are these criticisms? In this section we will take a look at evidence for the establishment of dynamic equilibrium, which in turn requires a close look at measurements of species turnover and assessment of whether it tends to be homogeneous (across all species) or heterogeneous (involving only a subset of species)?

Problems of measurement, pseudoturnover, and cryptoturnover

Coastal islands are notorious for their accumulation, at all seasons, of a staggering variety of migrant, stray, and sexually inadequate laggard birds. In the absence of specific information on reproductive activity, it is therefore unwarranted to assume tacitly that a bird species, even if observed in the breeding season, is resident.
(Lynch and Johnson 1974, p. 372)

A dynamic equilibrium, *sensu* the EMIB, requires that there is limited variation in species richness around a long-term mean value, with a measurable degree of compositional turnover. Testing for this ideally requires high-quality, standardized survey data enabling estimation of the rates of immigration of propagules and the loss of breeding populations per island. Yet recording the arrival of every viable propagule and monitoring the fate of every species population are each impractical, Herculean tasks. While there are other forms of test and inference that are possible (MacArthur and Wilson 1967; Valente et al. 2020), most studies of turnover have relied upon compilations of species lists carried out intermittently, by different observers and without standardization of sampling effort (e.g. Whittaker et al. 1989, 2000). Estimates of immigration and extinction derived from such crude 'presence' data fall far short of the original definitions in Box 5.1.

Two categories of measurement error have been described in such data. **Cryptoturnover** refers to species becoming extinct and re-immigrating between surveys (or vice versa), thus depressing turnover rate estimates: a problem that will increase with the interval between surveys (Simberloff 1976). **Pseudoturnover** occurs when censuses are incomplete, and information on breeding status inadequate, leading to species appearing to turn over when present as a breeding population throughout, or alternatively, when they have been observed to be on an island at some point but never properly colonized. It follows that estimates of turnover (and of *I* and *E*) are dependent on census interval (Diamond and May 1977; Whittaker et al. 1989) and survey effort/bias (Lynch and Johnson 1974), although the latter is itself generally poorly quantified.

In experimental re-surveying of plants from small islets in a Swedish lake, Nilsson and Nilsson (1985) showed that notwithstanding consistent survey techniques, their teams achieved at best only 79% efficiency per survey. The approach permits the estimation of pseudoturnover, potentially allowing correction of estimates of real turnover. However, such procedures cannot be applied with confidence to surveys by different teams many years apart, wherein the expertise, experience, special taxonomic interests or biases, methods, and time spent in surveying are poorly documented. If the islands in question are large or otherwise difficult to survey, the problems are multiplied. Whittaker et al. (2000) provide an illustration of just such a system, in their account of the build-up and turnover of plant species on the Krakatau Islands (Indonesia) following the sterilization of the islands in volcanic eruptions in 1883. They showed how their own earlier estimates of extinction were inflated by incomplete sampling efforts and by the inclusion of several early singleton records of species that almost certainly failed to establish a breeding population, thereby inflating both immigration and extinction estimates. Problems of pseudoturnover were also identified by Lynch and Johnson (1974) in a study of bird species turnover across a 50-year interval in the Californian Channel Islands. They also noted that some of the genuine turnover was attributable to specific causation, such as the impact of pesticide poisoning on several birds of prey, rather than being the sort of random lottery envisaged in the EMIB. The practical issues of sampling, resolution, and attribution noted here provide some cautions regarding the following sections.

The rescue effect and the effect of island area on immigration rate

The EMIB assumes that the effect of island area is solely on extinction rate, while isolation affects only immigration rate (Fig. 5.2). In practice, both E and I may be affected to some degree by both area and isolation. Given the power of modern statistical and modelling tools these complications are no longer analytically intractable.

Immigration rate refers to the arrival of members or propagules of a species not already present on an island. But it is logical to assume that alongside propagules of new colonist species, a near-shore island will continue to receive additional arrivals of resident species (e.g. see McCollin 2017). These arrivals will boost flagging populations and boost genetic heterozygosity (e.g. Parrish 2002), saving at least some island populations from extinction. As propagule arrival generally declines with isolation, the moderation of extinction by such supplementation should also diminish with isolation. This was termed the **rescue effect** by Brown and Kodric-Brown (1977), who inferred it from the study of a model system of arthropods on individual thistles in Arizonan shrubland. The rescue effect has been invoked in studies of real and habitat islands and, in a few cases, direct evidence has been produced to support it (Lomolino 1984, 1986; Van Schmidt and Beissinger 2020).

Similarly, immigration rate may be affected by area. A large island presents a bigger target for random dispersers, as shown by the correlation between beach length per island and propagule interception rates for 28 Australian reef islands (Buckley and Knedlhans 1986). Larger islands also present a greater range of habitats and attractions to purposeful dispersers, such as some birds, elephants swimming across a stretch of sea, or mammals crossing ice-covered rivers in winter (Lomolino 1990; Lomolino et al. 2017). Both this **target-area effect** and the rescue effect may be detectable in the same study system. For example, Toft and Schoener (1983) undertook a two-year study of 100 very small Bahamian islands, in which they recorded numbers of individuals and species of diurnal orb-weaving spiders. Extinction rate was positively correlated with species number and inter-island distance, and inversely correlated with island area. Immigration rate was positively correlated with area and negatively with both species number and distance. The significance of both rescue and target-area effects are likely to vary depending on the scale of variation in area and isolation, as well as the behavioural ecology of the taxon concerned (see e.g. Si et al. 2014).

The path towards equilibrium

Very shortly after the publication of the equilibrium theory, a classic set of experiments was carried out by Simberloff and Wilson (1969, 1970) on tiny mangrove islets in the Florida Keys (Simberloff 1976). Having first surveyed the arthropods on individual trees, they eliminated them by fumigation to monitor subsequent recolonization. The data demonstrated return to approximately the original species richness. Intriguingly, there was a slight overshoot, suggesting that the islands could support more than their 'equilibrium' number of species while most species were rare, but as populations approached their carrying capacities, competition, and predation eliminated the excess species, producing the slight dip sketched at T2 in Box Figure 5.2b. A similar study by Rey (1984, 1985) used patches of the salt-marsh grass *Spartina alterniflora*, varying in size from 56 to 1023 m^2, but structurally simpler than the mangroves. Once again, fumigation was used, and arthropod recolonization was monitored on a weekly basis. Initially, the colonization rate was slow because extinction rates were high. As the assemblages built up, populations persisted longer and extinction rates fell. While these experiments were broadly supportive of the EMIB, neither fitted perfectly and both suffer from the limitation of being highly simplified 'model systems'. It is important therefore to look for larger-scale study systems.

MacArthur and Wilson (1967) in fact provided the first demonstration of the development of a larger island system, from bare slate to apparent equilibrium with the example of Krakatau. Sterilized by volcanic eruptions in 1883, the colonization curve for land birds up to c. 1930 appeared to have reached equilibrium as the forests returned, although plant diversity remained on a steeply upward course. It later became apparent that the ecosystems were in a phase of successional turnover from open to closed habitats, following which the bird species diversity began to climb again (Bush and Whittaker 1991). We return to this system in Chapter 7 as it provides illustration of some of the additional ecological detail that we need to add to EMIB.

What causes extinctions?

Equilibrium theory is not prescriptive on the causes of extinction, but rather argues that as richness increases, extinction rate will increase until balancing the forces of species gain (immigration plus, for remote islands, *in situ* cladogenesis) (Box 5.2). Given our knowledge of SADs, we know that as species density increases there are going to be increasing numbers of rare species. Extinction rates will therefore rise with richness as competition for resources increases. As the loss of species X may not be attributable solely to competition with a particular species Y, we may think of this as diffuse competition. There are of course other reasons for a species population failing, which can include: (i) predation; (ii) the lack or loss of any, or all, of optimal habitat, resources, or mutualisms; (iii) a significant disturbance event (hurricane, volcanic eruption), or pronounced environmental variation (e.g. drought); or (iv) demographic and/or genetic stochasticity.

Trends in colonization and turnover on the Krakatau Islands observed over the first century post-sterilization include some non-monotonic rate changes and, once pseudoturnover is accounted for, are largely attributable to: (i) successional turnover of communities, (ii) loss of specific habitats through geomorphological change in the archipelago, and (iii) the impact of further volcanic disturbance (Whittaker et al. 1989, 2000; Whittaker and Jones 1994a; Thornton 1996). Krakatau also supplies an example of extinction of bird species through predation. This occurred during the build-up of the ecosystems on Anak Krakatau (the new island that emerged within the archipelago in the 1930s) (Thornton 1996). It was observable owing to the very small size of the vegetated area and thus of the colonizing populations. Once an island's ecosystems have attained greater biomass, the ability of particular predators to cause the total loss of a prey species may be reduced. Indeed, the role of top-down consumers in regulating island diversity is typically not readily distinguishable when ecosystems are undisturbed. However, it can be starkly apparent in cases where we add to (or remove from) an island particular categories of top consumer (Section 14.6; Russell and Kaiser-Bunbury 2019).

Understanding extinction drivers and risk better is crucial for conservation interventions (Chapters 12 and 15) and for understanding compositional dynamics on islands (Chapter 6). In the process of a species declining (perhaps in fluctuating fashion) to extinction, its population must, for some period, be small. While the size of the population may be a good proximal predictor of extinction, we may lack understanding of why the population is small in the first place. The explanation could lie in the trophic status of the organism; in predation, disease, or disaster; in resource shortage; in competition with other species; or synergies between several of the above. A focus simply on ecological interactions with competitors or even the place of a species in a food web may miss the point that some larger-scale process of habitat loss or change has reduced the population to a parlous point in the first place. It follows from the above that there is no reason to assume a single pathway to extinction. This remains a key area for further research.

Evidence of equilibrium

Long-term, high-quality data resampling the same system are scarce, and some tests of such data have produced equivocal results that fail to provide strong support for the envisaged dynamic equilibrium (e.g. Golinski and Boecklen 2006). However, 69 years of breeding land bird census data from Skokholm, a small Welsh island, provides more encouraging results. McCollin (2017) demonstrates that when analysing census intervals of up to six years, the turnover is dominated by repeated colonization and loss of rare species, but that over longer intervals of 12–24 years, 11% of the species both obtained sizable populations (some above 50 pairs) and experienced either colonization or extinction during the study. For example, the stock dove *Columba oenas* established a breeding pair in 1967, built to 62 pairs within eight years, remaining at this density for six years, before declining to extinction three years later. Skokholm is a small island, just 4 h km off the mainland, and McCollin notes that this and other near-shore islands often show trends in richness and turnover that are correlated with mainland population trends. Larger, more distant island systems are needed to continue our evaluation.

There are other means of testing for dynamic equilibrium and recent efforts to do so have employed molecular phylogenetic methods to reconstruct long-term patterns of accumulation of bird species on oceanic islands. In a series of papers, Valente et al. (2015, 2017, 2020) have used this approach to estimate colonization and speciation timing for island avifaunas in 41 oceanic avifaunas. Although not every case demonstrates attainment of saturation/equilibrium, their results provided a remarkably strong degree of support for the MacArthur–Wilson (1967) theory, including: (i) declining colonization and increasing anagenesis (speciation without *in situ* radiation) with isolation from mainland; (ii) increasing rates of cladogenesis (radiation) with area—accentuated with increased isolation; and (iii) the expected decrease in extinction rate with island area. It may be notable that these findings are for analyses undertaken at the level of archipelagos, rather than individual islands. We will provide fuller detail of these studies in Chapter 7, where we will consider the argument that because of the intersection of geo-environmental and biodynamics in remote island systems, analyses cast at the level of entire archipelagos may be more likely to comply with the assumptions of dynamic equilibrium than the individual oceanic islands within them.

$\Delta s = M + G - D$

The literature we have synthesized in this chapter is vast and our approach to it has been selective. We hope, however, to have provided a narrative structure that explores some of the more important themes in island macroecology, and which establishes the theoretical basis for the next few chapters. In summing up progress, we must return once more to the seminal place of the MacArthur–Wilson theory. The statement in Box 5.2, $\Delta s = M + G - D$, is of course a truism. The number of species on an island at a given point in time *has* to be a function of the numbers previously recorded, gained, and lost in the interim. The extent to which systems are,

in practice, at or close to equilibrium condition no longer seems so critical to us as we once considered to be the case. We now have analytical and modelling tools to permit the partition (if desired) of the elements of a fauna or flora that approximate equilibrium and those that don't. Recent developments in mechanistic modelling applied to island macroecological problems hold particular promise for exploring the causal processes underpinning the variation in rates of immigration, speciation, and extinction and the resulting emergent diversity patterns (e.g. Hubbell 2010; Rosindell and Harmon 2013; Chisholm et al. 2016; Cabral et al. 2019a, b).

It is apparent, but scarcely surprising on reflection, that the relationships with island area—and especially with isolation—turn out to be rather more complicated than shown in the core EMIB, exhibiting various forms of scale-dependency. Similarly, the extent to which particular islands can be regarded as being in a dynamic, equilibrium condition is also debatable. Yet, the widely shared view that the theory was largely inapplicable to many remote islands and archipelagos deserves reassessment. Recent advances in statistical and modelling tools, combined with efforts to compile environmental, species, and genetic data, have permitted new tests of the driving factors of island diversity in space and time. The work reviewed above reveals new support for many of the ideas set out by MacArthur and Wilson over half a century ago and demonstrate the vitality of island macroecology as an approach to understanding the ecology, evolution, and biogeography of islands. In the next two chapters we go on to explore compositional patterns on islands (Chapter 6) and efforts to integrate biological dynamics and geo-environmental dynamics (Chapter 7), thereby building a bridge between the subject matter of Part II ('Island Ecology') and Part III ('Island Evolution').

5.11 Summary

Island macroecology (the study of the emergent statistical properties of island systems) gained a lasting research focus with the publication in 1967 of Robert H. MacArthur and Edward O. Wilson's equilibrium theory of island biogeography. The theory stipulates that species number tends towards a dynamic equilibrium via opposing rates of immigration and local extinction, which decline with isolation and area, respectively. There is much more to their theory, including (i) the increasing contribution of cladogenesis with distance from mainland source pools and (ii) that equilibrium values should increase over evolutionary time from those initially established.

Oft referred to as an ecological law, the species–area relationship (SAR) comprises multiple phenomena and two distinct classes: species accumulation curves (SACs), built from nested samples, and island species–area relationships (ISARs), which are not. ISAR models are typically regressions of species number present per island as a function of island area. While multi-phase models—for example, flat–steep (the small-island effect) and sigmoid models (especially for larger ranges of island area)—may be preferred for particular datasets, the overall winner is one of the simplest, the power model, which describes a convex upwards curve, rising steeply initially and gradually levelling.

The log version of the power model linearizes the ISAR, permitting systematic comparison of slope (z) and intercept (log C) for multiple archipelagos. Slope tends (i) to decrease with the area range of the islands but (ii) increase with isolation (alongside declining log C values). These trends are noisy, partly reflecting the confounding effect of the clustering of islands within archipelagos, interfering with the predicted distance from mainland effect. Speciation in remote archipelagos contributes to steeper archipelago species–area relationships (ASARs) than ISARs and steeper 'endemics ISARs' than found for non-endemic native species. Multivariate models highlight important contributions from properties such as elevation, habitat diversity, climate/energy flow, and island age, alongside area and isolation. Which are significant in a particular study depends largely on the scale of variation for each variable and frequently varies between taxa for the same study system, reflecting different scales of interaction with the environment.

Underpinning the emergent island diversity patterns, most communities (insular or not), have strongly skewed species abundance distributions (SADs), such that a high proportion of species are rare at community and landscape scales within any island. Recent work has shown that the form of SADs on islands is controlled by many of the same variables that underpin island species richness, such as isolation, area, and the increased impact of disturbance on small islands. Models of population processes are needed (and some are now available) that link their dynamics to island-level diversity and turnover.

Species turnover plays a central role in island theory and in the final section we examine the challenges of estimating the vital rates. While island area and isolation are theorized to affect only loss and gain of species, respectively, there is evidence that area can influence colonization rates and isolation can influence extinction, complicating but not invalidating the EMIB. System behaviour appears to vary between equilibrial and non-equilibrial, and between high and low biodynamism, showing variable fit to the MacArthur–Wilson theory. However, the sophistication of modern analytical and modelling tools now allows us to draw value from all such forms of behaviour, without discarding the basic island biogeographical framework. It is noteworthy that recent analyses, based on molecular phylogenetic methods, demonstrate equilibrium dynamics at the scale of the archipelago, operating over evolutionary timescales. The macroecological approach has thus generated numerous insights and a flexible theoretical platform, on which we build in the next two chapters.

CHAPTER 6

Assembly rules for island metacommunities

6.1 Hidden tramlines and the detection, attribution, and resolution problems

In the previous chapter we took a macroecological approach to analysing island biotas and species were treated largely as exchangeable units. Here we take the next step, seeking to understand governing processes through the analysis of compositional patterns across island systems. Ecological and evolutionary theory posits that species within the pool of potential colonists are distinguished by meaningful trait differences and that as island systems assemble, these differences filter the potential colonists (Fig. 1.2). Within the island, interactions among species influence the composition of the communities that develop. Hence, the discovery of repeated emergent patterns in composition implies the existence of assembly rules, hidden tramlines guiding island assembly. The null hypothesis is that the assembly of island biotas is a matter simply of chance and that any apparent emergent patterns are largely illusory. The challenges arising are, therefore first, to develop methods to be sure that island biotas truly depart from being random draws from the species pool, and second, given that such patterns are identified, to attribute causation.

Our start point is Jared Diamond's (1975) theory of assembly rules for land birds on islands around New Guinea. This work, largely based on analyses of species distributions, illustrates that the challenges of pattern detection and attribution are both significant. A third key problem that emerged was to find agreement on the groupings of species within which to test for positive or negative associations; that is, should assembly rules be sought for whole faunas or within limited ecological guilds? The latter implies a focus on subsets of species that are most likely to compete for resources (Box 6.1). We will call these three challenges the detection, attribution, and resolution problems. They turn out to be recurrent issues in analysis of occurrence and co-occurrence patterns across island systems.

In the following sections, we first review where we have reached in describing and understanding island species assembly through tools such as incidence functions, checkerboard distributions, and nestedness. We then consider more recent developments involving: (i) species relatedness via phylogenetic analyses and metrics of phylogenetic diversity; (ii) functional traits and functional diversity of species; and (iii) interaction networks (e.g. between pollinators and plants, or seed dispersers and plants). Exploration of these topics provides further insights into the multiple levels of interaction involved in structuring island **metacommunities** (Box 6.1).

The work reviewed in this chapter demonstrates that non-random patterns are frequently observed in island metacommunities. As Roughgarden (1989, pp. 217–219) put it, 'studies of island biogeography show that a community is not simply a collection of all those who somehow arrived at the habitat and are competent to withstand the physical conditions in it . . . a community reflects both its applicant pool and its admission policies'. Island metacommunities may be shaped by a variety of ecological and environmental properties. They may reflect legacies from the non-random arrival of species drawn from a source pool, and sometimes the non-random loss of species following changes such as increased

Island Biogeography. Robert J. Whittaker, José María Fernández-Palacios, and Thomas J. Matthews, Oxford University Press.
© Robert J. Whittaker, José María Fernández-Palacios, and Thomas J. Matthews (2023). DOI: 10.1093/oso/9780198868569.003.0006

Box 6.1 Guilds, communities, and metacommunities on and across islands

Some studies restrict analysis to a particular **ecological guild;** that is, species of similar niche, or more specifically 'groups of species that exploit the same class of environmental resources in a similar way' (Root, 1967, p. 346) and which co-occur on an island. An example would be Diamond's (1975a) studies of the guild of forest fruit pigeons found on islands around New Guinea. Other studies considering broader ecological groupings use the term **community** for their level of analysis. An example is Heinen et al.'s (2018) analyses of extinction-driven change in the vertebrate frugivore communities across 74 oceanic islands. Their analyses are for all birds, mammals, and reptiles that have fruit in their diet, at the level of entire islands, which given the size of many of the islands, implies a rather coarse-scale use of 'community'.

When referring to the entire set of species found on an island, we have in the past preferred terms such as island floras, faunas, or biotas, which are neutral about the level and extent of interactions across communities or habitats. This is because the ecological community was traditionally defined as the set of species populations *interacting* in a local area within a particular habitat or ecosystem. By such a definition, larger islands (with few exceptions) actually contain multiple communities distributed across their habitat template. However, the traditional approach to community ecology was critiqued by Robert Ricklefs (1987, 2008), who argued that 'the local community is an epiphenomena that has relatively little explanatory power in ecology and evolutionary biology'. He advocated for a community concept 'based on interactions between populations over a continuum of spatial and temporal scales within entire regions'

(Ricklefs 2008, p. 741). The concept of the **metacommunity** has since taken hold to describe 'a set of local communities that are linked by dispersal of multiple potentially interacting species' (e.g. Leibold et al. 2004; Brown et al. 2011). This concept was originally inspired by the idea of metapopulations, referring to the equivalent idea at the single species level (Section 12.3). The conceptual shift to metacommunities embraces the importance of processes of interaction, of varying strengths, across multiple spatial scales (Brown et al. 2011; Whittaker et al. 2018).

In sum, the search for non-randomness within island biotas has variously been restricted to subsets of species that are expected to overlap strongly in niche (guild level), or to broader ecological or taxonomic groups, such as 'land birds', or 'all higher plants' (island level). The terms used to describe the level of analysis include guild, community, flora, fauna, biota, or metacommunity, and they are not always consistently applied. Differences of view on how best to operationalize these concepts provide plenty of scope for disputation. However, as more evidence accumulates of non-random patterns within whole island faunas and floras (and especially as this often appears to arise from interactions among species), the more reasonable it seems to deploy the term metacommunity in discussing them.

The metacommunity concept has been developed in a number of directions, including within Hubbell's (2001) unified neutral theory of biodiversity and biogeography, which took inspiration from the MacArthur–Wilson theory and has been applied to island biodiversity patterns (e.g. Rosindell and Harmon 2013).

---

island isolation. It follows that, where possible, a good approach to studying assembly and disassembly is to scrutinize temporal ('longitudinal') studies of island systems: this we do in the final part of the chapter.

6.2 Jared Diamond's assembly rules

Since at least the mid-19th century, island biotas have been regarded as distinctive in their composition, comprising 'disharmonic' (Section 4.3) suites of native and endemic species, ultimately sourced from mainland source pools but filtered and shaped

by insular rules of assemblage. Jared Diamond's work (e.g. 1975) on the avifauna of islands surrounding New Guinea represented a step change in analysing such compositional patterns within the broad framework of MacArthur–Wilson theory. His assembly rules were derived principally from analysis of the insular distributions of species within guilds (see Box 6.1). The approach became mired in the detection, attribution, and resolution problems described above. As we shall see, these challenges also tend to apply to many more recent studies of island assembly, yet with improved data, analytical, and modelling capabilities, progress is being made in solving them.

Incidence functions, checkerboards, and supertramps

Incidence functions show the frequency with which a species occurs in a set of islands as a function of, for example, their species richness, area, or isolation (Diamond 1975; Watson et al. 2005). Diamond (1975) used richness-based incidence functions for birds of the Bismarck Archipelago, categorizing them into those that are highly sedentary and only occur on the richest (also largest) islands (52 species), through four intermediate categories (26, 17, 19, 14 species, respectively), to the supertramp species (13 species), which occur exclusively on comparatively species-poor islands. Exemplars are shown in Fig. 6.1.

Within the study system: some islands were regarded as at equilibrium; former land-bridge islands (connected during Pleistocene sea-level lows) as supersaturated; and some disrupted by volcanic action (e.g. Long and Ritter Islands) were considered displaced below equilibrium (Fig. 6.2). The deviations from the ISAR and some aspects of the species incidence functions were together interpreted as indicative of long-term lags linked to historical events. In addition, Diamond hypothesized roles for ecological controls such as competition, territorial requirements, habitat controls, and reliability of food supply (Table 6.1).

The gradient from sedentary to supertramp species corresponds with the continuum from *K*-selected (late successional) to highly *r*-selected

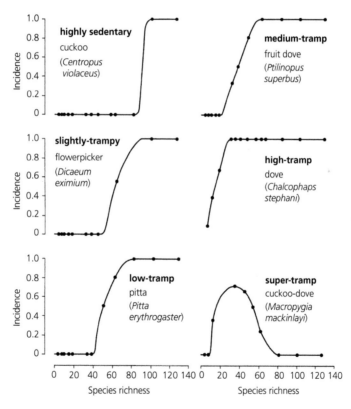

Figure 6.1 Incidence functions for six of the bird species of the Bismarck Archipelago as a function of island species richness. Each point represents mean incidence of occurrence (0 = no islands, 1 = all islands) based on between three and 13 islands of a given narrow band of richness, except the two largest values, which each represent a single island. Species occurring on successively less rich islands show increasing degrees of 'tramp'-like behaviour, with the supertramps apparently unable to succeed on the richest islands.

Modified from J. M. Diamond (1975) in *Ecology and Evolution of Communities* (ed. M. L. Cody and J. M. Diamond). Copyright © by the President and Fellows of Harvard College. Reprinted by permission of Harvard University Press.

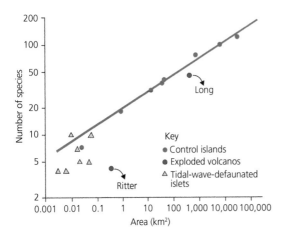

Figure 6.2 The island species–area relationship for resident, non-marine, lowland bird species on the Bismarck Islands (redrawn from Diamond 1974). The line of best fit is for the so-called control islands. The explosively defaunated islands of Long and Ritter are significantly below the line of best fit, which Diamond interpreted as demonstration of incomplete succession and failure to re-attain equilibrium in the period since their volcanic disturbance.

Table 6.1 Factors invoked in explanation of non-random co-occurrence of island birds in relation to Diamond's assembly rules.

Negative	Positive
Competition	Shared habitat
Differing distributional capacities/strategies	Shared distributional strategies/capacities
Differing distributional origins	Shared geographical origins
	Single-island endemics on same island

(pioneer) species discussed by MacArthur and Wilson (1967) (Section 5.2). The restriction of habitat specialists, those of large home range and those that rarely fly across large open expanses, to large islands and/or former land-bridge islands, is easy to grasp. More intriguing is the limitation of some species at the supertramp end of the spectrum to small, remote islands and to species-poor islands in the process of recovery from past volcanic disturbance. Where found on an island, these supertramps appeared to be broad in their habitat use and were characterized as good colonizers but poor competitors, which 'breed, disperse, tolerate anything, specialize in nothing' (Diamond 1975, p. 381).

Diamond also described how several ecologically similar congeneric species had mutually exclusive

but interdigitating distributions. Diamond suggested that these checkerboard-like distributions could be explained either (i) by subtle ecological advantage allowing each species to win on an island-by-island basis, or (ii) through priority effects, whereby an earlier-established species prevents a later colonist establishing by building up large populations before the competitor can gain a hold. For example, the flycatcher *Pachycephala melanura dahli* occurs on 18 islands, and its congener *P. pectoralis* on 11, but they do not co-occur on any island. Similarly, the cuckoo-dove *Macropygia mackinlayi* (mean body weight 87 g) occurs on 14 small islands, its congener, *M. nigrirostris* (86 g) on six, including the four largest islands occupied by either one: but they do not co-occur (Fig. 6.3). *M. mackinlayi* is one of the supertramp species, implying that it may be a good disperser but a weaker competitor than its congener.

Combination and compatibility—assembly rules for cuckoo-doves

Although not co-occurring with *Macropygia nigrirostris* (above), *M. mackinlayi* (97 g) does co-occur with larger cuckoo-doves, sometimes with *M. amboinensis* (149 g) and sometimes with a species of *Reinwardtoena* (297 g). It was suggested that a weight ratio of 1.5–2.0 between members of the same guild may be necessary to provide the niche difference enabling coexistence (Diamond 1975). Of the 15 possible combinations of the four cuckoo-doves (AMNR, indicating the species) found in the Bismarcks, just six were observed on the 26 islands that had at least one cuckoo-dove species, which Diamond concluded from a simple modelling exercise represented a departure from a random process of assembly.

The interpretation of the combinational patterns (Table 6.2) begins with the **incidence rules**, which determine that certain ('forbidden') species combinations do not occur. Second, so-called **compatibility rules** are exemplified by the allopatric *M. mackinlayi* and *M. nigrirostris*, which are so ecologically similar in their resource requirements that they are presumed to be unable to co-occur other than on the most temporary basis. Diamond

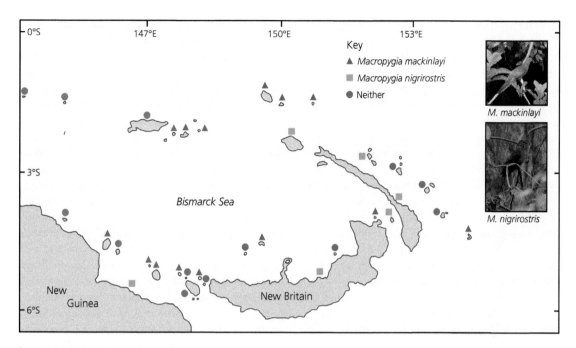

Figure 6.3 Checkerboard distribution of cuckoo-doves (*Macropygia*) in the Bismarck region. Of those islands for which data are available, most have one, but no island has both species.

Modified from J. M. Diamond (1975) in *Ecology and Evolution of Communities* (ed. M. L. Cody and J. M. Diamond), Fig. 20. Copyright © by the President and Fellows of Harvard College. Reprinted by permission of Harvard University Press. The photograph of *M. mackinlayi* is by David Cook and is used under licence (https://creativecommons.org/licenses/by/2.0/deed.en). The photograph of *M. nigrirostris* is by Marcel Holyoak and is used with permission.

Table 6.2 Island assembly rules for birds of the Bismarck Archipelago, as set out by Diamond (1975, p. 423). Based largely on distributional data and focusing on guilds rather than the entire avifauna, the rules were interpreted in relation to evolutionary change, habitat controls, and successional processes of recovery from disturbance, but placed particular emphasis on competition.

---

1 If one considers all the combinations that can be formed from a group of related species, only certain ones of these combinations exist in nature.

2 Permissible combinations resist invaders that would transform them into forbidden combinations.

3 A combination that is stable on a large or species-rich island may be unstable on a small or species-poor island.

4 On a small or species-poor island, a combination may resist invaders that would be incorporated on a larger or more species-rich island.

5 Some pairs of species never coexist, either by themselves or as part of a larger combination.

6 Some pairs of species that form an unstable combination by themselves may form part of a stable larger combination.

7 Conversely, some combinations that are composed entirely of stable subcombinations are themselves unstable.

---

suggested that the two may be the product of fairly recent speciation from within a former super-species. Thus, the combinations MN, MNR, AMN, and AMNR 'cannot' occur. The final rule type, **combination rules**, in effect mops up the unknowns. For instance, Diamond argued based on their incidence functions that the combination AMR should occur frequently on islands with a medium species number. It is allowed by the compatibility rules that had been deduced, yet it was not found to occur within the dataset. This

was taken to imply additional combinational rules preventing apparently reasonable combinations occurring with the frequency to be expected by chance. As most changes to an island's avifauna are likely to be produced by single species additions or losses, recombinations requiring two or more gains and/or losses simultaneously (to avoid a 'forbidden' combination) might be anticipated to have low transition probabilities. Hence, some combinations might be viable, but difficult to assemble from the 'permitted' combinations.

The niche relationships among the species of *Ptilinopus* and *Ducula* fruit-pigeons provide a classic example of resource segregation. Diamond first described community structure within the 'mainland' lowland forests of New Guinea, where the larger pigeons forage preferentially on bigger fruits—providing one element of resource segregation. Where large and small species feed within individual trees, the lighter pigeons are able to feed on the smaller, peripheral branches, better able to support their weight. Thus, some degree of niche separation can be maintained within a guild of co-occurring species. Yet, within a single locality in New Guinea, no more than eight species will be encountered, forming a graded size series, each species weighing approximately 1.5 times the weight of the next smaller species. There are eight size levels filled by combinations drawn from 18 species. On satellite islands off New Guinea, subsets are drawn in such a way that on smaller or more remote islands, size levels are emptied as the guild is impoverished according to a consistent pattern. Level 1 (the smallest birds) empties first, followed by levels 2 and 5, followed by level 8. It is intuitively sensible that the smallest birds should be excluded first as they will be restricted to the smallest fruit types, for which they would have to compete with slightly larger pigeons that can also make use of larger fruits. This form of structuring could also be identified on those islands regarded as undersaturated, suggesting that the time required to adopt such structure is shorter than that required to attain equilibrium in species numbers.

Criticisms and responses

The assembly rules became the subject of long-lasting and at times pointed debate (see e.g. Simberloff 1978; Connor and Simberloff 1979; Gilpin and Diamond 1982; Diamond and Gilpin 1983; Sanderson et al. 2009) centred on the detection, attribution, and resolution problems described above.

Detection: Determining whether the patterns observed might arise as a function of a random draw from the species pool may seem a simple concept, but requires agreement on the appropriate species pool, and how to constrain the randomization exercise, to take account of existing patterns of richness and incidence. This turned out to be a topic of prolonged disputation. Although perhaps not clear at the time, these problems are general to comparative analyses of this kind, are hard to resolve, and have resurfaced in numerous contexts since (e.g. Section 6.4; Weiher and Keddy 1995).

Attribution: The detection problem was at the forefront of criticism of Diamond's assembly rules, but closely allied to this was a concern that the importance of competition was over-emphasized. Critics suggested that ongoing processes of evolutionary change could be at work in creating some of the patterns, as lineages evolved and species formed and spread into new territory (or failed to). A further problem, which besets many analyses in island biogeography, is whether the patterns being analysed have been distorted by anthropogenic change and especially extinction: a key factor shaping contemporary oceanic island avifaunas (Steadman 2006; Heinen et al. 2018). Indeed, based largely on evidence of extinctions from islands in Remote Oceania, Steadman (2006) argues that anthropogenic extinction provides a more parsimonious general explanation for the failure of multiple congeners to co-occur on islands than does interspecific competition and that it has affected small islands in particular (Box 5.4), thus also distorting incidence functions (but see below).

Resolution: By this we refer to the challenge of determining which groups of species to include in analyses. Diamond (1975) restricted analyses to limited guilds, such as the cuckoo-doves (above). Some of the challenges to his conclusions were based on analyses of much larger suites of species, across multiple guilds, within which competitive effects are likely to be clouded by habitat differences. Subsequent work has confirmed that how broadly or tightly the groups are defined for 'checkerboard' analysis can be critical to whether evidence is found for negative associations between species groups (e.g. Stone and Roberts 1992; Sanderson et al. 2009).

Efforts to test for assembly rules in this and other systems continued for many years without a sense of clear progress. In retrospect, the main reason that the debate proved so difficult to conclude was that computational power and statistical approaches were, until recently, inadequate for the task of resolving the detection problem.

While the construction of suitable null models will always be a matter for debate, these problems no longer seem so intractable. We can now identify non-randomness within island biotas with a greater degree of confidence (e.g. Gotelli and McCabe 2002; Ulrich and Gotelli 2007) and indeed, rather than relying on a single null model, researchers can compare results from multiple models (see Strona et al. 2017).

In a landmark study covering both the original Bismarck system of 41 islands and the nearby Solomon archipelago of 142 islands, Sanderson et al. (2009) defend the validity of the distributional data now available and they provide an analysis of pairwise co-occurrence that confirms non-randomness of island assembly. Their approach involved the construction of vast numbers (10^6) of randomly drawn 'null' communities to permit identification of 'unusual' negative and positive pairings, broadly corresponding to the conventional statistical thresholds for significant departure from random expectations. They found that, in some cases, positive associations between species pairs arise not because of similarities in habitat use or ecology, but because both species favour islands of a similar range in richness, an example being provided by the pairing of a mound-builder (*Megapodius freycinet*) and a parrot (*Eclectus roratus*). They also argue that when considering complementary distributions (i.e. negative associations, tending towards checkerboard), it is important to consider how broader distributional and incidence patterns may constrain outcomes. The species pairs showing true checkerboards tend to involve one of the previously identified supertramps, restricted to species-poor islands, with a paired species restricted to species-rich islands, such as the *Macropygia mackinlayi* and *M. nigrirostris* pairing of Fig. 6.3.

A rather different example is provided by the pairing of an aquatic duck (*Anas superciliosa*) and an insectivorous flycatcher (*Monarcha cinerascens*). The duck and flycatcher are found, respectively, on islands of more than 33 species and fewer than 29 species. Using null matrices unconstrained by incidence functions, they will be identified as negatively associated. However, given their contrasting ecologies, we can rule out competition for resources: neither has excluded the other from colonizing any

island. To address this 'resolution' problem (above), the authors chose in this instance to avoid using guilds, the boundaries of which may be subject to disagreement, but instead to compare species pairs (i) among congenerics, (ii) among confamilials (excluding congenerics), and (iii) across the entire set (excluding confamilials). This provides a gradient of pairs of generally very similar to less similar autecology. In this analysis they found (Fig. 6.4 top panel) a marked prevalence of negative relationships in pairwise comparisons in both archipelagos but only at the within-genus level. They also show that when restricting analysis to pairs with broad overlap in incidence, the excess of observed complementary relationships persists: indeed, it can be seen at both within-genus and within-family levels (Fig. 6.4, lower panel). Hence, while some complementary relationships reflect differences in habitat use, that explanation can be ruled out for many congeneric pairings based on their traits and within-island distributions and behaviour. They also found that for those congeneric pairs found in both archipelagos, those showing complementary distributions in one tended to do so in the other.

Sanderson et al. (2009) did not extend their pairwise analyses to three or more species sets, as the computational requirements for doing so remained challenging. With this minor caveat, their study provides strong support for the existence of assembly rules (as per Table 6.2). The authors emphasize that extending analyses to entire island avifaunas involves the 'duck/flycatcher effect', a dilution of expected competitive effects by inclusion of the bulk of the species pairings for which, for a variety of reasons, direct competition between species is not to be expected. However, by filtering the analyses to subsets of pairs of closely similar ecology and by bringing improved computational science techniques to bear, they have resolved both the detection and resolution problems. By doing so, they have made a compelling case that interspecific competition influences the distributions of particular subsets of species within the avifauna of these archipelagos (see also Diamond et al. 2015).

As we gain improved resolution on the evolutionary relationships of the species involved and are able to place them in the context of the earth system dynamics of the region, it is likely that

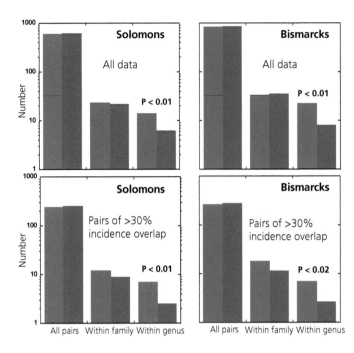

Figure 6.4 Pairwise comparison of distributional overlap of land birds on the Solomon and Bismarck islands. Observed (red) and expected (blue) number of 'unusually negative' relationships: top, among, all species pairs, and, bottom, among those which overlap > 30% in incidence (the range of island richness over which they are found). In each panel, analyses are provided for pairs within a genus, pairs within a family (excluding congeners), and pairs within all birds (excluding confamilials).
Modified from Figure 2 of Sanderson et al. (2009).

our understanding of Diamond's assembly rules will continue to develop. For example, within the Bismarck Islands, the whistler *Pachycephala citreogaster* is found on larger islands, while the congeneric *P. melanura*, identified as a supertramp, is restricted to smaller islets (although it is occasionally found in coastal locations on the larger ones). The different island populations of *P. citreogaster* have recently been shown to be genetically distinct from one another and to be around three times as old as those of *P. melanura*, which are undifferentiated across their small islands, consistent with their supertramp status (Jønsson et al. 2017). By way of contrast, Ó Marcaigh et al. (2022) provide data on another of Diamond's supertramps, the island monarch, *Monarcha cinerascens*, which sit less easily with the supertramp narrative as originally conceived. The island monarch is a widespread species, occurring on islands in both Melanesia and Wallacea. If the supertramp strategy is the result of *r*-selection, there should be minimal population structure across the species' range; that is, the high dispersal ability of the species should mean there is little genetic divergence between island populations. However, genetic data showed that there was in fact significant population structure within the island monarch's range (Ó Marcaigh et al. 2022).

This implies that the supertramp strategy may not be permanent. Rather, it could represent an early stage of an evolutionary cycle, where relatively good dispersers have 'high colonizing potential' but successful colonizers differentiate from other populations linked to a post-colonization evolutionary reduction in dispersal ability (see also Le Pepke et al. 2019; and Section 9.2 on taxon cycles).

6.3 Exploring incidence functions

Incidence functions, as per Fig. 6.1, are simple tools that provide a form of direct gradient analysis for a single species. Providing only that there are enough islands and enough incidences of a species, logistic regression can be used to fit and explore relationships with system properties such as richness, area, and isolation. We will review a few studies exploring the effects of one and sometimes two potential controlling variables. For a fuller review with particular focus on habitat islands, see Matthews (2021).

Biedermann (2003) provided an interesting analysis of area–incidence relationships of 50 species of vertebrates and invertebrates in which he demonstrates that area requirements increase essentially linearly with increasing body size on a log–log scale. This contrasts with Hanski's (1986) analyses of three shrew species on Finnish lakes, in which the largest species, *Sorex araneus*, had a low extinction rate from islands larger than 2 ha, but the two smaller congeners (*S. caecutiens* and *S. minutus*) were absent from many in the 2–10 ha range as their populations are more susceptible to environmental stochasticity. Variability in area–incidence functions was also evident in a longitudinal study of birds in woodland habitat islands in England by Hinsley et al. (1994), showing that population crashes in harsh winters changed their form.

When combined with population monitoring or other observations, the form taken by incidence functions can sometimes be linked to differential propensity to extinction (above), or immigration. As an example of the latter, Lomolino (1986) showed that small mammals active in the winter were able

to colonize islands in the St Lawrence River (New York State) across the ice and this was reflected in their insular incidence. It is consistent with island theory that area and isolation should be key determinants of turnover and of species incidence, and Lomolino (1986) identified five most likely scenarios: random distribution, minimum area thresholds, maximum isolation thresholds, a combination of area and isolation thresholds, or a trade-off between area and isolation (Fig. 6.5, panels b–f, respectively).

Applying this approach to 10 mammal species on islands in the St Lawrence River, Lomolino (1986) found that six exhibited minimum area effects, three compensatory effects, and one, the vole *Microtus pennsylvanicus*, showed no area or isolation effects. The carnivorous shrew *Blarina*

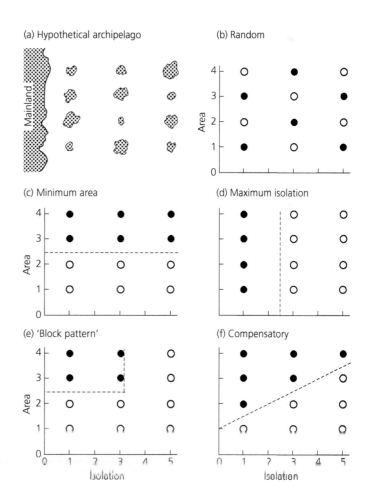

Figure 6.5 Patterns of insular occurrence of a hypothetical species on islands of a hypothetical archipelago (a). Presence and absence in (b)–(f) are indicated by filled and open circles, respectively. The units for isolation and area are arbitrary. Redrawn from Lomolino (1986), his figure 1.

brevicauda exemplifies the compensatory pattern, consistent with independent evidence that it is a relatively poor disperser. This study system also illustrates the potential of predation to shape distributions and abundance. The shrew preys mainly on immature voles and, in the absence of the shrew on more remote islands, the vole was found to undergo ecological release, occurring in habitats atypical of the species (Lomolino 1984). When introduced into islands, the shrew caused drastic declines in the density of the vole, restricting it to its optimal habitat and, in at least one case, causing its extinction. As the vole is the better disperser, the two species exhibit a negative distributional relationship across the islands studied. Lomolino's studies thus demonstrate the significance of recurrent arrivals and losses and also of ecological controls such as predation in structuring the mammalian communities of these islands. We might call this the vole/shrew effect, to set alongside the contrasting duck/flycatcher effect (above).

Incidence functions have also been used more directly to analyse predator impact. The Falkland Islands (Islas Malvinas) are an archipelago of several hundred islands roughly 460 km east of southern South America. They have been impacted by both the loss of native vertebrate predators (the Falklands wolf, *Dusicyon australis*) and the introduction of non-native predators: the Norway rat (*Rattus norvegicus*) is particularly widespread and problematic. Tabak et al. (2014) demonstrate the varying impact of rat presence on the incidence functions of several vulnerable passerine birds (including endemic subspecies). Most species were negatively affected by rat-infestation, which also therefore reduced the intercept of the ISAR. However, some species were unaffected and two, *Cistothorus platensis* and *Melanodera melanodera*, had higher incidences on small islands and lower incidences on large islands in the presence of rats, hinting at the importance of multi-species interactions in determining incidence (Fig. 6.6; Box 6.2).

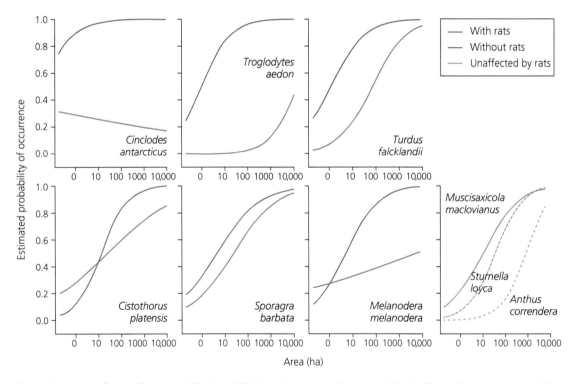

Figure 6.6 Incidence functions for selected Falklands land birds based on presence/absence on 44 islands with rats (*Rattus norvegicus*) and 59 rat-free islands. Three species statistically unaffected by rat presence/absence are represented by a single function each.
Redrawn from figure 3 of Tabak et al. (2014).

Box 6.2 Rats, seabirds, and the functional ecology of atolls

Many island communities are highly dependent on the surrounding seas as a source of food and, from an ecological perspective, the flows and linkages between land and sea are crucial. It is thus a little odd, at first glance, that so much island biogeographical analysis sets these linkages aside; for example, focusing—as many studies do—on land birds to the exclusion of seabirds. Graham et al. (2018) take the opportunity of a rare 'natural experiment', to demonstrate the importance of the intimate links between land and sea, in their study of the northern atolls of the Chagos Archipelago (Indian Ocean). These islands have been uninhabited for four decades and are protected from fishing. Some have long been infested with black rats (*Rattus rattus*), which predate seabirds, but others have escaped infestation.

They studied six infested and six rat-free islands, which were found to have mean seabird densities of 1.6 and 1243 birds per ha, respectively (for biomass values see Box Figure 6.2a). The nutrient transfer from the oceans to the land via guano (which is rich in phosphorus and nitrogen) deposited by the seabirds was found to scale accordingly, with an estimated nitrogen input from the guano of 0.8 and 190 kg ha^{-1} yr^{-1}, respectively. They quantified these transfers by sampling soils and plant foliage on the islands, sponges and macroalgae on the adjacent reef flats, and turf algae and damselfish tissue on the more distal reef crests. They found that not only were the islands themselves nutrient-enriched, but the effect remained detectable, if lower in amplitude, even as far from the atoll as the reef crests, some 230 m offshore. The effects of the deposition of so much natural fertilizer, derived from the open ocean, on these small islands, extended through the food chain. Not only did fish growth rates and biomass show a clear response (see Box Figure 6.2b) but the trophic structure was changed, with increased proportions of the reef being grazed by parrotfish, whose bioerosion functions are important to the dynamics of the substrates, feeding back into the geomorphological dynamics of the atoll.

This study hints at the widespread impact of humans in spreading rats throughout the world's islands and provides strong evidence of the cascading ecological impacts that are likely to follow rat eradication, or for that matter, further rat introductions (see also Brooke et al. 2018).

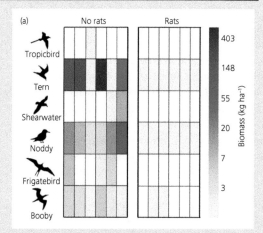

Box Figure 6.2a Biomass heatmap of seabirds per family, for each of six rat-free and six rat-infested islands in the Chagos archipelago. Each row is one island.
Modified from figure 1b of Graham et al. (2018).

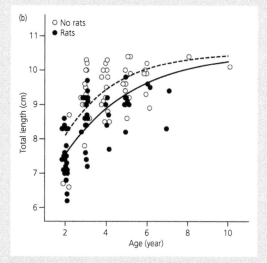

Box Figure 6.2b Growth of damselfish (*Plectroglyphidodon lacrymatus*) on rat-free and rat-infested islands in the Chagos archipelago.
Figure 3a of Graham et al. (2018).

Incidence functions can also be used to highlight the impact of matrix effects on species occurrence within habitat islands, as shown by Watson et al. (2005) for woodland birds in habitat islands embedded in three contrasting landscapes clustered around Canberra, Australia. The landscapes were: (i) an agricultural landscape, used for both pastoral and arable farming; (ii) a peri-urban landscape featuring pastures, hobby farms, and small urban areas; and (iii) an urban landscape, namely the city of Canberra. They tested for the patterns shown in Fig. 6.5, finding considerable variability in the incidence functions between the three landscapes and among species. Nineteen out of 27 species showed area and/or isolation sensitivity, but only four showed the same pattern of sensitivity in all four landscapes. Eastern yellow robin exemplifies this variation in response, showing area sensitivity in two landscapes and a compensatory area/isolation effect in a third (Fig. 6.7a–c). It is interesting to note that some species fare better in response to area reduction in woodlands embedded in agricultural landscapes, but others do better in woodlands embedded in urban landscapes (Fig. 6.7d–f).

Hence, in addition to the original richness-based incidence analyses, incidence functions can provide snapshots of the sensitivity of a species across a series of isolates to area, isolation, predation, and other factors. Such analyses typically reveal that at least some proportion of species display structured distributions across isolates but these responses can be temporally and spatially variable across the distribution of a species.

The importance of multi-species interactions with both competitors and predators (and by extension mutualists, pathogens, etc.), and of environmental differences and stochasticity, mean that such simple tools of analysis can only get us so far, providing a limited basis for developing predictive models of species responses to island environments.

6.4 Nestedness

As stated above, the null hypothesis of island assembly is that island biotas are randomly drawn from a regional species pool. Were that so, then it follows that within a given archipelago there should be predictable numbers (and proportions) of species shared between each island. Incidence functions, checkerboards, and the rest of Diamond's assembly rules describe apparent or actual departures from such an expectation. Evidence of these forms of non-randomness tends to be clearer when analyses are restricted to ecologically related species within guilds or genera (above). A **nested distribution** is simply another form of non-randomness, where smaller insular species assemblages tend towards being proper subsets of the species found at all other sites possessing a larger number of species (Patterson and Atmar 1986). Nestedness metrics (like incidence functions) provide descriptive tools that could be applied to a narrow group but are typically calculated for broad groups of species (e.g. higher plants, or land birds). This is relevant to note as, hypothetically, there might be significant patterns of complementarity of distribution within a particular guild of birds, while the avifauna analysed as a whole exhibits nestedness.

Just as nestedness is a departure from randomness in one direction, there is the possibility of 'anti-nested' patterns (e.g. Santos et al. 2010; Matthews et al. 2015a). Unlike nestedness, however, there are varied forms of 'anti-nestedness', as illustrated in Fig. 6.8, reflecting different patterns and drivers of distribution (Almeida-Neto et al. 2007).

It was not until the 1980s that statistical tests for nestedness were developed. Early work indicated it to be a common feature of insular systems and indeed of many other species–sites datasets, in part, it now seems, because the methods used were biased in that direction. Re-analysis of large numbers of datasets using more conservative algorithms, have shown a minority of datasets to be significantly nested, perhaps as low as 10% (e.g. Ulrich and Gotelli 2007; Matthews et al. 2015a). It should be noted that there can be a (big) difference between 'significant' nestedness, which denotes a dataset nestedness value (calculated using a given metric) that is significantly larger than expected based on a given null model, and 'perfect' nestedness, which describes the aforementioned nestedness definition involving perfect proper subsets (Matthews et al. 2015a). The former often involves many departures from perfect nestedness, which can have important implications if nestedness analysis is being used

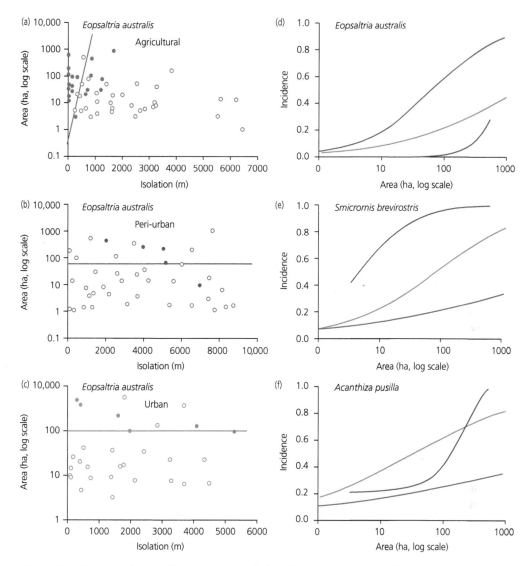

Figure 6.7 Incidence functions as determined by logistic regression for birds of woodland habitat islands from three nearby landscapes in south-eastern Australia: agricultural (blue circles and lines), peri-urban (orange), and urban (red). Panels a–c, presence (closed circles) and absence (open circles) for the eastern yellow robin *Eopsaltria australis* in relation to patch area and isolation; with the fitted lines showing the threshold where the probability of occurrence reaches 50%. Panels d–f, area-incidence functions for three species, selected to illustrate different patterns of sensitivity in the different landscapes. Note that panel d also features *Eopsaltria australis* for comparison with panels a–c.
Modified from Watson 2004; Watson et al. 2005; Whittaker et al. 2005.

in a conservation setting (Section 12.6). Previous synthetic analyses of nestedness have tended to include a mix of islands, habitat islands, and other datasets. The frequency of nestedness for real island systems thus remains unclear, with earlier literature an uncertain guide (Ulrich and Gotelli 2012, Matthews et al. 2015a). We review the evidence and

implications of nestedness in habitat island systems in Section 12.6 and here focus mostly on nestedness in true islands.

Wright et al. (1998, p. 16) observe that 'nestedness is fundamentally *ordered composition* ... Any factor that favours the "assembly", or disassembly ... of species communities from a *common pool*

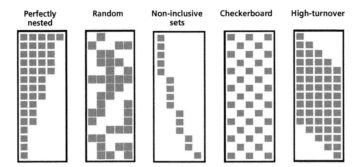

Figure 6.8 Nestedness and anti-nestedness. Each column represents an island, each row a species, with shaded cells indicating species presence and unshaded cells absence. Species–sites matrices may display tendencies towards nested distributions, they may be indistinguishable from random, or they may show various 'anti-nested' patterns: non-inclusive sets, checkerboards, or high-turnover systems. Nestedness analyses are not restricted in scope to island datasets, but many nestedness studies have employed island or habitat island data.
Redrawn from Almeida-Neto et al. (2007).

in a *consistent order* will produce nested structure' (italics in original). They identify four factors that could influence tendencies towards nested structure and that we briefly consider in turn: area, distance, passive sampling, and habitat nestedness. We also consider temporal turnover and variation in source pools.

First, considering area, many of the first tests for nestedness were deployed on land-bridge islands, for which the dominant hypothesis was the creation of a nested pattern through differential extinction following isolation and area loss as sea level rose after the last glacial. The expectation was that the variation in species' area–incidence thresholds (above) would lead to predictable and repeated sequences of species loss (disassembly) across an archipelago to generate a nested distribution (Fig. 6.9a). Second, where nestedness corresponds better to island isolation than to area, this might indicate that differential dispersal capacities are driving nestedness (Fig. 6.9b). Variants on the original practice of ordering island-incidence matrices by their richness have included ordering the matrix by area or by isolation, permitting more direct tests of the role of these variables on compositional overlap (e.g. see Roughgarden 1989; Matthews et al. 2015a). Alternatively, correlations between the row ranks in the maximally nested island–incidence matrix and different variables (or their ranks), such as island area or isolation, can also be used to test alternative hypotheses regarding the drivers of nestedness in a given system (e.g. Wang et al. 2010).

Third, passive sampling provides a null hypothesis for nestedness. Assuming that potential island propagules are drawn randomly from a pool of species in which the abundance of species varies, for example, following a lognormal species abundance distribution (see Section 5.9), then simulations suggest this would generate a strong degree of nestedness across a series of islands. This would be true even if the islands were all the same distance from the mainland pool (Wright et al. 1998). Fourth, many species have tight habitat requirements. If islands also tend to accrue habitats in a highly predictable pattern (**habitat nestedness**), this could produce a tendency towards nestedness of species composition (e.g. Honnay et al. 1999). Fifth, a temporal gradient of successional stage across a series of islands might also contribute towards a nested distribution if, once again, the order of species assembly over time tends to be highly predictable and extinction during the succession was minimal. Sixth, many archipelagos are known to receive species from more than one source pool (Chapter 4). It follows that extending analyses to systems spanning increasing geographical extents will mean different islands sampling from different source pools, inevitably reducing compositional nestedness.

These six scenarios may be neither a complete list nor mutually exclusive. For example, increasing island area may correlate with the accumulation of habitats, while differential sampling of a source pool could be linked to successional patterns and to isolation effects. Indeed, there is good reason

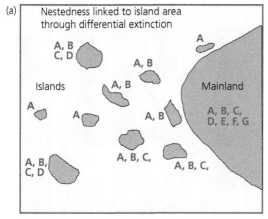

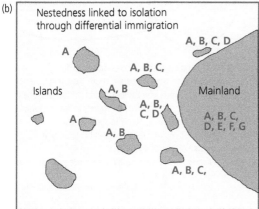

Figure 6.9 Hypothetically, nested insular subset patterns drawn from mainland species A–G may be generated by several mechanisms, including (a) ordered extinction, with minimum area thresholds determining the order of extinction predicted to occur across a set of islands of varying size following, e.g. the severing of land-bridges through sea-level rise, and (b) differential colonization ability, leading to nestedness linked to variation in island isolation.

to think that the imprint of different drivers, past and present, may be detectable within the distribution and diversity patterns of a single archipelago. For that reason, it is desirable to test a range of hypotheses rather than solely testing for one pattern such as nestedness. Equally, the fact that many of these drivers may be acting in tandem may result in reduced nestedness if they are acting in different directions; for example, if area-dependent extinction and distance-dependent colonization are both operating in an archipelago but the larger and less isolated islands are not the same.

In their study of 93 bird species across 42 islands within the Thousand Island Lake, China, Wang et al. (2011) tested for four types of assembly rules. These were: (i) Diamond's assembly rules 1, 2, and 5 (Table 6.2); (ii) the so-called 'favoured states' model, whereby species joining richer systems are drawn from complementary functional groups; (iii) constant body-size ratios, as previously exemplified by Diamond's New Guinea fruit-pigeons (above); and (iv) nestedness. They conducted analyses both for all bird species and for each of seven ecological guilds. They found no evidence for assembly rules of types (i)–(iii), but they did find significant nestedness, both at guild and overall metacommunity level. The nested patterns appeared to be explicable by a combination of selective extinction and habitat nestedness and could not be related to isolation or to passive sampling. Area/extinction effects were also found to be the preferred explanation for nestedness in lizards and butterflies in the Zhoushan Archipelago, China, by Xu et al. (2017). We have previously mentioned the hypothesis that there may be long-term legacies demonstrated for land-bridge islands in the form of extinction-driven nestedness. However, contemporary extinction events have also been invoked in some cases. In the Falkland Islands study cited earlier, Tabak et al. (2014) reported that infestation by non-native Norway rats both impacted on species incidence functions (Fig. 6.6) and also strengthened the degree of nestedness.

6.5 Partitioning beta diversity into turnover and 'nestedness' components

The terminology of diversity metrics is fairly impenetrable to the new reader. The use of the terms gamma, beta, and alpha diversity was inspired by R. H. Whittaker's separation of diversity components into richness ('alpha') and difference ('beta') components *and* scales of assessment (from fine to coarser: point, alpha, beta, gamma, delta, epsilon) (see Whittaker 1977). Here we are concerned with just two terms, gamma and beta, which in the present context can be described as: (i) gamma diversity, the richness of an archipelago of islands, and (ii) beta diversity, the degree of compositional difference between each pair of islands. Many different beta diversity metrics have been developed, with some focused on the approach pioneered by R. H. Whittaker of partitioning gamma diversity into alpha- and beta-components, while others involve analysis of the full site–species matrix. Recently there have also been efforts to partition two distinct elements of beta diversity: (i) that resulting from turnover in membership; and (ii) that resulting from nestedness/richness differences between sites (or islands). [Readers should be aware that debate on the validity of these alternative approaches is ongoing (see e.g. Baselga 2010, 2012; Carvalho et al. 2012; Cardoso et al. 2014; Matthews et al. 2019b), and the insights derived from their application may be subject to revision, as previously seen for topics such as nestedness and assembly rules (above).] As an illustration of the meaning of turnover in this context, we might find that the cuckoo-dove *Macropygia mackinlayi* occurs on island A and its congener *M. nigrirostris* occurs on island B. Regarding richness differences, and as discussed in the previous section, if a set of islands of increasing richness shares the same species apart from the richer island of each pair having a few additional species, we term this a nested distribution. Hence, the richness-related part of beta diversity is linked to nestedness, albeit imperfectly (Fig. 6.10).

With this caveat in mind, we exemplify the approach of beta diversity partitioning by reference to one of many studies from the Thousand Island Lake system in China, an inundated system of former land-bridge islands created in 1959 by dam construction (Box 6.3). Si et al. (2015) provide beta-partitioning analyses for bird and lizard metacommunities in this system. They expected to find that birds, being more vagile, would display reduced tendencies towards nestedness and thus lower overall beta diversity than lizards. Their results vindicated the hypothesis that lizards would show a lower degree of nestedness. For both taxa, they found that spatial turnover contributed more to beta diversity than did nestedness. Isolation (which is slight for all islands) was not found to be an important driver of the compositional patterns, which were largely attributable to island area and habitat diversity. Similar results were found for spiders within the same system, with turnover primarily responsible for beta diversity variation (Wu et al. 2017). Differences were found between frequent

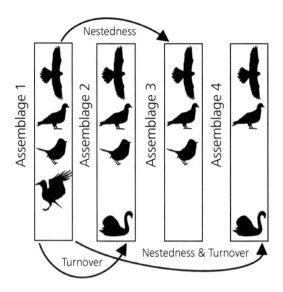

Figure 6.10 Illustration of the turnover and nestedness/richness-difference partitioning of beta diversity. Focusing on a target assemblage (Assemblage 1), the composition of this assemblage can differ from that of other assemblages due purely to species turnover (Assemblage 1–Assemblage 2), purely to species richness differences/nestedness (Assemblage 1–Assemblage 3), or a combination of both (Assemblage 1–Assemblage 4).
Bird vectors are from phylopic.org and are in the public domain; image used with permission of Joseph Wayman.

ballooners and less dispersive occasional or non-ballooners, with the latter contributing more to patterns of dissimilarity across islands. The finding that turnover tends to represent a larger proportion of overall beta diversity than nestedness/richness differences is not unique to these two studies, with the majority of beta-diversity partitioning studies reporting similar findings, including studies undertaken in island (e.g. Wayman et al.'s 2021 study of birds in 100 km² grid squares across Britain; and Borges et al.'s 2018a analysis of woody plants in plots in Terceira and Réunion) and mainland systems (e.g. spiders in Iberia; Carvalho et al. 2020).

Analyses of beta diversity for island metacommunities may seem a long way removed from the island assembly rules of Diamond that we began with; however, they continue the theme of searching for and testing structure in island metacommunity composition, whether this is explicable through processes that are dominated by colonization, extinction, or speciation, or perhaps

Box 6.3 Islands of the Anthropocene by K. A. Triantis and R. J. Whittaker

As humans are dramatically changing the planet in the Anthropocene epoch, they are creating new archipelagos of islands. The formation of isolates (habitat islands) through fragmentation of natural habitats led conservation scientists to turn to island theory in the search for general scientific guidance on how these new systems would function (Chapter 12). Hence, island theory became influential in new scientific disciplines and sub-disciplines, such as conservation biology, landscape ecology, and others (e.g. Haila 2002). But human activities have also created additional types of islands and archipelagos, which provide new opportunities for: (i) experimental tests of island theory, which have been limited owing to the difficulty in experimental manipulation of islands; (ii) the development of new theories—since the initial conditions are known, allowing for inferences; and (iii) appropriate spatial and temporal scales for informing land management and conservation planning.

Dam construction schemes are a particularly important means by which new land-bridge archipelagos have been formed. As of April 2020, there were c. 58,713 large dams (more than 15 m high, or of 5–15 m height and impounding > 3 million cubic metres) globally (from www.icold-cigb.org/GB/world_register/general_synthesis.asp), created as reservoirs and/or for hydroelectric power generation. Many contain newly created islands. We will provide brief illustration of the value of such systems in island biogeography.

Barro Colorado Island (BCI) sits within Gatun Lake in the central section of the Panama Canal. The island was formed in 1913 and has been the focus of ecological research for a century. The 50-ha forest dynamics plot established on BCI c. 1982, and within which every free-standing tree and shrub is recorded every five years, has been the basis for studies of species abundance distributions, forest dynamics and maintenance of diversity, processes of extinction, and effects of extreme events such as El Niño (Willis 1974; Condit et al. 2004). BCI has also been instrumental in the development of the *Unified Neutral Theory of Biodiversity and Biogeography* (Hubbell 2001), which provides a null model for assessing the role of adaptation and natural selection and contributes to the understanding of dispersal limitation, speciation and ecological drift in natural systems (Rosindell et al. 2011).

The construction of the Guri dam and hydroelectric plant from the 1960s onwards led to the flooding of the central Caroní Valley (Venezuela), forming an impoundment of

c. 4300 km^2. Flooding fragmented the formerly continuous dry forest of the mainland, leading to the isolation of hundreds of hill tops, forming islands ranging in area from less than 1 ha to about 760 ha. Island theory (Chapter 5) predicts a strong extinction signal in subsequent dynamics and distributions. Surveys of 'hill tops – turned islands' just four years after their isolation showed that about 75% of all vertebrates present on the nearby mainland had already disappeared from islands < 12 ha. Pollinators and seed dispersers were either absent or significantly underrepresented. This and other similar study systems have shown how trophic cascades can operate following fragmentation and isolation of small forest patches (Terborgh et al. 2001; Terborgh 2010).

In 1959, the Xinanjiang River in China was dammed to create the Thousand Island Lake, within which there are some 1078 land-bridge islands of > 0.25 ha. It has become a classic model system, generating insights into species–area relationships, nestedness, turnover, extinction, functional and phylogenetic diversity, and environmental filtering (e.g. see main text and Wang et al. 2011; Si et al. 2015, 2017; Wu et al. 2017; Lu et al. 2019; Zhao et al. 2020).

Humans create many other islands and archipelagos. For example, city parks have been seen as islands and key sites in the development of urban biogeography (Faeth and Kane 1978; Fattorini et al. 2018). Apart from providing novel grounds for testing island biogeography predictions (e.g. Fattorini 2014; Fattorini et al. 2018), city parks can also help us understand evolutionary dynamics. In illustration, the populations of a lizard species in four parks showed significant differentiation from non-urban populations, in a number of morphological traits, within a time window of less than 150 years (Littleford-Colquhoun et al. 2017). Tree plantations may also be framed as *de novo* islands, in which initial conditions may be controlled to a greater degree than most 'natural experiments'.

It has been argued that the biogeography of true, geographic islands, will increasingly be determined by the economic isolation/connectivity of their human populations rather than largely by their area, isolation, and evolutionary time (Helmus et al. 2014). Meanwhile, some anthropogenic islands will provide tractable systems for recording initial conditions and temporal change, with the possibility of intervention to modify the 'experiments' underway.

through stochastic processes. In regard to the latter, Almeida-Gomes et al. (2022) have shown that random placement/passive sampling models, similar to those used in analyses of nestedness (Section 6.4), do a good job at predicting the observed species dissimilarity patterns of generalist but not specialist frog assemblages in Atlantic Forest habitat islands, likely due to the broader environmental tolerances of generalist species. Further work on beta diversity and the role of stochastic processes, on birds from the Thousand Island Lake system, suggests that analyses of beta diversity patterns provide a means to distinguish between species equivalent (neutral) and non-neutral models and thus a means of developing improved mechanistic models of island assembly (Lu et al. 2019).

6.6 Trophic hierarchies and ecological networks within island metacommunities

Whereas most island biogeographical analyses tend to focus on a single taxon or ecological group, it is intrinsic to general ecological theory that there is interdependency among members of different ecological and trophic groups. This must have a bearing on island assembly (e.g. Bush and Whittaker 1991). In general, as energy flows through ecological systems, there is less available to sustain each subsequent layer, leading to generally smaller population sizes. The **trophic theory of island biogeography** recognizes this and predicts that species at higher trophic levels should be more strongly regulated by island area than those of lower levels (e.g. Holt 2010; Gravel et al. 2011; Roslin et al. 2014; Ross et al. 2019; Holt et al. 2021). Such effects are likely to be most clearly seen when including islands of a large range in area and, in particular, when that large range includes very small islands (Section 5.8). Analysis by Post et al. (2000) demonstrates this effect by showing that food-chain length increases with lake size, as larger lakes sustain both larger bodied fish and more intermediate layers in the ecosystem.

Relationships across trophic levels that may be important to determining colonization and persistence include predation, parasitism, pollination, seed dispersal, and the habitat-forming and modifying impact of vegetation. For species that are involved in specialist interactions, the presence of a specific interaction partner can be a pre-requisite for their establishment and persistence on an island. For example, the presence of the right fig tree species will determine the capacity of its pollinating fig wasp to colonize, while the perpetual absence of its fig wasp will prevent a colonist fig species from reproducing *in situ* (Compton et al. 1994; Mawdsley et al. 1998). Equally, if a species goes extinct on an island, any species depending solely on it will also go extinct. In a series of studies, Holt and colleagues (reviewed in Holt et al. 2021) explored these ideas analytically using incidence functions (Section 6.2). These incidence function models predict the percentage of islands occupied by a given species as a function of island area and trophic rank. In the models, a species of trophic rank i can only persist on an island if its resource species (of rank $i–1$) is also present; however, the species can still be absent even if its resource is present. Given this incidence model architecture, it is then possible to define the (conditional) incidence function of a given species as the probability it will be present on island, given that its resource is present, as a function of island area, or indeed any other variable of interest (Holt et al. 2021). It should be noted that these particular incidence function models characterize systems best when trophic dynamics are predominantly 'bottom-up'; when predators can extirpate their prey (i.e. top-down effects) the dynamics become more complex to model. Based on these incidence function models, Holt and colleagues (e.g. Holt 2010) theorized that ISAR slope should increase as one goes up the trophic pyramid, particularly if the focus is on specialist food chains (Section 5.8).

Jacquet et al. (2016) argue that the trait distribution sampled by an isolate should be a sub-sample of the regional pool, filtered in different ways depending on the area and isolation of the patch. They generated a mechanistic model to explore this proposition and tested their ideas using data for body-size distributions of fish from 134 tropical reefs. They found that the proportion of large-sized species was higher for small and isolated reefs compared with larger and better-connected reefs: this seemingly counter-intuitive result might be linked to increased diet generality, permitting larger-sized species to persist on small and isolated reefs.

Another illustration of trophic interactions is provided by Hanna and Cardillo (2014)'s analysis of extinction within 934 mammal populations on 323 Australian islands in response to introduced mammal species. Their analysis is also one of many that tackles a recognized problem of comparative analysis, namely phylogenetic non-independence. This occurs where differing degrees of shared common ancestry among the species undermine assumptions of statistical testing (Sakamoto and Venditti 2018). Hanna and Cardillo used generalized linear mixed models to partially control for phylogenetic non-independence and decision trees to explore interactions. They found that larger mammals had higher extinction probabilities than smaller species and their risk of extinction increased with island isolation. The presence of non-native black rats was the principal predictor of extinction for smaller species. Interactive effects consistent with meso-predator suppression (cf. Section 12.4) were also detected as, for example, extinction probabilities of smaller species were lower when black rats occurred alongside a larger non-native predator than on rat-infested islands lacking a larger predator. Their analyses thus demonstrate emergent outcomes from quite complex interactions of multiple non-native predators, prey traits, and island geography (cf. Box 6.2).

An obvious step in exploring island assembly rules across trophic levels is to quantify ecological networks, such as plant-seed dispersal and plant–pollinator networks by systematic observation of interactions in the field (e.g. Traveset et al. 2015; Vizentin-Bugoni et al. 2019). Whereas many of the analyses reviewed in this chapter have focused exclusively on native species, network analyses of this type inevitably span both native and non-native species and they have proven interesting.

Traveset et al. (2015) collected four years of data on flower visitation by birds from 12 of the Galápagos islands, recording observations of feeding on flowers and/or transporting pollen for 19 native bird species and 106 plant species, of which around 30% were endemic, 41% native but not endemic, and 29% non-native. Most bird species were found to be highly generalized, each pollinating around 22 plant species (with a range from 1 to 77). All the species-rich networks within islands were found to be both nested and modular. Nestedness we have already encountered at the island level, but here the analysis is being conducted between sites within islands and refers not to species composition per se but to species interactions (generally in the context of bi-partite networks). A nested network is one where the species with few interactions (specialists) interact with subsets of the species that generalists (those with many interactions) interact with. Modularity refers to the tendency of species interactions to form groups (modules), within which the members are strongly interconnected but between which connections are weaker (Olesen et al. 2007). The analyses showed that there was variation between islands (unsurprising as there are different species present) and between lowland and highland networks within islands (although they were coupled) and between seasons. Overall, there was only a slight preference for native plant species among pollinators and two of the most visited plants were invasive non-natives (*Psidium guajava* and *Impatiens balsamina*).

A three-year study based on faecal samples quantified seed-dispersal networks involving 21 bird species and 44 plant species (87% and 66%, respectively, being non-natives) on the Hawaiian Island of Oahu (Vinzent-Bugoni et al. 2019). Remarkably, there were no interactions recorded in the study between native bird and native plant species, so that the entire network involved novel (anthropogenic) interactions. Local networks were found to be non-nested, involved low-to-intermediate connectance (i.e. only a subset of potential interactions was observed), and demonstrated specialization and modularity. On an island-wide scale, networks showed lower connectance, moderate specialization, modularity (Fig. 6.11), and nestedness. We return to the topic of ecological networks and how non-native species can generate ecological cascades in Sections 13.8, 14.2, and 14.6. Here it is worth emphasizing that the structure (and stability) of these novel seed-dispersal networks on Oahu were found to be comparable with those of native-dominated communities sampled within islands and continents across the world. The non-native species have evidently become assimilated into well-integrated novel networks, despite a general low level of shared evolutionary or ecological

Modularity

Animals   Plants

Figure 6.11 Island-wide seed-dispersal interactions observed in a three-year study on Oahu Island, Hawaii. The figure shows non-random interactions between animal species (left) and plant species (right), which cluster within three modules (blue, orange, green), with comparatively fewer interactions across module boundaries (grey links). Thickness of lines indicates frequencies of interactions. From figure 1B of Vizentin-Bugoni et al. (2019). Reprinted with permission from AAAS.

history. Unfortunately, the introduced bird species often favour invasive non-native plant species and they therefore fail to replicate the role once played by the many extinct Hawaiian bird species in terms of seed-dispersal services of native plants. This is likely related to the fact that most of the introduced species are functionally different from many of the species that were driven extinct; this has been shown to be true both in Hawaii and

several other island systems (Carpenter et al. 2020; Sayol et al. 2021).

These examples illustrate, first, that analyses of island metacommunities can be conducted for units ranging from whole archipelagos down to the scale of samples within local communities across a single island. Second, they reveal that at each scale, we may detect non-random compositional patterns that have a variety of causes, from the geographical context (e.g. isolation), through environmental and habitat controls, to the biotic (e.g. competition, mutualisms, predation) and anthropogenic influences. Analyses at the finer scales of resolution considered here demonstrate important interactions occur not simply in competing for resources within communities and guilds, but also across trophic hierarchies.

6.7 Functional and phylogenetic diversity

Species richness has long been the standard metric of choice in biogeographical and ecological analyses. Recently, increased attention has been given to two complementary facets of the diversity of communities and metacommunities: (i) functional diversity (FD) and (ii) phylogenetic diversity (PD) (see Box 6.4). In general, the more species are sampled, the more evolutionary history and the more distinct functional traits will be represented in the community. As such, most FD and PD metrics quantifying the size/volume of the functional/phylogenetic space (e.g. FD richness, Faith's PD) have a tendency to increase simultaneously with species richness. The challenge is, therefore, to detect departures from a random sampling of a species pool in terms of the phylogenetic or functional characteristics represented by a particular set of communities or islands (e.g. Whittaker et al. 2014; Santos et al. 2016).

A commonly employed approach (reviewed in Münkemüller et al. 2020) involves using a null model in combination with FD and/or PD metrics to try and shed light on the processes that led to the assembly of a given set of species on an island or archipelago. For example, if a mainland species pool is defined for a given island, a null model can be used that randomly samples the observed number of species on the island from the pool and

---

Box 6.4 Functional and phylogenetic diversity: metrics

Functional traits are morphological, phenological, and physiological characteristics of species that determine the interactions between species and other species, and between species and the environment. **Functional diversity (FD)** metrics aim to quantify the range and value of functional traits in an assemblage (e.g. an island), and thus provide a theoretical link between diversity and ecosystem functioning (Petchey and Gaston 2006). To calculate FD for a given assemblage, a set of functional traits is collected for all species in the assemblage.

These traits are often split into two overlapping groups: (i) response traits—those that relate to species' responses to environmental conditions—and (ii) effect traits—those that control the effect a species has on ecosystem functioning. Traits can be categorical (e.g. reproduction mode), ordinal (e.g. size classes), circular (e.g. phenology), or continuous (e.g. body mass) in nature. Multidimensional trait datasets are often condensed into a smaller number of axes (e.g. using principal components analysis if all traits are continuous) due to the often inherent correlation between them. Traits are often standardized (if continuous) and then either used to build a multidimensional space directly (again if continuous) or converted into a dissimilarity matrix, before further techniques are then applied to either of these objects, allowing a range of FD metrics to be calculated (Mammola et al. 2021). If all traits are categorical there is a finite number of trait combinations, and the usual approach involves clustering species into functional groups or entities.

FD metrics can be based on incidence and/or abundance data, and can be used, broadly speaking, to measure one of three different dimensions of FD: (i) functional richness—the amount of trait space occupied by a set of species (e.g. by using convex hulls or probabilistic hypervolumes), (ii) functional divergence—the spread of species across trait space (often relative to the centroid of the space), and (iii) regularity/evenness—the regularity of species' distributions within the trait space (Villéger et al. 2008). A wide range of FD metrics, and approaches for dealing with different types of traits, exist and interested readers are directed to Villéger et al. (2008) and Mammola et al. (2021) as useful starting points.

Analyses of FD can complement traditional island studies of species richness. High levels of functional redundancy can result in relatively low FD in systems with high richness, and vice versa. By extension, two islands may have the same number of species but very different levels of FD. This could happen, for instance, if one island contains a full complement of native species, while the second has seen replacement of many functionally distinct extinct native species with similar numbers of functionally similar (and thus functionally redundant) introduced species (Sayol et al. 2021).

Phylogenetic diversity (PD) metrics are designed to reflect how much evolutionary history is represented by a set of species in an assemblage. These metrics are often calculated using a phylogenetic tree (phylogeny)—a branching diagram displaying the evolutionary relationships between species. One influential measure is Faith's PD, an estimate of the minimum total branch length required to span a set of species in a phylogeny (Faith 1992). Other commonly used metrics capture different aspects of the evolutionary relationships between species, including mean nearest taxon distance—the average shortest patristic distance (i.e. the distance between a pair of species in a branching diagram) between each species and all others—and mean pairwise distance—the average inter-species distances. Just as there is an array of traditional diversity and functional diversity metrics, incorporating richness and abundance, so too is there an array of alternative PD metrics, some of which are also weighted by abundance data (reviewed in Tucker et al. 2017). Many of these metrics can be placed in a framework analogous to the aforementioned richness–divergence–regularity framework used in the study of FD.

PD is often used as an integrative measure encompassing the overall similarity that stems from multiple traits when trait measurements are not available for a focal taxon but a phylogeny is. However, doing so relies on the assumption that phylogenetic relatedness is a good proxy for functional differences (i.e. there is phylogenetic signal in the traits under study), which may not always be the case (Mayfield and Levine 2010; Münkemüller et al. 2020).

---

calculates 'null' values of FD/PD using this sample (e.g. Santos et al. 2016; Si et al. 2022). This process is repeated multiple times and the observed FD/PD value for each island compared with each island's distribution of null model values to calculate a standardized effect size. If an observed island FD/PD value is larger than the majority of values generated via this null model, the island community is said to be over-dispersed, whereas if it is smaller it is said to be under-dispersed or clustered. These patterns may then be interpreted as evidence of island assembly rules operating.

Over-dispersion is often interpreted as indicating interspecific competition (invoking the concept of limiting similarity), while clustering is interpreted as indicating strong habitat filtering. Patterns not distinguishable from random are often viewed as resulting from neutral dynamics, although it is conceivable that a random pattern could result from habitat filtering and competition acting in tandem but cancelling each other out. However, all such interpretations are open to criticism. For example, given that competitive exclusion is theorized to relate to both niche differences and competitive ability differences, interspecific competition has been argued to result in both over-dispersion *and* clustering (Mayfield and Levine 2010). In addition, there is the issue of taxonomic 'resolution' (as Section 6.2), whereby evidence of interaction effects may be evident within a group of closely related species (e.g. pigeons) but diluted when broader taxonomic groups (e.g. all birds) are considered. In short, we encounter the same sorts of technical problems of determining the right focal group, the right source pool, the appropriate metric, and null models as evident for all the other approaches to island assembly rules that we have discussed earlier (see Münkemüller et al. 2020).

The following case studies exemplify the use of FD and PD metrics in island biogeography, demonstrating the nature of the inferences being drawn. We begin by looking at a study of island functional diversity–area relationships (IFDARs), as an alternative tool to the use of island species–area relationships (ISARs).

Figure 5.11 (in the previous chapter) compares the IFDAR with ISARs for the beetles and spiders of the Azores, subdividing the data by chorotype. Spiders are all predators, but beetles range from fungivores to herbivores and predators and thus have a lower mean trophic rank. Yet it is the beetles that have the steeper ISARs and IFDARs (Whittaker et al. 2014), contra trophic theory (above). It is notable that the introduction of non-native species, which now exceed the native species in richness, had a greater impact in altering the beetle IFDAR

than the ISAR, disproportionately boosting the functional diversity especially of smaller, less rich islands.

Ross et al. (2019) examined the interaction of trophic status and functional traits, by quantifying how diversity varies with area for birds of the Ryūkyū archipelago, Japan. They found that both species and phylogenetic diversity increased with area but that functional diversity did not. They also found that groups of different trophic rank (herbivores, intermediate, and apex predators) did not differ significantly in their ISARs, again contradicting the trophic theory (Section 6.6), although there were differences between granivores and frugivores. Intriguingly, there was evidence of trait-based functional assembly of intermediate predators, but not of apex predators or herbivores.

Schrader et al. (2021) employ a rather different analytical approach, of analysing community-weighted means of functional traits of woody plants based on transects from 28 small Indonesian islands. They reported that small islands of simple forest structure favoured woody species with small, light seeds and that larger islands featured more complex forest with taller trees possessing heavier seeds, lower leaf nitrogen concentrations, and higher chlorophyll content. These findings point to non-random selection of species from the regional pool by islands of differing characteristics.

In a further recent development, Si et al. (2017) combine a PD and FD metric into a composite functional-phylogenetic distance (FPD) metric and used it to assess non-random assembly of breeding birds from 36 islands in Thousand Island Lake. They found that island avifaunas were more clustered than null communities, with this effect declining with increasing island size such that the composition of the largest island was indistinguishable from a null model. In this case, therefore, environmental filtering is supported. A subsequent study of 97 species of ants on 33 islands from the same system reported a tendency for a switch from phylogenetic and functional clustering on smaller islands (implying environmental filtering), to over-dispersion (potentially through competitive effects) on larger islands (Zhao et al. 2020).

Two global-scale studies present further interesting findings. In their study of 300 islands, Weigelt et al. (2015) reported that the phylogenetic structure of palms and angiosperms exhibited a clustered pattern on most islands, but that this was not the case for ferns. Si et al. (2022) presented analyses for mammals on 212 islands. They reported an overall tendency towards phylogenetical and functional clustering. Phylogenetic clustering increased with island area and, to a lesser extent, isolation but functional clustering was weaker and unrelated to area and isolation. These results are consistent with limited colonization, followed by *in situ* diversification on the larger islands.

Illustrating the diversity of potential applications of phylogenetic metrics, Matthews et al. (2020) analysed the phylogenetic diversity of plants on 173 Aegean islands as part of an investigation of the small-island effect. They predicted that if the presence of a limited array of habitats on small islands drives the small-island effect, then only a specific subset of closely related well-adapted strand-line species should be able to occur, and thus phylogenetic diversity would be lower than expected (i.e. phylogenetic clustering) on small islands. As island area increases, the number of habitats is also predicted to increase and a wider range of species from a broader array of lineages would be able to establish, leading phylogenetic diversity to increase. The results of their analyses showed phylogenetic clustering to occur on smaller islands with PD increasing with island area, supporting the hypothesis (Fig. 6.12). Further support for this general mechanism has come from studies of butterflies on islands in the East China Sea (Chen et al. 2021) and insular tree communities in Indonesia (Schrader et al. 2021—above).

Both FD and PD have been extended to the study of compositional differences on islands through the analysis of functional and phylogenetic nestedness (Section 6.4) and beta diversity (Section 6.5) (e.g. Matthews et al. 2015b, 2020; Hébert et al. 2021; Zhao et al. 2021). These approaches aim to evaluate how communities differ; for example, in terms of the trait space occupied or the branches shared in a phylogeny. In their study of forest habitat islands, Matthews et al. (2015b) found that birds in several different forest habitat island datasets exhibited significant functional nestedness, with island area a likely driver. In terms of beta diversity, in a further

study of ants in the Thousand Island Lake system, an analysis of all three diversity components utilizing abundance and incidence data by Zhao et al. (2021) found that taxonomic and phylogenetic beta diversity were dominated by the turnover component, while turnover and nestedness contributed equally to functional beta diversity. All three beta diversity measures were found to increase with increasing (i) distance between islands and (ii) difference in terms of isolation from the mainland, both emphasizing the importance of dispersal limitation in structuring ant communities on these islands (Zhao et al. 2021).

In summary, deployed as tools alongside conventional taxonomic diversity, phylogenetic and trait data can provide nuanced insights into the classic macroecological patterns such as ISARs that we explored in Chapter 5. They also simultaneously provide tools to test for and quantify nonrandom patterns of island assembly/disassembly. Linking demonstrations of pattern to inferences of process has often generated debate and controversy in island biogeography. This may well turn out to be the case with these comparatively new analytical tools and approaches, but this looks like being a productive avenue for future research, especially as it allows these themes to be explored at multiple scales from local communities within islands and habitat islands up to whole archipelagos (Dias et al. 2020; Escoriza 2020; Zhao et al. 2020, 2021).

6.8 Longitudinal studies of island assembly and disassembly

A general limitation of many of the studies reviewed so far is that they necessarily infer processes operating over lengthy periods of time largely from snapshot data. We now turn specifically to 'longitudinal' studies (i.e. those monitoring system behaviour through time). We have selected systems both of disassembly in the form of recently created land-bridge islands, and of assembly on newly emergent or newly sterilized islands. Colonization and ecosystem development of a not-too-distant island arguably constitute just a special case of ecological succession, a process

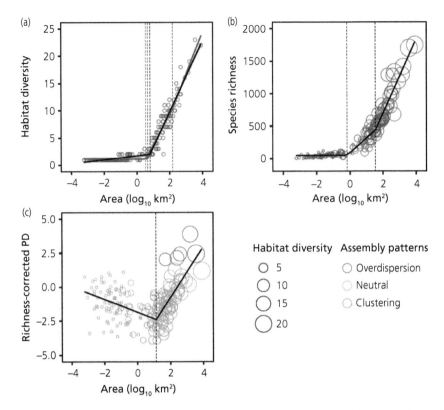

Figure 6.12 The relationship between (a) habitat diversity, (b) species richness, and (c) richness-corrected phylogenetic diversity (PD) and island area, for 3262 native vascular plant species on 173 Aegean islands. Coloured circles represent islands, and in (b) and (c) the size of the points is scaled by habitat diversity. Thick solid lines represent the predicted values of the best model(s) fitted to the data (black = best, red = second best, blue = third best model) and dashed vertical lines represent the threshold(s) of the best model(s). In (c), a null model was used to generate PD values corrected for island species richness. The different coloured circles represent different null model outcomes (community assembly patterns): yellow islands are those with less PD than expected (clustering), blue those with more PD than expected (overdispersion), and grey those where PD was not different from expected (neutral).
From Matthews et al. (2020).

(realistically many processes) that in complex ecosystems, dominated by long-living organisms, may be evident over hundreds of years, involving biotic and abiotic interactions that are critical to shaping island assembly from a regional species pool. We use the case studies of Surtsey (Iceland) and Krakatau (Indonesia) to illustrate some of these interactions and processes. We pay particular attention to Krakatau, which was the first such test system used by MacArthur and Wilson (1967). Both these study systems illustrate the importance of trophic hierarchies (Section 6.6) in the process of island metacommunity assembly over extended periods.

The dynamics of island disassembly

Land-bridge islands and especially those newly created by flooding areas through dam construction (Box 6.3) provide opportunities to study processes of island disassembly, by which we mean the process of partially or wholly dismantling a complex and diverse island metacommunity. Numerous such studies have been conducted but comparatively few involve both isolation by water and 'before' and 'after' data.

The Saint Eugène Fragmentation Project is an exception. Initiated in 1993 by the French Museum National d'Histoire Naturelle, the project involved sampling a relatively homogeneous area within the

Sinnamary River catchment, French Guiana, prior to flooding through the construction of the Petit-Saut Dam (Cosson et al. 1999a, b). Just three years after fragmentation by flooding, the vertebrate diversity (lizards, birds, small mammals, primates, and frugivorous bats) of the newly isolated communities was found to have shifted towards dominance by species that are: (i) of larger within-guild body size, (ii) generalists in habitat and food requirements, and (iii) wide-ranging occupants of undisturbed tropical forests. Understorey frugivorous bats appeared more sensitive to the fragmentation effects and changes were particularly noticeable on smaller islands (Cosson et al. 1999b).

A more typical study is that of Benchimol and Peres (2015), which lacks a formal 'before' baseline. They used an array of vertebrate survey techniques to assess persistence as well as evidence of movements between islands for a system of 3546 islands created by the Balbina hydroelectric reservoir in the central Brazilian Amazon. Their data showed that most terrestrial and arboreal vertebrate species were driven to extinction across the large majority of the islands some two decades after flooding. Only the largest of the islands were able to maintain a majority of the species, with dispersive species, adept at moving through or crossing water, having higher occupancy. Indeed, after three decades, islands of < 10 ha were found to be virtually empty of terrestrial vertebrates (Benchimol and Peres 2021). Gibson et al. (2013) provide another example of faunal disassembly following insularity without a strict baseline, although researchers had a good understanding of the fauna of the study area prior to isolation. Surveying small mammals across different time periods (5–7 and 25–26 years post-isolation) on islands isolated by the flooding of once continuous forest in 1986/1987 to create the Chiew Larn Reservoir in Thailand, the authors found almost complete defaunation took place in response to insularity. After 25–26 years, many islands only supported one small mammal species: the invasive Malayan field rat (*Rattus tiomanicus*). The rate of defaunation was found to be a function of island area, with the smaller islands (< 10 ha) losing the majority of small mammals within five years of isolation, and the larger islands (10–56 ha) within 25 years (Gibson et al. 2013).

The implications of such substantial changes to vertebrate fauna for the entire metacommunity can be profound. A classic demonstration comes from islands created by the damming of the Caroní Valley to form Lago Guri in Venezuela (Box 6.3; Terborgh et al. 2001). With diminishing island size, the island communities were found to collapse trophically, so that on the smallest islands, communities consisted of predators of invertebrates, seed predators, and herbivores. Consumer numbers increased in the absence of predators, greatly reducing densities of seedlings and saplings of canopy trees. Similar cascading trophic effects were recorded in the Balbina study (Benchimol and Peres 2021).

Trophic cascades connected with the loss of the top trophic tier appear to be a common feature in systems in which fragmentation and isolation have been imposed on formerly connected habitats. They have been demonstrated in experimental microcosms (e.g. Staddon et al. 2010), terrestrial habitat island systems (Chapter 12), and, as we have seen, in association with islands created by large-scale damming. However, there appear to be comparatively few longitudinal studies with temporal depth for islands in water where the baseline is as well established as in the Saint Eugène Fragmentation Project.

Trophic cascades and community disassembly can also arise from the introduction of species, especially where they represent the addition of a higher trophic level. A classic example being the impact of non-native rats on small atoll ecosystems (Box 6.2). We will return to the theme of island metacommunity disassembly in Part IV, where we consider the impact of human colonization of island systems in detail.

The dynamics of island assembly: Surtsey

Surtsey is a small volcanic island of *c*. 2.7 km^2 formed between 1963 and 1967 at the southern end of a string of such islands, off the south coast of Iceland. It experiences a mean annual temperature of just 5.0 °C and annual precipitation of *c*. 1580 mm. Plant recolonization has been painstakingly recorded by scientists since 1965 and other visitors prohibited from landing to ensure that the natural dynamics can be recorded undisturbed (Magnússon et al. 2009).

Initial colonization was dominated by shore plants and while species persistence was good over the first decade, vegetation cover remained low. The system showed relative stagnation in the second decade, until a gull colony established in 1985, marking a step change in system dynamics. The gulls represent an external—and especially a marine—subsidy, fertilizing the area of the colony, resulting in enhanced nutrient status and increased biomass. As the colony grew, other birds, including buntings and geese, also began to breed on the island. As a consequence, after the initial period where sea-dispersed species led the way, bird transport became the dominant mode of plant introduction to the islands. Over time, as the habitat has developed, more complex food webs have developed and wind-dispersed plant species have also managed to establish but biomass outside the gull colony remained significantly lower than within (Frontispiece to Part I; Magnússon et al. 2009, 2020).

The role of the seabird colony was thus critical in advancing the process of ecosystem development across Surtsey, providing a classic illustration of the subsidized island hypothesis (Box 5.5; cf. Box 6.2). Surtsey thus provides a high-latitude example of how island assembly can involve fairly distinct phases during which different ecological groups of plants and animals are able to establish as conditions allow, with key linkages to external ecosystems (i.e. marine subsidies) having a critical bearing on system development.

The dynamics of island assembly: Krakatau

We introduced Krakatau as a test system for equilibrium theory in Section 5.10. Here we provide richer detail, centred around our search for metacommunity assembly rules. Drawing in particular on work by Whittaker and colleagues (Whittaker et al. 1989; 2000; Bush and Whittaker 1991; Whittaker and Jones 1994a, b; Shilton et al. 1999; Whittaker 2004), we demonstrate some intriguing parallels with the Surtsey system, notwithstanding that Krakatau (6°S of the equator) experiences a radically more favourable climate (mean annual temperature

c. 27.8 °C; annual rainfall c. 2600 mm) and draws from a far richer source pool of colonists.

The Krakatau natural experiment commenced with a catastrophic sterilizing eruption on 27 August 1883, involving a massive reshaping of the three islands of the archipelago (Rakata, Sertung, and Panjang) and emplacement of pyroclastic flows tens of metres deep across their remaining land surface (Whittaker et al. 1989; Thornton 1996). No evidence for any surviving plant or animal life was found by a scientific team surveying the islands later that year, and in May 1884 the only life spotted by visiting scientists was a spider. The first signs of plant life, a 'few blades of grass', were detected in September 1884. Since then, biologists have worked intermittently on the Krakatau system to track the colonization of plants and animals and the development of its ecosystems. A fourth island, Anak Krakatau, emerged in the centre of the caldera between 1927 and 1930, adding further twists to the dynamics of the archipelago through damaging eruptions affecting both its own ecosystem development and that of the other islands over the following decades (Chapter 7).

Plant colonization of the Krakatau coastlines occurred swiftly through sea-dispersed seeds and fruits washed up on the beaches and comprised typical elements of the Indo-Pacific strand-line flora. For these elements, Krakatau is not an isolated location and so vegetation characteristic of other coastlines in the region swiftly accumulated. This strictly coastal flora featured a rapid decline in immigration rate over time and very little turnover as the limited strand-line habitats filled up (Fig. 6.13). In the interior, succession has been more complicated. Initially, colonists comprised wind-dispersed pioneering ferns, herbs (especially Asteraceae), and grasses. Colonization by animal-dispersed species, which ultimately provided the large majority of forest shrub and tree colonists, was initially much slower. Although we would expect nitrogen to have been limiting early on, nutrient limitation does not appear to explain delayed appearance of zoochorous plants. The more likely reason is that a barren island, or indeed one lacking in fruit, provides little to attract the frugivorous bats and birds moving through the Sunda Straits. Some of the strand-line plants are, however, diplochorous. They

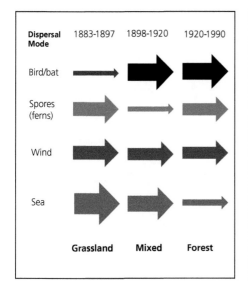
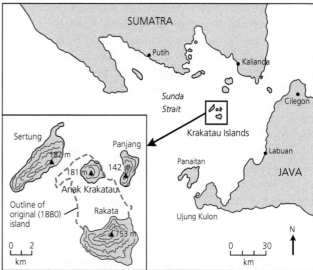

Figure 6.13 Plant recolonization data for Rakata, the largest island of the Krakatau group, following sterilization in 1883. Arrow widths are proportional to the increase in cumulative species number (thinnest = 0.14 species/year, thickest = 1.64 species/year) for three-time slices and four ecological groups: bird-dispersed, sea-dispersed, and wind-dispersed angiosperms, and ferns. The three phases correspond with vegetation of low stature (ferns, tall grasses), mixed savanna-parkland, and forest cover.

Data and map adapted from Whittaker and Jones (1994a). Accompanying map shows the islands *c.* 1983 and, in red, the pre-1883 outline of the larger island of which Rakata is the remnant.

produce fruits or seed that float in the sea and retain viability, accounting for their initial and early colonization. Once established and fruiting, they also provide a reward for birds or bats. A classic example is the sea almond, *Terminalia catappa*, a very early colonist on Krakatau, much loved by the large fruit bats *Pteropus vampyrus*, which visit and occasionally form roosts of several hundred animals on the islands.

Botanical observations record that the first fully zoochorous shrubs and trees to colonize and fruit were concentrated in and among the strand-line communities, highlighting that the frugivorous birds and bats which introduced them were indeed attracted to these localities. Once established and providing local fruit sources, these zoochores were then dispersed into the interiors. As they matured, they enabled the establishment of local populations of frugivores on the islands and continued to attract further visiting frugivores. With the principal exception of the wind-dispersed *Neonauclea calycina*, zoochorous trees became numerically dominant in the forest cover, which by *c.* 1930 clothed

the islands from coast to summit. It is this process (and the richness of the source pool in such species) that accounts for the uptick in arrival of animal-dispersed species, which continued through the second and third phases depicted in Fig. 6.13.

In turn, forest formation drove new waves of colonization by wind-dispersed forest epiphytes, including many ferns and orchids, as also reflected in the figure. The ease of dispersal of ferns, with their microscopic propagules, can be demonstrated by comparison of the size of the Krakatau flora (all islands 1883–1994) with the native flora of West Java. The ratio for spermatophytes is 1:10.1 and for ferns 1:4.2, indicating Krakatau to have now accumulated a remarkably rich fern flora (Whittaker et al. 1997) and demonstrating that the impediment to their colonization over the early decades was a lack of forest habitat.

As a function of these successional processes, the interior habitats of the islands initially, at the start of the 20th century, assembled animal communities characteristic of dense grassland 'savanna'-type ecosystems. As the forest formed up and these tree-less habitats were replaced with shaded forest, the birds and butterflies of the 'open' habitats swiftly declined in numbers. We therefore see a peak in immigration rate around the early 1920s, as forest habitat becomes available, followed by a peak in extinction rate as the grassland habitats largely disappeared. This was broadly repeated in data for plants, birds, and butterflies (Fig. 6.14). The irregular quality and frequency of survey means that the resulting estimates should be regarded as fairly crude approximations, but they serve to illustrate that the same sort of successional waves of assembly and turnover feature at the whole island metacommunity level as can be seen laid out within primary successional landscapes elsewhere (cf. Walker and del Moral 2003).

A lot of the species turnover observed in the Krakatau system has been driven by successional replacement of open and savanna habitats with closed forest habitat, or constitutes species lost due to the unconnected destruction of habitats (such as a lagoonal environment lost due to coastal erosion). Beyond these broad-scale loadings of the dice, analysis of the plant turnover data has revealed another unsurprising pattern: the species that both colonized and spread to become widespread at some point within the first 50 years, have generally persisted better than those that were always restricted to limited areas and/or small populations, typically within just one of the islands.

With the exceptions of around 24 diplochorous (sea-colonist) species and a few species introduced by humans, zoochorous species on Krakatau have been introduced by birds and fruit bats, mostly via gut passage. Bats are likely to have introduced only a few small-seeded plant species in this way, as they mostly do not swallow larger seeds, but their contribution includes some key early pioneering fig trees (*Ficus* spp.). Birds have doubtless introduced far more species, encompassing a greater range of seed sizes. In particular, it is likely that two larger fruit pigeons, *Ducula aenea* and *Ducula bicolor* (which is often seen flying between islands in sizable flocks), have played key roles in introducing larger-seeded trees and shrubs.

Hence, there are key distinctions to be drawn within the frugivore metacommunity as to (i) the

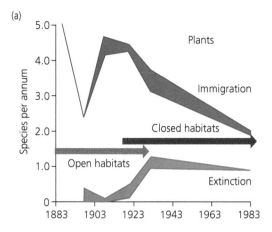

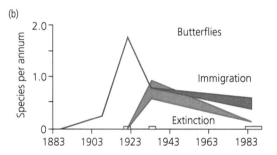

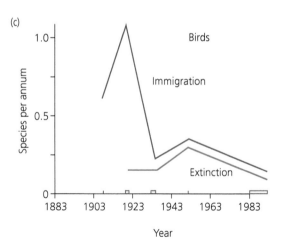

Figure 6.14 Immigration and extinction rates calculated in species/year for Rakata, the largest and best surveyed of the Krakatau island group: (a) spermatophytes 1883–1989; (b) butterflies; (c) resident land birds.
Data from Bush and Whittaker (1991) and Thornton et al. (1993). Shaded area indicates uncertainty within the estimates for Rakata. The successional transition from 'savanna' grassland to closed forest drove the linked peaks in immigration and extinction.

spectrum of plant species they introduce, (ii) their own status as residents or visitors, and (iii) their contributions to *arrival* as distinct from subsequent *spread* of plant species within the islands. There are, of course, numerous other ways animals interact with plants; for example, as seed predators and pollinators. In illustration, some 23 species of figs (*Ficus* spp.) have colonized Krakatau, each of which is dependent for its pollination on its own specific fig wasp species (family Agaonidae) also having colonized. The establishment of a breeding *Ficus* population thus requires multiple introductions of each fig and the assistance of their mutualist pollinators (Compton et al. 1994). It follows that not all colonization efforts succeed. Another cause of failure may be predation, with evidence both from Krakatau and from Ritter Island (a recovering volcanic island in New Guinea) of raptor predation of potential colonist bird species (Diamond 1974; Thornton 1996).

In comparison to Diamond's assembly rules, the description and interpretation of the Krakatau system places relatively little emphasis on competitive effects. Yet the analyses of plant turnover provide some suggestion that ecological success following colonization is not equally distributed. Two further mechanisms have been invoked in the Krakatau literature: (i) habitat controls and (ii) priority effects. (i) Some species may colonize habitat that is not particularly suitable for them, or through successional or other system changes, it becomes unsuitable, leading to their failure to expand their populations—or even to their disappearance. (ii) Other species may arrive in suitable habitat but find their ecological niche already occupied by a species of similar traits that arrived earlier and has built up a large population, thereby monopolizing available resources. These slightly-too-late arrivals may therefore be unable to expand on their initial foothold. Thornton (e.g. 1996) has suggested that differences in the prevalence of three lowland forest canopy dominants on three of the islands can be explained by priority effects, but the historical data are lacking to distinguish satisfactorily between this and other hypotheses (edaphic controls, disturbance effects). Similarly, the early inadvertent introduction by humans of *Rattus rattus* to Rakata Island and of *Rattus tiomanicus* to Panjang

and Sertung led to large populations forming that may well have prevented further populations of either species successfully colonizing an already occupied island. From the evidence available we are able neither to confirm nor refute these hypotheses for the Krakatau system, but it seems ecologically entirely plausible that these and other such mechanisms of biological interaction operate here as has been demonstrated more systematically elsewhere.

Ridley may well have had the early Krakatau data in his thoughts when he wrote in the preface to his classic text on plant dispersal (Ridley 1930, p. xii): 'An island rises out of the sea: within a year some plants appear on it, first those that have sea-borne seeds or rhizomes, then wind-borne seeds, then those borne on the feet and plumage of wandering seafowl, and when the vegetation is tall enough, come land birds bringing seeds of the baccate or drupaceous fruits which they had eaten before their flight.' The singular data series from newly constituted volcanic islands have provided invaluable, generative accounts of the processes of assembly. Yet, while the points of similarity between such systems as Krakatau and Surtsey are fascinating, we do not have the degree of system-level replication to permit the formulation of precise assembly rules from them. Nonetheless, we contend that these systems provide rich illustration of the importance of biotic and especially of hierarchical (cross-trophic-level) interactions in island metacommunity assembly that we have discussed elsewhere in this chapter. We shall return a final time to consideration of the Krakatau case study in Chapter 7, where we consider the impact of renewed volcanism within the archipelago on the unfolding biological dynamics.

6.9 Extending the time frame

As will have become evident, island assembly theory has long been rather mired in technical aspects of the detection, attribution, and resolution problems discussed above. However, there is a clear sense of progress in the complex business of describing and understanding the often loosely drawn rules by which island metacommunities are assembled, persist (or vary), and, sometimes, are disassembled or degraded.

Yet, there are a couple of elephants in the room. The first is that the natural environment doesn't remain constant over time. Systems of such complexity of process and large scale as Krakatau and Jared Diamond's New Guinea islands don't simply build themselves once and then remain constant. Rather they continually respond to fluctuating environments, to long-term climate change, associated sea-level fluctuations, and geo-environmental dynamics (uplift, subsidence, volcanism) (Whittaker et al. 2008, 2017; Jønsson et al. 2017).

The second elephant in the room is evolution, which we have acknowledged (especially in Section 6.7) but have yet to tackle in detail. Indeed, before leaving Diamond's assembly rules behind, we should note that there are parallels between Diamond's assembly rules model and E. O. Wilson's (1959, 1961) taxon cycle model (Section 9.2). Both theories seek to explain distributional patterns by means of competitive effects, habitat relationships, and past evolutionary processes. Although we have placed them for convenience in separate chapters, they are both evolutionary ecological models, both are dynamic models, and both invoke a key role for competition—indeed analyses of one may inform understanding of the other (Jønsson et al. 2017; Ó Marcaigh 2022). They differ principally in that the taxon cycle model focuses more on evolutionary change within the taxon as a biogeographical process, while the assembly rules ideas are concerned with identifying and explaining compositional regularities across a series of islands, invoking ecological dynamics linked to disturbance and ecological recovery.

In the final chapter of Part II we go on to examine the effects of island geodynamics on biodynamics, still largely through an ecological, or macroecological lens, before progression to island evolution in Part III.

6.10 Summary

The null hypothesis of island assembly is that islands draw species at random from a regional (mainland) species pool. On the whole, empirical evidence permits us to reject the null in favour of various forms of compositional structure, which may be manifest at a range of scales (from local communities within islands to archipelago memberships) and resolutions (from patterns of distributional co-occurrence within guilds up to functional or phylogenetic structure of entire biotas). Recognizing that islands and archipelagos typically possess local communities linked by dispersal of multiple potentially interacting species, we use the metacommunity concept to frame our review of island assembly rules.

We start by reviewing Jared Diamond's seminal analyses of land bird distributions on the Bismarck Islands, published in 1975. He employed simple graphical analyses of the incidence of individual species on islands of increasing richness, alongside their co-occurrence patterns within guilds that appeared to show mutually exclusive 'checkerboard' distributions, suggestive of competitive exclusion operating between species of closely similar niche. Based on these analyses and supported by ecological data and functional trait data, Diamond drew up seven descriptive rules highlighting non-random assembly of island bird communities. The approach drew pointed criticism, with debate focusing on problems of detection (were the patterns really different from random?), attribution (if so, was competition really the key driver?), and resolution (should tests incorporate whole biotas or limited guilds?). Limits to statistical and computational tools hampered progress in solving the detection and attribution problems for over a quarter of a century. Recent work has reaffirmed the existence of non-random patterns in avifaunas from the Bismarcks and Solomon Islands, providing evidence consistent with competitive effects in some species pairings, and of different causation in other pairings.

We go on to review evidence from incidence function analysis from a range of systems, highlighting that they are not always robust over time or across the range of a species. Next, we turn to nestedness, the condition where the species found in each of a series of islands of increasing richness form a proper subset of the next richest island. Nestedness metrics are typically applied to entire faunas/floras rather than to limited ecological guilds. Advances in statistical methods have shown nestedness to be less common than once thought and, indeed, a variety of forms of anti-nestedness can also be detected and may be as frequent (collectively) as nestedness. Recent work has begun to delve further into the patterns of compositional difference (beta diversity), distinguishing between nestedness/richness-difference and turnover components, highlighting how overall metacommunity-level patterns may comprise multiple ecological species sets with distinct distributional patterns responsive to different drivers.

Alongside competition, other important trophic relationships (e.g. predation, pollination, mutualisms) influence island communities, as demonstrated through the analysis of data for multiple taxa and through tools such as ecological network analysis. Recognition of nestedness and modularity within interaction networks provides another set of insights into island assembly and also processes of disassembly and the ecological cascades that have sometimes been generated by anthropogenic modification of island systems.

Analyses of taxonomic diversity patterns have recently been complemented by metrics designed to capture phylogenetic diversity (i.e. the evolutionary span of a set of species) and functional diversity (i.e. the ecological trait space occupied by a set of species). The application of such tools has provided further evidence of non-random selection of island species from regional species pools, variously demonstrating the operation of competitive effects

or environmental filtering, although the field lacks the maturity to provide robust generalizations at this stage.

Most of the work reviewed in the chapter leans heavily on inferences drawn from the analysis of the distributions of species across multiple islands, relying on statistical and modelling tools that may yet be found wanting. In the final section, we turn instead to a limited number of studies that provide a temporal series of island disassembly or assembly. Island disassembly studies are drawn from some recently created land-bridge islands that have resulted from flooding connected with dam construction schemes and they illustrate the ecological cascades that can follow from extreme area reduction and the consequent loss of top consumers. Finally, we turn to two descriptive accounts of island assembly in ecological time: Surtsey and Krakatau. These systems provide illustration of off-island subsidies and of the crucial importance of hierarchical links between plants and animals and especially the role of particular subsets of animals involved in plant dispersal. These accounts are illuminating but lack the replication of the many spatial comparisons discussed in the chapter. We conclude the chapter by highlighting that we have largely set aside to later chapters the consideration of island evolution and of the significance of island geodynamics for island biodiversity dynamics.

CHAPTER 7

Extending the timescale: island biodynamics in response to island geodynamics

7.1 The historical and the dynamic

Brown (1981) observed that MacArthur and Wilson's (1967) equilibrium theory had three attributes of a successful general theory of diversity.

- First, it is an equilibrium model, thus historical factors, climatic change, successional processes, and the like are acknowledged, but side-stepped. The theory explains rather the ultimate limits, the theoretical patterns of diversity.
- Second, it confronts the problem of diversity directly, number of species being the primary currency of the model. It also takes account of biological processes and some, at least, of the characteristics of the environment.
- Third, it is empirically operational, making robust, qualitative predictions that can be tested.

Perhaps the key attraction of MacArthur and Wilson's model in fact is that it is a dynamic model. The distinction between **historical** and **dynamic** **hypotheses** is based on the repeatability or probability of recurrence of a particular state or form. Historical (time-bound) knowledge refers to the analysis of complex states that have very small probabilities of being repeated (i.e. states of low recoverability). Physical or dynamic (timeless) knowledge refers to the analysis of states that have a high probability of being repeated (Schumm 1991). The search for dynamic hypotheses of species richness can thus be likened to the search for the general laws of ecology, from which historical contingency supplies the deviations (Brown 1999; Whittaker et al.

2001). Hence, the abiding attraction of MacArthur and Wilson's theory.

In its simplest form, it is close to being a neutral model, with species immigrating, speciating (on larger and more remote islands), and going extinct, at random, generating continual turnover of composition. As shown in Chapters 5 and 6, subsequent analyses of macroecological and compositional patterns in space and time have demonstrated that the null hypotheses of random assembly, turnover, or disassembly of island metacommunities can often be rejected, as anticipated based on general ecological theory. But not always. Moreover, the detection of structure within ecological subgroups or guilds, may sit alongside or within a largely random process of metacommunity assembly.

Thus far, our exploration of island ecological theory has been deliberately cast within the general assumption that once the stage is set for the ecological play, it remains largely unchanged. We have touched on ways in which this is not the case. For instance, we have considered evidence of responses to volcanic disturbances and to the fragmentation and isolation of land-bridge islands following anthropogenic dam construction (Sections 5.10, 6.8; Box 6.3). But we have yet to consider how to extend MacArthur and Wilson's dynamic model into longer time frames, explicitly incorporating island and archipelagic geo-environmental processes that are broadly repeated rather than entirely historically contingent. In this chapter we show that there is indeed scope to specify and test for general dynamic models that extend the macroecological

Island Biogeography. Robert J. Whittaker, José María Fernández-Palacios, and Thomas J. Matthews, Oxford University Press.
© Robert J. Whittaker, José María Fernández-Palacios, and Thomas J. Matthews (2023). DOI: 10.1093/oso/9780198868569.003.0007

approach into evolutionary time frames. Doing so requires us to make rather sweeping simplifications of how island geodynamics operate (for a reality check see Chapters 2 and 3).

We start by examining short-term environmental dynamics and consider their implications for island biodynamics and equilibrium. Next, we return to the Krakatau system, to consider how the arrival on the scene of a newly emergent volcanic island within the archipelago has impacted on island assembly processes. We then introduce and describe the general dynamic model of oceanic island biogeography (GDM). This model intersects MacArthur and Wilson's dynamics with a simplified island ontogeny (life cycle), generating predictions as to the resulting biogeographical patterns. We evaluate these predictions using macroecological tools of analysis in the current chapter, holding over consideration of the implications for phylogenies to Chapter 9. We also show how the analytical approach developed to test the GDM has been successfully applied to alternative models of long-term change in island configuration linked to Pleistocene

sea-level dynamics, again setting the scene for more explicit consideration of island evolution in the following chapters.

7.2 Extreme events and climate-driven fluctuations in carrying capacity

Over time, island carrying capacity may vary due to a wide variety of short-term geological, geomorphological, and climate-driven events and fluctuations and—since our emergence and spread as a dominant force—through human interventions (Table 7.1). Many of the natural changes in carrying capacity are inherently unpredictable at a fine scale, even if they may be explicable at a coarser scale and possible to model at particular resolutions. For instance, the general conditions for hurricane formation are well understood but the precise storm path and which islands will be hit are currently unresolvable other than at extremely short notice. Where they hit hardest, carrying capacities are inevitably temporarily reduced and ecosystems disrupted (Section 3.7). During post-hurricane

Table 7.1 How might the capacity for richness of an island vary through time? Left and right columns provide changes that may lead to increase/decrease in carrying capacity.

Factors leading to an increase	Factors leading to a decrease
Primary succession, increasing biomass, system complexity and niche space, in part driven by species colonization and enabling further colonization.	Late successional stages if passed through simultaneously across an island (or habitat island) might see competitive exclusion of suites of earlier successional species.
Arrival of 'keystone species' (e.g. fig trees on Krakatau).	Arrival of a 'superbeast' (e.g. first vertebrate predator) may over-predate naive prey species
Humans may increase habitat diversity, introduce new species, manipulate ecosystems to raise productivity, etc.	Humans may clear habitat, introduce new pests/predators, or highly competitive species, degrade habitats, and may introduce and release browsers/grazers to become feral (e.g. rabbits, goats).
Moderate disturbances might open up niche space, e.g. for non-forest species on an otherwise forested island.	Major disturbance (e.g. hurricane or volcano) may wipe out much standing biomass, massively reducing populations.
Increasing climatic favourability, increasing biological activity, increased NPP.	Climatic deterioration, lower NPP.
Area/habitat gain, e.g. through coastal deposition, island uplift, sea-level fall (e.g. during Pleistocene glacial periods), building phase of island ontogeny (as per the GDM).	Area/habitat loss, e.g. through coastal erosion, sea-level increase (associated with interglacial periods, global warming), or island subsidence; erosion and subsidence combining in declining phase of island ontogeny (as per the GDM).
Evolution, on remote islands, a similar effect to primary succession, principally operating through hierarchical links between organisms at different trophic levels.	*In situ* co-evolution is unlikely to generate reductions in richness, but as oceanic islands age, they subside and erode (Chapter 2), such that in parallel with the processes of speciation, carrying capacity will eventually decline (above).

GDM, general dynamic model of oceanic island biogeography; NPP, net primary productivity succession in the Caribbean, it has been found that the animal community may continue to change long after the initial 'greening-up' of the vegetation, through invasions and local extinctions as the vegetation varies through successional time (Waide 1991). Initially, nectarivorous and frugivorous birds suffer more after hurricanes than do populations of insectivorous or omnivorous species, a consequence of their differing degrees of dependence on re-establishment of normal physiognomic behaviour in the vegetation. Unsurprisingly, high-intensity storms can have long-term impacts on a variety of macroecological features of an island's flora and fauna (Section 5.9; Ibanez et al. 2020). Moreover, island ecologies and rates of immigration, evolutionary change, and turnover may be influenced by events of much lower magnitude than hurricanes.

In the late summer of 1984, 26 black-shouldered kites (*Elanus caeruleus*) took up temporary residence on San Clemente Island, 80 km from the Californian mainland, for a period of several months, although previously only one or two kites had been recorded over 19 autumn/winter cycles of bird observation. Scott (1994) suggested that the 1984 irruption may have been a consequence of an unusual pattern of Catalina eddies—which are seasonal cyclonic winds—possibly linked to an El Niño event. Each pulse of kites arriving on the island coincided with the first or second day of a Catalina eddy, suggesting that the eddy system acts like a door to the island that irregularly opens and closes. The same El Niño influenced carrying capacity for Darwin's finches in the Galápagos (Grant and Grant 1996a).

Abbott and Grant (1976) argued that the degree to which climate fluctuates about a long-term average value determines the extent to which species number does the same. 'In this sense, species number "chases" and perhaps never reaches a periodically moving equilibrium value, hindered or helped by stochastic processes' (Abbott and Grant 1976, p. 525). The notion that many islands spend substantial amounts of time not quite hitting equilibrium is captured in Fig. 5.20, which suggests there to be value in comparing equilibrium with non-equilibrium models.

Russell et al. (1995) reported that a non-equilibrium model provided improved predictions of observed turnover of birds for 13 small islands off the coast of Britain and Ireland. They suggested that turnover could be viewed in these islands as operating on three scales: first, year-to-year 'floaters' (trivial turnover); second, on a timescale varying between 10 and 60 years, an intrinsic component equivalent to that envisaged in the EMIB was observable; third, islands showed a change in numbers over time due to so-called extrinsic factors, including human impacts such as habitat alteration.

Environmental forcing, such as pronounced El Niño–Southern Oscillation (ENSO) events, hurricanes, and even volcanic eruptions, are features of island life and to some degree are keyed into island biogeographical and evolutionary dynamics. They may impact on native and endemic species populations (e.g. Miskelly 1990; Arendt et al. 1999), but on secular timescales they will only very rarely eliminate endemics altogether. This is because the species and ecosystems have evolved within the context of such disturbance regimes. Except on very small oceanic islands (such as cays and atolls and other similarly small volcanic islands) or where there are truly catastrophic eruptions, there will be sufficient refugia and sufficient resources somewhere within the island for the endemic species to persist.

The reduction in habitat availability that has occurred for many island species through human agency in the last few thousand years has, however, tipped the balance of the scales for many island endemics (e.g. Szabo et al. 2012), meaning that a hurricane or a large volcanic eruption that would otherwise be survivable may finally finish off an already endangered endemic. An example of a near miss is provided by the Montserrat oriole (*Icterus oberi*), which was very nearly eliminated by large volcanic eruptions of Chances Peak in the 1990s (Arendt et al. 1999). Another example is the Puerto Rican parrot *Amazona vittata*, which was reduced by habitat loss and hunting to a tiny remnant population in the El Yunque forest. This population was then further reduced by a succession of devastating hurricanes that hit the forest between 1989 and 2017. The parrot has only survived as a wild-living species through intensive captive breeding, translocation, and other conservation efforts (Paravisini-Gebert 2018). The Bahama nuthatch (*Sitta insularis*) and the Cozumel thrasher (*Toxostoma guttatum*) may not have been so lucky. The former was endemic to pine forest on

Grand Bahama Island and had suffered severe population declines due to habitat loss and introduced species, before two hurricanes in 2016 and 2019 are believed to have finally driven it extinct (Matthews and Triantis 2021). The latter is (was?) endemic to the Mexican island of Cozumel and was believed to have already been in decline due to the introduction of the boa constrictor before a series of hurricanes between 1988 and 2005 reduced the remaining population to likely unsustainable numbers; indeed, no confirmed sightings have been reported since 2006.

We have focused in this short section on a variety of short-term fluctuations in carrying capacity, driven by extreme weather and geological processes. Longer-term mechanisms of tectonics and of climate change are, of course, of greater importance. The processes of island genesis, building, decline, and disappearance and the shifts between glacial and interglacial conditions that have characterized the Quaternary period, separately and in combination, set the stage for the island biogeographical theatre. It is to these process regimes that we turn next, beginning with consideration of how volcanic processes have impacted on the biogeographical dynamics of the Krakatau Islands over the last several decades.

7.3 Island assembly interrupted: geodynamics and biodynamics of Krakatau

The recolonization of the Krakatau Islands following the sterilization event of 1883 provided MacArthur and Wilson with a first test of the return to equilibrium. The survey data for birds between 1883 and 1930 appeared consistent with the EMIB: species number reached an asymptote and further turnover occurred (MacArthur and Wilson 1963, 1967). However, the plant data from the islands revealed a poorer fit, as species numbers were continuing to rise steeply, which they hypothesized might reflect incomplete successional processes. Subsequent surveys, extending the data series by another six decades, revealed that in fact the 1920s represented a key successional transition phase as grassland habitats were replaced by rain forest communities (Fig. 6.13, 6.14), driving a pulse of

species losses of plants and animals associated with open habitats. Species richness increased again over the following decades as forest species accumulated. By the end of the 1980s, the data for land birds were compatible with an approaching (dynamic) equilibrium, but plant richness and butterfly richness continued to rise (Fig. 6.14). Data for other taxa (reptiles, mammals) were also of variable fit to the expectations of the EMIB (Table 5.1).

From 1883 until the late 1920s, Krakatau comprised three islands, and the best survey data, used in most of these analyses, are for the largest island, Rakata. Between 1927 and 1930, a new volcanic island, Anak Krakatau, emerged in the centre of the archipelago. It enriched this natural experiment in several ways, not least by violently disturbing the forests of the islands Panjang and Sertung, but to date having very limited impact on Rakata Island. How has this changed the trajectory of island assembly within the archipelago? And, more broadly, what role does the geo-environmental dynamics of the archipelago have in shaping the biodynamics?

Following the 1883 eruption, two thirds of the largest island had disappeared and a mantle of volcanic ash tens of metres thick had settled across the three remnant islands, extending the area of each island, mostly on the outer flanks (Whittaker et al. 1989, 2000). The early pace of erosion of the ash mantle was dramatic. It created a deeply dissected 'badlands' topography that remains geomorphologically highly dynamic. The extensive new territories around the coasts were also subject to extreme rates of attrition, and steep cliffs formed rapidly around much of the shrinking shoreline, increasingly restricting the extent of the shallow shelving beaches most favourable for drift colonists. Longshore drift created a lengthy spit at one end of Sertung Island, and for a while there was a lagoon near the base of the spit, which gathered its own specific colonists. Coastal erosion eventually led to the loss of this lagoonal habitat and with it several plant and animal species were lost.

After 46 years of inactivity, the new island of Anak Krakatau began to form in the centre of the 1883 caldera, finally establishing a permanent presence in 1930. Through intermittent and often violent activity it grew by the mid-1990s to an island of over 280 m in height and 2 km in diameter (Thornton et al. 1994). The three 1883 islands have always shown some differences in terms of the pattern of arrival and the unfolding succession of communities, with Rakata being distinctive through its greater elevation and in possessing upland species lacking from the other two islands. However, the differences became increasingly marked following the intervention of Anak Krakatau. Since 1930, both Panjang and Sertung have been repeatedly swept by damaging clouds of volcanic ejecta, often weighing down trees and foliage with ash, at times defoliating large areas. As early as 1931, Sertung was described as resembling a European wood in winter, while as recently as December 2018, Panjang Island was entirely defoliated, and Anak Krakatau itself was reshaped and sterilized once again (Gouhier and Paris 2019; Borrero et al. 2020). Botanical survey work in the 1980s and 1990s revealed that the forests of Panjang and Sertung were a mix of damaged and recovering forests. A relatively small number of forest trees dominated stands that formed a mosaic of compositionally distinct patches, possibly reflecting chance effects of the timing of the end of volcanic episodes in relation to fruiting patterns of the more commonly available gap-filling species (Whittaker 2004). The 2018 eruptions effectively reset Panjang to an early successional system and doubtless caused many extinctions, especially of animal species. It also caused tsunami damage to coastal habitats on all three of the older islands and to coastal areas around the Sunda Straits (Borrero et al. 2020).

We don't have biological survey data from these most recent eruptions, but we do have records of the earlier colonization history of Anak Krakatau. Colonists on Anak Krakatau have mostly comprised those plant, bird, reptile, and bat species which colonized the archipelago in the early post-1883 phase (Thornton et al. 1992), although differences were observed among the butterflies. This was probably due to the dependence of butterfly species on the availability of particular habitats and food plants, which differed sufficiently between old and new runs of the experiment to buck the broader trend.

Partomihardjo et al. (1992) analysed the succession of floras following, first, the island's appearance in the centre of the Krakatau caldera in 1930, and, subsequently, following wipe-outs of the vegetation in 1932/3, 1939, 1953, and damaging, but not sterilizing eruptions around 1972. Again, the results showed a strong degree of repetition, a temporal nestedness, with a core of constant species, which make it back each time and are added to. The first assemblage was essentially a seedling flora, a few members of which were not found in the second assemblage, but of 32 species identified from these two surveys, 30 recolonized subsequently. The eruptions in 1972 are presumed to have severely reduced the third flora, but 41 of its 43 species were recorded again between 1979 and 1991.

If the Anak Krakatau flora is broken down into arbitrary functional guilds, there are basically three sets of species:

1 *Strand-line species*: the largest set, representing 58 of 125 spermatophytes found between 1989 and 1991. Propagules of these species are sea-dispersed and are produced in large numbers locally on the other islands. This is a very consistent set. Forty-two of these species have been found on all four Krakatau islands.

2 *Pioneers of interior habitats*: wind-dispersed ferns and grasses, with a few composites (Asteraceae). They are locally available, are very dispersive, and produce large numbers of propagules. As with the strand-line species, they are in essence a subset of those found on the other islands post-1883.

3 *Second-phase colonists of interior habitats*: later successional species that only became established as forest developed post-1979. Comparisons with the other islands suggest that this group of species is less predictable in colonization sequence because of the more variable patterns of seed production and dispersal (mostly by birds or bats) within the archipelago. For instance, some of the most abundant tree species on the other islands of the 1980s (e.g. *Arthrophyllum javanicum*, *Timonius compressicaulis*, and *Dysoxylum gaudichaudianum*) were recorded on Anak Krakatau at an earlier stage of successional development than in the post-1883 sequence. Should Anak Krakatau experience a long enough phase without such damaging eruptions, these suggestions can be assessed against another run at the experiment post the 2018 eruptions.

Whittaker and Jones (1994a) condensed the Krakatau observations into an informal 'rule table' in which the dispersal, successional, turnover, and compositional characteristics of five broad functional plant guilds were described, should the opportunity arise to evaluate them elsewhere. Within plant successions there may be both autogenic 'facilitation' effects of habitat alteration and construction, driving species accumulation within the newly formed forest, and diffuse competitive effects between guilds, culminating in later successional communities shading out earlier ones. Alongside these processes, an abiding theme of the Krakatau story has been the significance of birds and bats in determining which species have colonized and when.

In addition to there being structure as to which species assemble and in what sequence, other sets of species may be effectively debarred from colonizing. Later successional species with poor dispersal adaptations, those possessing large, winged, wind-scattered propagules, those dispersed by terrestrial mammals, and large-seeded 'bat-fruits' (lacking diplochory) remain deficient on Krakatau, whereas highly dispersive forms such as ferns and orchids have been over-sampled (Whittaker et al. 1997). In sum, the Krakatau analyses suggest that the draw of early pioneers is fairly predictable as a function of their superior dispersibility and environmental fit, but as the succession advances, it becomes much less predictable which species of intermediate dispersiveness and later successional stage will join, while very poorly dispersing species can be seen as highly unlikely to colonize. Thornton (1996) argues that those guild members joining later have, in general, correspondingly reduced potential to build large populations and influence the unfolding play.

Ecological research on Krakatau has thus spanned themes within primary succession, plant–animal interactions, and island biogeography theory. It was this context that led Bush and Whittaker (1991) to argue that the length of time for the biological dynamics to reach conclusion exceeded the time available before the system was subject to major disruption. They proposed a modification of MacArthur and Wilson's classic model (Fig. 5.2) to recognize that, in such dynamic settings as Krakatau, the progression towards equilibrium may

be set back by waves of extinction, re-immigration, and community succession.

Krakatau is a particularly active spot volcanically, but instability of the platform is perhaps a core characteristic of volcanic islands (Chapter 2), as is their exposure to erosive attrition. This is exemplified by Surtsey, which we discussed alongside Krakatau in the previous chapter. This newly emerged Icelandic island grew to a maximum area of 2.7 km^2 as of June 1967, three years after its emergence, but by 2004 was reduced by coastal erosion to 1.4 km^2 (Magnússon et al. 2009).

7.4 The general dynamic theory of oceanic island biogeography: model description and properties

The general dynamic model of oceanic island biogeography, or GDM, represents an attempt to generalize the ontogeny of a volcanic island, setting aside the 'stop, reset, and restart' complexity of Krakatau by considering the entire life span of an island from its emergence, through to its eventual loss (Fig. 7.1a, Whittaker et al. 2008; review by Borregaard et al. 2017). As described in Chapter 2 (see e.g. Fig. 2.7), and considering, for simplicity, an island of hotspot origin, such as in Hawaii or the Canaries, each island goes through a building phase to reach and breach the ocean surface and ultimately to reach elevations of perhaps as much as 3000 to 4000 m. It is likely to reach its maximum area coincidental with the peak in elevation. As it ages and moves away from the hotspot, becoming less active, it will subside and erode: area and elevation will thus be reduced while initially, at least, topographic complexity may increase through a combination of downcutting and secondary volcanism. With the further passage of time, the island will eventually be reduced below sea level, although in tropical waters, it may persist through coral formation, as a low-lying atoll. Hence, the carrying capacity of the island must, over its life span, follow an arc, commencing at zero and ending at zero (or in the case of a persisting atoll, a low value consistent with its small size and low elevation). The GDM represents the intersection of this simplified

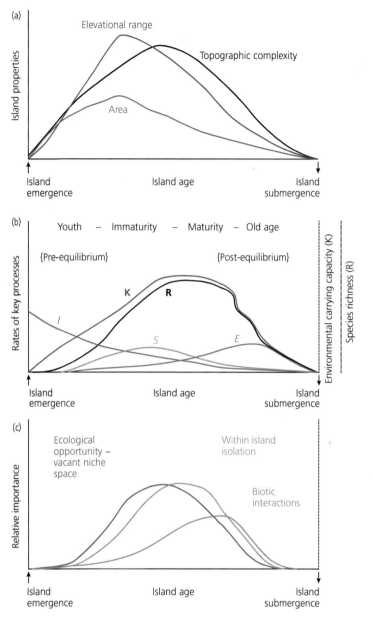

Figure 7.1 The general dynamic model of oceanic island biogeography. (a) A simplified model of island ontogeny, as might be found for a hotspot island as it emerges, builds, becomes isolated from its magma source, and eventually declines through subsidence and erosion. The relative heights of the three curves are arbitrary. (b) Environmental carrying capacity (K) is a hump-shaped function of island age. Species richness (R) is a product of the rates of immigration (*I*), speciation (*S*), and extinction (*E*), each calculated as species per unit time. (c) The inferred shift in relative importance of different modes of cladogenetic speciation anticipated within the GDM. Based on figures 3–5 in Whittaker et al. (2008). NB, panel b was first published as the 'Island immaturity–speciation pulse model of island evolution' by Whittaker et al. (2007, their figure 1), but the extinction curve was added only in the 2008 paper, in which it was incorrectly sketched as rising until island submergence, which is true only if expressing extinction as a function of extant richness rather than species per unit time. Although drawn as broadly symmetrical, time should be considered a log function, as the phase during which the island builds through volcanism to maximum area and carrying capacity, will typically be shorter than the decline phase.

ontogeny with existing island theory and in particular the arguments codified in MacArthur and Wilson's equilibrium theory (Fig. 7.1).

The model was initially developed as a response to a paper by Emerson and Kolm (2005a). Their paper analysed the relationship between the proportion of single-island endemics (SIEs) and abiotic and biotic properties of each island for arthropods and plants of the Canaries and of Hawaii. They found that the strongest statistical explanation for the number of SIEs per island for each taxon was the species richness of that same taxon. This might tell us, as inferred by Emerson and Kolm (2005a, b), that high species richness creates the conditions for high rates of speciation (e.g. through competitive interactions), but it could also be that the relationship is a by-product of circumstances whereby remote islands of high potential carrying capacity cannot be filled by immigration alone and so provide greater opportunities for diversification. Whittaker et al. (2007) developed this line of argument into a model they initially termed the island immaturity–speciation pulse model of island evolution. This model, relabelled a general dynamic model, was subsequently refined and developed into a broader dynamic theory for oceanic islands (Fig. 7.1b; Whittaker et al. 2008), incorporating additional theoretical arguments as to how a succession of islands of different stage fit together within an archipelago, generating predictions about emergent biodiversity processes and patterns.

There is broad support for the idea of an environmentally determined carrying capacity (K) for richness, largely determined by island area and reflected in the form of the ISAR, but young and recently disturbed islands may sit below the 'saturation' point formed by similar older islands (Chapter 5). Figure 7.1b argues that when a large remote volcanic island forms, the pace of immigration is too slow to saturate the potential carrying capacity. Moreover, as more dispersive colonists arrive first, immigration rate will decline over time. Hence, there will be prolonged opportunities for evolutionary diversification (cladogenesis) to occur, exploiting untapped resources (i.e. 'vacant' niche space). The pace of cladogenesis is initially expected

to pick up as more species colonize, providing for more interactions and specialization, but then to decline, as richness closes on the carrying capacity.

Taxa capable of rapid rates of speciation may perhaps so rapidly saturate even geologically young islands that a clear positive correlation can result between species richness and island area, as found for the highly volant Hawaiian *Plagithmysus* beetles. Other taxa, even if arriving early, may fail to respond with such rapidity. Hence, for example, the flightless Hawaiian *Rhyncogonus* weevils show a pattern of increasing species number with increasing island age and a decline with island area on the main islands (the largest Hawaiian island being the youngest) (Paulay 1994) and, in accordance with the GDM, the lineage persists in much lower numbers on the oldest and relatively much smaller north-western islands, with single species present on Laysan, Nihoa, and Necker (Hembry et al. 2021). This demonstrates that the propensity to diversify, and the rate of that diversification, is highly variable between different clades of invertebrates, plants, etc. The peak in speciation (cladogenesis) shown in Fig. 7.1b, c may be produced by rapid diversification within a comparatively small number of the original colonists to the island and some taxa may continue to diversify even as the carrying capacity and richness decline. Overall, however, extinction must become dominant as the island area, elevation, and habitat heterogeneity collapse.

As we will explore more fully in Part III, island speciation is not limited to contexts where cladogenesis occurs across an environmental gradient into unoccupied niche space, which we might broadly class as adaptive radiation, but can also occur through geographical isolation within the complex topography of oceanic islands. This isolation may occur through vicariance (subdivision of a formerly continuous distribution) or through dispersal between pockets of similar habitat, in each circumstance permitting allopatric speciation that may tend towards being non-adaptive. Hence, we might expect a tendency for speciation to shift from adaptive (across gradients) towards non-adaptive radiation (within isolates), while the signal from gradual specialization involving interactions (e.g.

between plants and pollinators) may become relatively more important later in the ontogeny, as the island declines (Fig. 7.1c) over time and extinction becomes increasingly dominant.

Figure 7.2 formalizes these arguments as a flow diagram and adds additional details that are lacking from Fig. 7.1 and which are relevant to the predictions that can be derived from the GDM. First, the figure recognizes that some colonist species become endemic through anagenesis: evolutionary change *in situ* but without diversification into separate species within the island. This process adds to *in situ* endemism without adding to species richness. Second, species that have become native to, or endemic within, our focal island may, over time, colonize other islands in their archipelago.

In particular, as the next youngest island emerges, they may act to seed that island with colonists (generating the island progression rule: see Wagner and Funk 1995). This may also serve to change the species status of the source island, as each single-island endemic that colonizes another island is no longer strictly endemic: thus richness remains unaltered but the proportion of single-island endemics is reduced. Hence, Fig. 7.2 codifies some of the arguments that allow the GDM to be extended from a single-island model to a representation of diversity dynamics across an archipelago. Based on this reasoning, Whittaker et al. (2008) list 10 predictions that can be derived from the GDM and which together are integral to the fuller general dynamic theory of oceanic island biogeography (Table 7.2).

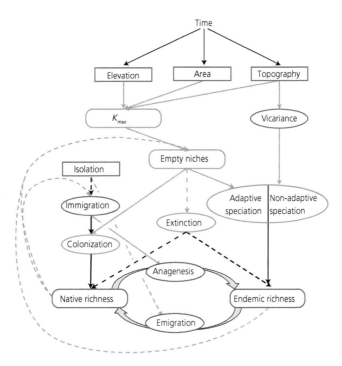

Figure 7.2 Causal relationships within the general dynamic model of oceanic island biogeography: (i) blue: processes of immigration, speciation, and extinction operate as functions of island isolation and area, as per MacArthur and Wilson's (1967) equilibrium theory; (ii) green: diversification is driven by unutilized ecological opportunity and within-island allopatry; and (iii) red: species carrying capacity and within-island allopatry ('vicariance') covary with the island's ontogeny. Thin arrows indicate causality; black arrows represent relationships that are either self-evident (e.g. speciation increases species richness and extinction decreases it) or empirically well established (e.g. island isolation decreases immigration rate); grey arrows indicate relationships that are more speculative. Dashed arrows are negative effects. Thick arrows depict the movement of species between categories. K_{max}, maximum carrying capacity. From Borregaard et al. (2016).

Table 7.2 Predictions derivable from the general dynamic model of oceanic island biogeography (GDM). Very slightly modified from Whittaker et al. (2008, 2010). Formal tests have since been carried for most of these predictions, especially focussing on predictions 1, 2, 3, and 6. Some predictions (e.g. 7, 8, and 10) are also reasonably well supported by evidence provided by previous and subsequent studies that did not set out specifically to test the GDM. See review by Borregaard et al. (2017).

1. Island species number and the number of single-island endemics (SIEs) should be a humped function of island age, and when examining snapshot data across an archipelago this will be combined with a positive linear relationship with area.
2. The amplitudes of the curves shown in Fig. 7.1 should vary in relation to the size of the island at maturity, with higher peak richness and SIE numbers on islands that attain greatest size (area and elevation) at maturity.
3. The relative amplitudes of the *immigration* and *speciation* rate curves should vary in relation to the effective isolation of islands: increasing isolation lowering the *I* curve and elevating the *S* parabola, thus also increasing endemism.
4. Lineage radiation (leading to multiple SIEs on individual islands) should be most prevalent after the initial colonization phase, in the period leading up to island maturity, coinciding with maximal carrying capacity (*K*) and the development of maximal topographic complexity.
5. Montane representatives on old, declining islands should gradually be lost because of loss of habitat, meaning that surviving montane forms are increasingly likely to be relatively old (i.e. basal) forms in relation to other members of an archipelagic radiation.
6. The proportion of SIEs should also be a humped function of island age as islands that decline to small size and carrying capacity should lose SIEs.
7. SIEs per genus should be higher on younger islands; intermediate-aged islands will have more lineages showing speciation than do young or old islands; SIEs per genus should decline on older islands so that as islands lose SIEs, there is a tendency towards monotypic genera, preserving maximal ecological spacing in the remaining endemics.
8. As islands age, some of their SIE species may colonize a younger island, so that they become multi-island species instead. Hence, the GDM also predicts that the progression rule (successively younger members of a lineage on each younger island [see Section 9.8]) should be a common/dominant phylogeographical pattern within an archipelago.
9. There should be a greater apparent tendency to anagenesis* on old, submerging islands as the dominant speciation signal, although in part resulting not from anagenetic speciation but from the collapse of former radiations (generated by cladogenesis).
10. Adaptive radiation will be the dominant process on islands at the point in time when the maximum elevational range occurs, as it generates greatest richness of habitats (major ecosystem types), including novel ones few colonists have experienced, whereas non-adaptive radiation will become relatively more important on slightly older islands, past their peak elevation, due to increased topographical complexity promoting intra-island allopatry. Similarly, composite islands (e.g. Tenerife, formed from two/three precursors), should have provided more opportunity than islands of simpler history for within-island allopatry, producing sister species that lack clear adaptive separation (e.g. Gruner 2007).

*Anagenesis and cladogenesis: we use the term anagenesis *sensu* Stuessy et al. (2006) to refer to geographical speciation that occurs within an island by reference to a mainland ancestral species; whereas cladogenesis (radiation) refers to splitting of lineages to produce two or more lineages *within* the island.

7.5 Evaluation of the general dynamic model: empirical tests and simulation models

Empirical tests

In a first test of the GDM reasoning, Whittaker et al. (2007) established that, as predicted (Table 7.2), the proportion of SIEs within the native arthropods and plants of the Canary Islands showed a humped relationship with island age. While this was an encouraging result it was far from conclusive, not least because there was no control for variation in the trajectory of area attained by each island over its life span. Indeed, it is important to emphasize that Figures 7.1 and 7.2 represent the combined geo-environmental and biodiversity dynamics of a single island over its life span and that it is only by combining such dynamics across multiple islands

Before moving on to tests and further developments of the GDM, we should mention that prior to publication of the GDM, the significance of the island life cycle for island biodiversity dynamics was recognized by several authors, such as Paulay (1994), Stuessy et al. (1998), Craig (2003), and Peck (1990, p. 375), who wrote: 'A relationship with island age should be expected, but it would not be a straight line ... Rather the relationship should be a curve which rises fast at first, reaches a peak or plateau, and then decreases as erosion destroys the island.' Moreover, Stuessy (2007) outlined a rather similar model for oceanic island floras, to that set out by Whittaker et al. (2007). The distinct contribution of the GDM was to integrate these ideas formally with island biogeographical arguments derived from MacArthur and Wilson's (1967) theory.

that the archipelagic pattern emerges. By analysing the proportion of SIEs across an archipelago, we are employing a chronofunction approach (ergodic reasoning), which assumes that the islands act as replicates. One obvious way that this assumption can be incorrect is if the islands attain very different dimensions at their maximum. A crude way to deal with this problem is to test models including a linear function of (log)island area (based on our knowledge of ISAR variation), alongside a humped relationship with island age (i.e. diversity = area + (time − time^2); or diversity = ATT^2), as shown diagrammatically in Fig. 7.3. This was the approach adopted by Whittaker et al. (2008) in using multiple regression to test models of species richness, the number of SIEs, their proportion, and finally a diversification index (number of SIEs/number of genera in which they occur) for 14 plant and invertebrate datasets from five oceanic archipelagos. Although the ATT^2 model was not always the most parsimonious model it was found to fit a higher proportion of datasets than alternative models (such as standard ISARs) and overall was found to be the best performing model.

Similar comparative analyses of the GDM alongside alternative models were carried out for Macaronesian spiders by Cardoso et al. (2010). They reported support for the GDM for the Canaries, and for Macaronesia as a whole, but not for Azorean spiders, where preferred models identified key roles for island area and remnant forest area (indicative of human impact). Another interesting test was carried out by Fattorini (2009), for 17 animal groups for the Aeolian Islands, in the central Mediterranean. This archipelago is both less remote and younger (oldest island c. 0.6 Ma) than the oceanic island systems evaluated by Whittaker et al. (2008). Once again, good support was found for the GDM. Fattorini (2009) also applied regression models to estimate expected number of species at equilibrium for a mature island (S_{eq}) and the time required to obtain this value (T_{eq}) and then used the ratio S_{eq}/T_{eq} to estimate an index of colonization ability (CAB). Using these three metrics he was able to characterize the biological properties of different animal groups quantitatively, finding that the differences were linked to biological traits such as body size and ecological specialization. In this relatively fast cycling set of volcanic islands, equilibration times across 17 groups of animals were reported to vary from 170,000 to 400,000 years for those groups well described by the GDM-derived model.

As pointed out by Bunnefeld and Phillimore (2012), the statistical power of these early tests of the GDM were limited by the need to fit multiparameter models to noisy datasets comprising

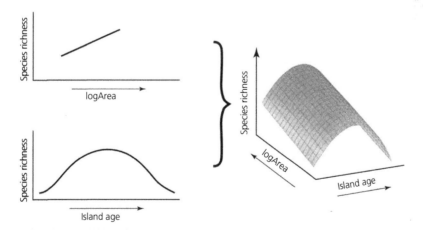

Figure 7.3 The GDM predicts that species richness and endemism patterns of hotspot-like archipelagos should exhibit a linear relationship with area (using log-richness and log-area) combined with a humped relationship with island age.
From Whittaker et al. (2008), their figure 7. For subsequent exploration of variants in the mathematical form of the relationship, see e.g. Fattorini (2009) and Carey et al. (2020).

(as most do) very few islands per archipelago. They also argued that the inclusion of multiple data points from a single archipelago violates the assumption that data points are entirely independent as, after all, the islands of an archipelago typically act as key source regions for each other. They therefore advocated use of a mixed-model approach, which they illustrated by applying it to the same SIE richness data compiled by Whittaker et al. (2008). Their models were built using 134 datapoints for four distinct taxa, spanning 39 islands, in four archipelagos. They found that the greatest amount of variance was attributable to archipelago, with a substantial taxon effect, highlighting the importance of the biogeographical context in which each island's biota develops. Their re-analysis nonetheless provided general support for the GDM, especially so for the Canaries and Hawaii. However, SIE richness on the Azores and Galápagos showed a positive relationship with island age, without evidence of a decline. Tests for other taxa for the Azores have generally found the same (e.g. Borges and Hortal 2009; Cameron et al. 2013), which probably reflects the young age of the archipelago; indeed, the maximum effective island age (at 4 Myr) is now thought to be even younger than used in these analyses (Hildenbrand et al. 2014; Ramalho et al. 2017).

Most subsequent studies of the GDM, and indeed of other similar island models, have followed Bunnefeld and Phillimore's (2012) lead in deploying mixed-effect models, permitting simultaneous fitting of models to islands from multiple archipelagos. An example is Cameron et al.'s (2013) analyses of native richness, archipelagic endemic richness, and SIE richness for land snails from the islands of eight oceanic archipelagos (Fig. 7.4). The ATT^2 [logArea + (island age − island age^2)] models were always within the group of best models and were often the single-best models. The humped relationship with island age was statistically supported for the data as a whole and for those archipelagos providing a fuller sequence of island ages (Canaries, Hawaii, Madeira, and Tristan da Cuhna), while a positive relationship was observed for the other four archipelagos (Fig. 7.4). Further analysis revealed that the emergent pattern for all land snails is underlaid by a degree of variation from family to family. The strength of the relationship for Hawaii was (unsurprisingly) found to be dependent on the inclusion of four small (older) islands, and it was reported that were the analyses restricted to just the larger six islands, thus reducing the age range, the humped relationship with island age would be weakened and be largely dependent on two of the richer families, the Achatinellidae and Amastridae. This serves to emphasize that a full range of island stages may be necessary to demonstrate the pattern predicted by the GDM.

More successful tests for the macroecological patterns predicted by the GDM have been carried out (see review by Borregaard et al. 2017, and e.g. Louiseau et al. 2018), although the model is not always the most parsimonious solution and sometimes is not supported. For example, a study of bryophyte diversity for 67 islands from 12 oceanic archipelagos found that island area and elevation provided adequate models, outperforming both time and isolation. This is consistent with the low level of endemism in the bryophytes of these islands, reflecting that even large distances fail to act as significant evolutionary barriers for the microscopic spores through which they spread (Patiño et al. 2013). Generally—and again unsurprisingly—the evolutionary effects predicted by the GDM appear to be weaker or undetectable if either or both (i) a limited range of island age is involved or (ii) the taxon involved is super-dispersive (Borges and Hortal 2009; Borregaard et al. 2017).

Steinbauer et al. (2013) also used a mixed-effects modelling approach using the same diversity indices but correcting for statistical weaknesses in the original tests undertaken by Whittaker et al. (2008). Their analysis again broadly reaffirmed the original results although fewer of the relationships were found to be significant. Their analysis showed that endemism peaked at later points in island life cycles than species richness and that the initial rise in species richness happens more quickly than the subsequent decline. Both Steinbauer et al. (2013) and Carey et al. (2020) have also explored the relative performance of GDM models using untransformed and log diversity as the response variable; the logged version often also involves taking the log of time (i.e. all variables are logged). Both studies reported improved model performance with the

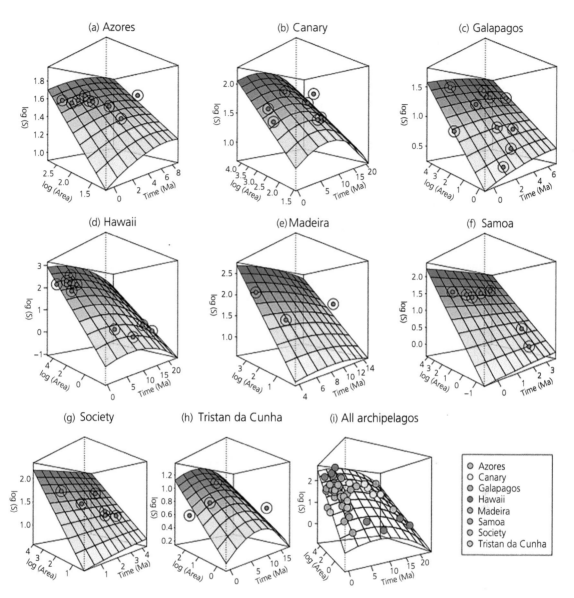

Figure 7.4 Fits for the logSpecies richness = logArea + (island age − island age^2) model (often termed the ATT^2 model, where A is island area and T = time since island emergence), designed to test the general dynamic model of oceanic island biogeography, as applied to a dataset of land snails from 56 islands (the datapoints) from eight archipelagos. The surface is the prediction from a linear mixed effect model including a random slope for log(Area) varying across archipelagos. For panels (a)–(h) species richness increases with the intensity of the colour shading. For panel (i) the slope corresponds to the grand mean fitted by the linear mixed effect model.
Cameron et al. (2013), their figure 2.

logged versions of the model, likely partly due to the logged versions modelling richness as a power function of area, which is known to be the best performing ISAR model (Chapter 5). Carey et al.

(2020) went further in exploring the mathematical properties of the various competing models, identifying that, when using the log version of the GDM, the humped time function (technically a lognormal function in time, with a steep rise and drawn out decline rather than symmetrical hump; a pattern observed in many empirical datasets) can be considered a replacement for c in the basic power law ISAR ($S = cA^z$), thereby providing a set of guidelines for further tests of the GDM and of alternative models (such as using a choros model function (*sensu* Triantis et al. 2003) to represent carrying capacity in place of area). They undertook comparative analyses for species richness for a range of different taxa from six archipelagos. Simpler area-only or time-only models were preferred as the best overall model in nearly 75% of cases, with models incorporating the expected humped-time function performing best for invertebrate datasets and also for the Canaries, where area is not a significant predictor owing largely to the large size, low elevation, and arid conditions of the two oldest islands (Lanzarote and Fuerteventura). They also reported that generally more dispersive taxa tended to reach peak diversity earlier than poorer ones.

Lim and Marshall (2017) developed an alternative model based on the same simple island ontogeny assumed by the GDM and which generates expected carrying capacity calibrated for specific clades across the Hawaiian Islands. The model generates the same humped trend in species richness as the GDM, increasing during island growth and declining as the islands are reduced by subsidence and erosion. When this model was contrasted with two alternatives employing a fixed carrying capacity, the island ontogeny model was found to be the best supported for 9 of 14 clades tested, with good support also reported for three other clades. The analyses supported a pattern of extremely rapid species accumulation in the early growth phase, followed by a decreasing rate of accumulation, prior to a switch to species losses from the older islands. In line with other studies we have considered, the peak in species richness typically lags behind the estimated peak in carrying capacity. A minority of lineages appear to accumulate more slowly, so that they have yet to reach carrying capacity, even on the oldest island in their analysis, Kauai. These findings are consistent with earlier work suggesting that those taxa capable of the fastest rates of speciation may so rapidly saturate even geologically young islands that a clear positive correlation can result

between species richness and island area (Hawaii being the largest and youngest island).

In their analysis of GDM model fits for oceanic island plants from 101 islands from 14 archipelagos, Lenzner et al. (2017) report that richness, endemic richness, and percentage of endemic species all show the predicted humped relationship with age of island. By generating models at the level of particular species-rich plant families they showed that the overall trends tend to be generated by just a few such families and within them by a few particularly diverse clades. Within those families conforming with GDM predictions, rapid attainment of peaks in SIE richness were provided by Asteraceae (within 3.6 million years), followed by Fabaceae, Crassulaceae, Poaceae, Lamilaceae, Euphorbiaceae, and Rubiaceae (within 7.7 million years). Orchidaceae and Brassicaceae were exceptions in showing no humped trend. These and other studies of island endemism and diversification (e.g. Cameron et al. 2013, above) demonstrate that in terms of evolutionary responses to island ontogeny, a minority of clades generate the pattern captured by the GDM.

As outlined earlier, Whittaker et al.'s (2007, 2008) arguments in the GDM were initially stimulated by questioning the expected form of the relationship between the proportion of SIEs (pSIEs) and island age. For reasons set out in Table 7.2, not only is overall species richness expected to show a humped relationship with island age by the GDM, but also the proportion of these native species that are endemic should also show a humped relationship over time (as e.g. Lenzner et al. 2017). In essence, the case is that much of the endemic diversity that accrues as an island matures will ultimately collapse, so that late-stage, low-lying islands shed SIEs, as generalist species of coastal habitats and wide dispersibility become comparatively more important in the diminished biotas. This effect may not be seen if small ancient islands are not available or not sampled.

As the above examples attest, the majority of tests of the GDM have focused on the model's predictions regarding diversity, particularly the richness of native and endemic species. However, recent studies have also tested a primary premise of the GDM, that environmental heterogeneity also exhibits a hump-shaped relationship with time (e.g.

Barajas-Barbosa et al. 2020; Kraemer et al. 2022). For example, in a study of vascular plants on 135 oceanic islands, Barajas-Barbosa et al. (2020) quantified environmental heterogeneity using 20 different metrics, encompassing both topographical and climatic variation, and related environmental heterogeneity to time using linear mixed-effects models. While, as expected, several of the metrics were found to be collinear, environmental heterogeneity in general (i.e. 16 of the 20 metrics) was found to exhibit an asymmetrical hump-shaped pattern with time, with environmental heterogeneity peaking relatively early in the island's life span before a drawn-out decline (Barajas-Barbosa et al. 2020).

An evaluation of trends in functional diversity (FD) in the context of the GDM has been undertaken by Kraemer et al. (2022) using data for *Naesiotus* snails in the Galápagos. They tested two broad hypotheses connected to island age: (i) FD may be driven solely by ecological opportunity and thus exhibit a humped relationship following the expected trend for species richness or (ii) competition between closely related species (with similar traits) on older and eroding islands may be high, leading to competitive exclusion of similar species, which in turn leads to the maintenance of high FD late into the island ontogeny cycle (Borregaard et al. 2017). While predictions of a hump-shaped relationship between richness and island age, and between a metric of ecological complexity and age, were found to hold, the trends for FD were more complex (Fig. 7.5; Kraemer et al. 2022). FD was very low on

the youngest islands and increased in a roughly linear fashion with island ontogeny, but with variable values on the older islands. These are intriguing but inconclusive findings: it is to be hoped further similar analyses will be undertaken for other taxa and island systems.

Simulation models

Another means of evaluating the GDM is via simulation modelling. A number of attempts, varying from comparatively black box to quite detailed mechanistic models, have been published that allow evaluation of GDM predictions (see Borregaard et al. 2017). Borregaard et al. (2016) developed a simulation model, running over 5000 timesteps, and providing a simple parameterized model of the form set out in Fig. 7.2. This provided a test of the ecological logic of the general dynamic theory and its capacity to reproduce the patterns set out graphically in Fig. 7.1. This was successful in qualitative terms, with one interesting modification: it was found that at initiation, colonization rate should be initially low, before building to an early peak and thereafter gradually declining. As we know from the Krakatau case study (and see also the Surtsey example), in the early stages following an island's emergence, colonization rate may not be maximal because the island has yet to stabilize and a successional process of habitat creation has to take place before many potential colonists can become established. Recently emerged islands

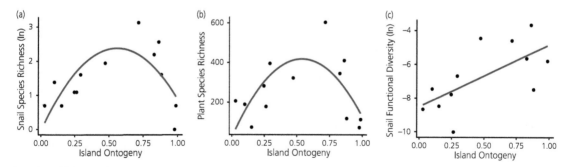

Figure 7.5 The relationship between island ontogeny and (a) snail species richness (log transformed), (b) plant species richness (used as a measure of environmental heterogeneity), and (c) snail functional diversity (log transformed). Blue lines represent the best fit linear model from a set of candidate models. Data reproduce *Naesiotus* snail species across 13 volcanoes/islands in the Galápagos.
Modified from Kraemer et al. (2022), their figure 2.

devoid of vegetation will also represent unattractive targets for potential volant seed dispersers such as birds and bats, again acting to reduce the colonization rate at this early stage. This modification to the spin-up phase of island biodynamics is thus ecologically realistic, but will only be evident if examining the early stages at sufficiently fine temporal resolution.

Valente et al. (2014) developed an approach that incorporated the GDM's island ontogeny, standard immigration–extinction dynamics, and phylogenetic birth–death models to provide a lineage-based simulation model, using it to explore predictions of the GDM. They implemented both a diversity-dependent version (as per the GDM) and a diversity-independent version to explore the importance of the role of carrying capacity encapsulated in the GDM arguments. The diversity-dependent version was found to produce more realistic results. Once again, their model replicated the humped trends in diversity as a function of island age and the peak in richness was found to lag behind the peak in carrying capacity. That is, island diversity only reaches the 'equilibrium' point when the island is already into the subsidence/erosion phase. This tallies with empirical findings of Bunnefeld and Phillimore (2012) and Cameron et al. (2013).

The Valente et al. (2014) simulation model has the added benefit of generating model phylogenetic trees, which may then be used to interpret empirical trees, providing certain technical challenges can be overcome. Their findings in this respect are broadly consistent with the analysis by Givnish et al. (2009) of the molecular phylogeny of Hawaiian lobeliads, which suggests a comparatively early peak in diversification, followed by a reduction of diversity within the lineage as islands attain old age and are reduced by erosion and subsidence. We discuss this and other evolutionary studies relevant to evaluating the GDM in Chapter 9.

The BioGeographical Eco-Evolutionary Model (BioGEEM) is a process-based model providing a spatially explicit representation of emergent patterns for an oceanic island integrating: (i) metabolic constraints, demography, dispersal, and competition processes; alongside (ii) mutation and speciation; and (iii) the geo-environmental dynamics as described within the GDM (Cabral et al. 2019a, b).

Plants were used as the model system for calibration purposes. The model generated key expected patterns such as the humped trend in species richness, with both differentiated (anagenetic) and radiated (cladogenetic) endemics peaking slightly later than richness and with trait richness also showing a humped pattern (Cabral et al. 2019a). In addition, the slope of the within-island species–area relationship was also found to be steeper during the island growth phase than during the erosion phase, matching the GDM's expectation of increased species packing with island age. A further interesting finding was that old, large, and isolated environments within the simulated islands accumulated greater proportions of endemics.

In a follow-up analysis exploring the impact of varying isolation within BioGEEM, Cabral et al. (2019b) reported that: (i) humped trends in species richness, and in richness and proportion of endemics generated by within-island cladogenesis, were robust to variation in isolation, while (ii) the proportion of endemism within the floras, and the importance of cladogenesis in generating it, both increased with isolation. Their models thus replicated several key patterns predicted by the GDM. They also generated some novel or less expected results, such as better persistence of endemics into late stages of the island ontogeny than suggested by Whittaker et al. (2008). In part this arises from BioGEEM considering island isolation only in relation to a distant mainland source area, rather than taking account of the existence and importance of biotic exchanges with other nearby islands. This issue has been addressed in another interesting simulation study of GDM-type dynamics, conducted by Aguilée et al. (2021), using an individual-based neutral simulation model. They found that, in an archipelago setting, model islands supported more species and more endemic species than an equivalent-sized island that occurs in isolation. This was because: (i) archipelagos allow for inter-island dispersal, which increases island-level colonization rates; and (ii) there is more opportunity for species to colonize islands and undergo anagenetic speciation. Their models also showed that the magnitude of these archipelagic effects should vary through time, as a function of changes to archipelago configuration through the birth and death of islands, which in turn generate variation in island diversity outcomes (Aguilée et al. 2021).

The GDM was originally set out via solely verbal and graphical arguments. The translation of these arguments into realistic mechanistic models is an important step and it is encouraging that the models support the internal theoretical logic of the GDM. There is clearly a role for further applications of simulation models in developing island theory. For instance, BioGEEM has the further benefit of generating spatially explicit output, thus enabling direct comparison with empirical patterns of species distributions of different chorotypes within islands (such as described below).

7.6 Downscaling the general dynamic model

Otto et al. (2016) analysed plant distribution per 1 × 1 km grid cell within three of the Canary Islands (Tenerife, La Palma, and El Hierro), using statistical models to test the power of terrain age, topography, and climate data to explain patterns of richness of SIEs. For each island, most SIEs were either restricted to, or were more frequently found on old terrain, with very few showing higher proportions of occurrences on young terrain. In each case, the best GLM (generalized linear model) for SIE richness, which explained some 40–50% of the deviance, found either geological age, or the interaction between age and slope, to be the best predictor(s), followed by precipitation. Figure 7.6 provides illustrative detail for Tenerife. The oldest massifs (shaded orange in the geological age map), Teno (north-west tip), Anaga (eastern tip), and Roque del Conde (south), possess high SIE richness and are characterized by the steep slopes associated with mature volcanic islands. There are also high concentrations of SIEs in some steep but younger areas; for example, around the southern side of the central caldera of Las Cañadas, in El Teide National Park (the crescent-shaped area of high richness in the centre of Tenerife). Nonetheless, these findings are consistent with those reported in the BioGEEM model by Cabral et al. (2019a), with endemics generally accumulating on older terrain within each island. As Otto et al. (2016) point out, the estimated divergence times for several calibrated phylogenies

of radiated lineages such as *Lotus*, *Echium*, and *Micromeria* are generally consistent with the ages of the older parts of the islands in question, supporting the idea that they began divergence at an early stage of each island's ontogeny.

In their subsequent geographical analysis of the distribution of endemics within plant communities across the Canary Islands, Fernández-Palacios et al. (2021b) demonstrated that the species within the diversified lineages of the Canaries are mostly rare at the community level within four of the major ecosystem types of the Canaries, the *Euphorbia* scrub, the thermophilous woodlands, laurel forest, and pine forest. It was only among the high-elevation mountain scrub and the rocky communities associated with steep slopes across the elevational gradient that species in diversified lineages provide significant components of the local vegetation. Combining these observations, it seems that for the Canarian flora, within-archipelago and within-island diversification have mostly produced relatively rare species in comparison with the rest of the native and endemic flora. The greatest opportunities for diversification come relatively early in an island's life span, or on steep slopes following rejuvenation of large areas via catastrophic eruptions, in relatively open terrain. We should note that 2000 years of human interference (e.g. de Nascimento et al. 2009, 2020; Castilla-Beltrán et al. 2021), land-use change, and the introduction of many non-native species (other plants, but also browsers and grazers), may mean that the current distribution of the native and endemic species is an unreliable guide to the underlying evolutionary processes that have unfolded over hundreds of thousands of years. Further empirical and modelling analyses for other island systems (e.g. Keppel et al. 2010) and taxa will therefore be needed before we can be sure of the robustness and generality of these findings.

A relevant analysis for Hawaii is provided by Craven et al. (2019), who assessed local-scale diversity patterns of woody plants in 517 sampling plots distributed across the archipelago. At the whole island-scale, the youngest island (Hawaii) hosted the fewest native and single endemic species, while the oldest island (Kauai) supported the most SIEs. At local scale, after accounting for island area and habitat diversity, it was found that the number of

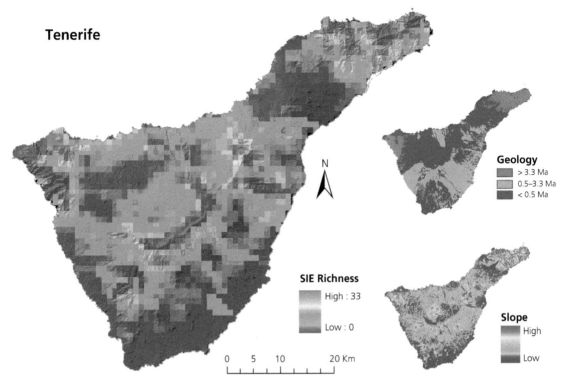

Figure 7.6 The species richness of single-island endemic (SIE) species of vascular plants per 1 km² grid cell, geological age of terrain, and mean slope for the island of Tenerife (Canary Islands).
Excerpted from figure 1 of Otto et al. (2016).

species per unit of area was greater on the older islands, highlighting the importance of macroevolutionary processes in driving local-scale as well as whole island-scale diversity patterns (Craven et al. 2019). Older islands were also found to (i) support a greater number of rare species, and (ii) have greater within-island beta diversity.

In an interesting extension of the logic of the GDM, Trøjelsgaard et al. (2013) tested whether interaction networks might also conform to the GDM. Using five of the Canary islands as a case study system, they assessed relationships with island area and age involving plants and their insect pollinators. Specifically, they used test sites within each island to quantify plant–pollinator interaction networks, including the numbers of 'single-island interactions' per island (ecologically equivalent to 'single-island endemics'). Sites were selected to include populations of the shrub *Euphorbia balsamifera*, to provide a degree of standardization of

habitat type. Both the richness of plant–pollinator interactions and the average degree of pollinator specialization showed humped relationships with island age. So too did site-based estimates of plant species richness. Intriguingly, the proportion of single-island interactions displayed a U-shaped relationship with age. This unexpected result appears to reflect changing patterns in the development of specialization with island age that differ between plants and their pollinators. Trøjelsgaard et al. (2013) provide some interesting ideas as to why this may be, providing new hypotheses that may be tested in further work on other archipelagos.

7.7 The general dynamic model as a bridge to a fuller theory

Several of the predictions in Table 7.2 concern phylogenetic patterns in island lineages and for that reason we will continue the evaluation of the GDM in Chapter 9. Thus far, we have reviewed studies that have explored the predictions of the model in terms of (i) emergent patterns of diversity for islands of different age, within and across archipelagos, and (ii) downscaled to within-island diversity patterns, particular clades, and even to interaction networks. In parallel with these empirical studies, there have been advances in the sophistication of the mathematical and statistical understanding of the model and how to test it more robustly. Overall, the GDM has been found to be useful in a substantial number of cases, either heuristically, or as the preferred model based on comparative model fitting exercises (Borregaard et al. 2017).

GDM applicability is strongest for oceanic island systems of hotspot origin and which include a full range of island stages. Of course, the ontogeny assumed by the GDM is a gross simplification of the geo-environmental histories of real islands and the pattern of variation in island properties across an archipelago can be confounded by both greater complexity of island geological histories and by other factors, such as variation in climate between islands (e.g. the western Canary Islands experience generally wetter climates than the two oldest, eastern islands). Whittaker et al. (2008) highlight Tenerife as a case in point. One account of its development is that it existed for several million years as two, or possibly three, separate islands, which became joined by an intense period of Quaternary volcanism to form a single island—certainly the central massif is far younger geologically and biologically than either the Teno or Anaga peninsulas (Fig. 7.6; Section 2.5). The placing of Tenerife in an age/stage gradient is thus challenging, as it really comprises a mix of young and mature island stages. Ultimately, the simplification of island ontogeny inherent to the GDM is perhaps both its strength as a general model and its greatest weakness when we drill down to detailed evaluation.

Several authors have suggested a need to develop alternative general island geodynamic models, for example, to take account of the differing dynamics of island arc systems (Heaney et al. 2013, 2016) and of land-bridge islands (Table 3.1). However, formal modelling attempts have so far mostly been limited to Pleistocene sea level variation (below). As both the Krakatau system and Section 3.7 remind us,

volcanic islands may also be influenced by intermittent and unpredictable forms of volcanism throughout their (active) life span, with important but messy impacts on emergent biogeographical patterns. One way forward is to move beyond the generalized ontogeny and to generate detailed reconstructions of changes in island configurations over hundreds of thousands of years and to use these island- and archipelago-specific models as the basis for locally calibrated tests of island biogeographical theories (see e.g. Ali and Aitchison 2014).

Notwithstanding the limitations we have discussed, the GDM appears to provide a bridge between ecological and geological/evolutionary time frames and to have a heuristic value to the further development of biogeographical theory, with potential wider applications to, for example, interaction networks (above), functional trait diversity (above; Borregaard et al. 2017), the island biogeography of marine organisms (Pinheiro et al. 2017; Ávila et al. 2019), and the incorporation of sea-level dynamics (Fernández-Palacios et al. 2016a; Ávila et al. 2019).

7.8 Incorporating glacio-eustacy as a key component of changing island configuration

The glacial/interglacial cycles of the Quaternary have driven repeated eustatic fluctuations in sea level—during the Last Glacial Maximum (LGM) reaching a global level c. 135 m lower than the present-day benchmark. However, we also know that relative sea-level changes can be regionally and even locally distinct due to a range of associated and independent isostatic adjustments (Section 3.9). It has long been understood that these changes in sea level have had a huge bearing on the number, size, and configuration of islands across the world and that this has profound importance to their biogeography (Wallace 1902). This is especially true of land-bridge islands: those continental shelf islands that are currently separated from mainlands by such shallow depths that they have been separated and rejoined reiteratively as a function of Quaternary climate change. These sea-level dynamics permitted extension of mainland species ranges to incorporate the hilltops that are now islands, followed by renewed isolation and loss of contiguous area as insular status was attained once again, leading to species extinction dominating their subsequent biogeographical dynamics. These extinctions are expected to be disproportionately severe on smaller islands (in line with the EMIB) and therefore should both lower and steepen ISARs for such

systems compared to their pre-isolation state: a process termed 'relaxation' of the ISAR (Section 12.4). Much the same, *but to a lesser degree*, should apply to true islands (those islands that never become connected to mainland), as their size should increase and their isolation diminish at sea-level minima, as set out by Fernández-Palacios et al. (2016a) in their **glacial-sensitive model of island biogeography** (and see Fig. 7.7).

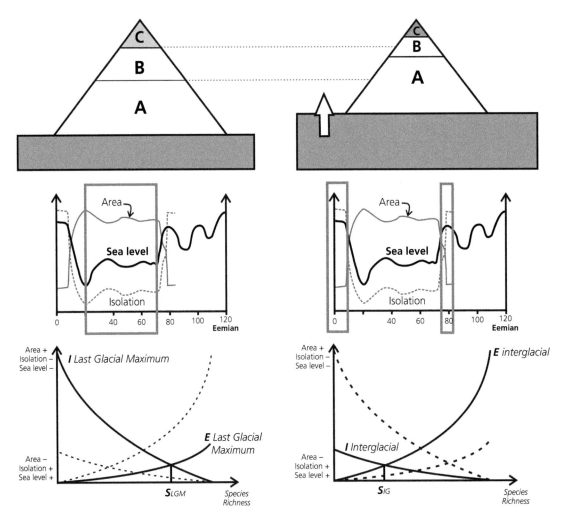

Figure 7.7 The glacial-sensitive model of island biogeography is a modification of the MacArthur–Wilson model focused on changing rates of immigration and extinction through time (bottom row), driven by glacial–interglacial fluctuations in sea level (middle row, showing the smoothed global curve since 120 ka), with consequent impact on the isolation, area, and distribution of major ecosystem types (A, B, C, in top row) of a hypothetical island. Grey shading and arrow in the top row indicate changing sea level. The left column shows the island at the Last Glacial Maximum, and the right column the same island during interglacial conditions. S = species richness, IG = Interglacial, LGM = Last Glacial Maximum, *I* and *E* are immigration and extinction rates, respectively.
From Fernández-Palacios et al. (2016a, their figure 6).

Changes in island configuration are, of course, only a part of the changing forcing factors of island biodynamics that are ultimately driven by profound changes in climate over the Quaternary glacial–interglacial cycles (Table 7.3). These changes undoubtedly varied in their relative significance with geographical location: high-latitude islands being subject to greater wholesale climate forcing than those at lower latitudes. Within low-latitude islands, or at least those of larger size and

elevational range, a far greater proportion of pre-glacial biodiversity has generally been able to persist by elevational adjustment over multiple glacial cycles. In so far as these processes are general and have a predictable geographical distribution, it should be possible to generate predictions for the impact of these events on emergent biodiversity patterns of true islands (including oceanic islands) and not just of land-bridge islands (Table 7.4; Fernández-Palacios et al. 2016a). However, focusing on the

Table 7.3 Summary of likely biogeographical consequences of changes in climatic and geographic parameters related to glacial cycles (modified from Fernández-Palacios et al. 2016a).

Shifts related to glacial cycles	Geographical effects on island/archipelago setting	Biogeographical consequences
Sea level	Area, elevation, and isolation shifts. Fusion/fission of nearby islands. Seamount emersion/ submersion, reducing system isolation.	Shifts in area availability and connectivity, affecting immigration and extinction rates. Genetic dilution/allopatric speciation. Availability/unavailability of stepping-stones.
Climate (temperature, precipitation)	Shifts in the vertical temperature/precipitation gradients. Shifts in the elevation of the cloud-sea influence zone.	Elevational shifts of zonal ecosystems. Contractions/expansions of species distribution ranges. Emergence/disappearance of summit ecosystems. Species extinctions, potentially repeated contributions to *in situ* allopatric speciation.
Marine currents and wind regimes (frequency, intensity, direction)	Changes in mainland–archipelago, between-archipelago, and within-archipelago connectivity.	Opening/closure of novel windows of dispersal. Back-colonization to mainland (boomerang events). Changes in marine species distributions.

Table 7.4 A selection of the predictions arising from the glacial-sensitive model of island biogeography drawn from Fernández-Palacios et al. (2016a; their box 1). See their box 1, the present chapter, and Chapter 9 for exemplification of several of these predictions.

(1): Predictions linked to shifts in area and isolation.
    (i) Palaeontological and palaeoecological research will make clear that several natural extinctions, caused by area reduction following sea-level transgressions, occurred before human arrival to oceanic islands.
    (ii) Some very rare extant species should present signs of having experienced a demographic collapse *c*. 15–12 ka resulting from range contraction due to sea-level rise.

(2): Predictions linked to fusion/fission of islands.
    The fission of one large island into two or more smaller islands due to sea-level rise will yield a high number of species shared among the once-attached islands, so that nearby islands subject to fusion/fission cycles will share more species than comparable island groups that were never physically linked.

(3): Predictions linked to seamount emergence/submergence.
    Fossil records from currently drowned seamount summits will reveal the presence (during glacial events) of terrestrial plant species belonging to coastal ecosystems and shared with other nearby islands.

(4): Predictions linked to climatic shifts.
    Some very rare species that are currently restricted to climatic refugia should present signs of having experienced a demographic collapse caused by range contraction resulting from the disappearance of their ecosystem during the last deglaciation.

(5): Prediction linked to shifts in marine currents and wind regimes.
    Shifts in marine currents and wind regimes have created new, bizarre dispersal windows, which should have left biotic signals in the form of retrocolonization (boomerang) events.

changes in island configuration represents a good first step.

While we know that island configurations change for all types of island as sea-level changes, for those substantial oceanic islands following hotspot-like ontogenies, we should expect sea-level dynamics to have much lesser impact (as per Table 7.3) than island stage. This is because the variation in carrying capacity over the island's life span, as represented by islands of radically different age across an archipelago, should be far more dramatic than the variation in carrying capacity and connectivity driven by eustatic change. Nonetheless, even in deep ocean, some volcanic islands become joined at sea-level low points—not to mainland—but to one another. This is true of the Maui-Nui complex (the islands of Molokai, Lanai, Kahoolawe, and Maui) within Hawaii, and of the Canarian islands of Lanzarote, Fuerteventura and islets surrounding them, which join to form the single island of Mahan at sea-level minima. The signal of these events can be detected in the distributions of species (e.g. within Galápagos: Ali and Aitchison 2014) and in the phylogeographic structure within them (e.g. within Hawaiian damselflies: Jordan et al. 2005). Eustatic sea-level reduction may also have resulted in the emergence of new stepping-stone islands between many archipelagos and the mainland, and thus a reduction in the isolation of many island groups.

Several studies have tested for the impact of changing island configurations through eustacy, alongside alternative models variously including island age, contemporary area and isolation variables (see e.g. Ali and Aitchison 2014, Weigelt et al. 2016, Norder et al. 2019; Barreto et al. 2021). An analysis of angiosperm diversity across 184 islands by Weigelt et al. (2016) reported that changes in island configuration, especially in area, post the LGM, have left a clear imprint. Both the number and proportion of endemics was found to be enhanced for islands that were considerably larger at the sea-level minimum, while native species richness was mostly a function of contemporary island configuration. Their model differed in structure from some of the others we have considered by using a spatial autocovariate built from model residuals, rather than assigning islands to archipelagos and including archipelago as a random effect.

The latter approach was used in a subsequent study by Norder et al. (2019). They used mixed-effect models to test alternative models of diversity of land snails and plants for 53 oceanic islands from 12 archipelagos, including several such as the Canaries, Azores, and Galápagos, used in studies of the GDM (above). They first reconstructed island configurations taking account of sea-level variation over the last 800,000 years, allowing testing of the explanatory power not only of the minimum (LGM) sea levels, which were comparatively short in duration, but also of intermediate glacial-era palaeo-configurations that describe configurations that were prevalent for much lengthier periods. Their response variables were native species, multiple-island endemics (MIEs) and single-island endemics (SIEs) for land snails, and natives and SIEs for plants. In total they evaluated six alternative models, one being based simply on current island area, a second current area and isolation, and four alternative palaeo-configuration models.

Consistent with expectations: (i) the richness of non-endemic natives was found to be poorly explained by glacial-era-configuration models for both taxa, the best models featuring either current area (snails) or area and isolation (plants) instead; but (ii) variation in SIE richness was best explained by models representing long-lasting (rather than the most extreme) glacial-era configurations (Norder et al. 2019). The form of the SIE models indicated that islands which gained most in area were richer in SIEs providing they remained separate islands, as those that had been joined together with other islands had lower SIE richness. This makes biogeographical sense as the physical connection would have allowed species to spread more easily between the formerly separated islands, thereby becoming MIEs when the islands became separated again. Intriguingly, the best MIE model for snails was also a contemporary area and isolation model, with no palaeo-configuration signal.

As with many of the variables commonly used as predictors of diversity in island biogeographical analyses, the magnitude of the effect of palaeo-configuration on the richness of a given taxon may be a function of that taxon's general dispersal ability. For example, in a global analysis of island mammals, Barreto et al. (2021) found that the richness of non-volant mammals on islands was more strongly driven by historical isolation patterns during the LGM, while volant mammal richness was more strongly affected by current isolation, indicating that, due to their relatively greater dispersal ability, the latter have overcome the historical isolation legacies of the LGM.

In many of the studies applying mixed-effect regression models to large compilations of islands, a large proportion of the variation is attributable to 'random' differences between archipelagos or regions (e.g. Weigelt et al. 2016; Norder et al. 2019; Barreto et al. 2021). This simply indicates that there are other archipelago-specific or regional factors generating differences in diversity additional to the factors of area, isolation, age, past changes in area and isolation, etc., that are coded in the particular model exercise. Likely important properties are those already reviewed in Chapter 5, such as their climatic characteristics, their net primary productivity, or differences in their geological and geomorphological properties. The archipelago identity does not, of itself, provide a quantitative metric that can be entered into the model, hence its inclusion as a qualitative 'random' effect. A more complete form of model may be obtainable if these other archipelago properties could be captured in a simple quantitative way. However, by adding in more variables, we run the risk of overfitting of models and given the inherently noisy nature of the data, in addition to the relatively small sample sizes, our alternative models of interest may become statistically indistinguishable.

While developing a more complete statistical model is always a reasonable ambition to hold it is not necessarily the driving motivation of every study. What the exercises reviewed here show is that it is possible to test and often find statistical support for particular theories, hypotheses, and models. Many of the papers reviewed here have recycled the same datasets from well-studied oceanic island groups such as Hawaii, Canaries, Azores, and Galápagos. It is notable that models representing alternative hypotheses can sometimes find support within the same datasets; that is, it may be possible to fit a standard ISAR, or the GDM, or a sea level adjusted model to particular archipelagos or to large multi-island datasets, demonstrating

that, for example, both island ontogeny and Quaternary sea-level changes have generated a legacy in contemporary patterns of island endemism. In fact, they present complementary hypotheses, addressing processes operating over different, yet overlapping time frames.

One way of building on and integrating these differing processes is to undertake highly detailed reconstructions of island geo-environmental history over lengthy periods. Ali and Aitchison (2014) provide just such a reconstruction for the last 700,000 years for the Galápagos, in which they reconstruct the configuration of the archipelago to account for thermal subsidence of the islands, eustatic changes in sea level, and changes in sea-floor loading associated with these changes. Their reconstructions show that much of the archipelago has undergone very significant changes in configuration, repeatedly connecting and isolating islands, generating well-specified and dated biogeographical hypotheses, which they were able to compare to empirical data for various vertebrate lineages. It was found that the distribution of tortoises was hard to relate to the geographical reconstructions, but the scenarios generated had explanatory power for the current distributions of the majority of reptile groups found within the core group of islands, in particular for snakes, lava lizards, leaf-toed geckos, and land iguanas.

The Aegean Sea provides another valuable island laboratory, which is increasingly being exploited for biogeographical research, combining a mix of land-bridge and permanent islands within a coherent (if complex) island region (Sfenthourakis and Triantis 2017; Sfenthourakis et al. 2018). This potential is nicely exploited by Hammoud et al. (2021) who used a reconstruction of relative sea-level change based on a detailed geophysical model, combining generalized sea-level curves with fine-scale isostatic reconstructions to identify islands that became joined together at: (i) the median sea level (MSL, for the last nine glacial–interglacial cycles), or (ii) the LGM sea-level low, or (iii) at no point within the Quaternary. Using generalized linear mixed models, they analysed data for plants, reptiles, butterflies, and centipedes and for chorotypes ranging from the native non-endemic species to SIEs. Their analyses demonstrated the importance of

LGM connections as endemics were clearly under-represented in land-bridge islands compared with true islands. Moreover, contrasting with the oceanic island data of Norder et al. (2019) the LGM models generally outperformed the MSL models. With the exception of the centipedes, the MSL land-bridge islands generally held more native nonendemic species as a function of contemporary island area in comparison to true islands. The authors concluded that the biogeographical impact of even very short-lived connectivity of Aegean islands to the mainland at the LGM had a more powerful impact on the emergent diversity patterns than similarly short-lived connectivity (between comparatively small landmasses) within oceanic island systems. These findings argue for careful separation of islands of differing ontogeny and especially mainland land-bridge islands from oceanic islands in island diversity modelling exercises.

As discussed for the GDM, simulation models have the potential to increase our understanding of the mechanisms through which eustatic sea-level change affects island biodiversity. Such models allow for greater flexibility in hypothesis testing, such as through the simulation of the entire sea-level history over a given time period. To take one example, Jõks et al. (2021) developed an agent-based simulation model that integrated past geological processes (e.g. volcanic island emergence and erosion) and eustatic sea-level change into a single model framework. Using data for multiple taxa from Hawaii, the Galápagos, and the Canary Islands to parameterize and test the model, it was found that, generally speaking, accounting for both eustatic and geological processes together resulted in improved correlations between predicted and observed native and endemic richness (Jõks et al. 2021).

7.9 Equilibrium and non-equilibrium dynamics across islands, archipelagos, and ocean basins

Throughout this chapter we have developed the argument that the geo-environmental dynamics of islands are crucial to understanding the biodynamics of island metacommunities. Island carrying capacity is not fixed and varies in response to island building, erosion, and subsidence, which together dictate the life span of each island's biota. It also varies in response to global climate change and associated eustatic (and isostatic) sea-level changes. Geological, geomorphological, and climate change processes are not constrained to fixed timescales, but oceanic islands typically last a few million years, while climate change has been particularly prominent over the last few tens and hundreds of thousands of years (Chapter 3). We have shown that the legacies of these processes are detectable in emergent macroecological patterns, specifically in the richness and proportional contribution of different chorotypes of species. They are also detectable in compositional patterns (cf. Chapter 6) and in phylogenies and in phylogeographical structure of island lineages (as will be evident in Chapter 9). Given which, where does this leave the debate on the value of equilibrium versus non-equilibrium models?

In his commentary on island theory, Haila (1990) differentiated four coupled space-timescales: (i) the individual scale; (ii) the population dynamics scale; (iii) the population differentiation scale; and (iv) the evolutionary scale (Fig. 1.3a). In the forerunner to the present volume, we argued that systems corresponding to the population dynamics scale—not too isolated, not too small, not too large—are most likely to show measurable turnover on secular scales and to best approximate the assumptions of MacArthur and Wilson's (1967) equilibrium theory. However, some recent work using molecular phylogenetic methods and based on compilations of archipelago-scale data serves to challenge this statement (Valente et al. 2015, 2017, 2020).

Applying novel phylogenetic methods, Valente et al. (2015) estimate rates of assembly (colonization and speciation) for the birds of the Galápagos archipelago. They report that Darwin's finches have diversified rapidly to reach carrying capacity but that the rest of the avifauna have failed to attain equilibrium, remaining in an ascending phase, as they continue to diversify (Fig. 7.8a). In a subsequent study, they develop this approach, by fitting the DAISIE model, a dynamic stochastic multi-archipelago model, to data for 91 taxa (both persisting and extinct) from the four main Macaronesian archipelagos. These analyses indicate

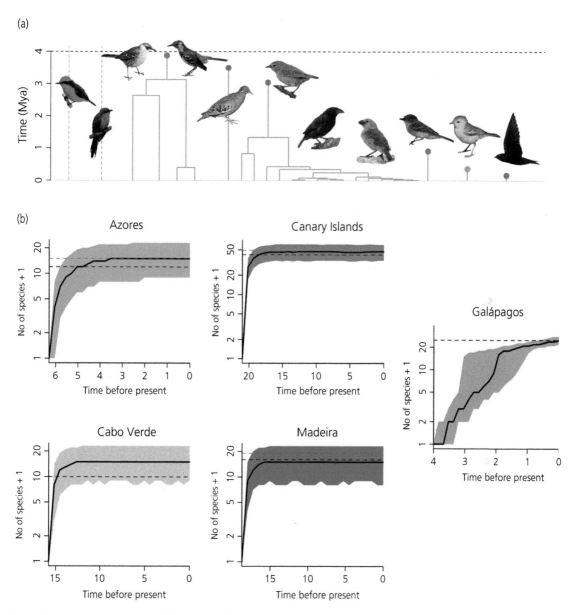

Figure 7.8 Apparent evolutionary equilibrium in island birds. (a) Estimated colonization and evolutionary branching times of Galápagos terrestrial birds. The circles represent estimated colonization points, blue = endemic lineages, green = non-endemic lineages. For two species (the vermilion flycatcher and dark-billed cuckoo) colonization time is poorly constrained. The horizontal dashed line represents the approximate age of the oldest island. (b) Estimated number of bird species over time (in millions of years) for four Macaronesian archipelagos and the Galápagos. Median estimates (black line) and 2.5th–97.5th percentiles (coloured shading). Grey dashed lines show pre-human diversity and black dashed lines show contemporary native species richness. Analyses were performed with DAISIE (Dynamic Assembly of Islands through Speciation, Immigration and Extinction), a dynamic stochastic island biogeography model, which models colonization, anagenesis, cladogenesis, and extinction.
(a) From Valente et al. (2015), their figure 1; Creative Commons CC BY. (b) From Valente et al. (2017), their figure 3.

that setting aside recent anthropogenic changes, each archipelago has achieved and maintained a dynamic equilibrium, which has persisted for several million years (Fig. 7.8b), and with similar biogeographical rates for three of the four archipelagos—the exception being that both speciation and colonization appear to have run faster in the Canaries.

It is intriguing that phylogenetic analyses suggest that most extant Macaronesian birds have colonized their archipelagos only within the last 4 Myr, notwithstanding that some extant islands date to 20–30 Ma (Illera et al. 2012). It may be that some part of the turnover reflects taxon cycles (Section 9.2) and/or the impact of late Pliocene–Pleistocene climate change, while the implications of varying archipelago size and subdivision over time (Section 7.8) are also largely set aside in these analyses. Nonetheless, at the archipelago scale, the case is made for biodiversity dynamics broadly conforming with dynamic equilibrium in bird richness.

In a further development of the approach, Valente et al. (2020) employ the DAISIE model to data for land birds from 41 oceanic archipelagos (including both Galápagos and Macaronesia), encompassing a total of 502 archipelago colonization events

and 26 *in situ* radiations. The resulting model provides a remarkably strong degree of support for a dynamic equilibrium *sensu* MacArthur and Wilson (1967) and for: (i) declining colonization and increasing anagenesis (species without *in situ* radiation) with isolation from mainland; (ii) increasing rates of cladogenesis (radiation) with area, accentuated with increased isolation; and (iii) the expected decrease in extinction rate with island area (Fig. 7.9). Intriguingly, the analyses produced no evidence of a minimum archipelagic area threshold being necessary for cladogenesis to take place. This doesn't mean that one does not exist for individual isolated islands, and indeed they did find the anticipated positive and interactive effects of increased archipelago area and remoteness on cladogenesis (point ii in the list), supporting the importance of untapped resource space ('vacant niches') for cladogenesis (as per the GDM).

Valente et al. (2020) noted that notwithstanding the attention generally paid to spectacular island radiations, their data reveal that most island endemic birds on oceanic archipelagos reflect speciation by anagenesis, with 231 of 350 endemic birds being the sole species of their colonizing lineage. This is a much higher proportion than the

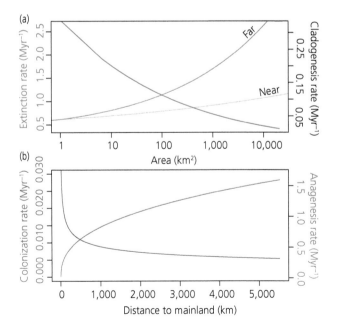

Figure 7.9 Estimates of key rates for 41 archipelagos derived from a dynamic stochastic model (DAISIE) applied to a global molecular phylogenetic dataset, showing: (a) extinction rate and cladogenesis as a function of archipelago area (note the interaction between area and isolation, wherein the lower line for cladogenesis is for near (50 km) and the upper line for far (5000 km) distances); (b) estimated colonization and anagenesis rates as a function of distance from mainland source pools. Extracted from figure 2 of Valente et al. (2020).

approximate quarter of endemic plants estimated as products of anagenesis by Stuessy et al. (2006) from analyses of 13 oceanic and continental island systems, reflecting the high rates of cladogenesis shown by plants on large, remote archipelagos such as Hawaii (93% cladogenesis), Canary Islands (84%), and Galápagos (57%). In the round, these observations indicate important differences of geographical scale of response between different taxa. While drosophilids and plants may attain geographical and reproductive isolation repeatedly within large, high islands (e.g. Wagner and Funk 1995), for birds, either (i) lineages are able to retain sufficient gene flow across an archipelago and/or (ii) enough colonization of other bird species takes place to limit opportunities for cladogenesis, or (iii) current taxonomy fails to recognize cryptic but valid allospecies within archipelagos (as e.g. Sangster et al. 2022). Following publication of the equilibrium theory in the 1960s, the earliest claims made for equilibrium on oceanic islands were based on Pacific bird data, but these claims were largely put aside as the scale of anthropogenic extinction of island birds became apparent. It is only now, with improved fossil data, the availability of well-specified molecular phylogenies, and of powerful analytical and modelling tools, that it has become possible to undertake the sorts of tests carried out by Valente et al. (2015, 2017, 2020).

Even so, we should temper our enthusiasm, by noting that these analyses remain reliant on assumptions concerning the completeness of the fossil record for extinct species, on island age estimates, on decisions about species status, and calibration of phylogenies (e.g. Steadman 2006; Duncan et al. 2013; Jønsson et al. 2017). We can illustrate some of these points with reference to Macaronesia, where the completeness of the fossil record remains in doubt (especially for the Canaries: Illera et al. 2012, 2016), as does how many extant species there are. For example, it has recently been claimed that instead of there being one species of robin, *Erithacus rubecula*, in the Canaries, there are actually three species, with Tenerife and Gran Canaria having cryptic SIEs (Sangster et al. 2022). Island age estimates are also subject to revision. For example, the maximum biological age of the Azores is probably around 4 Myr, rather less than shown in Fig. 7.8,

while the number of islands and overall area of the Azorean and Canarian archipelagos have each changed over this span of time (Hildenbrand et al. 2014; Ramalho et al. 2017). These issues affect many of the macroecological studies of oceanic islands reviewed in this chapter and mean that we can only draw tentative conclusions in relation to many of the issues we have been debating.

While many studies have failed to support or have refuted the dynamic equilibrium envisaged by MacArthur and Wilson (1967), the above analyses illustrate that support can be found for it even among remote oceanic archipelagos—the above caveats and concerns being noted but set aside! It is significant that the analyses by Valente et al. (2015, 2020) and the evidence of an evolutionary equilibrium for oceanic islands, are for birds, which are generally highly mobile, and at the level of the archipelago, rather than the level of individual islands within archipelagos. While different taxa certainly can show different diversity patterns across the same archipelago, it is worth pointing out in this context that the analysis of archipelagic species–area relationships (ASARs) by Triantis et al. (2015) demonstrated parallel scaling of species richness in plants, land snails, spiders, and land birds, for 14 oceanic archipelagos, indicating that at archipelago scales, species richness generally scales well with total available area (Fig. 5.12).

Yet, as we have seen in this chapter, island ontogeny and glacial/interglacial dynamics have left their signal on many island faunas and floras, especially so in taxa that are more species rich and have less active across-water dispersal capacity than birds. These signals are variably evident in data for different chorotypes (e.g. non-endemic natives versus SIEs), showing ongoing evolutionary adjustments on multiple individual islands in response to long-term geo-environmental dynamics. Indeed, the under-saturation of areas within islands and of whole islands at early stages of biological development following island emergence, or surface renewal, may exist within archipelagos that, analysed as a whole, can be well represented by equilibrium models. Indeed, the simultaneous existence of evidence for non-equilibrium and equilibrium dynamics within different elements of the Galápagos avifauna (Valente et al. 2015) highlights that both sets of models may have explanatory power.

While our focus has been on the search for dynamic models, we should not let this blind us to the importance of the biogeographical imprints of historical contingency and of how a deepening knowledge of both earth history and of phylogenies is increasingly providing powerful evolutionary narratives of the species of island regions. For example, Jønsson et al. (2017) demonstrate how it is possible to relate the biogeography and evolution of corvoid passerine birds (a group including nearly 800 species) to the complex geological history of the Indo-Pacific region, revealing New Guinea as a cradle, probably *the* cradle, for the evolution of this important group of passerines. Yet alongside the historically contingent, they also provide exemplification of how phylogenetic data may provide validation for general models, such as: (i) the confirmation of species as 'supertramps' (Chapter 6) and (ii) the applicability of the idea of taxon cycles having developed across the Indo-Pacific, involving repeated cycles of expansion and contraction of species distribution linked to evolutionary shifts in niche breadth and dispersibility (Chapter 9).

As we commented in Chapter 5, it no longer seems productive to dwell too long on picking a global answer to the question of whether islands are equilibrium systems, as we can find support for an array of emergent system behaviours (as e.g. Table 5.1), each of which may be regarded as analytically discernible. More interesting is to determine the scale dependency of key biogeographical processes, as they vary across major island types and contexts, and to continue the search for the general dynamic models that provide the theoretical focus to advance understanding. In the next part of the book, we will extend our focus from largely macroecological and compositional patterns, to evolutionary metrics and theories, and we will return to themes such as island ontogeny and Quaternary environmental dynamics through the lens of this evolutionary evidence.

7.10 Summary

The distinction between historical and dynamic hypotheses and models is that the former have explanatory value but limited predictive potential, while the latter set out to invoke the general processes and mechanisms that occur universally and which provide the basis for improved predictive models. In an island ecological context, the longevity of MacArthur and Wilson's equilibrium model arises from its focus on universal biological processes and their variation with simple geographical properties: island area and isolation. But its limitation in application to remote islands is that it pays no attention to island geo-environmental dynamics.

In this chapter we first consider the implications of fluctuations in island carrying capacity driven by extreme events and by climate variability before returning to the Krakatau system to explore how renewed volcanic activity within the archipelago due to the emergence and growth of Anak Krakatau has disrupted ecosystem assembly, repeatedly setting back ecosystem development and species accumulation on all but one of the Krakatau Islands. These disruptive events have been accompanied by ongoing geomorphological processes of erosion that have also influenced patterns of species turnover. Together these geo-environmental dynamics, when considered in relation to successional timescales, question the applicability of models based on simple monotonic trends in immigration and extinction that assume rapid establishment of dynamic equilibrium in secular, ecological timescales.

In contrast, the next model we consider aims to capture biodynamics in response to geo-environmental dynamics over the entire life span of volcanic oceanic islands, potentially extending to several millions of years. The general dynamic model of oceanic island biogeography (or GDM) describes the intersection of (i) the simplified ontogeny of a hotspot island with (ii) the immigration, speciation, and extinction dynamics of the MacArthur–Wilson equilibrium model. As an island progresses through building, mature and decline phases, this generates a humped trend in carrying capacity over the island's life span. In the building phase, large remote islands receive too few colonists to fill available niche space, generating a burst of cladogenesis, which slows as the island switches into the decline phase, and extinction becomes dominant. This model, when placed in the context of an archipelago of such islands, predicts emergent patterns in several diversity attributes. We explore the further development of the model, its translation into parameterized simulation exercises, and its empirical evaluation through various forms of multiple regression model.

Simulation models have been shown to validate the internal logic of the GDM, and can be used to generate novel predictions, including of the resulting geographical distribution of species derived from anagenetic speciation (where there is no *in situ* diversification) and cladogenetic speciation (where there is). Empirical testing through space-for-time substitution initially involved fits of a simple diversity = logArea + (time − time^2) format; that is, area combined with a humped function of island age. It advanced through application of a variety of mixed-effects models to permit data from multiple archipelagos (and even taxa) to be evaluated through a single model-fitting exercise. In general, these models were found to provide good fits and often the best fits within comparative model-fitting exercises, especially for oceanic island systems involving a full array of island ages/stages. It has been shown that the humped response in island endemism is typically driven by a small number of rapidly diversifying clades, with many lineages failing to exhibit any *in situ* cladogenesis at island or archipelago scales. The model has been extended in various ways, including via downscaling to within-island scales of diversity patterns and to analyses of plant–pollinator networks. Further developments of the GDM that may be envisaged include the specification of alternative island ontogenies both at the level of general models and as empirical reconstructions of actual island geodynamics, applications to marine systems, and additional work on ecological networks and functional trait space.

Quaternary climate change has driven profound changes in island configurations, and the glacial-sensitive model of island biogeography codifies the implications of these changes for vital rates and identifies a range of expected outcomes. Using similar models and often overlapping data used in testing the GDM, several authors have tested for legacies in contemporary island diversity arising from palaeo-configurations of island systems. Studies have shown that, for example, (i) legacies of past island configurations linked to intermediate glacial sea levels can be found in the richness of SIEs on oceanic islands, and that (ii) absolute sea-level minima have left legacies in endemism patterns on land-bridge islands in the Aegean. Further work is needed to establish the extent to which contemporary diversity patterns contain signals of Quaternary climate dynamics alongside those arising from longer-term island geodynamics.

We conclude the chapter by reviewing some exciting recent work applying novel phylogenetic methods to estimate key biogeographical rates for oceanic island birds analysed at the level of whole archipelagos. These studies indicate that: (i) within a single archipelago, the Galápagos, the finches fit equilibrium expectations while the rest of the land birds do not, while (ii) across a dataset of 41 oceanic archipelagos, there is remarkably good support for the equilibrium dynamics described by MacArthur and Wilson's equilibrium theory. All of which serves to highlight that both non-equilibrium/disequilibrium and equilibrium models have value, and that sometimes we can extract biogeographical signal from the same set of islands for different scales of geo-environmental processes; for example, there may be both a legacy of Quaternary eustatic sea-level dynamics and island ontogeny in the patterns of richness and endemism across a set of islands. Hence, the themes considered within this chapter serve as a bridge to the next part of the book: 'Island Evolution'.

PART III

Island Evolution

Frontispiece Double mutualisms involve animals acting as both pollinators and seed dispersers for the same plant species: (a, b) *Podarcis lilfordi* taking nectar (a) and fruit (b) of *Ephedra fragilis* in the Balearic Islands; (c, d) *Mimus parvulus* feeding upon the flowers (c) and fruits (d) of *Opuntia galapageia* in the Galápagos (excerpted from figure 2 of Fuster et al. 2019; photos (a) F. Fuster, (b) J. Rodríguez-Pérez, (c, d) R. Heleno).

CHAPTER 8

Colonization, evolutionary change, and speciation

8.1 Arrival and change

In Part II, we explored patterns and processes while treating species as essentially unchanging units of analysis. This is a useful simplifying assumption, but species are not homogeneous entities, and their populations may be subject to evolutionary change within islands, in many cases leading to speciation. However, it can be challenging to determine whether island forms are sufficiently distinct from one another and/or from mainland forms, to justify being designated as separate species. We therefore briefly review the nature of the species unit, before outlining a set of alternative frameworks within which we can organize ideas about island evolution, focusing on distributional, locational, mechanistic, and phylogenetic distinctions.

We then consider insular evolution from the moment of initial colonization of an island and both the chance (e.g. founder effects, genetic drift) and selective microevolutionary processes that may be at play, and which lead to changes in genotype and species characteristics. Initially, such changes represent population or subspecific levels of differentiation, but in time, they can contribute to the emergence of distinct island endemic species. The relevance of the distinctive community composition a species encounters on colonizing a remote island is profound and is discussed in relation to concepts of empty niche space, ecological release, density compensation and conversely, the role played by character displacement when ecologically overlapping species co-occur on the same island and compete for resources.

Most insular endemics arise by *in situ* evolutionary change, either with (cladogenesis) or without (anagenesis) *in situ* division into two or more species. However, islands may also gain endemics as an outcome of hybridization between two different species, or by evolutionary change in—or extinction of—the colonist species within the source region. We recognize and discuss these ways of gaining insular endemics alongside the various patterns and processes of cladogenesis. The chapter is intended as a simplified introduction to the topic and we reserve consideration of the most striking emergent patterns of island evolution to the following chapters.

8.2 The species concept and its place in phylogeny

It will be seen that I look at the term species, as one arbitrarily given for the sake of convenience to a set of individuals closely resembling each other, and that it does not essentially differ from the term variety, which is given to less distinct and more fluctuating forms.

(Darwin 1859, p. 108)

If we are to study distributions of organisms, let alone speciation, it is a prerequisite that we have a currency (i.e. units that are comparable). Hence, the most fundamental units of biogeography are the traditional taxonomic hierarchies by which the plant and animal kingdoms are rendered down to species level (and beyond). So, for example, the endemic Gran Canarian blue chaffinch, has traditionally been classified as follows:

Island Biogeography. Robert J. Whittaker, José María Fernández-Palacios, and Thomas J. Matthews, Oxford University Press.
© Robert J. Whittaker, José María Fernández-Palacios, and Thomas J. Matthews (2023). DOI: 10.1093/oso/9780198868569.003.0008

Kingdom	Animalia
Phylum	Chordata
Class	Aves
Order	Passeriformes
Family	Fringillidae
Genus	*Fringilla*
Species	*teydea*
Subspecies	*polatzeki*

According to which, it is a subspecies of *Fringilla teydea*. Recent work has, however, argued for its elevation to full species level with the name *Fringilla polatzeki* (below).

In most corners of biogeography, the species is the default unit of analysis, although we can point to analyses of higher (e.g. genus-level analyses of phylogenetic endemism by Veron et al. (2019)) and lower levels for particular purposes (phylogeographical analyses commonly focus on populations). There are in fact numerous operational definitions of the species unit. Singh (2012) lists 23 species concepts, while Lomolino et al. (2017) pick out six: (i) biological; (ii) recognition; (iii) phylogenetic; (iv) genealogical concordance; (v) cohesion; and finally (vi) the morphological or classical species concept. Traditionally, morphology was the principal basis for species recognition, but the second half of the 20th century saw a switch in preference towards Ernst Mayr's (1942, p. 120) biological species concept, which refers to 'groups of actually or potentially interbreeding natural populations [that] are reproductively isolated from other such populations'.

Unfortunately, whether two similar and related forms are reproductively isolated can be an unknown quantity, especially for geographically separated (allopatric) island populations. In such circumstances, populations may exhibit sufficient morphological differentiation to be considered subspecies, or even separate species, yet remain capable of interbreeding if placed into a common enclosure. Then again, species may be capable of interbreeding in the laboratory and have overlapping (sympatric) distributions but remain reproductively isolated in the wild because of behavioural differences. Sometimes a rider is added to restrict the definition to those individuals that can successfully interbreed to produce viable offspring. Yet this rule too can be broken, as species which are recognized as 'good' may interbreed successfully in hybrid zones, an example being the native British oaks *Quercus petraea* and *Q. robur* (White 1981). Hence, there can be uncertainty as to what constitutes a species and to whether the slightly different forms on different islands should be recognized as a full species: with the term allospecies being used to denote this boundary condition. Another term sometimes used to describe monophyletic groups of closely related and morphologically similar species of allopatric distribution is the superspecies.

The units below the species are a matter of considerable debate and strong opinions are held that they are either important ecologically and should be retained or that they are arbitrary and the whim of the taxonomist. Infraspecific units employed in the literature include 'varieties', 'aberrations', 'races', and 'forms'. However, the only infraspecific classification recognized by the 'International Code of Zoological Nomenclature' as a formal trinomial is the 'subspecies' (Braby et al. 2012). The subspecies concept was originally proposed as a means of categorizing variation within animal species across space; that is, the geographical subdivision of a species (e.g. Mayr 1963; see Phillimore 2010 for a review). They are often delineated based on traits indicative of reproductive isolation, such as vocalization and plumage in birds. Subspecies that are allopatrically divided are often viewed as representing incipient stages of the speciation process, but this may not always be the case (Phillimore 2010). Such allopatric separation of subspecies is quite common in island birds, where individuals have colonized an island and undergone divergence from the ancestral population that is insufficient for speciation to have run its full course. For example, while Kangaroo Island (South Australia) has no endemic bird species, it is home to 17 extant bird subspecies that have diverged from their mainland relatives since the island became isolated around 10,000 years ago following the end of the last glacial period (BirdLife Australia 2022).

The subspecies concept is probably less consistently applied and agreed upon than the species concept (Braby et al. 2012). This is not a trivial issue given that, within a given taxon, the number of subspecies can (far) outnumber the recognized full species. For example, there are roughly 10,000–11,000 recognized bird species (Tobias et al. 2022), but 17,000–20,000 recognized subspecies (Gill et al. 2022), while Groombridge (1992) estimated there to be *c.* 18,000 full species and approaching 100,000 subspecies of butterflies. These problems have particular force in relation to islands, where many closely related populations are allopatric. Taxonomic revisions can lead to elevation of such insular subspecies to full species level.

For example, the Canarian blue chaffinch (above), was traditionally regarded as one species comprising a Tenerife subspecies and a Gran Canaria subspecies, but Lifjeld et al. (2016) argued for their elevation to full species level (as *Fringilla teydea* and *F. polatzeki*, respectively) based on genetic data indicative of 1 million years of separation, plus differences between the two in sperm length, song, plumage, and body morphometrics. Most taxonomic sources now recognize the two as separate species. A second Canarian example is provided by the recent recognition of three (rather than one) robin species within the archipelago, two of which are single-island endemics (Sangster et al. 2022). The opposite process, whereby two previously recognized species are lumped together, can also occur, although it is less frequent (at least in birds). For some purposes—for example, diversity comparisons and conservation designations—the determination of whether an island population merits specific status, can be significant, but for other purposes—for example, phylogenetic and phylogeographical analyses—the precise status of a population is less important than understanding the overall patterns and timing of population divergence (below).

At higher levels, species are grouped into genera, then families, and so forth, representing increasingly deep evolutionary separation and divergence. In flowering plants such groupings of species are traditionally based principally on the evolutionary affinities of floral structure and so it is possible to find many different growth forms and ecologies within a single family. For instance, the Asteraceae (Compositae—the daisy family) all have a composite flower structure, but their growth forms range from annual herbs to long-lived trees. In general, there tends to be less variation in functional characteristics within a genus than between genera of the same family.

Some species are so distinct that they may be the only members of a genus, family, or even order, whereas others belong to extremely species-rich genera. Insular examples include, at the one extreme, *Amborella trichopoda* (Amborellales) and *Lactoris fernandeziana* (Lactoridales), endemic respectively to New Caledonia and the Juan Fernández Islands, both of which are the only extant representatives of their orders (Stuessy et al. 1990; *Amborella* Genome Project 2013). At the other extreme, on the Hawaiian Islands, there are approximately 60 named species of *Cyrtandra* in the widespread family Gesneriaceae (Kleinkopf et al. 2019). Large insular radiations occur across many taxa. As an example, 99 extant species are currently placed in the avian genus *Zosterops* (Tobias et al. 2022), roughly two-thirds of which are island endemics.

In recent decades, huge advances in genetic techniques accompanied by reductions in costs, have revolutionized both systematics and biogeography. Molecular phylogenies (or at least phylogenies incorporating molecular data for the majority of species) are now available for a wide range of taxonomic groups, including all birds (Jetz et al. 2012) and 32,000 land plants (Zanne et al. 2014). Molecular phylogenies are particularly useful in island studies as they can be used to: (i) explore the likely sequence of inter-island and intra-island speciation events within archipelagos, or between islands and mainlands (e.g. Ó Marcaigh 2022); (ii) provide information regarding the timings of island colonization; and (iii) estimate colonization, speciation, and extinction rates (Valente et al. 2015, 2018, 2020). For example, using the DAISIE model and incorporating data from dated molecular phylogenies, Valente et al. (2015) were able to show that, while the total species diversity of the terrestrial avifauna of the Galápagos has not reached equilibrium/carrying capacity, the Darwin's finches subset (Section 7.9) has reached a steady state. The increased availability of molecular data (e.g. through online databases), constantly increasing computational power, and reduction in costs of genetic sequencing, mean that the generation and analysis of molecular phylogenies in island biogeography will continue to increase, with the next frontier being the analysis of whole genomes for multiple species (e.g. see Leroy et al. 2021).

To maximize the information that can be harvested from molecular phylogenies, it is necessary to generate estimates of absolute divergence times between lineages. This enables divergence events to be interpreted in the context of major geo-environment events, such as island emergence, or the joining and splitting of islands due to sea-level changes. For this, we use molecular clocks, based on the proposition that the genetic distance between two species or populations is proportional to the time since divergence (Box 8.1; dos Reis et al. 2016). If the divergence time of a particular split can be constrained through fossil evidence, or from the timing of island emergence, the genetic distance can be translated into a rate of molecular evolution, which can in turn be applied to all nodes in the phylogeny to estimate absolute divergence times (Lee and Ho 2016; dos Reis et al. 2016). However, molecular clocks can involve false assumptions and may not always run to time, leading to incorrect inferences regarding the diversification history of a clade (Box 8.1).

Box 8.1 Molecular clocks, stem age, and crown age

Dating events in evolutionary biogeography is challenging because of the extended time periods involved. **Molecular clocks** provide a key solution to this problem. The idea, as developed in the 1960s, was based on the observation that proteins and DNA evolve at a fairly constant rate across very different species, allowing the extent of molecular divergence to be used as a metric for the timing of events within the development of a lineage. Advances in genomic sequencing and computational tools (particularly Bayesian methods, which enable better integration of uncertainties into the analysis), not to mention the increased availability of species-level genetic data, have since revealed that the rates of genetic change do in fact vary substantially across the tree of life (Lee and Ho 2016; dos Reis et al. 2016). As such, the 'molecular clock' is now used to refer to a broader group of models (including 'relaxed clock' approaches that model rate variation across branches of the tree, and 'partitioned clock' models where different portions of the genome are allowed to evolve according to different clocks) that can provide fine tuning for particular regions of the tree of life (Ho et al. 2015; Lee and Ho 2016; dos Reis et al. 2016). Nonetheless, molecular clocks inevitably involve assumptions and should be regarded as providing plausible hypotheses (subject to future revision) for the developmental sequence and timing of splits within monophyletic lineages (e.g. Box Figure 8.1).

Greater confidence can be placed in molecular clocks if they can be independently calibrated. Indeed, such calibration is essential if absolute rather than relative timescales are to be estimated. Typically, such calibration is undertaken using fossil data, but major geological events (e.g. the presumed sterilization of an island through volcanic activity) have also been used (Hipsley and Müller 2014; Ho

et al. 2015). For the many taxonomic groups that do not leave much fossil evidence (e.g. many invertebrate taxa), the latter is the only available approach. However, as Hipsley and Müller (2014) caution, the use of geological events as calibration points is often based on the assumption of vicariance as the driver of allopatric species distributions which, given the presumed role of long-distance dispersal in many island colonizations (Chapter 4), is not always the case. If over-water island colonization is presumed, island age can instead be used as a calibration point, which is occasionally done, but this works on the assumption that dispersal (and therefore lineage divergence) occurred close to the time of island formation (Ho et al. 2015). Violation of these and other assumptions (e.g. cases where island lineages are actually older than the island they now occur on), not to mention the difficulties in determining effective island age, or simply incomplete lineage sampling, will result in incorrect calibration and thus erroneous divergence dates (Heads 2011; García-Verdugo et al. 2019a). The use of fossil data is not infallible either as the age of the earliest fossil is not necessarily a good gauge of when the lineage originated (dos Reis et al. 2016).

As evident from this discussion, the time of island colonization, or TIC, is a critical piece of information for the calibration of phylogenies and for distinguishing between competing biogeographical hypotheses (see also Box 4.3). Within the literature, two key concepts linked to estimations of TIC are the stem age and the crown age (García-Verdugo et al. 2019a). The **stem age** of an island lineage is the estimated time of the split between the island lineage and its closest relative, whether that is a relative on another group of islands or from the mainland. As

Box 8.1 *Continued*

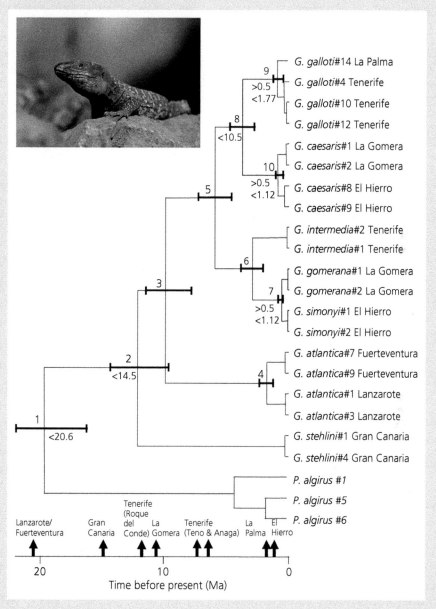

Box Figure 8.1 Hypothesis for the evolution of Canarian *Gallotia* in relation to the sister taxon *Psammodromus* showing the estimated divergence times (bars on the nodes) in relation to the approximate timing of island emergence (arrows). Dates within the tree represent maximal (<) and minimal (>) node constraints in Ma.

From Cox et al. (2010), their figure 3. Image of *Gallotia galloti eisentrauti*, which is found on the windward slopes of Tenerife, by Rubén Barone.

Box 8.1 *Continued*

García-Verdugo et al. (2019a) point out, the use of the stem age as a proxy for TIC may be affected by, for example, (i) failures to include relevant sister taxa/populations because of incomplete sampling or extinction; (ii) uncertainty over the colonization path involved in deeper splits (see discussion of Rand flora, Section 4.8). The concept of the **crown age** refers to the estimated date of onset of an insular radiation *within* the island or archipelago. The point at which the first evolutionary split has occurred *in situ* evidently cannot exceed and must—to a degree—underestimate the TIC. In archipelagos of a greater range of island ages, in which extinction of earlier products of the radiation may have occurred on older islands (whether extant or sunken to become seamounts), it is possible that the crown age may significantly under-estimate TIC, but in general it is argued to provide a better proxy than stem age (García-Verdugo et al. 2019a).

Keeping these issues in mind, we can exemplify the use of molecular clock dating with reference to Canarian endemic *Gallotia* lizards and their sister taxon *Psammodromus*. Carranza et al. (2006) constructed a dated phylogeogeny of the lacertid lizard *Psammodromus algirus* in the Iberian Peninsula and across the Strait of Gibraltar in North Africa. They sampled genomic DNA for three genes from 101 specimens of the subfamily Gallotiinae, mostly of the target species, but including specimens of

Gallotia species from the Canaries. These specimens provide a means of 'rooting' the phylogenetic tree, and also a means of calibrating it. This is because it is reasonable to assume that *G. caesaris caesaris*, endemic to the island of El Hierro, commenced divergence from its nearest relative, *G. c. gomerae*, endemic to the island of La Gomera, shortly after the formation of El Hierro (approximately 1 Ma).

The clock used by Carranza et al. indicated that diversification in the genus *Gallotia* began on the Canaries about 13 Ma, with a rather earlier data of 25 Ma for the first speciation events within the Iberian/North African genus *Psammodromus*. However, subsequent analysis of divergence in *Gallotia* using mtDNA and Bayesian techniques provide a somewhat different time frame, with initial colonization of the eastern Canary Islands *c.* 17–20 Ma, followed by a sequence of events constrained by assumptions linked to estimates of island subaerial ages (Cox et al. 2010). Their models suggest that cladogenesis was a response to the emergence of more islands within the group, with divergence in body size and multiple colonization events to particular islands (Box Figure 8.1). It is important to view such models as merely hypotheses of lineage development that are unlikely to be the final word, but they are sufficiently precise that future genetic analyses or fossil discoveries have the potential to provide meaningful tests of the models.

8.3 The geographical context of speciation and endemism

Distributional context

The geographical context of insular speciation/endemism can be viewed either in distributional or locational terms; that is, distinguishing the degree of overlap of populations involved in speciation events on the one hand, from the issue of whether in fact endemism has involved *in situ* or by change elsewhere (Table 8.1). **Sympatry** and **allopatry** denote, respectively, two populations (or species) that overlap in their distributions and two geographically separated populations (or species). A third condition, **parapatry**, refers to geographical situations of partial separation of two populations. Speciation events that occur where the two

diverging populations occupy the same geographical area constitute **sympatric speciation**. If taking place between two geographically separated populations, it is **allopatric speciation**, and where the separation is partial, and/or where there is a zone of contact (hybrid zone) between the two, then it constitutes **parapatric speciation** (Box 8.2).

The distinction between allopatric, sympatric, and parapatric speciation may be blurred for two reasons. First, the spatial scale we choose to describe the geography of the populations may not be concordant with the scale at which the members of the populations interact. For example, populations of a species on an island divided by a recent landslide may subsequently diverge in isolation, eventually resulting in the evolution of two new species. This speciation could be classified as allopatric at the

Table 8.1 A simplified framework for classifying speciation and the origin of endemism in insular systems.

Form	Pattern
i) *Distributional*	
Sympatric	Divergence occurs while populations overlap in space
Allopatric	Divergence occurs while populations are separated in space
Parapatric	Divergence occurs while populations are only partially separated
ii) *Locational (and historical)*	
Neoendemic	Change occurs and novelty arises within the island
Palaeoendemic	Island form is essentially relictual, with mainland ancestral population becoming extinct
iii) *Mechanistic*	
Allopatric	A combination of founder effects/drift and selection, in varying proportion
Competitive speciation	Sympatric speciation involving reproductive isolation developing through selection
Hybridization	Crossing between species that results in a novel species arising
Polyploidy (and apomixis)	Whole genome duplication arising from a single-parent species (autopolyploidy) or cross-fertilization between two (allopolyploidy)
iv) *Lineage subdivision*	
Anagenesis	Original colonist form evolves to become a recognizably different species without the lineage splitting within the island
Cladogenesis*	Lineage divergence (radiation) into two or more species within the island

*Stuessy et al. (1990) use the label anacladogenesis for cases where the lineage divides *in situ*, with the original colonist species persisting alongside new form(s), and the label cladogenesis for the subset of lineage divisions where radiation into two or more species has occurred and the original colonist form can no longer be found. This distinction has been less widely adopted and we do not use it herein.

within-island scale, but sympatric when described at a coarser scale, given that both species evolved on the same island. However, even at the within-island scale in this example, categorizing speciation mode is not straightforward given that there is no standard definition of how geographically isolated populations need to be to be considered truly independent. Diehl and Bush (1989) suggest that populations utilizing different, spatially segregated habitats should nonetheless be considered sympatric when all individuals can move readily between habitats within the lifetime of an individual. Determining this in practice will often be difficult and must be judged separately for different taxa as, for example, a within-island barrier for land snails may not be a barrier to bird populations. Indeed, the distinction between allopatric and sympatric *distributions* is something that varies substantially within higher taxa. In illustration, the Krakatau Islands, isolated from the mainlands by about 40 km, are beyond the apparent colonization limits of certain plant families (Whittaker et al. 1997), yet for some species of plants and their insect pollinators, the island populations can be considered as freely interbreeding (panmictic) with populations on the mainland (Parrish 2002).

Second, distributions change over time (e.g. as island configurations and climates change). A currently sympatric species pair may well have undergone cladogenesis while their populations were geographically isolated (allopatric) from one another. And, as shown by the presence on remote islands of many species that are native but not endemic, the existence of disjunctions in the distribution of a species does not have to generate rapid speciation. Even so, it will typically be matched by increasing genetic distance over time between the two (or more) populations, setting the populations on the path towards speciation (Ó Marcaigh 2022).

The endemic pigeons of the Canary Islands illustrate how two sympatric species from the same genus can arise where much of the evolutionary distance has been acquired in allopatry. Bolle's pigeon (*Columba bollii*) and laurel pigeon (*C. junoniae*) both inhabit broadleaved forests and can be observed in the same areas. They are also endemic to the same four islands. Using DNA nucleotide sequences from two mitochondrial genes, González et al. (2009) showed that *C. junoniae* occupies a basal position in the genus and based on fossil evidence appears to be derived from an ancient colonization of the archipelago, dating to c. 20 Ma, when only the island of Fuerteventura had emerged (Section 4.8). *C. bollii* was found to be more closely related to the common wood pigeon (*C. palumbus*) than to the other Canarian species and is estimated to have colonized perhaps as recently as 5 Ma (González et al. 2009). Thus, the two Canarian fruit pigeons are confirmed as an example of a 'double invasion' of the archipelago by two members of the genus that were already distinct at the point the second species colonized.

Many authors have assumed that allopatric speciation is by far the dominant mode of speciation. Indeed, sympatric speciation has been a relatively controversial idea ever since it was proposed (see review by Gavrilets 2014). However, we consider there to be compelling evidence that sympatric speciation can and does occur; for example, in ricefish (*Oryzias* species) in an ancient lake in Sulawesi and within several of the Hawaiian insect radiations (Hembry et al. 2021).

Box 8.2 Allopatry, sympatry, parapatry, and archipelago speciation: a matter of scale

We provide this simple summary as a quick look-up and clarification of the ideas involved with reference to geographical scale of application.

- **Allopatric speciation** is speciation occurring between two (or more) populations that are geographically isolated from each other. All island neoendemics exemplify allopatry in the sense that they are isolated from the mainland source pool. However, within an archipelago, populations may also be allopatric at the inter-island scale, or even within islands (microallopatry); for example, isolated within forest islands surrounded by lava (*kipukas*, Hawaii). Allopatric speciation can be split into two types: (i) vicariant (dichopatric)—where a single population is split by barrier formation (e.g. rising sea levels) and the resulting populations remain relatively large; and (ii) peripatric—typically involving dispersal by some individuals to form a new, typically small and isolated population. In oceanic island settings, speciation linked to initial colonization can be considered peripatric, but localized vicariance can also occur through barrier formation within an island or archipelago (e.g. Machado 2022; Ó Marcaigh 2022).
- **Sympatric speciation** is a term applied to cases where speciation occurs without geographical separation of the populations involved. In practice, the past roles of allopatry and sympatry can be difficult to distinguish at the intra-island scale, or even at the (inter-island) intra-archipelago scale when the islands are close to each other and the organisms involved have high-dispersal abilities. Theoretical modelling has shown that sympatric speciation demands certain conditions that limit gene flow. They include (i) the combination of disruptive selection (Box 8.3) with assortative/non-random mating (where individuals preferentially mate with those of similar phenotype), and (ii) a close association between trait(s) subject to disruptive selection and those that control assortative mating (Phillimore 2013; Gavrilets 2014; Lomolino et al. 2017). The beaks of *Geospiza* (a genus within the Galápagos finches) represent a possible case of association between traits constraining recombination. These beaks are subject to natural selection linked to seed handling, while beak dimension also affects male song, which is important to mate selection (Phillimore 2013). However, it should be noted that there is little evidence that any members of the genus have undergone perfect sympatric speciation (i.e. with no allopatric stage at all).
- **Parapatric speciation** refers to situations intermediate between the extreme cases of allopatry and sympatry, where the geographical segregation is incomplete; for example, the two separating forms are contiguous in space, with a hybrid zone in the area of contact/overlap (contrast with *peripatric*, above). Like sympatric speciation, there are also (theoretical) conditions for parapatric speciation to occur, although there is less necessity for an association between the traits subject to disruptive selection (Box 8.3) and those controlling assortative mating, given that in the non-overlapping parts of the range individuals are more likely to mate with other individuals from the same area, all else being equal.
- **Archipelago speciation** is a term used where speciation of island lineages has occurred mostly between islands rather than within them. This tends to imply an allopatric model, especially where inter-island distances are considerable, but may in fact involve both allopatric and sympatric phases. Sometimes, environmental differences across an archipelago such as the Canaries

Box 8.2 *Continued*

are great enough that **adaptive differences** are evident, but in other cases very little niche segregation is evident, in which case speciation might be considered **non-adaptive**. Archipelago speciation is a term that can encompass any mix of these conceptually distinct but often hard to separate forms of speciation.

• **Are these terms operational?** Given the complex geo-environmental dynamics involved over the life spans of volcanic islands and archipelagos, it can be very difficult to be sure that current distributions indicate the circumstances in which speciation occurred. Indeed, taking the Galápagos finches as a classic case study of island evolution, it appears likely that this adaptive radiation has involved allopatric, sympatric, and parapatric phases (Section 9.4), operating across changing configurations of islands (Ali and Aitchison 2014). Similarly, the distinction between adaptive and non-adaptive changes may not always be clear-cut: evolutionary change commonly involves a mix of stochastic and directional/deterministic mechanisms.

Locational and historical context

We have previously introduced the distinction between neoendemic and palaeoendemic species, a concept recognized as long ago as 1883 by the German botanist Adolf Engler. In an insular context, whereas neoendemics are species that have evolved their novelty *in situ* on the island, palaeoendemics are essentially relictual forms that have changed relatively little within the island but have gone extinct from their mainland range since island colonization (Box 4.2). Classic insular adaptive radiations provide numerous examples of neoendemics, but the identification of particular species, or traits, as 'relictual' can prove contentious (Box 4.3; Kondraskov et al. 2015; Médail 2022).

In their study of 4306 islands worldwide, Veron et al. (2019) used a phylogenetic endemism approach in combination with null models to assess the relative contribution of neoendemism and palaeoendemism at the genus level in monocots (grasses and grass-like flowering plants). Areas with significant amounts of palaeoendemism included Sri Lanka, Borneo, and islands in the Caribbean, while neoendemism was restricted to islands around Japan. Island regions containing significant amounts of mixed endemism (i.e. neither neoendemism nor palaeoendemism dominates) included Madagascar, New Guinea, and Japan. Several factors were found to drive or correlate with the different types of endemism, including island area, elevation, isolation, latitude, and climate stability.

The dichotomy between palaeo- and neoendemic is probably better thought of as a continuum, between older and sometimes clearly relictual taxa and rapidly diversifying lineages that demonstrate significant trait evolution within islands. While many lineages may defy easy classification at this stage, we are making progress in the placing of species and traits within increasingly robust phylogenies, allowing assessment of whether, for example, insular woodiness is basal or derived (Chapter 11) and also demonstrating examples of mainland species embedded within largely insular clades (Chapter 10).

8.4 Colonization and evolutionary filters

Potential island colonists must pass through a series of 'filters' to reach and become established on an island (Fig. 1.2; Table 8.2; Chapter 4). To recap and extend this idea, let us imagine a hypothetical volcanic island 100 km west of South America that has emerged, devoid of species, from the ocean. First, species have to be able to reach the island from the South American mainland, other nearby Pacific islands (such as the Galápagos), or even from a far distant continent. The surrounding ocean ensures that a dispersal filter constrains the chances of species reaching the island, favouring those with traits promoting long-distance dispersal. We should remind ourselves that traits can be imperfect predictors: much is down to chance. For example, a predatory bird might fly to our island shortly after eating a lizard, which had just eaten a fruit laden with seed; and on arrival it deposits those seeds in a viable state, thereby introducing a plant that has no obvious adaptation to long-distance dispersal through 'secondary dispersal'.

Of those species that are able to reach the island, many will arrive in such low abundance (e.g. a single non-reproductive individual) that they are unable to establish a breeding population, unless possessing specific traits, such as self-compatibility (Section 11.8) that enable them to beat the odds. They must also encounter the right environment, one that matches the fundamental niche of the species and in the case of plants, the establishment niche, which may be a matter not just of whether the island possesses appropriate climatic and habitat conditions in the first place, but whether the colonist is deposited in a 'safe site' for germination and establishment within that environment. The colonist may also be dependent on the presence of mutualists (e.g. mycorrhizal symbionts or

plant pollinators) or, for animals, the availability of appropriate food in sufficient quantity through the year that the population is able to persist and build in number (Weigelt et al. 2015; Russell and Kueffer 2019; König et al. 2020). Even with all these filters having been passed, some species may fail through a natural disaster (hurricanes, volcanic activity, island flank collapse, or tsunami) or less catastrophic shift in insular conditions (e.g. climate change or fluctuation), which we term a contingency filter in Table 8.2 (and see Chapter 4).

Having established and persisted, only a subset of colonists subsequently goes on to diversify into separate lineages within an island or archipelago. As per the radiation zone of MacArthur and Wilson (1967), diversification generally increases with isolation from source regions and across larger islands that are arranged in archipelagos permitting occasional inter-island movement of propagules (Fig. 8.1). The concept of a sequence of

Table 8.2 Successful establishment, persistence, and (in cases) evolutionary diversification on remote islands requires the potential colonist to pass through a series of 'filters' that we can conceptualize as sifting through the source pool to determine the winners. Properties of both the focal species and of the islands and of their existing biota influence the probability of passing through these filters.

Filter	Mechanism	Key factors
Dispersal	Possession of traits that enhance entrainment and long-distance dispersal	Island isolation; wind and marine currents
Breeding systems	Related to the ability to overcome inbreeding depression	Size of founder group; reproductive rate and flexibility
Environmental (abiotic)	Related to the existence on the island of the proper habitat where the species can thrive	Island characteristics (area, elevation, latitude, etc.)
Biotic interactors	Presence of key mutualists, appropriate food resources, etc.	Insular community characteristics
Contingency	Surviving volcanic activity, landslides, etc.	Chance, island geodynamics
Evolutionary	Related to existence of empty niches, species genetic plasticity, traits concerning dispersal powers, etc.	Traits possessed by lineage vs open niche space

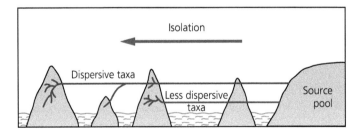

Figure 8.1 Dispersive taxa radiate best at or near to their effective range limits (the radiation zone), but only moderately, or not at all, on islands near to their mainland source pools. Less dispersive taxa show a similar pattern, but their range limits are reached on much less isolated islands. The increased disharmony of the most distant islands further enhances the likelihood of radiation for those taxa whose radiation zone happens to coincide with availability of high island archipelagos.

geographical, ecological, and evolutionary filters shaping the long-term dynamics of island systems is not a new idea and has sometimes been expressed in terms of geographical access, ecological access, and evolutionary access to resources (e.g. Stroud and Losos 2016).

8.5 Founder effects, bottlenecks, and genetic drift

Typically, a species immigrating to a remote island will have a tiny founding population, representing within it only a portion of the gene pool of the source population. This is known as the **founder effect** (Mayr 1954) and represents a **bottleneck**, a term used for this and any subsequent sharp reduction in population size that may result in the loss of genetic variation in the persisting population. While a founding population may be biased towards traits that have enabled colonization (e.g. high dispersibility), it may also just represent a chance selection of the potential gene pool.

Such effects become more significant when a population remains stuck in a **bottleneck**, remaining small in numbers for several generations, or when bottlenecks are repeated at short intervals. In such circumstances, **genetic drift**, the random change in allele frequencies from generation to generation (Box 8.3), may drive divergence from the source population. It has been suggested that this may lead to insular populations that become, to a degree, 'maladapted' *sensu* Farkas et al. (2015) and less able to cope with new or variable conditions. On the other hand, where a population expands in size following a bottleneck, it may rapidly gain genetic variability through mutation and re-sorting, and sometimes through mechanisms such as introgressive hybridization with a related species (Section 8.8).

Insular founder effects and genetic bottlenecks have been reported for a range of taxa and insular contexts, including: the ship rat (*Rattus rattus*) in the Guadeloupe archipelago (Abdelkrim et al. 2005), vascular plants in North Atlantic islands (Alsos et al. 2015), birds in Macaronesia (e.g. Spurgin et al. 2014), and a freshwater copepod in lakes within the South Orkney Islands (Maturana et al. 2020). Several studies have reported founder effects for the Hawaiian

Drosophila (Carson 1992; Kaneshiro 1995; O'Grady and DeSalle 2018), with founder effects inferred at three levels: in the colonization of the archipelago, of its constituent islands, and of patches of forest habitat (*kipukas*) within islands.

Random changes in genetic make-up of an insular population can be contrasted with those arising through selection (Box 8.3), although it is likely that evolutionary changes on islands (as elsewhere) reflect both chance effects and selection. For example, in their work on the genomic diversity of silvereyes (*Zosterops lateralis*) on South Pacific islands, Sendell-Price et al. (2021) found that genomic divergence between island populations was attributable to stochastic changes resulting from cumulative founding steps, but that some morphological differences (e.g. in body size) were the outcome of selective processes.

According to theories of **founder-effect speciation**, the narrowing of the gene pool during a bottleneck may serve to break up the genetic basis of some of the older adaptive complexes, liberating additive genetic variance in the new population (Carson 1992). Within a model proposed by Kaneshiro (e.g. 1995), an important element is differential sexual selection (Box 8.3). He found that a minority of male Hawaiian drosophilids perform the majority of mating events, and that females vary greatly in their discrimination, and acceptance, of mates. In normal, large populations the most likely matings are between males that have a high mating ability (the studs), and females that are less discriminating in mate choice. This provides differential selection for opposing ends of the mating distribution and a stabilizing force maintaining a balanced polymorphism. In a bottleneck situation, however, there would be even stronger selection for reduced discrimination in females, and if this was maintained for several generations, it could result in a destabilization of the previous co-adapted genotypes. In such bottleneck situations, with reduction in mate discrimination by females, the likelihood of a female accepting males of a related species increases, thus allowing introgressive hybridization to occur (see Section 8.8). Thus, although bottlenecks typically reduce the genetic base of a new colony, in some circumstances they may serve to provide *enhanced* genetic variability within a population, providing increased options on which natural selection (Box 8.3) may then go to work!

The strength of founder effects is likely to vary as a function of island/archipelago characteristics, especially their area and isolation. Alsos et al. (2015) provided a test of these ideas, using genetic data for plants that colonized North Atlantic islands following the last glacial period. Founder effect strength was inferred to be greater for smaller and more isolated islands, consistent with their having generally smaller founder populations and reduced subsequent immigration. Stronger founder effects were also inferred for insect-pollinated species than for wind-pollinated species, which is consistent with the limited availability of potential pollinators on islands leading to increased inbreeding in insect-pollinated insular plants (see Chapter 11).

Box 8.3 Evolutionary mechanisms: selection versus drift

Selection and drift are often viewed as competing evolutionary mechanisms. However, in most populations, both will be operating, simply to relatively varying extents, with drift comparatively more important when population sizes are really small. Following Charles Darwin's lead, selection can also be divided into natural and sexual selection, with contrasting viewpoints on whether sexual selection simply represents a mode of natural selection.

Natural selection

Natural selection occurs through the differential reproductive contribution to the next generation of individuals of different phenotypes. Genetic mutations and combinations that increase an individual's chance of survival and successful reproduction are passed on differentially and encode changes in phenotype over time. While natural selection is often seen as a process that purely increases adaptation, genetic correlations between traits (e.g. due to pleiotropy) can result in selection for a beneficial trait causing increased prevalence of a linked 'maladapted' trait, especially in small populations (Farkas et al. 2015).

Natural selection can be directional, stabilizing, or disruptive. Directional selection involves the selection for phenotypes at one extreme of the character distribution(s), shifting the mean value of this character in a population to higher or lower values (e.g. bill size in insular finches). Stabilizing selection is where an average phenotype is selected for, narrowing the character distribution (e.g. the clutch size of bird species). Finally, disruptive selection occurs when natural selection favours two or more phenotypes (i.e. the character distribution is multimodal), such as where there are two very pronounced food resource peaks, leading to selection for larger, heavier bills and conversely also for smaller, slighter bills, enabling more efficient tapping of the contrasting food types (Section 8.7).

Sexual selection

Sexual selection involves traits that provide an advantage in terms of securing a mate (e.g. plumage colour, horn size, display rate), and can be thought of as 'intraspecific reproductive competition' (Hosken and House 2011). Sexual selection can be 'multivariate' in that it may involve a combination of multiple traits (e.g. a preference for larger and more brightly coloured tails), and may be non-linear. Sexual selection can be stronger than natural selection in respect to certain traits, often resulting in trait values shifting beyond their optimum naturally selected value (Hosken and House 2011). The process is believed to have played a role in the diversification of certain classic insular radiations, including the Hawaiian Drosophilidae and the cichlid fishes of the African Great Lakes, as well as the putative Macaronesian chaffinch radiation (Recuerda et al. 2021; and see Section 8.8). Two mechanisms of sexual selection are often discussed: competition between members of the same sex (intrasexual selection), and situations where members of a given sex choose members of the opposing sex to mate with (intersexual selection).

Genetic drift

Genetic drift refers to the random fluctuation in frequency of a particular version of a gene (allele) within a population. Drift has been argued to be particularly important in very small populations, such as during founder events or subsequent population bottlenecks. The extent to which such neutral processes may drive speciation within islands remains poorly understood. It is entirely possible that both selection and drift may be involved in insular evolution of a single species (Sendell-Price et al. 2021).

Comparatively low levels of genetic variation have been observed in many oceanic island lineages, but Stuessy et al. (2014) caution that over millions of years, the signal of founder effects should be hard to detect, obscured by subsequent mutation, recombination, and selection, especially in lineages that have become endemic but have failed to radiate, raising the prospect that other factors may be at work in explaining low insular genetic diversity. However, others have argued for a role for repeated founder events/bottlenecks within oceanic archipelagos (and even within individual islands), in response to the active geodynamics that continually rework the space available (e.g. Carson 1992). For example, evidence for dispersal limitation of pollen and seed and the comparative prevalence of autogamic (self-fertilization) reproduction in many Canarian endemic vascular plants is consistent with repeated founding events having played a role in their evolution (Francisco-Ortega et al. 2000; and see Section 11.8), even if that role is typically not to provide the shortcut to speciation suggested in founder-effect speciation models (compare Carson 1992 with Sendell-Price et al. 2021).

Repeated bottlenecks

As a consequence of large distances between islands and archipelagos and the tendency towards reduced dispersibility found in many island lineages (Chapters 10 and 11), successful island colonists can experience multiple population bottlenecks as they spread from island to island. The chance element of genetic reduction can be accompanied by a selective component, especially where the colonizing populations encounter distinct physical and biotic conditions (see e.g. the impact of climate forcing on *Geospiza* in the Galápagos, below).

Returning to the example of Pacific silvereyes (above), the colonization of south-west Pacific islands from mainland Australia after AD 1830 provides an unusually well-specified quantification of founder effects. Single founder events appear to have had little detectable impact on genetic diversity, partly because founding populations have apparently been quite large (24–200 or so).

However, repeated island-hopping was accompanied by a gradual decline in allelic and genomic diversity (Clegg et al. 2002; Sendell-Price et al. 2021). A similar pattern, but developed over a far longer time frame, has been reported for Macaronesian chaffinch populations, with genetic diversity progressively decreasing along the path of archipelago colonization (Iberian mainland > Azores > Madeira > Canaries; Recuerda et al. 2021). Another example is provided by the Macaronesian endemic Berthelot's pipit (*Anthus berthelotii*). Genetic analyses indicate that the species colonized Madeira and the Selvagens from the Canary Islands independently and approximately simultaneously, probably in the early Holocene (Spurgin et al. 2014). Bottlenecks linked to the colonization events can explain differences in genetic diversity and morphology between populations. The general lack of evidence for selection in the traits examined was attributed to the similarity of habitats occupied by this species across the islands and the lack of competing insectivorous birds in those habitats (Spurgin et al. 2014). An interesting exception to this pattern is the mountain population of the species on El Teide, Tenerife, where increased body size is probably a selective response to low temperatures (consistent with Bergmann's rule; see Section 10.3).

The study of the long-term implications of bottlenecks has applied relevance in the context of the current biodiversity crisis (Part IV), in at least two ways. First, humans have reduced populations of many species to such small numbers that genetic bottlenecks are likely. Second, conservation translocations, which often involve just a handful of individuals, are increasingly used to move populations of (primarily vertebrate) species to different islands to protect them from anthropogenic extinction drivers, such as introduced species. While being the only realistic management option in many cases, one unplanned consequence of such translocations, when few individuals are involved, is the creation of a bottleneck and potential reduction in genetic diversity of the translocated population (e.g. see Ramstad et al. 2013, for an example of the little spotted kiwi in New Zealand, where one translocated population has effectively zero genetic diversity).

Sustained small population size—persistent bottlenecks?

As discussed above, the generally observed low genetic diversity of island populations could be a consequence of the founder effect, but it may also be the result of limited island size, constraining populations to small numbers of individuals over long periods (James et al. 2016; Leroy et al. 2021). Persistence at small population sizes may lead to a reduction in efficiency of natural selection and to the accumulation (and fixation) of slightly deleterious mutation. There is evidence from a number of studies in support of this 'nearly-neutral model of molecular evolution' for island lineages. For example, Leroy et al. (2021) report analyses of whole genome sequences of 25 passerines (14 insular and 11 continental species) that showed the island species to have lower nucleotide diversity and adaptive substitution rates compared with continental

species, supporting the idea that lower effective population sizes reduce the ability of natural selection to remove deleterious mutations (Fig. 8.2). However, not all studies have found such clear-cut results. James et al. (2016) used polymorphism and substitution data to compare the genetic diversity of 112 island–mainland species pairs across various taxa and island groups. Overall, island species were found to have significantly lower genetic diversity relative to mainland species, but this could be attributed to a subset of insular species that had recently undergone a bottleneck and for the rest there was no significant difference in either mutation rate or effectiveness of selection, despite the smaller average range size of the island species.

In conclusion, we have established in this section that the restriction of populations to an extremely small size, either on colonizing an island (the founding event), or subsequently in the history of a lin-

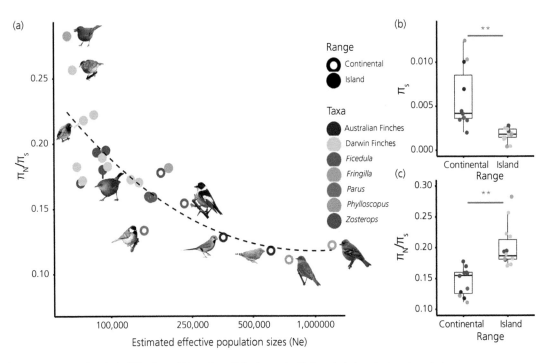

Figure 8.2 Genetic diversity in 14 insular and 11 continental island passerine bird species/subspecies based on whole-genome sequences. (a) The ratio between nucleotide diversity at synonymous (π_S) and non-synonymous sites (π_N) is taken as a proxy for the proportion of slightly deleterious mutations: it is shown to decline as a function of estimated population size. (b) Nucleotide diversity at synonymous locations is lower for island species than for mainland species, while (c) the ratio is higher for island species. From Leroy et al. (2021).

eage (temporary or persistent bottlenecks) may, in theory, produce a variety of effects, including both the loss of genetic diversity (reduced heterozygosity) and the addition of novel genetic combinations (typically during episodes of population increase). Empirical evidence supports a role for episodes of small population size, but that role can be variable. Thus, bottlenecks can variously diminish heterozygosity (a general indicator of fitness), or in contrast can serve to 'shake things up', generating a sort of mini-genetic revolution, contributing to rapid phases of evolutionary change in island populations. While we clearly still have much to learn concerning how these processes combine (Farkas et al. 2015), the occurrence of founder effects, bottlenecks, drift, mutation, and selection (natural and sexual) provide the building blocks to understanding the evolutionary pathways of founder populations.

8.6 After the founding event: ecological responses to empty niche space

Consider for a moment the arrival on a remote island of the first land bird, or perhaps the first seed eater or nectar feeder. In such situations the founding population can exploit, to varying degrees, hitherto largely untapped resources: in short, empty niche space. We should not, however, assume there to be a fixed architectural template of niches that constitute an equilibrium island biota for a given area/isolation combination, even if we accept the arguments from Chapter 6 that emergent assembly rules can be described for island systems. Indeed, historically, the **ecological niche** has been conceptualized as emphasizing the place of a species in the community, or alternatively, the array of physical and biological conditions within which a particular species can thrive—rather than attempting to identify templates for seemingly 'missing species'. Much ecological theory follows Hutchinson's (1957) depiction of the **fundamental niche** as an *n*-dimensional hypervolume of environmental axes describing the conditions permitting the occurrence of a species, within which **the realized niche** is the resource space actually occupied: the difference between the two reflecting the constraints imposed by dispersal limitations and biological interactions

(reviewed in Blonder 2017). In fact, biological interactions, such as pollinator mutualisms, can be key determinants of the capacity of a particular species to colonize and persist on an island, but the 'tightness' of such interactions is highly variable across species. Hence, when we turn from attempts to describe or quantify the niche of a particular species, to describing the array of 'vacant niches' provided by a recently emerged oceanic island, the concept becomes distinctly fuzzier.

The above points being registered, here we use the term **empty niche** to refer to the paucity of representation of a particular ecological guild or taxon, implying that a youthful island features an array of resources that are either untapped, or that are exploited by an unusually low diversity of species. Following the logic of the general dynamic theory (Chapter 7), the ecological and evolutionary opportunities for colonizing taxa to exploit such vacancies may be expected to peak not at the very earliest point of island emergence, but once an array of other interacting species have established. In time, the availability of incompletely filled niche space must reduce, eventually declining as a volcanically dormant island erodes and subsides.

Ecological–evolutionary responses to the differential occupancy of niches on remote islands compared with their source pools provide many interesting insights, which we will now review, starting with two general responses. First, the phenomenon of ecological release; that is, of niche expansion by colonist species, and second, density compensation, the term given for the general pattern of increased average density of island species that goes hand in glove with the lower richness (for their area) of island ecosystems.

Ecological release

Ecological release occurs when a species colonizing an island encounters a novel biotic environment in which particular competitors or other negatively interacting organisms, such as predators, are absent (Herrmann et al. 2021). The response is niche expansion (along one or more niche axes, e.g. macrohabitat, diet). Initially this may equate to a larger realized niche without change to the fundamental niche, but where the selective opportunity persists, it is often accompanied by **character release**, an increase in the variance of genetically coded features such as beak morphology. The increase in niche breadth following ecological release forms a key part in the reasoning of the competitive speciation model we consider later and is an important part of many scenarios for island evolution (e.g. adaptive radiation). Where the newly encountered resources are asymmetrically distributed, shifts in the mean value of relevant traits may occur (Herrmann et al. 2021). Island finches provide classic examples of both niche shifts and changes in niche breadth within radiating lineages. In illustration, granivorous finch species on remote islands such as Hawaii and the Galápagos are highly divergent in terms of beak sizes (and seed sizes utilized) compared with finches on continents and, plausibly, this can be related to an absence of competitor lineages on the islands (Schluter 1988; Section 9.4). Directional niche shifts following ecological release form a part of the taxon cycle model (Section 9.2), while the repeated colonization of new spaces within an archipelagic setting can provide opportunities for population divergence in traits reflecting key parameters of the niche.

Increases in niche breadth within an island lineage are not always morphologically apparent. The ancestor of the Darwin's finch of Cocos Island, *Pinaroloxias inornata*, colonized this very isolated island, within which 'empty' niche space and the lack of interspecific competition provided conditions for ecological release—but insufficient space for further cladogenesis. This endemic finch has diversified behaviourally and, while showing little morphological variation, exhibits a stunning array of stable individual feeding behaviours spanning the range normally occupied by several families of birds (Werner and Sherry 1987). This intraspecific variability appears instead to originate and be maintained year-round behaviourally, possibly via observational learning.

The changes in niche breadth may also be accompanied by a loss of defensive traits (reviewed in Chapters 10 and 11). The general interpretation offered for these changes is that organisms released from antagonist interactions no longer require such defensive characteristics, and so these features are gradually lost. Sadly, the loss of defensive traits

commonly involves a lack of fear of, and ability to flee from, humans and terrestrial vertebrate predators: undoubtedly a contributing factor to the demise of many island species, as reviewed in Chapter 14.

Density compensation

Assuming that carrying capacity is invariant, the lower area-adjusted species richness of island assemblages raises the expectation of increased mean population sizes per unit area relative to equivalent mainland habitats. This insular effect is termed **density compensation** (Fig. 8.3) and it is linked to competitive release and the expansion of the realized niche space discussed above. It was recognized by Crowell (1962) in a study comparing birds on Bermuda with those of similar habitats on the North American mainland. Density adjustment may occur to an apparently excessive degree, in which case it is termed **density overcompensation** (Wright 1980). This may occur where a species has undergone niche enlargement to occupy a wider range of habitats, foraging strata, techniques, or dietary components (MacArthur et al. 1972; Novosolov et al. 2016). In turn, these mechanisms may be underpinned by the compositional character of the island assemblage; for example, the

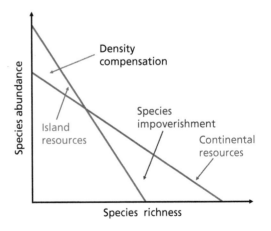

Figure 8.3 Density compensation in islands is a product of their lower species richness per unit area, which promotes the acquisition of available resources by the species present on the islands, allowing an increase in their abundance and thus density.
From Olesen et al. (2002), slightly modified.

absence of top predators, or of large-bodied competitors. In the former case, the reduced need for predator-avoidance behaviours may significantly improve the balance between energy expenditure and resource gain, while in the latter case, a similar bird biomass might be distributed among more numerous smaller-bodied bird species.

Insular density compensation has been demonstrated across taxa including birds, mammals, lizards, and fish. In their comprehensive evaluation of density compensation in lizards, Buckley and Jetz (2007) sourced data for 643 lizard populations (470 mainland and 173 island populations, encompassing 334 species) from around the world. Insular density compensation was found to be general, with island populations having average densities an order of magnitude higher than mainland population densities (1920 versus 128 individuals per ha). After statistically controlling for variation in energetic constraints, Buckley and Jetz (2007) found support for the role of ecological release from predation and competition in driving the observed density compensation. Due to high collinearity, it was not possible to determine which of the two was dominant. Novosolov et al. (2016), in their analysis of 346 lizard species on islands and mainlands, also found island lizard populations were denser. However, the subset of lizards occurring both on mainlands and islands had higher mainland densities than those restricted to the mainlands, which could indicate that mainland population density plays a role in island colonization, as (i) species with higher mainland population density will have more potential propagules, and (ii) the traits that facilitate high population density may enhance insular colonization success (Novosolov et al. 2016).

Several other mechanisms have been suggested that may influence density patterns in addition to those mentioned, such as the apparent tendency of islands to have less variable climates than mainland areas (Chapter 3), which may result in reduced mortality rates and thus higher population density (Novosolov et al. 2016; and for additional potential mechanisms, see MacArthur et al. 1972; Wright 1980). It is hard, therefore, to generalize on the combination of factors involved. Moreover, the occurrence and degree of density compensation is

variable, and a single species may exhibit higher, lower, or comparable densities on different islands as compared with the mainland. This is, on reflection, unremarkable, as particular species experience different conditions on different islands, not only in the habitats available but also in relation to the presence/absence of particular predators, pathogens, competitors, food resources, or mutualists.

The phenomena of ecological release and density inflation do not apply universally to all island populations, but there appears to be good evidence for their operation for many island lineages, providing selective environments that encourage selective genetic adjustments in the island populations. From first principles, we can anticipate that they will be particularly prevalent on remote islands at early stages of faunal and floral colonization.

8.7 Character displacement

Islands are classic systems for the study of **character displacement**, the evolutionary divergence of species that possess overlapping niches and traits upon coming into sympatry, where they compete for resources. It is commonly marked by changes in the mean and variance of particular morphological traits in the species populations involved. What distinguishes this form of niche shift is the causal interpretation: that it is competition between two ecologically similar varieties or species, which were initially allopatric but have attained sympatry, resulting in selection (in one or both) away from the region of resource overlap (Brown and Wilson 1956; Diamond et al. 1989; Stuart and Losos 2013). Darwin (1859) termed this 'divergence of character', and it has also been termed 'character coevolution' or 'coevolutionary divergence', although there is actually not one, but a suite of related theoretical ideas centred around this notion (Otte 1989; Pfennig and Pfennig 2020). It is a key concept in models of evolutionary diversification and adaptive radiation (Chapter 9). Generally, character displacement is theorized to involve genotypic change as a result of natural selection favouring individuals that are dissimilar from their sympatric competitors. However, in cases it may also involve, at least in the early stages, a role for phenotypic plasticity prior to divergent characteristics becoming genetically fixed (Pfennig and Pfennig 2020).

Arguably two of the most convincing examples of character displacement come from islands: Caribbean *Anolis* lizards and the Galápagos (Darwin's) finches. For the former, shifts in perch height and diameter were found to be consistent with competitive effects. Species of similar size were found to affect one another more than dissimilarly sized species, while larger species affected smaller ones more than the reverse (Schoener 1975; Losos 2009). With respect to the latter, character displacement has been used to explain the contrasts in beak size and shape observed in several species of Galápagos finch. As described by Lack (1947a) and

shown in Fig. 8.4, *Geospiza fuliginosa* and *G. fortis* showed similarity in beak size when occurring in allopatry on two separate islands, but when occurring in sympatry on two other islands they displayed non-overlapping beak size distributions, spanning a broader overall array of sizes. A further contrast is seen when three ground finches occur together: in this case, the range in beak sizes in *G. fortis* is attenuated compared with the two-species situation. Subsequent work demonstrated a strong link between beak size and seed diets, supporting inferences of competitive displacement linked to available food resources (Grant and Grant 2006). In 1982, the situation we have just described changed when *G. magnirostris* established on Daphne Major

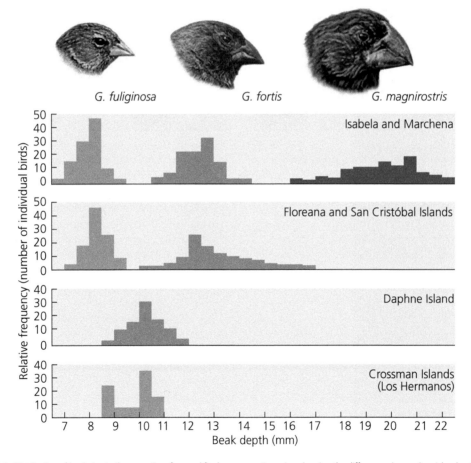

Figure 8.4 Distribution of beak size in three species of ground finches, genus *Geospiza*, showing the difference observed on islands where two or three of them occur in sympatry compared with islands where only one occurs.
From Lomolino et al. (2017), after Lack (1947a).

Island, bringing it into direct competition with *G. fortis* on that island (Grant and Grant 2006). Following the build-up of the population of *G. magnirostris*, a period of drought in 2003–2004 led to a significant reduction in resources, with declines in numbers of both finches and strong selection within *G. fortis* towards smaller beaks, directly attributable to interspecific competition with the larger-beaked congener. Moreover, genomic analysis indicated that selection for beak size was operating on a particular genomic region containing the *HMGA2* gene, with a different locus (ALZ1) being linked to variation in beak shape—beak size and shape being the key traits involved in the evolutionary radiation of the Galápagos finches (Grant and Grant 2006; Lamichhaney et al. 2016).

The fortuitous occurrence of 'natural experiments' on islands, such as reported above on Daphne Major Island, can provide some of the most convincing evidence for phenomena such as ecological release and character displacement. Another example is provided by Diamond et al. (1989), who reported a natural experiment demonstrating divergence within populations in sympatry for less than three centuries. In the mid-17th century, Long Island, off New Guinea, erupted violently and destructively, initiating a primary succession on both Long itself and the nearby islands of Tolokiwa and Crown. Thus, the bird populations of the three islands can be taken to have been founded roughly 300 years ago. The avifauna contains two different-sized species of *Myzomela* honeyeaters: *M. pammelaena* and *M. sclateri*. They are the only islands on which the two coexist. Allopatric populations on other islands can be identified as the probable sources for the founding populations of the Long group. Thus, the serendipitous 'experiment' has a maximum duration of about three centuries, for which two otherwise allopatric species have co-occurred. Furthermore, within the Long group, the two species exhibit microsympatry, occurring together in all habitats across the elevational gradient, often being found in the same flowering tree. Diamond et al. (1989) found that the Long populations are significantly more divergent from each other than are the allopatric source populations. The larger species, *M. pammelaena*, is bigger still on Long

than in the source populations, and the smaller *M. sclateri* is even smaller. Comparing samples of the two species from allopatric populations, the weight ratio was found to lie between 1.24 and 1.43, but for the sympatric Long Island populations the value was 1.52.

Dufour et al. (2018) provide an example of an anthropogenic insular 'experiment', arising from the introduction of *Anolis cristatellus* to the island of Dominica, where there is a functionally quite similar native lizard, *A. oculatus*. When the study was carried out, it was found that there were areas where the two species occurred in sympatry and others where they were allopatric. When occurring in sympatry, there was strong habitat divergence: *A. oculatus* using higher perches and *A. cristatellus* lower perches, compared with allopatric populations of each species. This habitat divergence was accompanied by morphological changes, but only in sites where the species had been sympatric for over a decade.

Most examples of character displacement relate to *ecological* character displacement arising from resource competition. However, *reproductive* character displacement has also been reported, whereby species diverge in reproductive characteristics, which may have the effect of minimizing deleterious reproductive interactions (Pfennig and Pfennig 2020). For example, there is some evidence for reproductive competitive displacement in Hawaiian *Laupala* crickets, where divergence in song is greater between coexisting populations than between allopatric populations (Otte 1989).

While the overall idea of character displacement is generally widely accepted, it can be challenging to distinguish it from alternative hypotheses (Pfennig and Pfennig 2020). Stuart and Losos (2013) suggested six criteria that need to be met to be confident of the diagnosis. Using these criteria in a review of 144 purported cases of character displacement, they concluded that only nine (including *Anolis* lizards and Darwin's finches) had strong evidence supporting ecological character displacement in favour of alternative explanations. We know that character displacement happens. It is reasonable to posit that it is far more prevalent on islands than this rightly cautious analysis reveals.

8.8 Mechanisms of speciation

When examining the mechanisms of speciation rather than focusing on the geographical circumstances, it becomes clear that there are many different processes at work. Rosenzweig (1995) summarizes these processes under the three frameworks: **geographical** or **allopatric speciation**, **competitive speciation**, and **polyploidy** (involving an increase in the chromosome number) and we also distinguish roles for **interspecific hybridization** and for **apomixis** as relevant phenomena to discuss herein.

Allopatric or geographical speciation

The essence of this model has been summarized by Rosenzweig (1995, p. 87):

- A geographical barrier restricts gene flow within a sexually reproducing population.
- The isolated subpopulations evolve separately for a time.
- They become unlike enough to be called different species.
- Often the barrier breaks down and the populations overlap geographically once again, but do not interbreed (or they interbreed with reduced success).

This is a highly generalized model and the more we learn of island evolution, the more variations we encounter. The first step of this model suggests a vicariant event (barrier formation). This scenario may apply to certain ancient continental fragment islands (e.g. Madagascar) and to former land-bridge islands. In such cases, the island population is assumed not to have experienced a bottleneck founding event. For true oceanic islands, the starting point is over-water dispersal, from another island, archipelago, or continent, typically involving a founding population of one or a few individuals, constituting a genetic bottleneck (above). In either scenario, the island population develops in isolation from the mainland source population. Subsequent environmentally forced bottlenecks, genetic-drift effects, and selective pressures in the novel island context then provide the engine for further differentiation from the ancestral form.

The failure of the isolating barrier may not be a frequent feature for oceanic islands, but can happen, for instance, as a function of sea-level falls connecting nearby islands, or providing additional 'stepping-stones' between a given island and the mainland. This can enhance opportunities for **double invasions**, where a second invasion of the original colonist or a closely related form occurs. Such repeat arrival events may lead to interbreeding and reduced evolutionary distance of the insular form, or there may be enough reproductive and niche separation to permit the two forms to persist as (mostly) non-interbreeding forms within the island. This was the hypothesis put forward by Lack (1947a) to explain the occurrence of the later arriving common chaffinch *Fringilla coelebs* alongside the endemic blue chaffinches (*Fringilla teydea* and *F. polatzeki*). Analyses of mitochondrial and genomic DNA have supported this idea (Fig. 8.5a, b), with the interesting twist that *F. coelebs* appears to have followed a circuitous pathway, first arriving on the Azores from (most likely) the Iberian mainland, before colonizing Madeira and then finally the Canaries (Recuerda et al. 2021), with genetic diversity progressively decreasing along the way, indicative of repeated founder effects. While all of these populations are currently considered to represent a single species, Recuerda et al. (2021) argue that each archipelago's population is phenotypically and genetically distinct, and thus diversification of *F. coelebs* across Macaronesia represents an example of a species-level allopatric (mostly non-adaptive) radiation (Fig. 8.5b). Populations have even diverged on individual islands within the Canaries, but these are more likely to represent subspecies than full species. For further examples of a range of colonization scenarios for other Atlantic island lineages see: Kvist et al. (2005), Stervander et al. (2015), and Sangster et al. (2022).

The barriers that exist between islands within an archipelago are undoubtedly important features of remote oceanic island groups. As Rosenzweig (1995, pp. 88–89) puts it: 'Suppose propagules occasionally cross those barriers, but usually after enough time for speciation has passed. Then the region and its barriers act like a speciation machine, rapidly cranking out new species.' The radiations that characterize many taxa on the Hawaiian Islands

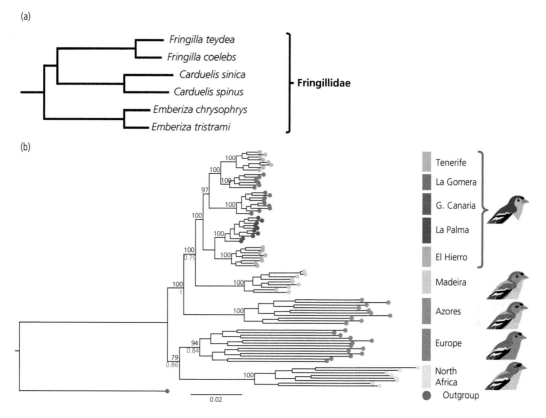

Figure 8.5 Phylogenies for Macaronesian chaffinches. (a) Excerpt from a phylogeny based on 33 species of Passeriformes, selected to illustrate the relationship between the Canarian blue chaffinch (*Fringilla teydea*) and the common chaffinch (*F. coelebs*) based on mitochondrial protein-coding sequences. Extracted from figure 4 of Marshall et al. (2013). (b) Phylogenetic tree based on genome-wide neutral SNP loci for *F. coelebs*, using *F. teydea* as the outgroup. Numbers in black are node support values, and in red are the average support values for species delimitation (using the mPTP method). Sketches on the right depict the main phenotypic differences between forms, with the Canary Islands represented by the La Palma form (figure 2 of Recuerda et al. 2021; see source paper for further details of methods).

reflect this, with interpretations for the relationships among the estimated 1000 species of Drosophilidae flies involving a number of phases of interisland movements as a part of this most spectacular of radiations (Carson 1992; Hembry et al. 2021). There are two major Hawaiian drosophilid lineages, which are sister to one another, *Drosophila* and *Scaptomyza*. Having both originated and diversified within Hawaii, *Scaptomyza* has also dispersed not only to other archipelagos across the Pacific but also to colonize all continents apart from Antarctica: the most dramatic example of 'back' and indeed *de novo* colonization to continents yet demonstrated for insects (Section 9.8; Lomolino et al. 2017; Hembry et al. 2021).

An illustration of the contrast between the archipelago context and the single remote island is provided by the Galápagos (Darwin's) finches. The Geospizinae, nested within the tanager family, is found in the Galápagos archipelago, which comprises nine islands larger than 50 km^2 (and a larger number of smaller islets), with predominantly desert-like conditions (Fig. 8.6). Most closely related to the dull-coloured grassquit (*Asemospiza obscura*) of the South American mainland, on the Galápagos they derive from a single colonist species and have radiated into between 15 and 17 species across four genera (the ground, tree, vegetarian, and warbler finches; Farrington et al. 2014). The more isolated islands support higher incidence of

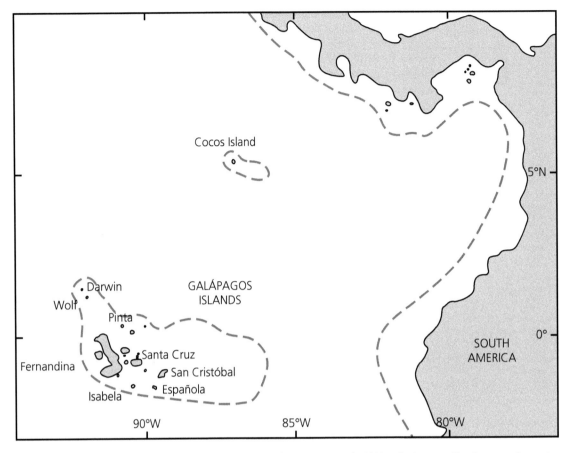

Figure 8.6 The Galápagos archipelago and Cocos Island. The dashed line approximates the 1800-m depth contour. Five degrees on the equator represent approximately 560 km.
Redrawn from Figure 9.1 of Williamson (1981).

endemic species and subspecies of the ground and tree finches, in part, it is thought, because during the Last Glacial Maximum, some of the central islands were connected whereas the peripheral islands have been constantly isolated, limiting intraspecific gene flow and driving population divergence (Farrington et al. 2014). The lineage is found also on Cocos Island, a lone, forested island of about 47 km², which is, just, the nearest land to the Galápagos (Fig. 8.6). Here there has been no possibility of the island-hopping that appears to characterize archipelago speciation, and the group is represented by a single endemic species *Pinaroloxias inornata*, the only member of its genus and one of just four species of land bird on the island (Valente et al. 2020). Recent analyses of genetic data indicate

that the Cocos finch was an early divergence within the radiation and is not a tree or ground finch as originally thought (Farrington et al. 2014).

Competitive speciation

Competitive speciation is an umbrella term used by Rosenzweig (1995) for a group of related models of sympatric speciation. The idea here is that a species expands its niche to occupy an unexploited ecological opportunity, followed by the sympatric break-up of that species into two daughters, one in essence occupying the original niche, and the other the newly exploited one. The expansion happens because of a lack of competition with other species, a feature common to remote island faunas and floras. But the break-up into separate species happens because of increased competitive pressure between those portions of the population best able to exploit the two different niches. The operation of these processes must take place over the course of many generations, and observation of the complete process in the field is thus not an option. Rosenzweig illustrates the idea by means of a simplified 'thought experiment', the essence of which, for an island bird radiation can be rendered as follows.

A colonist species will have a particular feeding niche, determined morphologically and behaviourally, and reflected in features such as bill size and shape (e.g. Pigot et al. 2020). According to theory, such characters may be expected to have a unimodal distribution of values. The environment of the island contains empty niches, unexploited ecological opportunities, which for the purpose of our experiment are taken to be distributed bimodally; that is, there are two resource peaks. The species, unconstrained by competition with the full range of continental species, expands from its original modal position somewhere along this resource space so that it occupies both of these niches. Individuals with genotypes that match best to one or the other of these niches become numerically dominant within the population, which has by this stage grown close to its carrying capacity. At this point, disruptive selection kicks in, as the mid-range phenotypes have lower fitness relative to those that are better suited to one or the other of the main feeding niches. Those individuals adapted primarily to a peak in the resource curve may also exploit the valleys either side of that peak. If they do so at all successfully, no opportunity remains for valley phenotypes between the two peaks. They will be few in number in the population, and their offspring will have fewer resources to tap. Although they may also reach up to the peaks, they will not be effective competitors in those portions of the resource continuum. Valley genes will thus have little success and will be bred out, as in time members of each 'peak' population that can recognize others of their type (and breed with them alone) will be at a selective advantage. Therefore, isolating behavioural mechanisms develop and at this point the lineage can be viewed as having split. By extension of this line of reasoning, if there are

more than two resource peaks, larger radiations may occur. This idea, of 'adaptive landscapes', analogous to topographically heterogenous landscapes, was originally introduced by Sewall Wright (1932) and has been incorporated into a number of different evolutionary models (see e.g. Futuyma 1986).

It is difficult to judge the importance of competitive speciation on islands. It can rarely be safely judged merely by investigating the degree of geographical overlap between sibling species in ecological time. This is because, given the probabilistic nature of dispersal and the magnitude of past environmental change, populations currently isolated from one another may once have been sympatric and vice versa (Box 8.2). Indeed, it may commonly be the case that **archipelago speciation**—diversification within archipelagos—involves a mix of allopatric and sympatric phases of subtly varying combination from one branch of a lineage to the next (Clarke and Grant 1996; Schluter 2000; Losos and Mahler 2010; Machado 2022). We can again approach this through a thought experiment.

Imagine that a bird species has once again colonized an archipelago with two available resource peaks (let's call then 'large fruits' and 'small nuts') but the ancestral population is highly specialized and adapted to only the large fruits. And, in this instance, interbreeding prevents natural selection from moving part of the population down into the 'valley' and up the other adaptive peak. Genetic drift, which may also operate to shift trait values, also fails to do so because the population size is too large and interbreeding too effective. However, the species successfully colonizes a second island in which the large fruits resource is missing but small nuts are abundant. Here, the population evolves to match the small nuts resource, with consequent changes in bill morphology and also gradual changes in reproductive behaviour. Subsequently, this population (re)invades the original island, successfully establishing a sympatric population specializing in the small nut niche. Depending on the degree of change that occurred in allopatry and the differences in the resource peaks and in the avifauna of the two islands, the two forms may evolve further once they have come into sympatry, via character displacement (Section 8.7; Aguilée et al. 2013). The distribution of the species in the original island may suggest sympatric speciation to have occurred even when a big part of the evolutionary divergence took place in allopatry. As both island configurations (through geo-environmental dynamics, sea-level changes, etc.) and species distributions can change over time, distinguishing allopatric from sympatric speciation is difficult from distributional data alone. The availability of detailed palaeo-geographical reconstructions, well-specified molecular phylogenies, and recently developed speciation models are increasingly permitting better resolution on such questions (see Ali and Aitchison 2014; Warren et al. 2015).

Hybridization

It has become increasingly realized that hybridization has played an important role in evolutionary dynamics, including in island settings (Givnish 2010; Caujapé-Castells et al. 2017; Meier et al. 2017, 2019; Lamichhaney et al. 2018). Hybridization is the production of offspring by parents of two different species, populations, or genotypes. Hybridization events involving two different species can lead: (i) essentially nowhere, as the individuals fail to contribute to future generations; (ii) to the incorporation of novel genes and/or alleles of one taxon into the gene pool of another; or (iii), in cases, to the formation of a new species, which can occur either with, or without, whole-genome duplication, respectively termed **homoploid hybrid speciation** and **polyploid hybrid speciation**.

Introgressive hybridization occurs when the initial F_1 hybrid is able to back-breed with one of the parent species, resulting in gene flow into one taxon from the other, increasing the genetic diversity on which natural selection can act. Within some island radiations, hybridization is more free-flowing, involving interbreeding between two previously isolated and genetically differentiated populations/taxa, in which case it is termed **admixture**. Admixture may often lead to the collapse of difference between two populations and thereby prevent insipient speciation from completion (Kleindorfer et al. 2014). Evidence of introgressive hybridization, admixture, and incomplete lineage sorting has been reported within the evolution of

particular insular clades, such as the spectacular radiation of 25 subgenera and 264 Macaronesian species and subspecies of *Laparocerus* weevils described by Machado (2022). Hybridization can result in **transgressive segregation** (Rieseberg et al. 1999), where the phenotype of the hybrid contains characteristics outside the range observed in either parent. Hybrids that result through transgressive hybridization may thus be able to exploit different niches to the parent species, or potentially even outcompete the parents in the ancestral niche. The process may also contribute to further speciation as it broadens the range of adaptations and genetic stock present in a lineage. Hybrid speciation facilitated by transgressive expression of ecological traits has been proposed as a diversification mechanism for some vascular plants on New Caledonia (e.g. *Codia*), some African rift-lake cichlids, and Sulawesi sailfin silversides (Givnish 2010).

White et al. (2018) provide an insular example of homoploid hybrid speciation from Tenerife, where two morphologically distinct members of the plant genus *Argyranthemum* (*A. lemsii* and *A. sundingii*) each derive from crosses involving the same two parental species. Another example of homoploid hybridization leading towards speciation has been described within the Galápagos finches. In 1981 an immature male finch, subsequently determined as belonging to *Geospiza conirostris*, was observed on Daphne Major, over 100 km from the nearest island on which the species occurs (Lamichhaney et al. 2018). Over the next 31 years, the survival and breeding patterns of its descendants were followed (e.g. Grant and Grant 1994), providing an unparalleled study of the process of homoploid hybrid speciation. The immigrant male initially bred with a *G. fortis* female, generating a hybrid offspring, as did one of its offspring. From the second generation onward, descendant individuals of this immigrant male behaved as an independent lineage relative to the other species on the island and, despite further inbreeding, possessed high fitness based on survival and reproductive output. The hybrid lineage (termed the 'Big Bird' lineage) is intermediate and slightly overlapping between the two parent species in terms of body mass. But in beak morphology, the Big Bird lineage is closer to *G. conirostris* (which does not occur on the island) and has undergone transgressive segregation, such that it occupies previously unoccupied morphological space relative to its sympatric competitors (Fig. 8.7). Given the changes in morphology and song relative to the parent species, the Big Bird lineage likely represents an incipient hybrid species that has become reproductively isolated from its parents in only a few generations (Lamichhaney et al. 2018).

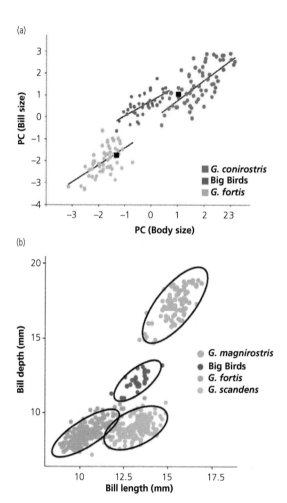

Figure 8.7 The distinctive morphology of individuals of the 'Big Birds' lineage of Galápagos finches resulting from an initial hybridization event between a colonist *Geospiza conirostris* and resident *G. fortis* on the island of Daphne Major, showing (a) principal components analysis of body size and bill size (black squares represent the two parents) and (b) a plot of bill length versus depth for coexisting competitor species (extracted from figure 3 of Lamichhaney et al. 2018).
Reprinted with permission from AAAS.

A further distinction can be drawn as to whether hybridization (admixture and/or introgression) occurs principally during or mainly prior to the radiation, respectively known as the syngameon hypothesis and the hybrid swarm hypothesis. The Galápagos finches provide an example of an important role for hybridization during a radiation (above; Lamichhaney et al. 2015). Whereas a hybrid swarm origin (where the hybridization of two distinct lineages provides the genetic variation which then facilitates the radiation) has been invoked in studies of the Hawaiian silversword alliance and *Laupala* crickets, in addition to some radiations of cichlids within the African rift lakes (Seehausen 2004; Givnish 2010; Meier et al. 2017, 2019; Chapter 9). The propensity for hybridization, when coupled with ecological opportunity, may explain why some lineages radiate extensively and others do not (Meier et al. 2017, 2019). Experimental work has shown that there is a 'sweet spot' in regard to the degree of divergence between hybrid parent species, which will vary across clades and taxa. Divergence needs to be above a minimum amount, such that the parents will have evolved novel and advantageous genotypes, but also below a maximum amount, beyond which genetic incompatibilities prevent admixture (see Gillespie et al. 2020).

With the introduction of so many non-native species to islands globally, the opportunity for hybridization between insular endemic forms and related but anthropogenically introduced species is increasing, with potentially important implications for insular conservation (McFarlane and Pemberton 2019; see Section 14.6).

Polyploidy and apomixis

One important class of sympatric speciation is through polyploidy, a condition linked to hybridization and to apomixis (below) that is comparatively common in plants and many invertebrates, but not in higher animals: it is unknown in mammals and birds, for instance (Grant and Grant 1989; Rosenzweig 1995). Polyploid species are those that have arisen by an increase in chromosome number, usually due to an error in the meiosis process. Polyploidy can result in instantaneous sympatric speciation as, although individuals of the new species can produce viable offspring with each other, reproduction involving individuals from the new species and the parent species results in sterile (triploid) offspring. Polyploids have been theorized to represent superior island colonists as they tend to have higher genetic diversity (relative to diploids) and an increased ability to survive with very small initial population sizes; this likely also explains the observation that polyploidy is relatively more common in invasive plants (Meudt et al. 2021). Polyploids have also been shown to have increased environmental tolerance and rates of vegetative reproduction, factors that will also improve their ability to colonize islands.

Two main forms of polyploid can be distinguished. **Autopolyploids** have twice the chromosome complement of a single-parent species and **allopolyploids** result from a cross between two species and have the chromosomes of both. It has been estimated that around a third of flowering plant species are of recent polyploid origin, with allopolyploids more common than autopolyploids (Rosenzweig 1995; Rice et al. 2019). Polyploidy increases with latitude on both continents and islands (if less clearly so on the latter), a pattern likely to be linked to declining temperatures and increased importance of perennial herbs in the flora (Rosenzweig 1995; Rice et al. 2019). In an analysis of the prevalence of polyploidy in the approximately 800 ecoregions globally, Rice et al. (2019) reported polyploidy percentages (of the species they were able to evaluate) of 32% for the Canary Islands (dry woodlands and forest; based on $n = 1098$ species), 44% for Madeira (evergreen forests; $n = 77$ species), 46% for the Galápagos (scrubland mosaic; $n = 26$ species), an average of 50% for Hawaii (across four tropical ecoregions; $n = 566$ species), an average of 46% for New Zealand (across 12 ecoregions; $n = 7833$ species), and an average of 23% for Madagascar (across 6 ecoregions; $n = 7154$ species). The Juan Fernández Islands are notable for having only around 3% polyploids in the flora evaluated to date (see Section 11.9; Rice et al. 2019; Meudt et al. 2021).

Although polyploids have been good island colonists and speciation through polyploidy evidently does occur on islands, it does not appear to be a key island speciation process in plants. On the other hand, there is some evidence that recent polyploids, when successful in colonizing islands, often show further success in diversification *in situ* (Section 11.9; Meudt et al. 2021).

Apomixis is uniparental asexual reproduction (i.e. production of seeds) without fertilization (Majeský et al. 2017). This form of asexual reproduction may enable isolated genotypes (such as newly arisen polyploids) to persist despite a dearth of mates or pollinators, allowing a new population to establish and grow, and thus subsequently to diversify genetically by the accumulation of mutations, or by the resumption of sexual behaviour (Koutroumpa et al. 2021). It was once presumed that apomictic plants, seemingly lacking adaptive variation, represented evolutionary 'dead-ends'. This view has been re-assessed given: (i) the genetic diversity often observed in apomictic populations, (ii) the discovery that apomixis is not an irreversible derived trait and transitions from apomixis to sexuality are more common than originally presumed, at least in certain clades, and (iii) observations that certain taxa are still able to reproduce sexually if apomixis fails, reducing genetic decay (Majeský et al. 2017; Koutroumpa et al. 2021). Given that asexual reproduction does not involve gene flow and apomictic taxa can hybridize with both sexual taxa and other apomictic taxa, resulting in novel genotypes and phenotypes, delineating species using many of the proposed species concepts (Section 8.2) is complicated.

The role of apomixis in island speciation remains poorly understood, but in some taxa it may be an important component in their diversification, especially in connection with hybridization and polyploidy. For example, Koutroumpa et al. (2021) found that apomixis was associated with accelerated diversification rates in Mediterranean *Limonium* (sea lavender), a cosmopolitan angiosperm genus with high diversity in the Mediterranean. Although extinction rates were either comparable or higher in apomicts (perhaps reflecting their lower genetic variation) than in sexual taxa, so too were speciation rates.

8.9 Lineage subdivision

We have one more framework to introduce, as we call upon it in the next chapter. It concerns where cladogenesis occurs, as represented in the phylogenies we construct. Following the natural colonization of an island by a species derived from a mainland source pool, a discrete number of options apply: (i) the colonization event fails; (ii) the species persists unchanged (evolutionary stasis), or changes insufficiently to be recognized as a new insular species or subspecies—that is, it becomes a native but non-endemic species; (iii) the previous scenario is followed but accompanied by extinction from the source pool to create an insular palaeoendemic; (iv) the insular population evolves gradually to the point that it is sufficiently distinct from the source population that it is recognized as an island endemic species—that is, it becomes an insular endemic; (v) it diversifies *in situ* to produce two or more distinct species, one or both of which are recognized as insular endemics.

Even on remote islands, many colonists fail to evolve to attain novel species status (option (ii) of our list). This can reflect recurrent colonization of particularly mobile island species, or that insufficient time has passed for evolutionary changes to reach a point warranting species or subspecies designation (Section 8.2). In many cases, detailed analyses reveal lengthy periods of geographical separation in the genetic structure of insular vs mainland populations (Box 8.4), so that at some level the 'native non-endemic' on the island has nonetheless become an 'evolutionarily significant unit' (ESU) and in time and with further scrutiny of genetic evidence and traits, such ESUs can be elevated to subspecies or species level (e.g. Ó Marcaigh 2022; Sangster et al. 2022).

With reference to options (iv) and (v), as a way of distinguishing and describing the amount of insular or archipelagic (if the analysis is done at that level) diversification involved, Todd Stuessy and colleagues (e.g. Stuessy et al. 1990, 2014; Takayama et al. 2015) promoted the use of the labels anagenesis and cladogenesis (Fig. 8.8). The application of this framework has proved useful, but it should be noted that: (i) classification of island species into this framework is challenging for groups lacking robust phylogenies, (ii) it assumes contemporary

---

Box 8.4   Haplotype network analysis and the Macaronesian buttercup

To understand evolutionary processes, we need to delve below the species level. Phylogeographical analyses permit the investigation of key island evolutionary questions at the level of populations. One such approach is the use of haplotype network analysis, an approach for analysing and visualizing relationships between individual genotypes.

The Macaronesian buttercup (*Ranunculus cortusifolius*) is one of a very few endemic angiosperms shared by the Azores, Madeira, and the Canaries. Until recently it was unknown whether this distribution resulted from one, or multiple, origins and what the relationships between the populations might be. Analysis of the haplotype data has now demonstrated that the Macaronesian populations probably derive from a single colonization event from the western Mediterranean during the late Miocene period (Williams et al. 2015). The data further show that none of the haplotypes were shared between archipelagos, indicating that the distances between them have been sufficient to promote a degree of diversification within the species across

Macaronesia (Box Figure 8.4). Tenerife and La Gomera in the Canaries harbour the more genetically diverse populations, with four similarly abundant haplotypes each, three of them single-island endemics, and the fourth (7H, red) shared with the rest of the Canaries. The dominant haplotypes of Gran Canaria (8H) and El Hierro (5H) are also unique, and only one mutational step away from the widely distributed 7H, indicative of a recent divergence. Madeira's populations have three haplotypes all exclusive to the island, whereas the Azores, with the exception of São Jorge (which shares haplotypes with other islands), have a single haplotype on each island. Grouping similar haplotypes indicates that distinct lineages of the species occur on each archipelago, with two or three distinct lineages being present on the Canaries.

Overall, these analyses provide an example of how haplotype network analysis can reveal patterns of cryptic genetic differentiation between populations of a species distributed within an island or across multiple island groups.

Box 8.4 *Continued*

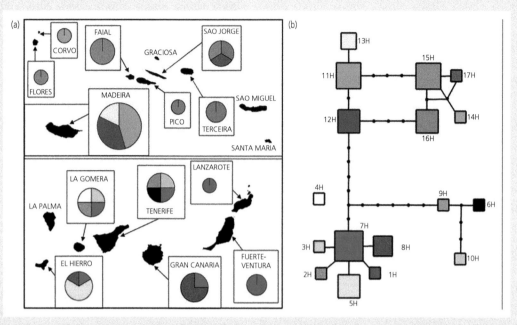

Box Figure 8.4 Analyses of plastid haplotypes of *Ranunculus cortusifolius* in Macaronesia, showing (a) contribution of haplotypes by island where pie-chart size indicates number of specimens, and (b) the haplotype network where each line indicates a difference of one base pair. Codes indicated haplotype labels.
Figure 1 from Williams et al. (2015).

patterns reflect developmental origins accurately (e.g. a group may radiate but collapse to one species over time), and (iii) this framework sets aside cases where lineages converge or cross, via hybridization (above). For a critique see Emerson and Patiño (2018).

Anagenesis refers to cases where the single insular species has evolved to become an endemic without branching *within* the island concerned (i.e. option (iv)). The process of anagenesis thus does not change the species richness of an island. Recent compilations for plants on oceanic islands indicate that the majority of insular colonists either remain native non-endemics, or they evolve to become an insular anagenetic endemic (as Fig. 8.8), especially when viewed at the level of individual islands rather than archipelagos. In the absence of systematic molecular data, a crude way of estimating the

proportion of speciation events fitting the anagenesis model is to count the number of cases where a genus is represented on an island/archipelago by a single endemic species. On this basis, Stuessy et al. (2006) estimate that for 2640 endemic angiosperm species from 13 island systems, about one quarter of speciation events match the anagenesis model. Based on more detailed analysis, it has been estimated that 88% of the endemic plants of Ullung Island have originated anagenetically, and about 36% of the endemics of the Juan Fernández Islands (Takayama et al. 2015). Anagenesis is particularly important on low-elevation islands of limited habitat diversity, such as Ullung Island (Stuessy et al. 2006; Igea et al. 2015; Valente et al. 2020) and on remote islands, although the relative contribution of cladogenesis also tends to increase with isolation (Phillimore 2013; Valente et al. 2020). NB Where

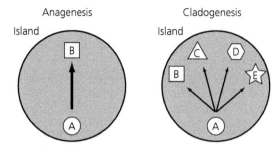

Figure 8.8 Island endemics can be classified as arising through anagenesis or cladogenesis. Anagenesis is where a colonist has evolved sufficiently after colonization to be morphologically and genetically distinct from the ancestral form in the source pool, but where there is only ever one island species. Cladogenesis is where the colonist diversifies into two or more separate species *in situ*.
Diagram from Takayama et al. (2015): their figure 1 (Creative Commons CC BY).

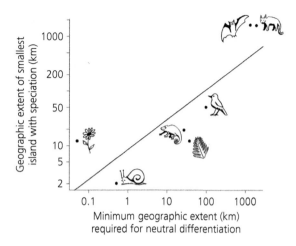

Figure 8.9 Meta-analysis of speciation from 64 island groups and from gene flow data, suggest the minimum island area requirements and distance required for neutral genetic differentiation within islands to permit *in situ* speciation (cladogenesis) for angiosperms, bats, birds, carnivores, ferns, lizards, Microlepidoptera, and land snails.
Figure from Givnish (2010), after Kisel and Barraclough (2010).

applying this framework at archipelago level, it may be possible to describe *archipelagic* cladogenesis within which speciation is anagenetic at the individual *island* level.

The process of cladogenesis, or *in situ* diversification (option (v)), can be adaptive (driven by natural selection and ecological opportunity), or non-adaptive (driven by founder effects and genetic drift). The prevalence of cladogenesis tends to increase with island area, elevational range, and habitat diversity and, of course, to increase with isolation of the island or archipelago (e.g. Losos and Schluter 2000; Valente et al. 2020; Burns 2022). Islands smaller than a particular size threshold fail to provide sufficient space and resources for populations to differentiate *in situ* and persist (Fig. 8.9). The minimum size limit permitting populations to differentiate and persist is imperfectly resolved but is known to vary with taxon characteristics that in turn link to population density and thus overall population size. For example, within-island speciation only makes an appreciable contribution to anole diversity on Caribbean islands > 3000 km^2 (Losos and Schluter 2000).

Generally speaking, even on the most remote islands, relatively few lineages experience explosive radiation and although we can generate statistical and process models that do well at replicating the emergent macroevolutionary patterns on islands, we have yet to fully understand what trait combinations control which lineages they are. It is to these

and other emergent patterns of island evolution that we turn in the following three chapters.

8.10 Summary

In order to understand the literature on evolutionary change, it is necessary to have a basic familiarity with the currency of evolution and knowledge of the problems associated with it. This chapter therefore begins by discussing the species concept, noting a gradual shift from morphological criteria to the use of the biological species concept (focused on reproductive isolation) towards increasing focus on genetic criteria. Advances in phylogenetic techniques and their application have revolutionized our understanding of evolutionary island biogeography. Molecular phylogenies are now available for many insular groups, enabling biogeographers to test competing hypotheses and models of island colonization, extinction, and speciation, although we stress that these phylogenies themselves should be seen as working hypotheses.

There are a number of alternative frameworks for describing how islands come to support endemic taxa. In this chapter, we draw a distinction between geographical, mechanistic, and lineage subdivision frameworks. Most island endemism can be attributed to evolutionary change following colonization, the main theme of this chapter, but we note that some insular endemics are relictual or palaeoendemic species that have changed relatively little *in situ* while becoming extinct from the source region.

Evolutionary filters of island life commence with the challenge of dispersing long distances to reach isolated islands, which leads to founder effects, the mostly random loss of genetic variation that occurs through very small founding populations. Over time, such population bottlenecks may persist or recur within the insular system, with cumulative impact on the genetic diversity of the population, although the scale and significance of founder effects remains a contested issue. Moreover, bottlenecks can pave the way for the accumulation of novel genetic combinations through mutation and recombination as populations expand (and sometimes hybridize).

Following the founding event, a colonist species may encounter an absence of key competitors and other interacting organisms, leading to niche expansion, which implies a reduction in the degree of resource/habitat specialization. Ecological release may be reflected in a general pattern of density compensation, arising from, for example, (i) there being fewer species to share resource space with, (ii) the use of novel resources, and (iii) the reduction in the costs of predator avoidance behaviours. Competitive effects can also be inferred to occur between ecologically similar (often related) species, through comparison of insular situations involving different combinations of species in space or time. Character displacement refers to niche shifts and changes of niche breadth attributable to competitive interactions and can be reflected in changing distributions of key functional traits. Such changes involve combinations of natural selection and sexual selection. While recognizing the importance of testing such ideas rigorously, we speculate that changes in competitive interactions are actually far more important in insular evolution than can be definitively shown with available data.

Such microevolutionary steps can contribute to species-level divergence. *In situ* speciation can occur in a condition of geographical isolation (allopatry) and/or in conditions of distributional overlap (sympatry). Moreover, these concepts can be applied on different scales: between island and mainland populations, within an archipelago, and within an island. Sympatric speciation may occur through more than one route. Perhaps the classic island examples are those that follow the competitive speciation model of niche expansion and break-up into daughter populations exploiting different resource peaks. Evolutionary change leading to species formation may be accomplished by a variety of genetic mechanisms, including hybridization and through polyploidy. Many insular plants are descended from recent polyploids but there is no evidence for polyploidy events happening on islands at a disproportionate rate. Certain evolutionary processes (e.g. apomixis), while not always leading to the generation of new species directly, can influence long-term diversification dynamics through increasing the likelihood of small, founder populations persisting on islands.

The final framework considered in the chapter—anagenesis and cladogenesis—concerns the geographic location of lineage subdivision. These terms are used, respectively, to denote: (i) the situation where the branching is between the source pool and the island, with no lineage splitting occurring within the island; and (ii) where lineage splitting occurs *in situ* within the island. Anagenesis is of unheralded significance, especially in insular plants. We reserve detailed consideration of these distinctions and of the other emergent phenomena of island evolution to the following chapters.

CHAPTER 9

Evolutionary diversification across islands and archipelagos

9.1 In search of general models of insular and archipelagic evolution

Having set out the basic tool kit of island evolutionary mechanisms in the previous chapter, we now examine how these processes combine to produce emergent patterns of diversification across remote islands. As we have noted, insular speciation can take the form simply of anagenesis, the gradual evolution of the island population to the point that it is sufficiently morphologically and genetically distinct from the ancestral form that it is considered a distinct species and an island endemic. Although anagenesis is an important contributor to island endemism (Section 8.9), we turn our attention in this chapter to patterns and processes of *in situ* insular diversification, turnover, and (largely inferred) extinction. Cladogenesis within an island or archipelago can range from a single *in situ* speciation event, up to hundreds of species (within multiple subgenera) from a single colonist. Insipient or partial speciation may give rise to multiple subspecies, varieties, or evolutionarily significant units within such radiations.

We begin by introducing and exemplifying three general models of island evolution: the taxon cycle (encompassing diversification and turnover of many taxa), adaptive radiation, and non-adaptive radiation. The studies reviewed show that it is often hard to maintain a strict separation between these general models, as large radiations span lengthy periods of time, interact with dynamic insular settings, may involve taxon cycle dynamics, involve a mix of vicariance and dispersal events, episodes of allopatry, sympatry, and parapatry, and a blend of stochastic and selective (natural and sexual) mechanisms—with increasing evidence of an important role for hybridization within many insular radiations. We illustrate these points by reviewing a range of case-study systems, in the process building up a broader picture of insular diversification processes and how they may relate to island geo-environmental dynamics.

9.2 The taxon cycle

The taxon cycle is a model of insular biogeography and evolution first outlined by E. O. Wilson (1959, 1961) to account for distributional and ecological differences among multiple lineages of Melanesian ants. It invokes several components and has been modified as it has been applied to different taxa (mostly ants, birds, and lizards) and insular regions (Wilson 1959, 1961; Roughgarden and Pacala 1989; Ricklefs and Bermingham 2002; Economo and Sarnat 2012). Within taxon cycles, there is typically a directional flow of species from a richer source region towards peripheral islands (Fig. 8.1). Colonization of an archipelago by an early arriving species is followed by the evolution of reduced dispersal propensity (see Chapter 11), accompanied by habitat/niche shifts. The isolation of these populations within different islands of the archipelago gives rise to a series of allospecies or subspecies (i.e. at this point lineage diversification has occurred). Competitive interactions between early and later arriving lineages effectively pen in populations descended from earlier colonization events to increasingly localized strongholds, with

Island Biogeography. Robert J. Whittaker, José María Fernández-Palacios, and Thomas J. Matthews, Oxford University Press.
© Robert J. Whittaker, José María Fernández-Palacios, and Thomas J. Matthews (2023). DOI: 10.1093/oso/9780198868569.003.0009

the implication that populations—representing more-or-less distinct species or subspecies—will eventually become extinct. Changing interactions with pests and pathogens may also be critical to changing fitness, abundance, and ecological success over time (Ricklefs and Bermingham 2002). We expand on and evaluate this model in the following sections.

Taxon cycles in ants

In his seminal study of Melanesian ants, Wilson (1959, 1961) recognized differences in the ranges of the ponerine ants as a function of their habitat affinities. Marginal habitats (littoral and savanna habitats) were found to contain both small absolute numbers of species and higher percentages of widespread species. These data were interpreted as a function of changes in ecology and dispersal capability as ants moved from the Oriental Region, and particularly its rain forests, through the continental islands of Indonesia, through New Guinea, then out across the Bismarck and Solomon Islands, on to Vanuatu, Fiji, and Samoa.

Wilson bundled the species into three stages. Stage 1 species dominate the marginal habitats: open lowland forest, grassland, and littoral habitats (Fig. 9.1), but also occur in other habitats and typically have a continuous distribution rather than being broken into locally distinct races. They tend to be trail-making ants, nesting in the soil. In stage 2, ants 'return' to and become restricted to interior and montane forest, within which they are more likely to be found nesting in logs or similar habitats. Those that succeed in adapting to the inner rain forests eventually differentiate to species level within Melanesia, forming superspecies or species groups (Fig. 9.1 stage 2/3, point E). As they differentiate, they are liable to exhibit reduced gene flow across their populations on different islands. Stage 3 species are those that have proceeded to the point where the species group is centred on Melanesia and lacks close relatives in Asia. We may also recognize a fourth stage (see section on birds, below), marked by reduction in distribution to one or a very few populations in the higher-elevation interiors of large islands. Progress within the cycle is thus marked by the increasing restriction

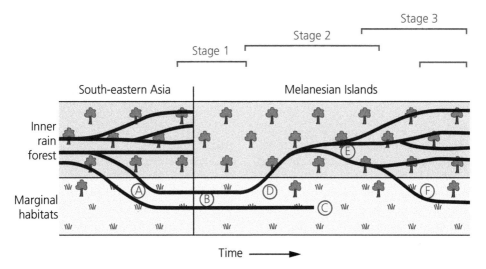

Figure 9.1 The taxon cycle, as hypothesized based on studies of ant species groups in Melanesia by E. O. Wilson (1959, 1961). The following sequence was postulated: stage 1—species adapt to marginal habitats (A) in the mainland source region (SE Asia) and then cross the ocean to colonize these habitats (B) in New Guinea. Some populations may become extinct in time (C). Other species enter stage 2—the ants invade the inner rain forests of New Guinea and/or surrounding islands (D). If successful in re-adapting to inner rain forest habitats they in due course diverge (E) to species level, resulting in multiple allospecies on different islands. Stage 3—diversification progresses within Melanesia, such that the lineage becomes centred in Melanesia. A few members of these lineages, especially those on New Guinea, may re-adapt to the marginal habitats (F), and expand secondarily.
Redrawn from Wilson (1959).

of individual species to narrower ranges of environments within the island interiors. Meanwhile, these earlier colonist lineages have been replaced by a new wave of colonists occupying coastal and disturbed habitats.

Integral to the cycle is the occasional arrival of new colonist species, which push earlier colonists from the open habitats. This is because the earlier colonists are likely to have become less specialized on these habitats (and thus less competitive) since they themselves colonized, as they spread into forest habitats. Wilson's model helps explain why species inhabiting interior forests of the oceanic islands can be more closely related to species of disturbed—than to interior forest—habitats of New Guinea. Although largely a unidirectional model, it was suggested that, occasionally, a stage 3 species may re-adapt to the marginal habitats, expanding its distribution to become a secondary stage 1 species (Fig. 9.1, point F).

Wilson's taxon cycle model, although compelling in its scope and vision, was based on rather provisional and unstandardized data. Half a century later, Economo and Sarnat (2012) provided a re-evaluation based on scrutiny of 200,000 Fijian specimens representing 177 species (all but six of the known ant fauna) from standardized surveys undertaken across all the major habitats of the archipelago. Consistent with the theory, they showed that older, 'deep' endemics (constituting a Fijian radiation) are largely restricted to undisturbed interior and especially upland forest, with endemic allospecies also weighted towards interior and less disturbed habitats (Fig. 9.2). The widespread native species, which are typically more recent colonists, are by contrast biased towards more disturbed environments of lower elevation. The numerous non-native species, mostly introduced in the last few hundred years, were found to be characteristic of human-dominated landscapes, perhaps representing a new, anthropogenic phase of the cycle.

A subsequent study of the exceptionally diverse and widespread ant genus *Pheidole*, indicates that it colonized the Old World from the New around 20 Ma, diversifying across the hemisphere to form some 500 species (Economo et al. 2015). Regionally, there is evidence of a faunal cascade from tropical Asia into the Pacific, via New Guinea, with no back-colonization. As predicted by the taxon cycle, widespread expanding species, which are distributed across the phylogeny, mostly occupy marginal habitats. In general, it is the first colonists of the lineage that have undergone phylogenesis (cladogenesis) while subsequent colonists typically have not radiated within these Pacific archipelagos. This is exemplified in Fiji by *Pheidole*, which is represented by a mix of non-native, native, and endemic species, derived from six separate colonization events (Economo and Sarnat 2012). A single early colonization event from Asia gave rise to all but one of the endemic species, with this early colonizing lineage diversifying into a range of specialized niches (Darwell et al. 2020). Consistent with the taxon cycle, ants shift following colonization, from occupation of widespread marginal habitats towards specialization within higher-elevation forests.

The *Pheidole roosevelti* group is particularly notable, being an early colonizing endemic group, the species of which are restricted to no more than three islands and in cases to one island, or even a single mountaintop (Economo et al. 2015). The lineage originated in the northern Fiji Islands, sequentially colonizing other islands. *In situ* evolution in this group has involved changes in habitat, alongside changes in morphology, foraging, and nesting behaviour and a loss of dispersal ability to the point that the within-archipelago barriers have permitted *in situ* diversification. There are parallels here with findings by Fernández-Palacios et al. (2021b), for plants within the Canaries: it seems that radiation (at first glance 'evolutionary success'), can be the mark of a lineage the members of which are increasingly trapped and localized within mountains ('ecological losers'). However, as also allowed by Wilson's original model (1961), the trend towards ecological specialization and population fragmentation is not inevitable, and some species within endemic radiations are able to enter phases of range expansion.

Phylogenetic analyses of Indo-Pacific *Odontomachus*, another of the ant genera originally studied by Wilson, has provided further support for the taxon cycle and especially the linkage between habitat preference and dispersal propensity

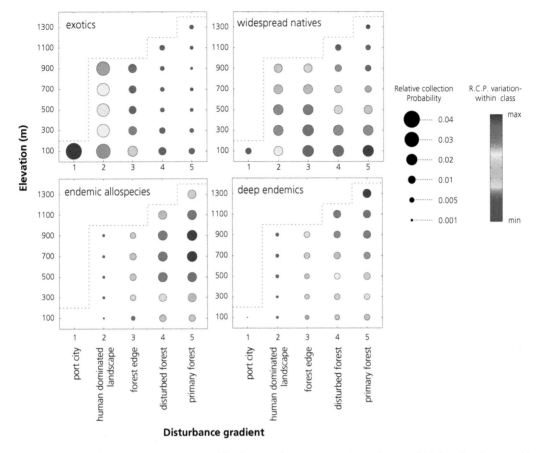

Figure 9.2 Response surfaces in relation to elevation and disturbance gradients for 177 species of Fijian ants divided into four chorotypes (the four panels), whereby the circle size indicates the collection probability across all chorotypes and the colour represents the likelihood of species detection within a chorotype.

The dashed lines represent the limits of potential distribution. Economo and Sarnat (2012), their figure 2.

(Matos-Maraví et al. 2018). One notable departure from Wilson's model, however, is that the group appears to originate from a New World lineage rather than from South-East Asia, a pattern shared with some other insect taxa and suggesting westward trans-oceanic dispersal by rafting. A further Fijian study provides somewhat mixed support for the taxon cycle, with evidence suggesting that a single colonization event of *Strumigenys* trap-jaw ants gave rise to a classic adaptive radiation, resulting in the filling of a large proportion of the morphospace observed for the genus globally and a lack of evidence of displacement of early by later colonists (Liu et al. 2020). In sum, the taxon cycle falls a little short of providing a one-size fits all model for Indo-Pacific ants, but it nonetheless provides a flexible model of heuristic—and some predictive—power.

Taxon cycles in birds

Ricklefs and Cox (1972, 1978) categorized the birds of the Lesser Antilles into a four-stage taxon cycle model (Fig. 9.3). Passerines with stage 1 distributions tended to occupy open, lowland habitats and to demonstrate habitat breadths and abundances rarely attained by mainland species, indicating that colonization of islands must involve some degree of ecological release. Late-stage species were largely restricted to montane or mature forest habitats.

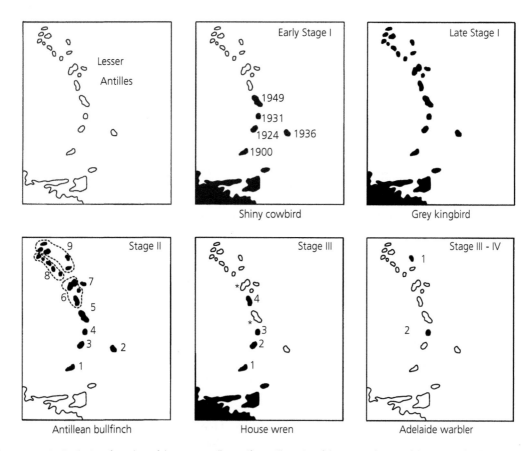

Figure 9.3 The distribution of members of the Lesser Antillean avifauna, illustrative of the proposed stages of the taxon cycle. Shiny cowbird, *Molothrus bonariensis*; grey kingbird, *Tyrannus dominicensis*; Antillean bullfinch, *Loxigilla noctis*; house wren, *Troglodytes musculus*; and Adelaide warbler, *Setophaga adelaidae*. The numbers for stage II–IV species indicate differentiated populations (subspecies). The house wren became extinct on the islands marked* within the 20th century. Dates refer to first colonization.
Redrawn from Ricklefs and Cox (1972), their figure 1.

Why should a new arrival be able to establish into an existing bird community? One reason put forward is that a small founding population may not only have left behind predators and close competitors but will also typically escape their parasites on colonizing an island, providing a significant fitness benefit. Over time, the evolutionary reaction, or 'counter-adaptation', of the pre-existing island biota to each established resident, would, it was argued, lower their competitive ability. The subsequent arrival of further immigrant species will then tend to push the earlier colonists into progressively fewer habitats and reduce their population densities.

The Ricklefs and Cox model was criticized by Pregill and Olson (1981), who argued that: (i) the four stages are simply criteria that define a set of distributional patterns, (ii) almost any species in any archipelago would fall into one of these categories without necessarily following the progression from stage 1 to 4, and (iii) insufficient attention was paid to the role of long-term environmental change as a driver of insular turnover. More recent work has confirmed departures from the simplifying assumptions of the taxon cycle, while tending to support the model as a generalized description. For example, yellow warbler *Dendroica petechia* has been found from mtDNA data to have colonized the Caribbean islands not once but multiple times and from different source regions of the mainland, meaning that some inter-island differences may be a consequence of founder effects rather than *in situ* differentiation of a single expansion event. In reviewing data for multiple bird lineages, Ricklefs and Bermingham (2002) noted other such discrepancies, but reaffirmed the taxon-cycle model as a fair representation of the temporal sequence followed by many species. Some 20 lineages colonized the Lesser Antilles between 10 and 7.5 Ma, with individual island populations typically persisting for about 4 million years, and with single-island endemics mostly among the oldest species, as required by the model. Notwithstanding that around 17 lineages appear to have colonized the Lesser Antilles between about 0.75 and 0.55 Ma (perhaps reflecting the onset of the most extreme Pleistocene climatic fluctuations), episodes of expansion and contraction did not appear to be correlated across species, nor to be linked in a simple way to Quaternary climate change cycles. A key role for pests and parasites in the proposed mechanism of counter-adaptation within taxon cycles appears plausible (Ricklefs et al. 2016) but remains hard to demonstrate empirically.

The taxon cycle has been found to provide a relevant framework for analysis of insular avifaunas elsewhere. For example, Kennedy et al. (2022) report that within New Guinea the youngest species are largely to be found in the lowlands and in outlying mountains, while the range-restricted, older endemics are concentrated in the montane forests of the central mountains, especially at mid-elevations, where range restriction is accompanied by ecomorphological specialization. Similarly, based on molecular, distributional, and ecological data, Jønsson et al. (2014) show that among Indo-Pacific *Pachycephala* (whistlers), earlier colonists (late stage of taxon cycle) tend to be found in higher-elevation interior habitats and to have little spatial overlap with more recent colonists that have expanded across multiple islands. Within the richest lineages, early colonization was followed by differentiation within islands, consistent with patterns found for Indo-Pacific ants (above).

Further support for taxon cycles in Indo-Pacific birds has been found in analyses of cuckooshrikes, cicadabirds, and trillers (the core Campephagidae) by Le Pepke et al. (2019). Once again, older species were found to inhabit higher-elevation habitats within relatively few, large islands. Intriguingly,

genetic data indicate that rather than continuing with an *r*-selected, fugitive lifestyle, some 'supertramp' species (such as the island monarch, *Monarcha cinerascens*, which occurs in small islands in both Melanesia and Wallacea but is not found in New Guinea and other large islands in between) display very little gene flow following colonization of small islands and their phylogenetic structure is more compatible with the taxon cycle than with the role assumed within Diamond's assembly rules theory (Le Pepke et al. 2019; Ó Margaich 2022; cf. Section 6.2).

Although originally developed to account for the biogeographical patterns of multiple species groups, the taxon cycle has also been invoked in the case of the Hawaiian honeycreepers (Pratt 2005; Ricklefs 2017), a monophyletic group undergoing adaptive radiation (Section 9.4, below). Pratt (2005) shows how particular species can be assigned to one of four stages of the taxon cycle based on their distribution, genetic, and niche differences. Analysis of the contemporary and reconstructed distributions of 87 known island populations, comprising 50 recognized species, 30 of which are now considered extinct, demonstrates a strong link between stage of the cycle and anthropogenic extinction likelihood (Ricklefs 2017). As shown in Table 9.1, extinctions of insular populations between Polynesian and European arrival were exclusively of stage 3 and 4 species, with persistence to the present day strongly biased against stage 3 species. The slightly better persistence of single-island endemics reflects the importance of avian diseases that are limited climatically to lower elevations.

How prevalent are taxon cycles?

The studies reviewed collectively suggest that taxon cycles in ants and in land birds occur repeatedly in insular regions where multiple islands and archipelagos provide the stage across which lineages are able to expand, diversify (mostly allopatrically)—and eventually contract in distribution—over timescales of millions of years. Although less frequently invoked for other taxa, the concept has also been discussed in work on other invertebrate and vertebrate taxa; for example, respectively weevils (Sequeira et al. 2000; Machado

Table 9.1 The status of 87 insular populations of species of Hawaiian honeycreepers as a function of the stage of the taxon cycle to which they were assigned based on distribution and differentiation, where: w = widespread, u = undifferentiated, d = differentiated, g = gaps in distribution, sie = single-island endemic; and prehistoric refers to extinctions between Polynesian and European colonization and historic to the post-European colonization phase. Data represent the number of insular populations in each category and, in italics, the percentage of the total number of populations that they represent.

Stage	Extant	%	Extinct historic	%	Extinct prehistoric	%	Total
1 (wu)	10	*58.82*	7	*41.18*	0	*0.00*	17
2 (wd)	5	*83.33*	1	*16.67*	0	*0.00*	6
3 (wd g)	6	*12.77*	15	*31.91*	26	*55.32*	47
4 (sie)	6	*35.29*	7	*41.18*	4	*23.53*	17
Total	27	*31.03*	30	*34.48*	30	*34.48*	87

Modified from Ricklefs (2017), his table 1.

et al. 2017) and lizards (Roughgarden and Pacala 1989; Richmond et al. 2021). As noted at the outset, a feature of this descriptive model is that the evidence supports or necessitates adoption of variants of the model in different implementations (taxa, or insular regions). Indeed, in their analysis of *Lepidodactylus* lizards, the distribution of which span the Philippines, New Guinea, Vanuatu, Fiji, and Tonga, Oliver et al. (2018) found evidence for biotic interactions driving disjunctions of distribution, consistent with the taxon cycle, but more strongly expressed on continental fringes than on islands, with habitat occupancy patterns at variance with the classic taxon cycle shift from open mainland into lowland forest habitats on islands. Their results indicate initial diversification of the group in the late Eocene and that persistence and transfer on geological fragments of a formerly more continuous arc of islands may be integral to understanding the contemporary distribution of these lizards and that peripheral habitats on island arcs have been particularly important for persistence and diversification of *Lepidodactylus*. Their work points to the need to integrate tests of models such as the taxon cycle with improved models of the geological history and dynamics of the region.

9.3 Adaptive and non-adaptive radiation in the context of island geo-environmental dynamics

The taxon cycle model is a framework of broad scope, both geographically (spanning multiple archipelagos) and in embracing patterns across multiple species groups. Adaptive and non-adaptive radiation, by contrast, refer to the diversification of groups that have derived from only a single colonization event of one species (i.e. they are monophyletic). Most students of biology have an idea what is meant by adaptive radiation, associating the term with classic insular examples, such as Darwin's finches, but non-adaptive radiation may be less familiar. In practice, this distinction follows on from the discussion of neutral versus selective mechanisms of change in Chapter 8 (e.g. Box 8.3), and applying it is to divide a continuum: indeed a single radiation may involve both adaptive and non-adaptive elements, as will become evident below. It is nonetheless a useful conceptual distinction. Our usage is that adaptive radiation involves genetically coded phenotypic diversification between members of a monophyletic lineage, expressed in combinations of morphological, phenological, and behavioural traits underpinning emergent niche differences. Non-adaptive radiation by contrast refers to diversification where the members do not appear to be ecologically differentiated to any meaningful degree and where the differences that have emerged have done so largely through processes such as founder effects and genetic drift (Section 8.5).

Opinions differ on the criteria for a system to qualify as adaptive radiation, and on the key drivers (Gillespie et al. 2020; Schenk 2021). In particular, there has been debate on whether the diversification needs to be rapid to be considered as an adaptive radiation. In practice, if you are looking for rapid diversification, remote islands provide plenty of examples that qualify (below) and this makes rapid diversification into novel trait space a key feature of insular adaptive radiations (see development of this point in Box 9.1). It is generally considered on theoretical grounds that as the trait space begins to saturate, diversification rate should decline towards an asymptote, leaving a characteristic signature in the structure of the phylogeny (Schenk 2021). Within certain limits, modern phylogenetic analyses and modelling techniques can permit these signatures to be dissected and inferences to be made about the evolutionary dynamics of the radiation. In the sections that follow, we review a small selection of many recent studies on these topics. Before doing so, we will return to a key theme of this book: the importance of island geological, geomorphological, and environmental dynamics.

In the context of short-lived oceanic islands, the general dynamic model (GDM) (Whittaker et al. 2008) argues for a decline in carrying capacity over time that will increase the extinction rate to the point that it exceeds the gain from further speciation (and from immigration) (see also Stuessy et al. 1998; Stuessy 2007). This too may be detectable in the phylogenetic structure and trait data of insular lineages, notwithstanding that detecting extinction purely from phylogenetic structure is challenging. Whether an entire radiation continues to increase, decrease, or remains at a constant size for a period of time across an entire archipelago or meta-archipelago will depend on (i) the interplay of a particular focal group of species with accumulating mutualists (e.g. plant/pollinator interactions) or antagonists (pests, pathogens, predators) and (ii) physical factors that determine the availability and distribution of suitable habitat. Crudely, is the archipelago growing or shrinking? While the evolutionary literature has discussed the first group of factors at length, it has in the past given insufficient attention to the second set of factors.

As a reminder and to keep to the essentials, the GDM predicts that the carrying capacity of an individual remote volcanic island will increase in the building phase of the island, with maximal opportunity for adaptive radiation on islands that are youthful–intermediate in stage (when elevational range and habitat diversity are at maximum), and declining opportunities thereafter as

the island subsides and erodes. Opportunities for non-adaptive radiation via intra-island isolation will peak with maximum topographic complexity, which may typically be attained at a slightly later point than opportunities for habitat shifts that drive adaptive radiation, and again will decline as the island declines. Co-speciation processes that benefit from increased availability of mutualists (Percy et al. 2008; Givnish et al. 2009) are expected to be of increasing relative importance with increasing island age and richness, but in time will also become limited with island decline. Composite islands, of more complex geological history (such as Tenerife, Chapter 2), should have provided more opportunity for within-island allopatry than islands with simpler histories, thereby producing additional sister species that lack clear adaptive separation. Hence, the theory provides predictions regarding the changing pace and cause of diversification and its eventual decline within an island (Whittaker et al. 2008, 2010; see also Lim and Marshall 2017).

Considering a hotspot archipelago as a whole, the GDM incorporates the progression rule; that is, that older islands predominantly act as key sources of colonists for successively younger islands, producing structured phylogenetic trees conforming with the island age sequence. In time, within the oldest islands, extinction should weed out earlier products of the radiation, particularly: (i) species that occupied habitats (such as montane ecosystems) which have been lost; and (ii) allospecies produced through non-adaptive processes that are insufficiently ecologically distinct to persist in sympatry as topographic barriers fail. Hence, the branch lengths to surviving members of early radiations should increase, as they become locally relictual (Whittaker et al. 2008, 2010). These features should leave signatures in the phylogenies of archipelagic radiation, at least for lineages that colonized concurrent with (or earlier than) the currently oldest islands. Whereas there have been multiple efforts to evaluate the GDM via macroecological analyses, the phylogenetic data required for evaluation of these evolutionary hypotheses are less conducive to meta-analysis (contrast e.g. Lim and Marshall 2017; Machado 2022). Nonetheless, in the sections that follow, it will become evident that a full understanding of evolutionary diversification across archipelagos requires attention to be paid to the geo-environmental dynamics, with various case studies providing at least partial support for the GDM.

Another important topic to consider in relation to evolution on islands is climate change and associated sea-level fluctuations, as encapsulated in the glacial-sensitive model of island biogeography (Section 7.8). The Quaternary period, commencing 2.588 Ma has featured multiple glaciation episodes, which have driven eustatic sea-level falls of up to *c*. 135 m, interspersed with warmer inter-glacials, sometimes featuring higher sea levels than present (Chapter 3). As many of the oceanic island radiations we consider have occurred either entirely or partly within this time frame, we would do well to keep in mind that the current configuration of islands—and of their constituent ecosystems—has been in place for only a few thousand years and may not reflect the conditions shaping events during lengthy periods in which the lineages we are studying have evolved (e.g. Fernández-Palacios et al. 2016a; Caujapé-Castells et al. 2017)).

In sections below, we begin with some iconic examples of adaptive radiation in insular birds, followed by illustration—through studies of land snails—of the concept of non-adaptive radiation to provide the counterpoint to adaptive radiation. We then go on to review further examples of insular radiations in a range of taxa, bringing further attention to biological processes such as hybridization and sexual selection, and to the role of insular geo-environmental dynamics.

Box 9.1 Is adaptive radiation a race towards specialization?

As we commented in the main text, rapid diversification into novel trait space is a defining feature of insular adaptive radiations. Does this mean that adaptive radiation is in essence a race towards specialization?

It is possible to view the distinction between specialist and generalist species as equivalent to the r–K gradient introduced by MacArthur and Wilson (1967) (Section 5.2). Thus, species that tend towards generalism may be regarded as more r-selected and specialists as K-selected. By this line of reasoning, we might expect generalist

(r-selected) species to have an advantage in island colonization because they will typically feature: (i) greater dispersibility, both in number of propagules produced per individual and in terms of propagule vagility; (ii) less restrictive abiotic (habitat) and biotic (niche) ecological requirements, enhancing the likelihood of finding the resources needed to establish and persist; and (iii) higher reproductive rate, reducing the likelihood of experiencing the inbreeding depression associated with the founder event.

Many of the trait shifts and syndromes we discuss in Chapters 10 and 11 can be viewed in the light of the outcome of a shift along the spectrum of r–K selection post-colonization on an island. Moreover, within insular adaptive radiations, a trend towards specialization may be expected for several reasons (Schluter, 2000):

First, the increase over time in the number (and density) of competing species should lead to an increasingly fine partitioning of available resources.

Second, as particular species evolve features that enhance their capacity to exploit specific portions of the resource spectrum, they may lose the genetic variation needed for exploiting other resources (but see Section 9.2 on the taxon cycle, which suggests that such a trend may not be irreversible).

Third, selection towards specialization may be expected to be more powerful than in the opposite direction, because an allele that confers an advantage in relation to a specific resource will spread more rapidly in a population specializing in that resource than within a population of generalists, in which it will have a more limited benefit.

Fourth, when sympatric speciation occurs through competitive speciation, as we outline in Section 8.8, intermediate genotypes are expected to be at a selective disadvantage, thus favouring mechanisms promoting reduced gene flow among specialists within the clade.

Finally, when considering predators, divergence in responses of prey towards the predators may favour trends towards specialization within the predators.

9.4 Adaptive radiation in island birds

Seeing this gradation and diversity of structure in one small, intimately related group of birds, one might fancy that, from an original paucity of birds in this archipelago, one species had been taken and modified for different ends.

(Charles Darwin, writing in 1939 about the beak sizes of *Geospiza* in the Galápagos, in Grant and Grant 2005).

We begin our examination of insular adaptive radiations with birds and in particular focus on the two best-known systems, the Galápagos (Darwin's) finches and the Hawaiian honeycreepers. Each is monophyletic and has radiated to occupy an expanded range of ecological roles, enabled by a remarkable degree of morphological disparification. Other impressive avian examples include the giant moas of New Zealand (Lomolino et al. 2021), white-eyes on islands in the Gulf of Guinea (Glor 2011), and the vangas of Madagascar (Jønsson et al. 2012). Indeed, for certain traits, the morphological variation exhibited by the Malagasy vanga radiation is even larger than that of its more famous Galápagos and Hawaiian counterparts (Fig. 9.4a), although they have had much more space and time (colonizing c. 25 Ma) in which to do so (Jønsson et al. 2012).

Galápagos birds and the Darwin's finch radiation

The Galápagos Islands are in the east Pacific, 930 km west of South America (Fig. 8.6; Ali and Aitchison 2014). Although equatorial, they are comparatively cool and average rainfall in the lowlands is less than 75 cm/year (Porter 1979). There are some 45 islands, islets, and rocks, of which nine exceed 50 km². Isabela, at 4700 km², represents over half of the land area and is four times the size of the next largest island, Santa Cruz. Isabela and Fernandina have peaks of some 1500 m a.s.l., but most of the islands are relatively low. They are true oceanic islands and remain volcanically active, many lava flows being recent and still unvegetated. Their history of emergence and change is now well documented and is integral to explaining evolution in the archipelago (Ali and Aitchison 2014).

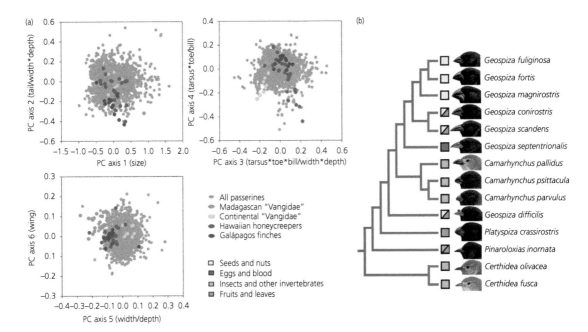

Figure 9.4 Adaptive radiation in three classic avian insular radiations: (a) principal components analysis summarizing the variation in seven morphological traits (as indicated in the axis labels), comparing three insular groups, continental relatives of the Madagascan vangas, and all passerines; (b) one of a number of alternative phylogenies of the Galápagos finch radiation, highlighting beak variation and dietary preferences. *Pinaroloxias inornata*, which is endemic to Cocos Island, is the single member of the radiation that occurs outside the Galapagos.

Panel a, figure 3 from Jønsson et al. (2012); panel b, extracted from figure 1b of Tokita et al. (2017) to exclude outgroup species used (CC-BY-4.0).

In addition to the finches, 15 species of land birds breed on the Galápagos, of which 10 are endemic, with one additional endemic species (the San Cristóbal vermilion flycatcher) known to have gone extinct (Triantis et al. 2022). The Galápagos mockingbird (*Mimus parvulus*) has also diversified into six endemic subspecies, each present on a different island or island group, although this radiation involves rather less morphological differentiation than seen in the finches (Reaney et al. 2020; Billerman et al. 2022). Genetic data indicate that the Genovesa Island subspecies (*M. p. bauri*) is of likely hybrid ancestry and is possibly undergoing speciation (Nietlisbach et al. 2013).

The Galápagos finches are nested in the tanager family (Thraupidae) and comprise between 15 and 17 species, allocated to four genera (*Camarhynchus*, *Certhidea*, *Geospiza*, and *Platyspiza*). There is also an additional (potential) hybrid species that might be recognized (Section 8.8) and one further species (*Pinaroloxias inornata*) endemic to Cocos Island, 630 km north-east of the Galápagos (Farrington et al. 2014; Lamichhaney et al. 2015). It is a young radiation, initiated only *c.* 1.5–1.0 Ma. As noted in Chapter 8, analyses based on molecular phylogenies suggest that the radiation may have reached equilibrium richness, in contrast to Galápagos land birds considered as a whole (Valente et al. 2015), although recent observations indicate that evolutionary change, including through hybridization, is ongoing. While many of the species look similar, they fill a wide range of ecological niches, as shown in Fig. 9.4b (which illustrates 14 of the species). Collectively, they feed on a remarkable diversity of food: insects, spiders, seeds, fruits, nectar, pollen, cambium, leaves, buds, the pulp of cactus pads, and blood from both seabirds and sealion placenta (Grant and Grant 2005). In illustration, the woodpecker finch (*Camarhynchus pallidus*) uses a twig, cactus spine, or leaf petiole as a tool, to pry insect larvae out of cavities. Small, medium, and sharp-beaked ground finches (*Geospiza fuliginosa, G. fortis*, and *G. difficilis*) remove ticks from tortoises and iguanas, and perhaps most bizarre of all, sharp-billed ground finches on the northern islands of Wolf and Darwin perch on boobies, peck around the base of the tail, and drink the blood from the wound they inflict.

Detailed study of the radiation was initiated in the 1940s by David Lack and more recently has been conducted over an extended period by Peter and Rosemary Grant. The following general summary is drawn from several sources (Grant and Grant, 2005, 2008; Lamichhaney et al. 2015, 2016, 2018; Stroud and Losos 2016; Rubin et al. 2022).

One of the Galápagos islands was colonized from the mainland in a one-off founding event, estimated at around 30–100 individuals. The founding population expanded quite rapidly, undergoing selective changes and/or drift. After some time, members of this population colonized another island in the archipelago, where conditions were slightly different. Further changes occurred through a combination of random genetic changes, drift, and selection (for different feeding niches). A degree of differentiation would then have been evident between the separate populations. At this point, individuals from the derived populations flew to islands already occupied by slightly differentiated populations. In some cases, selection would have favoured members of the two groups that fed in different ways from each other and so did not compete too severely for the same resources. For certain taxa, where resource use did overlap, divergence in beak morphology through character displacement took place, which was likely accelerated during periods of limited food supply (Section 8.7). This would be linked with the development of reproductive isolation through behavioural mechanisms. Female finches appear able to distinguish between acceptable and unacceptable mates based on beak size and shape and through song patterns (Grant and Grant 2005; Lamichhaney et al. 2018). This may have been the means by which females were able to select the 'right' mates, thus breeding true, and producing progeny that corresponded with the 'peaks' rather than the 'valleys' of the resource curves (Section 8.8). However, hybridization also has now been established to have played an important role in the radiation, in cases generating variants that had selective advantage in exploiting part of the resource spectrum (below). Multiple cycles of the above processes then resulted in the diverse adaptive radiation which we observe today.

As indicated, it is principally through changes in beak size and shape and associated changes in feeding skills and niches that the differentiation between the finches has come about (Grant and Grant 2005). A small number of loci are largely responsible for the variation in beak shape (*ALX1*) and size (e.g. *HMGA2*), with both traits being subject to strong contemporary selection (Lamichhaney et al. 2016; Rubin et al. 2022). Recently, phylogenetic analysis of whole genomes has revealed evidence of extensive interspecific gene flow and hybridization throughout the radiation, particularly among the ground (*Geospiza*) and tree (*Camarhynchus*) finches (Lamichhaney et al. 2015, 2018; Rubin et al. 2022). These processes have been found to be particularly prevalent on the peripheral islands in the archipelago, where high levels of admixture have been observed among *Geospiza* species. It is thought that this can occasionally generate novel morphological forms better adapted to the more extreme environmental conditions on these islands (Farrington et al. 2014). To sum up, the process of hybridization is thought to have (i) increased the genetic variation underpinning beak shape, a key adaptive trait in the radiation (Lamichhaney et al. 2015), and (ii) resulted in the evolution of several species of mixed ancestry, including a potential new species through a recent homoploid hybrid speciation event (Lamichhaney et al. 2018). However, Kleindorfer et al. (2014) has also reported that hybridization via disassortative mating appears to have caused the collapse of three closely related sympatric *Camarhynchus* species on Floreana Island, all present on the island in the 1900s, into two by the 2000s.

There has been a strong geographic component to the pattern of diversification, with volcanic activity occasionally generating new islands, and glacial–interglacial cycles resulting in the repeated coalescence and isolation of different central islands in the archipelago (Ali and Aitchison 2014). As a result, the more isolated peripheral islands support higher endemic species and subspecies richness of some clades than the central islands. Within the tree and ground finches, the deepest split in genetic structure corresponds to taxa on either side of the wide channel—which has been present for millions of years—that divides the core island group on the Galápagos platform in the south from the islands on a separate volcanic province to the north (Farrington et al. 2014).

To put these points together, the radiation of the lineage has taken place in the context of a remote archipelago, presenting extensive 'empty niche' space well distributed in space and time. The radiation has involved periods of allopatry followed by colonization and secondary contact, with evidence that natural selection, ecological character displacement, and introgressive hybridization have played important roles. Sympatric episodes of lineage development (Grant and Grant 1996b) allow us to invoke elements of the model of competitive (sympatric) speciation (Section 8.8) within the radiation.

Hawaiian birds and the honeycreepers radiation

The Hawaiian Islands have formed as a narrow chain from a remote mid-plate hotspot, which appears to have been operational since *c*. 85–74 Ma, although the oldest high island of the present group, Kauai, formed as recently as 5.1 Ma (Table 2.3). The archipelago is rich in endemic species (Chapter 4), including (prior to human colonization) around 100 land birds, 99% endemic (Triantis et al. 2022). However, there was a geologically recent period where there was no high island in the archipelago, and molecular clock data suggest that although some clades are older than Kauai, few Hawaiian lineages have founding ages > 10 Ma (Wagner and Funk 1995).

Avifaunal radiations on Hawaii include the native thrushes, which are placed in the same genus (*Myadestes*) as the solitaires of North and South America. Five species are currently recognized by Billerman et al. (2022), each occurring on its own island or island group, with two extinct and two currently threatened with extinction. However, unquestionably the most famous Hawaiian avian radiation is the honeycreepers. Prior to Polynesian settlement there were between 50 to 64 species of Hawaiian honeycreepers in this most impressive of insular avian radiations (Pratt 2005, 2014). Recent treatments have categorized the group as a single monophyletic tribe (Drepanidini), placed within the finch subfamily Carduelinae and sister to the

Eurasian rosefinches (*Carpodacus*) (e.g. Lerner et al. 2011; Zuccon et al. 2012). The ancestral colonist was probably a seed-eating Asian species with a typical cardueline finch-shaped bill. It radiated to fill seed-, insect-, and nectar-feeding niches, with a great variety of specialized beaks and tongues (Fig. 9.4a, 9.5; Pratt 2005; Tokita et al. 2017).

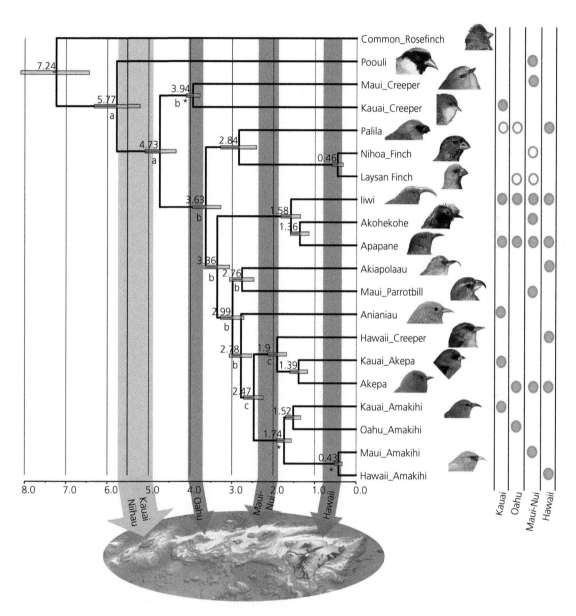

Figure 9.5 A phylogeny for the Hawaiian honeycreepers based on whole mitochondrial genomes, where point 'a' indicates divergence before Oahu formed, 'b' during/after Oahu formation, 'c' before/during Maui Nui formation. Bird images (by Jack Jeffrey) illustrate the range of beak morphology, grey circles indicate distributions, open circles known extinct distributions. Mean ages (Ma) and 95% interval are shown for each node. Asterisks indicate constrained nodes. Vertical shaded bars are the estimated shield building dates.
From Lerner et al. (2011), their figure 2.

They have been placed in over 20 genera, including: (i) *Psittirostra*, a monotypic (possibly extinct) genus with an unusually shaped beak, believed to allow it to eat the fruits of *Freycinetia arborea*; (ii) *Chloridpos*, an extinct genus of thick-billed seed eaters, and (iii) *Pseudonestor*, which uses its powerful beak to tear apart twigs to reach wood-boring beetles. However, the group is perhaps most famous for its many members possessing longer, narrower, and decurved bills, adaptations for feeding on insects (e.g. in the case of *Hemignathus wilsoni*) and on nectar (e.g. *Drepanis coccinea*). Indeed, the development of a tubular, brush-tipped tongue for nectar feeding, unique among passerines, appears to have turbo-charged the diversification (Pratt 2005). The development of long, curved bills within the honeycreepers is generally regarded as a key co-evolutionary process with key plant taxa such as *Hibiscadelphus* and the lobeliads (such as *Clermontia* and *Cyanea*), for which they provided pollination services in return for access to nectar.

Based on nuclear loci and mitochondrial genomes of several honeycreeper species and outgroups, and using island ages to calibrate DNA substitution rates, Lerner et al. (2011) determined that the ancestral colonist arrived in the archipelago around 7.2 Ma. The radiation had begun by *c.* 5.7 Ma, around the time of the formation of Kauai and intersecting in time with the slightly older and now much eroded Niihau. The genetic differentiation is unexceptional for the age of the radiation, but the morphological divergence is remarkable. It largely evolved after the emergence of Oahu (around 4–3.7 Ma) but before the emergence of the other large islands (Lerner et al. 2011; Gillespie et al. 2020). As new islands emerged, they were colonized in turn by honeycreepers, in line with the progression rule. Once populations had diverged, back-colonization became possible, with an important role for character displacement once populations came into sympatry (Pratt 2005). Thus, it appears that cycles of colonization, allopatric isolation, and insipient or full speciation linked to the emergence of islands were each critical to the evolution of the group. The exceptional morphological diversity exhibited by the tribe reflects the availability of empty niche space owing to the paucity of passerine colonists to Hawaii, and key innovations

(such as the development of the tubular tongue), natural selection, co-evolution (with mutualists), and episodes of sympatry marked by character displacement within the lineage.

Hybridization within the Galápagos finches and Hawaiian honeycreepers

Further comparison of the Hawaiian honeycreeper and Galápagos finch radiation is instructive, especially in relation to the role of hybridization. The richness and morphological variation exhibited by the honeycreepers exceeds that of the Galápagos finches (Fig. 9.4a; Tokita et al. 2017) but there are several examples of morphological convergence between members of the two radiations, including between *Geospiza magnirostris* in the Galápagos and two of the Hawaiian grosbeak species (*Chloridops* spp.), each with beaks enabling them to crack very hard seeds (Tokita et al. 2017). Covariation between the shape of the whole skull and individual skull elements (i.e. the degree of integration among skull modules) is stronger in Galápagos finches than in the honeycreepers and it is thought that this decoupling of skull modules in the honeycreepers has enabled greater evolutionary flexibility, allowing them to evolve a more diverse range of beak morphologies (Tokita et al. 2017).

Evidence of hybridization is considerably weaker in the honeycreepers. For example, in an analysis of convergent evolution in two Hawaiian honeycreeper taxa (the Hawaiian [*Loxops mana*] and Kauai [*Oreomystis bairdi*] creepers), Reding et al. (2009) found little evidence of past hybridization. They attributed the morphological and ecological similarities between these relatively evolutionary distant honeycreeper species to convergent evolution to fill the same resource niche on different Hawaiian islands, as observed in other classic insular radiations, such as the *Anolis* lizards of the Antilles (Losos 2009). However, more recently Knowlton et al. (2014) provided the first evidence of a potential hybrid between two honeycreeper species. An unusual looking honeycreeper was captured on the island of Hawaii in 2011. DNA marker analysis indicated the individual to be a hybrid offspring from a male *Himatione sanguinea* and a female *Drepanis coccinea*: two species that diverged *c.* 1.5 Ma.

Nonetheless, the prevalence of hybridization in the Hawaiian honeycreepers is evidently less than in Darwin's finches. Four hypotheses have been put forward to account for this (Grant 1994; Knowlton et al. 2014).

- Honeycreepers have been present in the Hawaiian archipelago far longer than have the finches of the Galápagos, and over this time have diversified further and have evolved prezygotic (behavioural) and/or postzygotic isolating mechanisms.

- The Hawaiian honeycreepers have evolved greater dietary specializations in the generally less seasonal and floristically richer and ecologically more diverse Hawaiian Islands. In an environment with more distinctive resource peaks, stabilizing selection for specialist feeding may have provided strong selective pressure against an ecological niche intermediate between the two parental species (Section 8.8).

- The limited observational evidence of hybridization in the Hawaiian honeycreepers could be due to the large number of recent anthropogenic extinctions effectively obscuring the evidence, although this is not evident in the museum specimens evaluated to date.

- A large proportion of congeneric honeycreepers (which are more likely to hybridize than other pairings) occur on different islands and thus potential parent species do not often come into contact.

The studies reviewed support the view that both allopatric and sympatric episodes can be involved in radiations of birds within remote archipelagos. Although events must vary from one lineage to another, the following is a general scenario to account for sympatric congeners (Grant and Grant 1996b):

- Following colonization of different islands there is an initial phase of rapid differentiation of allopatric populations.

- Subsequent interisland movements re-establish sympatry between certain species, following which further differentiation and character displacement takes place between the two populations, as behavioural mechanisms build and reinforce their reproductive isolation.

- Throughout the radiation, introgressive hybridization may occur to varying degrees, most likely declining over time and as species differences increase. In some cases, hybridization will inhibit speciation, but in others, the resulting recombinational variation plays a creative role, facilitating further divergence (e.g. Lamichhaney et al. 2015).

9.5 Non-adaptive radiation: island snails lead the way

We previously characterized adaptive radiation as involving phenotypic diversification between species, expressed in combinations of morphological, phenological, and behavioural traits and which (crucially) underpin emergent niche differences. However, some insular radiations involve trait, ecological, or niche differences that are so small (or so cryptic) that they have been labelled non-adaptive (e.g. Rundell and Price 2009; Czekanski-Moir and Rundell 2019). As discussed in Chapter 8, small population effects associated with founder effects and other bottlenecks, combined with mutation and genetic drift, may lead to the accumulation of genetic and phenotypic differences that are selectively more or less neutral among populations that are fragmented across similar environments within an island or archipelago. This produces an array of ecologically similar sibling species that share a common ancestor and which are allopatric (or parapatric) equivalents of one another ('allospecies').

In Chapter 8, we outlined mechanisms for sympatric speciation within islands, highlighting the importance of natural selection. However, it is also possible to envisage scenarios for sympatric speciation that involve a limited role for natural selection and instead focus on mechanisms such as polyploidy (which can essentially generate a new species in a generation), substantial genetic mutations, and sexual selection (Czekanski-Moir and Rundell 2019; Matsubayashi and Yamaguchi 2022). However, non-adaptive divergence is generally considered more likely to arise in conditions of allopatry, as outlined above, than in sympatry.

The concept of non-adaptive radiation can be traced back to J. T. Gulick's comments on Hawaiian land snails, published just 13 years after the *Origin of Species* (Czekanski-Moir and Rundell 2019) and land snails provide several more recently described, if largely anecdotal, examples. For example, (i) Cretan *Albinaria* land snails have radiated without evident niche differentiation, to occupy a narrow range of habitats with minimal co-occurrence of species (Gittenberger 1991), and (ii) non-adaptive radiation was argued to be responsible for the majority of the 47 endemic land snail species of Porto Santo (Madeira) (Cameron et al. 1996).

For non-adaptive diversification to occur within a single island, it requires large island size, complex topography, and a taxon of limited dispersal powers. However, when considering diversification across an archipelago or meta-archipelago, even comparatively dispersive taxa will hit the zone for effective genetic isolation at a given degree of isolation. Hence, phylogenetic structure within certain lineages of monarch flycatchers across Wallacea and northern Melanesia (Rundell and Price 2009; Ó Marcaigh 2022) reflect the importance of isolation by distance producing a suite of allospecies or subspecies that display only limited ecological differences.

In practice, classifying radiation events as adaptive or non-adaptive is difficult. It is impractical to measure all ecological characteristics of species, and thus important differences may be overlooked, leading to incorrectly identifying divergence as non-adaptive. Conversely, given that species arising from non-adaptive divergence are likely to be functionally similar, they may represent sets of cryptic species that are overlooked (Ó Marcaigh 2022) and thus non-adaptive radiations may in fact be more common than generally assumed. We think it likely that most of the larger radiations involve both adaptive and non-adaptive elements, as for example inferred within the radiations of Hawaiian *Tetragnatha* spiders (Cotoras et al. 2018; Kennedy et al. 2022) and Macaronesian *Laparocerus* weevils (Machado 2022, below). Another important mechanism for evolutionary change within radiations is sexual selection (Matsubayashi and Yamaguchi 2022), which has been invoked, for example, in insular studies of birds (Ó Marcaigh 2022) and cichlid fishes in African lakes. In regard to the latter, and particularly in the haplochromines, sexual selection has been identified as contributing a selective yet non-adaptive component within a spectacular radiation (see Box 8.3, Box 9.2).

It has been argued that allopatric non-adaptive radiation should typically be slower than adaptive radiation because populations are not experiencing strong selective pressures (Czekanski-Moir and

---

Box 9.2 The explosive radiation of cichlid fishes in the African Great Lakes

When considering freshwater organisms, lakes can be viewed as islands, and the African Great Lakes (Tanganyika, Malawi, Victoria, etc.), clustered in the Rift Valley, can be considered as an archipelago. They possess one of the most impressive examples of radiation (more accurately, multiple radiations) in the form of over 1400 species of cichlid fishes (Cichlidae), with some 240 in Lake Tanganyika, 600 in Lake Malawi, and over 500 in Lake Victoria (Turner 2007; Ronco et al. 2021). Incredibly, the Lake Victoria radiation is thought to have been generated in just the last 15,000 years (Meier et al. 2017). The precise number of species is hard to determine as: (i) there may be many cryptic species (e.g. distinguished by smell), and (ii) many sympatric cichlid species have been shown to be able to interbreed and hybridize in laboratory conditions, while reproductive isolation in the wild is maintained through divergent mate preference (e.g. based on colour) (Turner 2007).

The huge diversity in cichlids is accompanied by a wide range of ecological roles, including fish eaters, snail eaters, rock scrapers, zooplankton feeders, and one species that feeds on the parasites of catfish. Parallel evolution within the different lake radiations has meant that many of these different roles are found in multiple lakes (Turner 2007). Some lakes (e.g. Lake Mweru) contain multiple different adaptive radiations, with each radiation occupying distinct areas of morphospace, thus suggesting that sympatry limits the possible divergence within any given radiation (Meier et al. 2019). Cichlid radiation probability in African lakes is also related to lake depth, while the number of species in the radiation is linked to lake surface area.

Box 9.2 *Continued*

Many models have been proposed in explanation of these spectacular cases of radiation, including, for some of the big lakes, allopatric scenarios resulting from fluctuating water levels (Turner 2007). One of the more straightforward is the **three-stage radiation model** (Streelman and Danley 2003):

1 **Major habitat diversification.** In the first adaptive radiation stage, an initial divergence of the lineages takes place, resulting in the appearance of different clades related to the major available habitats (sandy and rocky shores, plus interior waters).

2 **Trophic diversification.** This second adaptive stage involves the rapid directional selection of the feeding apparatus, attaining resource partitioning between species. This process has yielded numerous endemic genera, featuring morphological differences in features such as jaw shape and tooth structure. Recent analysis of the species in Lake Tanganyika has shown that variation in key morphological traits (e.g. body and jaw shape) between species is strongly tied to variation in ecological roles (Ronco et al. 2021).

3 **Sexual selection.** The most intriguing step of the model is the third phase, a non-adaptive radiation process, involving sexual selection through mate choice by females. This relies on colour differences in males (and possibly also smell and sound) and is thought to occur only in transparent waters, where the colours are evident. As cormorants are thought to be able to hunt the colourful males more easily in transparent waters, it is assumed that bright colouration must come at a cost to fitness and thus can be considered essentially non-adaptive.

Recent whole-genome analysis by Ronco et al. (2021) of the Lake Tanganyika radiation has provided support for this model and for it having occurred entirely within the lake, commencing shortly after its formation. For four traits, they reported a pulse-like pattern of morphospace expansion through time, with the timing of pulses varying across traits: some undergoing rapid expansion early in the radiation, with others exhibiting a pulse (or multiple pulses) at a later stage. This variation in timing provides support for the idea that this adaptive radiation has progressed in discrete stages of macrohabitat use, trophic niche, and finally mate signalling. However, it appears that the most recent stage involves temporally overlapping pulses in both a mate signalling trait and an ecological trait (jaw shape), resulting in the observed tightly packed niche space (Ronco et al. 2021). Interestingly, this study also showed that the species richness of cichlid tribes in the lake was positively associated with genome-wide heterozygosity, supporting a role for hybridization in the radiation.

The analysis of genomic data from cichlid species in the Lake Victoria Region has shown (i) simultaneous divergence in macrohabitat and diet (McGee et al. 2020), and (ii) that the radiations there likely originated from a single hybrid swarm, involving two distantly related lineages (divergence estimated at more than a million years) that previously evolved in isolation in different river systems (Upper Congo and Upper Nile; Meier et al. 2017). This hybridization event provided the genetic variation which fuelled the subsequent adaptive radiations. Together, these results indicate that hybridization is likely a key process in cichlid radiations more generally (see also Meier et al. 2019; McGee et al. 2020).

Rundell 2019). Similarly, the GDM argues that adaptive radiation should peak in rate slightly earlier than non-adaptive radiation, and that both rates will decline towards the later stages of the life span of an oceanic island as intra-island barriers and environmental heterogeneity decline (Whittaker et al. 2008). It seems likely that the relative importance of natural selection, sexual selection, and neutral processes may change over time within the evolution of a particular insular radiation. For instance, populations initially separated through dispersal or vicariance and subject to founder effects and/or drift, may come into secondary contact at a later

point, allowing for introgression, admixture, and the formation of so-called hybrid swarms, or alternatively, leading to character displacement and differentiation through natural selection (Box 8.3, Section 9.4; Czekanski-Moir and Rundell 2019; Machado 2022). In sum, although the distinction between adaptive and non-adaptive evolutionary change has some heuristic value, it is liable to break down in the face of more complex realities, with both adaptive and non-adaptive (largely neutral, or even mildly deleterious) changes co-occurring in tandem within the same insular clades (e.g. Machado 2022).

9.6 Insular plant radiations

Considerable attention has been given to the case for labelling radiations as adaptive, which we characterize as establishing that the speciation process involved niche separation arrived at largely via natural selection (see Schenk 2021 for a full review). Valuable though such efforts are, as will have become clear, we regard it as pragmatic to consider that radiations of any size rarely arise through a single mechanism, or match perfectly with adaptive or non-adaptive labels. A case in point is the evolution of *Pericallis*, an Asteraceae genus endemic to Macaronesia. Jones et al. (2014) use multiple genetic tools to analyse the diversification of this group, finding that it has mostly diversified within the Quaternary, displaying three important habitat shifts and transitions to secondary woodiness associated with occupancy of laurel forest habitat, implying an adaptive element. However, the establishment of allopatric distributions appears more important than habitat shifts, as there are a number of single-island endemic species.

That a cladogenesis event occurs in allopatry doesn't rule out a role for natural selection in the separate populations of course. For example, there are animal radiations discussed in this chapter that have featured the repeated evolution of similar traits matching similar environments in allopatry (e.g. in some Hawaiian spiders and in Macaronesian *Laparocerus* weevils). Hence, determining that speciation has happened in allopatry is not equivalent to demonstrating that it is non-adaptive and just because we cannot detect trait or niche differences is not proof that there are none. Nonetheless, in *Pericallis* it appears that some part of the taxonomic structure has arisen allopatrically with limited ecological differentiation. In addition, hybridization was inferred to have a role in this genus, as it has in other Macaronesian plant radiations, including *Aeonium*, *Argyranthemum*, and *Sideritis*. Indeed, one emergent theme in recent work on insular radiations reviewed herein is that the application of multiple genetic markers and tools to phylogenetic reconstructions has revealed a far greater role for hybridization in diversification than previously suspected. Indeed, it seems that insular plant radiations are often (i) associated with the provision over time of newly available isolated territory, (ii) frequently involve taxa that are polyploids, and (iii) often involve a role for hybridization (Section 8.8; Schenk 2021).

Within remote archipelagos, only a small minority of plant lineages radiate spectacularly, but they are a major source of ecological diversity and new taxa. For example, in the native Hawaiian flora, the 20 largest clades, derived from just 7.6% of the initial colonists, have given rise to over half the native plant species. They typically display repeated shifts in their growth form, pollination biology and/or habitat, diagnostic of adaptive radiation (Givnish, 2010). Table 9.2 lists some of the notable plant radiations on remote archipelagos, most of which are oceanic archipelagos *sensu stricto*. Each of these radiations may be considered at least in part adaptive in nature and excepting two cases where relevant data are unavailable (*Pachycladon* and *Oxera*), each has involved hybridization events (Givnish 2010; Schenk 2021).

Estimates of diversification rates depend on accuracy of estimation of crown ages (Box 8.1) and on assumptions made about rates of species extinction and are thus subject to modification with further work. Nonetheless, it is evident that the more spectacular oceanic archipelagic radiations highlighted in Table 9.2 are exceptionally fast: on a per unit area basis, exceeding the pace of the most celebrated continental plant groups by orders of magnitude (Knope et al. 2012; Borregaard et al. 2017). Such speed of radiation tends to involve multiple mechanisms of speciation, as discussed elsewhere in this chapter, but the repeated founding and isolation of new populations on different islands (or within different massifs within one island) is a core feature.

The Hawaiian silversword alliance is another classic insular radiation. This is a monophyletic group of three genera (*Argyroxiphium*, *Dubautia*, and *Wilkesia*) and some 33 species, which derives from an herbaceous Californian species that reached Hawaii 4.9–2.0 Ma. Initial diversification appears to have been rapid and it has produced cushion form plants, rosette-shrubs, lianas, shrubs, and trees, between them occupying habitats from the lowlands up to 3750 m a.s.l. (Schenk 2021).

Table 9.2 Examples of radiation in insular plants that may be considered largely or in part adaptive, with estimates of number of species, crown age (point at which *in situ* diversification began), and range in diversification rate (assuming rates of both 0 and 0.95 species extinct per million years).

Family	Genus	Species no.	Crown age (Ma)	Diversification rate (sp./Myr)	Location
Asteraceae	*Argyranthemum*	24	2.5–3.0	0.30–0.99	Macaronesia
	Bidens	19	1.76–1.82	0.32–1.73	Hawaii
	Lipochaeta	25	0.73–1.79	0.43–3.46	Hawaii
	Scalesia	15	1.42–4.33	0.36–1.42	Galápagos
	Silversword clade	33	2.0–4.9	0.20–1.91	Hawaii
	Sonchus clade	34 (32)	?	?	Macaronesia
	Tetramolopium	11	0.6–0.7	0.54–2.84	Hawaii
	Tolpis	12	2.6–9	0.05–0.69	Macaronesia
Boraginaceae	*Echium*	28 (31)	9.9–20.7	0.04–0.27	Macaronesia
Brassicaceae	*Pachycladon*	11	2.2	0.35–2.19	New Zealand
Campanulaceae	Lobeliad clade	126	10.49–16.71	1.36–2.04	Hawaii
Caryophyllaceae	*Schiedea*	34	2.03–7.88	0.12–1.40	Hawaii
Crassulaceae	*Aeonium* alliance	61	13.25–17.15	0.11–0.26	Macaronesia
Cunoniaceae	*Geissois*	13	3.5–12.8	0.26	New Caledonia
Goodeniaceae	*Scaevola*	8	4.5	0.06–0.31	Hawaii
Lamiaceae	*Micromeria*	16	4.2–2.6	0.04–0.50	Macaronesia
	Oxera	33	2.7–6.92	0.2–1.04	New Caledonia
Plantaginaceae	*Plantago*	42	2.8–4	0.27–1.09	Hawaii
Violaceae	*Viola*	9	1.2–2	0.15–1.25	Hawaii

Extracted from Schenk (2021) and slightly modified (including alternative species numbers in brackets).

Argyranthemum is the largest endemic Macaronesian plant genus, comprising 24 recognized species of woody perennials, all but three being single-island endemics. They occur in a broad range of habitats spanning coastal to high-elevation shrubland and appear to have evolved from a widespread generalist to become more specialized and localized. Genotype-by-sequencing analyses by White et al. (2020) indicates that hybridization, habitat shifts, and geographical isolation have shaped the diversification of this group, with over half of speciation events occurring within an island and reflecting a combination of within-island geographical isolation and hybridization. The three multi-island endemics were found to be non-monophyletic and to constitute cases of morphological convergence on different islands from differing ancestral origins. For example, populations on La Gomera and on Tenerife, previously attributed to *A. broussonetii*, were found to comprise two distantly related taxa evolving within similar laurel forest environments on the two islands.

9.7 The trajectory of diversification within hotspot archipelagos

Large archipelagic radiations typically succeed in generating more species than there are islands. We may simplify how this occurs to a dichotomy of clades responding primarily to islands or to habitats. For groups that have limited dispersal capability, inter-island colonization is extremely rare and once it has occurred, there is time for *in situ* diversification to fill an array of available niches: this is what we mean by clades responding to islands. For more dispersive or perhaps for richer groups, there is a greater frequency of inter-island dispersal events, permitting several habitat (or other) specialists from an older island to colonize a younger island, where they subsequently evolve to become distinct allospecies, sharing a closest relative with the source island. This we can think of as clades responding to habitats. An even more dispersive species may be capable of maintaining sufficient gene flow across multiple islands that it remains a single species. When considering larger groups of taxa, such as flowering plants, or arthropods, examples of all three of these scenarios may be observed.

Studies by Rosemary Gillespie and colleagues on Hawaiian spiders provide interesting examples of repeated niche-filling on multiple islands, illustrating both inter-island and intra-island evolution (Box 9.3). For example, *Tetragnatha* spiders of the spiny leg clade have evolved the same three of four ecomorphs on each of the four high Hawaiian islands (e.g. Gillespie 2004), adapted to leaf litter (green), moss (maroon), twigs (small and brown), and bark (large, brown) substrates. Phylogenetic reconstructions indicate six evolutionary transitions from one ecomorph to another and eight cases of dispersal of an ecomorph between islands, thus indicating a mix of our two models. Similarly, *Laparocerus* weevils provide examples of both forms of response within Macaronesia (below; Machado 2022).

Box 9.3 Hawaiian spiders as model systems for studying diversification in a hotspot archipelago

Hawaiian spiders have proven to be excellent model systems for the study of archipelagic evolution, as revealed through a long-term programme of research by Rosemary Gillespie and colleagues. Spider morphology and hunting techniques provide traits that help define their niche differences and they have sufficient dispersibility (in the case of many smaller species through ballooning) to reach isolated islands, but they are not so dispersive as to be panmictic across the archipelago. This summary is drawn largely from Gillespie (2016) and Cotoras et al. (2018).

Analyses of their origins suggest that several Hawaiian groups have American origins (e.g. *Cyclosa*, *Orsonwelles*, and *Tetragnatha*—which colonized twice). Most spider groups arrived comparatively early and have followed a general pattern of older-to-younger island colonization within their radiations (the progression rule: Section 9.8), as exemplified by *Tetragnatha* and the endemic genus *Orsonwelles*, although the late arriving jumping spider genus *Havaika* provides an exception. Several groups show extensive diversification, classified into three main patterns by Gillespie (2016): (i) non-adaptive radiation, (ii) adaptive radiation featuring early ecological shifts, and (iii) adaptive radiation with repeated ecological differentiation.

(i) The 13 species of *Orsonwelles* exemplify non-adaptive radiation: there is little ecological differentiation and they do not co-occur, even when found on the same island. They display the pattern we describe as clades responding to islands, in that successful colonization events are rare, with subsequent diversification *in situ*. Diversity in this group increases steadily with island age.

(ii) *Mecaphesa* crab spiders colonized Kauai when it was the only high island. It was here that ecological differentiation into different niches occurred, with these differentiated taxa colonizing along the island chain, broadly in line with the progression rule. This is consistent with the concept of clades responding to *habitats*, although as the resulting ecologically differentiated species can be found co-occurring at the same site, it is really that they have differentiated to fill distinct *niches* and then colonized the islands in turn. In this genus, species numbers vary from 10 on the youngest island to 12 on each of the Maui-Nui complex and Oahu, dropping slightly to eight species on the oldest island (Kauai), although species density increases with island age.

(iii) The third pattern is shown by the remarkable radiation of *Teragnatha*: in particular, by the 17 species of the 'spiny leg' clade (Box Figure 9.3). This group had dropped the ancestral web-spinning behaviour and become cursorial predators. They have evolved four distinct ecomorphs, which are associated with distinct micro-habitats. They are maroon on moss, green on leaves, small brown on twigs, and large brown on tree bark. Two species on the oldest islands can change from one ecomorph to another as they age, but younger species appear to have become fixed in their ecomorph expression. Switching between ecomorphs has clearly occurred repeatedly in the evolution of different species within the lineage, exemplifying convergent evolution. Despite belonging to different (albeit closely related species), the representatives of each ecomorph are almost identical both morphologically and ecologically. It is thought that the colour matching is a result of natural selection due to bird predation. Species numbers and species density of *Tetragnatha* initially increase with island age and then decline on the older islands: a pattern shared with the genus *Ariamnes*.

Box 9.3 *Continued*

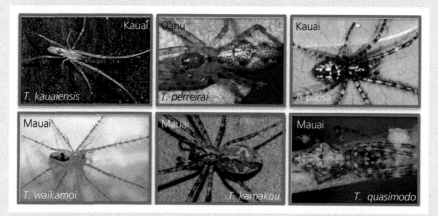

Box Figure 9.3 Members of the spiny leg clade of Hawaiian *Tetragnatha* spiders, illustrating three of the four ecomorphs. Members of different ecomorphs within the same island are more closely related than they are to the same ecomorph on a different island. From Gillespie (2016); her figure 2B; Creative Commons CC BY).

The distribution of ecomorphs of the spiny leg *Tetragnatha* have provided additional insights into the early phases of adaptive radiation (Cotoras et al. 2018). Multiple representatives (i.e. different species) of the same ecomorph do not generally co-occur, but at least on the older islands, one representative of each of the four ecomorphs can typically be found within the same area. However, within one volcano on East Maui, it is possible to find sites where combinations of two of three closely related species (*T. waikamoi*, *T. macracantha*, and *T. brevignatha*), representing the green ecomorph, co-occur, although over most of the island they do not do so. Although hybridization may have played a role at earlier stages of the evolution of Hawaiian *Tetragnatha*, admixture analysis revealed that these populations did not appear to be hybridizing, indicating that they have already attained reproductive isolation.

To explain their findings, Cotoras et al. (2018) suggest a scenario of multiple colonization events among East Maui and the other younger islands. Where species representing the same ecomorph (and that have evolved in allopatry) come into contact, one mostly excludes another. But in the context of the young—to intermediate-aged systems on East Maui, they may sometimes come into sympatry, where competition may be expected to shape further specialization. Cotoras et al. (2018) argue that these co-occurring populations represent a transient phase within the radiation, as co-occurring species of the same ecomorph have not been found on older islands. This phase may contribute to the intermediate-island age peak in diversity that has been observed in this and other taxa within hotspot archipelagos.

NB The repeated, independent evolution of the same ecomorphs within insular radiations is by no means limited to spiders and has, for example, been described in cichlid fish in Lake Tanganyika, Caribbean *Anolis*, and *Mandarina* land snails in the Bonin Islands (Chiba 2004; Losos 2009; Cotoras et al. 2018).

The most spectacular archipelagic radiations are of groups that have intermediate to low powers of dispersal, small body size, and rapid generational turnover, such as the Hawaiian drosophilids, Canarian *Laparocerus* weevils (below), and many plant taxa within both these archipelagos. Within these archipelagos at least, they also tend to follow the progression rule. However, to attain the richness seen in these exemplar groups, intra-island speciation has to assume comparatively greater importance: this is attained through combinations of niche shifts and within-island geographical isolation (Price and Wagner 2004; Machado 2022).

Hawaiian lobeliads: a spectacular radiation in a dynamic hotspot archipelago

The Hawaiian lobeliads (Asterales: Campanulaceae) provide a powerful illustration of archipelagic diversification in the context of a dynamic hotspot system. The single founding event may have occurred as early as 13 Ma on a former high island, eventually giving rise to the largest documented insular plant radiation. Within 3.4 million years, the group had given rise to six genera, which in time produced 126 species: around an eighth of the Hawaiian flora (Givnish et al. 2009). Invasion of the forest understorey was linked to a significant increase in diversification and, in turn, the lobeliads have enabled diversification in drosophilids and other invertebrate taxa that utilize them. The radiation has produced treelets,

trees, shrubs, woody rosettes, and vines and there have been multiple transitions in habitat occupancy and dispersal mechanisms, combined with pollinator partitioning, repeated as successively younger islands have been colonized. Differences in photosynthetic light responses, changes in floral characteristics, and a trend towards insular secondary woodiness (Section 11.2) are also evident (Fig. 9.6).

Saturation of habitat space with lobeliad species was attained within c. 1.5 Myr (Fig. 9.7) and a net attrition of lobeliad diversity per island can be inferred as older islands erode and subside, as few Hawaiian lobeliads occur below 200 m (Givnish et al. 2009). Another interesting feature is that *Cyanea*, which are bird-dispersed forest species, closely follow the progression rule and have

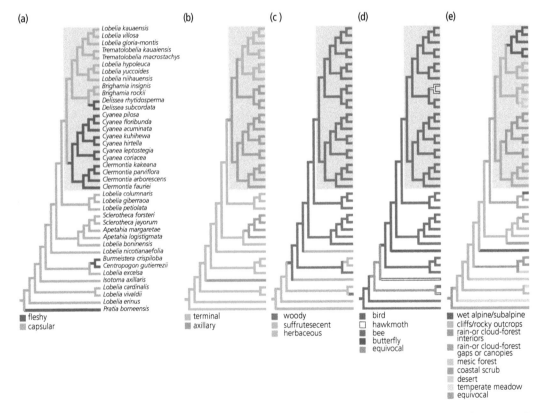

Figure 9.6 Trait and habitat evolution among members of the Hawaiian lobeliads (green shading) and close relatives. The inferred ancestral traits overlaid on the molecular phylogeny are: (a) fruit type, (b) inflorescence position, (c) herbaceous–woody habit, (d) pollination, and (e) habitat. *Lobelia erinus* and *Pratia borneensis* were grafted to the bottom of the in-group tree.
Figure 3 of Givnish et al. (2009).

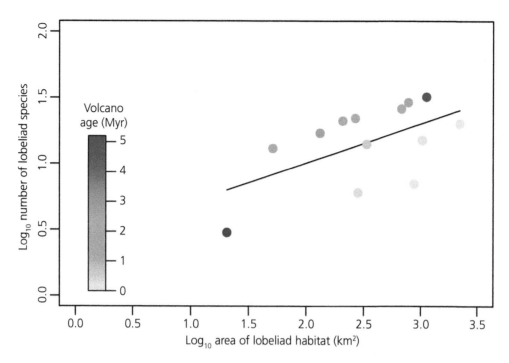

Figure 9.7 Species–area relationship of Hawaiian lobeliad species for individual volcanic mountains (incorporating the climate zones where most species occur), showing a linear relationship for volcanos > 1 Myr of age (with the exception of the old, degraded Ni'ihau volcano), while the richness of younger volcanos fall below the line.
Figure 6 from Borregaard et al. (2017).

diversified to a greater degree (76 species, nearly all single-island endemics), than have several wind-dispersed groups of more open habitats that appear to be more dispersive (Fig. 9.6). In sum, the molecular phylogeny, distribution, and traits of this large radiation provide strong support for several of the trends predicted by the GDM, including the comparatively early burst of diversification, general concordance with the progression rule, attainment of saturation on intermediate-aged islands, and the inferred attrition of diversity with island decline (Borregaard et al. 2017).

The diversification of Macaronesian *Laparocerus* weevils: varied responses to a dynamic archipelago

Laparocerus is a large genus of flightless Macaronesian weevils that has been the subject of extensive collecting and mapping of taxa, combined with morphological, taxonomic, and genetic analyses,

recently consolidated in a monograph by Antonio Machado (2022), from which the following account is drawn. The currently favoured scenario is of a founding event (from a mainland lineage that has since become extinct) into the Madeiran archipelago around 9.7 Ma. From Maderia, *Laparocerus* colonized the Canary Islands (and Selvagens), radiating to produce a swarm of around 217 species (261 species and subspecies), currently assigned to 25 subgenera.

Within the genus there are differences in habitat, feeding niche, plant preference, and seasonal prevalence, exemplifying both disparification of traits and cases of convergence of trait combinations. They are plant feeders, the larva feeding on roots and the adults favouring leaves, especially of shrubs, although species have specialized into different feeding niches, including herbs, trees, and leaf litter (Fig. 9.8). There are also some species that complete their life cycles below ground within the earth or in cavities such as caves. *Laparocerus* occupy all major

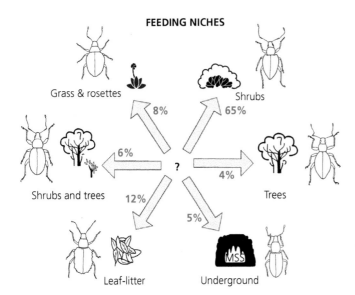

FEEDING NICHES

Figure 9.8 Diversification of Macaronesian *Laparocerus* weevils summarized by proportion of species (or subspecies) occupying each of six broad feeding niches. MSS = 'mileu souterrain superficiel'. From Machado (2022), his figure 453.

ecosystem types and can be found from just behind the beach to over 2000 m a.s.l., with a particular concentration of species in lowland shrubland and mid-elevation forests.

The radiation of *Laparocerus* has been favoured by their being comparatively generalist plant feeders. This means that as each island is added to the archipelago and becomes vegetated, it provides an abundance of suitable resource space, permitting a diverse range of niches to be re-filled once again, from relatively few founding events. The flightless nature of these weevils is also central to their diversification as it means that dispersal between areas separated by geographical barriers is infrequent and thus populations readily become geographically isolated within an island, even within one flank or massif.

The biogeography conforms well to the progression rule (Fig. 9.9), with most inferred colonization events being from older to younger islands, but with a few back-colonization events, including one from the eastern Canaries to the continent about 1.2 Ma, resulting in two species that are endemic to western Morocco. Within the larger Canary Islands, species richness peaks on intermediate-aged islands (Gran Canaria, Tenerife, La Gomera) with fewest species on the youngest (El Hierro) and oldest islands (Lanzarote, Fuerteventura). Within the oldest islands, it

is inferred that there were once larger radiations that have been trimmed back by habitat loss as the islands have aged.

Genetic and morphological data point to the importance of introgressive hybridization and of admixture in the evolution of *Laparocerus*. Within closely related species or subspecies, it is possible to find examples that are allopatric, parapatric, and sympatric. Where overlapping in distribution, some of the species or subspecies have unclear boundaries, suggesting incomplete lineage sorting. Hence, to account for the explosive radiation of *Laparocerus* it is necessary to refer to multiple speciation scenarios including: (i) adaptive changes within lineages, (ii) hybridization within areas of distributional overlap, and (iii) non-adaptive changes within lineages that became allopatric through either (a) dispersal and colonization events, or (b) via barrier formation (linked e.g. to volcanic eruptions, or climate change/sea-level change). The importance of within-island barriers to movement is evident in the presence of many intra-island subspecies, although some variation may be clinal.

Mega-landslides are a repeated feature of the Canary Islands (Section 3.7), occurring on average once per 135 k years and it is conceivable that they may generate both colonization and vicariance events. García-Olivares et al. (2017) find genetic

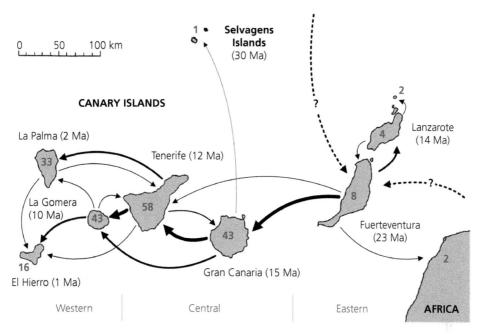

Figure 9.9 The species richness and possible colonization patterns of *Laparocerus* weevils for the Canary Islands and Selvagens, indicating possible source colonization direct from Africa, or via Madeira. Arrows indicate direction and arrow weights indicate proportional importance of colonization events. Approximate island ages are shown in parentheses.
Slightly modified from Machado (2022), his figure 449.

evidence that is supportive of the colonization of La Palma Island from Tenerife by four female lineages of the *L. tessellatus* species complex as a direct result of the Orotava mega-landslide. The hypothesis is that large rafts of material displaced from the flank of Tenerife were carried and deposited inland on the flank of La Palma by a tsunami associated with the collapse event. These events involve vast volumes of material, may occur incredibly swiftly, and within the Canaries are detectable in the form of tsunami deposits emplaced on other islands scores to hundreds of metres above contemporary sea level (Pérez-Torrado et al. 2006). Areas stripped by such flank collapse events often feature secondary infill by renewed volcanism, presenting youthful landscapes that in time are recolonized by plants and animals and where closely related taxa can come back into sympatry, and hybridize, as suggested for example for *L. holiaari* and its sister species *L. boticarius* in connection with the Icod mega-landslip, also on Tenerife (Machado 2022).

Moreover, large-scale eruptions, such as sterilized the majority of Gran Canaria *c.* 3.5 Ma, appear to have provided opportunity for renewed diversification within the affected area, as well, most probably, as causing the loss of localized endemics unknown to science.

Reviewing the findings for *Laparocerus* provides broad support for key tenets of the GDM. Across the genus, diversification rate declines with age of the lineage, likely reflecting a combination of reduced opportunity and increased extinction over time. The fauna of La Palma Island shows evidence of an initial burst of diversification, followed by a settling down over time indicative of 'a reduced set of winners tuned by extinction – including competition – or dilution after secondary sympatry' (Machado 2022, p. 581). However, it is also evident that the evolutionary biogeography of *Laparocerus* reflects the complexities of these islands and of their geological development, which is only crudely matched by the highly simplified colonization assumed within the

GDM. In particular, the long duration of volcanic activity within the Canaries is linked to periodic terrain destruction/renewal cycles, generating opportunities for colonization of young habitats on old islands, sometimes from older habitats on younger islands. In such contexts, the heuristic value of a simplifying model such as the GDM will exceed its predictive power.

9.8 The island progression rule, back-colonization, and onward colonization

The progression rule

Colonists for newly formed islands within oceanic archipelagos are mostly drawn from within their archipelago, disproportionately so from the nearest islands. Within hotspot settings this gives rise to the island progression rule (Funk and Wagner 1995), whereby older islands predominantly donate colonists to the next youngest island emerging in the sequence. This pattern can be detected in the phylogeography of a lineage, and in rapidly speciating groups is evident in the distribution of species and subspecies (Box 9.3; Fig. 9.5, 9.9, 9.10).

The island progression rule pattern is particularly evident for Hawaii, an archipelago in which the islands are arranged in a tidy linear age sequence (Funk and Wagner 1995; Percy et al. 2008; Knope et al. 2020a; Hembry et al. 2021). Examples include plant (silversword alliance, *Cyrtandra, Hesperomannia, Hibiscadelphus, Kokia, Metrosideros, Psychotria, Remya,* and *Schiedea*) and animal taxa (e.g. *Orsonwelles* spiders, *Laupala* crickets, *Hyposmocoma* moths, and flycatchers). The Hawaiian bark louse genus *Ptycta* (Psocidae) also exemplifies the pattern, with molecular dating indicating a colonization event to a previous high island *c.* 7.14 Ma, followed by a split into two main clades on the island of Oahu (then the youngest island) around 3.2–2.6 Ma (Bess et al. 2013). Inter-island colonization has mostly followed the progression rule, leading to the production of 51 species, mostly single-island endemics. There is evidence suggesting the loss of members of a once larger radiation within one of the two clades on the oldest large island, consistent

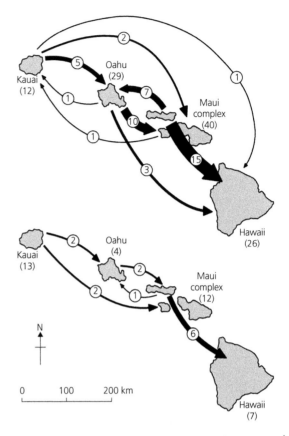

Figure 9.10 The inter-island dispersal events inferred for Hawaiian picture-winged *Drosophila* flies (above) and silverswords (tarweeds) (below). Arrow widths are proportional to the number of dispersal events, and the number of species in each island is shown in parentheses.
Redrawn from Cox and Moore 1993, Fig. 6.9, after an original in G. D. Carr et al. (1989). Adaptive radiation of the Hawaiian silversword alliance (Compositae—Madiinae): a comparison with Hawaiian picture-winged *Drosophila*. In *Genetics, Speciation and the Founder Principle* (ed. L. Y. Giddings, K. Y. Kaneshiro, and W. W. Anderson), pp. 79–95. © Oxford University Press, New York.

with the GDM. Groups largely conforming to the progression rule elsewhere include Macaronesian plants (e.g. *Olea*), lizards (*Gallotia*), and beetles (*Pimelia, Hegeter, Laparocerus*), Galápagos scarabs and weevils, and Austral Islands crab spiders *Misumenops rapaensis* (Whittaker et al. 2010; Machado 2022).

There are, of course, exceptions to the progression rule, even within clades that largely conform, and some groups fail to follow it. This is particularly evident in late-arriving colonists, in more dispersive taxa, and in taxa that invade an archipelago more than once. Macaronesian land birds provide illustration, exemplified by a recent reconstruction of the much-contested colonization history on the Canaries of the Afrocanarian blue tit (*Cyanistes teneriffae*) based largely on single nucleotide polymorphisms (Gohli et al. 2015). This analysis indicated that different islands were colonized from different mainland sources, with the youthful western island of La Palma colonized early and a more recent colonization event (or events) from Africa to four other western islands, followed by the recent arrival of north-west African birds into the two eastern islands. The biogeographical progression of *Argyranthemum* among the Canaries, Madeira, and Selvagens Islands remains tentative but also fails to conform with the progression rule, demonstrating some old-to-young (west-to-east) inter-island colonization events within a late-colonizing group in which the radiation appears to be correlated with an intense aridification episode within the Pleistocene (Hooft van Huysduynen et al. 2021).

Back-colonization, boomerangs, and surfing syngameons

The progression rule as just described refers principally to inter-island colonization. However, the idea is closely linked to the view of mainland to island relationships, which has traditionally viewed islands as evolutionary blind alleys; that is, the flow of colonization events was considered essentially unidirectional from continents to islands, implying a generally lower fitness of island endemic forms. Evolutionary models such as the taxon cycle and insular syndromes of 'slowing down' of life history, loss of dispersal, and of defensive traits (reviewed in Chapters 10 and 11) explain why colonization of continents from islands is comparatively rare. However, occasionally island species do manage to colonize continents (Fig. 9.11). The first reference we have found of this was in an 1878 publication by John Ball and Joseph Dalton Hooker (Fernández-Palacios and Whittaker 2020). Recently, as detailed molecular phylogenies have become available for many taxa, we have

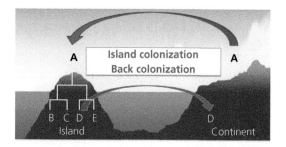

Figure 9.11 Schematic of island colonization by species A, followed by *in situ* evolution and eventually a single back-colonization event (boomerang) of one of the descendant taxa (species D) to the continent.

seen increasing evidence of island–mainland back-colonization (examples in Table 9.3). Such cases are sometimes termed boomerangs, although it is often a descendant species within a radiation that completes the return journey to the continent.

As we know, founder effects should in theory produce reduced genetic variation in island colonists compared with mainland populations, and indeed many insular radiations are marked by relatively limited genetic variation. However, Caujapé-Castells (2011) highlights that in the Canaries, the endemic flora features radiations in which relatively high levels of population genetic variation are observed, notwithstanding that the sequence markers indicate monophyly. He proposes an explanation for this termed the surfing syngameon hypothesis. A syngameon is a group of species that lack strong reproductive barriers, and which frequently hybridize when occurring in sympatry, producing multiple hybrid combinations. Caujapé-Castells proposes that the eastern Canary Islands, which are only around 95 km from Africa, and were only around 65 km from Africa during Pleistocene sea-level minima, may have been colonized by multiple related species, quite possibly with subsequent secondary contact events, generating the genetic variation in the founding members of radiations and producing monophyly that is more apparent than real.

The applicability of the surfing syngameons model is yet to be more widely evaluated and aspects of the model would appear intrinsically less likely to hold broad relevance to really isolated

Table 9.3 Some putative back-colonizations (boomerangs) suggested by molecular evidence for the Macaronesian flora. (A = Azores, CV = Cabo Verde, E = Europe, EA = E-Asia, I = Iberia, Mac = Macaronesia, Ma = Madeira, Me = Mediterranean, mu = Multiple, NEA = NE Africa, NWA = NW Africa, NW = New World, SA = South Africa).

Taxon	Family	Origin	Target
Plants			
Aeonium balsamiferum	Crassulaceae	Mac	NEA (*A. leucoblepharum*)
Andryala pinnatifida	Asteraceae	Ma	Me (*A. ragusina, A. integrifolia*)
Arbutus canariensis	Ericaceae	NW	Me (*A. andrachne, A. unedo*)
Convolvulus spp.	Convolvulaceae	Mac	I (*C. fernadezii*)
Helichrysum gossypinum	Asteraceae	Ma	Me (ancestor of several species)
Ilex perado	Aquifoliaceae	mu	EA (*I. leucoclada, I. latifolia, I. rugosa*)
Kleinia neriifolia	Asteraceae	EA	NWA (*K. anteuphorbium*)
Lotus campylocladus	Fabaceae	NWA	NWA (*L. assakensis*)
Matthiola bolleana	Brassicaceae	NWA	NWA (*M. bolleana, M. longipetala* ssp. *viridis*)
Sideroxylon mirmulans	Sapotaceae	EA	CV (*S. marginata*), SA (*S. inerme*)
Teline stenopetala (ssp. *micropauci*)	Fabaceae	NWA	NWA (*T. monspessulana*)
Tinguarra montana	Apiaceae	NWA	NWA (*T. sicula*)
Tolpis azorica	Asteraceae	A	NWA (*T. virgata, T. barbata*)

Extracted from Caujapé-Castells (2011).

insular systems. However, Caujapé-Castells et al. (2017) list 28 Canarian plant genera where there is robust evidence for hybridization (e.g. *Aeonium, Argyranthemum, Echium,* and *Euphorbia*)—and the figure could be as high as 60 genera—to some of which the surfing syngameon hypothesis may apply. One additional prediction of the surfing syngameons model is that as islands age and topographical complexity begins to decrease, there may be an increase in population genetic diversity as previously allopatric populations come into secondary contact (Caujapé-Castells et al. 2017). Hence, the ideas captured in the surfing syngameons model are linked to the episodic production of new territory via island ontogeny, the turnover of available land surface through volcanism, mega-landslips, etc., and to climate change, especially that linked to the onset of glaciation within the Quaternary. In this way, the model links to another model, presented in Section 7.8, the glacial-sensitive model of island biogeography (Fernández-Palacios et al. 2016a).

As described by the glacial-sensitive model, sea-level change driven by the glacial–interglacial cycles have multiple impacts that may generate signal in phylogenies and not just in numbers of species or of endemics (Fig. 7.7). The impacts include (i) the changing configuration of major ecosystem types within islands and continents, (ii) the changing area of the islands, (iii) the emergence of additional stepping-stone islands, and (iv) changes to major ocean and atmospheric current systems (Fernández-Palacios et al. 2016a). These factors may each combine to influence the rhythms of immigration and extinction. When all combine in tandem, the upshot may be dramatic and has been described as akin to opening and closing colonization 'windows' (Carine 2005).

According to Caujapé-Castells (2011), roughly 35% of Canarian plant genera originate from the Mediterranean, 25% from North Africa, 18% from within the Macaronesian archipelagos (including the Canaries themselves), and the remaining 22% from East or Southern Africa (Rand flora elements) or the New World. It seems evident from the work reviewed in this volume that these contributions to the flora have not been constant in proportion through time. We might therefore extend the idea of Pleistocene-era windows of opportunity 'opening and closing', to other sets of windows opening and closing over varying timescales during the long history of Macaronesia (including Palaeo-Macaronesia): as the Atlantic widened, as the Tethys Seaway closed, and as the islands have experienced changing climate over the last few million years (Caujapé-Castells, 2011; Grehan 2017; Caujapé-Castells et al. 2017).

Meta-archipelagos and upstream colonization to continents

Most colonists to remote archipelagos derive from a mainland source pool, but archipelagos also gain colonists from other (mostly) nearby archipelagos. Where this happens frequently, we may think of the islands existing within a meta-archipelago (Fig. 4.7; Whittaker et al. 2018). Within Macaronesia, for example, interchange between the Canary, Selvagens, and Madeira archipelagos is evident in multiple cases (e.g. in chaffinches, *Laparocerus* weevils, and in *Argyranthemum*), while this concept also applies well to some lineages of birds in the Melanesian and Wallacean regions (Ó Marcaigh 2022).

There are also cases of lineages that appear to have originated in one continent, spread across a series of islands, and colonized another continent, for which 'upstream' (or onward, or *de novo*) colonization would seem a more appropriate label than 'back-colonization' (Jønsson et al. 2011; Katinas et al. 2013; Lomolino et al. 2017). Even the remote Hawaiian Islands have produced species that have managed to colonize mainland regions, an example being *Scaptomyza* (a genus of fruit flies in the Drosophilidae), which have colonized several continents as well as islands as distant from Hawaii as Tristan da Cuhna (Lomolino et al. 2017). A spectacular illustration of onward/upstream colonization stemming from an insular region comes from birds, and specifically concerns the origin of the core Corvoidea, a group of over 700 species of passerines. Molecular phylogenetic data indicate an origin of the group as generalist feeders favouring open habitats, in the emerging proto-Papuan archipelago during the late Eocene/early Oligocene. Not only did multiple colonization events into South-East Asia subsequently follow, but also the group has since spread across all the continents (Jønsson et al. 2011, 2017). This is fairly remarkable when it is realized how effective and persistent the deep-water barriers (exemplified by Wallace's line) between the Sahul region (Australia and New Guinea) and South-East Asia have been to the spread of the great majority of birds (e.g. Dalsgaard et al. 2014; Ó Marcaigh 2022).

Dissecting mainland–island relationships in relation to climate change

It is interesting to speculate on what traits or circumstances enable a small minority of species to go against the general direction of taxon cycles and the associated continental to island flow. Are back—or upstream—colonizers simply a random draw of insular species? Do they exhibit hybrid vigour, or feature novel innovations (e.g. brain size in crow-like birds: Jønsson et al. 2011), that provide them with the necessary dispersibility and/or competitive advantage? Might some insular systems have provided climatically shifted environments (compared with nearby continents), which act as reservoirs (generators, or perhaps refugia) of pre-adapted species that are well positioned to invade the mainland at points where global climate change opens up new niche space in the mainland ecosystems? Testing such hypotheses robustly will require more than just anecdotal evidence from a few case studies (but see, for example, discussions in Jønsson et al. 2011; Caujapé-Castells et al. 2017; Hooft van Huysduynen et al. 2021).

When phylogeographical analyses of species distributed both in the continent and on islands show a pattern characterized by small, haplotype-rich insular populations vs. large, haplotype-poor continental populations, we may raise two contrasting scenarios in the first instance. On the one hand, a severe bottleneck may have had affected the continental population in the recent past (e.g. range collapse during a glacial period) eroding their genetic variance until an extreme where only one or few haplotypes survive. In such circumstances, the continental haplotype(s) may be absent from the insular populations. A possible fit is provided by *Cistus monspeliensis*, a widely distributed Mediterranean stone rose species with populations ranging from Cyprus to Iberia that features a single haplotype (Fernández-Mazuecos and Vargas, 2011). The species also occurs in the Canaries, where it is inferred to have colonized from the Mediterranean region in the Early to Middle Pleistocene (> 1 Ma). Within the Canaries it has followed a stepwise differentiation since 0.6 Ma, diversifying to produce 10 different haplotypes, in very local and fragmented populations, the majority restricted to a single island. As the single continental haplotype is basal to the Canarian ones, we can rule out a boomerang event.

On the other hand, where the few (or single) haplotypes characterizing the wider-range continental population are also present in insular populations that are richer in haplotypes, we may consider two further scenarios: (i) a recent island to mainland colonization event has occurred, followed by a continental range expansion, or (ii) a continental bottleneck has occurred, followed by both dispersal to islands and subsequent range expansion on the continent. Here, discerning the direction of the dispersal event is complex, although the geographical distribution of the haplotypes may help in the reconstruction. The tree *Erica scoparia* features four different haplotypes in the Western Mediterranean, all of them also present in Macaronesia. The Macaronesian populations also feature five additional haplotypes (one in the Canaries, one in Madeira, and three more in the Azores) exclusive to these islands. Only one of the continental haplotypes is widely distributed, the other three being limited to the Atlantic coast of Iberia. In this instance, a recent upstream colonization from the Azores to the Iberian Peninsula appears likely (Désamoré et al. 2012). Upstream colonization from Macaronesia to Iberia has also been inferred for multiple populations of bryophytes, suggesting that these islands have played important roles in replenishing continental diversity post the Last Glacial Maximum (Patiño et al. 2015). For an extension of this idea to angiosperms see Caujapé-Castells et al. (2022), who argue that the high level of floristic endemism in the Canaries is in part attributable to extinction of mainland relatives during the Late Pleistocene.

9.9 Insular diversification: concluding observations

The geo-environmental dynamism of insular regions appears critical to the ways in which archipelagos and meta-archipelagos act as speciation machines. We have focused mostly on oceanic island systems and particularly on hotspot

archipelagos, because of their *comparative* simplicity. However, continental fragment systems and mixed oceanic, continental fragment and continental shelf systems such as the Wallacean region also provide fabulous natural laboratories that can be tapped to develop and test evolutionary models (Wilson 1959, 1961; Michaux 2010; Dalsgaard et al. 2014; Rheindt et al. 2020; Ali and Heaney 2021; Ó Marcaigh 2022). Running as a thread through the rich literature on insular evolution, and perhaps particularly evident in a region such as Wallacea, is the importance of incumbency. Naturally, insular barriers are hard to breach. Yet, newly emerged islands attract colonists and while there is typically a predominant geographical source region, colonists can arrive from diverse sources from a great range of distances.

It seems almost self-evident, especially from hotspot archipelago systems, that it is in the spin-up phase of an island's life cycle where opportunities for colonization and the initiation of diversification are greatest. While the filters discussed in Section 8.4 limit the phylogenetic and phenotypic character of the colonists, there is a large degree of chance to the suite of species that arrive on a young, remote island. Founding effects, natural and sexual selection, and hybridization within closely related species all serve to generate novel genetic combinations, with some of the resulting taxa successfully radiating and spreading from island to island and archipelago to archipelago. Ecological opportunity presented by vacant niche space and the novel community mix that newly arrived species encounter leads to a combination of forms of speciation, associated with the exploitation of novel habitats or resources (e.g. mutualists).

Ecological opportunity as so defined is not enough, and it is also necessary for colonizing species to possess traits that enable them to utilize the relevant resources (Valente et al. 2015). The evolution of a certain feature or trait (a 'key innovation', e.g. toepads in *Anolis* lizards, the tubular tongue in Hawaiian honeycreepers) may open up a wider range of resources to a lineage, thus promoting further diversification (Pratt 2005; Losos and Mahler 2010). The arrival or evolution of a new type of resource (e.g. the evolution of species creating a new macrohabitat on an island) can also be critical, spurring co-evolutionary diversification and dis-parification (e.g. in plants and pollinators). As a further example, it has been postulated that within the lobeliad radiation on Hawaii, some species may have radiated within niche space created by other members of the radiation (e.g. shade-dependent shrubs within the microhabitat created by forest trees) (Givnish et al. 2009).

The propensity to hybridize has gained increasing attention as a driver of diversification. As we noted earlier, hybridization between weakly differentiated populations can retard the speciation process and early observations suggested that polyploidization within islands, which is strongly linked to hybridization (all allopolyploids are of hybrid origin) is not a key means of speciation. However, (recent) polyploid taxa do appear to radiate well once they have colonized. Furthermore, in taxa ranging from plants, through beetles to birds, and through a growing number of case studies, hybridization appears to have an important role in providing the genetic variation which fuels adaptive radiation (Lamichhaney et al. 2015; Gillespie et al. 2020). In particular, the reassembly of old genetic variation, derived through past hybridization events, into novel genetic combinations has recently been theorized to facilitate rapid adaptive radiation (Marques et al. 2019; Rubin et al. 2022). The importance of such 'combinatorial' mechanisms is partly evidenced through the observation that, in many cases, the genetic variation underlying reproductive isolation and speciation (e.g. the genetic variation underlying beak shape and size in Darwin's finches, below) is often older than the age of splitting events between species within a radiation (Marques et al. 2019). Classic insular adaptive radiations where hybridization is believed to have played an important role include Darwin's finches on the Galápagos, the silverswords of Hawaii, and the cichlid fishes of the Lake Victoria region.

Important genetic changes that result in high evolvability within a lineage, such as the decoupling of skull modules in the evolution of the Hawaiian honeycreepers (Tokita et al. 2017), may allow for greater evolutionary flexibility and ability to take advantage of ecological opportunities. The capacity of the members of an emerging radiation to diversify rapidly is also facilitated by their interacting

with the environment at a fine scale, having sufficient space to form multiple geographically isolated populations, and being able to experience cycles of allopatric and sympatric development. Among plants, smaller-sized, shorter-lived colonists have tended to radiate far more dramatically than have trees. For example, within the Juan Fernández Islands radiating lineages are typically herbaceous, but within the Canaries, radiations more commonly involve shrubby species (Takayama et al. 2018; Fernández-Palacios et al. 2021b). It has also been suggested that such traits as the breeding systems, tendency towards outcrossing, and pollination mechanism—all of which may influence rates of gene flow—should have a bearing on rates of diversification (Givnish 2010; and see Section 9.7 above). Within at least some animal groups, sexual selection appears to be an important mechanism that is linked to rates of radiation, and alongside competitive displacement and disruptive selection, can facilitate the attainment of reproductive isolation in sympatry (Rundell and Price 2009; Recuerda et al. 2021; Matsubayashi and Yamaguchi 2022).

Another theme running through this book and evident throughout the chapter is the importance of the geographical template and its dynamics. Insular radiations are more spectacular with increased isolation, on larger islands, and where those larger islands are clustered into archipelagos that allow occasional exchange of colonists. An interesting counterpoint is provided by Cocos Island, 630 km to the north-east of the Galápagos (and some 530 km from Central America. The Cocos finch, *Pinaroloxias inornata*, is the sole representative of the Galápagos finch clade found outside the Galápagos. This species has failed to undergo cladogenesis within Cocos Island but has evolved a remarkable diversity of foraging behaviors spanning the range normally occupied by several families of birds (Werner and Sherry, 1987). This intraspecific variation in behaviour appears to originate and be maintained year-round, possibly though observational learning. Cotoras et al. (2021) report a similar case in another endemic Cocos Island species, the spider *Wendilgarda galapagensis*, which has also diversified its feeding niche by deploying one of three different web types, without speciation having occurred.

These observations are consistent with the idea that islands below a particular threshold of area (and associated topographic variation) typically cannot generate and/or sustain many closely related endemics. For example, in the adaptive radiation of Greater Antillean anoles (Losos 2009), the ISAR is characterized by a breakpoint relationship (i.e. a threshold island area above which the ISAR steepens) (Losos and Schluter 2000). This in turn is seemingly driven by a similar breakpoint relationship between speciation rate and island area, perhaps due to (i) increased likelihood of allopatric separation on larger islands, and (ii) greater habitat diversity on large islands providing more ecological opportunity (Losos and Schluter 2000).

Responses to area vary greatly between taxa, as a consequence of body sizes and traits influencing mobility and dispersal (Fig. 8.9; Kisel and Barraclough 2010). For example, using laurel forest beetle species on the Anaga Peninsula, Tenerife, as a case study, Salces-Castellano et al. (2020) found support for a model of geographical isolation by niche conservatism of environmental and climatic tolerances across a topographically complex peninsula. Microclimatic factors, such as humidity and temperature, vary significantly across the peninsula as a function of elevation and aspect. As many laurel forest beetles have specific microclimatic niche requirements coupled with limited dispersal ability, this microclimatic variation results in a patchwork of suitable and unsuitable areas (e.g. humid conditions within the forest vs. drier intervening areas), creating the ideal conditions for allopatric isolation and subsequent genetic divergence between populations separated by only a few kilometres (Salces-Castellano et al. 2020).

It presents something of a paradox that taxa such as land snails and flightless *Laparocerus* weevils should be so effective at colonizing new territory and yet become so isolated within larger volcanic islands as to radiate so extensively, sometimes seemingly non-adaptively. It seems likely that we underestimate their frequency of dispersal, but that occasional strays rarely establish or make effective contribution given the propagule pressure provided by incumbent populations. We suspect that successful colonization events are often enabled by coinciding with a point where recipient ecosystems have been disturbed in some way, to create temporary resource availability.

As we have seen throughout the chapter, it is the combination of environmental drivers and the specific traits of the focal group that determine the response. As a further example, for cichlid fish radiations in the African Great Lakes, it is the combined effects of the intensity of sexual selection and environmental opportunity (in the form of lake depth, energy availability, and lake age), that best predict whether adaptive radiation will occur (Wagner et al. 2012). Finding that propensity for (adaptive) radiation is controlled by several factors helps to explain why only some taxa radiate in islands and archipelagos that feature great examples of it. Radiation may be predictable, but only when lineage traits and environmental factors are jointly considered.

Studies in Hawaii and the Canaries also associate the generation of novel diversity (and the emergence of hybrid swarms) with large-scale disturbance events within those archipelagos (e.g. Carson et al. 1990; Machado 2022; Salces-Castellano et al. 2020). The emergence, development, disturbance and rebuilding, and eventual decline of islands are represented within the GDM of oceanic island biogeography in highly stylized form. As we have seen, it captures the essence of the dynamics of hotspot archipelagos and how clades respond to them at a range of scales, but fine-scale evaluation of the role of geotectonics in island evolution inevitably leads to more complex descriptive models of lineage responses to island geodynamics (Machado 2022).

The signal of other forms of environmental change: on longer timescales, of the movement of terranes, and changing configurations of archipelagos and continents (e.g. Grehan 2017), and within the last few million years, climatic and associated sea-level fluctuation, is also evident but often harder to unravel, in the phylogenies of insular groups and their mainland relatives (Fernández-Palacios et al. 2016a; Caujapé-Castells et al. 2017; Machado 2022). Despite the complexities of the analyses we have reviewed, insular systems are increasingly tractable study systems because of the number of youthful endemic radiations (or radiations that are largely endemic, but with embedded upstream colonists), for which well-specified molecular phylogenies are available. Moreover, insular systems not only illustrate the overarching patterns and processes of taxon cycles and of radiations, but also serve as excellent natural laboratories for trait evolution, a theme we will expand further on in the following two chapters.

9.10 Summary

In this chapter, we introduce three general but non-exclusive models for insular diversification: the taxon cycle, adaptive radiation, and non-adaptive radiation. We elaborate on and evaluate these models through scrutiny of several case-study systems. We commence with the taxon cycle, a model introduced by E. O. Wilson to account for the biogeography of Melanesian ants.

The taxon cycle is notable for embracing evolutionary dynamics across multiple lineages simultaneously. It describes how colonist lineages of open habitats spread and diversify across an island region, while undergoing an evolutionary shift towards forest interior habitats, in the process becoming more specialized and restricted in their distributions, propelled in part by competition with later arriving species that again invade the more open and coastal habitats. More recent work applying modern phylogenetic tools and more systematic distributional data, largely reaffirms the role of taxon cycles in ants. Similar work on birds of the Lesser Antilles, the Indo-Pacific region, and even on Hawaiian honeycreepers (a classic example of adaptive radiation) shows the model to have similar explanatory value. The taxon cycle is notable for inferring a final stage of extinction, where early products of insular evolution are lost.

Adaptive radiation refers to the diversification of a monophyletic group of species to fill a range of distinct niches, typically involving trait disparification as well as instances of trait convergence in allopatric species. Non-adaptive radiation, by contrast, involves diversification with only limited trait shifts, lacking obvious selective benefit, largely acquired in allopatry through stochastic processes. We highlight that the geo-environmental dynamics, as represented in the GDM for oceanic islands, indicate that opportunities for radiation should show a humped trajectory through time, with extinction

dominant in old declining islands. The trajectory at archipelago level, however, depends on whether the archipelago is growing or shrinking and may be influenced by climate and associated sea-level change.

We focus on the Galápagos finches and Hawaiian honeycreepers as classic illustrations of adaptive radiation. The radiations feature diversification of traits in responses to untapped resource space, with episodes of change in allopatry followed by character displacement in sympatry. Hybridization is an important element of evolutionary change in the former, less so in the older and larger diversification of honeycreepers, in which the importance of new island formation and co-evolution with other taxa is more clearly evident.

Land snails provide classic examples of non-adaptive radiation, but the idea has relevance to numerous insular radiations, including in plants, birds, and spiders. In practice, it is better to view radiations as a continuum, or perhaps a blend between adaptive and non-adaptive components, within which sexual selection (as exemplified by studies of cichlid fish) also often has an important role. We illustrate the mixed nature of insular radiations through plant examples, highlighting once again the importance of hybridization as a source of novelty and potentially as a means by which non-monophyly occurring at the outset of some insular radiations is disguised.

We consider Hawaiian spiders, Hawaiian lobeliads, and Macaronesian *Laparocerus* weevils as case-study systems that reinforce the emerging model of insular radiations as involving multiple evolutionary mechanisms and processes, intrinsically linked to the dynamism of the archipelagos in which they occur. Repeated phases of intra-island disturbance, isolation, and mixing, in tandem with inter-island biogeographical dynamics, are crucial to understanding larger radiations within dynamic archipelagos.

Finally, we consider the relationships among islands and between islands and continental regions in more detail. The progression rule describes the general pattern of old-island-to-young-island colonization, dominant among larger hotspot system radiations. But even in those systems, some back-colonization events occur, and in late arriving taxa, neither pattern may be dominant. With the increasing availability of molecular phylogenies, evidence of back colonization (boomerangs), or onward (upstream) colonization of island taxa to continents is becoming increasingly clear. Although comparatively rare, over lengthy periods of time such events can be significant, as evident in the spread and diversification of originally insular core Corvoidea around the globe.

We do not yet have a general model of what traits and circumstances lead to back colonization or upstream colonization events, but a link with major shifts in climate is one plausible hypothesis put forward in relation to the Macaronesian flora. In our closing comments we highlight: (i) the need to pay attention to these dynamics, in addition to (ii) the intrinsic geo-environmental dynamics of volcanic island systems, in developing our understanding of insular diversification (and loss). And, finally (iii), we again emphasize the role of hybridization as a frequently key component in the early stages of insular radiations.

CHAPTER 10

Island evolutionary syndromes in animals

10.1 When do trait changes become a syndrome?

As Europeans began their transoceanic travels at the end of the 15th century, natural scientists were increasingly fascinated by many surprising forms of life on the islands they visited (Chapter 1). Among them were tree lettuces, tree sunflowers, flightless birds and insects, animals of unusually large or small size, and animals with no apparent fear of humans. The absence of many continental species groups, such as terrestrial mammals and amphibians, contrasted with high densities of other groups (such as lizards), usually much rarer in mainland communities. Somewhat later, with the discovery of fossil bones of extinct island species, natural scientists were again amazed to learn of the former existence of flightless island birds in excess of 3 m in height, dormice larger than cats, or insular elephants and mammoths the size of pets.

These observations have generated a host of specific questions that resolve into the following general ones. Are the differences between island and mainland forms real and general, or have we become over-influenced by a few extreme cases in combination with sampling biases? If there are real differences, how consistent are they and can we exploit variation in island properties such as area, isolation, and duration of isolation, to develop evolutionary models capable of explaining and predicting trait evolution?

While particular traits are often analysed individually, it is the combination of physiognomic, reproductive, or behavioural traits together into colour of insular syndromes that has excited most interest.

Notably, Adler and Levins (1994) identified a combination of changes in morphology and behaviour of rodents that they simply termed *the island syndrome*, combining around six distinct properties (including body-size increases, reduced aggressiveness, reduced reproductive output) and for which we adopt the label the **slowing-down syndrome** (*sensu* Schwarz et al. 2020). In practice, commonly investigated properties such as changes in reproductive output, or loss of flight, can themselves be considered syndromes, often involving trade-offs in more than one contributory trait. For instance, alterations in reproductive output may involve reduced clutch size but increased investment per offspring. By this line of argument, there are multiple island syndromes and Adler and Levin's model represents a sort of super-syndrome integrating several other syndromes we consider herein. It is with this 'super-syndrome' that we open the chapter.

Many of the island traits and syndromes were identified by the founding figures of the natural sciences (e.g. Darwin 1859; Wallace 1902) and an important contribution was subsequently made by the botanist Sherwin Carlquist (1965, 1974), through his books *Island Life* (1965) and *Island Biology* (1974) (see Traveset et al. 2016). Determining which of these putative syndromes are robust and repeated features of insular systems requires painstaking research efforts (Box 10.1). As in other fields of island biogeography, recent compilations of systematic datasets combining traits of extant and extinct species, phylogenetic relationships, and island properties and their analysis with powerful statistical tools have enabled more robust tests and have thus generated rapid advances in

Island Biogeography. Robert J. Whittaker, José María Fernández-Palacios, and Thomas J. Matthews, Oxford University Press.
© Robert J. Whittaker, José María Fernández-Palacios, and Thomas J. Matthews (2023). DOI: 10.1093/oso/9780198868569.003.0010

understanding. In this chapter, we explore island syndromes in animals, and in Chapter 11, we shift our attention to plants and to plant–animal interactions.

Box 10.1 Cautionary note on tests for insular syndromes based on comparative analysis of large datasets

Systematic comparative analyses of multiple datasets testing for particular insular syndromes are essential to establishing the level of support and for identifying the factors controlling the evolution of each syndrome. Issues that may need to be considered include the following.

a) **The need for phylogenetic correction.** Comparative analyses of large datasets should ideally be corrected for phylogenetic relatedness to satisfy statistical requirements for independence. Robust phylogenetic trees are not readily available for many taxa, especially when considering extinct species. Moreover, unexpected evidence of island lineages colonizing continents shows the danger of inappropriately rooted trees.

b) **Distinguishing island types.** Continental fragment islands isolated for tens of millions of years, volcanic oceanic (Darwinian) islands isolated from their moment of origin, and land-bridge islands isolated for just a few thousand years, present very different evolutionary contexts. Studies reviewed herein and in Chapter 11 highlight that these differences matter for detection of island syndromes, but such distinctions are frequently not made in comparative analyses.

c) **Distinguishing colonization bias from *in situ* evolutionary changes.** For particular island syndromes, it may be unclear whether the pattern has arisen through *in situ* evolutionary dynamics or through colonization and establishment biases. One way to tackle this is to undertake separate analyses restricted to endemic species in comparison to non-endemic natives, to sift out the contribution made through *in situ* speciation.

d) **Distinguishing different duration of island presence.** Some syndromes, including body-size changes, can take lengthy periods to develop and so analyses that are able to reliably estimate the duration for which particular taxa have been present on an island (either through molecular phylogenies or fossil data) may be able to provide improved resolution on the generalities of trends in animal (or plant) traits.

e) **Considering extinct species.** As noted in (a) and demonstrated herein, inclusion of data for extinct species can increase or diminish evidence for particular syndromes. While the catalogue of extinct species has expanded greatly, we cannot be certain what cryptic biases arise from unrecorded insular extinctions.

10.2 The slowing-down syndrome in rodents and lizards

Based on review of work mostly on mice and voles, Adler and Levins (1994) identified that island rodent populations tend to evolve higher and more stable densities, better survival, increased body mass and reduced aggressiveness, reproductive output, and dispersal (Table 10.1). They labelled this collective set of trait changes/syndromes in rodents, **the island syndrome**. Subsequent workers have described the changes as amounting to a 'slowing down' of life-history traits and behaviour (Schwarz et al. 2020), hence a more informative label is the **slowing-down syndrome**.

Island isolation not only reduces immigration rates, but it also reduces the prospects of emigration by individuals at times of resource shortage (the fence effect). This selects for more sedentary behaviour (Table 10.1). As islands have fewer species than equivalent continental areas, population densities per species for any given taxon should, all other things being equal, be greater on average for island than mainland populations. This effect is termed **density compensation** and it has been found to be a general (although not universal) insular pattern (Section 8.6), with density overcompensation (i.e. higher densities than expected for the reduction in richness) also being reported in some cases, particularly for lizards (Novosolov et al. 2016). The higher densities (the crowding effect, Table 10.1) of island populations can reflect the release not only from interspecific competition but also from predation (Section 8.6).

Where there is reduced predation and/or interspecific competition, intraspecific interactions become of greater selective importance. Put this together with a limited benefit of dispersiveness and we see why there will be insular selection of small mammals for larger body size, reduced

Table 10.1 The slowing-down syndrome: short-term and long-term changes in island rodents and proposed explanations.

Island trait	Proposed explanation
Reduced dispersal	Immediate constraint (short-term response) and natural selection against dispersers (long-term response)
Reduced aggression	Initially, reduced population turnover, greater familiarity with neighbours, and kin recognition. Long-term directional selection for reduced aggression
Crowding effect	Isolation ('fence effect' resulting from reduced dispersal) and reduced number of mortality agents such as predation, both of which result in crowding of individuals and consequently higher population densities
Greater individual body size	Initially, a reaction norm as a response to higher density. Long-term directional selection for increased body size in response to increased interspecific competition
Lower reproductive output per individual	Initially, a reaction norm as a response to increased density. Long-term, directional selection in response to decreased mortality
Greater life expectancy (higher survival probabilities for individuals)	Reduced number of mortality agents such as predation

From Adler and Levins (1994), their table 2.

reproductive output (but greater investment in individual offspring), and greater longevity. Adler and Levins (1994) go further in arguing that: (i) increasing island isolation should reduce diversity, generating selective benefits from reduction in aggressiveness, leading to greater population stability; and (ii) reduced island area should enhance density compensation effects and increased body size in the few species occurring on the island (Fig. 10.1). Hence, they provide not only a 'super-syndrome' synthesis of previous work on small islands but also a set of predictions for further testing.

Russell et al. (2011) studied introduced black rats on two isolated atolls in the Mozambique Channel to test predictions of this syndrome. Europa Island is comparatively natural in its ecology, but Juan de Nova has been altered by multiple species introductions to become more like a continental system. As predicted, rats on Europa Island had larger body sizes, greater densities, lower rates of reproduction, and lower population turnover than those on Juan de Nova. However, a study of deer mice (*Peromyscus maniculatus*) and red-back voles (*Myodes gapperi*) in six mainland and 10 island sites in a Canadian river found only mixed support for the syndrome (Juette et al. 2020). Insular members of both species showed less aggression towards

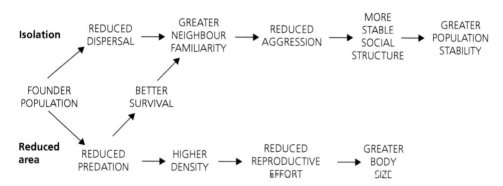

Figure 10.1 Schematic diagram showing the initial effects of island isolation and area on rodent populations. Long-term effects of insularity are directional selection for increased body size, reduced reproductive output, and reduced aggression.
Redrawn from Adler and Levins (1994), their figure 2.

predators, but aggressiveness of voles increased rather than decreased with island isolation. A separate study of deer mice from Saturna Island and nearby mainland sites (also in Canada) found through captive breeding and laboratory tests that while the increased body size of insular mice was heritable, there was no difference in aggressiveness in captive bred individuals from the two populations, which shows the reduction in aggression to be a plastic, behavioural response (Baier and Hoekstra, 2019).

Extending the scope of the syndrome, Novosolov et al. (2013) assembled data for 641 species of lizard, 100 of which were island endemics. Consistent with expectations, island endemic populations were on average four times denser than mainland populations and even after corrections for body mass and phylogeny (Box 10.1) the difference remained significant. The insular species also laid fewer eggs and produced larger offspring than did close mainland relatives, although endemics on larger islands laid more frequent clutches than mainland lizards. In a departure from Table 10.1, they hypothesized that larger offspring size in insular endemics may be selected for due to increased intraspecific aggression and cannibalism by adults.

These findings suggest that further tests for the **slowing-down syndrome** would be worthwhile. Recent work on evolutionary shifts in reproductive output in island lizards (Section 10.5, below) indicates that there remains much yet to be resolved about how insularity shapes these interlinked syndromes in different insular contexts and taxa.

10.3 The island body-size rule

Body-size changes are common within adaptive radiations, as exemplified by the lemurs of Madagascar (a large continental fragment island), and the *Gallotia* lizards of the Canaries (oceanic islands). In these cases, the insular lineages have exploited vacant niche space to diversify in size during cladogenesis (Hernández et al. 2000; Lomolino et al. 2017). When considering the island body-size rule, we are more typically considering cases of anagenesis so that what is being measured is the mainland–island body-size contrast of a pair of species/populations (e.g. Lomolino et al. 2013).

Mammals

The unusual body size of many island taxa, especially vertebrates, when compared with related mainland species, is one of the best-known features of island life. In a seminal paper, Foster (1964) tabulated size differences between mainland and insular mammals of different orders (Table 10.2). He noted that whereas island rodent species tend towards gigantism, the carnivores, cloven-hooved mammals (artiodactyls), and rabbits (lagomorphs) tend towards nanism or dwarfism. Based on analysis of a much larger dataset, Lomolino (1985, p. 312) subsequently described a 'graded trend from gigantism in the smaller species of insular mammals to dwarfism in the larger species'. It is this pattern that has been named the island rule, although a more precise label would be the **island body-size rule**, which we adopt here. Initially described for mammals, subsequent work has sought to expand its application to other vertebrate taxa and even to invertebrates and plants (Lokatis and Jeschke 2018; Chapter 11). We begin with mammals.

The data for insular mammals compiled by Lomolino (2005) and since updated (Fig. 10.2) demonstrates how the mammalian orders fit into a graded series, with pivot points for the switch between size increase/reduction occurring between *c.* 0.3 and 6.9 kg initial body weight (Lomolino et al. 2017). There is a lot of variation, with many individual cases bucking the trend, but the general trend appears striking. Nonetheless, it has been argued that the rule lacks generality (Meiri et al. 2006; Lokatis and Jeschke 2018) and, in particular, that data for carnivores fit poorly (Meiri et al. 2006). The

Table 10.2 J. B. Foster's tabulation of size differences between 116 insular races or species and mainland forms for six orders of mammals 'mostly living on the islands of Western North America and Europe' (Foster, 1964, p. 234).

Order	Smaller	Same	Larger
Marsupials	0	1	3
Insectivores	4	4	1
Lagomorphs	6	1	1
Rodents	6	3	60
Carnivores	13	1	1
Artiodactyls	9	2	0

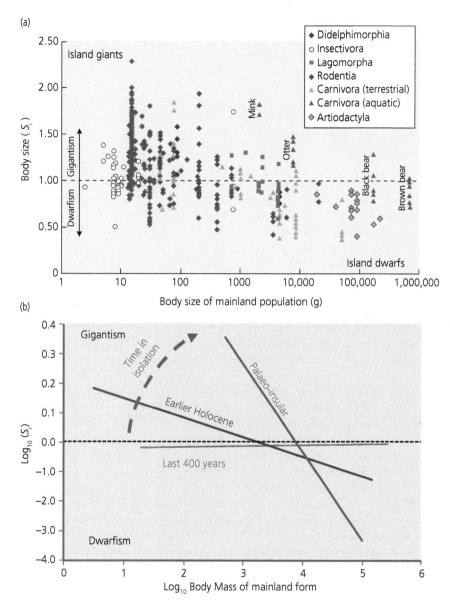

Figure 10.2 The island body-size rule in mammals. (a) Differences in body size between island mammal populations and those of their mainland relative or ancestor, expressed as S_i (the proportion of the mainland body size exhibited by the insular form). (b) Trend lines showing the increasing departure of relative insular body size (S_i) from that of the average mainland population body size with increased insular residence time. The lines of fit are for mammal species that: (i) were introduced within the last 400 years (not significant, but included for visual purposes); (ii) those established earlier in the Holocene; and (iii) those occurring before the Holocene and recovered from fossil data (palaeo-insular). The horizontal dashed line (black) indicates no change. The arrow indicates the expected change in the line of fit over time that generates the island body-size rule; that is, the graded trend from smaller species gaining body size to larger ones decreasing in size.

(a) Lomolino et al. (2017), their figure 13.46; (b) compiled from Lomolino et al. (2013), their figure 1 and van der Geer et al. (2018a), their figure 2.

controversy reflects differences in methodology, in assumptions as to how particular body features scale with body mass (cf. van der Geer 2018b), and perhaps in the outlook of different research groups (Lokatis and Jeschke 2018). Within carnivores, the data provided in Fig. 10.2a supports the role, but with an important distinction: those species largely dependent on terrestrial resources appear to follow one trend line, while those species that gain a large portion of their diet from aquatic resources (representing an insular subsidy) describe a separate and elevated trend line.

Syntheses of data for fossil bones of proboscideans (including the genera *Elephas*, *Mammuthus*, *Palaeoloxodon*, and *Stegodon*) from islands worldwide, demonstrates that the trend towards insular dwarfism in large mammals is stronger when considering extinct forms, many of which showed extreme dwarfism (Muhs et al. 2015; Lomolino et al. 2017). *Palaeoloxodon falconeri*, described from bones in Sicilian caves and dated to the Middle Pleistocene, stood around 1 m tall and had a body mass of about 250 kg, around 2% of that of its mainland ancestor *P. antiquus* (van der Geer et al. 2016; Romano et al. 2021). Turning to small mammals, Lomolino et al. (2013) estimate that the extinct hedgehog *Deinogalerix koenigswaldi*, known from the former island of Gargano (Italy) may have been 100 times larger than its mainland ancestor. It is one of many insular giants and dwarfs that have gone extinct naturally, through fluctuations in climate and sea level that reconnected land-bridge islands to mainland. However, most of the recent losses appear to have been anthropogenic. Faurby and Svenning (2016) show that inclusion of extinct island species in statistical analyses strengthens support for the island body-size rule in mammals, but inclusion of bats and of less isolated islands tends to weaken the trend. In sum, we take the island body-size rule in terrestrial mammals to be broadly supported as an emergent trend.

Based on comparison of contemporary measurements with fossil data, it is possible to address how consistently and how soon the pattern emerges (Lomolino et al. 2013; van der Geer et al. 2013, 2018a, b). van der Geer et al. (2013) compiled datasets for fossils of 19 non-volant mammals on four large Mediterranean islands with timespans of isolation

ranging from the Holocene to late Miocene. They found that small mammals tended to increase in body size and that this trend was often halted or reversed following the arrival of either competitors or predators. But not always. For example, the Cretan shrew, *Crocidura zimmermanni*, increased in size following the arrival of two other shrew species, perhaps reflecting elevational displacement towards areas of lower temperatures. In a subsequent paper, van der Geer et al. (2018a) show that the island body-size rule has yet to emerge in species resident for < 400 years, but can be detected in earlier Holocene arrivals and is strongest for those established by the Late Pleistocene or earlier (Fig. 10.2b).

The ancestor of Tenerife's endemic giant rat *Canariomys bravoi* is estimated from phylogenetic evidence to have colonized around 0.65 Ma (probably by rafting) (Renom et al. 2021). Its mean body mass of *c.* 1570 g is an order of magnitude greater than that of the ancestral African grass rat *Arvicanthis niloticus* (114 g), translating to an annual body mass increase of 0.0015–0.0023 g, and an evolutionary rate of 2.78–7.09 darwins, contrasting with typical rates for non-insular mammals of < 1.0 darwins. *C. bravoi* became extinct following human colonization, as did the closely related *C. tamarani* (endemic to Gran Canaria) and the lava mouse *Malpaisomys insularis* from Fuerteventura and Lanzarote.

Body-size reduction in woolly mammoth on Wrangel Island was once regarded as a classic example of very rapid dwarfism, but more recent work has cast doubt on the extent of dwarfing in this population (den Ouden et al. 2012). The reduction in body mass of the pygmy elephant *Palaeoloxodon falconeri* on Sicily to around 2% of the ancestral mass is considered to have taken 0.2–0.4 million years. The dwarfing of *Homo floresiensis* (the so-called hobbit) on Flores, Indonesia, to a height of 1 m, may also have taken a similar length of time, although this tantalizing example of dwarfing in an island hominin is marked by considerable uncertainty, including as to the identity of the ancestral species (both *H. erectus* and *H. habilis* have been proposed) (van der Geer 2018b).

Morphological changes across multiple mammal taxa have been estimated to occur a factor of 3.1 times faster for island than mainland populations:

the rate is fastest early on, from timescales of a few decades to a few thousand years (Millien 2006). Hence, we can conclude that: (i) rates of morphological change are variable, but (ii) in at least some cases the changes in insular populations are extraordinarily large and initially occur at a remarkably fast pace. This is testimony to the powerful selective force generated by isolation within a small area. Key to the pace of change and even its direction is likely to be the richness, ecology, and size distribution of other members of the insular mammal metacommunity (Faurby and Svenning 2016; Lomolino et al. 2017; Geffen and Yom-Tov 2019).

Extension of the island body-size rule to other animal taxa

Island species of unusual size are not limited to mammals. Large, flightless birds such as the dodo (Mauritius), elephant bird (Madagascar), and several species of moa (New Zealand) are emblematic of the loss of many extraordinary island species following human colonization. In the case of the moa, it has been argued that their extreme large size represented, in effect, an escape to an alternative body plan (*Bauplan*) and functional niche in the absence of non-volant mammals (e.g. Lomolino et al. 2017). Other giants survive. For example, the endemic Komodo dragon (*Varanus komodoensis*), at some 3 m length and 70 kg, is the largest extant lizard, while giant tortoises feature both in the Galápagos and in the Indian Ocean. But gigantism in tortoises may well be relictual and our focus here is on size evolution post-colonization.

Efforts to test for the island body-size rule in a wide array of animal taxa have been bedevilled by methodological issues and have shown rather inconsistent results. However, a meta-analysis by Benítez-López et al. (2021) constitutes something of a game changer. They compiled 2479 island–mainland comparisons (including 1166 insular and 886 mainland species) and controlled analytically for phylogenetic relatedness, sample size effects, and population variability (cf. Box 10.1). They found good support for the pattern as an emergent if noisy trend for mammals, birds, and reptiles, but insular amphibians mostly tend towards gigantism. The magnitude of change was shown to be greater for small and remote islands than for close and large islands (Fig. 10.3). Mammals and reptiles responded mostly to island area and isolation, but birds showed a greater relative influence of climate. The trends were widespread across sub-taxa, although some reptile and mammal groups were more prone to size changes than others.

What drives size changes in island vertebrates?

As we have seen, the transition from insular gigantism to dwarfism occurs at different body sizes for birds, reptiles, and mammals (and, within mammals, between orders): if there is an optimum body size for energy acquisition (*sensu* Damuth 1993) it is not uniform. At its simplest, this suggests that with the reduction in the array of interacting species, species colonizing islands evolve towards an optimal body size for their functionally distinct body plan (*Bauplan*) and insular context (Lomolino et al. 2017).

Food chain links can be important in selecting for size changes. For example, head size in adder *Viper berus* populations is larger in Baltic Sea islands than in nearby mainland populations (Forsman 1991) and this island effect is enhanced with increasing body size of the main prey, the field vole, *Microtus agrestis*. This may reflect stabilizing selection for head size within each population in response to the variation in body size of the main prey species and the small number of alternative prey species available on islands. Schwaner and Sarre (1988) also linked large body size in Australian tiger snakes (*Notechis ater serventyi*) to food resources. On Chappell Island they eat an abundant, high-quality resource, which requires little effort to locate, but which is highly seasonal: chicks of mutton birds. Here, larger body size might enable greater fat storage, thus enhancing survival of individuals through long periods of fasting. Hence, resource availability and its seasonal availability can drive selective changes in traits linked to body size. Predation pressure is also a key driver of morphological change, as shown in a study of insular lizards by Runemark et al. (2014), which showed results consistent with predation release on islets reducing escape behaviour and permitting increased body size.

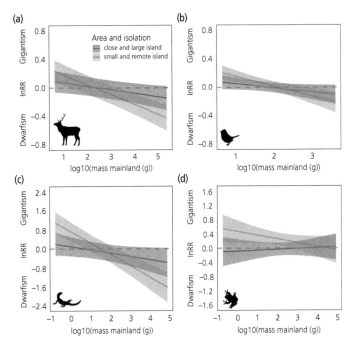

Figure 10.3 The island body-size rule assessed for 1058 mammal, 695 bird, 547 reptile, and 179 amphibian island–mainland pairs, using the log response ratio (lnRR); that is, the natural logarithm of the ratio between the mean body size of insular species and mainland relatives, fitted using phylogenetic multi-level meta-regression models, showing the effects of island size and remoteness, with 95% confidence intervals shaded.
Figure 4 of Benítez-López et al. (2021), reprinted with permission from Springer Nature.

Such observations have contributed to a range of hypothesized mechanisms connected with the island body-size rule. They include ecological release from predation pressure and from resource competition, and seasonality of resource availability. But as we have seen, factors such as climate can have a role in explaining variation in body size, alongside island area (which influences diversity and complexity of ecosystems), distance from mainland, longevity of isolation of the population, and functional and ecological differences between taxa. It seems likely that the morphological changes result from multiple processes that act in combination in insular contexts (e.g. Lomolino et al. 2012, 2017; Runemark et al. 2014; Benítez-López et al. 2021).

Figure 10.4 combines some of the more important drivers and processes into a conceptual model explaining the convergence in body size. Immigrant selection refers to the selective role hypothesized for more distant islands that may favour larger individuals successfully reaching them (e.g. better survivorship during drift dispersal), generating a founder effect that could initiate a shift towards larger body size of small vertebrates. However, this effect should not have a lasting impact unless larger body size is selected for within the island (Faurby and Svenning 2016). Bergmann's rule refers to the tendency of species populations to exhibit larger body sizes in cooler climates, meaning that they have a smaller surface area to volume ratio and better thermoregulation. The rule is often described as body size increasing with latitude, but this is only correct in so far as increasing latitude covaries with diminishing temperature, which it does within some geographical frames (e.g. within Western Europe) but not in others (e.g. within tropical latitudes). Islands can be temperature-shifted in relation to nearby areas of mainland (Chapter 3) and any such differences could again select for shifts in body size. The implications of limited insular space and resources (Fig. 10.4) are most likely to be evident in larger-bodied animals, which generally have larger area requirements. For these creatures, more agile forms that can breed at smaller sizes are favoured over larger ancestral forms, probably in part because they are no longer subject to predation. Resource subsidy from aquatic systems has already been mentioned and may have a role alongside seasonality of food resources.

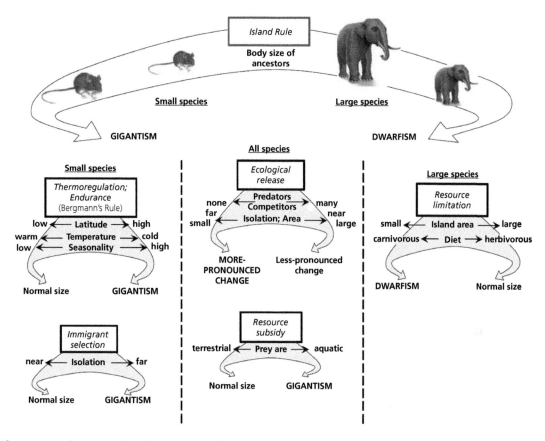

Figure 10.4 In this conceptual model, key hypothesized drivers of body-size change are set out in relation to the graded transition from small insular species evolving increased size over time and larger ones evolving decreased size.
From Lomolino et al. (2012), their figure 2.

Of all these factors and processes, both Lomolino et al. (2017) and Faurby and Svenning (2016) argue that the role of ecological release (from predation and interspecific resource competition) is probably key. Faurby and Svenning (2016) show the rule to be far better supported on non-land-bridge islands islands than land-bridge islands, on which island mammals have experienced only a few thousand years of ecological release following post-glacial isolation and subsequent relaxation of species numbers (Section 7.8). Similarly, they note that bats, which fail to support the rule, are subject to bird predation on islands and therefore also do not experience significant insular ecological release. A key role for release from predation in the evolution of terrestrial vertebrates on islands is also consistent with a subsequent enhanced risk of extinction following human introduction of predatory

mammals. Overall, we conclude that the island body-size rule is supported as an emergent tendency for an array of vertebrate taxa. There are exceptions to the rule that guide us towards important ecological and biogeographical factors contributing to shaping body-size variation (Lomolino et al. 2017).

10.4 Flight loss

Given that oceanic islands are evidently hard to reach, it presented a paradox to early natural scientists that they possessed so many species of reduced dispersal powers, including flightless birds, flightless insects, and plants lacking evident long-distance dispersal traits (Darwin 1859; Wallace 1902; Carlquist 1974). Ever since, the discovery of flightless forms of normally volant animals, or those with reduced wings and flight capacity, has continued to gain attention. Is loss of dispersal power indeed an emergent island phenomenon, rather than merely reporting bias and, in so far as it does occur, what causes it? We address these questions in turn for insects and birds, before briefly considering loss of dispersal powers in other animals.

Insects

Flightlessness occurs in several insect orders, including Coleoptera, Orthoptera, Lepidoptera, Diptera, and Hymenoptera. Many island forms have a reduced flight capacity or propensity—for example: (i) 18 of 20 endemic beetles on Tristan da Cuhna have reduced wings; (ii) 40% of native insects on Campbell Island have some degree of wing reduction; and (iii) 94% of New Zealand's Lepidoptera have limited dispersal powers (Gillespie and Roderick 2002). Surprisingly, an influential synthesis by Roff (1990) reported a *lack* of statistical support for insular flightlessness once latitude and elevation (i.e. climatic) effects were accounted for. However, he tested only for the complete loss of flight. For this reason and because the analysis used rather basic statistical tests, we should consider his analysis cautionary rather than definitive.

Immigrant selection should in general select for colonists of enhanced dispersal abilities on remote oceanic ('Darwinian') islands (e.g. Waters et al. 2020). However, insects can colonize islands passively (e.g. by drift, or by dispersal by birds), or in some cases may represent relict populations stranded through vicariance (e.g. on land-bridge or ancient continental islands). Hence, we should consider not just the overall dispersal profile but whether *in situ* transitions to flight loss have occurred once, followed by diversification of the flightless form, or multiple times in separate islands and lineages (Fig. 10.5).

Samuelson (2003) invokes bird-aided dispersal as the best explanation for the wide Polynesian distribution of *Rhyncogonus* weevils. All are flightless and they each tend to have very local distributions; for example, within the Hawaiian archipelago, 47 species have evolved from a flightless ancestor. An even more spectacular example is provided by the 264 species and subspecies of *Laparocerus*

weevils, all of which are flightless. *Laparocerus* is the largest Macaronesian genus and shows a pattern of colonization and diversification largely consistent with the age sequence of islands in the Madeiran and Canarian archipelagos (Machado et al. 2017; Machado 2022). However, there is also evidence of repeated evolution of flight-reduced insects on islands. For example, (i) flightless endemics have evolved within all 11 orders of alate species that have colonized Hawaii apart from the Odonata (the group including dragonflies), and (ii) transitions to flightlessness have occurred five times in Cratopini weevils in the Mascarenes (Gillespie and Roderick 2002; Kitson et al. 2018). In sum, many insular insect lineages have evolved reduced or lost flight *in situ* and there are also examples of spectacular diversification within flightless clades. We should revise our initial questions to ask *in what circumstances* are reduced flight and dispersiveness favoured on islands?

The prevalence of flightlessness is exceptionally high (47% against a global average of 5%) in Southern Ocean Island species, having evolved at least 62 times, and being especially common in indigenous Coleoptera, Diptera, and Lepidoptera (Leihy and Chown 2020). The proportion of flightless species is lower for two comparator groups: (i) species introduced to the same islands by humans (17%) and (ii) those occurring in five Arctic islands (8%). This suggests that there is something about these Southern Ocean islands, apart from simply their insularity, which repeatedly selects for this trait.

Darwin (1859) hypothesized that flightlessness in island insects might arise through the selective disadvantage of flighted forms being carried away on the wind. This idea has been criticized on the grounds that it is unlikely to apply within particular contexts such as for cave species (which are not exposed to the wind) or in general within larger oceanic islands such as Madeira (e.g. Roff 1990; Leihy and Chown 2020). Alternative hypotheses have been suggested, with varying relevance to islands. First, focusing on decreased benefits of flight, they include predator release, competitor release, habitat stability, and habitat complexity. Second, focusing on increased costs of flight, they include wind exposure, habitat fragmentation,

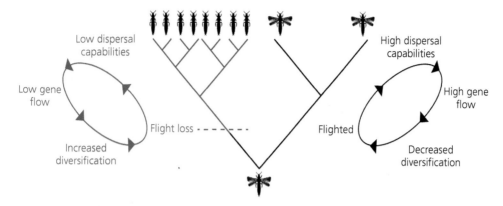

Figure 10.5 A scenario for the evolution of flightlessness in insects, placing emphasis on reduction in dispersal leading to low gene flow and thus repeated local isolation, leading to greater diversification in lineages that have evolved flight loss. Leihy and Chown (2020), who provided this graphic (their figure 1b), failed to find support for this scenario in their analyses of Southern Ocean island insects. It is important to consider such scenarios and to search for similar phylogenetic structure in evaluating hypotheses of island evolution.

low temperatures, and low air pressure. Leihy and Chown (2020) considered that variation in wind speed best explains their findings, with a subsidiary role for thermal seasonality. Noting that flight loss has evolved repeatedly on the windiest South Ocean islands, they suggest that this occurs not so much because dispersive forms are blown off the island, but through a more complex trade-off between the energetic costs of flight in windy environments and reproduction.

In other insular contexts a high prevalence of flightless species might arise through a higher tendency for less dispersive (including flightless) forms to become differentiated *within* an island or archipelago due to reduced gene flow (Fig. 10.5; Gillespie and Roderick 2002; Machado et al. 2017; Kitson et al. 2018; Salces-Castellano et al. 2020). Along these lines, Salces-Castellano et al. (2020) used standardized community sampling from Canarian laurel forests to compare 104 winged and 110 wingless beetle lineages. They found that species in wingless lineages have: (i) smaller range sizes at the archipelago scale; (ii) less representation in younger communities; (iii) stronger population genetic structure; and (iv) greater spatial structuring of species assemblages between and within islands. Their findings support the idea that dispersal limitation operates across multiple scales to shape insular metacommunities.

Arnedo et al. (2008) explore the idea of a link between loss of insect dispersal powers and taxon cycles (Section 9.2). Initially, islands may be colonized by fully winged species that exploit resources in relatively open lowland habitats. Subsequent competition with new colonist species will displace the original species towards higher-elevation forested habitats, where reduced wings are more likely to evolve. This scenario is proposed as a model for evolution in *Calliphona*, a small genus of crickets endemic to the Canaries. An alternative dynamic model is proposed by Kitson et al. (2018) for the radiation of Cratopini weevils across islands within the Indian Ocean. They find that flightlessness has evolved at least five times and that speciation events associated with flight loss occurred more recently than other speciation events. They speculate that flightlessness has evolved repeatedly but that flightless species have a higher probability of extinction, thus accounting for their relative youth in this system. In sum, it seems that: (i) selection on remote islands confers a tendency to reduced dispersal in many arthropod lineages relative to ancestral colonists; (ii) we do not fully understand the trade-offs involved, although in the Southern Ocean islands a role for wind has been identified, (iii) reduced gene flow in flightless lineages is often accompanied by high diversification rates relative to those of related flighted lineages; and

(iv) flightlessness associated with small range size and habitat specialization may lead to enhanced extinction risk.

Land birds

Some of the most iconic island birds are flightless, large-bodied, and extinct, including the dodo (Mauritius), the solitaire (Rodrigues), the moas (New Zealand), and the elephant bird (Madagascar). Flightless island birds survive, but they represent a fraction of those existing prior to human contact, as island bird extinctions have been strongly biased towards large-bodied and flight-reduced species (Boyer 2008; Boyer and Jetz 2014; Fromm and Meiri 2021): one estimate being that flightless species were 33 times more likely to go extinct than volant forms (Duncan et al. 2013; see also Sayol et al. 2020). The catalogue of extinct island bird species is far from complete (Section 15.1), hindering the task of assessing the strength of the insular trend towards flight reduction and loss (but see Fromm and Meiri 2021).

As Wright et al. (2016) point out, most island bird populations are not flightless, and some prolific island colonists, such as white-eyes, whistlers, kingfishers, and hummingbirds, have not produced any flightless species. Moreover, within the Columbiformes (pigeons and doves), in addition to flight loss in such classic cases as the dodo, transitions also occurred towards arboreal foraging, which led not only to increased speciation rates but also range expansions across multiple islands and back colonization of continents (Lapiedra et al. 2021). Nonetheless, flight loss clearly is a repeated insular phenomenon, as more than 1000 lineages of island birds have become flightless, including passerines, owls, pigeons, parrots, waterfowl, and most especially rails (Steadman 2006; Wright et al. 2016). Most Pacific islands that have been explored by palaeontologists have revealed the presence of an extinct flightless rail and we can be confident that many more await discovery (Steadman 1997). Indeed, extinct rails have also recently been identified from fossil bones on two North Atlantic archipelagos, one each from six Azorean and two Madeiran islands, all but one described to date being flightless (Alcover et al. 2015). Flight loss transitions have clearly occurred hundreds of times in rails colonizing islands across the world.

In addition, and as shown by Wright et al. (2016), there is also a strong tendency to reduced flight capacity in island birds that nonetheless remain volant. First, they showed that 38 island-restricted species have smaller flight muscles for their body mass than their continental relatives. Second, they examined forelimb and hindlimb dimensions (permitting estimation of flight muscle size) from specimens representing 366 bird populations from 80 Caribbean and Pacific islands. They found that birds on smaller islands, which tend to be less species rich, evolve longer legs and smaller flight muscles. The trend holds for nine avian families across four orders, for both regions and for both continental and oceanic islands. It is notable that the groups of birds we listed above as not producing flightless species (white-eyes, whistlers, kingfishers, and hummingbirds) nonetheless exhibit changes consistent with the trend towards flight loss.

Wright et al.'s (2016) statistical analyses identified the presence of mammalian predators and the richness of raptors to be the best predictors of their morphological indices, superior to island area, species richness, or isolation, although these factors tend to be interlinked. As raptor richness increases, it is more likely that the metacommunity includes specialist bird hunters, which maintains selection for rapid escape capabilities. The failure, thus far, to find evidence of extinct rails on the Canary Islands, despite their presence elsewhere in Macaronesia (above), might be explained by the presence of a richer predator community, including the now extinct endemic mice (*Malpaisomys insularis*), and giant rats (*Canariomys* spp.). This explanation is also consistent with the discovery of extinct flightless quails from three Macaronesian archipelagos—Cabo Verde (Cape Verde), the Azores, and Madeira—whereas a weakly flighted (but also extinct) species occurred on the Canaries (Rando et al. 2020). On the other hand, the Canaries do provide an example of another rarity, an extinct flightless passerine, the Canarian long-legged bunting (*Emberiza alcoveri*) (Rando et al. 1999).

We conclude that the trend towards reduced flight is a regular feature of avian evolution on islands.

This has a lot to do with the absence of predatory mammals from more remote islands and likely has something to do with reduced incidence of specialist bird predators among island raptor communities. As discussed for insects, the tendency towards flight loss may be connected to taxon cycles, as the reduction in dispersiveness reduces gene flow and traps the flightless form on a single island, ultimately increasing its vulnerability to extinction (Ricklefs and Bermingham 2002; Wright et al. 2016). The loss of flight on islands is connected to other traits, including gigantism, ground nesting, and foraging behaviours. In the most extreme cases, these traits and syndromes combine with a significant shift in ecological role, as island species experience a *Bauplan* shift (i.e. evolutionary transitions in body plan and in the proportions of different elements of the body, such as flight muscles, leg size, and body size), as they move into 'empty' feeding niches. Classic cases include the Malagasy elephant birds (Aepyornithidae, reaching up to 650 kg) and the New Zealand moas (Dinornithiformes, up to 250 kg). Examples of such transitions are a feature of other types of animals on islands, as exemplified by some species of New Zealand wetas (Anostostomatidae, Orthoptera, which weigh up to 70 g).

10.5 Evolutionary adjustments in reproductive investment

Lizards

Evolutionary adjustments in reproductive investment noted for island vertebrates include reduced litter/clutch size in small mammals, land birds, and lizards (Section 10.2). The expected trade-off is that more energy can be invested per offspring given the reduced predation pressure on eggs and young and the corresponding increase in intraspecific competition for resources. This is consistent with a comparison of clutch sizes of Canarian and Ibero-Balearic lizards, which showed that insular lizards tended to have smaller clutch sizes for a given body size (Siliceo and Díaz 2010). This is confirmed by a more extensive analysis by Novosolov et al. (2013), who reported that island-endemic lizards have reduced clutch sizes and larger hatchlings than comparable mainland species (see Section 10.2 for details).

Oceanic island endemics provide the most convincing evidence of adjustments in reproductive output, producing more frequent, but small clutches, giving rise to larger offspring than do similarly sized mainland species (Novosolov and Meiri 2013). Lizards on land-bridge islands evidence shifts in the same direction but these shifts were only significant for clutch size. Novosolov and Meiri (2013) caution that the shift to more frequent clutches on oceanic islands is in part due to the large number of *Anolis* lizards in their dataset and the difference disappears when the analysis is corrected for phylogenetic structure. In a subsequent study based on data for 2511 lizard species, Schwarz and Meiri (2017) highlight the distinction between those lizards that lay small invariant-sized clutches (e.g. geckos and anoles) and those that produce variable numbers of eggs. The former exhibit increased frequency of laying, but no difference in investment per egg, compared with mainland species. By contrast, insular species of variable clutch size lay larger eggs in smaller clutches, and their hatchlings are larger at emergence. The differences in the evolutionary responses shown by geckos and anoles may owe something to their physiology and behaviour that limits clutch sizes the female can produce.

Various ideas have been put forward to explain reduced clutch sizes in birds and lizards, including roles for reduced seasonality, greater resource predictability, and conversely environmental instability through hurricanes (particularly relevant to many islands hosting geckos and anoles) (e.g. Covas 2012; Schwarz and Meiri 2017). However, these factors do not seem to provide a sufficient general explanation for the emergent trends in reproductive investment, which surely reflect first and foremost the decreased predation and interspecific competition pressures and correspondingly increased intraspecific competition experienced by island species. Hence, we tend to see these effects emerging most clearly on the most isolated, Darwinian islands. This is not to exclude a role for other island characteristics. For example, Schwarz et al. (2020) find that within 31 populations of the gecko *Mediodactylus kotschyi* in the Aegean, variation in both predation pressure and island isolation influenced variation in life-history traits. Their findings suggested that increased species richness of predators on islands drives faster life histories (larger clutches, smaller eggs), but that the presence of one particular predator, the boa *Eryx jaculus*, pushes life-history changes in the opposite direction. It is not clear why the snake might have this effect: one hypothesis is that it may prey on rats, which in turn may eat gecko eggs.

Land birds

It has long been known that bird clutch sizes (i) increase with latitude, and (ii) decrease on islands when compared with continental areas at the same latitude (Lack 1947b; Ricklefs 1980; Covas 2012). Clutch sizes have also been reported to decrease with diminishing island area and with increasing isolation within the same archipelago (Higuchi 1976) and at least for some related bird species, clutch size also decreases with increasing endemism level of the island species. For instance, the Canarian endemic blue chaffinches (*Fringilla teydea* and *F. polatzeki*) have a mean clutch size of 2.0 eggs, the endemic subspecies of chaffinch (*Fringilla coelebs canariensis*) a mean of 2.6, and the northwest African subspecies of chaffinch (*F. c. africana*) a mean of 3.9 (Cramp and Perrins 1994; Martín and Lorenzo 2001). Canarian endemic laurel-forest pigeons (*Columba bollii* and *C. junoniae*) lay one egg, instead of the two eggs laid by the native but non-endemic rock dove (*C. livia*). The Canarian kestrel (*Falco tinnunculus canariensis*) has a mean clutch size of 4.41 eggs, whereas Moroccan kestrels at the same latitude have a mean clutch size of 4.80.

Clutch size is just one trait indicative of changing patterns in reproductive investment, as demonstrated by Covas (2012), who examined multiple life-history traits for subsets of 153 pairs of island birds and close mainland relatives. The islands were restricted to those of very small size (< 12 km^2) but included continental and oceanic islands. Her analysis confirmed smaller clutch sizes on islands for a given latitude and that the increase in clutch size with latitude occurs 4.5 times faster in mainland species than in island relatives. Alongside reduced clutch sizes, she reported an increase in egg volume in island birds, longer incubation periods, longer nesting periods, and a less clearly established tendency for longer post-fledging care. Consistent with

the increased density of island bird populations and of the shifts towards fewer eggs but greater investment per egg and hatchling on islands, there was also a significant increase in the proportion of island species undertaking cooperative breeding (7% of mainland and 33% of island species), in which often closely related individuals assist the parents in feeding the young. This is an intriguing finding that merits further study.

That the insular reduction in fecundity is exaggerated with latitude suggests a role for climate or food resources. The ocean's buffering effect means that islands are generally more benign in winter in temperate latitudes than continental interiors and they are less liable to extreme heat in the summer, thus favouring more stable populations (Covas 2012). Various other hypotheses have been put forward for the shifts in reproductive strategy following island colonization, including: (i) increased intraspecific competition due to reduced interspecific competition (as per Section 10.2) and (ii) reduction in predation pressure and parasite load with increasing island isolation (e.g. MacArthur et al. 1972; Ricklefs 1980; Covas 2012). Another factor is provided by Beauchamp (2021), whose comparative analysis of 154 island and 543 mainland bird species from across the globe showed a tendency for adult birds on islands to have greater longevity than mainland species. Increased adult survival sits well alongside reduced fecundity but increased investment per egg and hatchling, in a general shift towards *K*-selection. It would be interesting to investigate how reproductive investment varies across tropical islands along a gradient of increased exposure to tropical storm damage (cf. Chapter 3).

10.6 Island tameness and the loss of defensive behaviour

Charles Darwin (1839) famously commented on the extreme tameness of the birds of the Galápagos Islands, writing in his account *The Voyage of the Beagle* that 'In Charles Island, which had then been colonized about six years, I saw a boy sitting by a well with a switch in his hand, with which he killed the doves and finches as they came to drink. He had already procured a little heap of them for his dinner, and he said that he had constantly been in the habit of waiting by this well for the same purpose. It would appear that the birds of this archipelago, not having as yet learnt that man is a more dangerous animal than the tortoise or the *Amblyrhynchus*, disregard him, in the same manner as in England shy birds, such as magpies, disregard the cows and horses grazing in our fields.' Darwin argued that fear of predators, such as humans, was inherited and at the end of his section on island tameness commented: 'We may infer from these facts, what havoc the introduction of any new beast of prey must cause in a country, before the indigenous inhabitants have become adapted to the stranger's craft or power.' How common a trait is tameness in island species and is it learnt behaviour or inherited?

Although Darwin's comments were about birds, most work on this topic concerns lizards, in which flight initiation distance (FID), the distance between prey and predator at which the prey starts to flee, is one commonly used metric of defensive behaviour. The importance of antipredator behaviours is obvious where predators abound, but fleeing has an energetic cost and, where unnecessary, the behaviour should be selected against.

A study by Brock et al. (2015) of FID in Aegean wall lizards (*Podarcis erhardii*) in one mainland site and 37 land-bridge islands in the Cyclades provides relevant evidence. The islands range from < 0.05 km^2 to 1000 km^2 and in duration of isolation from under 10 years to 3.6 million, although most became isolated as sea levels rose around the start of the Holocene. Species preying on the lizards within the islands include birds, snakes, rats, and other mammals. It was found that: (i) very small islets (< 0.05 km^2) lack predators but as island area increases, their diversity increases; (ii) FID generally increases with increased diversity of predators; and (iii) evidence for a decline in FID with duration of isolation is inconclusive. In a separate study, Cooper et al. (2014) collated data for FID in 66 species of mainland and island lizards. They reported shorter FID in mainland than island populations and that FID among island populations decreases with island isolation. Both studies support island tameness in lizards as an emergent tendency.

In their study of *Podarcis erhardii*, Brock et al. (2015) examined rates of autotomy (self-severing

of the tail), another behaviour regarded as an antipredator defence mechanism in lacertid lizards. Autotomy is a far more expensive strategy than hiding or fleeing. Against expectations, they found rates of autotomy to be higher on islands that lacked predators than on islands with predators and there was no difference in rates as a function of increased diversity in those islands that did have predators. The higher rates of autotomy on predator-free islands appeared to be a result of intraspecific aggression and especially of battles between rival males (a contrast with the slowing-down syndrome: cf. Section 10.2). Brock et al. (2015) also highlight an earlier study of tail shedding in 15 Mediterranean lizard taxa from islands and mainland sites, which found that the presence of vipers best explained variation across populations and that neither insularity nor the presence of other predators had explanatory value.

In sum, these findings indicate that different defence mechanisms may show differing responses and that behavioural responses to insularity can be complex. It is likely that there are both learnt and inherited components to FID and other related behaviour (cf. Baier and Hoekstra, 2019; Satterfield and Johnson 2020). A loss of fear of predators in island species is one factor, alongside other traits such as ground nesting and flight loss in birds, which together have increased the risk of extinction of island endemics following the introduction of non-native predators by humans (Chapter 15).

10.7 Relaxation in territoriality

Although we have given counter examples, insular lizards, birds, and mammals often exhibit reduced situation-specific aggression towards conspecifics (Section 10.2). This relaxation in aggressive behaviour can be expressed in the form of: (i) reduced territory sizes, (ii) increased territorial overlap with neighbours, (iii) acceptance of subordinates on the territory, (iv) reduced aggressiveness, or (v) abandonment of territorial defence (Stamps and Buechner 1985). These changes are often associated with unusually high densities, niche expansion, low fecundity, and the production of few, but competitive offspring.

Stamps and Buechner (1985) outline two non-exclusive hypotheses for relaxation in territoriality. The resource hypothesis suggests that the absence of competitors leads to greater resource availability and thus reduced territory sizes and increased territorial overlap. The defence hypothesis focuses on the increased density of fledged conspecifics on islands, generating 'have-nots' that intrude and may contest with those that own territories. This will elevate the costs of defence for owners of territories, contributing to shifts towards rearing fewer but more competitive young (above) within reduced territories, as well as acceptance of territorial overlap. Acceptance of subordinates that engage in cooperative breeding (as reported above for birds) may also be involved in changes in territoriality in island vertebrates.

It remains unclear how general these putative island syndromes may be and whether they may differ across or within major taxa. A recent study of the breeding behaviour of the Hawaiian crow, *Corvus hawaiiensis*, provides a caution. This species became extinct in the wild in 2002, with surviving individuals maintained in captivity. Reproductive success was found to be negatively affected by high social density and by pairs being housed in aviaries in sight of conspecifics. Earlier records indicate that the crows once occupied large territories, separated by a kilometre per breeding pair, although whether this reflects their density prior to human contact is unknowable. Thus, insular adjustments in territoriality (and intraspecific aggression) appear to be another syndrome in need of further systematic research.

10.8 Herbivory in island lizards

Only around 3% of lizards consume significant amounts of plant food, although many more do so occasionally (van Damme 1999). Factors driving a shift towards herbivory may include larger body size (generating increased energy demands and perhaps less agility) and low levels (or inconsistency) of availability of arthropods. To shed light on this, van Damme (1999) collated data on diets of 97 lizard populations (67 continental and 30 insular) representing 52 species. Correcting for

phylogeny he found a significantly higher tendency for herbivory on islands than in continental habitats. Such dietary shifts were sometimes accompanied by a body-size change, but neither consistently, nor in a consistent direction. van Damme's classification of species as herbivorous does not exclude the consumption of arthropods but instead indicates that plant matter (e.g. pollen, nectar, fruits, seeds, leaves, flowers) formed a meaningful part of the diet based on studies of guts, faeces, or field observations. Reduction of predator pressure in island environments may well be a factor in the switch to more plant material in the diet, as it allows longer periods safely basking while digesting plant material.

The endemic *Gallotia* lizards of the Canary Islands provide examples both of island giants and island herbivory. Initial colonization took place in the older, eastern islands *c*. 17–20 Ma and *Gallotia*, which forms a monophyletic clade, in turn colonized the younger, western islands (Cruzado-Caballero et al. 2019), giving rise to a still uncertain number of distinct species and subspecies, ranging *c*. 0.4–1.2 m in length. All five western islands have (or had) giant forms. Analyses by Čerňanský et al. (2016) indicate that *Gallotia* derived from an ancestral European lizard lineage already of large body size. Following colonization, both size increases (leading to the extinct *Gallotia goliath* from a smaller ancestor) and, more commonly size decreases, have occurred. The fit of *Gallotia* with the island body-size rule (Section 10.3) appears from these analyses to be weak, as the genus has diversified in body size from a putative large colonist, to fill distinct ecological niches; for example, on Tenerife producing *Gallotia galloti* (small, currently abundant, and widespread), *G. intermedia* (large, rare, and localized), and *G. goliath* ('giant' and extinct) (Crowley et al. 2019). The likely mainland ancestor was faunivorous (Čerňanský et al. 2016) and it therefore appears that the transition to a plant-oriented diet has occurred *in situ* within the Canaries. All extant *Gallotia* rely heavily on plant matter and based on isotope analysis it appears that the same applies for at least one extinct giant species, *Gallotia goliath* (Crowley et al. 2019). We tentatively conclude that living on islands encourages a switch towards more plant-based diets in lizards.

10.9 Acquisition of 'low-gear locomotion' in large herbivores

Alongside, but distinguishable from insular dwarfing, insular mammals have often evolved differences in other morphological features, such as the skull, teeth, and limbs. In particular, many extant and extinct ruminants and some other large mammals (e.g. elephants) from islands developed shortened and thickened metapodials, sometimes involving bone fusions (Rozzi et al. 2020). These changes have been described as indicating a shift to 'low-gear locomotion' (Sondaar 1977), advantageous for stability in steep and uneven terrain and without selective penalty in the absence of predators on islands.

Rozzi et al. (2020) assess several metrics indicative of low-gear locomotion and test for their relationship to factors including body size, S_i (an index of relative change in insular body size, as used in Fig. 10.2), island area, elevation, and topographic complexity. They found that low-gear locomotion does not result simply from phyletic dwarfing and release from predators in insular ruminants but is also associated with reduced numbers of competitors, enhanced by increased terrain roughness. The most extreme cases were found on islands without mammalian predators. Interestingly, they found that mountainous terrain appeared to lead to slightly stouter metapodials in Bovinae (cattle, bison, buffalo) but the opposite trend in Antilopinae (antelopes). Nonetheless, they concluded that a tendency towards low-gear locomotion is a repeated emergent trend in insular ruminants in both families, with reduction in interspecific competition a key factor. It will be interesting to see these analyses extended to more taxa and more insular systems.

10.10 Colouration and song

A distinguishing feature of many island species, subspecies, and populations is differences in colouration from their closest mainland relatives. A tendency towards dull colouration is present in many island birds and in some insects, once again as noted by Darwin (1839) for the Galápagos. Indeed, insular melanism, having been observed in birds, reptiles, spiders, and insects (Table 10.3;

Uy and Vargas-Castro 2015), provides another candidate island syndrome. Plausible explanations for melanism include: (i) more effective thermoregulation (dark colours absorb more heat than light colours); and (ii) crypsis (the ability of an animal to avoid observation or detection by other animals).

Spectophotometric analysis of the colouration of 116 endemic island bird species from around the world and of their closest mainland relatives, revealed reduction in brightness and colour intensity in island birds of both sexes (Fig. 10.6; Doutrelant et al. 2016). Moreover, the number of patches of distinct colours in male birds increases with increasing diversity of confamilial bird species in the island metacommunity. These results support the evolution of dullness in island birds as a repeated pattern, probably reflecting a reduction of selection for species recognition. It has also been suggested that: (i) reduction of sexual selection on islands can occur because reduced genetic diversity in turn reduces the benefits of mate choice; and (ii) melanism may be closely linked to the production of sexual hormones, so that it arises from indirect selection through pleiotropy (where one gene influences two seemingly unrelated traits). This mechanism (essentially of reduced sexual selection) has been suggested for repeated evolution of melanism in insular populations of fairywrens *Malurus leucopterus* (Walsh et al. 2021).

Nonetheless, some insular species are more colourful than their mainland relatives. Taking a similar comparative approach for 110 island/mainland species pairs, Bliard et al. (2020) found that insular colouration increases as the number of insular predators decreases, although plumage brightness was found to be unaffected by predation pressure. These analyses in combination indicate that different aspects of plumage colouration may be differentially affected by diminishing competition and predation across island metacommunities.

Island bird song is often found to differ from that of the same or related species in mainlands (e.g. Lachlan et al. 2013). This is, of itself, unremarkable, but of greater interest is whether there are emergent trends in the evolution of island birdsong. A study of the chaffinch *Fringilla coelebs*, across 14 island and continental populations, showed that the syntax of

Table 10.3 Examples of insular melanism, indicating the overall distribution of the species.

Taxa	Distribution	Alternative colour morphs	Frequency of insular melanism	Evolutionary mechanism inferred
Birds				
Coereba flaveola	Central and South America	yellow and black	nearly fixed	thermoregulation?
*Malurus leucopterus**	Australia	blue and black	fixed	reduced sexual selection
Monarcha castaneiventris obscurior	Solomon Islands	chestnut and black	polymorphic	unknown
M. c. ugiensis	Solomon Islands	chestnut and black	fixed or nearly fixed	unknown, but evidence for selection
Myzomela spp.	Pacific Ocean	red and black	fixed	unknown
Riphidura fuliginosa	New Zealand	rufous	nearly fixed	unknown
Reptiles				
Elaphe quadrivirgata	Japan	striped	nearly fixed	thermoregulation
Nerodia sipedon	North America	striped	polymorphic	thermoregulation
Podarcis lilfordi	Balearic Islands	several	nearly fixed	unknown
Invertebrates				
Nephila maculata	Japan, Indonesia	yellow and green	polymorphic	prey capture and crypsis
Oedaleus senegalensis	Sahel, Macaronesia	green and brown	nearly fixed	unknown
Philaenus spumarius	Europe	several	polymorphic	thermoregulation and habitat

* See text and Walsh et al. (2021). Slightly modified from Table 1 of Uy and Vargas Castro (2015).

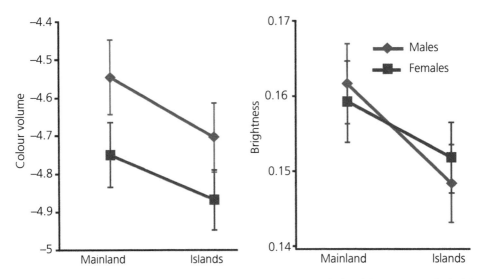

Figure 10.6 Contrast in colour intensity (volume) and brightness between 116 island endemic birds and their nearest mainland relatives. Doutrelant et al. (2016), their figure 2. See source article for further explanation of units.

the songs has become less structured as the species has sequentially colonized islands from Europe into the Atlantic and south to Gran Canaria (Lachlan et al. 2013). A larger study by Morinay et al. (2013) analysed song complexity in pairs of closely related island and mainland birds from around the world. They found that several features of song performance and complexity did not differ, but that island species produced fewer broadband and aggressive song elements (rattles and buzzes), suggesting that presence of competitor species may be important in birdsong evolution.

The possibility that changes in song might co-evolve with colour changes in island birds was investigated by Reudink et al. (2021) in analyses of island and mainland species or subspecies for three songbird families: Frigillidae, Meliphagidae, and Monarchidae. Controlling for phylogeny they found a trend towards melanism in the Meliphagidae (only) and found no evidence of changes in song structure between island populations and their mainland relatives in these three families. However, they did find that birds on smaller islands had simpler song structures than those of larger islands.

10.11 Niche shifts in island bats

Bats, whether insectivorous or frugivorous, are often the only land mammals that have naturally colonized remote islands. There are no entirely flightless bats, but New Zealand's short-tailed bat (*Mystacina tuberculata*) is one of the few bats in the world that spends a lot of time foraging on the ground (it feeds on insects, fruit, nectar, and pollen). While this species has morphological features that enhance its mobility on the ground, it is not unique in its ground foraging and agility. For example, it shares both traits with the common vampire bat *Desmodus rotundus* of South and Central America, which differs, however, in its diet (it is an obligate blood-feeder) (Riskin et al. 2006). While the absence of terrestrial vertebrate predators in New Zealand prior to human arrival may have played a role in the evolution of ground foraging in the short-tailed bat, there is no evidence that this is a general insular trend in bats.

Nocturnality in bats probably evolved largely in response to diurnal predation by birds and appears to be a well-conserved evolutionary behaviour (Russo et al. 2011). Some island bats have undergone a partial shift towards diurnal activity, very likely enabled by the absence or scarcity of avian predators. A few flying foxes (large Old World fruit bats) are known to fly in daylight on predator-free islands, one prominent being the Samoan flying fox (*Pteropus samoensis*), which can be seen exploiting

thermals for soaring. Diurnal activity has also been recorded in at least three predominantly nocturnal, insectivorous bats on islands, the Azorean bat (*Nyctalus azoreum*), the São Tomé bat (*Hipposideros ruber*), and Blyth's horseshoe bat, *Rhinolophus lepidus* on Tioman Island, Malaysia (Russo et al. 2011; Chua and Aziz 2019). The São Tomé bat is described as commonly active during the day, with males observed to be more diurnally active than females, suggesting an element of temporal (and thus feeding) niche partitioning (Russo et al. 2011). However, it would be premature to declare diurnal behaviour to be an island syndrome based on these few examples, especially as occasional diurnal activity has also been observed in some mainland bats (Chua and Aziz 2019).

10.12 Parthenogenesis

According to Hanley et al. (1994), **parthenogenetic** (asexually reproducing) lizard species are relatively common on islands (Hanley et al. 1994; Kearney et al. 2009), possibly reflecting: (i) an enhanced colonization ability (one founder being sufficient); (ii) the opportunity to flourish in a less diverse biotic environment in which genetic inflexibility is not too disadvantageous; or (iii) escape from competition with their sexual relatives. Hanley et al. (1994) examined the coexistence of parthenogenetic and sexual forms of the gecko *Lepidodactylus* on Takapoto Atoll in French Polynesia. The two forms, although closely related and hybridizing to a limited extent through the activities of males of the sexual form, are considered separate species. The asexuals were found to be extremely heterozygous relative to the sexual population, probably reflecting their hybrid origin. In this case, far from the asexual form suffering from competition with the sexual, it was the parthenogenetic form that had the wider distribution within the atoll, with different clones having overlapping, yet distinct, habitat preferences. The coexistence of the two forms appears to be facilitated by the specialization of the *less* successful sexual taxon to beach habitats.

At least some reptiles show a capacity to switch between sexual and asexual reproduction. For example, two captive female Komodo dragons (*Varanus komodoensis*), isolated from males in

separate zoos, produced offspring parthenogenet-
ically (Watts et al. 2006). One of the females
subsequently produced further offspring sexually,
demonstrating flexibility in reproductive mode.
While this capacity would seem to convey enhanced
colonization potential, in the case of this giant
lizard, its present distribution is narrow and seemingly relictual.

Parthenogenetic lineages of three genera of
geckos—*Lepidodactylus*, *Hemidactylus*, and *Nactus*—
are widespread throughout the tropical Pacific, but
only some are confined to islands, and they may
occur alongside sexual relatives (Kearney et al.
2009). While islands may indeed commonly host
parthenogenetic reptiles, they appear to be biased
in their distribution to relatively open and/or disturbed environments, rather than to islands per se
(Kearney et al. 2009).

Parthenogenesis occurs in many insect groups but
only one example is known within the Odonata, in
populations of the damselfly *Ischnura hastata* in the
Azores, where only females are found. The species
is native to South, Central and North America,
and to the Caribbean and Galápagos, but no other
parthenogenetic populations have been recorded in
either the mainland or island populations, apart
from in the Azores (Lorenzo-Carballa et al. 2017).
The species is a good flyer, but the Azores, lying
over 3300 km from the rest of the range, may well
have been reached only by a female (or females),
leading to a switch to parthenogenesis. It remains
uncertain how prevalent parthenogenesis is as an
island syndrome.

10.13 Concluding remarks: a plethora of island animal syndromes?

We began by outlining the slowing-down syndrome (originally formulated for rodents), a super-
syndrome of behavioural and physical changes. As
we have reviewed the separate components of this
syndrome and other putative island syndromes,
we have shown that the strongest of them can
best be described as emergent statistical tendencies, while other syndromes may not be particularly
biased to islands at all, or simply remain unresolved
(Table 10.4).

The causal mechanisms for the trends summarized in Table 10.4 vary, with multiple hypotheses
available for each. However, the common factors
and best supported factors are the simplification of
island communities, the reduction or absence of vertebrate predators (especially terrestrial mammals),
and the reduction in diversity of competitors, followed by aspects of the insular environment such
as the assumed increase or decrease (!) in environmental or resource stability.

It is perhaps important to emphasize that when it
comes to behaviour, as opposed to morphology (or
phylogenetic relationships), the question of the heri-
tability of the behaviour must always be considered.
For example, in birds, plumage colour is genetically
determined but to what extent does this apply to
song, or to escape behaviours linked to potential
predators?

This question brings us to another proposed
island syndrome. Comparative analyses for > 1900
bird species have shown an evolutionary trend
towards increased brain size in island species compared to their mainland relatives (Sayol et al. 2018).
Increased brain size carries an energetic cost and
involves a longer period of development, but this
in turn is compatible with the slowing-down syndrome and the linked patterns of reduced clutch size
and greater investment per individual. Larger brain
capacity may be of selective advantage as it permits greater behavioural plasticity, which is likely
to be beneficial where island environments are subject to unpredictability of conditions and resources
over time, favouring individuals who are able to
adopt novel foraging techniques or food sources
(Sayol et al. 2018). Some of these larger-brained
insular species, the core Corvoidea, having evolved
in the proto-Papuan islands, subsequently invaded
continents around the world (Jønsson et al. 2011,
2017)—in opposition to the 'islands as sinks' concept. Another recent bird study from the region has
shown hitherto cryptic niche segregation between
the sexes in the Sulawesi babbler (Ó Marcaigh 2022).
Sexual dimorphism was found to be increased in
insular as opposed to mainland populations, possibly reflecting an increased role for intraspecific
resource competition in the islands. Although this
is not a unique example, too few systems have been

Table 10.4 Island syndromes in animals. A summary of the syndromes considered herein, distinguishing those for which we have reported support from comparative statistical tests (or meta-analyses) involving multiple cases.

Syndrome	Description	Taxon	Evidence
Slowing down	Multiple traits involving slowed life history, reduced aggression, etc.	Mammals	Breadth of syndrome makes statistical evaluation challenging: there is support from case studies for the interrelationship of multiple traits proposed in this syndrome.
		Lizards	Support from comparative analysis notwithstanding the point above.
		Other	Invoked as relevant to birds.
Island body-size rule	Smaller bodied species evolving larger size and vice versa	Mammals	Contested but overall good support from comparative statistical analysis, strengthened when extinct species considered.
		Birds and reptiles	Recent comparative analysis supports the trend as a noisy but significant trend.
		Amphibians	Recent comparative analysis suggests a trend to gigantism.
Flight loss	Loss or reduction of flight	Insects	Flightlessness is a recurrent feature of (but is not unique to) islands; a role for windiness evident in some datasets.
		Birds	Comparative analyses show flight reduction to be widespread in island lineages; flight loss a repeated emergent trend, especially in ground feeding birds (key role for absence of vertebrate predators).
		Bats	Very few cases of flightlessness in island bats: not established as a trend.
Reduced clutch size	Fewer eggs but more investment per individual	Lizards	Comparative analyses demonstrate largest effects for oceanic islands and differences in nature of shifts between lineages of different ecologies.
		Birds	Comparative analysis supports an island effect, distinct from and additional to a latitudinal gradient.
Tameness	Loss of defensive behaviour	Lizards Birds	'Flight initiation distances' lower on islands lacking predator diversity. Anecdotal evidence strong for this on first human arrival; we are unaware of a comparative statistical analysis
Relaxation in territoriality		Mammals, birds, and lizards	A corollary of density compensation, invoked in the slowing-down syndrome and supported with case studies; we are unaware of a comparative analysis.
Diet switching	Herbivory in lizards	Lizards	Comparative analysis supports a tendency to increased use of plants in the diet of island lizards.
Low-gear locomotion	Stouter, shorter limbs	Mammalian herbivores	Comparative analysis finds it to be a repeated pattern, associated with isolation on mountainous islands.
Melanism	Loss of bright colouration	Birds	Comparative analysis supports a general but not universal trend; linked to competition and predation pressure (and reduced sexual selection?).
		Other taxa	Has been observed in reptiles, arachnids, and insects; not clear how strong an insular trend.
Insular songs	Changes in syntax and complexity	Birds	Complexity and syntax of songs can differ; emergent trend of reduction in seemingly aggressive song elements.
Diurnal bats	Increased day time activity	Bats	It occurs in certain insular fruit bats and insectivores; unclear if there is an emergent trend.
Parthenogenesis	Switch to asexual reproduction	Lizards	Claimed as common on islands, but not confined to islands and strength of association unclear.
		Insects	Comparative analyses appear to be lacking, unclear if this is biased towards islands.
Larger brains	Increased brain size	Birds	Comparative analysis (Section 10.13) indicated this to be an emergent island trend.

studied to establish whether this may be yet another insular syndrome.

Another way in which island syndromes can be identified is through metrics of functional and phylogenetic similarity (cf. Section 6.7). In illustration, Triantis et al. (2022) computed $MNTD_{TURN}$ (mean nearest taxon distance turnover) values to reveal patterns of morphological and phylogenetic convergence in the avifaunas of 18 Atlantic, Indian, and Pacific Ocean archipelagos. They compiled trait and phylogenetic data from each archipelago (including extinct species) and from their respective mainland source pools, to allow correction for the differences that would arise by random colonization from each source pool. They found clear signals of convergence in phylogenetic structure, consistent with island avifaunas being drawn non-randomly from the avian tree of life; in particular, from pigeons and passerines. This may reflect the dispersiveness of these taxa but also their colonization prowess. Morphological convergence was also evident in avian body plans (*Baupläne*), assessed via body mass-corrected wing, tail, tarsus, and beak length. By running the analyses for different subsets, they were also able to show that body-plan convergence is weaker (but still significant) in the endemic avifauna and has been increased by anthropogenic extinctions, which have tended to weed out some of the more morphologically distinctive species. Analysis of body mass showed convergence for 'all species' but not when restricting analysis to insular endemics, or to diurnal primary and secondary consumers. For a specific example of insular morphological convergence, in wing size of the globally distributed barn own complex (*Tyto* spp.), see Romano et al. (2021). Morphological convergence arguably deserves a place in Table 10.4 as another insular syndrome.

The above studies demonstrate how innovative comparative analyses are in the process of rapidly advancing our understanding of island syndromes. There is surely more to come from such approaches in building our understanding of the intertwined themes of island assembly rules (Chapter 6) and of island evolutionary syndromes. The distinction between the two themes being in the extent to which the biases we detect in island metacommunities are the result of filtering at the point of dispersal, of colonization, of persistence (i.e. avoiding natural or anthropogenic extinctions), or are the outcome of *in situ* evolutionary change (Fig. 1.2).

Some apparent island peculiarities are not unique to islands, and none are universal across all island species. However, we conclude that there are multiple evolutionary syndromes among island animals. That is, there are evolutionary trends that are broadly repeated across different lineages in multiple islands and archipelagos. As we have seen, analyses of how the trajectory of evolution varies, which syndromes co-vary across island systems to provide 'super-syndromes', and which characteristics of the islands, or of the taxa, explain this variation, are providing invaluable insights into evolutionary processes.

10.14 Summary

The peculiarities of many island endemics greatly intrigued Darwin and his contemporaries in the 19th century. Over time, generalizations and hypotheses were formulated describing such features of insular animals as loss of flight, absence or diminution of fear, the occurrence of dwarfism and nanism, reduction in clutch size, etc., which gradually have been codified into 'rules' and syndromes. However, not all peculiarities of island species constitute repeated syndromes biased to islands. Systematic, large-scale, comparative analyses are needed to evaluate each proposed syndrome.

The slowing-down syndrome (also known as *the island syndrome*) refers to a suite of morphological and behavioural changes that co-occur in dense island populations occurring in isolation from close competitors and predators, involving a slowing down of life history and reduction in aggressive behaviour. Initially provided as a conceptual model for insular rodents, it has been supported by comparative analyses of insular lizards and has also been invoked in studies on birds. We start with this model in illustration of how different traits may (potentially, at least) evolve in tandem in insular lineages, before going on to consider both the component parts of the slowing-down syndrome and other proposed island syndromes.

The island body-size rule (also known as *the island rule*) describes the graded change in body size in island vertebrates as initially small-bodied species evolve larger body size, while larger-bodied species experience the opposite. It appears best supported for mammals (for which it was first proposed) but also to be evident in birds and reptiles. Ecological release from predation and interspecific resource competition on islands appear to be key drivers. In the case of this and other generally well-supported syndromes, the exceptions and deviations from the general trend can be ecologically informative as to the factors regulating evolutionary shifts from ancestral forms and behaviour.

Reduction and loss of flight appear to be repeated tendencies in island insects and especially in birds, but not in bats. Causal forces are less well understood for insects (windiness is one factor) than for birds (reduced predation is again important). Shifts in reproductive strategy are evident in reduction of clutch size, accompanied by more investment per individual, in lizards and in birds. A reduction in defensive behaviour is known from foundational observations on island birds and has been quantified in island lizards, which are not hasty in fleeing from potential trouble. We are unaware of systematic analyses of the hypotheses of (i) reduction in territoriality in island species, or of (ii) increased incidence of parthenogenesis in island lizards and invertebrates, or of (iii) diurnal behaviour in bats, but there are instances of each of these changes. Rather better supported are trends towards (iv) herbivory in island lizards, (v) low-gear locomotion (shorter, stouter limbs) in large ruminants, and (vi) the loss of bright colouration, (vii) altered songs (including reduction in aggressive elements), and (viii) larger brain size, all in island birds.

Our understanding of island syndromes as emergent statistical tendencies in island animal populations is rapidly developing. We are gaining clarity as to which orders (in cases even which families) of animals follow specific evolutionary trends/syndromes, and what factors influence them. Fundamental to island syndromes in animals are the geographical isolation and limited area of islands. Closely intertwined with these key features of insularity are the absence of most, or all, terrestrial vertebrate predators and the reduction in interspecific competition, which appear to be the key drivers of most insular syndromes in vertebrates.

CHAPTER 11

Island evolutionary syndromes in—and involving—plants

11.1 The peculiarities of island plants

In this chapter, we extend our examination of island evolutionary syndromes to plants (following Carlquist 1965, 1974) and then to plant–animal interactions. Some proposed insular plant syndromes are similar to those formulated for animals, concerning changes in dispersibility, reproductive systems, size, and defensive traits. Others, such as the development of secondary woodiness, are more plant-specific, at least at first glance. Reflecting both the greater richness of insular floras, and the variability within and between species in plant traits, we so far generally lack the systematic comparative analyses now available for many island vertebrate syndromes (but see Burns 2019 for a detailed review).

In scrutinizing these proposed syndromes, the first challenge is to distinguish differences between island and mainland floras that carry both statistical and ecological significance. The second challenge is to distinguish those differences that arise because of differential colonization and establishment—and which are essentially forms of assembly rule (as Chapter 6)—from those arising through *in situ* evolutionary change.

11.2 Insular secondary woodiness

Darwin (1859) observed that islands often feature woody species belonging to groups that are herbaceous on the continents. He hypothesized that this reflected the difficulty of trees reaching islands and the subsequent selective advantage for herbaceous plants evolving greater stature in the absence of the usual array of woody species. Where an ancestral species is herbaceous and derived species are woody, this is known as secondary woodiness (as opposed to primary or basal woodiness) and where the evolutionary transition has occurred on islands it is termed insular secondary woodiness (or simply insular woodiness). Insular secondary woodiness is typically associated with increased plant stature and longevity (Lens et al. 2013; Nürk et al. 2019). We might therefore regard it as indicative of a shift towards *K*-selection, or a form of 'slowing-down syndrome' (cf. Section 10.2) for plants. However, the extent to which insular woodiness is derived *in situ* rather than basal has long been debated. Hence, we must first establish that the transition to woodiness is indeed an emergent island phenomenon. Where it does occur, a further key question is whether it typically occurs multiple times, or just once in a clade, followed by disproportionate diversification of the woody lineage.

In their recent, systematic compilation, Zizka et al. (2022) identified 175 evolutionary transitions to woodiness distributed across 31 archipelagos (375 islands), resulting in some 1097 insular woody species, belonging to 149 genera and 32 families (Fig. 11.1). Although island secondary woodiness was widespread across dicots, it was concentrated in the superasterids clade, especially in Gesneriaceae (327 insular woody species/30%) and Asteraceae (256/23%). The large number of insular woody species of Gesneriaceae owes much to the spectacular radiation of the genus *Cyrtandra* in Polynesia. Hence, the order of importance is different when looking at the number of transitions, which are led by Asteraceae (47 transitions/27%),

Island Biogeography. Robert J. Whittaker, José María Fernández-Palacios, and Thomas J. Matthews, Oxford University Press.
© Robert J. Whittaker, José María Fernández-Palacios, and Thomas J. Matthews (2023). DOI: 10.1093/oso/9780198868569.003.0011

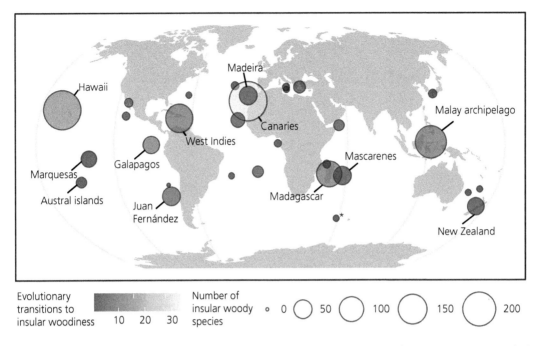

Figure 11.1 The minimum number of evolutionary transitions to insular woodiness for those archipelagos featuring at least one such shift. The asterisk indicates data lumped for several Southern Indian Ocean islands (Crozet, Kerguelen, Prince Edward, and Heard and MacDonald). Figure 3 from Zizka et al. (2022).

Amaranthaceae (15/9%), Brassicaceae (12/7%), Rubiaceae (10/6%), and Campanulaceae (7/4%). The Canary Islands (204/18% of the species and 33/19% of all transitions) and the Hawaiian archipelago (199/18% species, 17/10% transitions) emerged as centres of insular secondary woodiness, followed by the Malay Archipelago (129/11% species, 12/7% transitions), and the West Indies (98/9% species, 11/6% transitions). Based on 61 clades with calibrated phylogenies, stem ages ranged 19.7–0.1 Ma and only 10 clades had origins > 10 Ma. Many of the more recent stem ages were from the Canaries (Zizka et al. 2022).

Outstanding examples of insular woodiness are provided by the silversword alliance (*Argyroxiphium*, *Dubautia*, and *Wilkesia*) in Hawaii, tree lettuces (*Dendroseris*) in Juan Fernández, tree sunflowers (*Commidendrum*) in St Helena, and buglosses (*Echium*), sow thistles (*Sonchus*), daisies (*Argyranthemum*), and tree houseleeks (*Aeonium*) in Macaronesia (Givnish 1998; Lens et al. 2013; Nürk et al. 2019). All Macaronesian *Sonchus* (64 species; Asteraceae)

and most *Aeonium* (c. 44 species; Crassulaceae) are secondarily woody (Dulin and Kirchoff 2010). The Macaronesian clade of *Echium* (Boraginaceae) derive from a single herbaceous continental ancestor, which evolved secondary woodiness, diversifying to produce around 30 endemic species, within which there have been three separate reversals back to herbaceousness (i.e. secondary herbaceousness) and one to a monocarpic-rosette shrub habit (García-Maroto et al. 2009; Lens et al. 2013; Graham et al. 2021).

Within the Canary Islands, 46% of the insular woody species occupy arid coastal areas and just 13% occur in the mesic laurel forests, where they mostly occur within open areas (Hooft van Huysduynen et al. 2021). The evergreen laurel forests of Macaronesia feature around 30 tree species, most of which have changed little since colonization from the Mediterranean region, with several becoming extinct from the source region, making them relictual or palaeoendemic species (but see Kondraskov et al. 2015). Diversification in the Macaronesian flora appears to be primarily a feature of plants of lesser stature, including not only the foregoing examples of secondary woodiness but also (partially) woody plants that are most likely descended from *woody* continental ancestors, including *Descurainia* (Brassicaceae, 7 species), *Pericallis* (Asteraceae, 14 species), and *Tolpis* (Asteraceae, 11 species) (Dulin and Kirchoff 2010).

The Hawaiian lobeliads, with 126 species, constitute by far the largest radiation of plants derived from a single colonist found on any oceanic archipelago (Section 9.7; Givnish et al. 2009). All are woody. Although they appear to have a distant herbaceous ancestor, they evolved their woody habit long before they colonized any Pacific islands. Givnish et al. (2009) suggest that the woody habit, the possession of fleshy fruits and of axillary inflorescences, may be adaptations to forest understorey living. Within this lineage, transitions have occurred in features such as fruit type, pollination mechanism and habitat preferences, but there have been no back transitions to herbaceousness or partially woody (suffrutescent) status.

It therefore appears that many of the most spectacular radiations of plants on islands involve woody lineages, some of which have transitioned to woodiness post-colonization, some of which were already woody at the point of arrival. They tend to involve species of relatively small stature, up to sub-shrubs and shrubs, rather than canopy trees, and they tend to have comparatively rapid (but perennial) life cycles (Takayama et al. 2018). Secondary woodiness is also a feature of tropical alpine mountain ('sky island') systems (Hughes and Atchison 2015).

Not only have insular secondary woody lineages shown a propensity to diversify into multiple species (Carine et al. 2010), many such lineages also show divergence in growth forms and key functional traits ('disparification'). Nürk et al. (2019) consider four such radiations—Hawaiian silverswords, Macaronesian *Echium*, and Andean 'sky island' *Lupinus* and *Hypericum* (Fig. 11.2)—finding that secondary woodiness appears to be associated with dispersal to islands, accelerated rates of speciation, growth form divergence, and divergence in habitat occupied (see also Kim et al. 1996 on Macaronesian *Sonchus*).

Hypotheses for insular secondary woodiness include the following:

1 *Competition* (Darwin 1859). Trees are unlikely to reach remote islands owing to dispersal limitations, providing opportunity for successful herbaceous colonists to gain competitive advantage by growing taller. This would lead to selective pressure for increased woodiness, which would eventually lead to arborescence. Givnish's (1998) *taxon cycling hypothesis* provides an extension of this idea, highlighting that successful island colonizers are likely to be weedy species that establish easily in open/marginal habitats, where they can grow rapidly. Competition may then select for a shift toward persistence through woodiness.

2 *Longevity and promotion of sexual outcrossing* (Wallace 1878; Böhle et al. 1996). Woodiness might allow extended life spans, increasing the chance of sexual reproduction on remote islands where pollinators are scarce or unreliable.

3 *Seasonality release* (Carlquist 1974; Nürk et al. 2019). In comparison with the closest mainlands, islands typically have milder (and moister) climates, extending growing seasons, thus allowing a root system that can support more leaf growth and greater woodiness, permitting additional photosynthesis, promoting the development of herbs to larger stature.

4 *Herbivory release* (Carlquist 1974; Dulin and Kirchoff 2010). The absence of vertebrate herbivores on islands allows plants longer to complete their life cycles. Rather than being consumed, plants grow year-round, exploiting climate moderation, selecting for the evolution of perennial growth forms and increased woodiness.

5 *Drought stress resistance* (Lens et al. 2013; Hooft van Huysduynen et al. 2021). Many secondary insular woody groups occur in relatively dry environments, suggesting a link between secondary woodiness and increased resistance to hydraulic failure (embolism) in their vessels. For example, woody Canarian *Argyranthemum* daisies have been shown to be more resistant to embolism in droughted conditions than herbaceous mainland relatives (Dória et al. 2018). A high proportion of Canarian insular woody

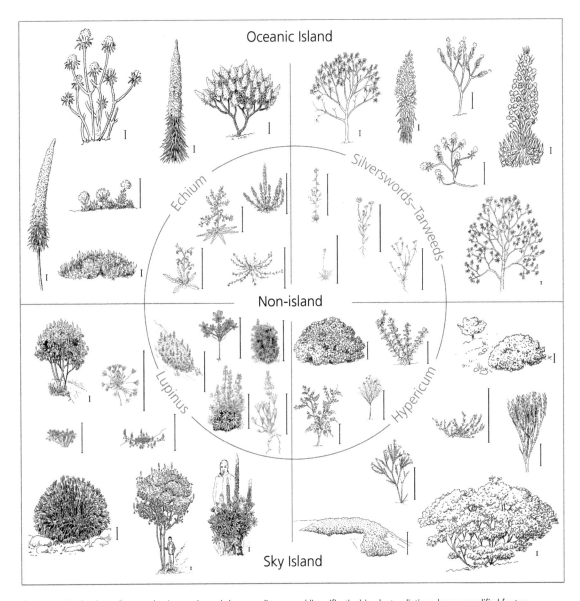

Figure 11.2 Island woodiness underpins accelerated character divergence (disparification) in plant radiations, here exemplified for two oceanic-island and two sky-island plant lineages: Macaronesian *Echium*, Hawaiian silverswords and tarweeds, Andean *Lupinus* and Andean *Hypericum.* Non-island members shown within the circle. Stature indicated by relative size of vertical bars.
Nürk et al. (2019), their figure 1; Creative Commons CC BY.

lineages originated within the last *c.* 3 Myr, during which the islands became subject to seasonal summer droughts (Hooft van Huysduynen et al. 2021).

6 *Co-evolution with leaf size* (Burns 2019). Burns (2019) argues that woodiness may be selected for as a by-product of increased leaf size. He observes that large leaves may perform better in wetter island environments (contrasting with (5)) and also in islands lacking vertebrate herbivores (as per (4)).

It is plausible that more than one of the above mechanisms is at work (Hooft van Huysduynen et al. 2021). This is supported by the recent synthesis by Zizka et al. (2022). They applied structural equation models to their dataset of 1097 insular woody species from 375 islands, to determine that across all islands (including continental shelf, continental fragments, and oceanic), reduced herbivory pressure by large mammals, increased drought and island isolation (in declining order of importance) were the key factors correlating with richness of insular woody species. When restricting the analysis to oceanic islands, which lack native large mammals, herbivory was replaced in the model by possession of a favourable aseasonal climate as the most important factor, with drought and island isolation again following as contributory factors. That both our list of hypotheses and these models invoke both aseasonal climates and conversely responses to drought indicates that there may be different combinations of circumstances that promote insular woodiness. We should also note that the transition to secondary woodiness is not simply an island phenomenon. Nonetheless, it is associated with an increased capacity to exploit the ecological opportunities provided by insular systems filtered at the point of colonization, and it features in multiple island radiations (Nürk et al. 2019). Secondary woodiness permits and is associated with longer life cycles, larger stature, and more prolonged reproductive effort, which are features that, as Darwin argued, are likely to be selected for in the absence of browsing pressure from vertebrate herbivores and in the context of particular climate regimes.

11.3 Loss of dispersibility

The inaccessibility of remote islands initially selects for plant species and groups (such as ferns) with excellent long-distance dispersal mechanisms (Section 4.4; Whittaker et al. 1989; Carlquist 2009; Schrader et al. 2021). Indeed, a high proportion of colonists are weedy in character. However, some insular species of remote islands lack evident adaptations to long-distance dispersal, which may reflect: (i) vicariant origins via past land bridge connections, (ii) chance 'non-standard dispersal' mechanisms, or (iii) post-colonization loss of dispersibility (Section 10.4). It is on the latter that we focus here. Early observations of island endemic plants

that have evolved greatly reduced dispersibility led to the identification of this trend of 'precinctiveness' as a classic island plant syndrome (Carlquist 1965, 1974, 2009).

One explanation is that dispersive propagules get disproportionately carried out to sea and are lost from the insular gene pool. However, we should bear in mind that plant diaspores are variously adapted to dispersal by water, wind, and animals. Strand-line plants typically have seeds that float and survive prolonged immersion in seawater: they are thus primarily sea-dispersed (thalassochory). Long-distance wind-dispersal (anemochory) is aided by small propagule size (fern spores, the dust-like seeds of orchids), or by wings or parachutes (many Asteraceae). Animal dispersal to islands can be effected by spurs that attach a seed capsule to fur or feathers (exozoochory/ectozoochory) or by providing a reward (fleshy fruit, or lipid-rich aril) that leads to ingestion by a bird or fruit bat (endozoochory). Some plants exhibit diplochory (e.g. long-distance thalassochory, followed by zoochorous spread inland). The trait changes involved in reduced dispersibility across these distinct dispersal syndromes must inevitably vary. We consider first exozoochorous and second anemochorous examples.

A classic example of dispersal loss in exozoochorous plants is provided by *Bidens* (Asteraceae), a cosmopolitan genus of around 230 species, which is polyphyletic within *Coreopsis* (Carlquist 1974, 2009; Knope et al. 2020b). The Polynesian radiation, which originates from South America, is understood to be monophyletic. It is a recent and rapid radiation (< 1.63 Ma), of at least 41 insular species, mostly single-island endemics. The Hawaiian species alone illustrate greater ecomorphological diversity than found in the entire genus across five continents. Pacific island *Bidens* owe their wide distribution to attachment of the dry indihiscent fruits to bird feathers via barbed awns. However, the insular species have reduced barbs and shortened awns compared to mainland species, decreasing their likelihood of becoming hooked on bird feathers and being moved among islands. This presents something of a paradox: if dispersibility is reduced, how have they managed to colonize so many islands? Examining the distribution of species within islands, the species retaining dispersal traits closest to ancestral types are those growing on open cliff-sites, where shore and seabirds are most likely to come into contact with them. Those species with reduced dispersal features mostly occur inland, where such contacts are less likely. The evolution of reduced dispersibility thus appears to accompany evolutionary shifts in habitat and other aspects of morphology, a feature repeated within other Hawaiian lineages (Knope et al. 2020b).

The Asteraceae provides further examples of reduced dispersibility in wind-dispersed island plants. Cody and Overton (1996) monitored populations of *Hypochaeris radicata* and *Lactuca* (*Mycelis*) *muralis*, on 200 small, near-shore islands in Barkley Sound (Vancouver Island, Canada). Significant shifts in diaspore morphology occurred within 8–10 years (around five generations) of population establishment. The diaspores comprise two parts, a tiny seed with a covering (the achene), which is surrounded by or connected to a much larger ball of fluff (the pappus). The clearest findings were for *Lactuca muralis*, the species with the largest sample size. Founding individuals had significantly smaller seeds (by about 15%) than typical of mainland populations, illustrative of a founder effect (**immigrant selection**). Within eight years, pappus volumes had decreased below mainland values while seed sizes had returned to those of mainland populations (Fig. 11.3). If the pappus is thought of as a parachute and the seed the payload, then the ratio between the two indicates the dispersibility of the diaspore: indeed it has been found to be a good metric for descent time (e.g. Fresnillo and Ehlers 2008). Cody and Overton's study thus demonstrates rapid evolution of reduced dispersibility in wind-dispersed higher plants on small islands. Turning briefly to lower plants, comparative analysis by Patiño et al. (2013) suggested a general pattern of reduced long-distance dispersal in insular bryophytes, evidenced in a greater relative importance of asexual diaspores, which are heavier compared to the spores involved in sexual reproduction.

Contradictory findings have been obtained in some other cases. For example, a comparative study of diaspore properties and descent times of three herbaceous plants on three small Danish islands and

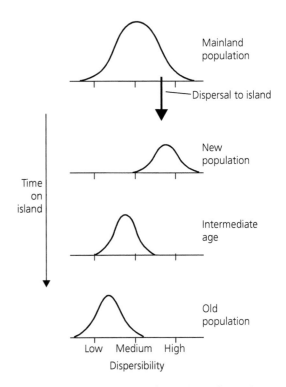

Figure 11.3 A schematic diagram of the evolution of reduced dispersibility of wind-dispersed plants on islands, as suggested by the changes recorded in *Lactuca muralis* (Asteraceae) on near-shore islands in British Columbia, Canada. Maximal dispersibility characterizes the youngest island populations, as the founding event(s) will be from the upper end of the dispersal range of the mainland population. Thereafter selection acts to decrease dispersibility of propagules on islands, one hypothesis being that more dispersive diaspores are lost to sea.
Simplified from Cody and Overton (1996), their figure 1.

in mainland Denmark, showed reduced dispersibility in the island populations of *Cirsium arvense* but the reverse in *Epilobium angustifolium* and *E. hirsutum* (Fresnillo and Ehlers 2008). Based on the compilation in Table 11.1, reduction in dispersibility was detected in 64% of cases, 36% involving a reduction in dispersal aids, and 73% an increase in seed sizes. We should qualify these statistics by noting that multiple species are involved in some comparisons (e.g. *Bidens*) and that there is in any case no reason to consider this a representative selection of insular species. Based on the limited hard data available, Burns (2019) concluded that reduced dispersibility remains of uncertain status as an insular plant syndrome (see also Waters et al. 2020). Where

Table 11.1 Case studies demonstrating changes in dispersibility in insular plants.

Species	Dispersal mode	Geographic comparison	Reduced dispersibility?	Reduction in dispersal aids?	Increase in seed size?	Source
Bidens spp.	Exozoochory	Polynesia vs. American mainland	yes	yes	no	Carlquist (1974)
Cirsium arvense	Anemochory	Islets off Danish strait	yes	yes	yes	Fresnillo and Ehlers (2008)
Epilobium angustifolium	Anemochory	Islets off Danish strait	no	no	yes	Fresnillo and Ehlers (2008)
Epilobium hirsutum	Anemochory	Islets off Danish strait	no	yes	no	Fresnillo and Ehlers (2008)
Fitchia spp.	Exozoochory	Polynesia vs. American mainland	yes	no (increase)	yes	Carlquist (1974)
Hibiscus spp.	Thalassochory	Ogasawara vs. Asian mainland	yes	yes	yes	Kudoh et al. (2013)
Hypochaeris radicata	Anemochory	Islets off British Columbia	yes	no	yes	Cody and Overton (1996)
Mycelis muralis	Anemochory	Islets off British Columbia	yes	no	yes	Cody and Overton (1996)
Periploca laevigata	Anemochory	Macaronesian and Mediterranean islands vs. African mainland	no	no	yes	García-Verdugo et al. (2017)
Rumex bucephalophorus	Mixed	Macaronesian and Mediterranean islands vs. European mainland	no	no	no	Talavera et al. 2012)
Sonchus spp.	Anemochory	Chatham vs. New Zealand	yes	no	yes	Burns (2019)

Modified from table 3.1 of Burns (2019) by inclusion of additional data.

reduced dispersibility is observed, it often involves increased seed size, which may indicate that a driving force is selection for enhanced establishment through larger seedlings.

Finally, as we stressed above, ecomorphological changes in island plant lineages can be substantial and may involve changes in *mode* of dispersal. For example, within the spectacular radiation of Hawaiian lobelliads (Fig. 9.6), *Cyanea* has diversified to produce 76 species (60% of all Hawaiian lobelliads), a high proportion being single-island endemics (Givnish et al. 2009). They tend to be forest species that are dispersed, mostly locally, by forest-interior birds. Those lobelliad lineages with wind-dispersed seeds have both lower species richness and a lower proportion of single-island endemics. Here the inference is that seed dispersal of forest understorey species by avian forest frugivores produces lower rates of gene flow than wind dispersal

of open habitat species and is contributory to greater precinctiveness and thus greater diversification (see *Bidens*, above). Testing these ideas quantitatively would be challenging, particularly in the context of the great reduction in native frugivores both in Hawaii and in many other oceanic island systems (e.g. Heinen et al. 2018; Carpenter et al. 2020). More generally, we currently lack an adequate framework by which to assess shifts in dispersibility in insular endozoochorous plants (Burns 2019). We also know little about changing dispersibility in thalassochorous plants, although Table 11.1 features one example, the loss of buoyancy in diaspores of *Hibiscus glaber* (which has colonized interior habitats), in relation to the widespread strandline progenitor *H. tiliaceus* (Kudoh et al. 2013).

In conclusion, reduced dispersibility in island plants: (i) is detectable in multiple lineages of exozoochorous and anemochorous plants, (ii) appears to be associated with some spectacular radiations of insular lineages, (iii) may be invoked where insular lineages show a switch in dispersal mode (as per Hawaiian lobelliads), (iv) may be a by-product of selection for traits such as increased seed mass rather than reflecting the outcome of the export from the gene pool of overly dispersive propagules, and (v) is of unknown frequency of occurrence.

11.4 Size changes in island plants

Changes in body size are a key topic in island animal syndromes, within which we noted that changing dimensions of body parts may not be isometric (e.g. Sections 10.4, 10.9). Island plants may also evolve differences in overall stature and in plant parts (e.g. in fruit, seed, or leaf size) relative to their mainland progenitors. While the often weedy nature of remote-island colonists sets the scene for a general trend towards both larger stature and diversification of stature within radiating lineages (exemplified above), island plants may also become smaller in size. Testing for emergent trends in plants is a challenging task given how variable properties such as plant height and leaf size can be within a population (for leaves, within a single plant). For example, conspecific trees growing on a valley floor on Krakatau are typically very much taller than those growing on an adjacent ridge, while the canopy height reduces with elevation more rapidly than on nearby mainland Java or Sumatra due to the compressed climate zones found on small islands (Section 3.4). Demonstrating that trait differences between island and mainland populations are genetically determined is needed for firm conclusions to be drawn. Such concerns are alleviated when examining large radiations such as discussed in Sections 11.2 and 11.3, in which there are numerous examples of island plants that have diversified in leaf, fruit, or overall plant sizes in relation both to continental progenitor species and to each other (Fig. 11.2).

The following studies give an indication of the variability of patterns reported. *Alyxia ruscifolia* (Apocynaceae) is a woody shrub native to Australia and to Lord Howe Island, 600 km east of the continent. Plants were found to become reproductively mature earlier on the island, to have smaller stature at maturity and to produce larger and more elongated seeds (Burns 2016). Another carefully executed comparative study of several closely related woody plants growing on New Zealand and nearby islands showed a trend to larger plant size and larger leaf size for a given stem size in the island populations (Burns et al. 2012). Analyses of populations of two wind-dispersed Canarian endemic plants *Kleinia neriifolia* and *Periploca laevigata* in comparison with sister species from nearby mainland areas also supported an increase in mean leaf size in the two island species, but the former produced 25% smaller seeds and the latter 23% larger seeds than the mainland sister species (García-Verdugo et al. 2019b).

Recently, initial efforts have been made to test for the plant equivalent of the island body-size rule (Section 10.3) (Burns et al. 2012; Biddick et al. 2019). Comparative analyses for 175 pairs of related species from 'mainland' New Zealand and 10 nearby archipelagos rather intriguingly demonstrated a graded trend from gigantism (in smaller plants) towards dwarfism (in larger ones) for both stature and leaf area, but seed size of insular plants tended to be larger than those of their mainland relatives (Biddick et al. 2019). Given that the multiplicity of differences between plants and vertebrates, it is not immediately apparent why an island body-size rule should be predicted for plants. Indeed, Brian and Walker-Hale (2019) have questioned the conclusions owing to the degree of heterogeneity displayed within the data, arguing that the methods used may have obscured an underlying trend of island lineages 'exploring more of trait space than their mainland relatives', an interpretation rejected by Biddick and Burns (2019). Evolutionary changes in plant traits, including in overall dimensions, are evidently common features of island plant lineages but thus far the theoretical and analytical frameworks for detecting emergent trends remains underdeveloped (Diniz-Filho et al. 2021).

11.5 Altered defensive adaptations to herbivory

Many plants have anti-herbivore traits. They include: (i) chemical compounds that reduce their palatability (terpenoids, alkaloids, flavonoids), or digestibility (tannins, phenols, phytoliths); (ii) spines, thorns, and hairs that protect to some degree against browsing; (iii) architectural features such as divaricate branching (below) and microphyly (the production of very tiny leaves); (iv) heteroblasty (significant functional changes, e.g. of leaves, occurring over the life span); and (iv) mutualistic relationships—for example, with ants that protect the plant against insect damage (Carlquist 1974; Burns 2019; Moreira et al. 2021). Following plant colonization of oceanic islands lacking mammalian herbivores, the selective advantage of such features may no longer outweigh the energetic costs of producing them and hence it is expected that the plants of remote islands should show a tendency to lose such traits over time, especially those related to herbivory by mammals. Indeed, the increased palatability of island endemic plants is often cited as a contributory factor to their endangerment and loss following introduction of non-native mammalian herbivores (e.g. Cubas et al. 2019). However, as we shall show, the picture is rather more complicated than simply a pattern of loss of defensive traits in insular plants.

The absence of top-down (predator) control of vertebrate herbivores introduced by humans onto oceanic islands is of crucial importance to their general impact on island vegetation, in the worst cases causing huge reductions in plant biomass (Chapter 14). They may also have selective impacts, as illustrated by a study of introduced rabbits (*Oryctolagus cuniculus*) on Tenerife (Fig. 11.4). Browsing damage was found to be lowest for non-native plants, increasing for natives with the level of endemism (Cubas et al. 2019). It appears that many of the secondary woody perennials of the more open, low-stature vegetation types on Tenerife are particularly susceptible to rabbit browsing. Exclosure experiments in the high-elevation summit scrub have shown the dramatic impact that rabbits are having on the vegetation dynamics of this zone, which is naturally dominated by neoendemic plants (Cubas et al. 2018).

Quantification of plant traits linked to defence against herbivory permits testing of island–mainland differences in closely related taxa. Bowen and van Vuren (1997) measured chemical defences (phenols and tannins), morphological defences

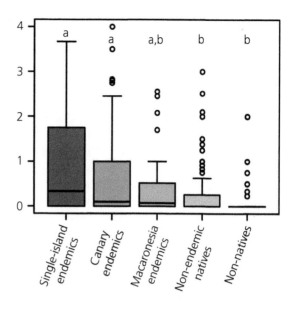

Figure 11.4 Rabbit-browsing damage was quantified on a scale from 0 (no damage) to 5, in 215 transects distributed across the island of Tenerife. Damage levels increased steadily from non-natives to localized (single-island) endemics. Letters indicate chorotypes statistically distinct from others ($p < 0.05$). A separate test confirmed that non-natives and native non-endemics (combined) were less damaged than insular endemics (combined) ($p < 0.001$). Chorotypes are non-overlapping, e.g. Macaronesian endemics are those Canarian species shared with at least one other archipelago but not found outside Macaronesia.
From Cubas et al. (2019), their figure 1b.

(leaf spines and leaf size), and structural features linked to digestibility for six species of insular shrubs from Santa Cruz Island (California Channel Islands) and their closest mainland relatives. They found island–mainland differences in at least one defensive mechanism for each species, nearly all in the predicted direction. Consistent with the presence of invertebrate herbivores but the absence of mammalian herbivores from the island (at least since the extinction of mammoths c. 11,000 ka (Muhs et al. 2015) until the introduction of sheep in the 19th century), morphological changes (larger leaves and reduced spinescence) were more evident than chemical traits. Structural traits were indistinguishable between island and mainland populations.

Some other studies have failed to show the anticipated pattern of insular loss of defensiveness. Chemical defences of the shrub *Periploca laevigata* were assessed via measurements of tannin levels in seedlings via a common garden approach (Monroy and García-Verdugo 2019). Populations were sampled from the Cabo Verde Islands, the Canaries, and both insular and mainland sites in the Mediterranean. Leaf tannin levels generally declined with increasing latitude (suggesting climatic control) and were, if anything, higher in insular populations, in contradiction to the hypothesis of reduced defences on islands. Similarly, in a comparative study of two archipelagos, Moreira et al. (2022) reported that insular species appeared to be better defended than the non-native flora, with Balearic Island plants tending to feature physical defensive traits and Canary Island plants featuring chemical defensive traits. On the same theme, a comparative study of spinescence of 18 insular and 18 mainland sites in Australia found no significant insular effect (Meredith et al. 2019; but see e.g. Burns 2016).

Although remote oceanic islands naturally lack browsing land mammals, many feature reptiles, sometimes including giant tortoises (e.g. Galápagos and Aldabra), which have largely plant-based diets and can act as key ecosystem engineers (Gibbs et al. 2010). Due to their shell structure and weight, they feed on leaves close to the floor, creating a relatively low browse line. In the Galápagos, the endemic *Opuntia* cacti are a favoured food of the endemic giant tortoises (*Chelonoidis* spp.), which can significantly reduce densities of these cacti (Gibbs et al. 2010). On those islands currently (or recently) supporting tortoises, juvenile *Opuntia* tend to produce long, rigid, sharp spines, but once above the reach of giant tortoises, they produce soft stems with bristle-like spines. On islands that never had tortoises (Genovesa, Marchena), *Opuntia* species tend to be smaller statured or prostrate and are significantly less spiny (Dawson 1966).

Another possible plant defence syndrome is divaricate branching, involving consistently wide branching angles that create wire-like plants, within which leaves are harder for browsers to access. Divaricate branching is comparatively common within the floras of Madagascar and New Zealand. Indeed, nearly 10% of native New Zealand woody plant species are divaricately branched and the growth form has evolved independently in 17 plant families (Burns 2019). Both island groups possessed

now-extinct giant ratite birds (elephant birds and moas, respectively). In New Zealand, divaricately branched tree species tend to be strongly heteroblastic. Younger plants are divaricately branched, with small leaves, until around 3 m high (roughly the reach of the tallest moas), above which they begin to produce larger leaves and shallower branching angles. These properties suggest specific adaptations in insular plants in response to herbivory by large-bodied indigenous birds and reptiles, distinct from those prompted by long exposure to herbivorous mammals. In turn these evolutionary responses could be linked to differences in mouth parts (toothless vs toothed browsers) and eyesight between these groups that influence browsing behaviour and impact. Although these large insular ratites are now extinct, experiments involving emus as analogues provide some support for divaricate branching as being a response to browsing by large avian herbivores (Burns 2019).

The hypothesis of evolutionary reduction of plant defences on islands is thus arguably based on an over-simplified assumption regarding the browsing regime of islands being substantially lower, when reptiles and birds have in cases filled some of this feeding niche space. To assess emergent empirical patterns, Moreira et al. (2021) applied meta-analysis to a compilation of around 200 comparisons of conspecific (90% of cases) or congeneric (10%) native plants in insular and mainland settings. While they found evidence for greater levels of mammalian herbivory on islands, there was no difference in herbivory by invertebrates. There was also no overall island–mainland difference in either physical or chemical defensive traits.

We may conclude that some case studies are consistent with the evolution of a loss of defences in insular plants driven by release from large mammalian browsers. But, equally, some studies indicate either no difference, or a pattern of increased defensive traits in insular populations. Although there are other driving factors, it is of course possible that anthropogenic effects are clouding the picture. For example, humans introduced goats to the Canary Islands over two thousand years ago (Chapter 14) and so it is possible that poorly protected species have gone extinct and that the persisting native flora has been selected for increased defensive traits over this period. Indeed, Moreira et al. (2021) suggest that such historical legacies of past introductions and extinctions may explain the failure of their study to provide clear support for the hypothesis of reduced insular defences.

The absence of clear support for a unidirectional island syndrome of reduced defensive traits in plants need not, however, be the end of the story. Islands provide contexts where vertebrate herbivore pressures are often distinct in nature, diversity, and intensity compared with comparator mainland sites. Vertebrate herbivory is often naturally reduced (many oceanic islands), but sometimes it is greater (e.g. high densities of introduced browsing mammals), and often it involves—or until recently involved—browsers of different characteristics and impact (e.g. larger flightless birds, giant reptiles) compared with mainland sites. The literature already provides instructive 'insular experiments' indicating that long periods of exposure to different forms and levels of herbivory on islands can produce detectable evolutionary changes in plant traits linked to herbivory, especially in morphological features, less so in chemical properties. More systematic testing of how plant traits evolve in response to changing browsing and grazing regimes on islands over varying time frames is likely to prove informative (Moreira and Abdala-Roberts 2022).

11.6 Reduced fire resilience

It has been argued—in something of a generalization—that islands experience relatively mild, humid climates, which should lead to a reduced natural incidence of forest fires. If this is so, we should expect to see reduced adaptations to fire in insular plants compared with their mainland relatives (Burns 2019). Relevant traits include thick bark, serotiny (e.g. retaining seeds in woody cones/fruit, released after fire), the capacity to resprout from epicormic buds (dormant buds beneath the bark), self-pruning of low branches, or seed germination induced by high temperature or smoke.

The proposition that islands have reduced fire regimes in the absence of humans in that is hard to evaluate, as we lack matched palaeoenvironmental

records from islands and directly comparable mainland areas free from human impact. Charcoal data series from island lake cores demonstrate that human colonization of forested islands throughout prehistorical and historical periods has been strongly associated with the use of fire as a clearance tool and very often have then featured sustained high frequencies of burning (Chapter 13). But we also know that fires were a natural feature (initiated e.g. by lightning strikes, volcanic eruptions) of many islands prior to human contact, albeit typically at a lesser frequency.

Within Macaronesia, for example, fire regimes have been characterized through charcoal data for sites within islands in the subtropical wet Azores, within at least some ecozones in the Canaries and even within the comparatively arid low-latitude Cabo Verde Islands (Connor et al. 2012; Nogué et al. 2013; Castilla-Beltrán et al. 2021; Ravazzi et al. 2021). Within the higher Canarian islands, fire is rare in the moist laurel forest zone but is an important feature of the more arid, higher-elevation Canarian pine forest zone. Canarian pine (*Pinus canariensis*) features multiple fire-resistant traits and is probably more effectively fire-adapted than its closest mainland relatives (Climent et al. 2004). It possibly holds the status of a relictual or palaeoendemic species, suggesting that it was largely pre-adapted to active fire regimes, although its ancestry is not settled and it may also have developed increased fire-resistance post-colonization (Burns 2019). This again illustrates the challenge in assessing plant syndromes based on limited case studies. As a further complicating factor, humans have used fire as a primary land-clearance tool throughout history; hence, it is possible that fire-susceptible island plants have been eliminated prior to their scientific description.

Notwithstanding these limitations, Burns (2019) assesses a small number of studies showing features such as reduced bark thickness and serotiny in conifers on Isla Guadalupe (North America) and reduced serotiny in island conifers in glacial lakes in Canada and in paperbark trees on Lord Howe Island, in each case in relation to mainland relatives. Other studies he reviews are less supportive of such an evolutionary trend. While the negative effects of anthropogenic fires on ecosystems that have evolved essentially without fire can be dramatic and such effects can indeed be found on islands (e.g. Carrera-Martínez et al. 2020), the available literature provides limited support for loss of fire adaptations as an island syndrome.

11.7 Tufted-leaved (*Federbusch*) growth

Plant species in several families have evolved shrub forms featuring repeated branching, with more or less upright branches and leaves only in rosettes in the branch tips. Termed tufted-leaved growth or *Federbuschsträucher* (feather duster plants), Schimper (1907) suggested this growth form provides resistance to frequent, strong winds. Meusel (1952), on the other hand, emphasized the benefits of these traits for the production of new autumn leaves following the shedding of foliage during summer droughts. Finally, Beyhl (1995) attributed this growth form to their benefits for taking up atmospheric water from nocturnal dew and/or diurnal fog.

Federbusch growth is considered globally rare, but it is a feature of some islands. In the Canaries such plants are common, especially in the drier coastal zones. They include some species within the genera *Aeonium* (Crassulaceae), *Bupleurum* (Apiaceae), *Carlina, Kleinia* and *Sonchus* (Asteraceae), *Echium* (Boraginaceae), and *Euphorbia* (Euphorbiaceae). In addition, similar physiognomy (dracoid growth) can be seen also in certain monocot plants, such as the dragon tree (*Dracaena draco*). The growth form is less frequent elsewhere in Macaronesia and where it occurs may reflect colonization from the Canaries (e.g. of *Echium, Euphorbia*, or *Aeonium*). In the Pacific, the Juan Fernández Islands also feature some shrubs or trees of similar growth form (e.g. *Dendroseris, Robinsonia*) (Moreira-Muñoz et al. 2014). The climate of these islands being cooler and lacking a pronounced dry period, the frequent occurrence of rosette shrubs does not appear explicable through summer leaf loss. While these growth forms are striking where found, *Federbusch* does not appear general enough to qualify as an emergent island syndrome.

11.8 Reproductive syndromes in island plants

Self-compatibility in insular plants: Baker's law

H. G. Baker (1955) proposed that self-compatible (SC) species capable of uniparental reproduction should have an advantage in colonizing remote islands over self-incompatible (SI) species, because a population could potentially be founded by a single propagule. Baker's law, as it became known, has been disputed, in part because of the high frequency of dioecy—where male and female gametes are produced by different individuals—in insular floras (Pannell et al. 2015). However, Baker's law has recently received support from two systematic analyses of global datasets. Before reviewing them, we should note that, as formulated, Baker's law is essentially a form of assembly rule and that it would be necessary to show a tendency for the SC/SI ratio to change via *in situ* evolution for it to qualify as an insular evolutionary syndrome. This is tricky to assess, especially as the distinction between SC and SI is not always clear-cut, with some mostly SI species showing a degree of pseudo-self compatibility (PSC) (Crawford et al. 2011).

Grossenbacher et al. (2017) compiled data for 1500 species of insular and mainland representatives of three families: Asteraceae, Brassicaceae, and Solanaceae. As many species do not comply perfectly with the SC/SI dichotomy, species that were not consistently reported as either SI or SC were screened out within their analysis. Overall, the prevalance of SC was 66% for islands against 44% for mainland species, with a similar pattern found for each family (e.g. Fig. 11.5). Discounting island endemics from the comparison did not alter their findings. However, it is worth noting that some of the more successfully diversifying island plant lineages are members of the Asteraceae that are ancestrally either SI or PSC, including *Argyranthemum* and *Sonchus* in the Canaries and the Hawaiian silversword alliance (Crawford et al. 2011).

Razanajatovo et al. (2019) compiled data for 1752 species, representing 161 higher plant families, to test for an insular bias in the prevalance of

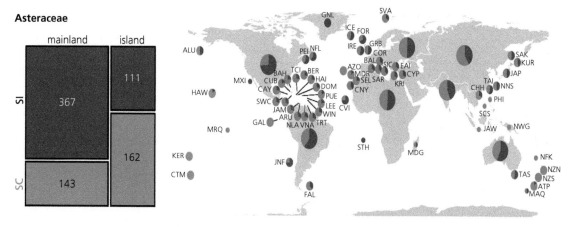

Figure 11.5 The numbers of self-compatible (SC) and self-incompatible (SI) Asteraceae species in the floras of continents (mainlands) and islands globally. The 'mainland' category comprises species only found on continents, but the island category includes both island endemics and island native species shared with mainlands. Pie charts show proportions, where small pies = 1 or 2 study species; medium pies = 3 or more species; large pies = mainlands. Letters permit areas to be identified.
Excerpt from figure 1 of Grossenbacher et al. (2017).

self-compatibility and of autofertility (self-fertility in the absence of pollinators). Their results indicated a significant (if slight) bias of island colonists towards SC and that this tendency is greater for species that are also autofertile. They found the same pattern for naturalized non-native species, indicating that introduced species with selfing ability are also more likely to succeed in becoming an established part of insular floras. Interestingly, they found no difference in selfing ability of island endemics compared to mainland species, which might be taken to indicate that *in situ* evolution has tended to favour reduced selfing ability. However, noting the small number of endemics considered in their study (57 species), we concur with the authors that more rigorous testing is needed before a conclusion can be drawn. In particular, we need more studies that quantify the propensity for SC/SI ratios to shift within island floras as a function of *in situ* evolutionary transitions or differential rates of diversification.

Sexual dimorphism in plants

Around 90% of flowering plants are hermaphroditic (defined as individuals acting as both females and males), with around 6% being dioecious (individuals are either female or male) (Barrett and Hough 2013). However, in some well-studied insular floras, the proportion of dioecy is markedly higher (Table 11.2). There are a number of additional variants, which may be grouped into monomorphic sexual systems (comprising hermaphroditism, monoecy, andromonoecy, and gynomonoecy) and dimorphic sexual systems (comprising dioecy, androdioecy, and gynodioecy), but as can be seen from Table 11.2, the stand-out feature is the bias of the island floras towards dioecy and away from hermaphroditism (Crawford et al. 2011; Schlessman et al. 2014). Crawford et al. (2011) are clear in their review that insular dioecy/sexual dimorphism can be either basal (mostly the case for the Galápagos: Chamorro et al. 2012) or derived. As before, the key questions are thus: is there a general island bias towards dioecy, and if there is, is it biased colonization and establishment (i.e. an assembly rule), or *in situ* evolution that is responsible?

We cannot be certain how representative the data in Table 11.2 may be. However, high values of dioecy have been indicated for other archipelagos, including 15–20% for La Réunion, 16% Tonga, and 13% Guam and Ogasawara (Abe 2006), 9% Juan Fernández (Bernardello et al. 2001), and 16% for Galápagos (Chamorro et al. 2012), although these estimates are subject to amendment as more data become available. All lie above the global mean,

Table 11.2 Species richness, endemism, and frequencies (%) of sexual systems among the angiosperm native floras of New Caledonia, Hawaii, New Zealand, and the world.

	New Caledonia	Hawaii	New Zealand	World
Land area (km^2)	19,103	16,887	268,680	148,900,000
Angiosperm spp.	3051	971	2066	c. 350,000
% endemicity	78	89	84	–
Total (% endemic) dioecious spp.	640 (89)	143 (95)	225 (?)	c. 14,620
% world dioecious spp.	4.4	1.0	1.5	100
Dioecy	21	14.7	12–19	4–6
Gynodioecy	0.1	3.8	2	7
Androdioecy	0.1	0	NA	NA
Total dimorphic	**21.2**	**18.5**	**14–23**	**10**
Hermaphroditism	61.2	62.4	NA	72
Monoecy	12.8	7.6	9	5
Andromonoecy	3.7	4.5	NA	1.7
Gynomonoecy	1.0	3.9	NA	3
Total monomorphic	**78.8**	**78.4**	**77–86**	**90**

Source: Schlessman et al. (2014) and references therein. NB Inconsistencies in sub-totals and variation in estimates stem from the combination of estimates from different sources. NA = not available.

although the Canaries provides a counter-example, with an estimated 3% of dioecy (Francisco-Ortega et al. 2000).

Plant sexual systems can be messy to categorize and as with the SC/SI distinction, it is possible that deviation from consistent dioecy may enable population establishment from a single individual colonist in a dioeceous species. An example of 'leaky' dioecy is the Canarian endemic *Withania aristata*, males of which can occasionally produce fruits. Another association that has been noted is that many insular sexually dimorphic species possess fleshy fruits, associated with bird dispersal, raising the possibility of multiple individuals being introduced in a single colonist event (Crawford et al. 2011; Razanajatovo et al. 2019). Such features may explain successful colonization by sexually dimorphic species.

Given support for Baker's law as an assembly rule weighting insular establishment towards SC species (above), the high incidence of sexual dimorphism (dioecy in particular) in insular floras, rather implies a role for *in situ* evolution. Crawford et al. (2011) provide the following illustrative examples: (i) whereas 11% of Hawaiian plant colonists were dimorphic and gave rise only to dimorphic species, around one third of dimorphic species derive from monomorphic colonists; (ii) within the

Juan Fernández Islands, *Robinsonia*, the largest dioecious genus, appears to have a hermaphroditic ancestor; and (iii) dioecy in Canarian *Bencomia* has arisen twice *in situ*. The switch to dioecy favours outcrossing, thus avoiding the inbreeding depression that may disadvantage hermaphrodites in the long run, especially in small populations (Carlquist 1974). Beyond true doiecy, another common feature of oceanic floras is the presence of functional dioecy; that is, mechanisms promoting outcrossing, such as unisexual flowers, protandry, protogyny, herkogamy, heterostyly, or allelic self-incompatibility (Carlquist 1965, 1974; Barrett 1996). As we noted above, the Canarian flora has few true dioecious speices (3%), but around half of the Canarian endemics have mechanisms promoting outbreeding (Francisco-Ortega et al. 2000).

The data demands for inventory of sexual systems of insular floras are particularly challenging, often requiring detailed manipulative experiments in greenhouses (Anderson and Bernardello 2018). Unsurprisingly, more research is needed to establish the extent to which the proportions of different sexual systems in insular floras reflect both non-random colonization (assembly rules) *and in situ* evolution. From the data available, it appears: (i) that the proportions of dioecy may be positively linked with tropical distribution, woody habit, fleshy zoochorous fruits and plain flowers (Vamosi et al. 2003); and (ii) that *in situ* evolution favours increasing richness of sexual dimorphism in insular floras, but the limited evidential base currently available makes this a tentative conclusion.

Floral traits and pollination

Island plants often display a loss of flower showiness, featuring small, actinomorphic (radially symmetrical), white or green non-showy flowers with simple bowl-shaped corollas, instead of the large, tubular, zygomorphic (bilaterally symmetrical), bright flowers frequently displayed on the continents (Carlquist 1974). This trend has been noted for Hawaii, Galápagos, Juan Fernández, and Ogasawara (Bonin) floras (Crawford et al. 2011) and is evidenced in a comparison of Australian and New Zealand floras. It probably reflects the generalistic and promiscuous nature of island pollinators (Barrett 1996), as well as their smaller sizes when compared with many continental pollinators. Inoue and Kawahara (1996) demonstrated a direct correlation between the flower size of Japanese *Campanula* species and of their pollinators. Flowers were larger on the mainland, where they were visited by larger pollinators (bumblebees and megachilid bees) lacking from the islands. Pollination in the island populations was carried out by smaller pollen-collecting halictid bees, which were shown experimentally to be indifferent to flower size, in contrast to the preference for larger flowers shown by the nectar-feeding megachilid and bumblebees on the mainlands.

It has been suggested that anemophily (wind pollination) is more prevalent on islands (Carlquist 1974). Possible explanations for why wind pollination might be favoured despite it being a less targetted mechanism than animal pollination include: (1) gaining independence from impoverished pollinator services (Traveset 2001), (2) that windy conditions prevalent on islands favour anemophily, and (3) outcrossing fitness benefits yielded by wind pollination (Barrett 1996). The most obvious problem for animal-pollinated species is at the point of colonization, especially if they possess highly specialized pollination systems, such as found in *Ficus*,

in which each species has a unique pollinating fig wasp species (Compton et al. 1994).

Based on data from 42 study sites, it has been estimated that anemophily occurs in 11.5% of angiosperms globally (Ollerton et al. 2011). But values for the 42 study sites vary from zero to 67%, with a tendency for lower values in the tropics and higher values in colder climates. Hence, although values above the global mean have been estimated for particular islands, e.g. 20% for Hawaii (Sakai et al. 1995), 26% for Ogasawara (Abe, 2006), 29% for New Zealand (Webb et al. 1999), and 47% for Juan Fernández (below), we cannot be sure that this represents departure from source pool values. A systematic comparative analysis will be needed before an insular bias can be confirmed and this is hampered by the paucity of knowledge of pollination systems and how well syndromes predict actual pollination mechanisms. For example, some plants may retain features such as nectar rewards while becoming in effect wind pollinated (Bernardello et al. 2001; Crawford et al. 2011). Detailed studies of specific insular systems do, however, provide evidence of switching of pollination mode.

Probably the best-studied oceanic system from the perspective of pollination syndromes is the Juan Fernández Islands (Bernardello et al. 2001; Crawford et al. 2011). Around 50% of colonists are inferred to have been insect pollinated, 40% wind pollinated, with a handful of bird- and mixed-pollination system species. Within the contemporary flora 47% are wind-pollinated, 9% hummingbird-pollinated (the island has two hummingbirds, one of which is endemic), one species (*Libertia chilensis*) appears to have retained insect pollination and the pollination mechanisms of the rest are unknown. These data are consistent both with a colonization bias (assembly rule) favouring wind pollination and with subsequent evolutionary gain of wind pollinators. Although the large number of species for which data are lacking necessitates caution, field studies indicate a minimal role for native insects in pollination. Bernardello et al. (2001) suggest that at the point of colonization, the following genera were probably insect pollinated but now are wind pollinated: *Azara, Drimys, Juania, Pernettya, Robinsonia*, and *Ugni*, while a few likely insect-pollinated founders have switched to hummingbird pollination, including species of *Centaurodendron*, *Eryngium*, *Dendroseris*, and *Rhaphithamnus*.

Switching of pollination mode is known from other insular systems. For example, the 34 species of the monophyletic endemic genus *Schiedea* on Hawaii variously exhibit wind-, insect-, or bird-pollination, or are autogamous (self-fertilizing). Diversification has involved expansion into drier habitats associated with a switch to dimorphic breeding systems and wind pollination: this combination accounting for nearly a third of the species in the genus (Wagner et al. 2005). Another facet of pollination switching that appears to be a recurrent feature on islands is from insect pollination in ancestral species towards mixed vertebrate-insect pollination systems, involving birds and/or lizards (Navarro-Pérez et al. 2013; Alarcón et al. 2014). Macaronesia provides examples of lineages considered to be bird-pollinated at the point of colonization (e.g. *Canaria* and *Lavatera*) and those where it appears to have arisen *in situ* (e.g. some species of Campanulaceae, *Lotus* and *Echium*). Navarro-Pérez et al. (2013) provide a further example in *Scrophularia calliantha*, which has an inferred wasp-pollinated continental ancestor, but which is now pollinated by endemic *Gallotia* lizards in addition to birds and insects. We return to the theme of animal pollination in Section 11.12, where we look at plant–animal syndromes in more detail.

11.9 Polyploidy on islands

Polyploidy, or genome duplication, is an important form of plant speciation, with around 35% of flowering plant species being of recent polyploid origin (Rice et al. 2019). Globally, the proportion of polyploids increases with latitude, reflecting the influence of decreasing temperature, through its effects on plant lifeform and (reduced) species richness (Rice et al. 2019). There are two forms, **autopolyploid**, where the derived form results from a single parent species and **allopolyploid**, resulting from cross-fertilization between two species (Section 8.8). The reproductive barriers with the diploid parents is sufficiently strong, especially in allopolyploids, as to represent effectively instantaneous speciation.

Based on those species analysed to date, Hawaii, with 50%, Galápagos and New Zealand, each with 46%, are comparatively rich in polyploids, whereas the Canaries with 32% are unexceptional and Juan Fernández, with 3%, are below average (Rice et al. 2019; Meudt et al. 2021). Meudt et al. (2021) deploy an innovative path analysis for 150 lineages to explore the contribution of polyploidy to *in situ* diversification (Fig. 11.6). Lineages from larger source pools and that changed ploidy levels close to the point of colonization have higher insular polyploidy levels. Lineages that were shown from dated phylogenies to have been on islands longer (greater stem age) also had higher ploidy levels and this directly and indirectly contributed to greater diversification, although the effect of stem age on ploidy levels was reversed for Hawaii. Further differences in the models for Hawaii and Juan Fernández were also evident (denoted as 'island effect' in the figure). Direct effects of source pool size and repeat colonization on endemic diversity were not evident in the model.

While these analyses are for subsets of data from just four archipelagos and leave much unexplained variation, they suggest: (i) that polyploids may have advantage in long-distance dispersal and establishment (constituting an assembly rule); and (ii) that polyploidy also contributes to insular diversification (see also Linder and Baker 2014). *In situ* or neopolyploidization events can certainly occur, as shown for the Balearic Islands by Rosselló and Castro (2008), but more typical are Hawaiian lineages such as *Santalum* and the silversword alliance, where polyploidization occurred shortly before insular colonization, with diversification (sometimes involving further hybridization) then following (Keeley and Funk 2011). Once again, more systematic work is needed to establish how environmental factors, such as climate, influence the role of polyploidy in insular evolution and how it varies with archipelago age and isolation and with timing of lineage colonization.

11.10 Mycorrhizal symbionts

Delavaux et al. (2021) compiled data for 515 mainland regions and 313 islands, using family-level

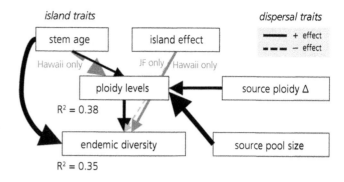

Figure 11.6 Fitted path analysis model of the influence of polyploidy on endemic insular plant diversity based on data for 150 lineages from New Zealand (66% of the data), Hawaii (15%), Canaries (15%) and Juan Fernández (3%, JF). Arrow size indicates standardized effect sizes for significant pathways ($P < 0.05$) and those in grey are significant only for Hawaii or JF, as indicated. Figure 3 of Meudt et al. (2021).

assessments to classify plants based on evidence of fungal symbionts into arbuscular mycorrhiza (AM), ericoid and ecto-mycorrhizal (EEM), orchid-mycorrhizal (ORC, i.e. all orchids), or non-mycorrhizal. AM fungal symbionts are considered particularly effective at improving plant access to inorganic phosphorous and nitrogen, while EEM symbionts are better able to access organic nitrogen, which enables EM plants to dominate in cold climates. Mycorrhizal fungi disperse independently of their plant hosts, raising the question of whether their absence from islands might limit plant colonization.

AM fungi are of particular interest as they lack adaptation for long-distance dispersal and cannot grow independently of their hosts, suggesting poorer ability to colonize islands. On the other hand, AM fungi have lower host specificity than the other types, while more AM plants are merely facultatively dependent on their symbionts. Consistent with these prior observations, it was found that AM plant richness declines more steeply with island isolation than any other category, suggesting colonization limitation (i.e. an assembly rule) for this category of plant. However, consistent with the lower host-specificity of AM symbionts and the benefits of the mutualism, the contribution of AM plants to increasing insular endemism was found to be positively related to island isolation. It should be cautioned that the data are noisy and that other environmental variables such as island area, elevation, and climate also play important roles. However, the analyses support a role for

mycorrhizal association type in structuring the assembly and *in situ* evolution of insular floras.

11.11 Plant syndromes on islands: a synthesis

Throughout this review, we have stressed the challenges involved in distinguishing island assembly rules from *in situ* evolutionary dynamics. In essence, assembly rules refer to non-random assembly of island biotas via differential dispersal and/or establishment of lineages possessing particular traits. Evolutionary syndromes, on the other hand, arise from *in situ* trait changes in particular lineages, which can involve combinations of genetic changes within particular species, instances of hybridization (whether or not involving polyploidy), and differential rates of diversification across taxa possessing different trait combinations. As we have seen, many of these traits are much messier and less easily quantified than in animals. The reader will also have appreciated that many of the traits examined may be loosely or tightly bound up with other traits and syndromes that we have examined.

The potential importance of mycorrhiza in plant establishment on islands (Section 11.10) has been understood for some time, but to the best of our knowledge, it hasn't hitherto been proposed as the basis of an insular syndrome (Table 11.3). It seems unlikely that we have exhausted the list of plant traits that may be invoked as potentially important for insular assembly rules and evolutionary syndromes (see Ottaviani et al. 2020). In addition to the

Table 11.3 Proposed island evolutionary syndromes for plants, grouped (left column) as 'probable', 'possible', and 'doubtful' according to Burns (2019), see his table 7.1, and 'not listed' by Burns, with (right) summary comments synthesized from the present chapter.

Syndrome	Summary comments arising from our review
Burns' *probable* components	
Defence displacement	There are some intriguing indications of evolutionary responses to distinct vertebrate grazing regimes, but there is *a lack of clear support* for general loss of insular plant defences.
Baker's law (self-compatibility)	*Supported* as an emergent tendency in **island assembly** (rather than an *in situ* evolutionary syndrome), with tantalizing but inconclusive suggestion of a counter-balancing trend generated by *in situ* evolution.
Seed gigantism	Seed size increase *is one means* by which dispersibility is reduced in some insular plant lineages.
Leaf gigantism	Insular radiations often involve diversification in plant traits including leaf size, but the generality of directional size changes *remains to be established*.
Burns' *possible* components	
Leaf heteroblasty and spinescence	*Possibly favoured* on islands browsed in the past by large/giant birds and reptiles.
Loss of protection mutualisms	*Not reviewed herein*. Burns (2019) provides examples of insular species losing their extra-floral nectaries in the absence of ants but concludes it has been too rarely documented to stand as an emergent insular syndrome.
Divaricate branching	Appears to be a feature of certain continental fragment islands featuring giant ratites, but *by no means clear* that this is a common/emergent insular trend.
Seed-size constraints on dispersal	Although we comment only briefly on this, plant species lacking long-distance dispersal traits (e.g. many large-seeded species) are largely excluded from colonizing islands (i.e. *this is an assembly rule* cf. Chapter 6).
Island rule in plant stature	Evidence in favour of an island body-size rule for plants has been presented by Burns and colleagues but others have argued that data heterogeneity is problematic; *no clear conclusion*.
Insular woodiness	Insular secondary woodiness *is an emergent trend* and a key factor in adaptive radiation in island plants.
Reduced incidence of heterostyly	Heterostyly is a mating system that facilitates outcrossing. *Not reviewed herein*, but Burns' review showed *insufficient evidence* to draw a conclusion.
Anemophily	Anemophily as a feature enhancing insular colonization remains unproven. *Switching away from insect pollination* towards either wind or vertebrate pollination *appears to be a feature* of a range of insular systems and lineages.
Generalized floral morphology	Generalized floral morphology *is supported* by data for several oceanic archipelagos.
Loss of fire-adapted traits	There is *limited support* for this as an island evolutionary syndrome.
Burns' *doubtful* components	
Loss of dispersibility	*Of uncertain generality, but is a repeated evolutionary feature*, sometimes involving a switch in dispersal mode; it is associated with some spectacular insular radiations.
Insular dioecy	There is *support (albeit tentative)* for *in situ* evolution favouring increased richness of sexually dimorphic species, countering to some extent Baker's law/assembly rule.
Not listed by Burns	
Tufted-leaved growth	*Not general enough* to qualify as an emergent island syndrome.
Polyploidy	Recent polyploidization *appears to be an advantage* for insular colonization (i.e. island assembly) and *in situ diversification*.
Mycorrhizal limitation	AM symbiotic relationships appear to constrain plant colonization with increasing island isolation, with *in situ* evolution subsequently favouring diversification in AM lineages: *tentative* as based on a single study, albeit of a large numbers of islands.

pervasive challenge of so far inadequate data, key remaining challenges include: (i) to identify how the various traits considered in this chapter sum to emergent syndromes; and (ii) to determine which traits may be secondary or largely correlative, as opposed to primary determinants of insular success. Table 11.3 lists the traits considered by Burns (2019) in his recent synthesis, alongside summary comments of our own. As indicated in the table,

we are more supportive of the two features that Burns labels 'doubtful' than we are of several of his 'possible' components. We also identify Baker's law as a form of island assembly rule rather than an evolutionary syndrome. But our conclusions remain tentative.

A key challenge emerging from our review is whether there is such a thing as a unitary insular syndrome for plants? Our view is that it may be more productive to search for multiple syndromes, suites of traits that together are selected for in particular insular contexts, regulated by environmental features such as climate, elevation and soil properties, and by biotic properties such as a paucity of terrestrial vertebrates, reduced availability of specialist insect pollinators and (at least initially) paucity of mycorrhizal symbionts. Insular subtropical cloud forests evidently do not select for the same combination of plant traits as semi-desert environments, even on the same islands. However, this is not to argue for an approach restricted to considering each plant trait independently. For example, the multiple features of plant reproduction, from the various forms of monomorphic and dimorphic sexual expression, through flower structure, to pollination mechanisms, clearly also need to be analysed as systems. Moreover, when analysing dispersal filters and subsequent evolutionary outcomes *in situ*, properties of the fruits and seeds resulting from these systems are also part of the piece. For example, in considering the relative success of two plant families on islands, Takayama et al. (2018) note that Asteraceae and Orchidaceae are both highly dispersive, but the former is more successful as a contributor to insular floras (both in colonization and diversification). However, as they comment, we cannot say whether orchids are limited principally by overspecialist pollination requirements, or perhaps by their reliance on mycorrhiza, by a combination of both, or indeed by other traits.

11.12 Island syndromes involving plant–animal interactions

Given that islands tend to be species poor and disharmonic, a key challenge for new immigrant species is to find the biological resources they need to survive and complete their life cycle. This can include pollinators for plants, specific plant resources for many insects, and particular host animals for parasites or parasitoids (Santos et al. 2016; Hembry et al. 2021). The new communities into which the immigrants arrive may simply lack suitable resources, in which case the colonization fails. On the other hand, the island may present an abundance of untapped resources, with some plant and animal taxa disproportionately abundant

and lacking a full array of pests or parasites. It is these circumstances that have selected, within remote oceanic islands, for the exploitation of alternative hosts, new food resources and shifts in habitat (e.g. Hembry et al. 2021). We have already provided numerous examples in earlier sections. Here we focus on two ecologically crucial forms of plant–animal mutualism on islands: plant–pollinator and plant–disperser relationships.

We highlight two features. First, it has been suggested that species-poor island pollination networks tend to involve supergeneralists, often endemic species, which have experienced ecological release to become involved in multiple pollination interactions (Olesen et al. 2002). Second, pollination and/or dispersal relationships can appear bizarre in relation to those found on continents, because the functions are carried out, in the absence of more common continental pollinators or dispersers, by groups of animals not frequently involved in such functions (Section 11.8; Traveset 2001). We will delve a little deeper into both ideas.

Endemic super-generalists within pollination networks

Traditionally, pollination studies tended to focus on single plant species or, at most, on guilds of related plant species and the insects pollinating them, reflecting the view that pollination mutualisms are the result of highly co-evolved relationships (Traveset and Santamaría 2004). Recent work has shown that pollination processes are typically more generalized than previously thought, with many insect and plant species interacting, and thus yielding weak pairwise relationships. As the degree of linkage within networks has also been found to vary between biomes and to diminish in species-rich communities (Olesen and Jordano 2002), it follows that species-poor insular communities should have relatively higher proportions of connectance. Following this line, Olesen et al. (2002) hypothesized that island networks typically involve a few supergeneralized animal pollinators serving many plant species, and a few super-generalized plant species that are visited by many pollinators. To explore this, they quantified connectance (C) as the proportion of potential pollination links that were observed and thereby identified the following as endemic super-generalists: *Bombus canariensis* (bee, Canaries, $C = 48\%$), *Halictus* (bee, Azores, $C = 60\%$), *Phelsuma ornata* (day gecko, Mauritius, $C = 71\%$), *Xylocopa darwini* (bee, Galápagos, $C = 77\%$), *Aeonium holochrysum* (plant, Canaries, $C = 80\%$), *Azorina vidalii* (plant, Azores, $C = 75\%$), *Gastonia mauritiana* (plant, Mauritius, $C = 62\%$), and the native (rather than endemic) *Cordia lutea* (plant, Galápagos, $C = 45\%$).

In a subsequent comparative analysis of 23 continental-island, 18 oceanic-island, and 11 mainland pollination networks, Traveset et al. (2016) confirmed that oceanic-island networks tend to be less species rich, posses fewer links, and a lower density of interactions, consistent with their possessing super-generalists. They also noted that whereas plant niche overlap was significantly higher on oceanic islands compared with the mainland and continental-island networks, this was not the case for pollinators, which are typically depauparate in comparison to plants on islands. However, super-generalism in island endemic insects is strongly supported in analyses of a Fijian bee *Homalictus fijiensis*, which was found to utilize a much wider range of plants than introduced bees (including the honeybee *Apis mellifera*), including in their network many non-native plants (Draper et al. 2021). Support for high connectance in pollination networks involving insects was also provided by Wang et al. (2020) in a study of Yongxing Island, in the South China Sea, in which it was also observed that plant floral traits poorly predicted the composition of pollinators visiting the plants.

Although more comparative studies are still needed, cases of insular super-generalism have been reported for endemic lizards, insects, and birds. A spectacular illustration of pollination generalism in the Galápagos avifauna is provided by Traveset et al. (2015), who collated data for 500 h of observations of 19 native birds (all but two endemic) across 12 islands. Nearly all Galápagos land bird species were found to be involved in pollen transport, with an average of 22 plant species visited per bird species. Many of the interactions involved non-native plants, reinforcing previous findings that insular endemic super-generalists readily incorporate non-natives into insular networks, facilitating

their establishment and spread (and see Hayes et al. 2019).

Shifts in pollination syndromes

The absence of many common continental pollinators (bees, wasps, hummingbirds, etc.) from remote islands has provided opportunities for animals that do not usually perform these functions to acquire or extend these roles (above). Olesen (1992) suggested that various island–mainland switches may occur: (1) from insect pollination on mainland to wind, rain, or self-pollination on islands; (2) from a high to a low diversity of flower visitors; (3) from insects to vertebrates; (4) from social to solitary bees; (5) from long-tongued to short-tongued or nectar-robbing bumblebees; or (6) from specialist insects to generalized flies or bumblebees.

Beginning with shifts within insect communities, Hiraiwa and Ushimaru (2017) quantified insect pollinator services in sand dune communities in an oceanic archipelago (the Izu Islands) compared with similar plant communities in mainland Japan (Honshu). Their study of 211 pollinators and 80 plant species showed that in the oceanic archipelago, pollinator niches shifted, such that diverse short-tongued species increasingly visited long-tubed flowers. These shifts in realized niche compensated for lower availability of long-tongued pollinators, although with a cost to the plants in terms of reduced seed set.

We have already assessed the evidence for transitions from insect pollination towards wind and vertebrates from the plant perspective above (Section 11.8). We now turn to the implications for insular vertebrates of the paucity of pollinating insects. First, it implies reduced insect-resources to feed on, and second, it implies reduced competition for plant resources (nectar), driving niche expansion in some vertebrate species towards a role in pollination (Olesen and Valido 2003). In particular, there is evidence of expanded pollination roles for lizards (saurophily), birds (ornithophily), and bats (chiropterophily) on islands, as we review next.

Lizards have adopted roles as pollinators almost exclusively on islands, including New Zealand, Mauritius, the Balearics, and Tasmania. Olesen and

Valido (2003) reported that of 37 lizard species then known to visit flowers and/or act as pollinators, only two species were from mainland areas, both peninsulas (Baja California and Florida). Insular examples are known from: (i) the Balearic Islands, where the endemic *Podarcis lilfordi* is a possible pollinator of some 23 plant species; (ii) from New Zealand, where *Hoplodactylus* geckos have been observed visiting flowers of several species; (iii) from the Canaries, where endemic *Gallotia* lizards contribute to pollination of a number of species; and (iv) from Indian Ocean islands, where *Phelsuma* day geckos have a role as pollinators as well as consuming nectar, pollen, and fruit (Olesen and Valido 2003; Rodríguez-Rodríguez 2015). As these examples indicate, flower visitation may be largely to feed on pollen and nectar and doesn't necessarily equate to the provision of an effective pollination service. A recent intensive study of endemic Galápagos lava lizards (*Microlophus* spp.) showed that at least 7 of the 9 lizard species were involved in pollen transport, but that only 8% of lizards captured were transporting pollen, with just 10 pollen taxa recovered, implying a relatively modest role in plant pollination (Hervías-Parejo et al. 2020).

Bird pollination is far from being restricted to islands and it is important to note that the two bird species involved as pollinators in the Juan Fernández studies cited in Section 11.8 are hummingbirds, which are particularly important in plant pollination in mainland Neotropical mountains (Dellinger et al. 2021). However, the Galápagos study by Traveset et al. (2015) highlights an expanded role for birds and the involvement of a remarkably high proportion of the native avifauna in pollination. In crude terms, birds and insects shared one third of the host plant species, half were exclusively ornithophilous, and just one sixth were exclusively entomophilous. Thus, although birds constituted only 8% of the total pollinator fauna, and 36% of all links, they contribute to the pollination of 85% of the native flora. Along similar lines, a more restricted study of three Canarian plants with traditional bird–flower syndromes, *Canarina canariensis*, *Lotus berthelotii*, and *Isoplexis canariensis* was undertaken on Tenerife by Ollerton et al. (2008). Notwithstanding possessing floral features normally associated with specialist

pollinators (lacking from the current Canarian avifauna), these plants received effective pollination services from an array of generalist (i.e. non-flower specialist) bird species, in particular Canarian chiffchaff (*Phylloscopus canariensis*). More systematic observational studies, such as that carried out in the Galápagos, are required to determine the extent to which plant–bird pollination mutualisms may constitute a co-evolved insular trend, but such investigations are hampered by the extent of insular extinctions, habitat change, and non-native species introductions.

Plant pollination by fruit bats can be observed on both mainlands and islands. Around 62% of Old World pteropodid fruit bats (flying foxes) are insular and they have successfully colonized islands across the Indian Ocean and much of the tropical Pacific (Fleming et al. 2009). In the New World, only around 12% of phyllostomid bats (leaf-nosed bats) are restricted to islands. Bat pollination services thus differ between islands in the Old and New World, with insular pteropodid bats visiting flowers in 53% of Old World bat-pollinated plant families, and the corresponding figure for New World (mostly Caribbean) islands being just 18%. Most plant families featuring bat pollination were around before the evolution of nectar-feeding bats, implying that bat pollination is a derived condition. Although bats are viewed as keystone pollinators and seed dispersal in some Pacific island systems, the extent to which bat pollination has co-evolved in insular settings is hard to assess.

Reptiles as dispersal agents: an island phenomenon?

Island lizards represent one third of lizard species but two thirds of fruit-eating lizards (Valido and Olesen 2019). Frugivory is part of a tendency towards more plant-based diets in insular lizards (Section 10.8), alongside a number of other putative syndromes discussed in Chapter 10. This insular characteristic is regarded as an example of a form of interaction release *sensu* Traveset et al. (2015). Valido and Olesen (2019) suggest three key elements. First, islands have reduced food resources through lowered species richness and abundance of insects. Second, they feature reduced predation and interspecific competition, which in combination leads to the third element, density compensation. This implies a greater degree of intraspecific competition and selects for expansion of the feeding niche towards alternative resources, such as fleshy fruits. Density compensation in island lizards appears to be a general, global phenomenon. Mean insular densities are reported to be more than an order of magnitude higher than on mainlands (a mean of 1920 vs 128 individuals/ha) (Buckley and Jetz 2007), with *Sphaerodactylus macrolepis* in the Virgin Islands being recorded at a remarkable density in excess of five individuals/m^2 (Rodda et al. 2001).

Examples of partially frugivorous lizards acting as seed dispersers include: (i) 18 New Zealand species, among them the gecko *Hoplodactylus maculatus* and the skinks *Oligosoma grande* and *Cyclodina alani*; (ii) the Balearic Islands lizards *Podarcis lilfordi* and *P. pityusensis*, which feed on the fruits of at least 26 species (Olesen and Valido 2003); and (iii) the endemic Canarian lacertids (seven *Gallotia* species) and skinks (three *Chalcides* species), which are all known to feed on fruit. *Gallotia* lizards eat fruits of at least 40 native plant species, which is slightly more than the half of the Canarian plant species bearing fleshy fruits, and they also feed on the fruits of at least 11 introduced species (Valido and Nogales 1994).

Analysis of seed dispersal effectiveness (SDE) in the Canaries, showed that although *Gallotia* lizards are highly frugivorous and have mean population densities around 20 times higher than those of birds, both lizards and birds dispersed a similar quantity of seeds, reflecting the slower metabolism and lower seed intake of the lizards (González-Castro et al. 2015). Whereas birds disperse more seeds within woodlands, lizards were the predominant dispersers in shrublands and open sites. Lizards provided higher SDE for 7 of 11 plant species and, with a single exception, the disperser with the greater quantitative contribution also provided the best quality in terms of seedling emergence/establishment. Comparative analysis of SDE for a lizard-dispersed shrub, *Neochamaelea pulverulenta*, on three of the Canary Islands has demonstrated that the extinction, or

near extinction, of larger-bodied species of *Gallotia* following human colonization of the Canaries has resulted in diminishing SDE for at least some zoochorous plant species (Pérez-Méndez et al. 2015). It should be noted that the relatively sedentary behaviour of lizards means that long-distance dispersal of zoochorous species, such that might lead to foundation of new populations or contribute to maintain gene flow across islands and archipelagos, relies upon members of the avifauna, on occasion involving seeds ingested by a lizard subsequently killed, eaten, and then deposited by a predatory bird (Nogales et al. 2012).

Frugivory and seed dispersal is common in chelonians (turtles and tortoises) and there is no evidence to suggest that this behaviour is an island syndrome (Falcón et al. 2020). Nevertheless, islands tortoises are (Galápagos, Aldabra), or once were (Bahamas, Balearic Islands, Barbados, Cabo Verde, Canaries, Madagascar, Mascarenes, Seychelles, etc.), among the largest native terrestrial frugivores of remote islands. Many have gone extinct since human colonization, although, for example, the disappearance of the giant tortoises in the Canaries long pre-date human arrival (Rhodin et al. 2015). Where still extant, tortoises assume the role of dispersers of several large-seeded endemic plant species, such as in the Galápagos (Racine and Downhower 1974) or Aldabra Atoll (Hnatiuk 1978). Giant tortoises represent low-risk, high-impact taxon substitutes in island rewilding strategies (Hansen et al. 2010) and several insular rewilding efforts are underway involving their deployment. For instance, the Aldabran species (*Aldabrachelys gigantea*) has been introduced on several islets off Mauritius where goats and rabbits have recently been eradicated, as a functional substitute for the extinct Mauritian tortoises (*Cylindraspis* spp.), which are considered to have been the original dispersers of the endemic Ile aux Aigrettes ebony (*Diospyros egrettarum*). After a pilot study in a fenced enclosure, viable fruits of the ebony were found in tortoise faeces at a distance from the parent trees, demonstrating the theory in action (Zavaleta et al. 2001).

Double mutualisms on islands

A study of the critically endangered endemic Mauritian woody liane, *Roussea simplex*, showed that both pollination and seed dispersal was carried out solely by endemic diurnal geckos (*Phelsuma cepediana*), which therefore constitute a 'double mutualist' (Hansen and Müller 2009). An intensive multi-year study of pollination and seed-dispersal interactions in the Galápagos enabled the construction of a database of 479 interactions involving 21 bird species (19 native) and 108 plant species (one third, introduced) (Olesen et al. 2018). Overall, 48% of the birds and 18% of the plants were involved in double mutualisms (see Frontispiece to Part III) and although double mutualisms constituted just 5% of the links, as against 73% and 22% of single-pollination and single-dispersal mutualisms, respectively, the species involved in double mutualisms were also typically highly connected to other species via single mutualisms. Hence, the double mutualists constituted the core of the pollination/dispersal network.

To test for geographical bias in the occurrence of such double mutualisms, Fuster et al. (2019) compiled records of 302 cases, from 17 insular and 16 mainland study areas. With the exception of one ant species, all other cases involved vertebrates (in declining order birds, mammals (mostly bats) and lizards). Most plants involved were woody (85%). These double mutualisms have most commonly been reported from tropical regions and from islands. Across all taxa—and for birds and lizards separately (but not for mammals)—significantly more double mutualisms were insular and a significantly higher proportion of insular cases involved double mutualisms with multiple plant species.

Although there is a danger of a reporting bias towards better-studied insular systems, there are several reasons why we might expect double mutualisms to have a tendency to develop on islands. First, it is consistent with the phenomena of interaction release and niche expansion described for many insular vertebrates (above). Second, it is consistent with the shifts described above towards more generalism in pollination networks and towards plant pollination by vertebrates in particular. Third, as Fuster et al. (2019) highlight, there may be a role for improved palatability of plant parts, although the jury is still out on whether reduced chemical defences are common enough to constitute an insular syndrome (above).

Plant–animal interactions: further evidence for insular evolutionary syndromes?

In so far as they are insular biased in the first place, a key question hanging over from the above subsections is to what extent these particular plant–animal interactions (pollination and dispersal mutualisms) represent insular evolutionary syndromes as opposed to merely reflecting colonization bias? There is no simple answer to this and no single best way of approaching it. Table 11.4 provides a summary of our discussion of four specific forms of plant–animal interaction.

In addition to switches in feeding niche, pollination systems, dispersal, and other interspecific relationships across trophic levels, island life is replete with numerous examples of other forms of ecological shift, including in elevation occupied, from above-ground to below-ground (lava tube caves), from aquatic/riparian to terrestrial habitat (and vice versa), many of which are repeated multiple times within and across different archipelagos (e.g. Parent et al. 2008; Hembry et al. 2021; Machado 2022). Hence, it is entirely plausible that additional animal, plant, and plant–animal insular syndromes may be identified and confirmed in the future. However, as we have stressed throughout, while identification of a particular syndrome as strongly and simply associated with degree of insularity may be especially satisfying for island biogeographers, it is ultimately more useful to view islands as experimental systems that permit dissection of the factors that drive patterns in metacommunity assembly and evolution and which therefore can contribute to improved mechanistic understanding of the often intricate, intertwined, and complex processes involved.

Table 11.4 Possible island syndromes involving plant–animal interactions.

Evolution of super-generalism within pollination networks	There is *support for in situ evolution of generalism* of endemic plants and animals (insects, birds, and lizards) within insular communities, driven by colonization limitation (especially of specialist pollinating insects), although more comparative studies are needed.
Shifted pollination syndromes (see also 'Anemophily' in Table 11.3)	Following from the above, *there is also evidence for an emergent directional tendency* away from specialist insect pollination mechanisms, including towards wind, bird, lizard, and sometimes bat pollination, i.e. suggesting co-evolutionary changes across insular plants and animals.
Plant dispersal by reptiles	Frugivory in endemic lizards *appears to be an emergent island syndrome*, with evidence of effective seed-dispersal services being provided to a range of island endemic plants; giant tortoises also have important roles where present, but frugivory in chelonians is not specifically insular behaviour.
Double mutualisms involving the same species in both pollination and seed dispersal	The limited comparative data available suggest double mutualism to be comparatively more important in insular metacommunities, especially involving endemic vertebrates, but *too few studies are available to establish the generality* of this pattern.

11.13 Summary

Continuing the theme of the previous chapter, we first review a selection of proposed insular syndromes in higher plants and second turn to plant–animal interactions, focused around pollination and dispersal systems. The first challenge is to decide whether trait profiles for insular plants actually differ from a random sampling of mainland source pools. The general variability within potentially relevant plant traits, their multiplicity, and the lack of systematic comparative data is such that this first task is currently slightly less well advanced than for animal syndromes. The second challenge, where differences are detected, is to determine the relative contribution of dispersal and establishment filters versus *in situ* evolutionary change.

We begin with insular secondary (derived) woodiness, which is associated with increased stature and longevity and is akin to the 'slowing-down syndrome' in animals. Phylogenetic data have provided confirmation of this evolutionary syndrome, which is a feature especially of the superasterids clade (e.g. Asteraceae and Gesneriaceae) and is well exemplified within the floras of Hawaii, Galápagos, and Macaronesia, as well as some sky-island systems. Loss of dispersibility, often involving increased seed mass, (i) is found in many insular lineages of exozoochorous and anemochorous plants, (ii) is a feature of some particularly large insular radiations, and (iii) can involve a switch in dispersal mode. However, systematic comparative analyses of frequency of occurrence of these

trends are currently lacking. Diversification in plant dimensions (plant height, seed or leaf size) during insular evolution has been documented within many plant radiations, but recent, intriguing, suggestions of a 'body-size' rule (small plants increasing in size while large ones get smaller), remain subject to debate.

Reduced physical and chemical defences in the absence of terrestrial mammals on remote islands has long been considered an insular plant syndrome, but recent tests have produced contradictory findings, while in some insular floras there is evidence of features such as divaricate branching, spinescence, and heteroblasty that suggest a legacy of browsing regimes dominated by giant tortoises and large flightless birds. The occurrence of reduced fire tolerance and of tufted-leaved (*Federbusch*) growth forms are each too infrequently documented to qualify as emergent insular syndromes.

Changes in plant reproductive systems within insular plants are much better supported. Baker's law describes the tendency for self-compatible species to be slightly favoured in colonization (making it an assembly rule). There is evidence of a trend away from self-compatibility and a trend towards sexual dimorphism during insular evolution, but both conclusions are tentative. Insular plants also often feature a loss of flower showiness alongside a tendency for switching from specialist insect pollination towards either wind—or vertebrate—(especially bird) pollination as lineages diversify on islands. The frequency of polyploidy (genome duplication) and hybridization within insular plant lineages appears to be slightly elevated, with polyploidization shortly before (sometimes shortly after) colonization appearing to be associated with more rapid insular diversification. Finally, we describe a recent global-scale study indicating that plants with arbuscular mycorrhizal symbioses may be limited in colonizing more distant islands, but that once established, they may contribute disproportionately to insular endemism: a finding requiring confirmation in further work.

The final section considers plant–pollinator and plant–dispersal interactions and the possibility of co-evolutionary syndromes spanning plant and animal networks. We find support for: (i) increased levels of generalism in insular pollination networks, (ii) a shift in insular vertebrates towards increased use of plant resources, linked to (iii) enhanced roles in pollination, and of (iv) *in situ* evolution of tendencies towards frugivory (and, hence, seed dispersal) in insular lizards. We also review cases of double mutualism, in which endemic vertebrates are involved in both pollination and seed dispersal for the same plant species, often in fact for multiple plant species, although we caution that relatively few mainland (or insular) networks are sufficiently well studied to be sure of the generality of this tendency as an island evolutionary pattern.

Although our conclusions are tentative, there is tantalizing evidence of a range of plant traits and syndromes being selected for (or against) in island systems, many of them owing to interactions with— and featuring some level of co-evolution with— members of the disharmonic animal metacommunities found on islands.

PART IV

Human Impact and Conservation

Frontispiece: The Moai of Ahu Nau Nau on Anakena Beach, Rapa Nui (Easter Island). Anakena Bay is believed to be where Ariki Hotu Matu'a landed with his people to establish the first settlement of Rapa Nui.
Photograph by Nigel Purse, 9 February 2014.

CHAPTER 12

The application of island theory to fragmented landscapes

12.1 Habitat islands

Humans have had a profound influence on the ecology and biodiversity of the planet, altering the composition and functioning of ecosystems across the globe (Watson et al. 2018). We have done so through habitat alteration, hunting, agriculture, pollution, and the introduction of non-native species (Maxwell et al. 2016), the impact of which on islands we detail in the subsequent chapters. Here we focus on the implications of the ongoing fragmentation and reduction in area of natural habitats, which have together long been regarded as key drivers of terrestrial species extinctions, whether on islands or continents. The remaining areas of more-or-less natural habitats are, in many regions, becoming mere pockets within a matrix of profoundly altered habitats. In short, they are **habitat islands**. The reduction of contiguous area should, according to island theory, imply losses of species, but can we predict how many species should be able to persist in a landscape of habitat islands? This is the general problem that we address in the present chapter.

The implications of increasing insularity of remaining 'natural' habitat were recognized in the seminal works of Preston (1962) and MacArthur and Wilson (1963, 1967), with Preston astutely observing that in the long run species would be lost from wildlife preserves, for the reason that they constitute reduced areas in isolation. MacArthur and Wilson's dynamic (EMIB) model provided a theoretical basis to develop this argument (Chapter 5), allowing conservation scientists to explore the outcomes of differing configurations of protected areas, assuming them to act as virtual islands in a 'sea' of radically altered and thus essentially hostile, empty habitats (Fig. 12.1). Hence, with this assumption of 'binary habitats', island theory became an early cornerstone of the emerging field of conservation biology during the 1970s and 1980s.

An early but remarkably unyielding question posed was: 'given the opportunity to put a fixed percentage of land into conservation use is it better to opt for a "single large or several small" reserves?' The focus of the 'SLOSS' debate was on the entire regional assemblage of species and how to maximize the carrying capacity of the protected area (i.e. habitat island) system. The answer provided by island theory (EMIB) appeared to be that larger reserves are better and this mantra became embedded within conservation policy. Some argue that this was misguided (see Wintle et al. 2019). We will come back to this key debate a little later, as we have organized this discussion to start with the smallest units: populations of individual species.

The most basic question concerning populations in isolates is: 'how many individuals are enough to ensure the survival of a single isolated population?' We will consider this first. Many species of conservation concern are not, however, restricted to a single locality—or at least, not yet. Rather, their survival in a region relies on a network of island-like habitats or habitat islands. Most of this book has been concerned with real islands in the sea. Is it realistic to expect island-like habitats within large landmasses to behave according to the same principles as real islands? Instead of the barrier of saltwater, a forest habitat island might be separated from another patch of forest by a mixture of meadows, hedgerows, arable land, roads, and

Island Biogeography. Robert J. Whittaker, José María Fernández-Palacios, and Thomas J. Matthews, Oxford University Press.
© Robert J. Whittaker, José María Fernández-Palacios, and Thomas J. Matthews (2023). DOI: 10.1093/oso/9780198868569.003.0012

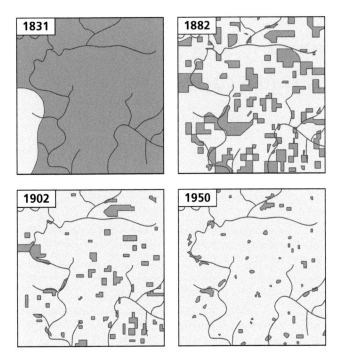

Figure 12.1 The first figure in MacArthur and Wilson's (1967) *The Theory of Island Biogeography* is a map of an area (Cadiz Township) in Wisconsin, reproduced from an earlier source: it was selected to illustrate the process of habitat loss and fragmentation, highlighting that the principles of island theory should be applicable to habitat islands.

urban areas. The implications for species movements between patches might be rather different (Itescu 2019; Matthews 2021).

Another important consideration in addressing the impact of habitat subdivision on the movement of species is the scale of isolation. Mountain habitat islands ('sky islands') within continents, but surrounded by extensive arid areas, can be more effectively isolated than real islands that happen to be located within a few hundred metres of a mainland. However, habitat islands are often (perhaps typically?) sufficiently close to one another that their populations are in effect linked: they form metapopulations, the second theoretical element that we will consider. Having done so, we will then consider the whole-system implications of fragmentation and how 'island' approaches have contributed to their understanding.

When island theory was first applied to forecasting species losses, the implicit assumption was that the base state of a 'natural' area is a state of balance, or equilibrium. Hence, following piecemeal habitat destruction, the newly isolated fragments are cast out of equilibrium. Subsequently, they must shed species as the system moves to a new equilibrium,

determined by the area and isolation of each fragment. This way of thinking is consistent with the **balance of nature paradigm**, which sees the world in the absence of human intervention as a self-regulating, balanced system (Pickett et al. 1992).

We have stressed the importance of considering physical environmental factors in island biogeography, recognizing that ecological responses to environmental forcing factors are often played out over too long a time frame for a tightly specified dynamic equilibrium to be reached. The notion that ecological systems may only tend towards equilibria, is consistent with an alternative viewpoint, the **flux of nature paradigm**, which stresses the openness of ecological systems and the role of episodic events (Pickett et al. 1992).

In some respects, both these paradigms of nature are problematic as it has become increasingly clear that few areas of the planet, if any, are pristine, untouched by human hand. It follows that most 'natural ecosystems' that we are reducing and fragmenting feature the incompletely resolved imprint of past human activity and land uses (Willis et al. 2004; Watson et al. 2018). If the starting assumptions concerning the prior equilibrial state of the system are flawed, some of the theoretical island ecological effects may, in practice, be relatively weak. These issues are addressed in the present chapter in the applied, conservation setting as we review the contribution of the island paradigm to conservation science (Matthews 2021).

12.2 Minimum viable populations and minimum viable areas

How many individuals are needed?

The **minimum viable population** (MVP) is the estimated number of individuals required to ensure long-term survival of a population. It is often formalized in terms such as 'the number of individuals providing 95% probability of persistence for 1000 years'. Efforts to model the trajectories of single populations and to estimate MVPs are termed **population viability analyses** (PVA). Important elements that may be relevant to PVA include demographic stochasticity, key traits of the species, genetic erosion in small populations, extreme events such as hurricanes or fires, and human actions (both negative and positive) (Fig. 12.2).

Simulation models suggest that demographic stochasticity of small populations can lead to species loss from small or unsuitable habitat islands. Where populations do persist for several generations at very small sizes they are deemed to have passed through a **bottleneck**, which can result in the loss of genetic variation through **inbreeding** and **genetic drift** (Section 8.5). These populations may then lack the genetic flexibility to cope with either the normal fluctuations of environment or an altered environment. They may also accumulate deleterious genes. In short, they lose **fitness** (Bouzat 2010). It has been estimated that to avoid these effects requires a population size of around 50 to ensure short-term fitness, with 500 individuals recommended to avoid genetic attrition. These commonly cited numbers arose as rough estimates derived, respectively, from data from animal breeders and studies of bristle number in *Drosophila*. Reviews have argued on the one hand for increasing the values to 100 and 1000 (or even 5000), respectively, or for retaining the original 50/500 figures to avoid writing off populations that sit below the higher targets (Jamieson and Allendorf 2012; Frankham et al. 2014). The recovery of the Mauritian kestrel from a single pair, following assisted breeding interventions (Bouzat 2010), provides exemplification of why these rules of thumb should be seen as highlighting species for action rather than abandonment. In any event, wherever possible, species-specific PVAs (e.g. Saunders et al. 2018) are preferable to rule-of-thumb targets.

The term **effective population size (Ne)** acknowledges that in many species only a subset of the *adult* population participates in breeding. The ratio between the effective and actual population sizes varies. For the isolated population of grizzly bears

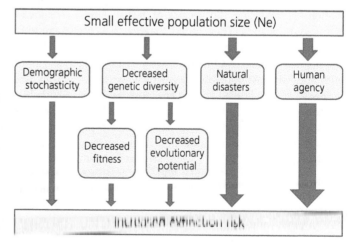

Figure 12.2 Small effective population sizes (Ne) increase the risk of extinction by several routes. The impacts of genetic attrition and of demographic stochasticity are easier to model but probably less important on the whole than natural disasters (hurricanes, fires, etc.) and human agency. The relative importance of each factor, as we perceive them, is illustrated by the varying width of arrows. Human impacts may, as shown, be negative (e.g. illegal hunting/collecting, poisoning, introduction of predator population); but via conservation responses can also be positive (e.g. supplementary feeding, predator eradication, *ex situ* breeding and reintroduction, etc.). Often, different factors are involved in the initial reduction in Ne and the eventual extinction event. Original.

in the Yellowstone National Park, the Ne has been estimated to have varied historically between around 80 and 110, with a resulting inbreeding rate of 2.3%, which sits above the recommended maximum 1% inbreeding rate, suggesting that the population is experiencing slow genetic attrition (Miller and Waits 2003).

In practice, putting a population through a bottleneck of just a few individuals may not be particularly damaging to the genetic base, providing it is allowed to increase in numbers again fairly quickly, or occasional reinforcement takes place. In illustration, the great Indian rhino was reduced within the Chitwan National Park in Nepal to an effective population of only 21–28 animals in the 1960s. After a short period (in relation to the generation time) the population recovered to about 400 individuals, within which genetic variation was found to be remarkably high (Tudge 1991). A further complication is that where a species is split into numerous separate populations in fragmented habitats, there may be multiple bottlenecks involved. This may result in reduced variation *within* each population but increased genetic differentiation *between* populations (Leberg 1991). This may be of significance to lengthening persistence in metapopulation scenarios (below). In sum, genetic attrition in small populations is likely to increase the probability of extinction but it is hard to determine by how much: typically, other threats provide far more immediate causes for concern (Fig. 12.2; Reed 2007; Bouzat 2010).

Pimm et al. (1988) analysed turnover of 355 populations representing 100 species of British land birds on 16 islands. They found that the risk of extinction decreased sharply with increasing average population size and that extinction risk varied with body size. At population sizes of seven pairs or fewer, smaller-bodied species are more liable to extinction than larger-bodied species, but at larger population sizes, the reverse is true. This can be explained as follows. Imagine that both a large-bodied and a small-bodied species are represented on an island by a single male. Both will die, but the larger-bodied species is liable to live longer and so, on average, such species will have lower extinction rates per unit time. On the other hand, if the starting population is large, but it is then subject to heavy losses in an extreme event, the small-bodied species may climb more rapidly back to higher numbers as they tend to have higher intrinsic rates of natural increase. The larger-bodied species is likely to remain longer at low population size, and thus be vulnerable to a follow-up event. They also found that migrant species are at slightly greater risk of extinction than resident species. Numerous factors might be involved in migrant losses, such as events taking place during their migration or in the non-breeding season range (e.g. Runge et al. 2014; Bairlein 2016).

It is difficult to determine how to incorporate extreme events such as hurricanes, volcanic eruptions, or illicit introduction events of non-native predators into PVA. Yet, such haphazard events may have a crucial impact on an endangered bird species. In systems at risk of such catastrophic disturbance, much larger populations are needed to ensure survival (Williamson 1989b; Trail et al. 2007; Oppel et al. 2014). In view of the likelihood of failure of small populations, it is unsurprising that very small islands tend to lack their own endemic species (see Larrue et al. 2018). On larger islands, disturbance events have to be of large magnitude and extent if they are to cause the global extinction of an insular endemic vertebrate, unless the species concerned has already been reduced in range (cf. Mortensen and Reed 2016). The endemic Bahama nuthatch (*Sitta insularis*) is an example of an island species that may have finally been extinguished by large hurricanes in 2016 and 2019 (Matthews and Triantis 2021). Another example is provided by Hurricane Hugo, which in 1989 caused extensive damage to El Yunque National Forest, the home of the last remaining wild population of the Puerto Rican parrot (*Amazona vittata*). In 1975 the parrot population had reached a low point of 13 wild birds due to habitat destruction, capture for pets, and hunting. Given protection, the population built up to 47 individuals, from which it was reduced by Hurricane Hugo to 22 birds. Unfortunately, Puerto Rico is rather prone to hurricanes, and the area was badly affected again by Hurricane George in 1998 and by Hurricane Maria in 2017. The El Yunque population stood once again at 13 individuals when it took a direct hit from Maria, decimating or scattering this tiny population (Paravisini-Gebert 2018). By this point, fortunately, captive breeding and release programmes had established a secondary population in the north-east of the island: without these efforts, the species would now be extinct in the wild.

A further example, attributable to volcanic action, is provided by the Montserrat oriole (*Icterus oberi*), which although confined to only 30 km² of forest within the island of Montserrat, was not considered globally threatened until the eruption of the Soufrière Hills volcano began in 1995. The eruptions continued for two decades, but were particularly damaging in the early phases, devastating much of the forest habitat. It is estimated that the population declined from several thousand individuals before the eruptions to fewer than 700 birds and with only limited sign of subsequent recovery the species remains of conservation concern (Oppel et al. 2014). The sterilization (near or absolute) of the Krakatau Islands in the eruptions of 1883 provide further exemplification of the power of tectonic forces to cause species extinctions from islands, the distinction being that it is unlikely that the Krakatau Islands held any endemic species (Thornton 1996).

Models have been developed to explore the implications of such major events on species populations, an example being a study of the viability of short-tailed albatross (*Phoebastria albatrus*). This species was chosen because 80% of the global population of around 2500 birds breed on Torishima Island, which features an active volcano that has erupted three times since 1900. The model took account of the possibility that all birds on the island could be killed in an eruption but that not all birds would be on the island at the time. In fact, although the model showed an increased likelihood of extinction if a new, intensive phase of eruptions were to be initiated, it also showed that sensitivity to a sustained increase in the background mortality rate (e.g. from fisheries bycatch, plastics, or disease) was of at least equal concern (Finkelstein et al. 2010).

To capture some of the complexity of such scenarios, it is desirable to include data on age structure, catastrophes, demographic and environmental stochasticity, and inbreeding depression in PVA. Reed et al. (2003) use this approach to derive MVP estimates for 102 vertebrate species, using a working definition of MVP as 'one with a

99% probability of persistence for 40 generations'. Across this data set, mean and median estimates of MVP were 7316 and 5816 *adults*, respectively: much larger than obtained when considering only genetic effects. In a subsequent meta-analysis, including 212 species and using the same MVP thresholds, Traill et al. (2007) reported a median MVP of 4169 individuals. In many cases, the area required for a large vertebrate population of this size to be contained in a single reserve may be unobtainable, requiring conservation strategies that either allow population exchange to occur between major reserves, or that are not dependent on reserve systems.

Fortunately, most species that are of conservation concern do not exist in just a single population. The challenge is then to estimate extinction risk across the set of populations. Multiple population viability analysis (MPVA) is a recently proposed statistical approach designed for exactly this purpose (Wenger et al. 2017). Although insufficiently tested to determine its reliability, initial application for a fish species unsurprisingly indicates that survival is likely at smaller population sizes than if assuming a single population (cf. Section 12.3).

How big an area?

How does the MVP estimate translate into the **minimum area requirement** (MAR) or **minimum viable area** (MVA) for a given species? In general, we would predict that the larger the body size and/or higher up the trophic chain, the larger the area needed. It has been estimated that a single pair of the very likely extinct ivory-billed woodpeckers (*Campephilus principalis*) would have required 6.5–7.6 km² of appropriate forest habitat; that the European goshawk (*Accipiter gentilis*) has a home range of about 30–50 km²; and that male mountain lions (*Felis concolor*) in the western United States may have home ranges > 400 km² (Wilcove et al. 1986). Pe'er et al. (2014) undertook a systematic survey of the literature to compile MAR estimates for 216 animal species (mostly birds and mammals) derived from 80 studies (Fig. 12.3). MAR values were estimated by several distinct methods, mostly linked to PVA, producing values varying from 0.1 ha (a butterfly and a froghopper) up to 3.5 m ha (wolverine). As expected, there is a general increase in MAR as

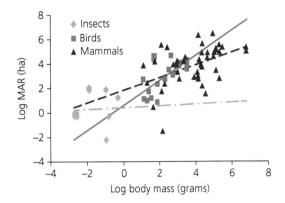

Figure 12.3 Minimum Area Requirements (MAR) for a subset of 73 of 216 animal species collated by Pe'er et al. (2014), from their Fig. 3: these being species for which estimates were based on PVA. Blue diamonds, insects; red squares, birds; green triangles, mammals. Lines of fit follow the same colour scheme.

a function of body size and some variation between taxa and/or feeding guild, although the picture is noisy (Fig. 12.3). Even some plant and insect species may need surprisingly large areas; for example, the MAR for some species of strangler fig may be as large as 200,000 ha because of their low population densities (Mawdsley et al. 1998). Thus, for many species, reserves must be really rather large if their purpose is to maintain a MVP entirely within their bounds.

One proviso attached to the calculation of MAR (in common with MVP) is that the approach assumes that the area concerned acts rather like a large, enclosed paddock in a zoo, with freedom of association within it, but no exchange with any other paddocks or zoos. Unless the reserve does indeed contain the only population of the species, or they are completely immobile creatures, there is the possibility of immigration from another population. The problem then becomes one of estimating survival across a network of patches while estimating chances of reinforcement between them (see Section 12.3).

Applications of incidence functions

Incidence functions are graphical or statistical models of the occurrence of a species in relation to, for example, the species richness, area, or isolation

of a series of habitat patches (Section 6.3; Wilcove et al. 1986; Watson et al. 2005; Matthews 2021). They are typically based on snapshot surveys of presence/absence, and they provide a simple way of exploring the role of single factors and of the potential interaction between two or more of them (e.g. area and isolation), in determining which patches a particular species may be capable of occupying. The information returned is not equivalent to PVAs, as incidence functions simply describe the properties of islands in which a target species currently occurs, not those in which it may persist in the long term. We next exemplify this approach with examples from both real and habitat islands.

Simberloff and Levin (1985) generated incidence data for indigenous forest-dwelling birds of a series of New Zealand islands, and for passerines of the Cyclades archipelago (Aegean Sea). In both systems they found that most species occur remarkably predictably, with each species occupying all those and only those islands larger than some species-specific minimum area. A minority of species did not conform to this pattern, possibly because of habitat differences among islands and because of anthropogenic extinctions.

Moving to habitat islands, Hinsley et al. (1994) quantified the incidence functions of 31 bird species in 151 woods of 0.02–30 ha in a lowland arable landscape in eastern England over three consecutive years (Fig. 12.4a, b). Small woods were found to provide poor habitat for specialist woodland species, although some breeding occurred even in the smallest of the woods. Only the marsh tit (*Parus palustris*), nightingale (*Luscinia megarhynchos*), and chiffchaff (*Phylloscopus collybita*) failed to breed in woods of < 0.5 ha in any year of the study. Variables describing the landscape around the woods were important in relation to woodland use by both woodland and open-country species. For instance, long-tailed tits (*Aegithalos caudatus*) were found to favour sites with lots of hedgerows around them, whereas yellowhammers (*Emberiza citrinella*) preferred more open habitats, small woods, and scrub.

A recent, innovative use of incidence functions was to quantify patterns of use by frugivorous birds of isolated fig trees in agricultural landscapes in Assam, with increasing distance from a large intact block of forest. While species richness and

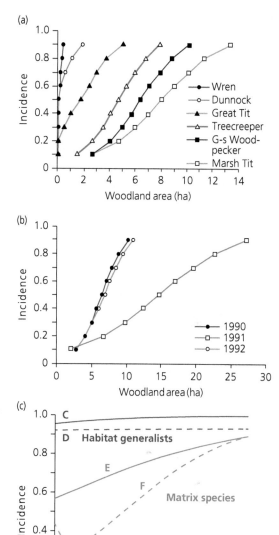

Figure 12.4 Incidence functions. (a, b) The probability of breeding as a function of woodland area, based on 151 woods of 0.02–30 ha area, in a lowland arable landscape in eastern England. (a) Data representative of widespread and common woodland species (wren, dunnock, great tit) compared with that for more specialist woodland species (treecreeper, great spotted (G-s) woodpecker, marsh tit). All relationships are for 1990, except that for the marsh tit, which is for 1991. (b) Interannual variation in the incidence functions for the great spotted woodpecker, reflecting a period of severe weather in February 1991. Small woods were reoccupied in 1992. (c) Logistic regression models for observations of frugivorous birds in 122 isolated fig trees in agricultural landscape in Assam, as a function of distance from an intact forest block, showing examples of forest dependent species (A) great pied hornbill *Buceros bicornis* and (B) blue-eared barbet *Megalaima australis*; habitat generalists (C) red-vented bulbul *Pycnonotus cafer* and (D) coppersmith barbet *Megalaima haemacephala*; and matrix specialists (E) great myna *Acridotheres grandis* and (F) black-hooded oriole *Oriolus xanthornus*.

Parts (a) and (b) redrawn from Hinsley et al. (1994), part (c) redrawn from Cottee-Jones et al. (2015).

functional diversity (Section 6.7) were each relatively unaffected by distance over a range of 32 km, the incidence functions demonstrated rapid replacement of forest-dependent forms with species favouring more open habitats (Fig. 12.4c; Cottee-Jones et al. 2015).

To establish the broader reliability of incidence functions as indicators of species requirements, they need to be shown to be reproducible through time and across the range of a species (Matthews 2021). In Hinsley's study (above) it was shown that a harsh winter could significantly alter the incidence functions of some species, with specialist wood-land species being more likely than generalists to disappear from small woods. Recovery of these populations could take over a year (Fig. 12.4b). Incidence functions calculated by Watson et al. (2005) for birds of woodland habitat islands in and around Canberra, Australia, demonstrated a mix of area-sensitivity, isolation-sensitivity, and compensatory area–isolation effects. They also reported significant variation in the form of the incidence functions between three distinct landscapes, pointing to the importance of the character of the habitat matrices in which the woodland systems were embedded. Prugh et al. (2008) analysed occupancy data for 785 species of animals from over 12,000 habitat patches from 89 studies distributed across six continents. They found that area effects were stronger than isolation effects but that together they accounted for a median of only 25% of the deviance in occupancy. This may have reflected the inclusion of many species for which the scale of area and isolation in the database was inappropriate to capture a meaningful response, but again matrix effects may have been involved. They concluded that while both factors do indeed have importance, there are likely to be greater conservation returns from manipulating the character and quality of the matrix habitats in which the fragments are embedded than by a focus solely on habitat patch configuration (see also Ramírez-Delgado et al. 2022). Collectively, these findings suggest that incidence functions may not be reproducible, or that they are influenced by other important factors and therefore that more complex, multi-factor models might be needed to accurately predict species incidence.

Some recent work in this area has applied more powerful statistical tools such as generalized linear mixed-effect models. Exemplifying this approach, Keinath et al. (2017) analysed occupancy of 1559 species across 3342 patches from studies across the globe. Their analyses reinforced the importance of patch area but also help identify other important factors determining sensitivity, not least, the degree of habitat specialization of the species involved, the habitat type of the system and the taxon involved. They found that reptiles are particularly sensitive to patch area, while for birds, mammals and reptiles, forest species are generally more affected than grassland species.

In conclusion, the analysis of species occupancy in fragmented habitats can generate useful insights, particularly if multiple and repeat surveys are available to establish spatial and temporal variability in their responses. Notions of threshold population sizes and area requirements need to be offset against dispersal efficacy, habitat requirements, and specific factors influencing mortality. A common theme emerging from the analysis of incidence is that the 'island–sea' analogy is imperfect, and that we should be paying greater attention to the ecology of the matrix through a landscape ecology lens. As more datasets accumulate and more sophisticated analytical tools are employed, more nuanced understanding of responses to habitat characteristics across complex landscapes are becoming possible.

12.3 Metapopulation structure and source–sink dynamics

Imagine that you have a collection of populations of a particular species, each existing on patches of suitable habitat. Each patch is separated from other nearby habitat patches by unsuitable terrain. Although the populations each have their own fairly independent dynamics, as soon as one crashes to a low level, or indeed disappears, that patch will provide relatively uncontested space for individuals from one of the nearby patches, which will soon colonize. This describes a **metapopulation**, a network of separate patches occupied by a species (although some suitable patches may be empty), which are interconnected by occasional movements of individuals and/or gametes (Fig. 12.5). Thus, within a metapopulation, member sub-populations may change in size independently but their probabilities of existing at a given time are not independent of one another, being linked by mutual recolonization following periodic extinctions, on timescales of the order of 10–100 generations (Harrison et al. 1988). The first metapopulation models were developed by Richard Levins in papers published in 1969 and 1970 and the concept has since been developed and extended, in large part through

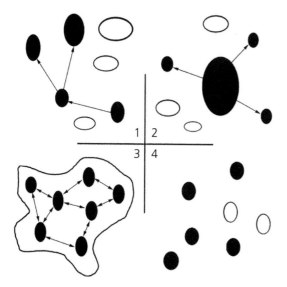

Figure 12.5 Metapopulation and related scenarios. 1 = a classic metapopulation in which patches fall empty stochastically and are resupplied from other patches; 2 = core–satellite or source–sink structure, where very large or high quality patches acts as stable source populations supplying individuals to resupply or maintain populations in small or low-quality patches; 3 = where patches are so close together that they act as a single population; and 4 = where patches are too far apart and cease to resupply other patches in the short run, such that metapopulation models no longer apply.
From Triantis and Bhagwat (2011), their figure 8.7.

Table 12.1 Problems that have been identified with classic metapopulation models, some of which have been addressed in more recent models (see text).

1. Occurrence of core–satellite relations
2. The rescue effect is not dealt with adequately
3. Distance effects and varying dispersal abilities
4. The occurrence of 'sink' habitats and of source–sink dynamics
5. Metapopulations are very difficult to replicate for study
6. Much extinction is deterministic
7. Patch vs matrix assumptions do not always apply for particular species or times

seminal contributions by Ilkka Hanski and collaborators (Ehrlich and Hanski 2004; Hanski 2010; Schnell et al. 2013; Ovaskainen and Saastamoinen 2018).

While the general idea is compelling, relatively few systems have been studied long enough and in sufficient detail to fully describe their dynamics (Ovaskainen and Saastamoinen 2018). Often, empirical tests have suggested the metapopulation models to be over-simplistic (Table 12.1). However, empirical case studies have been described (below) that show: (i) a degree of independence of population dynamics between the isolates, combined with (ii) a demonstration of recolonization, favoured by proximity of occupied patches, thereby fulfilling the essential requirements of metapopulation theory (e.g. see Fig. 12.6; Vögeli et al. 2010).

The classic metapopulation models assume equivalence in size and quality of patches, with all patches susceptible to extinction and all occupied patches able to act as re-supply centres (Fig. 12.5, panel 1). There are two common problems with such a simplistic approach (Dawson 1994). First, if one patch is big, and the others small, then they may have a **core–satellite relationship**, in which the core is effectively immortal and the smaller patches are like satellites resupplied from the big patch (Fig. 12.5, panel 2; Table 12.1). For example, in a study of orb spiders on Bahamian islands, larger populations were found to persist, whereas small populations repeatedly went extinct and re-immigrated from the larger areas and then become extinct again (Schoener and Spiller 1987), thus contributing little or nothing to the chances of persistence of the whole metapopulation. Second, even seemingly isolated populations may be saved from extinction through continued supplementation by individuals migrating from the core population. In other words, the 'rescue effect' (Section 5.10) may operate: a modification that has been built into later generations of metapopulation models (e.g. Eriksson et al. 2014), as have distance effects (e.g. Vögeli et al. 2010; Eriksson et al. 2014).

The metapopulation concept has similarities to MacArthur and Wilson's (1967) EMIB. However, whereas the emphasis in the EMIB is on species number as a dynamic equilibrial function of area and isolation, in metapopulation studies the focus is on the status of populations of a single species of strong habitat specificity: extinction and subsequent recolonization are seen principally as a function of (temporary) alterations in the carrying capacity of the system. It is consistent with the metapopulation approach although not necessarily with all metapopulation *models*—that rates of **movement**, and of population gains and crashes, may vary

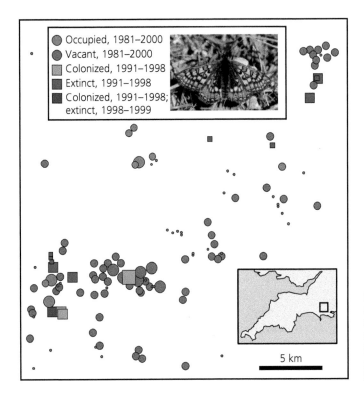

Figure 12.6 Changing occupancy of patches within a 625 km² grid cell in southern England (inset) by the marsh fritillary butterfly (*Euphydryas aurinia*): circles show no change and squares change in occupancy between 1981 and 2000. Symbol sizes are scaled by patch area.
Modified from Bulman et al. (2007).

greatly over time. For a system to function as a metapopulation requires just the right range of inter-patch isolation so that the dynamics of the islands are to some extent interdependent (Haila 1990). If the patches were much more isolated, their population fluctuations would be entirely independent and then they would not constitute a metapopulation (Fig. 12.5, panel 4). If, on the other hand, a more dispersive species was being considered, the entire system might constitute a single functional population, albeit one spread across a fragmented habitat (Fig. 12.5, panel 3). Thus, metapopulation models form a bridge between the study of population ecology and island theories of the EMIB form. In addition, the metapopulation idea has also been extended to the community level via the concept of the **metacommunity**, which refers to a set of local communities linked by the dispersal of multiple potentially interacting species (Leibold et al. 2004; Leibold and Chase 2018).

Intensive studies of the checkerspot butterfly (*Euphydryas editha bayensis*) in the Jasper Ridge Preserve (USA) provide a classic depiction of metapopulation dynamics (Harrison et al. 1988; Ehrlich and Hanski 2004). The checkerspot butterfly is dependent on food plants found in serpentinite grasslands. The study area of 15 × 30 km comprised one large patch (2000 ha) supporting hundreds of thousands of adults—in effect a permanent population—and 60 small patches of suitable habitat. A severe drought in 1975–1977 is known to have caused extinctions from three of the patches, including the second largest patch, and so was assumed to have eliminated all but the largest population. By 1987, eight patches had been recolonized. Small patches over 4.5 km from the 'mainland' patch were found to be unoccupied. Harrison et al. (1988) showed that the distribution of populations described an apparent 'threshold' relationship both to habitat quality and distance. Patches had to be both good enough habitat and near enough to the 'mainland' in order to be inhabited at that time. Similar data for the silver-studied blue butterfly in sites in North Wales feature in Fig. 12.7.

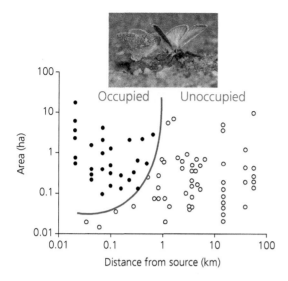

Figure 12.7 Occupancy of suitable habitat by the silver-studded blue butterfly (*Plebejus argus*) in North Wales in 1990. Most patches larger than 0.1 ha were occupied (filled circle), provided that they were within about 600 m of another occupied patch. Beyond this distance, no patches were occupied (open circle), regardless of patch size. This response to area and isolation suggests either a 'block pattern' or 'compensatory' incidence function as per Figure 6.5. Redrawn from Thomas and Harrison (1992).

The application of metapopulation models to conservation problems has been central to their development. As an extension of the idea of minimum viable populations, Hanski et al. (1996) coined the term **minimum viable metapopulation** (MVM) for the notion that long-term persistence may require a minimum number of interacting local populations, which in turn depend on there being a minimum amount of suitable habitat available (exemplified for the marsh fritillary butterfly *Euphydryas aurinia* by Bulman et al. 2007). Equally important is that the habitat patches be within reach for the species in question. There are many species of relatively limited powers of dispersal, which, like the checkerspot butterfly, may show distance limitation of patch recolonization on a scale of a few kilometres. For them to survive in increasingly fragmented habitats may require conservation intervention.

The most celebrated example of a conservation strategy leaning on metapopulation modelling is that of the Interagency Spotted Owl Scientific Committee. In 1990 they proposed the creation and maintenance of large areas of suitable forest habitat in proximity, so that losses from forest remnants will be infrequent enough and dispersal between them likely enough that numbers of the northern spotted owl (*Strix occidentalis caurina*) will be stable on the entire archipelago of forest fragments. However, the plan proved controversial, in part because of increased levels of predation of young owls in cleared areas. Over time it became evident that the decline of this owl is only in part determined by reduced habitat availability and that a factor external to the metapopulation models, competition with the invasive barred owl (*Strix varia*), presents a much greater threat to recovery of the northern spotted owl than originally appreciated (USFWS 2011; Wilk et al. 2018).

Metapopulation studies have served to highlight the importance of habitat quality. Indeed, a habitat patch may contain a breeding population with slightly negative growth rate, making it a **sink population**, incapable of sustaining itself in the absence of immigration. Thus, while some suitable habitats will be unoccupied because of their isolation from **source populations** (those with positive growth rates), other poorer-quality habitats will be occupied because of their proximity to source populations. Although relatively few systems have been quantified in detail, source–sink dynamics are suspected to be commonplace (Furrer and Pasinelli 2016). This implies that some proportion of populations of conservation interest may occur in 'sink habitat', either because of aspects of the physical environment, competition with other species (as with the northern spotted owl example), or because of predation, as exemplified by invasive non-native trout feeding on native fish in New Zealand rivers (Woodford and McIntosh 2010).

A further caution on the use of metapopulation models is that colonization and extinction of particular species from habitat patches can be driven by a range of deterministic factors, sometimes having little to do with the insular nature of the patch (Table 12.1)—for example, succession following disturbance, climate change, regional range expansion/contraction, hunting by humans, or altered survivorship of migrant species in the winter range or in passage. Success or failure of a species may therefore reflect the ability of a species to cope with regionally pervasive change in environment (Thomas 1994). In cases, quite subtle changes in habitat may produce a species extinction that might be mistakenly classified as 'stochastic'. One classic example being the demise of the large blue butterfly *Maculinea arion* in southern England, which occurred despite conservation efforts, because of changes to the sward height in its grassland habitats caused by altered grazing regimes. Metapopulation models therefore need to be superimposed on an environmental mosaic, which in many cases will itself be changing. Efforts have therefore been made to extend metapopulation models to account for patchy disturbance phenomena and for specific extinction drivers in dynamic landscapes (e.g. Ovaskainen and Saastamoinen 2018).

In sum, when applied at the landscape scale, metapopulation models have had some success in predicting system behaviour, but in other cases and especially at more local scales of application, complicating factors, such as interactions with other species, limit their predictive power (Driscoll and Lindenmayer 2012). The original metapopulation concept and models were highly simplified and therefore easy to find fault with. However, as the study of metapopulation patterns and processes has advanced, greater realism has been added into the models, thus providing improved tools for analysing the spatial structure and temporal interdependency of networks of local populations (Ovaskainen and Saastamoinen 2018).

12.4 Habitat fragmentation, extinction debt, and species relaxation

The term 'habitat fragmentation' is often used inconsistently and as a broad umbrella for many patterns and processes that accompany landscape change. This has made it a panchreston or an explanation or theory used in such a variety of ways as to become meaningless.

(Lindenmayer and Fischer 2007, p. 127)

Island theory founded on MacArthur and Wilson (1967) predicts that the progressive conversion of contiguous areas of habitat into smaller and more isolated patches of habitat should simultaneously reduce immigration rates and increase extinction rates, thereby reducing the equilibrium richness of each patch (Fig. 12.8a). At the point of habitat conversion, the newly created habitat island is therefore **supersaturated**; that is, it contains too many species (it may even gain fugitive displaced populations) for its newly downsized and isolated state, creating an **extinction debt** (Ewers and Didham 2006; Halley et al. 2014). It follows that each habitat island

should in time undergo a phase of extinction (fast at first, gradually declining) until it reaches the new, lower equilibrium point (Fig. 12.8b). Given knowledge of isolation and area of patches, it should be possible to estimate the richness of each member of the resulting new network of habitat islands and, as each element of the network sheds species, the logSpecies–logArea curve should re-adjust (relax) to a steeper form with a lower intercept. As an historical note, the original use of the term relaxation in the island biogeographical context referred to the time taken to adjust to any displacement from an existing equilibrium condition, whether that involved increase or decrease in richness on an island (see Diamond 1972), but the term **species relaxation** has mostly since been used to describe the expected shedding of species from a network of islands or habitat islands following isolation.

Providing we assume that a landscape was close to ecological equilibrium prior to habitat fragmentation, then the expected adjustments in the form of the species–area relationship provides a basis for predicting species losses. There are several caveats to this, starting with the possibility of past disturbance having occurred that invalidates the equilibrium assumption. A second caveat is that there is normally a lot of scatter around the ISAR regression line, with points lying above and below the line of best fit because of variation in habitat diversity and quality (e.g. Matthews et al. 2016a). Where a newly created habitat island lies above the line it doesn't necessarily follow that it is supersaturated and destined to lose species, or that losses will follow swiftly (e.g. see Halley et al. 2014). Third, it is important to recognize that there is a key difference between predicting the number of species that each new isolate may hold and how many are held across all of them collectively: this depends on the extent to which different species win through in different patches (Matthews et al. 2015a; May et al. 2019). Fourth, the underlying assumption that populations on habitat islands are distinct and isolated may itself be flawed (Section 12.3; Lindenmayer and Fischer 2007).

Given that island theory, as set out in Fig. 12.8, is fundamental to efforts to predict the biodiversity effects of habitat loss and fragmentation (Ladle 2009), it is important to establish: (i) the extent to which species relaxation operates, (ii) if it is selective in the species that win and lose, as well as (iii) the

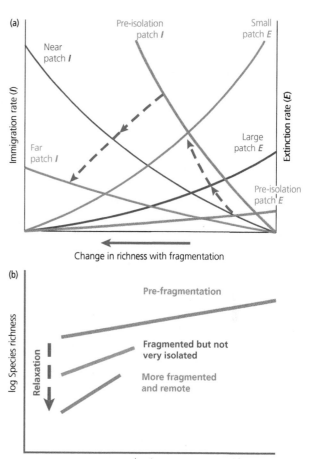

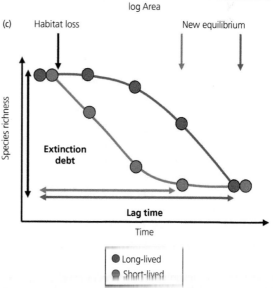

Figure 12.8 Implications of fragmentation of habitat for an area within a large block of habitat, progressively reduced to smaller and more isolated patches. (a) Some turnover is expected even in non-isolated habitat but immigration rate declines and extinction rate increases as the fragmentation process continues, driving equilibrium species richness down. The process of adjustment may not be instantaneous, leading to patches being supersaturated and to the existence of an extinction debt. (b) The corresponding adjustment of the island species–area relationship from the pre-isolation state results in steeper ISARs, with lower intercepts and, as shown, greater disparity for smaller and more isolated patches. The matching island data points are omitted to avoid clutter. (c) Lag times to paying the extinction debt may vary with generation time, but not necessarily in the manner suggested, as other traits such as body size, trophic level, and mobility may be as or more important; the relaxation time is the time taken for payment of the entire extinction debt.

Original compilation, drawing on various sources; panel c from figure 1a of Kuussaari et al. (2009).

time lags involved before extinction debts are finally paid. If, for instance, the same sets of species consistently lose out, that would imply different regional extinction rates from a situation where losses were randomly distributed across fragments. While, if lag times span several decades, there may be more time than anticipated for conservation measures to avoid extinction.

Species turnover is rarely entirely random. Some species tend to be stable in their distribution across habitat islands and a high proportion of turnover events involve 'ephemerals', species marginal to the habitat, or successional change (Chapters 5 and 6). The metapopulation literature also provides evidence of the tendency for large populations to persist, while smaller satellite populations may come and go, without jeopardizing the metapopulation. Nonetheless, species of conservation concern do disappear from habitat islands, and from entire landscapes, and in cases the loss will mean final extinction of a species. The recent global extinction of several bird species in the heavily fragmented Brazilian Atlantic Forest, provides disheartening exemplification (Develey and Phalan 2021). It is therefore important to ask, how strong is the relaxation effect, how soon does it operate and in what ways is it selective?

One of the classic examples of relaxation on ecological timescales is that of birds lost from Barro Colorado Island (BCI), in Panama. BCI was formerly a hilltop in an area of continuous terrestrial habitat, but it became a 15.7 km² island of forest when the central section of the Panama Canal Zone was flooded in 1914 to form Lake Gatun. There were around 150 breeding species of birds shortly after isolation, declining to *c.* 100 by 1980 and 85 by 1996 (Lomolino et al. 2017). The ocellated antbird, *Phaenostictus mcleannani*, disappeared after a particularly harsh dry season in 1969. Although maintaining populations in forests near to BCI, it failed to recolonize during the next four decades (Touchton and Smith 2011). Following its loss from BCI there appears to have been a doubling in the density of the smaller and usually subordinate spotted antbird (*Hylophylax naevioides*)—a possible case of density compensation. Several of the other species that failed are thought to have occurred at low density at the outset, so their extinction could

easily be accounted for by chance demographic fluctuations, or temporary adversity. Other losses are attributable to quite specific ecological processes that were initiated by insularization. For example, the abandonment of farming initiated a process of secondary succession that led to the loss of open habitats and some associated bird species. In addition, some of the birds lost were members of the guild of ant followers, which have been found to be vulnerable to fragmentation elsewhere (below).

Also in Lake Gatun, top carnivores, such as the puma, with large area requirements, became extinct from smaller islands formed within the lake. This, in turn, led to increased numbers of smaller omnivores and predators—a process termed **mesopredator release** (Soulé et al. 1988). The knock-on effects included the loss of certain ground-nesting birds. Similar cascading effects have been documented in other island/habitat island systems (Fig. 12.9; Ritchie and Johnson 2009; Laurance et al. 2017).

In a simple sense, the Lake Gatun islands provide evidence of relaxation, with a mix of deterministic and stochastic losses, combined with some interesting ecological twists reflecting changing habitat, competitive and food-web relationships. Such knock-on effects through the ecosystem, if pronounced enough, are termed **ecological or**

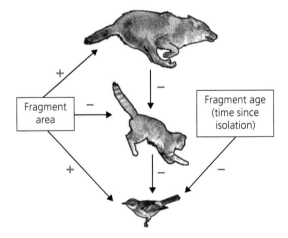

Figure 12.9 Model of the combined effects of trophic cascades and island biogeographical processes on top predators (e.g. coyote), mesopredators (domestic cat), and prey (scrub-breeding birds) in a fragmented system. Positive effects are indicated by a plus sign and negative effects by a minus sign.
From Crooks and Soulé (1999).

trophic cascades and they tend to be most acute on islands when a whole trophic level or role (e.g. top predators) is lost, or gained (Rayner et al. 2007; Terborgh 2010). These studies also show the imprint of differing barrier-crossing propensity, even among volant species (see also Van Houtan et al. 2007). These newly formed lake islands have thus proven to be useful study systems, but we need to consult other studies to determine how representative they are of habitat islands in other contexts.

Extinction debt

Even where remnant patches are left during habitat conversion, some species may immediately be lost from a landscape: those not occurring in the remnant patches at the time of conversion. The term **extinction debt** applies to those species that survive temporarily within the remaining fragments, while the full ecological consequences of habitat fragmentation feed through and the systems relax to a new equilibrium (Ewers and Didham 2006; Kuussaari et al. 2009; Figueiredo et al. 2019). The delay or **lag-time** involved may relate to successional changes in the remnant habitats, trophic cascades initiated by the loss of top predators, the spatial configuration of remnant patches and thus the extent to which metapopulation processes may operate, the generation times of the species of interest, etc.

While Fig. 12.8c indicates that longer generation times will lead to longer persistence times, in practice, the evidence for this is mixed, as multiple factors influence reaction times. In their review, Ewers and Didham (2006) note a case in an experimental microcosm of extinction debts in microarthropods being settled within a year (speedy, as expected), but analyses of extinction debt for the arthropods of the Azores at a massively larger spatial scale, suggest system lag times of well over 100 years (Triantis et al. 2010). A similarly long period of adjustment was also inferred by a comparative analysis of forest plant data from two sites in the UK and Belgium (Vellend et al. 2006).

A key challenge in estimating extinction debts and lag times is a general lack of systematic longitudinal (through time) data monitoring actual turnover in fragmented systems. A suite of comparative, SAR-based and modelling approaches

has therefore been applied to the problem. Analyses of the form of 207 habitat island datasets by Matthews et al. (2016a) showed that the power model provides the best general fit. They reported that values for z (slope) for the log–log power model are typically lower for habitat islands ($n = 135$ significant ISARs; median $z = 0.22$) than for true islands ($n = 460$; median $z = 0.29$), and that mountaintop and urban habitat islands tend to exhibit steeper slopes than do other habitat island types (e.g. forest patches). Slope and intercept values varied in part in relation to the ratio of largest to smallest island. Hence, it appears that the slope and intercept of ISARs may be more sensitive to the range in area of isolates sampled than generally realized. This in turn may impact on our assessment of the relative richness gain from having larger-sized reserves: although it should again be noted that simply being able to predict the richness of each isolate does not reveal the overall richness of the system.

A wide-ranging review of around 100 studies (not all connected specifically to habitat conversion) found evidence of extinction debts being sustained over a span of 5 to 570 years, and indeed with estimates of as much as 1000 years to settle debts in specific cases (Figueiredo et al. 2019; see also Halley et al. 2014). In a more tightly focused review of the results of habitat fragmentation experiments spanning 35 years, multiple continents, and ecosystems, Haddad et al. (2015) found evidence of significant changes to ecosystem function, alongside levels of biodiversity loss ranging from 17% to 75% and spanning several decades.

Fragmentation, relaxation, and the habitat amount hypothesis

One of the basic assumptions derived from island theory is that losses are generated not only by the reduction in contiguous area but also by the degree of isolation of remaining habitat. However, for a variety of reasons discussed throughout this chapter, these effects have turned out to be hard to detect and quantify. Fahrig (2013, 2017) goes so far as to claim that fragmentation has little distinct effect and that the loss of biodiversity of a focal ecosystem within a landscape is a function of the overall habitat remaining (see also Rabelo et al. 2019). Her **habitat amount hypothesis** has, however, been contested on both theoretical and empirical grounds. Notably, Saura (2021) provides a critique of the logic of the habitat amount hypothesis, arguing that it has been misinterpreted and fails to support the dismissal of fragmentation and isolation as factors influencing species richness of habitat island networks. Fahrig's claims have also been contested by Haddad et al. (2017), who use data from two experimental systems (one a moss microcosm and the other a manipulated plantation system) to show that fragmented habitats have steeper SAR slopes than unfragmented habitats and that the difference increases over time. Backing this perspective, Horváth et al. (2019) present an analysis of changing invertebrate species prevalence in a series of ponds in eastern Austria, affected by six decades of habitat loss, which demonstrated that the losses exceeded those explicable as a function of the amount of habitat lost and indicated a clear role for the loss of connectivity that stems from habitat fragmentation.

As with other topics in island biogeography, the outcome of fragmentation is almost certainly sensitive to such issues as system nestedness (Section 6.4), taxon sensitivity to the scale of isolation and range of habitat sizes involved, and how confounding factors are treated in analysis (Fletcher et al. 2018). Moreover, the technicalities of analyses, particularly those concerning inferences drawn from species–area relationships, are in practice something of a source of confusion in understanding the impacts of habitat fragmentation (Saura 2021; Box 12.1).

Threshold responses

There is evidence of threshold responses to reduction of isolate area, often involving hierarchical links that generate ecological cascade effects, such as mesopredator release (Fig. 12.9; Terborgh 2010). Below a certain size, an isolated habitat island cannot support populations of large predatory vertebrates, with the consequence that higher densities of their prey species may be maintained. Thus, the densities of medium-sized terrestrial mammals are between 8 and 20 times greater on BCI, which lacks top predators, than in a comparable 'mainland' site in which they occur (Terborgh 1992). Terborgh (2010) summarizes similar findings for a system of hundreds of small islands created by the formation of Lago Guri in Venezuela in 1986. Within four years, islands of < 12 ha had lost 75% of their vertebrates, including predators of vertebrates. Persisting species showed increased levels of abundance, and in particular herbivores were favoured on smaller islands. The impact on plant recruitment over time led to significant shifts in the composition of the vegetation, especially of the smaller islands, changes which were still ongoing when the project ended in 2003. Interestingly, there was also evidence of compositional differences between islands, arising either from initial differences in composition or from variation in which species became extinct from island to island.

There are relatively few longitudinal studies of changing diversity and composition for habitat islands comparable with the above 'lake island' studies. They tend to be focused on a small number of famous long-running habitat fragmentation experiments (e.g. the Biological Dynamics of Forest

---

Box 12.1 Does island theory provide a basis for the use of species–area regressions to predict extinction threat?

Species–area relationships have long been used as a basis for forecasting extinction levels. The general idea being that with continuous habitat being fragmented, the characteristic species accumulation curve will give way to an island species–area relationship of steeper slope (cf. Ewers and Didham 2006; He and Hubbell 2011; Matthews et al. 2021a). As discussed in the text, there are numerous problems with this and in Box 5.3 we highlight one of them,

Box 12.1 *Continued*

the incommensurability of different types of SAR. This is exemplified by Matthews et al. (2016b) who show that when comparing the same set of islands, the parameters of the species accumulation curve (SAC; how total richness rises with increased island area) and of the island species–area relationship (ISAR; richness of each island considered separately) differ.

In a similar vein, He and Hubbell (2011) have argued that mishandling of SAC methods has given rise to systematic over-estimation of extinction debt. Their argument is that many estimates have relied upon working their way down SACs to smaller areas, but in fact much larger areas of habitat have to be removed to ensure the removal of the last individuals of a species. This controversial claim was contested by other authors, who provided arguments and evidence to show that their findings are not general (e.g. Pereira et al. 2012; Rybicki and Hanski 2013). Some of the confusion certainly stems from using different types and methods of constructing SARs. However, much also depends on the spatial pattern of the residual network.

Put crudely, it is often assumed that an estimate can be made of species losses based on the regional figure for habitat loss. But, this is to assume that the configuration of the patches remaining, how big they are and how scattered, is unimportant. For example, in a study cited in the main text, Triantis et al. (2010) estimated extinction debt for each of the Azorean islands based on the total forested area of the islands and how this has changed over time. Yet, the remaining forests of the Azores are not clustered into a single patch per island, but rather into variable numbers and sizes of patches: an important detail that was set aside as a simplifying step and which limits the power of the analysis. Similarly, Grelle et al. (2005) use a SAR model to predict expected losses of species of vertebrates due to habitat loss in Rio de Janeiro state. That is, the authors assume that the whole of Rio de Janeiro state is acting as a single fragment, which has lost 81% of its area and which is therefore travelling along a trajectory towards its lower equilibrium point. However, in practice, this is a large region in which forest habitat persists in a very large numbers of patches, scattered across the region in varying configurations. This loss of habitat is certainly bound to vastly reduce the number of individuals of forest species across that region, but whether this results in particular species falling below their minimum viable population in all isolates is not knowable simply from the regional figure for habitat loss.

This brings us back to the debate on whether it is area lost regionally or the combination of area loss and fragmentation that drives losses (see discussion of the habitat amount hypothesis in the text). The answer to this question has to depend in large measure on the scale parameters of the system. Consider a thought experiment where a number of small woodland patches are distributed within a few hundred metres of one another. For most vertebrates the populations within can function as a single population, or at least a functioning metapopulation. Yet distribute these same woods a few score kilometres apart, and they become truly isolated, and many of the constituent forest-dependent species will fail to persist in the network. Finally, consider a network of far larger woodlands, each capable of containing viable populations of many woodland species, but distributed over increasingly large regions. The further apart these woodlands are, the fewer species they will hold in common, as each will hold the imprint of increasingly different climates and biogeographical histories. Such factors of scale (system extent and grain) are in fact detectable in the variation recorded in ISAR parameters (e.g. Triantis et al. 2012; Matthews et al. 2016a, b), as discussed in Chapter 5.

It should also be recognized that even where species–area analyses are linked phenomenologically to island theory, they provide only a coarse stochastic model of the impacts of fragmentation. The actual drivers of extinction may be structured, and in some instances avoidable by appropriate management. For instance, hunting and collection of birds and mammals may be important drivers of species declines that are worsened by the road networks that typically accompany fragmentation (Peres 2001; Ribon et al. 2003; Benchimol and Peres 2013) but they can be addressed through specific mitigation measures within those same landscapes.

Hence, in conservation science applications, the use of rather simplistic assumptions about the parameters of species–area relationship to generate forecasts of extinctions should be understood as an inherently crude and broad-brush approach to the problem. Basing predictions of regional losses on SAR models that take no account of habitat fragmentation and that make no effort to estimate realistic parameters for the scale of isolation and of patch size, is problematic. Given these issues, several studies have proposed extensions to simple SAR models that account in various ways for the characteristics of fragmented landscapes, including the matrix calibrated, edge effect calibrated, countryside, and lost-habitat SAR models (reviewed in Matthews 2021). However, wherever possible, the use of species-level assessments may be preferable to the analysis of species–area fits. Such assessments enable attention to be given to species most at risk, and necessarily involve an assessment of what factors threaten particular species.

Fragments Project in South America). However, it is possible to test for non-linearity in response to fragmentation by means of piecewise regression of habitat island ISARs (Fig. 12.10). Several studies have claimed to show the so-called 'small island effect' occurs quite often in habitat island systems. As discussed in Section 5.4, this effect takes the form of limited or no diversity response to increasing area until the threshold area is reached, beyond which a rapid increase occurs, generating a two-phase shallow-steep species–area relationship. However, the opposite form can also be found, where richness increases rapidly until an upper threshold is reached, beyond which little additional gain in richness occurs as area increases. Furthermore, as shown by Matthews et al. (2014b), the detection of threshold effects is influenced by which data transformation is used prior to analysis, with even the same datasets able to suggest flat–steep and steep–flat slope transitions (see also Fig. 12.10b, c). In sum, these analyses demonstrate that threshold responses are statistically detectable but that they can take very different forms, with breakpoints ranging over five orders of magnitude of variation in area (although virtually all occurring at < 50 ha) and spanning the classic small-island effect, the opposing steep–flat relationship, and a number of other less ecologically intelligible forms.

By reference to the dataset shown in Fig. 12.10b and comparing the two graphs, we might reasonably ask whether the apparent decline in richness above around 250 km^2 (left panel) is a meaningful response to increased area, or more likely an artefact of some confounding variable(s)? The lack of response to area within the half-dozen smallest patches shown in the right-hand panel also suggests that other, confounding factors are at work (see Watson 2003). The data shown in Fig. 12.10c are in some ways more typical, in that we see plenty of scatter, and less convincing lines of best fit. Noisy and indeed non-significant relationships are not uncommon. Depending on transformation used, a third to a half of the datasets examined by Matthews et al. (2014b) failed to produce a significant fit by any of the six models used. This reminds us that other factors are always at work in determining richness of patches and indeed this was explicit in the sampling and analyses of many of the source papers used by

Matthews et al. (2014b). These factors can include isolation, elevation, local climate, rock type, disturbance history, and others. The quality of the matrix may also be critical to the richness of the habitat islands embedded within complex landscapes (Prugh et al. 2008; Frishkoff et al. 2019; Matthews 2021; Ramírez-Delgado et al. 2022). Detecting such interactions and developing refined understanding of the impacts of altered configuration of habitat for overall system diversity requires that analyses consider more than just the form and slope(s) of the ISAR (Box 12.1; Prugh et al. 2008; Matthews et al. 2019a). We can only conclude that the detection of threshold effects in habitat island ISARs should be accompanied by careful analyses of other potentially important variables and at best provides only one piece of the story.

Winners and losers

In assessing the ecological impacts of habitat fragmentation and loss, it is important to recognize that there may be winners (relative or absolute) as well as losers. The net changes in bird species richness in habitat remnants typically disguise a switch towards generalists and away from habitat specialists (Matthews et al. 2014a; Cordeiro et al. 2015), a trend that may manifest also at a regional level. Fragmentation of tropical forest appears to have strongly negative effects on particular guilds of forest insectivores, especially those that assemble in mixed-species flocks (Ribon et al. 2003; Van Houtan et al. 2007). This is exemplified by a study of fragmentation in the forests of the East Usambara Mountains in Tanzania where richness and size of mixed-species flocks declined, with forest specialist insectivores losing out. One species, the square-tailed drongo, *Dicrurus ludwigii*, was found to play an important role as a 'nuclear species' in the assembly of mixed flocks: with its loss or gain from a forest area having knock-on consequences for the presence of other bird species (Cordeiro et al. 2015).

Although more work has probably been carried out on birds than any other taxon, it is clear from reviews and meta-analyses that the non-random nature of sensitivity to habitat loss and fragmentation is general. Keinath et al. (2017) report the results of analyses designed to distinguish sensitivity to

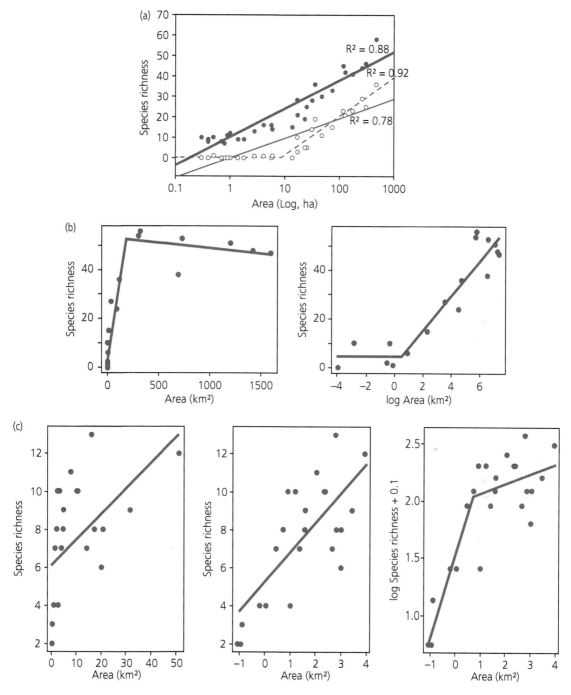

Figure 12.10 Comparison of linear and piece-wise regression models for habitat island ISARs. (a) Bird species from littoral forest fragments, south-eastern Madagascar. Closed circles (and bold, solid line): all bird species; open circles: forest-dependent bird species, fitted by linear regression (solid line), and by break-point regression (broken line) (from Watson et al. 2004b). (b) Breakpoint regression models of Watson's (2003) dataset for birds in habitat islands in Mexico, showing (left) a steep–shallow response using untransformed richness and area data and (right) a shallow–steep response when fitting the same system using log transformation of area. (c) Best models for the same dataset (Kitchener et al. 1980) for mammals in Australian habitat islands were found to be linear for untransformed Species v Area and Species v logArea analyses, but a steep–shallow response for logSpecies v logArea. Panels b and c are illustrative examples from Matthews et al.'s (2014b) analyses comparing the performance of a range of piecewise and linear models applied to 76 habitat island datasets.

fragmentation per se for amphibians, birds, mammals, and reptiles for 1559 species across 3342 patches, distributed across the globe. They found sensitivity to be a function of habitat specialization, trophic status, and body size (large species being mostly losers), with evidence of interactions between these traits. Reptiles were reported to be particularly sensitive to habitat fragmentation.

12.5 Reserve configuration and the 'single large or several small' (SLOSS) debate

Given a finite total area can be set aside for conservation as a natural landscape is being converted to other uses, what configuration of reserves should conservationists advocate? This question had crystallized by the late 1970s as the SLOSS debate, posing a hypothetical choice between a single large or several small reserves amounting to the same area but scattered across the landscape (e.g. Diamond and May 1981). The framing of the debate stemmed from efforts to apply the logic of MacArthur and Wilson's (1967) EMIB, as they themselves had clearly envisaged (Fig. 12.1).

As described above, the argument is that reduced immigration and elevated extinction rates in small fragments generates extinction debts and an ongoing process of relaxation to a new and less diverse equilibrium state in each patch (Fig. 12.8). On this basis, arguments were put forward favouring single or few large rather than several small reserves, circular rather than elongated reserves (reducing edge effects), closely spaced reserves, and the use of corridors to improve their connectivity (e.g. Diamond and May 1981). Although highly influential in conservation planning (Wintle et al. 2019), the theoretical and empirical answer to the basic SLOSS question has in fact been disputed and for good reason: in our view there is not a single general answer to it, for much the same reasons of scale and context dependency rehearsed already in this chapter. Initially conceived of as a source of general guidance in making urgent conservation decisions in the absence of adequate biogeographical data, it has, despite its failings, provided a heuristic framework of lasting value in thinking about

conservation landscapes as, over time, research efforts have focused on evaluating the role of edges, corridors, and matrix effects, while various practical considerations have been added to the mix, as shown in Fig. 12.11 and Fig. 12.12.

Although inspired initially by the EMIB, that theory fails, for several reasons, to provide a clear answer to the SLOSS debate. One reason is that the EMIB is a stochastic model, which, in its simplest form, assumes that membership of a series of islands is a random draw from a larger species pool. Yet island and habitat island species composition are not entirely random, but are, to varying degrees, determined by the physical environmental properties of the isolates and by biotic interactions and ongoing community dynamic processes (e.g. responses to past habitat disturbance). Hence, the EMIB provides insufficient resolution on the compositional overlap or beta diversity encompassed within a particular set of isolates—crudely predicting the species number per island (Fig. 12.8b) but not the richness of the whole system—and it fails to identify which species may be at risk from a particular selection of areas (Worthen 1996; Tjørve 2010). At its simplest, a strongly non-nested series of small reserves may hold more species than a single large reserve, but where the system is perfectly nested, the largest reserve will hold most of the species (below).

While there may not be a simple, general answer to the SLOSS debate, the questions posed nonetheless retain importance in conservation science (e.g. Tjørve 2010; Lindenmayer et al. 2015). Intuitively, we know that larger reserves are desirable, as they can, at a minimum, hold larger populations of species dependent upon them, but when traded off against an increased number of reserves the equation is less straightforward. First, a set of smaller reserves may incorporate more different habitats (i.e. they may capture beta diversity across a landscape). Second, competition may lead to the exclusion of species of similar niches from any given reserve and so it may be good to have several reserves so that different sets of species may 'win' in different reserves. Third, there is an epidemiological risk inherent in having 'all your eggs in one basket'.

Overall, the substantial literature of modelling and experimental studies has seen a gradual shift

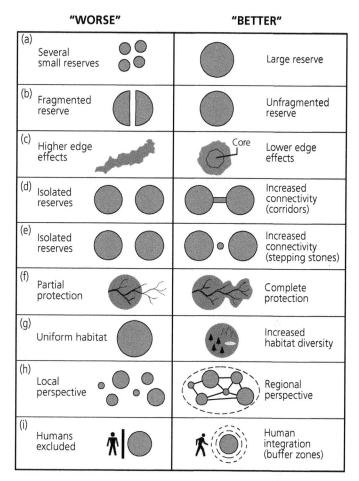

"WORSE"   "BETTER"

(a) Several small reserves — Large reserve

(b) Fragmented reserve — Unfragmented reserve

(c) Higher edge effects — Core — Lower edge effects

(d) Isolated reserves — Increased connectivity (corridors)

(e) Isolated reserves — Increased connectivity (stepping stones)

(f) Partial protection — Complete protection

(g) Uniform habitat — Increased habitat diversity

(h) Local perspective — Regional perspective

(i) Humans excluded — Human integration (buffer zones)

Figure 12.11 Design guidelines arising from the 'single large or several small' (SLOSS) debate, each of which has some measure of support but several of which are contentious, or fail in particular circumstances. Indeed, this applies to the first question, whether a single large reserve is preferable to several small reserves of the same total area. Some of these suggestions were put forward based directly on consideration of island theory in the 1970s, but several have been added subsequently and based on other forms of argument.
Modified from figure 8.2 of Triantis and Bhagwat (2011).

away from the overriding emphasis on the 'single large' end of the gradient that became embedded in much conservation policy towards the end of the 20th century. The following case studies are of illustrative value. Lindenmayer et al. (2015) report results for an experimental study in a *Eucalyptus regnans* forest in south-eastern Australia, in which four distinct treatments were used, involving clear-felling to create habitat islands of different size. They report no sensitivity to island size for birds or for small mammals and that the overall outcome was neutral with respect to SLOSS, most likely because the scale of isolation (around 100 m or so) and range in area (0.5 ha vs 1.5 ha) involved were too small. A study by Rösch et al. (2015), this time on plants and several invertebrate taxa in habitat fragments in central Germany, reported that a set

of 14 small fragments (totalling 4.6 ha) contributed 85% of the overall richness whereas the two largest (totalling 15.1 ha) reached just 37%. Similarly, Ulrich et al. (2016), in their study of birds in a dozen cloud forest fragments in the Taita Hills, Kenya, found that the three largest fragments (of 90 to 179 ha) featured no more than 75% of the bird species distributed across the smaller fragments (each < 15 ha), despite constituting more than twice the combined area of smaller patches.

A more extensive analysis is provided by Wintle et al. (2019), who gathered data on the size, isolation, and conservation value of habitat patches from 31 studies across four continents. Their focus was not so much on the richness of the patches, but their overall contribution to regional biodiversity. Contrary to the generalizations in Fig. 12.11, they found

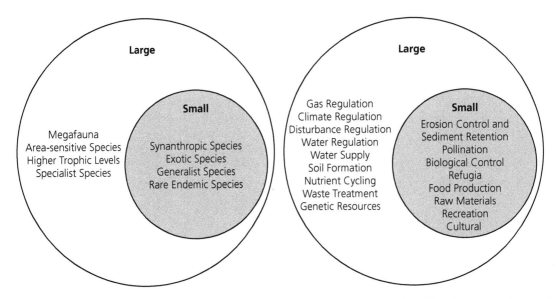

Figure 12.12 Conservation value of large versus small reserves, as summarized by Volenec and Dobson (2020) based on a systematic review of 58 taxon-specific studies. Left: species sensitivity; right: ecosystem services.

that relatively isolated habitat patches of small size (even of around just 1 ha) and often of high complexity of shapes, contributed better value than similar sized patches within intact areas of habitat. Such data do not provide assurance of continued persistence but do highlight that comparatively small areas can be of conservation significance and should not be ruled out of consideration for conservation action (but see e.g. Van Houtan et al. 2007).

Based on systematic review of 58 papers (screened from nearly 2000 papers), Volonec and Dobson (2020) also find support for small reserves (< 100 ha), noting that although smaller reserves often disproportionately feature generalist species, they often harbour high proportions of regional diversity, sometimes including specialist and endemic bird species. They point out that reserves of < 100 ha provide over half the global Protected Area estate and in addition to contributing to regional species diversity, these reserves also provide important ecosystem service value (recreational use, pollination services, etc.), again arguing for greater attention to be paid to the conservation of smaller reserves (Fig. 12.12).

Nonetheless, it is worth cautioning that there are circumstances for which larger reserves of more or less continuous habitat are essential (e.g. where dealing with highly diverse rainforest and larger megafauna). Species requiring large areas are often the ones most threatened by us and in need of protection (e.g. top carnivores, primates, and rhinoceros). They tend to be the larger vertebrates, species needing a large area per individual, pair, or breeding group, and/or requiring 'undisturbed' conditions. Although some may be absent from smaller habitat patches, predators undoubtedly have crucial roles to play in fragmented landscapes, not just in terms of predation within the habitat island, but also within the matrix. Simberloff (1992) notes that in the Manaus fragmentation study the bat falcon (*Falco rufigularis*) has repeatedly been seen to chase birds flying over cleared areas from fragment to fragment. It may therefore be unsurprising that most understorey birds will not willingly cross gaps of even 80 m. Similarly, in the USA, the northern spotted owl (*Strix occidentalis caurina*), of metapopulation theory fame, previously was a species which rarely left closed forest. In the newly fragmented landscapes, many yearlings are killed by great horned owls (*Bubo virginianus*) and goshawks (*Accipiter gentilis*) as they disperse over cleared areas.

Whereas top carnivores, such as big cats, may require large territories and disappear from small habitat islands, it might be possible to keep the highest diversity of butterfly species by means of a number of small reserves, each targeted to provide particular key habitats (Ewers and Didham 2006), providing that they are not too small or isolated that they consist principally of 'sink' populations (above). We have also to think about how species use their habitat space. For instance, many seabirds and some birds of prey have large territories incorporating both good and poor habitat. This doesn't necessarily mean large reserves are required, providing that their breeding sites are protected. But it may require a range of measures taken to discourage and prevent killing of the birds away from the nests within a mixed-use landscape. The 'reserves' required to protect such nest sites can be small compared with the ranges of the individuals.

The optimal configuration of patches for conservation protection is likely to depend on the traits of the target taxon, the scale of area, and isolation encompassed, and a number of other system-specific properties—biotic, abiotic, and societal/political. Key questions may include: what are the costs involved in establishing and maintaining protected areas? Is it necessary to patrol to prevent poaching or encroachment into the reserve? How many wardens do you need? Do you have to equip and manage park headquarters? In many situations, fewer, larger reserves make more sense from these practical, financial, and inevitably political perspectives. Moreover, when considering coarse scales and large regions and where the data are available for analysis of actual distributions, a superior approach is to use systematic conservation planning tools to select the particular set of areas for incorporation into protected area networks (e.g. Ladle and Whittaker 2011). However, the *implications* of different reserve configurations still demand attention, irrespective of the outcome of the SLOSS and other such theoretical debates.

12.6 Habitat island nestedness and its relationship to system diversity

A series of habitat islands exhibits nestedness when there is a significant tendency for the species present in species-poor patches to be found also in successively richer patches. There are other ways of ordering a matrix of species occurrences and other ways of defining nestedness (Section 6.4), but this captures the essence of the idea. Why might this be useful information?

Consider the scenario of a continuous block of habitat being reduced to a series of irregularly sized habitat islands in which the richness of the individual patches is largely a function of area. Providing we can predict the tendency towards nested distributions, we might be able to use this information to design more biodiverse landscape options, taking account of the subsidiary roles of isolation, linking habitat corridors and so forth. Now imagine that we have five patches and that we are told that in order of increasing area they contain 3, 6, 9, 15, and 40 species of birds from a regional species pool of 100. How many species do we have in total in our five patches? Should the system be perfectly nested, with each smaller set being a perfect subset of the larger one, then we would have 40 species. Were the system perfectly anti-nested, each patch containing entirely different species, then we would have 3 + 6 + 9 + 15 + 40 (i.e. 73 species). Should each patch represent a random draw from the pool of 100, we would have a species total somewhere between 40 and 73. The challenge statistically is to determine how to set up the null model to determine the random pattern and how to measure the significance of departure from random. Since the first attempts to test for nestedness in the 1980s, a range of different metrics have been developed, promoted and, in many cases, subsequently challenged as unsatisfactory (e.g. Ulrich and Gotelli 2012). Caution is therefore required in evaluating older papers that claim to show significant patterns of nestedness.

Nonetheless, whether or not departure from randomness is deemed significant or not, the fact remains that our imagined system of habitat islands (above) could return anywhere between 40 and 73 species and thus differing species accumulation curves, despite having a constant per island richness and thus a single ISAR. In the messier real world, empirical analyses of actual archipelagos of islands and of habitat islands have shown essentially the same thing (Santos et al. 2010; Matthews et al. 2015a). That is, ISARs typically over-predict archipelago richness when the system tends towards nestedness and under-predict archipelago richness when the system tends towards anti-nestedness.

In theory, a tendency towards nestedness may be caused by: (i) non-random extinction from habitat islands following their isolation, which is expected to be driven by the differential loss of species of larger habitat requirements; (ii) non-random colonization of islands/habitat islands, which is expected to be generated by differential failure of less mobile species to reach more isolated patches; or (iii) by strong nestedness of habitat types with increasing area, permitting occupancy of nested sets of species linked to those habitats (Section 6.4). A tendency towards anti-nestedness implies: (i) a high proportion of species unique to particular patches, which may reflect a lack of nestedness of habitat types, each with their own specialist species; (ii) that the habitat islands draw species from different bioclimatic or biogeographical species pools; or (iii) that habitat islands contain their own endemic species that have evolved *in situ*, or perhaps are relictual. Complicating matters further, these and other drivers (Matthews et al. 2015a) may interact with one another within dynamic ecological landscapes, as land-use change, other environmental change, and lag effects play out over time.

Expanding a little on the above, we should also expect nestedness, overall beta and total diversity of a system of habitat islands to exhibit a form of scale dependence, which we can illustrate by another thought experiment. Imagine creating two networks of a dozen habitat islands from a contiguous area of habitat, each network identical in terms of the starting richness of each island, but one more widely spaced than the other. We can highlight two expectations. First, the widely spaced system will exhibit a lower frequency of species populations mutually reinforcing one another, increasing the likelihood of extinctions. Second, the widely spaced system is more likely to sample different habitats, perhaps even sampling different biomes or biogeographical source pools. Thus, a widely scattered system of reserves may capture more beta (differentiation) diversity (Whittaker et al. 2005); but whether it can retain this diversity in the long run is another question. Hence, returning to the subject of Section 12.5, the answer to whether it is better to have a single large or several small reserves may be sensitive to the grain (in this case, size range of habitat islands) and extent (the area described by a convex hull around all the habitat islands) of the system.

At coarse scales of analysis, considering huge reserves across whole nations or sub-continental regions, the relevance of island theory is less clear. Here, the 'island effects' are subsumed in larger scales of biogeographical pattern and process. Indeed, the field of systematic conservation planning has been developed to tackle such problems, using large databases of distributional data in digital form to find optimal solutions to species representation on a coarse scale. For instance, computer algorithms have been developed to achieve goals such as finding the 10% of sites/grid cells that maximize representation of rare or endemic species (Ladle and Whittaker 2011). Such analyses provide a far more direct approach to conservation planning than the original SLOSS guidelines and effectively supersede them at coarse scales of analysis. Hence, in exploring other aspects of the SLOSS debate and of the application of island theory to conservation problems, we will continue to focus predominantly on the consideration mostly of comparatively small habitat islands embedded within landscapes that, from a suitable vantage point, a human might gaze across and encompass by a turn of the head.

Nestedness was once considered a common property of habitat island datasets, but more recent work, based on more conservative metrics, has overturned this generality, both for real islands and habitat islands (Ulrich and Gotelli 2012; Matthews et al. 2015a; Gibson et al. 2017; Section 6.4). Based on analysis of 97 datasets for a range of taxa drawn from previously published studies from around the globe, Matthews et al. (2015a) used NODF (nestedness metric based on overlap and decreasing fill) and assessed significance using the so-called proportional–proportional algorithm (Ulrich and Gotelli 2012), finding evidence of significant nestedness in only 9% of the habitat island datasets examined. Indeed, rather more, 16% were found to be significantly less nested than expected by chance (i.e. they were anti-nested). The remaining datasets failed to depart significantly from random. In a further step, Matthews et al. (2015a) analysed nestedness tendencies for 16 bird datasets, dividing the species into habitat specialists (those more dependent on the woodland patches) compared with generalist species. Overall, the specialists tended to be more nested than the generalists, but again, some datasets showed a tendency towards antinestedness. The general weak support for nestedness implies something of a win for the 'several small' school of thought. As Matthews et al. (2015a) conclude: 'These findings highlight that, in almost all instances, multiple islands of different size are required to represent all species.'

Notwithstanding the failure to generate clear conservation guidelines from nestedness analyses, data collated for such analyses hold further potential for analysis of species prevalence within fragmented landscape. For example, they can be used to solve the so-called minimum set problem; that is, to determine the set of islands/habitat islands that provide representation at least once, of each species (Matthews et al. 2015a; Gibson et al. 2017). The value of such data can be further enhanced if not only presence/absence data but also abundance data are recorded and surveys are repeated over time. The concept of nestedness may also be developed in other ways; for example, see Matthews et al. (2015b) for a test of nestedness of functional traits of bird species in forest habitat islands.

12.7 Edge effects

Much of the world's forest now occurs in fragments of < 10 ha, with perhaps half of the global woodland estate being within 0.5 km of a forest edge (Haddad et al. 2015). Such a startling statistic emphasizes the importance of examining edge effects in habitat islands, especially between forest and non-forest habitat (Fig. 12.13). The boundary zone, or ecotone, between two habitats, being occupied by a mix of the two sets of species, may have a higher density of species than either of the abutting core habitat types. For those taxa capable of exploiting features of the two contrasting habitats (e.g. roosting in trees but feeding in farmland), possession of a large expanse of such edge habitat can be beneficial. However, if the additional species supported by the edge habitat are those of the matrix that are not dependent on

the habitat of interest, then they may be discounted in evaluating the biodiversity benefits of the habitat island system (Matthews 2021). Moreover, if such edge effects penetrate a long way, then small habitat islands may be all edge and no core, in which case they will lack viable populations of habitat specialist species.

Edge effects have been documented for a range of taxa and across many different parts of the world (e.g. Schneider-Maunoury et al. 2016; Laurance et al. 2017). Watson et al. (2004a, b) describe responses of birds in littoral forests and surrounding habitats in south-eastern Madagascar. Core forest locations were found to be richer than edge or matrix habitats, with some 68% of the forest-dependent species found to be edge-sensitive. Frugivorous species and canopy insectivores (including six endemic vanga species) were generally edge-sensitive, while in contrast, sallying insectivores were edge-preferring. At least part of the edge-sensitivity recorded was attributable to changes in vegetation structure at the remnant edges. Forest-dependent species were generally lacking from fragments < 10 ha, thereby demonstrating a small-island effect (Section 5.4), masked in analyses of all bird species richness (Fig. 12.10a).

As illustrated by the foregoing study, a range of possible responses of species to forest/matrix transitions may be hypothesized. In their study of 51 species of amphibians and 82 species of reptiles, Schneider-Maunoury et al. (2016) found that 93% of species responded to the forest/matrix gradient, but they did so in varied ways, corresponding with five distinct response profiles that showed peak abundance at different points from well within the matrix to well within the forest. Edge effects were found to be detectable at least 250 m into matrix habitats and 2 km into forest habitats. More than half of forest-dependent species showed declines in abundance extending over 250 m into the forest patches, implying that for a majority of the species considered, a circular patch of 78 ha (approximately 10 soccer fields) is too small. One explanation for this is increased nest predation near forest edges (Wilcove et al. 1986).

The variability in edge effects is exemplified by an experimental study by Burkey (1993), who quantified egg and seed predation with distance from

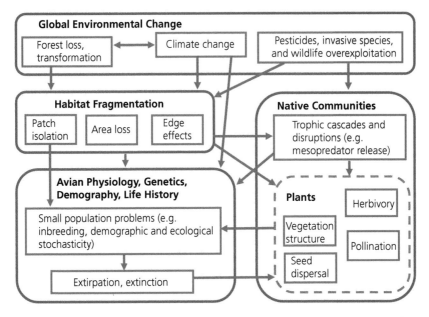

Figure 12.13 The network of factors involved in the decline and extinction bird-species populations invoked in a study of tropical forest fragments. Dashed-line enclosed community properties subject to alteration due to the inputs shown and which in turn feed in to changes in bird populations. This schematic captures many of the effects discussed throughout this chapter.
Slightly modified from figure 1 of Robinson and Sherry (2012).

the edge of a patch of Belizean rainforest. Egg predation rates were higher in a 100 m edge zone, but, conversely, seed predation rates were found to be higher 500 m into the forest than close to the edge. This illustrates that the relationship between a habitat patch/reserve and its surrounding matrix is not subject to easy generalization. There are species that share both zones, and just as there are matrix species which may impact negatively upon core reserve species, there may also be reserve species (dependent on it for breeding and cover) which utilize the matrix habitat (see also Levey et al. 2005; Mortelliti and Lindenmayer 2015; Matthews 2021). Favourable matrices can provide resources that many habitat island species can exploit, while more hostile matrices may act as 'traps' with elevated risk of mortality, as well as 'sources' of various agents of disturbance (Ramírez-Delgado et al. 2022). Hence, matrix permeability varies and influences species movements between habitat islands, affecting processes such as island colonization and rescue effects (Chetcuti et al. 2022). In this way, the matrix can be seen to modulate both the 'effective' area and isolation of habitat islands, thus complicating the

application of classic island theory to such systems (Matthews 2021).

12.8 Landscape effects, isolation, and corridors

The case for and against habitat corridors

It is generally assumed that habitat connectivity is beneficial to long-term survival, as it enables gene flow within populations and metapopulations. Habitat connectivity might be achieved by having stepping-stones, or corridors, of suitable habitat linking larger reserves together. In practice, habitat corridors act as differential filters, enabling the movement of some species but being of little value to others. All sorts of linear landscape features, including hedgerows, rivers, roads, and railways, may act as conduits for movement. Equally, they and other features, such as power cables, or even wind farms, may represent hazards or barriers (below; Benítez-López et al. 2010). Landscapes with great topographic relief channel their kinetic energy via dendritic tributaries, main channels, and distributaries, meaning that connectivity is highly structured across such landscapes. Much the same may apply to the distinction between a minor country road and a motorway. Studies of features such as hedgerows have not always found them to be as effective in connecting up woodlands as was hoped, with some species moving as well through surrounding fields and others simply not moving well along the hedges.

The relevance of corridors within changing landscapes is demonstrated by studies of Carnaby's cockatoo (*Calyptorhynchus latirostris*) in the Western Australian wheat belt, an area of 140,000 km^2, 90% of which has been cleared for agriculture (Saunders and Hobbs 1989). The Carnaby's cockatoo is one of Australia's largest parrots and was once the most widely distributed cockatoo in the region. It breeds in hollows in eucalypt trees and eats seeds and insect larvae from plants in the sand-plain heath. It congregates in flocks, both to nest and forage. Individuals may live 17 years and breeding status is not attained until at least 4 years of age, each pair rearing a single offspring at a time. The widespread clearance of the land has removed extensive areas of the vegetation type in which they feed, replacing it with annual crops that are useless to the birds. In some developments, wide verges of native vegetation have been left uncleared along the roads. These act to channel the cockatoos to other areas where food is available. Cockatoos have declined in areas of earlier clearances, carried out without these connecting strips, as once they have run out of a patch of acceptable habitat it takes a long time for the flock to find another patch of native vegetation. The more patchy the vegetation, the less successful the birds are at supplying adequate food to rear their nestlings. Furthermore, the narrow road verges of these early clearances result in a higher incidence of road deaths. The example illustrates that the degree and nature of connectivity of different landscape elements—in this case of the breeding and feeding habitats, and of the vehicular hazards—are critical to the survival of this species, which has continued to decline within these relatively recently fragmented landscapes (Williams et al. 2015).

Given the foregoing, it is important to establish the evidence for corridor efficacy via systematic review. This has been attempted in two studies, respectively analysing work published 1985–2008 (Gilbert-Norton et al. 2010) and 2008–2018 (Resasco 2019), both reporting benefits. In the latter case, 32 studies met the criteria for inclusion, providing data for a range of taxa and forms of evidence. Although not universally effective, corridors generally were found to benefit populations and communities by enhancing connectivity across the landscape. As ever, scale of the features is an important factor. Small corridors such as hedgerows may be of limited value for large mega-fauna yet have a key role for smaller-sized taxa. Some forest birds show strong preference for moving within forested corridors (e.g. Haddad, 2008; Şekercioğlu et al. 2015). For others, forested corridors act as 'drift fences', intercepting individuals moving through the matrix and redirecting them into desirable habitat. There is also evidence that some bird species effect seed dispersal within fragmented landscapes while tracking habitat corridors, preferentially following the edge, rather than flying within the corridor itself (e.g. Levey et al. 2005). Even fairly isolated trees within a landscape can function in a similar way, attracting birds and thereby generating patterns of seed dispersal within a landscape, although some species swiftly decline in prevalence with distance from continuous forest (Cottee-Jones et al. 2016a). The development of improved, miniaturized telemetry systems is continuing to add insights into the ways in which even quite small animals use corridors (e.g. Haddad 2008; Şekercioğlu et al. 2015).

There are also contexts in which *isolation* of populations of a target species might be desirable (Simberloff et al. 1992); for example, where there is a threat from disease. Young (1994) evaluated 96 studies of natural die-offs where large mammal populations had crashed by > 25%. He found that where the crashes are produced by disease epidemics, population subdivision may be beneficial, and the creation of linking corridors, and translocation efforts, may be harmful. An epidemic need not eliminate all the individuals in a population for it to have a crucial role in causing the eventual extinction of a population from a reserve. Nonetheless, there are relatively few studies that have claimed to show pronounced negative effects of habitat corridors. A systematic review and partial meta-analysis by Haddad et al. (2015) found little evidence for negative effects in relation to such factors as invasive species, spread of fire between reserves, or synchronized population crashes.

Overall, the case for favouring habitat corridors and improved connectivity of fragmented landscapes appears persuasive. Connectivity of habitats is not crucial for all species, but can be for many species of conservation concern, including both resident and migratory species. Ongoing concerns about 21st-century climate change are serious enough to imply that species range shifts should be anticipated and, in the face of the massive degree of habitat change generated by human societies the world over, connectivity of wild spaces seems highly desirable.

Reserve systems in the landscape

Over time, debates about SLOSS and habitat corridors have broadened to a consideration of how a whole range of differing natural, semi-natural, and artificial features are configured within a landscape (Figs. 12.11, 12.12). While conservation habitat corridors may be beneficial, other types of anthropogenic linear features often aren't. Countless birds, bats, terrestrial vertebrates, and insects are killed each year through collision with vehicles (Benítez-López et al. 2010; Fensome and Matthews 2016). Deaths through collisions with overhead power cables and wind turbines are also significant but can be greatly reduced by consideration of important flight paths during construction, by design features of the gantries, and by attaching a variety of objects to overhead power cables to enable birds to sight them (Alonso et al. 1994). Paved roads and tracks can play an important role as novel habitat corridors favouring species movement between previously disconnected systems, sometimes enabling the introduction and spread of non-native species (Delgado et al. 2007). On the other hand, wide roads and, especially, fenced highways, may act to impede the movements of large terrestrial animals, thus fragmenting populations and interfering with migration. Such information needs to be integrated into improved management of whole landscapes. In short, a landscape ecological framework is needed.

In practice, there may be greater opportunity for dispersal between two distant reserves linked by a river and its adjacent riparian corridor, than between two similar but closer reserves separated by a mountain barrier of differing habitat type. For butterflies, differences in land use as subtle as switches from one woodland type to another can significantly influence movement of individuals from one favoured habitat patch to another (Ricketts 2001). An early attempt to place island theories into realistic landscape contexts was provided by Harris (1984). Recognizing that there were limits to the amount of land that would be put over to reserve use, he advocated a series of reserves placed within the landscape as dictated by geographical features such as river and mountains, to encourage population flows between reserves. Most reserves would remain in forestry management. Within such reserves, there should be an undisturbed core of forest that is never cut, providing habitat for species requiring undisturbed, old-growth conditions, around which commercial operations might continue, following a pattern of rotational partial felling. This would provide a mosaic of patches of differing successional stage, maximizing habitat diversity, while maintaining income. His strategy aimed to satisfy the requirements of island theories, patch dynamic models, and economic realities. The implementation of such a policy requires concerted action from a wide variety of agencies and is thus easier to sketch out than to bring to a realization. In some regions of the world, such thinking has in any case been overtaken by different priorities linked, for example, to carbon capture or rewilding (Pereira and Navarro 2015).

Migratory and nomadic species

For birds, most attention in the habitat island literature is paid to resident or breeding species, but nomadic species and migratory species in their non-breeding ranges present particular challenges for conservation. Despite their mobility, such species may be particularly vulnerable to habitat loss (Runge et al. 2014). They may require conservation action in widely separated localities, spanning different states. Species that carry out a regular annual migration cycle may follow particular favoured routes and can be dependent on very specific stop-over sites, as well as on conditions in 'summer' and 'winter' ranges. Such information is beginning to be incorporated into both terrestrial and marine conservation planning strategies, but large knowledge gaps remain (Runge et al. 2014; Bairlein 2016; Cottee-Jones et al. 2016b). Nomadic species that follow resource peaks in regions of high inter-annual climatic variability are even more challenging, as conventional static protected area networks fall short (e.g. Cottee-Jones et al. 2016b). Conservation of such systems and species requires an approach to conservation planning that transcends political boundaries, considers habitat connectivity at a range of scales and focuses not merely on providing 'good' habitat but also mitigating threats outside of the protected area network (Box 12.2).

Box 12.2 Land-sharing versus land-sparing

The initial framing of the land-sparing vs land-sharing dichotomy is attributed to a seminal paper by Green et al. (2005), although the debate has deeper origins. It has long been understood that highly intensive agriculture is bad for wildlife. Crops are bred or 'designed' to provide high yield for the agriculturalists and low yields for pests (and pathogens). These crops are often sustained by systems involving pesticide and fertilizer applications to produce large-scale monoculture agriculture. The economic argument for this is that it is the mode of agriculture with the best prospects for viable farming livelihoods, and that it is thereby key to food security. The conservation (as opposed to economic) argument in favour of high-intensity agriculture is that the high yields created allow more land to be spared from agriculture (**land-sparing**). Wildlife-friendly farming, or **land-sharing**, on the other hand refers to systems which enhance wild populations on farmland, with the preference for higher yields being traded against the wildlife benefits of less intensive agriculture. Hence, following Green et al. (2005), land-sparing implies more land left unconverted to agriculture, including as formal protected areas, versus more extensive but less intensive agriculture.

Essentially then, land-sharing vs land-sparing is a debate about how to balance conservation and agricultural production (Fischer et al. 2014). This is to set aside other aspects of the debate centred on implications for human diets, power dynamics involving large agribusiness corporations, etc. Although the dichotomy between land-sharing and land-sparing may initially seem straightforward, in practice it can be hard to apply to particular agricultural systems and to determine what scale of set aside distinguishes sparing from sharing (Fischer et al. 2014; Lichtenberg et al. 2017). For example, we may view land-management systems and incentives that are designed to support wildlife through setting aside field margins and other areas, thereby providing connectivity of semi-wild habitats within the landscape as land-sharing. The counter argument is that the negative effects of intensive agriculture are too pervasive, leading to spill-over pollution in run-off and herbicides, fungicides, and pesticides that too often reach non-target individuals and areas, reducing populations of insects, birds, and other wildlife within the landscape; that is, resulting in lowered biodiversity benefits than anticipated (e.g. Melchett 2017).

Moving on from the simplicity of the land-sparing vs land-sharing dichotomy, systematic analyses of multiple data sets point to the desirability of coherent land-use policies that create extensive areas where farm-management schemes are based around organic farming and diversified agricultural production systems, both within fields and across landscapes. Such systems bring evidence of higher biodiversity and of the ecosystem services associated with pollination and biological control of pests (Lichtenberg et al. 2017). Although the biodiversity conservation case for such practices is strong, to achieve them requires strong regulatory frameworks and access to markets willing to pay a premium for organic produce.

12.9 Further implications of habitat fragmentation

Fragmentation of habitats generally happens in a selective fashion reflecting such physical features as geology, drainage, and topography as well as a range of socio-economic factors. Given the choice, farmers take the best land and leave or abandon the least useful. In much of southern England, this has meant that most remaining native woodland is on very heavy, clay-rich soils, with the lighter soils in agricultural use. The species mix in the resulting fragments may not be representative of the pre-fragmentation landscape. To give just one example, the short-leaved lime (*Tilia cordata*) does not favour the heavier soils, and it is now known as a species of hedgerows. In the past it was probably an important forest tree. Similarly, transport networks must link urban areas, industrial facilities, and access to raw materials, and collectively carve up landscapes in structured ways. The outcome is that habitat loss and fragmentation in actual landscapes take an infinite varieties of forms. Similarly, drainage and irrigation systems are profoundly altered by human endeavour. All of this has implications for conservation because it happens in textured landscapes rather than abstract planes. How landscape elements are connected can be crucial to biodiversity properties and conservation outcomes. As a simple example, many amphibians require both terrestrial and aquatic environments to provide for all stages of the life cycle: their conservation may not be attained by a focus on a single habitat type (e.g. Zimmerman and Bierregaard 1986).

Habitat fragmentation, especially of forested areas, can lead to significant immediate and ongoing impacts on the fluxes of radiation, wind, water, and nutrients across the landscape (Laurance 2002; Laurance et al. 2017; Reis et al. 2018), summed up by Laurance (2002, p. 595) in the idea of **hyperdynamism**: 'an increase in the frequency and/or amplitude of population, community, and landscape dynamics in fragmented habitats'. The following examples are illustrative.

- **Radiation fluxes**. In south-western Australia, elevated temperatures in fragmented landscapes reduced the foraging time available to adult Carnaby's cockatoos (*Calyptorhynchus latirostris*) and contributed to their local extinction.
- **Wind**. When air flows from one vegetation type to another it is influenced by the change in height and roughness. At the edge of a newly fragmented woodland patch, increased desiccation, wind-damage, and tree-throw can occur. Increased wind turbulence can affect the breeding success of birds by creating difficulties in landing due to wind shear and vigorous canopy movement. Wind-throw of dominant trees can result in changes in the vegetation structure and allow recruitment of earlier successional species.
- **Water and nutrient fluxes**. Removal of native vegetation changes the rates of interception and evapotranspiration, and hence changes soil moisture levels and nutrient dynamics (Haddad et al.

2015). In parts of the wheat belt of Western Australia, the new agricultural systems can bring stored salts to the surface, and this secondary salinity has caused problems both to agriculture and the remnant patches. In the fenlands of eastern England, drainage for agriculture historically led to peat shrinkage and a drop of 4 m in land level in 130 years, with knock-on consequences for remaining uncultivated ecosystems.
- **Fire regimes**. Modified landscapes can be a major source of surface fires; for example, from burning of adjoining pastures. Penetration of such fires into fragment interiors can increase plant mortality, disturb the fragment boundaries, and in time cause an 'implosion' of forest fragments (Laurance 2002).

The physical effects of edges thus have important knock-on effects for the biota, especially soon after forest fragmentation (Table 12.2) and some of these effects may penetrate far into the remnant patches (Reis et al. 2018). In time, the system adjusts to the new physical conditions, and the woodland edge fills and becomes more stable. Yet, just as there is a wide range of disturbance regimes across real islands, so must there be for habitat islands, and when habitat alteration is extensive, these processes can involve regional-scale feedbacks. Land-use of the matrix in which the habitat islands are embedded typically continues to change (Laurance et al. 2017; Reis et al. 2018). In general, in such contexts, smaller habitat islands will be subject to greater disturbance impacts than larger islands, thereby making such sites less suitable for particular species, often causing their local extinction.

Humans may also interfere directly in events within reserve systems. Increased fragmentation and road construction often increase the exposure of forest fauna to markets, driving increased bush meat extraction and pushing larger vertebrate populations towards local extinction, especially from smaller habitat patches (Peres 2001). Such effects are a consequence of habitat fragmentation but require different forms of information and a different analytical approach to understand, such as agent-based modelling (Iwamura et al. 2016), rather than classic island models.

Table 12.2 Classes of edge-related changes triggered by the process of forest fragmentation, as informed by the Minimum Critical Size of Ecosystem project. The first-order effects may lead to second-order and, in turn, third-order knock-on effects (based on Lovejoy et al. 1986 and Laurance et al. 2017).

Class	Description of change
Abiotic	Temperature increase
	Relative humidity decrease
	Penetration of light increase
	Increased exposure to wind throw
Biological	
First order	Elevated tree mortality (standing dead trees)
	Treefalls on windward margin
	Leaf-fall
	Increased plant growth near margins
	Liana infestation
	Altered balance of shade-tolerant/light-tolerant trees
	Reduced forest-dependent bird populations near margins
	Crowding effects on refugee birds
Second order	Increased insect populations (e.g. light-loving butterflies)
Third order	Disturbance of forest interior butterflies, but increased populations of light-loving species
	Altered abundances of insectivorous species

12.10 Ecological change and rewilding in fragmented landscapes

The fragmented world of biological communities in the future will be so different from that of the past, that we must reformulate preservation strategies to forms that go beyond thinking only of preserving microcosms of the original community types.

(Kellman 1996, p. 115)

The processes of land-use change that create fragmented systems frequently initiate successional changes in the habitat island remnants. This was apparent in the Barro Colorado Island study (above). Another example is provided by Weaver and Kellman's (1981) analysis of plants in newly created woodlots in southern Ontario, Canada, in which species relaxation involved the successional loss of a particular subset of species. With appropriate management, the observed losses could be avoided. If occurring at all in the absence of disturbance and succession, turnover appeared to be very slow-paced among the vascular flora. In this system, the EMIB effects were clearly weak and subordinate to other ecological processes.

In practice, protected area management often pays considerable attention to ecological succession, particularly in small reserves. Failure to do so often leads to the loss of desirable habitats. This applies to many of the lowland heath reserves of southern England, which have long been anthropogenically maintained by combinations of grazing, cutting, and use of fire. Without continued management, most areas suffer woodland encroachment. Simplified representations of woodland dynamics view a stand of trees as going through phases of youth, building, maturity, and senescence, each of which may support differing suites of interacting species. Reserves should therefore be large enough (or managed) to ensure that they contain enough habitat patches at different stages of patch life cycles to support a full array of niches. A related and often contentious issue is how fire regimes are managed. In fire-prone regions there is often a cyclical pattern of post-burn succession and fuel accumulation, leading to an increased likelihood of fire, repeating the cycle. The maintenance of a particular fire regime and patch mosaic structure may be of crucial relevance to species diversity in a reserve system, but fire regimes are often poorly understood (Carrera-Martínez et al. 2020; Ravazzi et al. 2021). Fire is also politically contentious because of the threat it poses to people and property.

Rewilding is an idea that has gained a lot of traction in recent conservation discourse. It takes a variety of forms but broadly refers to conservation strategies promoting self-sustaining (self-managing) ecosystems, typically via the reintroduction of missing animal species, especially larger vertebrates, which have important roles in regulating ecosystem processes and which therefore may have cascading impacts on other trophic layers (Pereira and Navarro 2015; Pedersen et al. 2019). The general ethos is that land managers should permit free movement of these larger animals across landscapes, which can involve removing fencing (although often the outer perimeters remain fenced). Rewilding thus implies the deliberate reduction of directed management, and hence also the blurring of management distinctions between habitat islands and the matrix of other habitats surrounding them. Intensively managed and highly fragmented landscapes containing small natural habitat patches are typically not considered great targets for a rewilding approach (Pedersen et al. 2019). However, sometimes, larger landscapes do contain important habitat patches that have high conservation value for specific taxa (e.g. Lepidoptera). In such contexts, it may well remain necessary for conservation managers to continue directed management of relatively small habitat islands to conserve the habitat structure on which these rare and localized taxa depend, notwithstanding that a less controlling approach is being pursued in the wider landscape (Chapter 6 in Pereira and Navarro 2015).

Much active management of nature reserves is about keeping a mosaic of different successional stages and habitat types to maximize habitat diversity and/or the provision of particular desired states, and not allowing an entire area to march through the same successional stage simultaneously. Moreover, continuing changes in the matrix can also be extremely important to the fate of populations of native species. When viewing the whole landscape, it becomes apparent that changes in the matrix, whether ongoing agricultural intensification, phases of agricultural abandonment and natural regeneration, corridor creation, set-aside, or deliberate rewilding, can be important in re-evaluating conservation priorities within habitat islands embedded within complex landscapes (e.g. Daily et al. 2003; del Castillo 2015; Pereira and Navarro 2015).

12.11 The island paradigm in conservation biogeography: assessment and prospect

It is commonly assumed that at some stage the remnant will re-equilibrate with the surrounding landscape. It is, however, questionable whether a new stable equilibrium will be reached since the equilibration process is liable to be disrupted by changing fluxes from the surrounding matrix, disturbances, and influx of new invasive species. The final equilibrium can be likened to an idealized end point that is never likely to be reached, in much the same fashion as the climatic climax is now conceptualized in succession theory. Management of remnant areas will thus be an adaptive process directed at minimizing potential future species losses.

(Saunders et al. 1991, p. 23)

The application of the island 'paradigm' to conservation science has produced many insights over the last half century and retains heuristic value. Nonetheless, it will be apparent to the reader that the guidelines for conservation to be derived from 'island approaches' remain rather heavily context dependent. For one thing, it is obviously an oversimplification to regard habitat patches as akin to remote islands, with the relevant processes occurring between patches and a mainland source pool, uninfluenced by the intervening landscape. The importance of land use within the matrix and of linear features connecting or subdividing the landscape in which habitat islands are embedded has repeatedly been stressed in this chapter.

The relevance of equilibrium assumptions to habitat islands can be debatable. Lag times involved in finding revised equilibrium points following habitat alteration are such that other dynamics may be as or more important: agricultural and climate change, episodic fires, hurricanes, or high-intensity storms, for example. Moreover, where species relaxation is evidenced, there is a need to determine the extent to which this results in selective loss of particular species (or guilds of species) as ecological processes unfold within the remaining habitat patches (Fig. 12.13). And, as so much 'fragmentation research' has concerned forest island patches, there is arguably a need to invert the focus and to consider the losses of populations and species typical of more open habitats, many of which thrived in cultural landscapes but have declined spectacularly in the era of chemical/industrial farming. In essence, while taking note of the information returned from island approaches to conservation problems, we also need to pay attention to how whole landscapes function ecologically. This doesn't in fact necessitate abandoning equilibrium theory and models but may mean contesting some of the early generalizations derived from island approaches in favour of a broader array of considerations and with a realization of the importance of scale effects in relation to questions such as SLOSS.

In this respect, interactions across trophic levels clearly demand more attention than evident in much of the theory discussed in this chapter. In particular, this may involve considering the roles of plant–pollinator and plant–disperser interactions, as well as the role of animals as grazers and browsers. Plants (collectively or individually) determine, to varying degrees, the ability of particular animal species to occupy a given habitat island. Hence, successional changes in plant communities can drive turnover in bird or butterfly communities (above; Section 6.8; Thomas 1994). Equally, the activities of animals as dispersers and pollinators of plants (and as predators) may be crucial to plant communities, both in contexts of ecosystem maintenance and of the recolonization of disturbed areas.

Forest fragmentation has been shown to impact negatively on frugivore-mediated plant dispersal (e.g. Martínez-Garza and Howe 2003; Levey et al. 2005). Different animals disperse different sets of plants. For instance, studies of cleared areas in the Neotropics have suggested that bats may be more significant in seeding open habitat nearby forest than are birds, while birds generally disperse a greater range of seed types. Bird dispersal can be positively influenced by the availability of some woody cover in an otherwise open area (Guevara and Laborde 1993). Even a low density of scattered trees can act as focal points for bird activity, thereby encouraging seed dispersal across landscapes (Cottee-Jones et al. 2015, 2016a).

The practical implications of such findings are worth noting. Efforts towards wildlife-friendly agriculture and land use can produce dividends and should form a part of so-called reconciliation ecology initiatives (Rosenzweig 2003). There can be many aims and purposes for shaping conservation management, ranging from the aesthetic, through the scientific, to the economic. We may wish to conserve systems or species which are 'representative', 'typical', rare, diverse, 'nice to look at', of recreational value, or provide economic return (Ladle and Whittaker 2011), and in cases we may wish to emphasize functional properties of assemblages in relation to wider ecosystem service provision (e.g. Hatfield et al. 2018). Such multiplicities of purpose require requisite tools; the island theories have their place in the toolkit, but they should not always be the first to be reached for.

We consider the evidence reviewed in this chapter to point to the need for what we might call a greedy conservation strategy. There is clearly conservation value in maintaining habitat islands and in maintaining and improving their connectivity (Haddad et al. 2015). We should also continue to advocate protected areas in which nature conservation is a key goal of management. But, it is evidently not enough and there appears to us a strong case for greater priority to be given to *extensive* alongside *intensive* conservation; that is, for environmental management policies to encourage the persistence (or safe passage) of many species outside of the more closely protected reserves systems (cf. Daily et al. 2003; Rosenzweig 2003). In short, to work to prevent habitat islands and reserves becoming more and more like real islands—except in those biogeographical contexts where insularization is actually beneficial to survival prospects! We have seen in this chapter that issues such as size, shape, and configuration within a landscape are important to reserve success, not just in terms of how many species will be held within a protected area, but also which sets of species. The number of species held in a reserve (or reserve system) is actually less important than to conserve those species which cannot survive outside the remnants.

Some recent efforts have been made to move beyond an exclusive focus on (forest) fragments towards understanding the role of such habitat islands within mixed-use landscapes (reviewed in Matthews 2021). Such arguments are framed by reference to a variety of labels; for example, matrix effects, reconciliation ecology, countryside biogeography (Rosenzweig 2003; Watson et al. 2005; Frishkoff et al. 2019). The common element is a realization that effective conservation must include consideration of what happens outside reserves, as the way we shape the countryside—whether we farm intensively or extensively, whether we retain hedgerows and trees within mixed landscapes—can have profound implications for regional diversity and for the abundance of wildlife (e.g. Frishkoff et al. 2019).

Conservation requires pragmatic decision-making. As we continue to fragment landscapes, island effects may inform such decision-making, but should not be oversimplified. There is no single message, and no single island effect; indeed, insularity of habitat patches may, in at least a minority of cases, bring positive as well as negative effects. Island effects may be weak or strong. The implications of insularity vary, depending on such factors as (i) the types of organism involved, (ii) the type of landscapes involved, (iii) the nature of the environmental dynamics, (iv) the biogeographical setting, (v) the nature of human use and involvement in the system being fragmented, (v) the scale and degree of isolation, and (vi) the range in area of resulting habitat patches.

Looking to the future, the development of increasingly affordable and practical systems for monitoring the incidence and movement of individuals of even quite small-bodied animal species within complex landscapes will continue to improve the evidence base on the effects of habitat alteration in fragmented landscapes (e.g. Şekercioğlu et al. 2015). The application of standardized monitoring systems for a range of organisms and habitat types is another important component in improving the evidence base (e.g. Borges et al. 2018b). The combination of the increasing evidence base and the use of techniques of meta-analysis and key evidence syntheses (e.g. Haddad et al. 2015; Keinath et al. 2017; Wintle et al. 2019; Volonec and Dobson 2020; Ramírez-Delgado et al. 2022), hold the prospect of stronger, more nuanced scientific advice for policy makers.

12.12 Summary

The conversion of more-or-less wild habitats to other land-uses is fragmenting, reducing, and isolating 'wild' areas across much of the globe. Here we examine the implications of this insularization from the scale of individual populations up to whole landscapes.

The minimum viable population (MVP) is the smallest number of individuals required to ensure long-term population persistence. We know that MVPs vary from species to species, and that as typically some individuals are not involved in breeding, the effective population size is smaller than the actual population size. Recent work suggests figures of the order of 4000 to 6000 adults may be necessary, although species reduced to a handful of individuals have been rescued from extinction by management intervention. Population loss may be due to stochastic demographic and/or genetic effects, or to environmental disturbance and

change—and has both natural and anthropogenic drivers. MVPs therefore need to be established separately for different types of species, and management regimes must be responsive to changing circumstances.

The area required to support the MVP is termed the minimum viable area (MVA). Estimates for MVA can be generated in various distinct ways and range from < 1 ha to > 1 m ha, with variation linked to traits such as body size and trophic level. Incidence functions provide one way of estimating area (or habitat, or isolation) requirements, although the patterns revealed may be confounded by multicausality. In addition, incidence functions can be inconsistent over time and space. Hence, they may fail to predict changes that follow from increased habitat fragmentation.

Where geographically separated groups are interconnected by patterns of extinction and recolonization, they constitute a metapopulation. Within a metapopulation, each patch can follow its own internal population dynamics, but when patch populations crash to extinction, they are repopulated from another patch within the metapopulation: a form of rescue effect that prolongs estimates of population viability. However, habitat patches often appear to describe a source–sink relationship rather than a mutual support system. The smaller or lower-quality sink habitats may have little relevance to overall persistence-time of the metapopulation.

Island biogeographical theory has provided a long-lasting framework for studying the impacts of habitat fragmentation on species diversity of the resulting habitat island archipelagos. The core expectation is the relaxation of diversity to new, lower values of richness, with smaller patches disproportionately impacted, although the time taken for these processes to play out can be lengthy, generating widespread extinction debts that can take many decades to settle. Losses can be strongly ecologically structured, involving ecological cascades across trophic levels.

The significance of isolation effects resulting from habitat fragmentation as a driver of species loss, over and above that resulting simply from regional area loss, has recently been subject to debate. Disentangling threshold effects, scale dependency, and the role of matrix effects from the insularity effect is technically challenging. Nonetheless, evidence reviewed herein strongly supports isolation as having an impact for particular species, guilds, and contexts. For example, tropical forest insectivores participating in mixed-species flocks appear often to be losers from forest fragmentation.

Another long-running controversy arising from the application of the island paradigm is provided by the SLOSS debate. Given opportunity to save a given amount of natural habitat, is it better to advocate a single large or several small reserves? Early arguments for the former have been challenged by analyses demonstrating support for the latter. The demonstration that compositional nestedness is far rarer across habitat island systems than previously thought provides one such line of evidence. In practice, however, the optimal reserve configuration for one type of organism, landscape, or scale of study system may not be optimal for another. An initially theoretical debate has generated a large literature, providing an increasingly pragmatic set of considerations involved in designing optimal solutions.

Fragmentation is so widespread that perhaps half of the global forest estate lies within 0.5 km of an edge. Strong edge effects are particularly associated with forest/non-forest boundaries and can sometimes be detected at distances of the order of 0.5 km, reducing the suitability of small habitat patches for forest specialists. Recent literature demonstrates that (i) habitat corridors play a diverse range of roles in connecting habitat islands within landscapes, and (ii) the environmental properties of the matrix of habitat types and land-uses, within which habitat patches are embedded, can also be hugely important to the biodiversity of the patches themselves. Such work points to the need to move beyond the island paradigm to encompass a broader set of theoretical and practical conservation arguments, such as bound up in land-sharing vs land-sparing, landscape ecology, countryside biogeography, systematic conservation planning, and 'rewilding' literature. Nonetheless, the heuristic and applied value of the substantial 'insularization' literature is reflected in the increased application of techniques of systematic review and meta-analysis to provide quantitative assessment of the effects of fragmentation.

CHAPTER 13

The human transformation of island ecosystems

13.1 Arrival

Commencing in the Late Pleistocene, human societies of diverse origins succeeded in reaching and settling on islands all around the globe, in late pre-history reaching even the most remote inhabitable islands of the Pacific. Within each oceanic region there is considerable variation in the length of human occupancy, often with very early and quite late settlement of islands that are in the same ocean basin (Table 13.1; Fig. 13.1). Hence, when Europeans began their major phase of expanding across the world around the start of the 16th century, they encountered islands with varying antiquity of human occupation as well as some that had never been occupied. We choose to distinguish three groups: (1) palaeoinhabited islands, settled several millennia or tens of millennia before European expansion (e.g. New Guinea, the Solomons, Tasmania, the Antilles); (2) neoinhabited islands, those discovered and settled for just one or two millennia (or even a little less) (e.g. the Canaries, Madagascar, Marquesas, Hawaii, New Zealand); and, finally, (3) previously uninhabited islands (e.g. the Azores, Madeira, Cabo Verde, St Helena, Tristan, Mascarenes, Galápagos, Juan Fernández). This rather Eurocentric way of looking at history is adopted for two reasons: first, European colonization (and colonialization) marked a key turning point in the human history and biogeography of islands, and second, it also, in most of the world, marked the start point of the historical documentation of island life, through the collection of specimens and scientific descriptions of island plants and animals.

Advances in palaeoecology, palaeoenvironmental reconstruction, genetics, and archaeology increasingly enable reconstruction of the dynamics of island systems over longer time frames, putting the European expansion into a fuller perspective alongside the impacts of earlier island colonists. The improved resolution of these reconstructions confirms the scale of anthropogenic extinctions of island life, which amount to thousands of endemic species and, within the historic period, accounts for 60% of global extinctions (Whittaker et al. 2017). It also provides a more nuanced understanding of the sustainability (or otherwise) of indigenous cultures on islands. The sense of sadness about the loss of endemic species and ecosystems that we feel as biologists is increasingly counterposed with admiration for the extraordinary achievements, innovations, and adaptability of so many indigenous island societies (Dawson 2014; Braje et al. 2017).

The ever-increasing extent and resolution of our (still fragmentary) knowledge of human impacts on islands means that we can accommodate only a selective review (see also Leppard et al. 2022). Those systems we have chosen provide case studies from palaeoinhabited (the Caribbean), neoinhabited (Polynesia, Canaries within Macaronesia), and historically colonized systems (parts of Macaronesia). We show that each cultural wave to reach remote islands has its impact and that they become layered upon one another so that each island increasingly supports non-native rather than indigenous food webs and ecosystems, with the scale of impact typically massively ramped up following European conquest.

Island Biogeography. Robert J. Whittaker, José María Fernández-Palacios, and Thomas J. Matthews, Oxford University Press.
© Robert J. Whittaker, José María Fernández-Palacios, and Thomas J. Matthews (2023). DOI: 10.1093/oso/9780198868569.003.0013

Table 13.1 Approximate date of first human colonization in relation to European contact dates from the historical period for several archipelagos around the world (from several sources; and see text). There remains uncertainty over many of the dates for pre-European establishment (see Fig. 13.1 and text).

Island or archipelagos	Pre–European establishment	European contact (AD)
Palaeoinhabited		
New Guinea	40–50,000 BC	1526/27
Tasmania	c. 33,000 BC	1642
Cyprus	> 10,000 BC	–
Crete	> 10,000 BC	–
Greater Antilles	5000–4000 BC	1492
Neoinhabited		
Fiji, Samoa	c. 1050 BC	1643
Canaries	c. 400 BC	c. 1400
Madagascar	AD 750–900	1500
Marquesas	AD 1190–1290	1595
Rapa Nui (Easter)	AD 1190–1290	1722
Hawaii	AD 1190–1290	1778
New Zealand	AD 1190–1290	1643
Previously uninhabited		
Madeira	–	1420
Azores	–	1432
Cabo Verde (Cape Verde)	–	1456
St Helena	–	1502
Galápagos	–	1535
Juan Fernández	–	1574
Mascarenes	–	1598
Tristan da Cunha	–	1812

More recently, a high proportion of the world's islands have been swept up in the 'Great Acceleration'—the rapid transformation of the human relationship with the natural world that began shortly after WWII (Steffen et al. 2015). Hence, in the next chapter, we build on these selected narratives by presenting a synthesis of anthropogenic extinctions on islands and review the status of island biotas in the light of contemporary threats to island biodiversity. We reserve to the final chapter a deeper look at contemporary challenges to island societies and island nature, alongside threat-mitigation, conservation strategies, and actions.

13.2 The first islanders

In some parts of the Old World, modern humans were not the first human island colonizers. For example, *Homo erectus* reached Socotra, the Greater Sunda Islands and the Philippines during the Middle Pleistocene, demonstrating a repeated capacity to cross sea passages. They reached the island of Flores around 1 Ma, most probably providing the ancestors from which the dwarfed hominid *Homo floresiensis* (dubbed 'the hobbit') evolved *in situ* (Bellwood 2017). Fossils of another South-East Asian insular dwarf have also been found on Luzón (*H. luzonensis*). However, it was the arrival of anatomically modern humans, *Homo sapiens*, on the islands of the world that marked an irreversible threshold-crossing moment. This is an event that has been staggered in time across and within different island regions over a span of thousands of years, reflecting the pattern of emergence, cultural development, and dispersal of our ancestors out of Africa and across the world (Fig. 13.1; Lomolino et al. 2017). It is significant that this process began in the Late Pleistocene, in many regions coinciding with glacial conditions when the world was very much colder and sea levels much depressed (Chapter 3). One corollary is that many

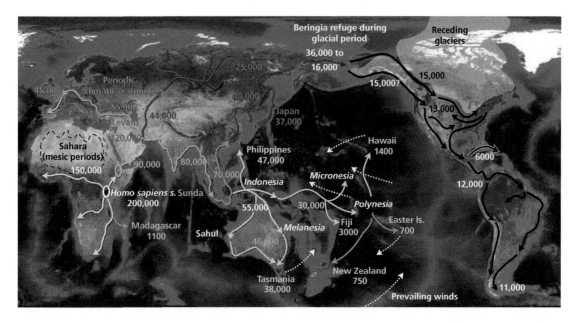

Figure 13.1 The geographic range expansion of anatomically modern humans (*Homo sapiens sapiens*) out of Africa was strongly geographically structured, as shown by this reconstruction from Lomolino et al. (2017). The islands of South-East Asia and the Mediterranean were colonized a great deal earlier than those of the Atlantic, Caribbean, and Pacific. Such remote groups as the Azores (central Atlantic) and New Zealand were among the last islands of size to be reached and colonized by humans. The timing(s) of first colonization of the Americas is underestimated by the dates shown here, with plausible (but contested) claims of earlier human presence in Central and Southern America ranging *c.* 21–37 ka: see text.

potentially important coastal-fringe archaeological sites are now submerged (e.g. Cunliffe 2012).

At an early stage of our expansion during the Late Pleistocene, modern humans migrated along the coastal regions of the Indian Ocean into tropical South-East Asia and Wallacea, reaching Australia and New Guinea before 50 ka, but failing to spread further into the central Pacific until the last 5 ka (Fig. 13.1; Bellwood 2017). Colonization of Mediterranean islands appears to have begun rather later than in South-East Asia, in a sequence partly reflecting area and distance from the mainland (Table 13.2; Dawson 2014; Cherry and Leppard 2018). Intriguingly, while 75% of the Mediterranean islands of > 300 km² were colonized by the 4th millennium BC, the Balearic Islands in the west of the basin, including the large island of Majorca, had yet to be settled (Leppard et al. 2022).

Turning north to Britain, there is evidence of the presence of earlier species of human *c.* 0.5 Ma, and there is intermittent evidence of human activity in periods of milder climate between that point and the

first arrival of modern man *c.* 12,000 BC (*c.* 14 ka), when there are signs of activity in a few scattered localities (Cunliffe 2012). This period coincided with the warming at the end of the last Ice Age, when Britain was an extension of the continent, which it was to remain until around 5000 BC (*c.* 7 ka) when the final land connection (via Doggerland) was severed by rising sea levels. A regionally intensive cold spell, the Younger Dryas (*c.* 12.0–11.7 ka), saw the return of glacial conditions across Britain and NW Europe, and it was only at the end of this cold snap that evidence of hunter-gatherer communities reappear in the archaeological record within SE England. Initially, and for several thousand years, these people were able to move across the land connection to populate Britain, through territory now long-since drowned.

Abundant archaeological evidence reveals that whereas the eastern side of Britain long exhibited strong connections to northern France, Belgium, and the Netherlands, the western coastal fringes maintained cultural and trading connections

Table 13.2 First evidence of human colonization/presence on larger Mediterranean islands commenced around the end of the Pleistocene: 75% of these islands had been colonized by the 4th millennium BC, in a sequence partly reflecting island size and location (based on Dawson 2014; Cherry and Leppard 2018).

Island	Western/Eastern Med.	Date (millennium cal BC)	Size (sq. km)	Rank size
Sicily	W	> 10	25,708	1
Sardinia	W	> 10	24,089	2
Cyprus	E	> 10	9251	3
Crete	E	> 10	8259	5
Naxos	E	> 10	430	18
Corsica	W	9	8722	4
Thasos	E	6	380	20
Lefkas	E	6	303	22
Hvar	W	6	300	24
Lesbos	E	5	1633	7
Rhodes	E	5	1400	8
Chios	E	5	842	9
Kephalonia	E	5	781	10
Corfu	E	5	593	12
Solta	W	5	588	13
Samos	E	5	477	17
Andros	E	5	380	21
Karpathos	E	5	301	23
Lemnos	E	4	478	16
Zakynthos	E	4	402	19
Mallorca	W	**3**	3740	6
Menorca	W	**3**	702	11
Ibiza	W	**3**	572	14
Jerba	W	1	568	15

through the Atlantic fringe to the Iberian Peninsula by means of the early and continuing use of small boats. The rich archaeological literature reviewed by Cunliffe (2012) illustrates the cultural connectivity that early human societies in this part of Europe were able to maintain via maritime links: these movements undoubtedly involved the translocation of at least some domesticated and commensal species, particularly as hunter-gatherer lifestyles later gave way to agropastoralism.

The timing of human colonization of the Americas remains controversial, as does the extent to which humans were subsequently responsible for the megafaunal extinctions (Lomolino et al. 2017; Lesnek et al. 2018). Climate change undoubtedly had a huge role to play in changing patterns of diversity during the Pleistocene, and specifically was important in the period spanning the Last Glacial Maximum (*c.* 21 ka) and the early

millennia of post-glacial conditions of the Holocene. Yet mounting evidence from around the world supports a leading role for humans in the Late Pleistocene extinction of most of the largest land vertebrates occurring across the Americas, and also across Eurasia and Australia (reviewed in Lomolino et al. 2017). There is a consensus for human population movements into North America from Beringia (a Pleistocene land-bridge connecting north-east Siberia with Alaska), via a deglaciated north Pacific coastal corridor *c.* 18–15 ka (Fig. 13.1; Lesnek et al. 2018; Bush et al. 2022), but it seems likely that this does not represent the earliest human arrival. Recent archaeological studies have provided estimates from sites scattered across New Mexico, Mexico, Uruguay, and Brazil of human presence and activity for dates ranging *c.* 21–37 ka (Rowe et al. 2022). Such claims have hitherto been received with caution, ranging to scepticism (e.g. Potter et al. 2022).

Based on more cautious estimates of first arrival *c.* 18–15 ka, the colonists must have rapidly spread through the length and breadth of the Americas. On the west coast the California Channel Islands were occupied by 13 ka. On the east coast, excavations at the Page-Ladson site on the north-western landward end of the Florida peninsula reveals that *c.* 14.5 ka humans of pre-Clovis culture had butchered a mastodon in a riverine site. The remains were subsequently buried by several metres of sediment during the Late Pleistocene marine transgression (Halligan et al. 2016). Undoubtedly much of the record of early human occupation of the coastal fringes of the continent (and indeed of other regions of the world) suffered a similar fate. Notwithstanding such gaps in the record, there appears to have been a significant delay between occupancy of the coastal fringe of this part of the continent and subsequent colonization of the Caribbean (Fig. 13.1; Siegel et al. 2015). Should the earlier dates for human occupation of Central and Southern America be accepted, then the delay in occupation of the Caribbean would seem even greater—reflecting that these events largely involve colonization of islands beyond sight of land (Napolitano et al. 2019).

In the Greater Antilles, first occupation of Cuba, Hispaniola, and Puerto Rico spanned *c.* 7.0–6.0 ka according to Keegan and Hofman (2017), but began a little later, *c.* 5.8 ka, according to a more recent account by Napolitano et al. (2019). Settlement of Jamaica appears to have occurred much later, with the first permanent occupation no earlier than 2.7 ka. Turning to the 15 major and numerous smaller islands of the Lesser Antilles, the earliest evidence of occupation is for Trinidad, just 10 km from South America, *c.* 8.4–7.3 ka (Napolitano et al. 2019), with a confusing picture of later and perhaps initially patchy colonization of other islands. Some islands in both the northern and southern Lesser Antilles (e.g. respectively, Nevis and St Lucia) were colonized in a later phase *c.* 1.8–0.5 ka. In common with many other island systems in the prehistoric era, there is abundant evidence of multiple cultural waves for many islands, bringing new tools, cultivars, and techniques to Caribbean islands over the millennia of the pre-Columbus era (Keegan and Hoffman 2017). Movements appear to have been initiated from various mainland and insular points,

reflecting still only partially understood networks of human mobility and cultural exchange across the region.

In the Pacific, the islands around New Guinea (near Oceania) were occupied from around 40 ka. Settlement of the more distant islands (remote Oceania/Polynesia) was greatly delayed, being initiated around 3.1 ka following the emergence of a distinct Lapita culture in the Bismarck Archipelago around 3.4–3.2 ka (West et al. 2017). Alongside archaeological and other evidence, analysis of genomic data for the commensal Pacific rat (*Rattus exulans*) supports a scenario of the colonization of central east Polynesian archipelagos from west Polynesia, followed by rapid expansion northwards to Hawaii, south-east to Rapa Nui, and south-west towards New Zealand, reflecting the extraordinary mastery of techniques of navigation attained by the Polynesians. Recent analysis of human genomic data from 21 Polynesian populations from across the region reassuringly provides a similar scenario, differing only a little in detail, showing a pattern of expansion originating in Samoa, and spreading via the Cooks, to the Society Islands, the western Austral Islands, Tuamotu, and then outwards to the remote northern outposts of Hawaii and the southern Marquesas, and to the south-east, to Rapa Nui (Ioannidis et al. 2021).

Even here, the precise timing of events is yet to be settled. The estimates shown in Fig. 13.1 suggest a much earlier northwards extension of settlement to Hawaii than to the south-east and south-west, but based on analysis of 1434 radiocarbon dates, Wilmshurst et al. (2011) propose a scenario where the Society Islands were colonized *c.* AD 1025–1120, with a further major pulse incorporating all remaining islands (including e.g. Rapa Nui, Hawaii, New Zealand) *c.* AD 1190–1290; that is, around 0.72–0.78 ka (Table 13.1; Fig. 13.2; and see West et al. 2017). These estimates shorten the time frame of human influence for some islands compared with some previous estimates. Also consistent with late colonization dates, many of the smaller, low-lying Pacific atolls (e.g. in the Tuamotu Archipelago and northern Cook Islands) were unsuitable for human settlement until sea levels declined from the mid-Holocene hydro-isostatic highstand sufficiently to allow their stabilization, which occurred

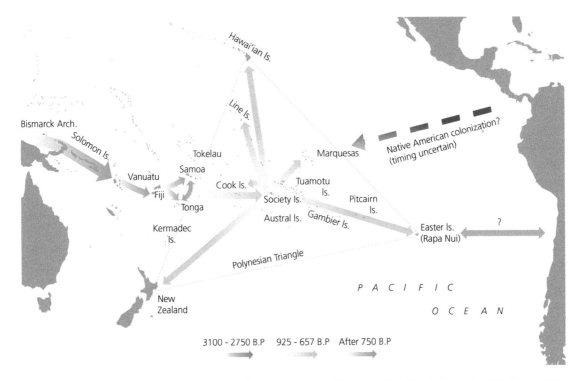

Figure 13.2 An approximate chronology for colonization of Polynesia from around Taiwan, arriving in Rapa Nui around AD 1200. The possibility of some contribution from Native Americans to Rapa Nui is shown in two scenarios—one a direct voyage and the other joining a returning visit with some Polynesians: these links are contested (see text). Figure modified from an original in West et al. (2017). For an alternative scenario based on genetic interrelationships among Polynesians and differing in some details, see Ioannidis et al. (2021).

around 1.0–1.5 ka (Dickinson 2009; Ioannidis et al. 2021).

Colonization of the Indian Ocean also spanned a great range in time, involving both mid-Pleistocene arrival of *Homo erectus* on Socotra and within South-East Asia, and later waves of colonization by modern humans (above). The range of proposed dates for first settlement of Madagascar ranges from *c.* 10.5 ka to *c.* 1 ka, with Hansford et al. (2020) providing a detailed defence of the earliest of these dates, noting that multiple human migrations to the island followed, with a major pulse of megafaunal losses accompanying rapid expansion of human impact *c.* 1 ka.

The North Atlantic islands of Macaronesia have a relatively recent history of human colonization (Fig. 13.3; Fernández-Palacios et al. 2016b). The Canary Islands may have been colonized as early as 2900 ka, although de Nascimento (2020) suggest between 2400 and 2000 ka as a better supported time

frame. The first inhabitants were of Berber affinity from Northern Africa. There may have been a period of continued connectivity with the Mediterranean for a while, following which the islands and their people were isolated from external contact for around a thousand years. Exploratory voyages from the Mediterranean to this part of the North Atlantic resumed in the 14th century, with conquest of the Canaries spanning the 15th century, during which the other unoccupied Macaronesian archipelagos were colonized. This is not to rule out earlier human visits to Madeira (Rando et al. 2014) or even the Azores (cf. Raposeiro et al. 2021; Elias et al. 2022). Radiocarbon-dated evidence of house mice remains indicates their arrival on Madeira *c.* AD 1036. Genetic evidence suggests their accidental introduction from Scandinavia or northern Germany, perhaps by an undocumented Viking voyage (Rando et al. 2014). The Portuguese took possession of Madeira in 1419. Some decades later Christopher

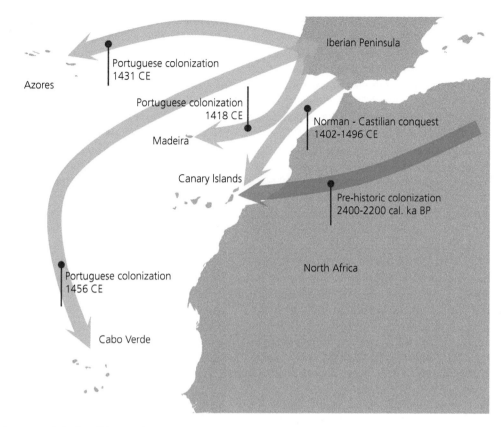

Figure 13.3 Human colonization of Macaronesia. Modified from Fernández-Palacios et al. (2016b).

Columbus spent some time living on the Madeiran island of Porto Santo. This was prior to his epoch-defining first transatlantic voyage, departing from the Spanish port of Palos in 1492, sailing via the Canary Islands, before making landfall on San Salvador in the Bahamas 33 days later. This momentous event initiated the European conquest of the Americas and marked the start of European expansion across the oceans of the world.

13.3 The prehistoric island survival package

Island colonization in pre-history typically followed corridors of accessibility along coastlines dotted with islands and via stepping-stone routes (Wilmshurst et al. 2011; Cunliffe 2012; Siegel et al. 2015). But as we have seen, this did not always

result in a simple pattern of spread from a particular continental origin point or by shortest geographical distances. In the Pacific, those master mariners the Polynesians developed sophisticated navigation techniques, based on the study of the ocean (water temperatures, colours, currents, wave patterns) and sky (e.g. clouds, stars, birds), to detect land beyond the horizon. Hence, they were able to spread with remarkable rapidity and efficiency across what became Polynesia, voyaging in outrigger and double-hulled canoes. When setting out on voyages of settlement, they took with them a more-or-less standard package of the tools, plants, animals, and techniques of resource use needed not just for the voyage but also to establish new colonies.

Indeed, notwithstanding the undoubted diversity of cultures and timings of island colonization, it has been argued that Neolithic packages carried by colonizing voyages of islands in the Mediterranean, Pacific, and Caribbean tended to involve broadly similar elements and thus to generate common pathways of ecosystem change (Fig. 13.4a; Braje et al. 2017; Leppard 2017). Where the people were agropastoralists, initial colonization would typically involve a set of commensals (e.g. rats, mice), domesticated plants, and animals. Evidence of rat-gnawed seeds, and of the arrival of plant and animal commensals, are increasingly being used to refine estimates of arrival and ecological impacts (e.g. Prebble and Wilmshurst 2009). Landscape clearance involving the use of fire is another very common element (Fig. 13.4a). Indeed, spikes in charcoal and the detection of *Sporormiella* fungal spores—associated with the dung of domesticated animals (pigs, goats, sheep)—provide classic means of identifying human colonization through analysis of lake sediment cores (Nogué et al. 2017). Land clearance, in turn, is associated with increased erosion (e.g. Morales et al. 2009; de Nascimento et al. 2009; 2020), and therefore increased sediment deposition rates in lakes is also typical.

Many prehistoric societies adopted mitigation practices such as terracing and lithic mulching, while around coastlines marine resources were often manipulated (e.g. by modification of coastal pool systems). In cases, transformations were comparatively slow, in others (such as New Zealand and Rapa Nui), extensive and rapid (Wilmshurst et al. 2011). Indeed, in many parts of New Zealand, forest cover was largely eliminated by fire within 200 years of colonization by the Polynesians (McWerthy et al. 2010), while Madeira was said to have 'burned for seven years' during first settlement and land clearance by the Portuguese *c.* 1420/1425. Palaeoecological analyses of vegetation change from a range of oceanic islands show that the rapidity of vegetation change has typically been greater for islands colonized in the historic period than in prehistoric times (Nogué et al. 2021).

Exploitation of native food resources, including 'naïve' resident land birds and seabird colonies, led to declines and local or global extinction of many native animal and some plant species from islands. Simultaneously, introduced species also contributed to the loss of native species. Prime examples in the prehistoric period included commensal rodents, while in the historic period a rather larger array of rodents, mustelids, and ungulates have been introduced. Whereas remote archipelagos and many individual islands possess their own endemic species, the introduction of similar sets (or bundles) of commensals leads, through each wave of human colonization and increased connectivity with the outside world, towards **biotic homogenization**, as part of a general process of environmental convergence (*sensu* Leppard 2017; Fig. 13.4). The package of translocations in the Pacific and Indian oceans, in addition to ubiquitous rats, commonly featured coconut, taro, yam, banana, breadfruit, pigs, dog, and domesticated chickens. With the exception of maize, the translocation of plant products during early colonization of the Caribbean is less clear, but it appears that armadillo, agouti, guinea pig, peccary, and opossums were introduced, possibly as sources of non-domesticated protein (Leppard 2017).

13.4 Each contact leaves a trace

Each distinct human cultural wave to reach an island system is likely to bring with it some new impact connected to the exploitation of natural resources and the introduction of new species. The time taken for many of these changes to fully manifest can be very lengthy, and so it can be difficult to unravel the impacts of past cultural shifts and more recent ones, and indeed to distinguish human impacts from ongoing natural changes in climate and environment (but see e.g. McWerthy et al. 2010; Nogué et al. 2017, 2021). Geographers sometimes use the analogy of a palimpsest to describe such layered and confused effects.

Nonetheless, for many island regions, the expansion of European influence around the globe from the 15th century onwards marked a very significant gear change in human impact, with lasting biogeographical legacies (e.g. Lenzner et al. 2022). Continuing to the present day we may conceptualize this process as one of de-insularization and globalization, often starting with highly exploitative phases of resource extraction, human translocation (e.g. the slave trade impact in the Caribbean, in Macaronesia, and in parts of Polynesia), and

(a)

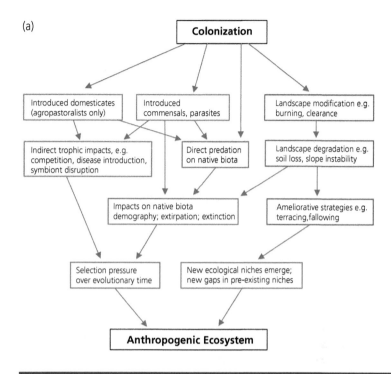

(b)

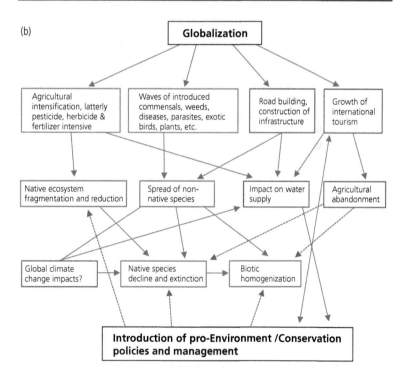

Figure 13.4 The influence of humans on island ecosystems tend to have involved a common set of characteristics, (a) commencing with the initial wave(s) in pre-history, and (b) continuing alongside modern features of industrialization and globalization of recent decades. Blue boxes: drivers of landscape change; green boxes: changes to biota and ecosystems; and orange boxes: island biodiversity responses.

Part 'a' modified slightly from figure 2 of Braje et al. (2017); part 'b' original.

ecosystem transformation (Fig. 13.4b). The most recent phase of the cultural transformation of many remote island systems in the decades up to *c.* 2020, especially in warmer climates, has been driven by the phenomenon of international tourism, boosted during the global post-WWII 'Great Acceleration' (Steffen et al. 2015) by low-cost flights and the boom in the cruise industry.

13.5 The Caribbean

As true also of much of South-East Asia, the arrangement, geological, and sea-level history of the Caribbean arc (Section 2.5) permitted natural colonization and subsequent *in situ* evolution of non-volant land mammals to a degree not found in more distant oceanic archipelagos. Within the Caribbean, since European arrival (AD 1492) at least 29 non-volant species have become extinct (Turvey et al. 2017) but other losses occurred earlier. The Greater Antilles possessed at least 116 species of mammal at the start of the Holocene, of which just 56 survive (Upham 2017). Whereas 75% of bats remain extant, 80% of the non-volant mammals (sloths, shrews, rodents, and primates) are extinct. Hispaniola has lost 23 out of 25 non-volant mammals, Cuba 13 of 23, Jamaica 5 of 6, and Puerto Rico all 7 of its species. Work continues to refine timing and attribution of these losses, with climate change and associated sea-level rise providing the obvious natural alternative to human impact. However, the majority of losses appear to follow human arrival, with attribution studies pointing the finger of blame mostly at humans and their commensals (rats, dogs). In particular, the initial wave of extinction involved losses of endemic radiations of sloths and large-bodied rodents ('giant hutias') (Turvey et al. 2017).

On the largest island, Cuba (109,884 km^2), sloths, bats, rodents, and shrew-like mammals survived well into the Holocene, beyond initial human arrival and in many cases beyond initial European contact (Borroto-Páez and Mancina 2017). The first European colonists arrived in 1509 (17 years after Columbus' first voyage brought black rats to the island), in short order bringing house mice, cats, pigs, cows, horses, and additional breeds of dogs to the island (Fig. 13.5). All had become feral and invasive within the first 50 years, transforming habitats,

spreading new diseases and parasites, and competing with and preying on native species. Directly and indirectly, these animals became key agents of the decline and extinction of native non-volant mammals.

The introduced species were assembled from around the world, but because of the Spanish connection and the sailing routes, many earlier introductions were sourced from the Canary Islands and/or the Iberian Peninsula. While many of the most influential ecosystem transformers were introduced early in the 16th century, the mongoose (a particularly problematic small predator) was not introduced until the late 19th century, and since then there has been a steep rise in the rate of introductions (Fig. 13.5), the impacts of which are surely yet to be fully manifested. In illustration, the endangered endemic vertebrate almiquí (*Solenodon cubanus*) was on the menu of the Amerindians prior to 1492, has been affected by habitat transformation, and is known to be predated by feral dogs and cats. Dogs and pigs destroy their burrows and black rats infest them. The almiquí is now an endangered species and greatly restricted in distribution. Many island endemics are similarly threatened by multiple and often interacting threats (Chapter 14).

European colonization and conquest of the Caribbean, as more broadly in the New World, had enormous impacts on the native peoples and the ecology of the region (Braje et al. 2017). The transfer of pathogens from the Old World led to the deaths of a high proportion of the native inhabitants. The introductions of non-native plants and animals, and widespread land clearance for agriculture, transformed the ecology of the islands. Disease, warfare, and forced labour following initial conquest led to such declines in human population that the Spanish began the enforced movement of Africans to their territories as slave labour as early as 1518 (Benn-Torres et al. 2008). The slave trade reached a peak in the 18th century, with abolition following in the 19th century. The biological component (the transfer of plants, animals, and diseases) of the European age of expansion and conquest was so significant to the process that it has been dubbed a process of *ecological imperialism* (see Crosby 2004).

Across the Caribbean, broad differences emerged depending on the controlling colonial power.

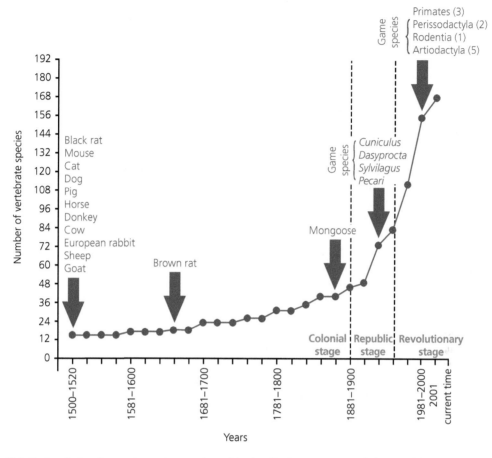

Figure 13.5 The introduction of non-native vertebrate species to Cuba since first European contact, which occurred during Columbus' first voyage in 1492. Since then, 168 species have become established on the island, sourced from around the world. Redrawn from figure 4 of Borroto-Páez and Mancina (2017).

Spanish islands for a long time served largely as the gateway to the mainland beyond, whereas the British focused heavily on sugar production within the islands of the Caribbean. Sugar was highly labour intensive, with around two million enslaved Africans being transported to British-controlled islands in the 18th century alone. One result is that the ethnic composition and genetic mix of African, European, and Native American ancestry is varied, with typically higher persistence of Native American ancestry in Spanish than in other islands.

Barbados was a classic sugar island. Following the arrival of the Spanish in 1518 the island was swiftly depopulated and in time abandoned, with the British arriving in 1627 and converting 80%

of the land area to sugarcane within two decades (Watts 1970). This practice was highly dependent on imported slave labour. A recent study of individuals of African descent from Barbados revealed levels of ancestry of around 89% West African, 10% European, and < 1% Native American (Benn-Torres et al. 2008). Elsewhere in the Hispanic Caribbean, contemporary populations with a > 50% indigenous component have been documented in less accessible regions. In Barbados, following initial clearance, a history of extreme boom-and-bust agriculture followed (typical of many islands, e.g. see Garfield 2015), driven by market vagaries and amplified by the reliance of small island economies on a limited range of commercial crops. The introduction of non-native plants and animals, and the cycles of clearance and abandonment of the colonial era, compounded the impacts of the first human colonists. As Braje et al. (2017, p. 7) state, '... Competition by European powers for commercial and strategic control accelerated environmental impacts to such a degree that many Caribbean islands are ecological shadows of their pre-Columbian states'.

The multiple and interacting pressures on island ecosystems and on island endemic species described above appear rather typical. Across the Caribbean as a whole, 15 surviving endemic land mammal species are known, of which at least 13 are classed as threatened by a combination of habitat loss, invasive species, and hunting (Turvey et al. 2017). Endemic birds and lizards have also been lost, with many surviving species now considered endangered (e.g. Szabo et al. 2012; Devenish-Nelson et al. 2019). Altogether, over 60 bird extinctions have been reported from the Caribbean (53 in prehistoric, and 9 in historic, times; Matthews et al. 2022). Since European contact, 55% of the endemic parrots of the Caribbean have become extinct, while a third of the endemic forest-dependent birds have been classed as endangered owing to habitat loss, over-exploitation, and interactions with non-native species (Devenish-Nelson et al. 2019). As forest-dependent species constitute 80% of the endemic birds and forest cover has been reduced to an average of < 30%, it is easy to see the scale of the problem. Specific examples from Puerto Rico and Montserrat were described in Section 12.2.

13.6 Macaronesia

The Atlantic archipelagos of the Azores, Madeira, Selvagens, Canaries, and Cabo Verde (Cape Verde) comprise the biogeographical region of Macaronesia (Chapter 4). The Canaries were settled first, followed well over a thousand years later by the other archipelagos—although the tiny Selvagens have never sustained a permanent population (Table 13.1; Fig. 13.3).

The Canaries

The Canaries were probably settled sometime between 2400 and 2000 cal. years BP by people of Libyan-Berber stock from North-West Africa, although a slightly earlier arrival date cannot be ruled out (above; de Nascimento et al. 2020). There is some intriguing evidence of links between them and Phoenician and Roman trading activity, indeed it is possible that the first settlers may have been in effect a vassal people (Camacho 2012). But, as the Mediterranean empires retracted, contact ceased and the people of the Canaries remained isolated for well over a thousand years, until explorers from Europe reached the islands in the first part of the 14th century. The adventurer Jean de Bethencourt initiated the process of conquest on behalf of the Crown of Castile, landing on Lanzarote in 1402. It was not until 1496 that the conquest was completed, with the fall of Tenerife.

The first inhabitants of the Canaries generally go by the name of the Guanche, although strictly speaking this name only applies to the inhabitants of Tenerife. They were pastoralists, who practised a form of transhumance, grazing animals in drier coastal areas in winter and upland areas in summer. They brought with them goats, sheep, pigs, and dogs and also exploited coastal food resources (fish and shellfish) and grew barley, wheat, beans, lentils, peas, and figs. As the case on islands around the world, increases in charcoal and sedimentation in lake sediments provides early evidence of forest clearance through burning (Fig. 13.4; de Nascimento et al. 2020; Nogué et al. 2021; Ravazzi et al. 2021), although the palaeoecological record suggests varying levels of ecological impact. Woody vegetation cover was reduced most in the low/mid-elevation thermophilous forest belt and in the upland (dry but cooler) pine forest belt. The mid-elevation cloud forest belt of broadleaved evergreen (laurel-dominated) woodland largely persisted and provided a rich source of useful plants (for medicines, timber, etc.) exploited by the Guanche. At least two species of extinct trees, an oak (*Quercus*) and hornbeam (*Carpinus*), are abundant in the pollen record from Tenerife, but within which forest association they grew is uncertain: both had been eliminated by the time the Castilians arrived (de Nascimento et al. 2009, 2020). Similarly, on Fuerteventura, charcoalized timber of three species of the laurel forest (*Laurus azorica*, *Persea indica*, and *Arbutus canariensis*) has been recovered from a cave site in layers dating from the 3rd to the 7th century. This formation no longer occurs on Fuerteventura, and these species are absent from the cave record from the 9th century onwards (Morales et al. 2009). Alongside forest clearance by cutting and burning across the archipelago, the grazing stock undoubtedly had a big influence, selectively foraging the most palatable native species. Indeed, the clearance through fire of forest cover in favour of open formations more suitable for grazing goats and sheep appears to have been a very early symptom of human arrival on these islands (e.g. Ravazzi et al. 2021).

The ecological transformation initiated by the Guanche resulted in the loss of several species of native vertebrates through a combination of habitat transformation, but especially hunting and the activities of commensal predators (which included the house mouse, pigs, dogs, and cats) they brought with them. Losses included two species of giant rat (*Canariomys tamarani* and *C. bravoi*), giant lizards (although three islands possess surviving single-island endemic species), a lava mouse, a flightless bunting (a rarity in being a flightless passerine), and two other passerines of limited flying ability. Other species (including monk seal) were greatly reduced in population size but probably became finally extinct following the Castilian conquest, during which two non-native rat species were introduced, alongside cats. Since initial human arrival, at least 12 bird species, including seven endemics, have been extinguished from the archipelago. Alongside these archipelagic losses, there have been other species (sometimes distinct subspecies) lost from individual islands and many greatly reduced in extent and population size.

There is no question that the Guanche initiated massive ecological change on the Canaries, including wide-scale reduction of tree cover, increased soil erosion, reductions in populations, and extinctions of endemic species, but the evidence also shows that their environmental impact was of slower pace than followed the conquest in the 15th century, or indeed over the past century. This is reinforced by Nogué et al.'s (2021) comparative palaeoecological analyses, which show that across oceanic islands generally, vegetation change following prehistoric colonization was much slower than the ravages that typified islands colonized for the first time

by Europeans in the historic period. That some species and resources (e.g. monk seal populations) were depleted is evident (Morales et al. 2009), but the inhabitants also developed an intimate knowledge of the plant and animal resources at their disposal, adapting their lifestyle to the environment and resources they encountered, and despite the over-exploitation of particular resources, there is no cause to conclude that the prehistoric societies were unsustainable at the point at which contact was re-established.

European colonization in the 15th century was initiated by small expeditionary forces by adventurers in loose association with the Crown, but conquest of the larger islands required more men and resources and became an enterprise of the Crown of Castile. From the first contacts in the 14th century and spanning over a century, a combination of trading, religious conversion, local alliances, and the classic package of 'guns, germs, and steel' (the title of a 1997 book by Jared Diamond) led to the overthrow of the indigenous rulers and their people. Enslavement, transportation, and land grabs followed. The Canaries are in some way a model of what was to follow (see Crosby 2004). Colonization and overthrow were brought about by small forces with superior weaponry (the Guanche had no metal, for example), the advantages provided by the horse, and by the weakening of the defenders by disease introduced from the continent, which devastated the defending forces on Tenerife before its final fall in 1496. Nonetheless, genetic analysis demonstrates a continuing legacy of the Guanche heritage, especially through the maternal line, and to a large extent we should assume that essential knowledge of the natural resources built up over centuries of occupation would have been maintained and passed on.

In 1492, Christopher Columbus journeyed to the Canaries from Palos de la Frontera, Spain, pausing for a while in the deep harbour of La Gomera, before setting sail to discover the New World. Thereafter the pivotal position of the archipelago in relation to the trade wind system, as well as its agricultural produce, ensured continued connectivity of the islands as an essential Atlantic hub. The following centuries were marked by massive land-use change, forest loss, introduction of non-native animals and especially plants, wide-scale ecosystem transformation, and continuing losses of native species. We will pick up on the theme of the recent human impact and conservation responses in the Canaries in the final two chapters.

The Azores, Madeira, and Cabo Verde

The Azores, Madeira, and Cabo Verde archipelagos were each colonized by the Portuguese during the 15th century, although the possibility of earlier human arrival and failed settlement cannot be ruled out (Section 13.2; Rando et al. 2014; Raposeiro et al. 2021; Elias et al. 2022). On each of these archipelagos, the Portuguese settlers went through the processes shown in Fig. 13.4a and 13.4b in short order. On their discovery, Madeira and the Azores were largely clothed in dense laurel forest (closely related to that found at mid-elevations in the Canaries; Section 4.8). The settlers swiftly embarked on changing that to release land for agriculture. The steep terrain of Madeira and lack of indigenous population presented huge challenges, met by the use of fire, terracing, and the importation of slaves, initially from the Canaries (Crosby 2004). It is said that in the early years of settlement (from 1420), Madeira burned for seven years as the lower parts were converted to agricultural use. Nonetheless, large areas of laurel forest persisted in the steepest terrain of the interior, especially in cloud-drenched areas unsuitable for agriculture. Extraordinary systems of water channels (levadas) were constructed to transport the precious water from these forests to feed agriculture in the lowlands.

A similar speed of conversion was evident in the Azores, with widespread clearance of lowland areas in the 100 years following colonization in 1432 (Connor et al. 2012). Forest loss continued into the 20th century and today around 95% of the original native forest has gone, replaced for the most part by cattle pasture and, to a lesser extent, plantations of exotic trees, notably *Cryptomeria*. Around 70% of the vascular plants and 58% of the arthropods of the Azores are now exotic, many of them invasive (Triantis et al. 2010). The combination of habitat loss and interactions with non-native species has generated numerous extinctions of vertebrates and invertebrates, and many species are today considered threatened (Terzopoulou et al. 2015).

Although the picture remains fragmentary, losses from the Azores and Madeira since human colonization are known to include at least the following extinct bird species: from the Azores, six endemic flightless rails (three of which are still undescribed), a Scops owl, a giant bullfinch, a species of thrush, probably two endemic quails, and an enigmatic pigeon, known locally as *pombo da terra*; and from Madeira, three endemic rails, three flightless quails, an endemic owl, and undescribed, likely endemic, species of thrush and finch (Alcover et al. 2015; Rando et al. 2020; Matthews et al. 2022).

The Cabo Verde islands, far to the south of the Canaries, are hot and arid, supporting far less woody cover prior to human settlement. Nonetheless, the same processes of vegetation burning, introduction of grazing animals, and introduction of non-native species followed human colonization in the mid-15th century, with the most fertile and productive land having since been converted to agriculture (Castilla-Beltrán et al. 2020). Palaeoecological data from sediment cores again attest to the unprecedented speed of vegetation change following human settlement. Recently, the first case of an extinction event recovered from fossil evidence was reported: a quail (Rando et al. 2020).

The degree of economic development has varied greatly across Macaronesia, with the Canaries and Madeira now oriented strongly towards international tourism but the Azores still heavily dependent on agriculture. And while the ecological transformation of each archipelago has been profound, it is in the Canaries and Madeira, with their high proportions of endemic species, that the impacts have been greatest in terms of (known) biodiversity loss. It perhaps deserves repetition that the scale of impact of the European colonization of Macaronesia from the 15th century onwards, rather dwarfs that of the prehistoric colonization of the Canaries.

13.7 Polynesia

Polynesia is a roughly triangular area of the Pacific Ocean encompassing over 1000 islands bounded by New Zealand in the south-west, Hawaii to the north, and Rapa Nui (Easter Island) to the east (Fig. 13.2). These islands (and a few others outside this triangle) were colonized within the last 1000 years by people of the Lapita culture (Austronesians), whose origins can be traced back to Taiwan. This is not to imply that all islands were colonized, and there were some, dubbed the Mystery Islands, which were colonized and then abandoned (Leppard 2016). We have picked the three corners of the Polynesian triangle for consideration, focusing especially on Rapa Nui because its disputed story is both fascinating and instructive. We return to the Mystery Islands at the end of the section.

Rapa Nui (Easter Island)

The incongruity between the small and apparently impoverished Rapa Nui population that early European travellers encountered and the magnificence of its numerous and massive stone statues has fed a deep fascination with the island.

(Puleston et al. 2017)

In the two precursors to the present volume, we presented the story of Rapa Nui (Easter Island; 163 km^2, 27° 9′ S, 109° 26′ W) as both an enigma and a warning of the folly of unsustainable resource exploitation. At the start of the 18th century, Rapa Nui was one of the most isolated pieces of inhabited land in the world. Palaeoecological investigation has shown that prior to human colonization, the island possessed extensive forest, dominated by a large palm (possibly endemic). Following Polynesian colonization, the forest declined such that it had essentially disappeared by the time of European contact and the palm, all indigenous land birds, and several endemic land snails were extinct (Bahn and Flenley 1992; Rull and Giralt 2018). Here was an island, cut off from the rest of the world, where the indigenous people had been capable of constructing the remarkable and now iconic giant moai (see Frontispiece to Part IV), which so astonished the first Europeans to reach the island. Yet, whose inhabitants had apparently carelessly squandered the natural resources of their island, causing an ecological and cultural collapse (Diamond 2005). This ecocide narrative has been shaped as a warning to us all, of the dangers of over-exploitation: as captured in the title of the book by Bahn and Flenley (1992), *Easter Island: Earth Island*. More recently, counter-arguments have gained prominence, claiming that rather than arising from intrinsic failures of the first settlers, the collapse was a direct consequence of European contact and the demise of the Rapanui

was closer to a case of genocide. What happened and what lessons should we take from it all?

The essence of the ecocide narrative is that the Rapanui allowed their own populations to grow too large, over-exploiting the resources of the island (including seabird populations). They reduced the forest by a combination of burning, felling, and indirectly by a collapse in regeneration owing to seed predation by the commensal Pacific rat (*Rattus exulans*), prompting soil erosion and a general loss of fertility of the island's ecosystems. Supporting the ecocide narrative, reference is sometimes made to the scorched or burnt appearance of the vegetation, the lack of trees, the poor state of the people, and the inadequacy of their canoes, as recorded by Jacob Roggeveen at the point of European discovery of the island on Easter Day (5 April) 1772. The inference being that the collapse was underway. When James Cook arrived in 1774, he reported a reduced population, abandonment of former cultivated land, and that many moai had been overthrown, signalling a population responding to their own failures by abandonment of the old gods.

The following (translated) excerpt from Roggeveen's log provides a rather different impression of the condition of the island at first contact:

Nor can the aforementioned land be termed sandy, because we found it not only not sandy but on the contrary exceedingly fruitful, producing bananas, potatoes, sugar-cane of remarkable thickness, and many other kinds of the fruits of the earth; although destitute of large trees and domestic animals, except poultry. This place, as far as its rich soil and good climate are concerned, is such that it might be made into an earthly Paradise, if it were properly worked and cultivated; which is now only done in so far as the Inhabitants are obliged to for the maintenance of life.

**(*Source:* www.easterisland.
travel/easter-island-facts-and-info/history/ship-
logs-and-journals/jacob-roggeveen-1722/,   visited
25 September 2019)**

Recent analyses from archaeological sites suggest that perhaps half the protein in the diets of the Rapanui was from marine resources (Jarman et al. 2016). Additionally, $\delta^{15}N$ values in human collagen were higher than expected from background levels in the environment, consistent with evidence that the people were consuming crops grown in carefully managed and manipulated soils.

What is not at issue is that the Rapanui transformed their environment, largely eliminating forests from the island and driving population reductions and extinctions of many native species. However, this does not mean that their resource use was on an unsustainable trajectory in relation to a fixed carrying capacity and that their culture was at the point of collapse when Roggeveen's ship hailed into view. Rather, it was contact with the outside world that was the decisive factor. The death of around a dozen Rapanui in a shooting incident on first landfall was in some respects a signal of what was to follow, although as elsewhere, it was the combination of guns and germs that was decisive. Diseases to which the Rapanui had no prior exposure, or immunity, took their toll, as did subsequent slave raids over the next century and a half. In 1862/63 alone, slave raiders forcibly removed over 1407 people from Rapa Nui, perhaps one third of the population at the time. They were transported to Peru, alongside other Pacific islanders, and sold at auction to work in guano mines and other menial jobs. Most soon died, but under international pressure, Peru repatriated 15 survivors to Rapa Nui in the following year (Ioannidis et al. 2020). The returning survivors brought with them another smallpox epidemic. By 1872, the population stood at only 111 people.

Just 16 years later the Chilean government annexed the island, and it became in effect a sheep ranch, with the remaining islanders largely confined to a single village as forced labour. The abandonment of the indigenous cultivation practices, such as lithic mulching alongside the introduction of grazing animals, led to accelerated soil erosion and a further phase of environmental transformation. Hence, it is that Hunt (2006) and other scholars have concluded that genocide, rather than ecocide, is the more appropriate term for the demise of the Rapanui. Hunt cites the French ethnographer Alfred Metraux, who visited the island in the 1930s, as describing what happened as 'one of the most hideous atrocities committed by white men in the South Seas'.

We will now proceed to put some detail to the account of the Polynesian settlement and

subsequent transformation of Rapa Nui (Fig. 13.2, 13.6): for a fuller updated account see Rull and Stevenson (2022a). Most Polynesian islands were incorporated into trade networks with other islands—even quite distant ones—but the island of Rapa Nui was very much out on a limb and there seems to be little evidence of ongoing connectivity following initial colonization (Puleston et al. 2017). Archaeological evidence points to a time frame for colonization of *c.* AD 1200–1253 (Wilmshurst et al. 2011; Ioannidis et al. 2021), although some authors do not entirely dismiss a slightly earlier arrival (Rull and Giralt 2018). Remains of the Pacific rat *Rattus exulans* are associated with the earliest archaeological evidence of human occupation, indicating that it arrived alongside the Rapanui.

There is intriguing evidence for an Amerindian connection with Easter Island in the presence of the South American sweet potato *Ipomoea batatas*, which was found across east Polynesia prior to European presence. One scenario is that Polynesians reached South America and then brought this valuable food plant back to the islands (Fig. 13.2). There is also some suggestion of pre-European admixture of Native Americans into the Rapanui people (Rull and Giralt 2018). Contact with Native Americans and their culture could also potentially explain both the presence of the bottle gourd *Lagenaria siceraria* and the construction of the moai and ahu (the platforms on which some moai stood).

Recent genome-wide analyses based on sampling across Polynesia claims conclusive support for an alternative scenario. There was a single contact of Polynesians with Native Americans from present-day Columbia, possibly a population who had themselves established on the island of Fatu Hiva in the South Marquesas. This occurred around AD 1150, thus allowing for incorporation of Native American genes, culture, and plants within the Polynesian expansion, prior to settlement of Rapa Nui (Ioannidis et al. 2020, 2021).

Acknowledging the increasing consensus on Polynesian arrival within the window AD 1200–1253, it appears that only around 550 years elapsed prior to European arrival in AD 1772. Rapanui culture initially developed around the construction and erection of the > 950 moai and the associated stone platforms, or ahus (see Part IV Frontispiece),

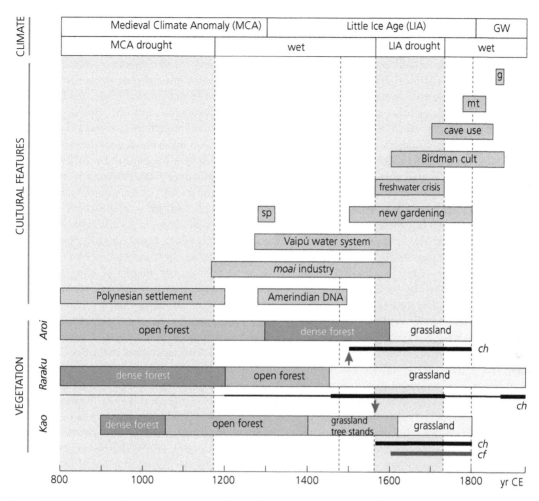

Figure 13.6 A chronology of vegetation, cultural, and climate change on Rapa Nui. The figure shows vegetation reconstructions from three palaeoecological sites, with indication of charcoal layers (ch) and spopres of coprophilous fungi (cf) and human population movements (red arrows). Also indicated are the timings and duration of key cultural phases, including the Vaipú water system (a system of hydraulic installations and terraces), the introduction of sweet potato (sp), the toppling of the moai (mt), the genocide following European contact (g), and the timings of significant climate episodes and the only partially corresponding periods of comparative dry and wet climates, with the final phase (GW) indicating the period of anthropogenic global warming.

Image provided by Valenti Rull (8/11/22, pers. comm.), based on figure 24.3 of Rull and Stevensen (2022b), which was in turn a modification from figure 1b of Rull et al. (2016).

of which there were around 300. But the moai cult was replaced by the so-called birdman cult, involving very different social, religious, and political structures around AD 1600 (Fig. 13.6). Rather than the previous dynastic hierarchy, within the birdman cult the highest authority was elected annually from among clan chiefs, based on whose champion managed to return with the first egg of sooty terns from an offshore islet. The ecocidal narrative has

this transformation occurring through overpopulation, leading to a collapse in the island's carrying capacity, followed by conflict and warfare within the island and the overthrow of the moai cult. But more recent research rejects this narrative, arguing that the human population never rose to really high levels, but built up to around 3000, remaining around this figure until European arrival (Hunt 2006; Rull and Giralt 2018; Rull 2019), although a slightly higher peak (around 6000), followed by a decline cannot be ruled out (Rull and Stevensen 2022a, b). Moreover, rather than a swift island-wide deforestation event, the latest palaeoecological research supports a spatially and temporally patchy forest decline, with some areas converted much earlier than others and some tree cover remaining even when the Europeans arrived (Rull 2019).

The notion that large numbers of trees must have been felled to provide rollers for moving moai around the island is another part of the eco-cide narrative that has been challenged, as it has been shown that statutes could have been 'walked' across the island by teams of people working with ropes. Both the diary evidence from Roggeveen and the dietary analyses mentioned above support the notion that Rapanui society had proved to be resilient to the environmental changes they had experienced (both those they initiated and regional climate change), developing local-scale horticultural practices, resource extraction methods, and cultural practices that provided for a largely stable population and a sustainable way of life. Cultivation techniques involved small garden features, *manavai*, in which rock walls were used to protect crops and soils, while lithic mulching was used to reduce water loss, and they also practised limited use of terracing. Manavai were widely distributed in the lowlands, across up to 20% of the island's surface, and provided a system that continued until the time of European arrival (Rull 2019). Nonetheless, it has been hypothesized that the shift from the moai cult to the birdman cult may have reflected a combination of deforestation and drying (reflecting regional climate changes) of the area around Rano Raraku, the site from which the moai were quarried, with a shift of population and power to the other end of the island, around Rano Kau, where fresh water and forests remained available (Rull and Giralt 2018).

Rapa Nui may never have had a large indigenous flora, as only around three-dozen native species are known, alongside 14 introduced by the Polynesians. But the loss of trees (including the now extinct palm) and other plant species is matched by the loss of bird species (Steadman 2006). There were probably six native land birds, including a heron, two rails, two parrots, and an owl. None survive. Whereas there were once 25 nesting seabirds, only around 12 breed on the island today (Flores et al. 2014). The losses are thought to include an endemic breeding seabird.

The economy of the island today relies on tourism, aided by the opening of an airport near the settlement of Hanga Roa. The island's mixed population is predominantly of Polynesian descent and at the turn of the century stood at around 3300, much in line with the estimated pre-European population size we give above. But by 2017 numbers had increased to over 7000, reflecting recent immigration and the growth of tourism, which focuses around the unique history of the island and the remarkable moai. The people of Rapa Nui have made increasing efforts to reclaim their heritage and to wrest greater say over their island from Chile.

How to sum up the enigma of Rapa Nui? The increasingly large literature remains hard to resolve into a confident narrative. Yet we can say that human colonization by the Polynesians utterly transformed the ecology of the island, driving many native and endemic species to extinction. The land birds of the island today, such as the house sparrow (*Passer domesticus*), common diuca finch (*Diuca diuca*), and rock pigeon (*Columba livia*) are introduced species and reflect a pattern of biotic homogenization repeated across many of the world's islands (e.g. Li et al. 2020). The unique culture of Rapanui people developed *in situ* and in isolation over only a few hundred years, had responded to the changes of environment for which they were largely—but not wholly—responsible, showing resilience in their capacity to adapt. All this changed in 1772, when the outside world came calling.

Hawaii

Hawaii has featured prominently in this book, reflecting its significance within island biogeography. While we stop short of a full account of human impact on Hawaii, a brief account of the scale of extinctions is surely essential. We concentrate, for illustration, on birds. Fossil data have shown that a wave of bird-species extinctions followed Polynesian settlement of the archipelago, carrying away some 56 species of the native non-migratory land—and freshwater—bird species, with losses biased towards flightless, larger-bodied, and ground-nesting species, many of which were hunted (Boyer 2008). Testifying to the broad impacts of the Polynesian settlers and the destruction of much of the dry forest habitat, many smaller granivores and frugivores also became extinct as part of this initial pre-European extinction wave, which eliminated all but 55 native species. Flightlessness, a common trait in certain insular bird lineages (especially in rails), was a fatal weakness in the face of newly introduced terrestrial predators. A further 27 endemic species are known to have become extinct since European arrival (Matthews et al. 2022), not including the Hawaiian crow, currently classified as Extinct in the Wild, despite various reintroduction efforts. Reflecting the continued transformation of Hawaiian nature, especially the introduction of additional predators, mosquito-borne diseases and non-native birds (acting as alternative hosts as much as competitors), there has been a disproportionate loss of granivores and frugivores in the historic period (Boyer 2008). Thus, over 80 endemic bird species are known to have gone extinct on the Hawaiian Islands since human colonization, including the giant Hawaii goose, over 30 Hawaiian finches (Section 9.4), an extinct genus of flightless ibis, an extinct genus of stilt owls, an extinct genus of large, flightless ducks, and an entire family of honeyeaters.

Today the islands are infested with somewhere between 50 and 100 non-native species (e.g. Sayol et al. 2021). The differential effects of extinction on vertebrate feeding guilds (and dispersers and pollinators) in prehistoric and historic times have had huge implications for plant-community dynamics, with the many non-native species failing to provide like-for-like replacements for the ecological roles of the extinct endemic avifauna (Boyer and Jetz 2014; Heinen et al. 2018; Sayol et al. 2021).

New Zealand

The Polynesians discovered New Zealand around AD 1190. They brought with them six species of plants and two species of mammals. Polynesian settlement led to the destruction of half of the lowland montane forests (mostly via burning), widespread soil erosion, and the loss or reduction of much of the vertebrate fauna. During the phase of their expansion, New Zealand lost its sea lions and sea elephants, and numerous bird species, including the nine species of moa (Lomolino et al. 2021), which, in the absence of terrestrial mammals, had evolved to occupy the browser/grazer role normally occupied by animals such as kudu, bushbuck, or deer. By about AD 1400, the major New Zealand grazing systems had ceased to exist, their place being taken by unbrowsed systems: the removal of the moas thus had significant effects on the structure and species composition of the vegetation, although the role of burning in landscape clearance was also a key driver of ecological change (McWerthy et al. 2010). As an indication of the scale of changes, since humans first set foot in New Zealand, around 40 endemic bird species have become extinct (Duncan et al. 2002; Garcia-R and Di Marco 2020; Lomolino et al. 2021).

An interesting story is attached to the loss of one of the smaller New Zealand birds, the Stephens Island wren, named *Traversia lyalli* after the lighthouse keeper (Lyall) who discovered it. Stephens Island is a tiny island in Cook Strait, off South Island. The wren was unknown to science until in 1894 the lighthouse keeper's cat brought in a few specimens it had killed. The exotic predator met the island endemic: end of story. At least, this was the conclusion reached at the time, prompting a correspondent to the *Canterbury Press* to suggest that in future the Marine Department should see to it that lighthouse keepers sent to such postings should be prohibited from taking any cats with them, 'even if mouse-traps have to be furnished at the cost of the State'. Although cats have been claimed to have by far the worst record in exterminating New Zealand birds and indeed to have been largely responsible for 14% of recent island extinctions of birds, mammals, and reptiles (Medina et al. 2011), we should not place all the blame on the lighthouse keeper and his cat. Before the arrival of the Polynesians, the wren was a widespread mainland resident. Its loss from mainland New Zealand was probably due to a combination of habitat changes and the introduction of rats by the Polynesians; the loss from Stephens Island was but the final step to extinction.

The Mystery Islands

The Polynesians did not colonize all islands within the Polynesian triangle and some that they did colonize were subsequently abandoned. These so-called 'Mystery Islands' comprise two types: (i) young, high islands in the subtropics (Necker, Nihoa in the Hawaiian chain; Pitcairn and Henderson in far eastern Polynesia; Norfolk and Raoul in the Kermadecs, north of New Zealand); and (ii) low-lying coralline islands such as Kiritimati, Fanning, and Palmerston in the central Pacific (Leppard 2016).

On Henderson Island, settlement was accompanied by the introduction of cultigens and tree species, and by burning for crop cultivation. Marine molluscs, birds, and turtles were heavily predated and five endemic land bird species, several ground-nesting seabirds, and at least 6 out of 22 land snail species became extinct. Sustained occupation of what was a fairly marginal environment was probably only possible through interaction with Pitcairn Island and, 400 km to the west, with the island of Mangareva, where settlements also failed (Benton and Spencer 1995; Weisler 1995; Wragg 1995).

It seems likely that the precise reasons for abandonment may vary between the different types of island, but that like Henderson, they are all typically marginal in some respect, either climatically, and/or because they lack reliable marine resources. Alongside susceptibility to fluctuations in climate, Leppard (2016) also suggests that many of them were also marginal in terms of trading networks. He also speculates that there may have been enough contacts to permit the introduction of human pathogens, as subsequently happened to such devastating effect with the arrival of the Europeans.

There are mystery islands elsewhere, in the Caribbean, Mediterranean, and in Macaronesia—islands where there is evidence of human arrival and impact but no extant human population at the point historic European expansion was initiated *c.* AD 1500. But the sustainability of island populations in pre-history is not linked in a simple way to the extent of their impact on biodiversity (Leppard 2016). Similarly, island abandonment has occurred in the historical period. For example, some Scottish isles have been abandoned in the last few years and decades, not because of insular degradation but simply reflecting the many challenges of making a living and sustaining a viable society in economically marginal environments.

13.8 Micronesia

Micronesia is one of the three main island regions of Oceania, alongside Polynesia (above) and Melanesia (which comprises islands to the north and east of New Guinea, including the Bismarcks, Solomon Islands, New Caledonia, Vanuatu, and Fiji), with which it has shared cultural history. Micronesia is the area west of Polynesia and north of Melanesia and comprises thousands of small islands, notably including the Caroline Islands, Mariana Islands, Marshall Islands, Nauru, and Wake Island.

The Marianas, Guam, and the brown tree snake

The Marianas archipelago may have been the first colonized in Remote Oceania, *c.* 3.2–3.0 ka (Petchey et al. 2017). Intriguingly, and contrary to the general picture for Polynesia, humans colonized the Marianas without introducing the Pacific rat (*Rattus exulans*), which arrived over two thousand years later, around 1.0–0.8 ka. Moreover, chickens, dogs, and pigs were also absent throughout pre-history. These absences and especially the late arrival of rats, explain the relatively good persistence of native land birds, including Guam's flightless rail, until recent times (Fig. 13.7).

Guam is the largest (*c.* 550 km^2) and most populated (164,000 inhabitants) island of the Marianas and deserves our attention as a notorious and recent example of the impact that the accidental introduction of a predator can have on an oceanic island. Analyses of bones from sites across the island reveal that there were at least 24 species of native land birds prior to human arrival, with 18 persisting to the historic period, of which just five were described as surviving by Pregill and Steadman (2009). A more recent assessment of Guam's native forest birds states that only Mariana grey swiftlets and Micronesian starling survive in the wild. Guam kingfishers are held within captive breeding programmes and the Mariana crow has been extirpated from Guam (it survives in the wild only on the island of Rota)

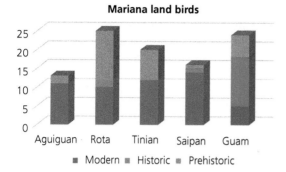

Mariana land birds

■ Modern ■ Historic ■ Prehistoric

Figure 13.7 Species richness of resident land birds of the Mariana Islands in modern times, in the historic period (based on historic records or studies of bones from the past two centuries), and the prehistoric period (based on bird bones). Anthropogenic extinctions have significantly altered bird richness across the archipelago, but the pre-human arrival richness figures undoubtedly still reflect knowledge deficiencies as well as variation in such properties as island area and elevation. The area, elevation, and number of bones available per island are as follows: Aguiguan (7 km^2, 166 m, 944 bones), Rota (85 km^2, 491 m, 475), Tinian (102 km^2, 170 m, 647), Saipan (122 km^2, 466 m, 12), and Guam (544 km^2, 406 m, 448). Based on data in Pregill and Steadman (2009).

(Engeman et al. 2018). The Guam rail has been successful re-introduced to Cocos Island (1 mile off the southern tip of Guam), subsequent to the eradication of rats from the island, leading to it being only the second bird to be downgraded by the IUCN from Extinct in the Wild to Critically Endangered.

A native of the Australasian region, the brown tree snake (*Boiga irregularis*) was introduced accidentally to Guam, possibly as a stowaway in military cargo, soon after the end of the Second World War. The snake was first sighted in the 1950s, and by 1968 it had spread all over the island. Its nocturnal activity, its ability to live in close proximity to humans, and the wide range of prey it consumes (lizards, rats, fruit bats, bird), enabled a demographic explosion, with densities of 12,000 to 15,000 snakes per square mile being reported (Patrick 2001). Snakes can be abundant while remaining relatively cryptic, and the devastating role of this non-native in the decline of native birds only emerged through detailed investigation.

Before the arrival of the brown tree snake, the only snake on the island was a small, blind snake that lives in the soil and feeds on ants and termites. The vertebrates of Guam, having evolved

for many millennia without such native predators, lacked the usual defensive behavioural mechanisms of continental fauna, and this 'naivety' is thought to have been an important contributory factor to their demise (Rodda et al. 1999). Brown tree snakes are considered primarily responsible for the extirpation from Guam of almost all its breeding population of seabirds, as well as the majority of native forest birds, including: Guam rail, white-browed rail, white-throated ground-dove, Mariana fruit-dove, Guam kingfisher, nightingale reed-warbler, Guam flycatcher, rufous fantail, cardinal honeyeater, and bridled white-eye (Rodda et al. 1999). Two out of the three species of native mammals, the Pacific sheath-tailed bat and the little Marianas fruit bat, as well as 5 of the 10 native species of lizards, have also been extirpated. The Mariana fruit bat (*Pteropus mariannus*) has also been greatly reduced in numbers by a combination of hunting and brown tree snake predation (Engeman et al. 2018). In addition to the snake, habitat loss, rats, and feral cats (again) have also had roles in historical species losses (Rodda et al. 1999).

The ecological (and socio-economic) consequences go beyond this immediate biodiversity loss. A trophic cascade can be considered to be under way, due to the loss of birds and bats that acted as pollinators and dispersers of trees, with impacts detected on trees and even spiders (Wald et al. 2019). A comparative analysis of seed rain along forest transects on Guam and the nearby island of Saipan—which still has a relatively intact vertebrate frugivore community—revealed a number of striking findings. For example, in degraded forest in Saipan, roughly 1.66 seeds per 26 days landed in a square metre of forest, the equivalent number in Guam being zero (Caves et al. 2013). Thus, forest regeneration in Guam is likely to be severely retarded due to the near complete loss of frugivorous vertebrates. Concerns have also been raised that there may be consequent outbreaks of insect populations that are no longer subject to predation by insectivorous birds and lizards (Rodda et al. 1999). Around the end of the 20th century, large populations of mosquitos contributed to outbreaks of dengue fever among the people of Guam, possibly linked to these ecological cascades (Pitts and Leasman-Tanner 2001). Economic damage due to the brown tree snake was reported to have included electrical outages, at one stage occurring every third day, including island-wide blackouts, costed at $1–4 million a year. It is also a recognized hazard to humans, especially children. The central role of Guam in the Pacific transportation network highlights the risk of the further extension of the snake's range to other oceanic islands in the Mariana archipelago and elsewhere. Mitigation, control, and education efforts are essential to managing these dangers (Kimball et al. 2016; Engeman et al. 2018; Wald et al. 2019).

Across the Marianas, over 20 bird species (including several undescribed species) endemic to the archipelago are known to have gone extinct (Steadman 2006; Matthews et al. 2022), with many more likely to still be discovered. The timing and degree of decline of bird species on the islands of the group has been variable. This variation reflects still imperfect understanding of the pre-human baseline, as well as differences in habitat diversity, island area, and human interaction with the natural environment, of which the snake is one prime example (Pregill and Steadman 2009). The picture emerging is nonetheless again one of waves of anthropogenic impact, with particular moments marking the irreversible crossing of new thresholds: the first humans, the arrival of rats, first European contact in 1521, and the post-WWII arrival of the brown tree snake.

13.9 Summary

Human colonization of islands commenced during the Late Pleistocene, with many remote islands occupied only within the last two millennia, blurring the distinction between prehistorical and historical periods. We show that significant differences are apparent in the timing of colonization and the extent of impact among and between archipelagos across the world, even between those in the same broad sector of each ocean. In both historical and prehistoric eras, islands have often been visited (with the visitors sometimes leaving a calling card of introduced rodents) prior to being colonized and in cases we are unable to distinguish between an early visitation and a failed colonization attempt.

Prehistoric colonization in most cases afforded a degree of continuing connectivity, trading, and exchange among islands or with nearby mainland, and frequently involved broadly similar toolkits or packages of plants, animals, and cultural practices. In some cases, however, island societies became cut off, sometimes for hundreds of years. Prehistoric societies, who were typically agropastoralists (as well as utilizing other food acquisition strategies), transformed their island environments, initiating sustained changes in vegetation (with fire a key tool) and introducing non-native commensals such as rats, dogs, and pigs, and sometimes goats or chickens also. In general, as exemplified here for the Caribbean, Polynesia, Micronesia, and the Canary Islands, large numbers of species extinctions followed. However, these societies typically displayed adaptive management strategies such that their occupancy was sustainable at the point of European contact. There are islands where prehistoric colonization failed, not because resource extraction methods and impacts on biodiversity were substantially worse than elsewhere, but mostly because the islands were environmentally marginal and often poorly connected within trading networks.

The age of European colonial expansion that began towards the end of the 15th century brought a new and continuing level of upheaval to island ecosystems, typically proving devastating to the indigenous inhabitants (through military means, disease, and enslavement). The ongoing introduction of non-native plants, animals, and diseases was a feature both of previously inhabited and previously uninhabited islands. However, the pace of vegetation and ecological transformation of islands first colonized in the historic period has typically been faster than that of the prehistoric period. We present a series of case-study systems from the Caribbean, Macaronesia, Polynesia, and Micronesia, which reveal subtle differences in the degree of transformation to native biodiversity and resulting biotic homogenization brought about by the different cultural waves that have swept over the world's islands.

CHAPTER 14

Anthropogenic extinction on islands: a synthesis

14.1 The scale of the losses

Human transformations of islands across the globe provide some of the most dramatic and convincing illustrations of prehistoric and historic anthropogenic species extinctions (Szabo et al. 2012; Duncan et al. 2013; Sayol et al. 2021; Matthews et al. 2022; Soares et al. 2022); indeed, Russell and Kueffer (2019) contend that throughout the history of human settlement of the globe, most recorded extinctions have happened on islands. The notion that we live in a new period in earth system processes in which humans influence the environment on a global scale is embodied in the much-debated concept of the Anthropocene epoch, for which a range of start dates from within the 20th century to thousands of years ago have been posited. For insular systems, rather than a single global Anthropocene we have innumerable local ones of radically different timings (Tables 13.1, 13.2) and they have consisted not of a single moment of humans gaining dominance, but of successive waves of transformation (Fig. 13.4), a form of anthropogenic ratchet effect.

Wood et al. (2017) distinguish three phases, although some islands show evidence of only one or two of these phases. The first, 'visitation', refers either to visits during exploratory voyages or to failed colonization events. Such visitations have led to the accidental introduction of non-native rodents to some islands, including Madeira (Rando et al. 2014) and the Mascarenes, in the latter case likely causing the extinctions of a skink and at least four bird species (Wood et al. 2017). Other animals left on islands prior to settlement have included goats and sheep, introduced as potential foods sources for future visits. The second and third phases recognized by Wood et al. (2017) are the prehistoric (pre-European) and historic (mostly European) phases. Notwithstanding the commonalities highlighted earlier (Fig. 13.4), the relative degree of transformation brought about by each cultural wave to crash over remote islands has varied very considerably from archipelago to archipelago and even island to island within an archipelago.

In moving from specific case studies to a synthesis, we begin with a simple question, which is, however, difficult to answer (Box 14.1). How many species have become extinct as a direct or indirect result of human settlement of islands? The previous chapter attests to the level of losses on particular oceanic islands and suggests as yet uncatalogued further losses from across the world's islands, for many of which we have very limited data on pre-human baselines. Not only have many islands lost numerous endemic species prior to scientific inventory, but also many undoubtedly hold extant species currently scientifically unknown that are waiting to be discovered and described, a knowledge deficit termed the Linnean shortfall (e.g. Ladle and Whittaker 2011).

In illustration, the cataloguing of the Canarian biota is generally considered to have begun c. 1724, when the French priest Louis Feuillée began to describe and classify endemic plant species—although he was far from the first collector. Despite three centuries of scientific attention, species new to science continue to be discovered, in recent decades at a rate approximating one species every six days

Island Biogeography. Robert J. Whittaker, José María Fernández-Palacios, and Thomas J. Matthews, Oxford University Press.
© Robert J. Whittaker, José María Fernández-Palacios, and Thomas J. Matthews (2023). DOI: 10.1093/oso/9780198868569.003.0014

(one *animal* species every 10 days), including two large lizards (*Gallotia intermedia* and *G. gomerana*) and two trees (*Morella rivas-martinezii* and *Dracaena tamaranae*) (Izquierdo et al. 2004; Machado 2022). Many new species have escaped detection until recently because they are small and comparatively cryptic, or they persist only in small populations, others are known local forms upgraded by taxonomic work to species status (e.g. Lifjeld et al. 2016; Sangster et al. 2022; Section 8.2). Hence, no sooner are they discovered than many are categorized as in danger of extinction.

A further problem is the attribution of the extinction event. Species extinctions from islands are part of the natural biogeographical dynamics of islands, although extinctions of endemic species in the absence of humans occur infrequently. For some island systems (especially land-bridge islands), certain extinctions have been attributed to pronounced environmental change, such as associated with the Late Pleistocene climatic and sea-level fluctuations (Wood et al. 2017). Some islands were colonized early enough by humans for both sets of processes to coincide. However, the majority of remote islands were colonized at a late stage of the Holocene (Chapter 13), during a period of relatively stable climate and limited sea-level variation: in such cases attribution of extinctions to humans or to particular natural causes (e.g. the eruption of Krakatau) is generally unproblematic.

The group that provides the most insight into the scale of anthropogenic extinction on islands is undoubtedly the birds (Fig. 14.1). Focusing on all known extinctions (including discovered but not yet fully described species), recent work has shown that almost 600 bird species have gone extinct globally in the last 125,000 years, with 477 of these (i.e. ~80%) being island endemics (Sayol et al. 2020; Matthews et al. 2022). Using a baseline of AD 1500 is to begin in the middle of the movie, but is often done due to the increased quality of available data on extinct species. Of the 477 known island endemic bird extinctions, 149 are recognized by the IUCN as having occurred since this AD 1500 baseline, with evidence that an additional 66 island endemic subspecies have also gone extinct in this period (Matthews et al. 2022). These historical losses were initially concentrated on islands in the Mascarenes, Caribbean, Hawaii, New Zealand, and French Polynesia, especially connected with the arrival of new cultures and technologies on islands (Fig. 14.1; Szabo et al. 2012; Matthews et al. 2022).

These *known* island extinction events include many extraordinary birds, including the giant flightless moas of New Zealand, Haast's eagle (the largest known eagle and predator of the moa), the elephant birds of Madagascar (one of which, *Vorombe titan*, is believed to have been the largest and heaviest bird to have existed), the majority of the Hawaiian honeycreeper finches (Section 9.4),

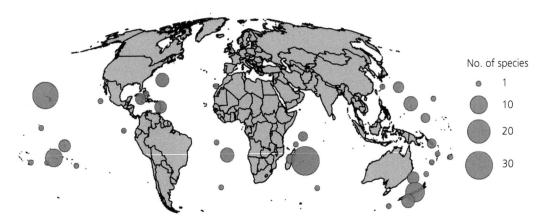

Figure 14.1 Hotspot map of extinct island endemic bird species, according to the IUCN Red List. Only species classified as Extinct by the IUCN are included (i.e. species that went extinct since AD 1500). Antarctica was cropped out of the map to save space (no extinct species were located on Antarctic islands). A Mollweide projection was used.
Adapted from Figure 4 in Matthews et al. (2022).

as well as petrels, prions, pelicans, ibises, herons, swans, geese, ducks, hawks, eagles, megapodes, kagus, aptornithids, sandpipers, gulls, pigeons, doves, parrots, owls, owlet-nightjars, and many types of passerines (Hume 2017). However, it is almost a certainty that many other island birds have gone extinct that have yet to have been discovered, with estimates of perhaps 2000 species lost from the world's islands before European contact alone (Steadman 1997). Virtually every Pacific island that has been examined palaeontologically has been found to have had one or more rails of such reduced flight capabilities that they must have been endemic species (Steadman 2006). Extinct rails have now been found also from islands in two Atlantic (Macaronesian) archipelagos (Alcover et al. 2015; Rando et al. 2020), providing further support to the suggestion that overall, several hundred island endemic rail species were lost from the world's islands following human arrival and before AD 1500. Going beyond the rails, Duncan et al. (2013) report on a form of mark–recapture analysis of non-passerine land birds for 41 Pacific islands from which fossil bones have been collected and which were colonized by people within the last 3500 years. They comprise birds typically of larger body size commonly targeted by human hunters. They found that two thirds of the populations went extinct following human colonization and prior to European contact, leading to estimates that initial colonization of remote Pacific islands may have led to the extinction of around 1000 species of non-passerine land birds alone, far in excess of the known island bird extinctions.

Before moving on from the birds, it is worth highlighting that around 30% of persisting island endemic bird species are currently threatened with extinction (Fig. 14.2a, b), a much higher proportion than for continental species (Fig. 14.2b), with 39 of these having global population sizes of less than 50 individuals (Matthews et al. 2022). Several species (e.g. the Makira Moorhen, endemic to the Solomon Islands and last recorded in the 1950s) have not been definitively seen for decades. Taken all together, one particularly sobering statistic is that almost 50% of known island endemic bird species have become extinct or threatened with extinction since human colonization of the islands of the world.

A similar picture holds for other taxa (Fig. 14.2c), with island extinctions representing over 60% of documented extinctions across plants, invertebrates, and birds in the historical era (Whittaker et al. 2017). Among vertebrates, the losses are skewed to birds, in large part reflecting the paucity of mammals (apart from bats) on oceanic islands. Among invertebrates, land snails have fared particularly badly, thanks to a combination of habitat loss and the introduction of non-native snails that have preyed on endemic snails, but the data are poor for most invertebrate taxa on most islands, and detailed analyses of Hawaiian snail extinctions (which have been accelerating over time) suggest that invertebrate extinctions from islands remain hugely underestimated (Régnier et al. 2015). Plant extinctions are comparatively rare considering the massive extent of vegetation transformation on so many islands, but many species persist in greatly reduced and endangered populations (e.g. Marrero et al. 2003, 2019). While the trend in extinctions shifted somewhat to continents in the second half of the 20th century, islands continue to be the sites of most extinction events and the rate has increased in the last century (Fig. 14.2d).

14.2 Ecological cascades

When one species goes extinct, any evolutionary interactions involving that species must also cease. If this involves very tight mutualisms (e.g. concerning pollination services), then the surviving co-evolutionary partner, or 'widow' species (*sensu* Olesen and Valido 2004) may in time follow the same fate, unless it is able to find similar services from another locally available species. In illustration, the loss of avian pollinators has been suggested to have contributed to the extinction of more than 30 Hawaiian plant species (Boyer and Jetz 2014). Such losses may then multiply, as other mutualisms are interrupted, in a fashion analogous to the mesopredator release mechanism (Section 12.4). This process is termed a **trophic cascade**, which implies an effect propagating across trophic boundaries as an ecosystem unravels. There are some spectacular examples of ecological cascades driven by 'ecosystem transformer' species (e.g. Rodda et al. 1999; O'Dowd et al. 2003; Terborgh 2010; Laurance et al.

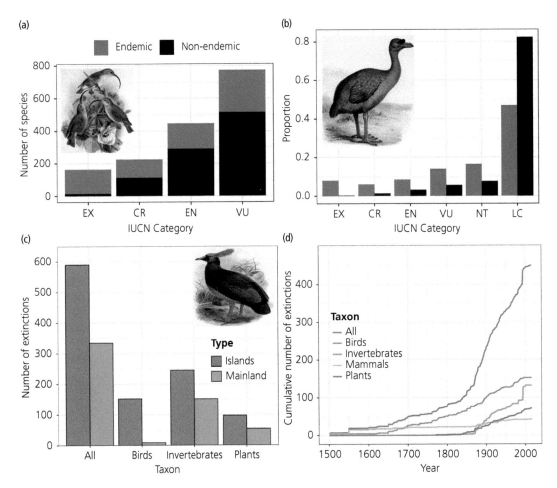

Figure 14.2 Threatened and extinct island diversity. (a) The number of bird species in four different IUCN Red List categories (EX = Extinct, CR = Critically Endangered, EN = Endangered, and VU = Vulnerable), split by island endemism status. (b) The proportion of bird species in each category, plus those classified as Near Threatened (NT) and Least Concern (LC). The proportion of island endemic bird species in each category was calculated relative to the total number of island endemics (1856), while the non-island-endemics proportions were relative to the total number of non-island-endemics (9256); both totals included extinct species. In both panels (a) and (b), blue columns = island endemics and black columns = all other species (i.e. those of mainland, or mainland plus island distribution). (c) Island species extinctions in the historical period (since AD 1500) by taxon in comparison to continental extinctions ('all' being the sum of birds, invertebrates and plants). (d) Island extinctions over time, for those species for which an estimate of date of extinction is available.

Inset images are all of extinct island bird species: in (a) a Maui Nui 'akialoa (drawn by Lionel Walter Rothschild and John Gerrard Keulemans); in (b) a Rodrigues solitaire (drawn by Frederick William Frohawk); and in (c) a Choiseul crested pigeon (drawn by John Gerrard Keulemans). (a) and (b) from Matthews et al. (2022), and (c) and (d) are modified from Whittaker et al. (2017).

2017), but it is by no means clear how often and how far such cascading effects operate. Many ecosystems in practice have a degree of ecological redundancy (Olesen and Jordano 2002), such that a cascading 'ecological meltdown' may be relatively uncommon. However, given the scale of species losses on many oceanic islands, knock-on effects are

inevitable. For example, there is evidence that the near wipe-out of land birds on Guam has seen economic impacts on crop production owing to loss of bird predation (notwithstanding increased densities of spiders) and declines in both pollination and seed-dispersal services (Caves et al. 2013; Boyer and Jetz 2014).

Box 14.1 The varied meanings of extinction

It is often said that extinction is for ever. But the term extinction is used in a number of ways and levels, such that it may refer to the local and temporary loss of a population of a species from a habitat patch or island, or to the death of the final member of a unique species from the globe. The following table (inspired by, but greatly modified from Ladle and Jepson, 2008) refers to forms of extinction some of which may not, in fact, be forever.

Type	Definition	Insular relevance
Linnean extinction (Centinelan)	Extinctions of undiscovered species inferred to have once existed through extrapolation methods.	The high prevalence of bones of extinct flightless rail species in well-studied islands leads to the inference that hundreds more once occurred. See text.
Wallacean extinction	Species that have not been documented for many years but in which extinction is uncertain because populations might survive in areas that have not been surveyed within the potential distributional range.	An example is provided by the Cebú flowerpecker: once thought extinct it was eventually rediscovered. See text.
Phoenix extinction	Extinct in wild but genetic material available in the form of stored material or closely related conspecific or congeneric variety/breed/hybrid, allowing the possibility of a future reintroduction of the same or a functionally equivalent form.	Something close to this means of 'avoiding' extinction has been raised as a possibility for the Floreana giant tortoise within the Galápagos following the death of the last known survivor (lonesome George): (Poulakakis *et al.* 2008)
Ecological extinction	Extinct in the wild but with extant captive bred population, or present in the wild at such low densities that it is considered functionally extinct.	Examples include around 12 species[*] of *Partula* land snails from Tahiti, maintained in a captive breeding programme, but extinct in the wild because of predation by introduced snails.
Local extinction	Extinct in the wild within a clearly defined geographic area but with extant free-living populations outside that area. Also known as extirpation.	This form of local 'extirpation' is invoked as a fundamental part of the MacArthur/Wilson Equilibrium Theory
Historical extinction	Documented extinction of a species 'known to science' (i.e. voucher specimens available), whereby thorough survey over time has established beyond reasonable doubt that the species is extinct in the wild and no captive-bred population or genetic material available.	Sadly, there are numerous clear-cut island cases, including the dodo, numerous Hawaiian honeycreepers, endemic *Partula* land snails from the Windward Islands (Haponski *et al.* 2019), etc.
Prehistoric extinction	Extinction of species in the prehistoric era, following human colonization of an island, where the species is known only from e.g. analysis of fossil bones and there is no authenticated record of the species from the historic period.	Many cases mentioned in the text, including bird species from the Canaries, Hawaii, etc. Care is needed as many island endemics have evolved and become extinct before humans ever arrived on islands.
Phyletic extinction (pseudo-extinction)	When a species evolves within an area to become distinct from its ancestor, whether diversifying into one 'new' species or more than one and where the original form no longer exists there, then the original species (in name at least) has become extinct in that area. This is a fairly abstract concept.	It is relatively uncommon to find representatives of the ancestral species, persisting on islands alongside the descendant island endemic species. All "anagenesis" speciation involves pseudo-extinction and most cases of cladogenesis involve pseudo-extinction of the original colonist.

---

[*] The precise number of *Partula* extinct in the wild but maintained in captivity is subject to taxonomic uncertainty (Haponski *et al.* 2019)

Comparative analysis of the loss of frugivorous vertebrates (birds, mammals, and reptiles) following human colonization catalogued extinctions from 45% of 74 islands (from 20 archipelagos) analysed and showed that one third of native species had become extinct (Heinen et al. 2018). The patterns of recorded loss may in part reflect sampling differences, but also show geographical differences, with much of the variation being linked to the archipelago: several Pacific and Indian Ocean islands lost more than half their frugivore species (Fig. 14.3). Across islands, larger bodied species fared worst, resulting in a 37% decrease in body size of persisting frugivore species relative to the original fauna. As these larger species often have special importance in dispersing plants of larger seed or fruit sizes, the data imply a general targeted impact of the loss of these key dispersal mutualists on the plant community dynamics and intra- and inter-island movement of plant species. The dodo may have been a very peculiar island species, but in being an extinct large-bodied island frugivore it is far from being alone. Within the sample of islands examined, smaller, remote islands of high elevation showed greater proportional losses. The loss of large-bodied species appears general across the world's islands, as exemplified by the extinction of New Zealand's moas (giant flightless ratites; Lomolino et al. 2021) and the elephant birds of Madagascar, which were accompanied into oblivion by two giant tortoises, pygmy hippo, and at least 14 species of lemur, most of which were larger than any surviving species (Groombridge and Jenkins 2002).

In their analysis of 41 Pacific islands, Boyer and Jetz (2014) report a significant impact of species extinctions on the functional trait profiles of the persisting avifauna. This reflected a pronounced overall loss of ground-level foragers, granivores, and herbivores. Substantial reductions of functional diversity appear to be commonplace for remote-island avifaunas, with some islands in Hawaii and East Polynesia having lost as much as 80% of their pre-contact functional diversity (Fig. 14.4a, b; Sayol et al. 2021; Soares et al. 2022; Matthews et al. 2022).

Alongside the loss of endemic and other native land birds from many oceanic islands, the human interest in birds has led to the introduction of many

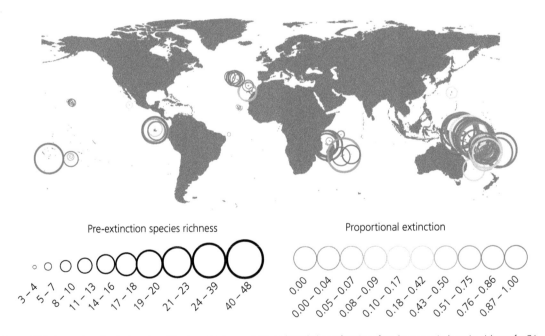

Figure 14.3 Proportion of extinct species of frugivore (birds, mammals, and reptiles) as a function of pre-human arrival species richness for 74 islands.
From Heinen et al. (2018).

(a)

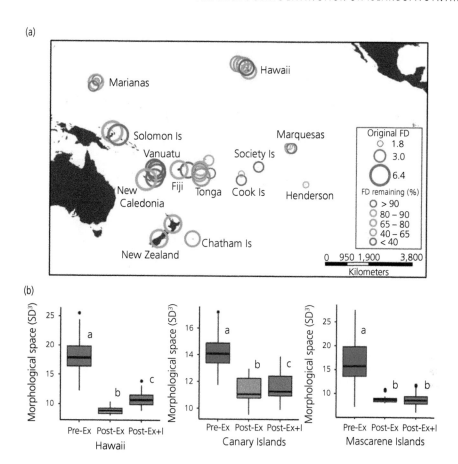

(b)

Figure 14.4 Functional diversity (FD) changes in island avifaunas. (a) Original (pre-human) and remaining functional dispersion (based on data for body size, foraging niche, diet, and activity period) for 44 Pacific islands. (b) Morphospace occupied for Hawaii, Canaries, and Mascarene Islands for: native pre-extinction (Pre-Ex) avifauna, native post-extinction (Post-Ex) avifauna, and post-extinction and alien species introductions together (Post-Ex + I), coloured in red, yellow, and blue, respectively. Letters (a, b, c) indicate whether scenarios differed in a paired samples Wilcoxon test (at $P < 0.05$).

(a) Figure 1 of Boyer and Jetz (2014). (b) Reprinted with permission of AAAS, from figure 3 of Sayol et al. (2021); © The Authors, some rights reserved; exclusive licensee AAAS. Distributed under a CC BY-NC 4.0 License (http://creativecommons.org/licenses/by-nc/4.0/).

non-native birds as part of a process of insular biotic homogenization (Li et al. 2020; Otto et al. 2020). These introductions include accidental escapes of caged birds but also the deliberate introduction of a selection of colourful birds and songsters by 'acclimatization' societies. Not all such introductions become naturalized, but many do, often in rough proportion to the species driven to extinction. Indeed, the island species–area relationships of non-native species are often found to have similar slopes to those for indigenous species, an intriguing feature of several analyses of different taxa

and archipelagos (Section 5.6). However, while the number of introductions is often similar to (or even exceeds) the number of extinctions, the functional diversity represented by introduced species on a given island is generally much lower than that lost through extinction (Fig. 14.4b; Sayol et al. 2021; Soares et al. 2022), likely because extinct species were often relatively functionally distinct whereas introduced species tend to be relatively more functionally alike/redundant. In addition, while the naturalized species may have become integrated into the insular food webs, they may nonetheless present

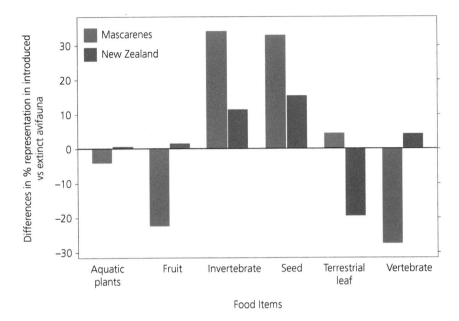

Figure 14.5 The difference in the proportional representation of different feeding guilds in the naturalized non-native avifauna in comparison with the extinct species of birds for the Mascarenes and for New Zealand. Note that while invertebrate and seed consumers have increased in relative representation in both archipelagos, reflecting the enhanced extents of open habitats brought about by land-use change since first colonization, trends in other feeding guilds vary between the two systems.
Redrawn from Wood et al. (2017).

very different functional roles, as illustrated for New Zealand and the Mascarenes in Fig. 14.5 (see also Carpenter et al. 2020).

Radical human transformation extends down to even small and low-lying atolls. Dramatic evidence of this comes from the Chagos Archipelago, where uninhabited atolls have provided a form of inadvertent ecological experiment. Graham et al. (2018) studied six rat-infested and six rat-free atolls. Rat predation on the infested atolls resulted in massive reductions in nesting seabirds (1.6 as against 1243 birds per ha). As a result, the nitrogen input from seabird droppings was only 0.8 kg per ha per year on the infested atolls as against 190 kg per year on the rat-free atolls. Run-off from the islands was correspondingly poor and rich in nutrients, respectively, leading to similarly stark differences in the biomass and structure of the coral reef ecosystems surrounding the atolls. Hence, the rats, by massively reducing densities of seabirds, were acting to disrupt energy and nutrient flows between the pelagic, insular, and island-adjacent

reef ecosystems: a classic illustration of cascading impacts transcending ecosystem boundaries.

14.3 Anthropogenic extinction: the causal factor complex

Assigning species extinctions to a single cause is, as all those who attempt it warn, over-simplistic. Clearance of native vegetation for agriculture, cutting of trees for timber, fragmentation of remaining native vegetation, introduction of non-native species, and hunting can all impact on the population levels of a particular native species. Often the impacts of two or more processes are multiplicative (i.e. synergistic) rather than simply additive. As a first example, it is thought that the giant Haast's eagle in New Zealand became extinct as a consequence of a combination of habitat loss *and* the extinction of its prey, which included the even larger flightless moa (Prebble and Wilmshurst 2009). As a second illustration, the introduction of non-native competitors and predators typically accompanies the clearance of native woodland. The extinction of the palm that once covered much of Rapa Nui is clearly attributable to human colonization but the relative roles of seed predation by the introduced Polynesian rat and deforestation by the settlers remains hard to resolve and sits within the broader ecocide/genocide debate (Section 13.7). Recent syntheses of palaeoecological data suggest that deforestation was not synchronous across the island and reflects a combination of deliberate clearance, climatic droughts, and synergistic feedbacks between land-use change and climate (Rull and Giralt 2018; Rull 2019), notwithstanding that rat predation of the seeds was also occurring. Synergisms between driving factors can thus be put alongside the disruption of native food webs that is intrinsic to the ecological cascades discussed above.

As evident earlier in the chapter and borne out by numerous analyses, the combination of the introduction of non-native commensal mammals (especially small rodents), hunting for the pot, and land clearance for agriculture (both arable and pastoral) were the predominant reasons for anthropogenic island extinctions in the prehistoric phase of island settlement. For example, analyses of bones from middens in New Zealand showed that those birds most intensively hunted typically became extinct at an early stage and that large-bodied birds were selectively targeted (Duncan et al. 2002). Similarly, the role of non-native rodents in the prehistoric era is underlined by evidence of their impact on islands visited but not settled by humans, in which the introduction of the rodents was the key distinguishing change agent involved in species losses of lizards and birds (above). Another classic example, from the historic period, is the introduction of the rabbit to Porto Santo Island (Madeira) by the first settlers in the 1420s. The rabbit multiplied to huge numbers and devastated the native vegetation cover, caused greatly increased erosion and opening the island to the spread of introduced weeds. (It is a historical note of interest and somewhat symbolic that Christopher Columbus married the daughter of the man who introduced rabbits to Porto Santo, and that he spent some time living on the island prior to planning his voyage across the Atlantic.)

Although remote oceanic islands mostly lacked native mammals (other than bats), there have been numerous extinctions of mammals from islands (Section 13.5; Fig. 1.4). Based on analysis of seven large and 31 small islands, Lyons et al. (2016) concluded that prehistoric extinctions from the former tended to filter out species of larger weaning mass and from the latter those of longer life spans, both being traits suggesting reduced capacity to recover swiftly from population depredations. Extinctions in the historic era continue this general trend, but less strongly, perhaps indicative of a wider array of extinction drivers.

Szabo et al. (2012) assessed the key driving forces of extinction of island birds in the historic period as being the introduction of non-native species, hunting, and agriculture (Fig. 14.6). A subsequent assessment, of 4350 extinct and threatened species from eight higher taxa in 15 island regions, attributed causation to the same three principal drivers: biological invasions, wildlife exploitation, and cultivation (Leclerc et al. 2018). In their study, Szabo et al. (2012) found climate change and extreme weather events as of very minor significance. Extreme events can be important in particular contexts and in Section 12.2 we gave examples of the reduction in population sizes of the Montserrat oriole due to volcanic eruptions and of the Puerto Rican parrot via hurricane damage. However, these examples also show that the impact of the natural event depends very substantially on the existing level of threat due to anthropogenic agencies.

In the contemporary period, the array of processes at work on islands has increased and often intensified (Fig. 14.7). Air and water pollution and the application of artificial fertilizers and pesticides were largely novel features of the last century, sometimes alongside industrial scale mining of phosphate deposits, extensive road building, and urbanization. Alongside these changes, in many cases, have also come mitigation measures, including the establishment of protected areas, restrictions on hunting, management of non-native species, etc. (Chapter 15). The upshot being a gradually shifting balance of anthropogenic extinction threats, varying in detail from island to island (Russell and Kueffer 2019). In addition to the largely 'within island' threats there are the twin problems of global climate change and associated risk of sea-level increase in the 21st century, for which Courchamp et al. (2014)

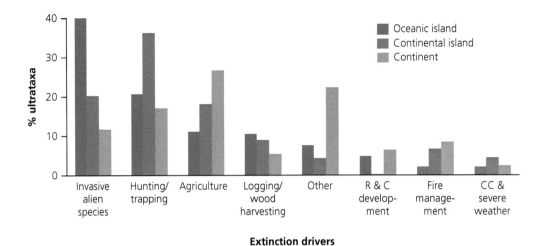

Figure 14.6 Extinction drivers for the loss of bird species and subspecies (collectively ultrataxa) on islands and continents since *c.* AD 1500. The category 'other' includes, for example, mining, transportation corridors, etc.; R and C refer to residential and commercial development, and CC to climate change.
Redrawn from Szabo et al. (2012).

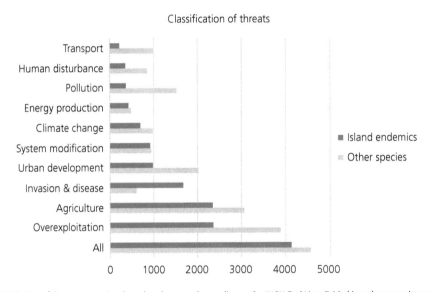

Figure 14.7 Attribution of threats to species classed as threatened according to the IUCN Red List, divided into those species endemic to islands and all other species.
Data from Russell and Kueffer (2019).

cite a range of 0.26–2.3 m as realistic, imperilling low-lying islands with inundation and threatening many hundreds of species of island endemics with extinction (Wetzel et al. 2013; Storlazzi et al. 2018).

Even in the absence of sea-level increase, many island species are likely to be at risk from future climate change given that the finite area of islands limits the ability of their species to track suitable climate conditions via dispersal (Leclerc et al. 2018). However, developing models for species responses to global climate change on islands is especially challenging because of the complex topography and compressed climate zonation of the most diverse islands (Chapter 3). These points are acknowledged by Pouteau and Birnnaum (2016) in their study of vulnerability of 469 tree species to climate change in New Caledonia, in which they nonetheless report that around 90% are likely to decline in range size, over half by greater than half their range, with up to 15% facing extinction. Climate change impacts are not simply about mean temperatures but are also about extremes of climate and, often, about altered incidence of fire, which may have significant impacts on the habitat of endangered island species (Moreno et al. 2018). In addition, there is evidence that warming on high mountains on islands may be accelerated compared with lower elevations. On Tenerife, warming has been occurring at *c.* 0.14 °C/decade at *c.* 2200 m elevation in the Teide National Park, twice as fast as in the lowlands (Martín-Esquivel et al. 2020). Within the park, rabbit populations appear to have increased thanks to the reduction in the severity of winter conditions and are currently preventing the regeneration of the dominant shrub *Spartocytisus supranubius* (Teide broom), alongside other palatable endemic species, suggesting that a phase shift may be underway in the vegetation structure and composition as a synergistic outcome of the non-native rabbit and climate change.

14.4 Habitat degradation and loss

Habitat alteration and the replacement of pristine native vegetation with altered ('cultural') vegetation landscapes typify areas occupied by modern humans. As we have seen, from the earliest stages of island occupancy, fire has been used to clear and manage landscapes, alongside agricultural activities, both pastoral and arable, with the relative significance and timing varying from system to system (Nogué et al. 2021). Through these means, colonist societies have typically massively changed the balance of major ecosystem types on islands. The loss of woodland habitat has been an important driver of the loss not only of certain tree species making up the forests, but also of many vertebrate and invertebrate species that are woodland dependent. For example, seven of 55 species of forest-dependent endemic beetles of the Azores are thought to have become extinct since around 1850, as 97% of the forest has been removed (Terzopoulou et al. 2015). These losses have been biased towards larger-bodied habitat specialists that are endemic to

single islands. Around 50% of surviving endemic beetles are considered at serious risk of extinction based on species–area estimates (Triantis et al. 2010).

A specific example cited by Whittaker and Fernández-Palacios (2007) was the loss of the four-coloured flowerpecker (*Dicaeum quadricolor*), which was considered extinct following the almost complete deforestation of the island of Cebú in the Philippines. However, this incredibly rare bird had in fact been rediscovered in 1992, in a woodland patch of around 2 km^2, where it co-occurred with four of the five other extant endemic bird species on the island (Dutson et al. 1993). Although such rediscoveries are encouraging, the authors noted that this was the only patch of closed canopy native forest remaining on the island. The flowerpecker remains critically endangered.

The extent to which islands were naturally covered in woodland varied as a function of climate and local microclimate and environmental controls. In many cases, closed woodland cover was not ubiquitous or was a feature only of the most favourable conditions within an island. It follows that comparatively open vegetation types constitute important native ecosystems, possessing many endemic species. This is the case for the Canaries, Cabo Verde, and Porto Santo in Macaronesia, and much of the Galápagos Islands, as well as many high-latitude islands. Such landscapes have also been subject to massive transformation owing to agricultural conversion for crops and the introduction of animals such as goats and rabbits. As well as the outright loss of particular ecosystem types and the reductions in carrying capacity linked to over-grazing and soil loss that we have described, it is fair to point out that, within many islands, adaptive agricultural practices were developed leading to the creation of productive cultural landscapes. Within the agriculturally productive areas of these landscapes, elements of the native vegetation are now integrated with many non-native species.

14.5 Predation and collection by humans

Some of the earliest human colonists of islands were hunter-gatherer societies. Even later colonists who practised agriculture (e.g. the Guanche in the Canaries and the Polynesians) relied also upon hunting of native fauna, and marine resources, including fish and shellfish but sometimes also marine mammals. Evidence, including charred bones of extinct birds from middens in Hawaii and other Polynesian islands, provides clear evidence of hunting of birds for food, often targeting larger prey (Duncan et al. 2002; Boyer 2008). Hunting has continued into the contemporary period, of course. An example is provided by the Madeiran endemic Trocaz pigeon, which almost went extinct as recently as the 1980s owing to a combination of forest loss and hunting. The species then recovered following the introduction of conservation measures and the banning of hunting, with the species being added to the European Birds Directive in 1986. Numbers increased from < 3000 to 10,000–14,000. Paradoxically, this pigeon is now subject to limited but officially sanctioned culling due to their propensity for raiding crops (Hance 2017).

Present-day trade in fruit bats (flying foxes) exemplifies how hunting for *ex situ* markets can reduce oceanic island populations to crisis point. The critical ('keystone') role that these bats play as pollinators and as seed dispersers means that their loss will have significant knock-on impacts (Vitousek et al. 1995). Hunting for collections also continues to be problematic. Today, the trade is driven mostly by individual collectors, but in the past hunting for museum collections was common and, for example, contributed to the extinction of the spectacular New Zealand bird, the huia (*Heteralocha acutirostris*) (Hume 2017). Another sad example of how collecting brought a species to the brink of extinction comes from the Canaries. The endemic Gran Canaria blue chaffinch (*Fringilla polatzeki*) was first discovered only in 1905, and was only raised to full species status in the last decade. Immediately after its discovery, natural history museums from all over Europe sent naturalists to Gran Canaria to gather specimens, with over 100 individuals being collected in just a few years, significantly depleting the population (Rodríguez and Moreno 2004). With an estimated population size today of around 430, the Gran Canaria blue chaffinch is considered by the IUCN to be one of the most endangered European bird species.

14.6 Non-native fauna, flora, and diseases

Predators, grazers, browsers, transformers

Species introduced to an area by humans are variously called exotic, alien, or non-native species. Of the vertebrates and plants, some species persist simply as domesticated (animal) or cultivated (plant) species, while others become naturalized; that is, they form self-sustaining populations within modified habitats (Henderson et al. 2006). Of these, some become feral or invasive, expanding into intact or semi-intact habitats. Of this subset, a few species cause serious ecological impacts. These species are termed ecosystem transformers (Henderson et al. 2006). Island examples abound.

Bellard et al. (2017) reported that over 70,000 islands worldwide support native vertebrates that the IUCN Red List has recorded as being threatened by invasive species. Due to the fact that certain types of species are more sensitive to invasive species than others (e.g. flightless birds), it is inevitable that future losses will have substantial impacts on functional diversity (Section 6.7), with knock-on effects on ecosystem functions (Section 13.8) (Marino et al. 2022). Although many of the most serious change agents have been vertebrates, we have either purposefully or inadvertently introduced to islands a remarkable array of seriously problematic invertebrate and plant species and disease organisms (Table 14.1).

We have noted how prehistoric and historic island colonists brought with them an array of both pest and/or useful vertebrate species, such as rats, dogs, cats, chickens, rabbits, pigs, sheep, and goats, some of which had the capacity to radically transform native vegetation either through ecological cascade effects, or directly via grazing and browsing. Sometimes these introductions preceded successful human colonization (sheep in the Azores, mice on Madeira); sometimes they were contemporaneous with first colonization (rabbits on Porto Santo, rats on many islands). Typically, introductions have continued following first colonization (Fig. 13.5).

Table 14.1 Examples of the worst invasive species on islands. The examples given here demonstrate that problematic non-native species that have invaded islands across the world are drawn from many higher taxa, and originate from all parts of the globe. Although most of these problematic invasives are continental in origin, some are drawn from other islands. NB We have omitted feral cats from this list for no good reason: they have been blamed for 14% of recent bird, mammal, and reptile extinctions on islands (Medina et al. 2011).

Common name	Scientific name	Origin	Examples of invaded islands
Insects			
Black twig borer	*Xylosandrus compactus*	Asia	Madagascar, Mauritius, Seychelles, Java, Sumatra, Fiji
Crazy ant	*Anoplolepis gracilipes*	West Africa	Hawaii, Seychelles, Zanzibar, Christmas Island
Little fire ant	*Wasmannia auropunctata*	Tropical America	Galápagos, New Caledonia, Solomon Islands, Hawaii, Vanuatu
Worms			
Triclad flatworm	*Platydemus manokwari*	New Guinea	Hawaii, Guam, Marianas
Snails			
Rosy wolfsnail	*Euglandina rosea*	SW United States	French Polynesia, Hawaii, Bermuda, Bahamas
Giant African snail	*Achatina fulica*	NE Africa	Hawaii, Samoa, Philippines, Sri Lanka, Tahiti
Golden apple snail	*Pomacea canaliculata*	South America	Japan, Philippines, Borneo, New Guinea
Amphibians			
Cane toad	*Bufo marinus*	Central America	Hawaii, Fiji, Samoa, Guam, Marianas, Caroline and Solomon Islands
Caribbean tree frog	*Eleutherodactylus coqui*	Puerto Rico	Hawaii, Virgin Islands
Reptiles			
Brown tree snake	*Boiga irregularis*	Australasia	Guam
Birds			
Indian myna	*Acridotheres tristis*	India	New Zealand, New Caledonia, Fiji, Samoa, Hawaii, Cook
Red-vented bulbul	*Pycnonotus cafer*	South Asia	Fiji, Samoa, Tonga, Hawaii
Mammals			
Goat	*Capra hircus*	Iran	Many, worldwide
European hedgehog	*Erinaceus europaeus*	Western Europe	New Zealand
Long-tailed macaque	*Macaca fascicularis*	South-East Asia	Mauritius
Small Indian mongoose	*Herpestes javanicus*	South Asia	Mauritius, Fiji, West Indies, Hawaii
Ship rat	*Rattus rattus*	India	Worldwide
Rabbit	*Oryctolagus cuniculus*	Iberian Peninsula	Worldwide
Plants			
Mile-a-minute weed	*Mikania micrantha*	Central America	Philippines, Solomon, Sri Lanka, Mauritius, Rarotonga
Kahili ginger	*Hedychium gardnerianum*	Himalaya	Micronesia, French Polynesia, Hawaii, New Zealand, Réunion, Jamaica, Azores, Madeira
Elephant grass	*Pennisetum* sp.	Africa	Hawaii, New Zealand, Sri Lanka, Guam, Galápagos, Canaries
Yellow sage	*Lantana camara*	Tropical America	Hawaii, Cabo Verde, Galápagos, New Caledonia, New Zealand
Diseases			
Avian malaria	*Plasmodium relictum*	unknown	Hawaii

Sources: Cronk and Fuller (1995); IUCN/SSC Invasive Species Specialist Group (2004).

Goats (*Capra hircus*) have historically been a leading cause of many insular endemic plant extinctions. A classic example is on the 121 km^2 island of St Helena. Following their introduction in 1505, almost 150 years before the first permanent human settlement, the native vegetation was devastated by vast herds of feral goats, leading directly to the extinction of several plant species (Lambdon

and Cronk 2020). Similarly, in Guadalupe Island (area 244 km^2), located off Baja California Peninsula, Moran (1996) records that from 156 native vascular plant species, 26 had been driven to extinction (or extirpated) by goats, which at one point reached a population size of *c.* 100,000 individuals. Campbell and Donlan (2005) estimated that feral or unrestricted goat grazing was the primary threat to 26%

of the IUCN threatened insular plant species, especially those located in tiny oceanic islands, which typically have small and scattered populations. An example is *Musschia isambertoi*, a critical endangered species found only in Deserta Grande (Madeira) (area 10 km^2). It has only recently been discovered and, due to goat grazing, only 10 seedlings survive in the single known population (Menezes de Sequeira et al. 2021).

As importantly, the human management of the species we have introduced has also changed over time, so that their roles as agents of vegetation change, soil erosion, and other ecosystem processes have also varied (Crosby 2004). The impact of managed domesticated populations of goats, dogs, donkeys, for example, is typically very different from the same species that have become feral.

It is the introduction of vertebrate predators, such as cats, rats, dogs, and mongooses, which have caused some of the worst problems (Keast and Miller 1996; Spatz et al. 2017, 2022). Feral cats are a feature of innumerable islands. They are opportunistic predators, eating what is most easily caught and they have extinguished numerous island species (Medina et al. 2011). Cats were introduced to the sub-Antarctic Marion Island in 1949 and at one stage it was estimated that there were about 2000 of them, responsible for killing > 400,000 birds per annum, mostly ground-nesting petrels. Given these losses, only by eradicating the cat from the island could the long-term survival of the native bird species be ensured (Leader-Williams and Walton 1989). Cats have been present on another sub-Antarctic island, Macquarie, for a lot longer, and in that case have been responsible for the loss of two endemic bird species. The combination of cats and mongooses on the two largest Fijian islands has resulted in the local extinction of two species of ground-foraging skinks of the genus *Emoia*; they survive only on mongoose-free islands. In the Lesser Antilles, three reptiles have been extirpated from St Lucia within historical time, coincident with the introduction of the mongoose. Domestic dogs can also be devastating, and feral populations have been responsible for the local extinction of land iguanas in the Galápagos. The introduction of predatory fish into lakes can also be highly damaging (Box 14.2).

Box 14.2 The decline of Lake Victoria's cichlid fishes

In this chapter we have concentrated on oceanic islands, but there are other types of islands that have similar characteristics of high endemism and which have experienced similarly alarming losses. Lakes are 'negative islands'; that is, they are largely isolated freshwater areas surrounded by a hostile land matrix, although they can be connected to other lakes through rivers. Long-lasting lakes (e.g. Lake Baikal in Siberia, the Great Lakes of the African Rift Valley, and Lake Titicaca in South America) have provided similar evolutionary opportunities to oceanic islands, and in the case of the African Rift Valley lakes (Tanganyika, Victoria, and Malawi), outstanding illustrations of radiations of fish species (Box 9.2).

Lake Victoria, shared by Kenya, Tanzania, and Uganda, is the largest tropical lake and the second largest freshwater lake in the world, comprising some 3000 km^3 of water. It is, or rather was, the unique habitat of c. 500 or more endemic cichlid fishes belonging to the genus *Haplochromis*, which are thought to have diversified within the last 15,000 years since the lake recharged (Box 9.2). Unfortunately, the majority of these species have disappeared since Nile perch (*Lates niloticus*) was introduced in 1954 with the aim of increasing the size of the fishing catch.

Important additional changes to the lake also co-occurred with the faunal collapse, including increasing pollution and sediment load, eutrophication, and associated oxygen depletion. Recent analysis confirms the hypothesis that the fundamental cause of the extinctions and of the trophic reorganization was indeed the introduction of the Nile perch, which is a voracious predator. The haplochromines did in time undertake a limited recovery in biomass, but most of the biodiversity appears to have been lost, with just 72 species recorded in 1979–1989 and rather fewer, 27 species, in surveys in 2006–2008 (Marshall 2018).

Whittaker and Fernández-Palacios (2007) reported that in the case of the extinct Hawaiian mamo (*Drepanis pacifica*), the famous royal cloak worn by Kamehameha I would have required some 80,000 birds to be killed. However, narrative accounts indicate that birds were caught and then released following harvesting of a few feathers, suggesting that the number of birds killed for cloak making would have been much lower than this estimate (wikipedia.org/wiki/'Ahu_'ula, visited

12 August 2020) and that other factors were key to its demise. Alongside hunting and habitat change, introduced commensals (pigs, dogs, and the Pacific rat) were important agents of extinction on Hawaii in this period. Following European arrival, the causal agents have been expanded to include continued deforestation, urbanization, additional introductions of non-native predators, alongside competition with and disease carried by introduced species, especially avian malaria (also introduced). Two bird species extinctions principally attributed to the introduction of the rabbit include the Laysan rail (*Porzanula palmeri*) and Laysan millerbird (*Acrocephalus familiaris*), both of which disappeared during the early 20th century because rabbits denuded their island. The Hawaiian rail (the smallest flightless bird) survived into the historic period but has since succumbed, probably to a combination of predators introduced by Europeans. The extinction sequence in Hawaii indicates that larger birds and ground-nesting birds initially fared worst, with rats playing an important role in some lowland areas. More recent extinctions have hit insectivores and nectarivores disproportionately (Boyer 2008).

As highlighted in Table 14.1, there are many cases of highly damaging introductions of seemingly insignificant invertebrate species, such as ants on Hawaii and the Galápagos. Some of the worst involve snail-on-snail predation. For example, the snail *Euglandina rosea* was introduced to Mauritius, in the hope that it would control introduced giant African snails of the genus *Achatina*. As in some other cases of failed biological control schemes, the introduced predator ignored the target species and instead gorged on native snail species (Griffiths et al. 1993). Some 30% of Mauritian snails are now extinct. As most losses preceded the introduction of *Euglandina*, they have been attributed principally to habitat destruction, but the exotic predator now represents one of the major threats to the remaining endemic snail species. Thus far, on Mauritius, primary forests do not appear to have been penetrated by the exotic species, and so habitat conservation remains key to the conservation of the native species. On Hawaii, the introduction of *Euglandina* has been implicated in the decline of 44 species of endemic *Achatinella*; and it has been a major factor in

the extinction of numerous *Partula* species in French Polynesia, perhaps as many as 19 endemic species having been lost from Raiatea alone (Haponski et al. 2019). For further examples, see Leclerc et al. (2018).

Modified pollination and dispersal networks

Non-native species can have varied ecological impacts. They include alterations to pollination and dispersal networks, hybridization, and the development of extensive thickets of non-native plant species, which transform habitats for other species (Meyer and Florence 1996; Henderson et al. 2006).

Species dispersing spontaneously or being introduced by humans to remote islands leave behind their usual web of interacting species, which may include pollinators, dispersers, competitors, food plants, prey, predators, parasites, and pathogens (Olesen and Valido 2004). If they cannot withstand the loss of evolutionary 'partners', they may fail to establish on the island, but at least a fraction of introduced species find a new web of interacting species they can interact with successfully (Traveset et al. 2015). Indeed, as discussed in Section 11.12, there is some evidence that pollinator networks tend to be more generalist on remote islands, involving super-generalist endemic species that assimilate non-native species into their networks, thus aiding their establishment (Olesen et al. 2002).

The introduction of non-native pollinators, such as the honeybee (*Apis mellifera*), continental bumblebees (*Bombus* spp.), and wasps (*Vespula* spp.), can impact both positively and negatively on insular plants and pollinators. If the introduced pollinator is more effective in transporting pollen than the native pollinators, then native (but also exotic) plants will increase their fruit and seed production in relation to those native plants not pollinated by the exotic species. Conversely, if it is less effective in pollinating a particular plant than a displaced native pollinator, then this may result in reduced seed set. For instance, introduced honeybees have been found to compete with the endemic Canarian bumblebee (*Bombus canariensis*), resulting in diminished pollen transfer success for several Canarian endemic plants (such as *Echium wildpretii*) (Valido et al. 2002).

The introduction of predators or disease organisms that knock out native pollinator networks can be expected to have significant negative consequences for native plant species. For instance, the introduction in Hawaii of predatory insects such as *Vespula pensylvanica* and of diseases, like avian malaria, have extinguished entire groups of pollinators, such as 52 endemic species of *Nesoprosopis* bees (Vitousek et al. 1987a), and many endemic bird species, respectively. The disappearance of exclusive pollinators is considered responsible for the extinction of many plants pollinated by them, including some 30 endemic Hawaiian Campanulaceae species (Cox and Elmqvist 2000).

The introduction of exotic dispersers can have positive consequences for native (but also for exotic) plant species. For example, the Balearic endemic plant *Cneorum tricoccon* was originally dispersed by *Podarcis lilfordi* lizards, which occupy elevations of less than 500 m asl. Following the introduction to Majorca of an alternative disperser, the European pine marten, *C. tricoccon* began to colonize elevations up to 1000 m asl, driven by the wider elevational range exploited by the marten (Traveset and Santamaría 2004). Conversely, the introduction of less-effective dispersers can result in negative consequences for native species. For example, two introduced mammal species (rabbit and Barbary ground squirrel) now compete for the fruits of the endemic shrub *Rubia fruticosa* in Fuerteventura (Canaries) with the indigenous dispersers (native birds and lizards). The seeds are dispersed less effectively by the non-native mammals and have a lower viability. In this case there are thus negative consequences both for the native plant (poorer dispersal services) and for the native frugivores (competition for a food resource) (Nogales et al. 2005).

Similarly, the introduction of an exotic plant can alter the insular networks in varying ways. The new plant can offer the native dispersers new resources that will produce positive effects for both species, but simultaneously have negative consequences for the native plants that have to compete with the exotics. For instance, the introduction of the prickly pears (genus *Opuntia*) on the Canaries has provided native dispersers, including ravens and endemic lizards, a new fleshy fruit to feed on,

with a much larger year-around availability than the native resources. The spread of *Opuntia* has thus been aided by native lizards and ravens, so that *Opuntia* species have successfully invaded large areas of Canarian coastal scrub rich in endemic *Euphorbia* and other plant genera (Nogales et al. 1999, 2005; Valido et al. 2003). Finally, it is often the case that interactions between two or more exotic species are crucial to the spread of problematic invasives. For example, within the Galápagos, introduced mammals are entirely responsible for the spread of the introduced guava (*Psidium guajava*), one of a number of ecosystem-transforming non-native plant species (e.g. Henderson et al. 2006).

Hybridization with native species

The introduction to an island of an allopatric congener, which is not fully reproductively isolated from the native taxon, can lead to a process of introgressive hybridization and the genetic dilution and eventual disappearance of the native form. This introgression process is currently affecting the Canarian endemic palm *Phoenix canariensis*, the plant symbol of the Canaries, due to the introduction of its closest relative, the date palm *Phoenix dactylifera*. The problem of introgression has recently been increased, following the importation of large numbers of date palms to decorate the new tourist resorts and paved roads across the archipelago (Morici 2004). Hybrid swarms are now found all over the archipelago, especially in the tourist honeypot areas on the eastern islands. In addition to hybridization, the endemic palm is also threatened by a lethal pest, the weevil *Rhynchophorus ferrugineus*, introduced with date palms imported from the Middle East. As a result of these pressures, the native palm faces an uncertain future. Genetically pure stands of the Canarian palm populations remain in the interiors of Gran Canaria and La Gomera, far away from the resorts (González-Pérez et al. 2004), potentially providing seed for future conservation programmes. A similar process of introgression is under way on the Cabo Verde islands, between the endemic palm *Phoenix atlantica* and the date palm (Morici 2004).

Disease

The problem of disease is closely associated with exposure to non-native competitors and, indeed, the separation of exotic microbes as a category from other introduced organisms is largely arbitrary. The most striking exemplification of the impact of disease on insular populations has, without doubt, been among the native peoples of islands, particularly on first contact with Europeans. This factor, in combination with more purposeful persecution, led to the decimation of many island peoples, such that in some cases little or no trace of them remains (Chapter 13; Crosby 2004).

The most commonly cited example of exotic disease afflicting native island birds is avian malaria, which is caused by mosquito-borne protozoan parasites of the genus *Plasmodium* (McClure et al. 2020). Avian malaria, along with another introduced disease, avian pox (*Avipoxvirus*), is perhaps the single most important driver of the extinction of native bird populations from areas of Hawaii below 1200 m since the start of the 20th century, although continued transformation of lowland ecosystems also plays a role. Carried by introduced mosquitos and able to sustain reservoirs in populations of many of the 54 naturalized non-native birds introduced to the archipelago, the diseases have had devastating effects on the fitness of the native bird species. Apart from two species of endemic honeycreeper, the lowland bird communities are, as a result, almost entirely composed of introduced bird species. As the introduced tropical mosquito *Culex quinquefasciatus* slowly adapted to cooler elevations, the die-off of native bird species advanced up the mountains, reaching 600 m by the 1950s and 1500 m by the 1970s, sweeping away several mid-elevation species in the process, and restricting the survivors to ever higher refugia. This process is ominously described by Pratt (2005) as 'the third and ongoing wave'. Interestingly, one species, the Hawaii 'amakihi, has seemingly started to evolve some resistance to the disease (Atkinson et al. 2013), pointing to a possible evolutionary escape route for some of the surviving species. However, scientists remain concerned that continued local adaptation of *C. quinquefasciatus* and ongoing climate change may lead to the further spread of avian malaria into the higher elevations of the islands: the final strongholds of most persisting native and endemic bird species.

14.7 Why have islands been so badly affected?

It has become axiomatic to represent oceanic islands as fragile ecosystems, which, because of their evolution in isolation from continental biotas, are particularly susceptible to the impact of anthropogenic change and especially the collision with continental floras and faunas that we have caused. Are island systems really particularly fragile, or have humans just hit them especially hard? There is evidence to support both propositions and several factors that have been identified as leading to increased impacts on islands. Here we briefly consider six points raised by Cronk and Fuller (1995).

- **Species poverty**. This may mean that there is more vacant niche space and less competition from native species. Although some island communities exhibit low invasibility where not subject to other forms of disturbance, species poverty is a contributory factor to island vulnerability, especially where niches are entirely vacant. However, species poverty needs to be analysed in tandem with the effects of evolution in isolation.

- **Evolution in isolation**. The evolution of flora and fauna in isolation from grazing, trampling, or predation by land mammals, has led to the loss of defensive traits and behaviours in many oceanic island endemics, especially birds (Table 10.4; Table 14.2; Anton et al. 2020). In his review of island plant syndromes, Burns (2019) concluded that a loss of defensive traits following colonization of islands has occurred repeatedly in plants, taking the form of the loss or reduction of both chemical and physical defences (such as spines). Although we have cast doubt on how general this syndrome is (Table 11.3), it is a feature of at least some island plant species. Plants that have lost defensive traits are at a competitive disadvantage and tend to disappear following the introduction of mammalian browsers, as was the fate of many endemic plant species on St Helena following the introduction of goats shortly after the island's

Table 14.2 Examples of hypothesized links between island evolutionary features in plants and animals and increased vulnerability to anthropogenic extinction drivers (modified from Fernández-Palacios et al. 2021a). The degree of support for the evolutionary syndromes mentioned is summarized in Tables 10.4 and 11.3.

Evolutionary feature	Implications linked to vulnerability
Insular secondary woodiness in plants	Logging targets for firewood, tools, etc.
Trend towards secondary dioecy/ functional dioecy in plants	Difficulty of mating in precarious demographic conditions
Loss of plant defences against herbivory	Vulnerable to predation by introduced herbivores
Gigantism in small vertebrates	Bigger reward for hunting, source of meat, generally slower rate of population replacement
Dwarfism in large vertebrates	Decrease of fierceness, facilitating hunting and/or predation by introduced predators
Flightlessness in birds	Easily hunted by humans and introduced predators
Diminution of defensive behaviour, tameness	Naivety against hunting/ predation
Reduction of clutch size	Slower recovery potential during disturbance
Lack of contact with diseases, pathogens	Vulnerability against imported diseases

discovery in AD 1502. The case of *Morella faya*, transported from one set of oceanic islands, in Macaronesia, to another, Hawaii, illustrates that the 'continentality' of the landmass from which a species hails may not always be the decisive factor (Box 14.3). For *Morella*, it appears to be its nitrogen-fixing capacity that has been crucial to its success in Hawaii.

Turning to animals, the loss of flight or a reduced tendency to flight characterizes many island birds (Wright et al. 2016) and, prior to the anthropogenic extinction of numerous island birds (Section 14.1),

characterized many more (Sayol et al. 2020). Island naivety—by which is meant behavioural vulnerability, such as a lack of fear of humans— is also a characteristic of many island birds on first contact, as remarked by Charles Darwin in his account of the Galápagos. The literature we have reviewed provides more formal analyses illustrating that to varying degrees of certainty, large body size, late maturation, small clutch size, and other indicators of low intrinsic rate of reproduction, characterize many extinct and endangered island birds (Table 10.4). Such observations support the notion that island faunas and floras have

---

Box 14.3 The invasion of *Morella faya* in the Hawaiian Islands

Over 13,000 plant species have been introduced to Hawaii, of which around 900 have become naturalized and 100 have become serious pests (Cox 1999). Among the more problematic tree species are guava (*Psidium cattleianum*), also a serious pest in the Galápagos, miconia (*Miconia calvescens*), tropical ash (*Fraxinus uhdei*), and, especially, the fire tree (*Morella faya*, formerly *Myrica faya*). It is a native to Macaronesia, where it is an early pioneer tree within the native laurel forest, and has been introduced from there into Hawaii, where it has become a pernicious forest weed (Cronk and Fuller 1995).

In the Canaries, *Morella faya* is the only nitrogen-fixing tree species of the community, in symbiotic association with

the actinomycete *Frankia*, and it exhibits a persistent pioneer strategy (Fernández-Palacios et al. 2019). It is a dioecious species some 20 m in height, possessing medium-size fleshy fruits dispersed by indigenous birds, such as the blackbird (*Turdus merula*). *Morella* maintains a local soil seedbank, and indeed its seeds germinate only after the appearance of a large canopy gap. Thus, *Morella* plays an important successional role in the return of forest to previously cleared areas, but unlike some other Canarian woody pioneers (*Erica* spp.) also persists after the gap is filled through its ability to re-sprout via suckers from the base.

Morella faya was introduced to Hawaii as early as AD 1880 and it was used extensively in reforestation, especially during

---

Box 14.3 *Continued*

the 1920s and 1930s. In Hawaii its fruits are consumed, and seeds dispersed, by a large number of both native (*Phaeornis obscurus*—Hawaiian thrush) and introduced birds (e.g. *Zosterops japonicus*—the Japanese white-eye, and *Carpodacus mexicanus*—house finch), as well as feral pigs. Between the 1960s and 1980s, aided by its ability to fix nitrogen, *Morella* rapidly invaded native forests of *Metrosideros polymorpha*, which it out-competed in reaching the forest canopy and forming pure stands (Vitousek and Walker 1989). *Morella* became invasive within the Hawaiian Volcanoes National Park in 1961 and has since spread over more than 35,000 ha. The new *Morella-Metrosideros* forest extends across a belt of 400–1200 m ASL on the southern slopes of Hawaii (the Big Island), particularly on recent (< 1000 years) lava flows (Mueller-Dombois and Fosberg 1998), and it is a well-established and problematic invasive across the archipelago.

Cronk and Fuller (1995) listed the following ecological characteristics as possible reasons for its success as an invader: prolific seed production, long-distance dispersal by birds, nitrogen fixation, possible production of allelopathic substances that may inhibit potential native competitors (specifically *Metrosideros polymorpha*), and the formation

of a dense canopy under which native species are unable to regenerate. The impacts of *Morella* on soil properties are particularly important, with knock-on consequences for ecosystem form and function. Vitousek et al. (1987b) calculate that the nitrogen availability in an unaltered Hawaiian forest is about 5.5 kg/ha per year (mainly through the fixation ability of the native *Acacia koa*). In stands invaded by *Morella* this may be increased up to 23.5 kg/ha per year. By substantially altering edaphic conditions, in combination with the activities of feral pigs, *Morella* has enabled the invasion of many continental plant species. *Morella* has varying degrees of ecosystem impact in different ecosystems and its patterns of spread are modulated by interactions with other non-native plants as well as animals (D'Antonio et al. 2017).

Options for the control of *Morella* infestation have been the subject of considerable research efforts, particularly in the search for a safe biological control agent. However, the most effective means of control at present appears to be cutting and the use of foliar applications of herbicide, both of which are only practical for dealing with newly establishing populations. The species continues to spread throughout the archipelago.

---

intrinsic vulnerability to certain kinds of introduced organism (such as terrestrial mammalian predators or grazers). Although many continental systems have also shown vulnerability to introduced species of plants, animals, and diseases (indeed most ecosystems are invasible to a degree), island evolutionary syndromes serve generally to enhance the fragility of island species and ecosystems to non-native species and in some respects at least to other change agents (D'Antonio et al. 2017; Burns 2019; Anton et al. 2020).

A large-scale analysis of 257 tropical and subtropical islands by Moser et al. (2018) has highlighted how, controlling for island area, elevation, and climate, the richness of *non-native* species increases with island isolation for ants, reptiles, mammals, and plants and that this pattern is not simply a function of economic activity or trading links. This trend is accompanied by a decline in richness of native species with isolation, which was evident in all taxa.

These results serve to confirm how with increased isolation, native richness declines and ecological naivety through *in situ* evolutionary change leads to a general increase in invasibility.

- **Exaggeration of ecological release**. Non-native species generally arrive on islands without their natural array of pests and diseases, and this provides them with an advantage over native species. A similar ecological release mechanism is proposed as part of the taxon cycle (Section 9.2), and it is therefore a logical extension to apply this idea to invasive non-natives on islands (e.g. Moser et al. 2018). The operation of this mechanism is, however, likely to be confounded with other factors, such as differential habitat specificity among native and non-native elements and variation in the disturbance mechanisms at work in different habitats (e.g. Kitayama and Mueller-Dombois 1995). Further systematic work to confirm what factors control ecological release among non-natives would be of considerable value to conservation management.

- **Small scale**. The extent of non-native species occurrence and of our transformative impact on oceanic island ecosystems mean that we put them under higher pressure. Everything is concentrated in a small area, within which there are few physical features large enough to prevent exploitation, disturbance, and introduction on an island-wide basis. Some 40% of the global human population lives within 100 km of the sea, and islands take their share of this, holding 10% of the world's population, while also supporting around 20% of plant, reptile, and bird species (Delgado et al. 2017). Islands also often sustain surprisingly large human populations. For example, Cronk and Fuller (1995) note that Mauritius, with a population of 530 people/km^2, may be compared culturally and historically with India, which at the time had 'only' about 240 people/km^2. As Delgado et al. (2017) point out, as humans encroach on island ecosystems, there is typically not much suitable habitat to be displaced into, hence many island species with restricted ranges have been or are being displaced.
- **Early colonization**. As we have seen, the history of island colonization over the Holocene has been complex, involving multiple waves of impact (Leppard et al. 2022). However, many islands were at the forefront of European colonial expansion and their natural resources were often plundered early on in a particularly rapacious fashion, sustained only by their connectivity to mainland markets. But it doesn't seem to be the length of contact and exploitation that is the key, so much as its scale and nature and the specific change agents introduced (above).
- **Crossroads of intercontinental trade**. Particularly in the days of sailing ships, islands have often been used as watering points, re-supply posts, and staging posts for trade and for plant transport and acclimatization. As some islands have also changed hands between different colonial powers and trading networks have changed over time in any case, they have, moreover, had opportunities to receive introductions from

differing networks of nations and biogeographical regions (cf. Lenzner et al. 2022).

Sometimes the introduction of a single exotic species has caused the extinction of numerous native island species. Examples include goats on several islands (above), the brown tree snake (*Boiga irregularis*) on Guam, the shrub *Miconia calvescens* on Tahiti, rats in the Mascarenes, and the carnivorous snail *Euglandina rosea* on several Pacific islands. Moreover, in the case of the snail, and of *Miconia*, the negative effect operates within the same taxon (Meyer and Florence 1996). However, the number of species extinctions attributable to a single extinction driver is very much in the minority.

Hence, notwithstanding that remote islands boast numerous endemic species, shaped by natural selection *in situ* and which therefore should be well adapted to the physical environment of their island homes, the deluge of multiple anthropogenic disturbance agents, notably including waves of non-native species, has swept vast numbers of them away. As we have described, among the survivors, many endemic island forms are threatened with extinction and are Red-listed; that is, they are highlighted as conservation priorities by the IUCN (Fig. 14.7; Matthews et al. 2022). The effects of, and synergisms between, multiple anthropogenic change agents make modelling, predicting, and mitigating these threats enormously challenging. In the final chapter, we will build on this analysis of past human impact to focus on the contemporary problems of islands and to highlight some of the many conservation efforts underway to address them.

14.8 Summary

As we saw in the previous chapter, the timing of first human arrival on islands across the world's oceans has spanned many thousands of years, from the Late Pleistocene to the Middle Ages. Our initial impact on species extinctions has sometimes preceded settlement via introductions of non-native vertebrates during brief visits, providing the first of many turns of the anthropogenic extinction ratchet. Efforts to synthesize the overall impact in terms of species losses have shown that species extinctions in the Holocene have been hugely biased towards islands, with thousands of species lost in the prehistoric era—many still unknown to science. Staggeringly, >60% of anthropogenic global species losses since AD 1500 have been insular.

Islands exemplify the idea of ecological cascades, where the loss of one or more species at one trophic level (often through introduction of a predator) cascades through interacting networks of species at other trophic levels. Dramatic illustration comes from comparative studies of small rat-infested and rat-free islands and from the devastating consequences of the introduction of the brown tree snake to Guam.

Extinction drivers have diversified over time, to include pollution, urbanization, climate change, and sea-level rise, alongside the long-standing problems of hunting, land-use change, and the pernicious impacts of many non-native species on island endemics. Distinguishing the role of specific processes in driving species to extinction is generally challenging, reflecting that many species populations are adversely affected by multiple processes, and often by synergistic interactions between them. Notwithstanding these complexities, we review some of the primary anthropogenic extinction drivers, commencing with habitat alteration, human hunting (both for food and, for example, for museum collections), and non-native organisms.

The importance of non-native species, both in the past and currently, can scarcely be overstated. The role of non-native predatory vertebrates among island faunas that have evolved in their absence has been matched in many islands by the impact of introduced mammalian herbivores on island vegetation. Even introduced non-native invertebrates have caused the extinction of endemic island species. The loss of native and endemic island species has occurred in tandem with the additions of numerous comparatively benign non-native species, greatly altering plant–pollinator, plant–disperser, and other interaction networks. Often the non-natives in these networks particularly assist other non-natives, which can result in further negative impacts on surviving native species. Sometimes non-native species provide surrogates for lost native ecosystem functions. Hybridization between non-native and endemic species/populations is another way in which distinct island species and genetic varieties may, in effect, be lost. The introduction of disease organisms provides a further extinction driver, exemplified by Hawaii, where introduced birds and mosquitos act as hosts for introduced disease organisms that have taken a heavy toll on endemic birds over the last century. Such processes represent a continuing threat to island endemics around the world.

Finally, we ask why islands have been so disproportionately affected by anthropogenic extinction? We find support for the conventional view that this is in large part due to 'ecological naivety' arising from evolution in isolation and that island syndromes have resulted in marked vulnerability to humans and their commensals. However, the physical context—the geographically confined nature of the systems, small distribution ranges of the species, and the lack of refugia remote from human interference—also comes into the picture.

CHAPTER 15

Meeting the conservation challenge

15.1 Island societies face distinct challenges

In the previous two chapters we described the history of anthropogenic transformation and biodiversity loss on islands, and we outlined the high levels of threat to many surviving island endemics. In further illustration, it has been estimated that a third of the endemic forest-dependent birds of the Caribbean are endangered (Devenish-Nelson et al. 2019), 2000–2800 island endemic plant species are 'on the verge of extinction' (Caujapé-Castells et al. 2010), and half of all Red-listed species are island endemics (Fig. 14.7; Russell and Kueffer 2019). In this final chapter, we develop the case that environmental and biodiversity conservation in remote archipelagos presents distinctive combinations of challenges, which arise from their insular character and from the socio-economic and political situation of the island societies (Table 15.1; Fig. 15.1; Nunn 2004; Caujapé-Castells et al. 2010). We go on to explore some of the conservation solutions promoted for remote islands. All of this involves asking fundamental questions about the goals of conservation and what we wish to conserve.

We previously illustrated some of the common elements of the problems afflicting island biodiversity: habitat loss/degradation and the role of introduced species foremost among them. Taking a complementary approach, we start this final chapter with a short selection of case studies drawn from different regions of the world, chosen to illustrate the varying circumstances and issues of concern in particular archipelagos. We include examples of islands where the loss of endemic species is not the central concern but where the biodiversity impact is manifest in terms of the loss of valued ecosystem goods and services.

15.2 The Maldives: in peril because of climatic change

The Maldives (mean elevation 1 m; total land area 300 km^2) form an extended equatorial archipelago of 1192 low-lying coral islands grouped into 26 atolls (from the native *atholhu*), extending 850 km across the Indian Ocean. Together with the Lakshadweep Islands to the north and the Chagos Islands to the south, the Maldives form part of a vast submarine mountain range, on the crest of which coral reefs have grown. Being small, low islands, they have a low rate of endemism and there are just five endemic plant species (all in the genus *Pandanus*) within a native flora of 277 species (Davis et al. 1995).

Human settlement of the Maldives dates back at least 2000 years. Latterly they fell under European influence, becoming a British protectorate in 1887 and regaining independence in 1965. Today they have a population of around 310,000. Projected sea-level rise resulting from 21st-century global climatic warming threatens the inundation of much of the archipelago, including of Malé Atoll, the capital, and most populated island. While over-washing can sometimes result in sedimentation and upward adjustment of elevation, flooding is not acceptable for inhabited islands with modern infrastructure and the projected range of sea-level rise this century represents an existential threat to this island nation, as it does for many other similarly distinctive societies on low lying islands (Nunn 2004). Mitigation

Island Biogeography. Robert J. Whittaker, José María Fernández-Palacios, and Thomas J. Matthews, Oxford University Press.
© Robert J. Whittaker, José María Fernández-Palacios, and Thomas J. Matthews (2023). DOI: 10.1093/oso/9780198868569.003.0015

Table 15.1 Islands as model systems for human societies. Phenomena that have been associated with islands (with varying degrees of evidence) and that have a bearing on the sustainability of their societies, economies, and ecosystems. They will largely be familiar from Chapters 13 and 14. We provide further illustrations in the present chapter.

Phenomena	Key ideas
Easter Island: Earth Island (ecocide)	Isolated, ill-informed human societies lacking adequate knowledge and governance structures overexploit their resource base leading to collapse {ecocide narratives}
Fragile to outside contact	Cultural naivety, susceptibility to disease and to devastating introductions of non-native species, insufficient power/inadequate *in situ* resources; at its worst, populations ravaged by disease and enslavement {genocide narratives}
Fragile to loss of contact	Trading inter-dependency with other islands and mainlands such that collapse in those links makes the way of life unsustainable. NB The tension between balancing the need for outside contact and protecting from it was illustrated during the Covid-19 pandemic in 2020–2021 (e.g. Davila et al. 2020)
Lack of suitable land for agriculture	Small area, limited water and/or water storage; fragility to climatic variation/change
Boom and bust	Narrow agricultural base, necessity of focus on limited natural products: victims of market vagaries
Trade route hubs	Crossroads of cultural, population, and biological exchange, and sometimes therefore of innovation (e.g. the grapefruit originated in Barbados as a hybrid from two introduced Asian citrus species)
Exacerbated costs of insularity	Archipelagic states need to replicate indispensable infrastructure (hospitals, schools, energy power stations, police, etc.) in different islands or assume the additional costs, e.g. of transporting sick people between islands by sea or air
Small-state crises	Susceptible to poor governance by a limited elite; states small, land area scattered
Resource limitations	Dependence on imports of fossil fuels and other natural resources (minerals, metals, etc.) not present within the island
Heavy dependence on the seas	Fine for a subsistence economy, but labour-intensive culture fails in modern settings; inshore marine resources often threatened by combinations of pollution, sedimentation, invasive species, and over-fishing of large fish
Rising sea level linked with climate change	Low-lying island nations susceptible to global sea-level rise linked to global climate change; impact of climate variation and change on islands at sensitive points in global circulation systems
Susceptibility to natural hazards	Hurricanes, earthquakes, volcanic eruptions, and tsunamis can present enormous threats with overwhelming per-unit-area and economic impact on small island communities and economies
Resource raiding	Exposure to overwhelming global trade forces, e.g. guano mining, fruit bat trading, offshore fishery depletion by foreign factory fishing fleets, timber extraction, and historically slave raiding
Victims of flag planting/geopolitics	Powerful nations claiming or gaining dominant influence (e.g. European expansion phase; bombardments and occupation in wartime; contemporary super power spheres of influence), or even constructing islands (e.g. South China Seas) to claim strategic resources
Tourism hotspots	Susceptible to the imposition of a large external ecological footprint from mass tourism via air travel or cruise ships, which may also provide vital economic inputs
Regulatory avoidance	Tax havens, potential routes of money laundering, poorly regulated trade or smuggling (including human trafficking); corrupting influences connecting to other themes in this table

options include the construction of artificial islands and sea walls, both of which have been undertaken at Malé Atoll. An artificial island called Hulhumalé was constructed from the late 1990s on adjacent reef flat. It was designed with an increase in sea level of up to 1 m by 2100 in mind, but current projections suggest a more rapid rise and continuing acute management challenges (Brown et al. 2020).

The economy of the Maldives is reliant on fisheries and tourism. The expansion of the tourist economy in recent decades has been heavily dependent

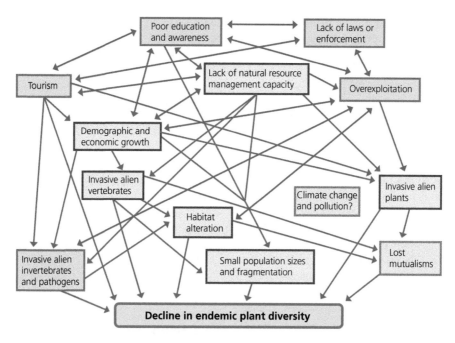

Figure 15.1 The complexity of societal, biological, and environmental factors and their interactions highlighted in a qualitative study of threats to endemic plant diversity for nine archipelagos. At the time, factors in blue boxes were judged as of greatest importance, those in orange boxes of intermediate importance, and green boxes, least importance. Important two-way interactions are shown in red; predominantly one-way relationships are in blue.
Redrawn from Caujapé-Castells et al. (2010).

on the reef ecosystems, but they are increasingly threatened by elevated seawater temperatures and the associated problems of coral bleaching (an extrinsic threat not amenable to local mitigation measures) and by siltation arising from dredging activities and by a lack of satisfactory waste disposal systems (Jaleel 2013). Further problems include invasions of the crown of thorns starfish (*Acanthaster planci*), which was first recorded in 1963 and which has since spread, despite efforts to remove them. While the Maldives have institutional and legal frameworks in place to address these issues, the level and nature of the threats provide huge ongoing challenges for a small state to address.

Threats to island reef ecosystems in the Maldives, as elsewhere, stem from a combination of extrinsic, global-scale anthropogenic drivers and intrinsic island-specific problems. Extrinsic factors include: ocean acidification; ocean surface warming; sea-level increase and impacts of intensified

hurricane activity (all of which can be linked to anthropogenic CO_2 emissions and climate change); the accumulation of plastics in ocean food chains; marine pollution, including from large cruise ships; disease outbreaks; and the introduction of invasive non-native fish and invertebrate species (e.g. Donner and Potere 2007; Green et al. 2012; Storlazzi et al. 2018; Andersson et al. 2019). Intrinsic factors include: local failures to regulate fishing techniques and methods; physical damage to reefs linked to the same; pollution and sedimentation derived from the island; and the interruption of natural ecosystem processes, such as can be caused by rodent infestation on small islands (Box 6.2; Graham et al. 2018). In small islands with limited agricultural land, fisheries are both the primary source of protein and a key (or only) source of income, with small-scale subsistence fishing comprising 80% of the catch in parts of the South Pacific, highlighting the significance of these threats to island societies (Donner and Potere 2007).

15.3 Okino-Tori-Shima and the South China Sea disputes: the strategic importance of rocky outcrops

As islands disappear under the sea through erosion, subsidence, or sea-level rise, so do their Exclusive Economic Zones (EEZs), the 200 nautical mile extension around a landmass, within which a nation retains rights of use to undersea resources, primarily fishing and seabed mining. This aspect of international law accounts for the interest countries have in claiming and holding on to tiny and desolate islands.

Okino-Tori-Shima (also known as Parece Vela), some 1750 km south of Tokyo, is Japan's southernmost island. The island is a largely submerged atoll, which at high tide consists of just two rocky outcrops jutting 60 cm above the surface of the Philippine Sea in the western Pacific Ocean (Chang 2000). Typhoons rip through the region annually, and their destructive impact, alongside regular wave action, has threatened to submerge the atoll. As it is the basis for an EEZ of 400,000 km^2 (larger than Japan itself), of very rich fishing waters (alongside mineral rights), Japan spent millions of dollars in 1988 constructing iron and concrete breakwaters to prevent its disappearance. A further $100 million investment was announced in 2016. China, South Korea, and Taiwan have disputed Japan's EEZ claim, arguing that Okino-Tori-Shima doesn't qualify under the relevant UN convention for technical reasons.

There are a number of other such territorial disputes ongoing far to the south-west, in the South China Sea, notably surrounding China's island-building actions in the area of the Spratly and Paracel island regions, in which it has been engaged since 2013. While most attention is on the geopolitics of the disputed regions, models suggest that the combination of island-building works, destructive clam harvesting, and overfishing within the affected archipelagos will be impacting the recruitment of reef components and diminishing the resilience of the reef ecosystems (Wolanksi et al. 2020).

By contrast, two uninhabited islands, Tebua Tarawa and Abanuea (ironically meaning 'the beach which is long lasting'), part of the island nation of Kiribati, were lost in 1999. Should sea level continue to rise as forecast over the 21st century, then territorial losses will require the recalculation of Kiribati's EEZ, which currently constitutes over 3.5 m km^2, almost 5000 times the land area, and equivalent to over a third of that of the USA (Chang 2000). Kiribati's population has increased from 30,000 to 120,000 between 1950 and 2020 and is predicted to double again by 2100. The pressures on fish stocks and the environment generate huge challenges to factor alongside climate change and the loss of land area implied by rising sea level (Cauchi et al. 2020).

These examples serve to illustrate the disproportionate importance of the loss (or gain) of land area to island nations and to nations claiming jurisdiction over islands. They also serve to highlight that sea-level rise is not a simple issue to assess, as it plays out alongside other change drivers pertaining to small islands and their marine resources.

15.4 Nauru: the destructive legacy of colonial resource extraction

Named Pleasant Island in 1798 by John Fearn of the British ship Hunter, the Pacific island of Nauru is located about 500 km north-east of New Guinea. It has a total land area of 21 km^2 and a population of almost 11,000 in 2020. The coast is fringed by a ring of sand, surrounded by a protective reef, but the interior is formed of guano deposits that have mixed and solidified with a coral limestone base. Early in the 20th century, the major powers began to take an interest as it became evident that Nauru happened to have the largest reserves of this easily accessible, high-quality source of phosphate. The mining of these deposits has devastated the island environmentally and has created severe financial, legal, and cultural problems for its people. The story is, once again, one of exploitation by the outside world (Gale 2019).

The island was initially ignored by the colonial powers but, following a carve-up of 'spheres of influence' between Britain and Germany in 1886, the Germans annexed Nauru in 1888. The export of phosphate (mostly to Australia) began shortly after this, with Australia taking advantage of the First World War to grab possession. Over the following decades, Australia and New Zealand governments, in cahoots with the British, manoeuvred to ensure continuing control over access to this rich resource. The supply lines from Nauru were successively targeted by the German and Japanese navies during the Second World War and Japan took the island in 1942, holding it until 1945. By 1947, the island was once again essentially under Australian control via the United Nations Trusteeship Council. Gale (2019) calculates that the depletion of the rich guano deposits of Nauru, along with additional supplies from Banaba (the nearest island to Nauru) and from Christmas Island in the Indian Ocean, were crucial to the agricultural and economic development of Australia and New Zealand over the 20th century. Between World War II and Nauru's independence in 1968, the mining operation produced two million tons yearly, nearly all destined for Australia and New Zealand.

Phosphate mining has been concentrated in the central plateau. It has left behind deep pits and tall pillars within a moon-like barren wasteland, useless for agriculture and covering over 80% of the island area, with the residents restricted to a narrow coastal strip. Following independence from Australia, the Nauruans began to experience the financial benefits of phosphate mining for the first time. In 1989, the new Republic filed suit against Australia with the International Court of Justice and five years later a settlement was reached amounting to over 100 million Australian dollars (Bambrick 2017).

The money was soon depleted, along with most of the remaining phosphate reserves, while pollution from the mining has devastated the local fisheries. The traditional culture, agriculture, and fishing practices have gone, leaving the Nauruans reliant on imported pre-packaged foodstuffs, while problems of obesity, diabetes, cardiovascular and other associated diseases have fuelled a 12-year reduction in life expectancy (Bambrick 2017). Nauru has become dependent on outside assistance and on revenues for hosting Australian asylum-seeker processing centres and, according to some commentators, in desperate need of new solutions to avoid becoming the Pacific's first 'failed state'. Clifford et al. (2019) sketch out one possible pathway for the island, which paradoxically proposes that restoration of the island must necessarily start by further resource extraction from the remaining phosphate and limestone reserves, providing that this is done in a far more environmentally controlled means.

15.5 The Canaries: overdependency on mass tourism?

The Canarian Archipelago is one of the most biodiverse territories within the European Union, possessing more than 12,500 terrestrial and 5500 marine species in, or around, a land area of only 7500 km^2. There are some 113 endemic genera and 3800 endemic species (Izquierdo et al. 2004), among them several spectacular radiations of animals (e.g. *Laparocerus* weevils, > 200 species) and plants (e.g. succulent rosette-forming members of the Crassulaceae, > 60 species).

The Canaries were first settled around 400 BC by people of Berber stock and subsequently conquered by the Castilians (Spain) during the 15th century (Section 13.6). Ecological transformation has thus spanned over 2000 years, accelerating in the last 500 years, leading to extensive deforestation of the mesic zones, goat grazing of the less productive land, and varied agricultural use of whatever land could be put into production. As a result of cheap air travel, the pattern of exploitation of natural resources began its third major phase c. 1960–1970, shifting from an agricultural society to a mass tourism model, especially on Tenerife and Gran Canaria.

The tourist boom of the last half-century has driven a huge increase in population and abrupt socio-economic and cultural changes (Table 15.2), with profound consequences for the natural and semi-natural ecosystems of the archipelago. Tourism increased from around 70,000 visitors per annum in 1960 to 2.23 million by 1980 (around the time of the political transition in Spain from the Franco dictatorship to the new constitution), and to 12 million by 2000. As of 2019 (prior to the Covid-19 pandemic of 2020), visitor numbers stood at around 15 million. Hence, human population densities have very recently risen far above historical norms (Table 15.2), especially on Gran Canaria (> 550/km^2) and Tenerife (450/km^2, overwhelmingly concentrated into one third of the land area). The increased resident and transient populations have placed huge demands on space

Table 15.2 The shift of the economic development model (1960–2018) on the Canaries (updated from Fernández-Palacios et al. 2004 using publicly available data). Oil consumption refers to within-island consumption, excluding aircraft and shipping.

Property	1960	1970	1980	1990	2000	2010	2018
Population (M)	0.94	1.17	1.44	1.64	1.78	2.12	2.11
Number of tourists (M)	0.07	0.79	2.23	5.46	12.0	10.4	15.6
Population density (inhabitants/km^2)	130	155	189	206	231	284	283
Cultivated area (K ha)	95	68	60	49	46	41.5	39.8
Oil consumption (K oil equiv. ton.)	–	827	1442	2473	3155	3721	3634
Electric energy consumption (GW)	–	890	1680	3423	6292	8232	8326
Concrete consumption (M ton.)	–	0.76	1.22	1.57	2.65	0.82	0.59
Number of cars (M)	0.02	0.08	0.28	0.5	1.08	1.62	1.73
Active population in agriculture (%)	54	28	17	7	6	3.6	2.5
Active population in services (%)	27	46	55	62	70	81.6	82.3
Unemployment (%)	2	1	18	26	13	28.7	20.1
Female life expectancy (y)	65	75	77	80	82	84.3	84.6
Literacy (%)	36.2	–	91.7	95.7	96.4	> 97	> 98
Per capita income (K dollars)	4.3	8.8	11.4	16.4	17.2	23.7	24.9

for houses, hotels, roads, and infrastructure, with increasing demands for food, goods, water, and other resources, while simultaneously producing more waste (e.g. García-Falcón and Medina-Muñoz 1999). It has been estimated that on a daily basis the average Canarian resident contributes more than 20 kg CO_2 to the atmosphere and produces 5 kg waste (Fernández-Palacios et al. 2004).

The Canarian pattern of development is clearly unsustainable in terms of the indigenous resource base (Fernández-Palacios et al. 2004). On Tenerife, large parts of both the arid south and of the best agricultural land in the north of the island have been gobbled up by urban expansion. Coastal ecosystems have been fragmented by buildings, harbours, roads, and golf courses. The water table has been depleted and water has been 'mined' from the central volcanic massif. Pressures have also increased on fisheries in the inshore waters. As a by-product of tourist-driven urbanization, about half the agricultural area (50,000 ha) has been abandoned. Due to the steep topography, agriculture in the Canaries is labour-intensive, with cultivation concentrated in small, terraced fields, not amenable to large-scale mechanization: easier and seemingly more attractive livelihoods are to be found in the urban areas.

The upshot has been a wholesale shift in the nature and geography of the human impact on Canarian landscapes, alongside a shift from net exporter to net importer of food. Significant

between- and within-island differences have emerged, as, for example, the island of La Gomera, and the rural park zones of Tenerife (Anaga and Teno) have seen declining populations over the last half-century, reflecting the move away from small-scale agriculture. On Tenerife, the mountain regions, which were once devastated by overgrazing and cutting, have now largely been handed over to replanting (especially in the endemic Canary Island pine belt). The mid-elevation laurel forest belt, although reduced to 10% of its original area, now has protected status. Some areas once in cultivation in the lower reaches of the laurel zone and below are gradually being spontaneously recolonized by a mix of native and exotic plants (Box 15.1), or else are built on. As mass tourism favours the sunniest environments, some of the biggest tourist developments have been built on the most arid areas, previously only sparsely populated.

Over-dependence on international tourism is a particular risk for island economies, which often gravitate towards over-reliance on a particular product or narrow product range ('boom and bust', Table 15.1; Garfield 2015). Butler's (1980) tourism area life cycle model is one lens for viewing island tourism that may find application, especially on the larger islands of Tenerife and Gran Canaria, which have proceeded furthest along a low-cost mass-tourism model (Rodríguez et al.

Box 15.1 Invasive plants in the Canaries: the incorporation of prickly pear (*Opuntia*) into the landscape

During the 19th century, several prickly pear species (*Opuntia*, Cactaceae) of Mexican origin were introduced into the Canaries, principally for the natural red-purple colorant (carminic acid) produced by the females of the parasitic insect *Dactylopius coccus* (Coccoidea, Homoptera), also introduced from Mexico, which feed on them. True cacti are not naturally found outside the New World. *Opuntia* rapidly naturalized and became invasive within the natural and semi-natural ecosystems of the Canarian lowlands (Otto et al. 2006), where they found suitably arid, warm conditions. Their large, sweet, fleshy fruits provide reward for both native and non-native dispersers year around (Section 14.6). They also regenerate and spread vegetatively, from falling blades taking root: behaviour not found in the native Canarian *Euphorbia*, which are the indigenous dominants of this scrub vegetation. Another clue to their success is that whereas all but one species of *Euphorbia* drop their leaves in summer drought conditions, *Opuntia* are still able to photosynthesize, using the crassulacean acid metabolism (CAM) photosynthetic pathway.

Although several prickly pear species grow wild today on the Canaries (*O. maxima*, *O. dillenii*, *O. tomentosa*, *O. tuna*, *O. robusta*, and *O. vulgaris*) (Izquierdo et al. 2004) only two of them (*O. maxima* and *O. dillenii*) are widely distributed. *O. dillenii* form part of the subdesert coastal scrub, whereas *O. maxima* thrives in the disturbed zones at mid-elevation, where it forms dense, almost monospecific patches. Their succulent blades were eaten by goats, providing them with a valuable water intake, whereas their fleshy fruits are not only eaten by humans, but also by endemic *Gallotia* lizards (Valido and Nogales 1994) and several native birds, especially the raven (*Corvus corax*).

Nogales et al. (1999) found *O. maxima* to be the most common seed in the pellets regurgitated by ravens, and both *O. maxima* and *O. dillenii* showed improved germination after passing through raven guts. Furthermore, starling (*Sturnus vulgaris*) has been observed feeding on the parasitic insect *Dactylopius coccus* growing on the prickly pear (Martín and Lorenzo 2001). *Opuntia* thus provides an illustration of how non-native plant species can compete successfully with native and endemic species, while becoming integrated into food networks of both native and exotic animal populations.

Alongside *Opuntia*, there are many other introduced plants on the Canary Islands, with estimates of the minimum number of non-natives ranging from 700 to 964 species, of which 82–117 are considered invasive (Fernández-Palacios et al. 2023). Despite their number and the presence of a few locally problematic species (e.g. *Pennisetum setaceum* and *Arundo donax*), invasive non-native species contribute <10% cover to the majority of Canarian ecosystems and compared with many other oceanic island systems, the Canarian flora appears to be unusually resilient (see discussion in Fernández-Palacios et al. 2023).

Today, after several centuries of presence in the Canaries, prickly pears are considered part of the Canarian landscape and apart from within very specific conservation sites, no attempt has been made to eradicate them. In contrast, a lot of money has been invested in efforts (of limited success) to control some other exotics, such as the grass *Pennisetum setaceum*, which is a problematic invasive on Tenerife as elsewhere in the archipelago and indeed on other archipelagos (Da Re et al. 2020).

2008; Inchausti-Sintes et al. 2020). Investment in the upward phase of the life cycle and competition between resorts for development can lead to saturation of the market and trimming of margins and, especially when accompanied by short-termism from investors, can lead to the build-up of ageing infrastructure that fails to generate sufficient margin to pay for reconstruction and renewal. At that point, a downward spiral can kick in, with reduced tourist demand and still tighter margins. In response to evidence of just such effects, efforts

were made by the Tenerife tourist board to improve the quality of tourism by developing a range of zones and brands, which comprised Tenerife Golf, Tenerife Natural, Tenerife Select, Tenerife and the Sea, Tenerife Convention Bureau, and Tenerife Film Commission. Along with a general raised awareness of the importance of environmental sustainability, Tenerife Natural picked out nature-based tourism as an important plank of the strategy. However, short-term economic imperatives play a powerful part in hindering efforts to reduce the ecological footprint of the mass tourism model and to put in place meaningful long-term conservation measures (Fernández-Palacios and de Nascimento 2011). We return to the Canaries to look at conservation measures in more detail below.

15.6 The Galápagos: a threatened evolutionary showcase

Comprising 11 large and over 100 small islands and islets, 1000 km west of Ecuador, the Galápagos are famous for their high degree of endemism, which includes iconic species such as marine iguana, Galápagos penguin, Gálapagos fur seals and sea-lions, flightless cormorants, and Darwin's finches, alongside 180 endemic plant and over 700 endemic insect species. The basic conservation problem of the Galápagos is the ever-accelerating breakdown of their isolation as a result of dramatic changes in human population size, activity patterns, and mobility. The islands were first settled in the 19th century and the human population of the archipelago has increased from just 120 in 1883 to 25,244 in 2015. Introduced species constitute the greatest threat to terrestrial species. Toral-Granda et al. (2017) report that 1579 non-native species have been recorded on the archipelago, 1476 of which have established, with at least 59% becoming naturalized. They include 821 terrestrial plants, 545 terrestrial insects, and 50 vertebrates. Particularly problematic non-natives include commensal and feral mammals (goats, pigs, cats, rats, dogs, cattle, and donkeys) and aggressive alien plants (e.g. *Lantana camara*, *Cedrela odorata*, *Cinchona succirubra*, *Rubus* spp., and *Psidium guajava*). The possibility of the introduction of new, virulent diseases, as has afflicted Hawaii's honeycreepers (Section 14.6), is of particular concern, with at least three viral diseases already recorded.

From 1979 to 2015, tourist numbers increased from just under 12,000 per annum to a quarter of a million (Toral-Granda et al. 2017), providing the engine for population and economic growth, driving increased deliberate and accidental introduction of non-native species (Fig. 15.2), urbanization, habitat fragmentation, agricultural encroachment, over-exploitation of native woody species, and fires (Jackson 1995; Cruz et al.

2005). Spin-off forms of enhanced mortality now include, for example, measurable levels of roadkill of birds in particular localities (García-Carrasco et al. 2020) and a heavy toll of feral dogs on Galápagos tortoise and land iguana populations, with both domestic and feral cats also predating endemic birds, reptiles, and even insects (Jackson 1995). Although only five islands have human settlements, the increased mobility within the archipelago associated with tourism, can only accelerate the risk of rapid spread of invasive non-natives. The Galápagos Islands are particularly sensitive to ENSO events (Section 3.7), which can bring significant fluctuations in precipitation levels, leading to phases of biomass reduction that in turn can be followed by sudden increases in distribution and abundance of particular invasive non-native plant species. Based on such observations, it seems likely that global climate change will generate further synergistic threats to native communities and species (Hamann 2011; Salinas-de-León et al. 2020).

The problems extend to the surrounding waters. Overfishing (e.g. for sea cucumber and lobster) and illegal fishing (especially for shark) by the artisanal fishing fleet has caused the erosion of coastal fisheries and generated considerable tension between fishermen and conservationists (Eddy et al. 2019). In addition to the local fleet, there have been recent incursions of very large fleets of as many as 200 or 300 Chinese-flagged boats, including huge, refrigerated container ships, which operate (for the most part) just outside the EEZ zone and appear to do so with scant regard for international conventions (Alava and Paladines 2017; Collyns 2020). The hugely increased levels of resource extraction from the seas around the Galápagos in recent decades, combined with the other change processes described, has already led to the probable extinction of endemic marine species and presents ongoing threats to many others (Edgar et al. 2010).

Given the totemic significance of the Galápagos in the development of evolutionary theory, it might be anticipated that there would be a matching conservation focus and effort. Indeed, the Galápagos National Park covers 97% of the archipelago and the islands have also been designated a World Heritage Site, a Biosphere Reserve, and a 'flagship' area for conservation (Toral-Granda et al. 2017). Yet, as

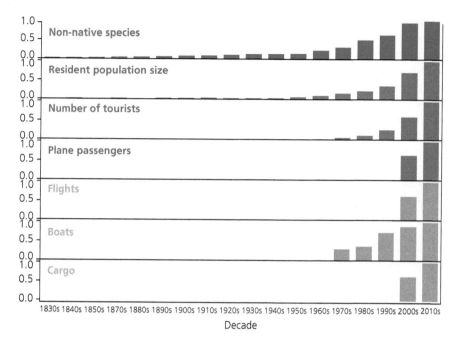

Figure 15.2 Numbers of non-native species, of residents, and of transportation vectors (except cargo: weight) for the Galápagos Islands, where all values are rescaled from 0 to 1.0.
Redrawn from Toral-Granda et al. (2017).

emphasized in Chapter 14, synergism among the many change agents can cause cascading ecological impacts, presenting unpredictable management challenges that call for concerted scientific research efforts. Research efforts in the Galápagos are coordinated by the Charles Darwin Research Station and the National Park Service and the management of the park aims to preserve the biological uniqueness and intactness of the islands and surrounding seas, through regulating resource use. Serious mitigation efforts have been made, including the establishment in 2012 of a dedicated biosecurity agency, but there are concerns that it remains under-resourced. A comprehensive zoning system has been adopted, which is central to conservation efforts in both terrestrial and marine realms.

The threats to the archipelago's biodiversity are many and varied and—given that many of the forces at work are international—present complex and at times seemingly overwhelming challenges to the authorities. It is something of a paradox that many if not most Galápagos tourists would count themselves environmentalists, but it is their interest in the islands that has stimulated the local economy and boosted immigration. This population growth has, in turn, brought additional pressures to bear on the environment and biota, fuelling conflicts between pro-sustainability and pro-development and extraction interests (Davis et al. 1995; González et al. 2008).

Combating non-native species in the Galápagos

In May 1989, seven feral goats were observed on Alcedo volcano on Isabela: the first confirmed report of goats in the habitat of the most intact race of Galápagos tortoise. By 1997, there were around 100,000 goats, and the habitat of the tortoises was collapsing. Feral pigs are equally problematic, as they consume plants, invertebrates, the eggs and hatchlings of endemic tortoises, lava lizards, and Galápagos petrels, among other species. Thankfully, efforts to remove feral pigs and goats have seen increasing success. The last pig was eliminated from the large island of Santiago in 2000, at the end of a 30-year campaign. Recent campaigns to eliminate feral goats have been increasingly sophisticated, well resourced, swift, and efficient. Over the first decade of this century, using a combination of ground and aerial hunting and Judas goats (Box 15.2), feral goats were removed from Santiago and Isabela and indeed from all but three of the Galápagos Islands. Over 140,000 goats were removed from Isabela alone, at the cost of US$10.5 million (Carrion et al. 2011). These successes show that with funding and political support, eradication of some key non-native species is technically feasible. Unfortunately, fishermen have also carried out

goat reintroductions as a food source or as a political tool within disputes with the National Park authority. Such reintroductions are costly to mitigate, particularly if not detected at the earliest stages, which serves to demonstrate the importance of gaining broad stakeholder support for conservation action (Carrion et al. 2011).

The destruction brought about by large populations of feral and commensal animals can involve the crossing of irreversible thresholds, particularly when native and endemic species have been lost and non-native species introduced over the

---

Box 15.2 The arrival and eventual removal of ecosystem transformers: feral goats, rabbits, and Judas goats

Goats and rabbits in the Canaries

Although there were once giant tortoises (*Centrochelys burchardi* and *C. vulcanica*) on the Canaries, these animals have been extinct since the Late Pleistocene, and otherwise, the Canarian flora evolved without large grazers until humans arrived and introduced goats (*Capra hircus*) in the first millennium BC (Section 13.6). For two thousand years, goat grazing played a central role in the transformation of the vegetation cover of the islands, leading to the loss of endemic populations of plants and animals (Marrero-Gómez et al. 2003; Morales et al. 2009; de Nascimento et al. 2020), as on many other archipelagos.

Following the declaration of the summit region of Tenerife as the Teide National Park in 1954, goats were removed from the uplands of Tenerife and, as the economic development model shifted from agriculture to mass tourism, also from most of the rest of the Canarian landscape. The last wild goats were successfully extirpated from the uninhabited 10 km² island of Alegranza in 1980, marking the end across most parts of the archipelago of an influence that lasted for more than two millennia. In response to the removal of the goats, in those areas escaping development, a trend of vegetation recovery can be seen. However, the goat is not the only non-native herbivore in the Canaries.

Rabbits were introduced to the Canaries in the 15th century and with the removal of goats they have become the predominant herbivore of open habitats. Their numbers have increased recently within the Teide National Park, perhaps in response to warming climate. Exclosure experiments have shown that rabbits are having a crucial role in modifying soil properties and vegetation dynamics, driving a switch

between dominance of the leguminous shrub *Spartocytisus supranubius* and the smaller *Pterocephalus lasiospermus* (Cubas et al. 2018). Selective grazing by rabbits constitutes a serious, continuing threat for many other endemic species in the Canaries (Cubas et al. 2019). Conservation efforts in the Teide National Park have focused on assessing the key life-history phases and germination requirements of highly threatened plant species, followed by efforts to re-establish wild populations (e.g. see Marrero-Gómez et al. 2003) but so far the only options to tackle the rabbit in this context are perpetual control measures combined with more and larger exclosures (Cubas et al. 2018). Rabbits have been introduced to > 800 islands worldwide: it rarely goes well.

Judas goats

Goats have been introduced on numerous islands around the world, sometimes to provide a food resource on uninhabited islands for visiting sailors, other times by settlers. Like rabbits, they are highly effective ecosystem transformers, which present huge conservation problems in many islands. A recent systematic review reported that as of 2016 there had been 147 successful goat eradication programmes and at least 90 successful rabbit eradication programmes from islands (Schweizer et al. 2016). Modern technology has now been harnessed to enable swift and effective eradication of a variety of invasive mammal populations from islands, but different species present distinct challenges. Common to all is the need for systematic monitoring to ensure that eradication has been 100% successful and that reintroductions do not occur, or if they do occur, that they are swiftly detected and addressed.

Box 15.2 *Continued*

One technique for goat eradication is the use of so-called Judas goats. Females are trapped, sterilized, and radio-collared before being re-released, upon which (being social animals) they typically find and remain with other goats. They are then tracked and the other animals accompanying the Judas goat are shot with high-powered rifles. The Judas goat may then be released again, and the process repeated. Although goat eradication programmes can involve both ground-based and aerial hunting (which can lower costs substantially), Judas goats can be a key means of rounding up the last members of the feral population (Carrion et al. 2011). Although some environmentalists and animal rights protestors raise principled objections to such culling programmes, they can be the most effective action towards protecting the native biota of many oceanic islands (Campbell and Donlan 2005; Cruz et al. 2005; Carrion et al. 2011; Schweizer et al. 2016).

same time frame (Chapter 14). However, vegetation recovery following removal of feral pigs, goats, and donkeys from Santiago Island was swift, with detectable impacts on ecological networks. These effects are not always straightforward. For example, there was a decline in the adult population of Galápagos hawks, probably as a result of reduced hunting success as vegetation thickened followed goat removal. At the same time, there was a switch in their diet towards black rats (*Rattus rattus*), from 20% before eradication to 73% of prey biomass after goat eradication (Jaramillo et al. 2016). These findings illustrate the cascading and often unpredictable outcomes of non-native species introduction, expansion, and removal, highlighting the need for continued scientific monitoring efforts and for systematic analyses of such conservation management interventions (e.g. D'Antonio et al. 2017).

15.7 What do we wish to conserve?

The compelling case for improved island conservation measures is recognized by a host of national and international governmental and non-governmental institutions and actors. Islands are included and often prioritized within major strategic conservation planning initiatives, even when the small scale and/or remoteness of island systems requires some adjustment to the original planning approach. Major schemes in which islands have figured significantly include Conservation International's biodiversity hotspots (Box 4.1), Birdlife International's Endemic Bird Areas, and targeted

islands schemes such as the IUCN Global Island Partnership.

We have shown in Chapters 13 and 14 just how much humans have transformed island environments and biotas, massively changing the extent and distribution of major ecosystem types, replacing natural biomes with cultural 'anthromes' (Russell and Kueffer 2019), driving many native and endemic species to extinction (or close to it) and inserting a profusion of non-native species into the mix. In innumerable islands across the world, the option to reset to a pre-human contact point is not available. And, in any event, we may value elements of the cultural landscapes we have created on islands and not wish to eliminate or lose them all (Nogué et al. 2017; Russell and Kueffer 2019). What then should the goals of island conservation be?

Avoiding extinction and making choices

For most conservationists, avoiding future extinctions of distinct island taxa (whether species or varieties/subspecies) holds a high and unquestioned importance, reflecting a range of ethical and other values-based concerns, alongside more utilitarian or pragmatic concerns for the loss of ecosystem goods and services (Russell and Kueffer 2019). However, in practice, not all island species and populations can be saved, as the resources and societal and political support are insufficient, and/or the threats are impractical to mitigate entirely. For these reasons, choices have to be made about which species, communities, or landscapes to invest in.

Often priority is given to more charismatic species (those which excite and motivate societal support), such as many bird species, rather than to less-charismatic species, such as many plant or invertebrate species. Sometimes attention is given to species that have a perceived ecological importance—those with so-called keystone roles (including ecosystem engineers)—as protecting them from extinction may be of great benefit to other species. Similarly, some species or taxonomic groups may be regarded as surrogates for unknown biodiversity in other groups, or as umbrella taxa, the conservation of which will serve to protect other species (Ladle and Whittaker (2011). Each of these concepts, while not specific to islands, may find application within insular contexts (e.g. Sibarani et al. 2019).

Within this volume, we have mostly focused on the traditional taxonomic unit, the species, as the unit of interest and on species richness (or richness of endemics) as the key diversity metric. However, within biodiversity science there has been an increasing interest in the last decade on two relatively new concepts—phylogenetic diversity and functional diversity (Section 6.7; Fig. 14.4)—as potentially important tools in conservation prioritization (Doxa et al. 2020), including within islands (e.g. Sibarani et al. 2019).

Phylogenetic diversity metrics aim to incorporate the evolutionary distinctiveness of each species, enabling prioritization of sets of species that maximize the representation of the tree of life. The upshot is that, for example, a monophyletic group such as the Hawaiian silverswords, Galápagos finches, or the approximately 264 species or subspecies of flightless weevils of the genus *Laparocerus* on the Canaries (Section 9.7; Machado 2022), being closely related phylogenetically, will be downweighted in relation to the many native (or endemic) species that have failed to diversify within the same islands and thus represent a greater portion of the evolutionary tree for plants, birds, or coleopterans, respectively. In short, although potentially emphasizing palaeoendemic relictual taxa on some islands, phylogenetic diversity metrics can be expected to downweight the significance of highly diversified neoendemic lineages, which may well run counter

to societal valuation of these showcase examples of island evolution.

Functional diversity metrics are conceptually similar to phylogenetic diversity metrics, but instead of being based on evolutionary relatedness (similarity), they assess the distinctive contribution of each species to overall diversity of so-called functional traits: such characteristics in animals as body size/mass, bill shape in birds, wing characteristics in birds or beetles, and in plants such properties as growth form, height, seed mass, and wood density. The presumption, as for phylogenetic diversity metrics, is that sites with a greater functional diversity are ecologically more valuable (or perhaps in better ecological health) than those with equivalent species richness that have a lower functional diversity.

As we have seen, anthropogenic extinction has differentially weeded out many endemic species characterized by island syndromes such as reduced flight and large body size, in the process reducing the overall phylogenetic and functional diversity of native island avifaunas. As importantly, key ecological functions, such as seed-dispersal services, have been reduced (Heinen et al. 2018). Pragmatically, we would do better to focus on the particular ecological roles of threatened species, in so far as we can determine them, as opposed to the summary statistics provided by the diversity metrics. Hence, while there may be a place for phylogenetic diversity and functional diversity metrics in conservation prioritization decisions, the implications of such analyses for ranking species for conservation action on islands will require careful assessment.

Composition or function?

Notwithstanding the importance of doing so, the conservation movement is not and never has been solely concerned with avoiding extinctions, rather it seeks to emphasize and promote a range of values concerning the human–nature relationship. These values include the aesthetic and intellectual contemplation of nature and of natural monuments, the health and well-being to be derived from access to nature, and the emotional connection between indigenous biota and/or landscapes and a sense of belonging (e.g. Ladle and Whittaker 2011). In more prosaic terms, people need access to wild spaces for relaxation and exercise, ecotourists value beautiful natural landscapes, and people will travel to see particular (often charismatic) fauna or flora, rock formations, or landscapes (including cultural landscapes) (Newsome et al. 2013). What should these wild spaces and landscapes look like and what biota should they comprise?

It has been pointed out that conservation decision making is generally based on an understanding of ecosystems that is temporally shallow, rarely calling upon empirical data series of even 50-years depth (Gillson and Willis 2004). While we lack long-term monitoring data for island ecosystems, we do have increased availability of palaeoecological records from lake sediment cores, for island systems across the world (Nogué et al. 2017, 2021). Such data may provide some quantitative guidance on

how vegetation cover has changed in the period before and since human colonization. We also have fossil, archaeological data, and historical accounts. We may use such data to identify at least some elements of the pristine pre-human **baselines** for island systems (Fig. 15.3).

Approaches for managing areas for conservation have been divided by Callicott et al. (1999) into compositionalist and functionalist. In a useful if over-simplifying analysis they argue that an emphasis on composition tends to be based around a worldview that regards humans as separate from nature and acting to defile and destroy the pristine. Compositionalist approaches, according to this view, would seek to remove the imprint of humans as much as possible and to restore to a pre-contact baseline. They characterize functionalist approaches on the other hand, as based around a worldview in which

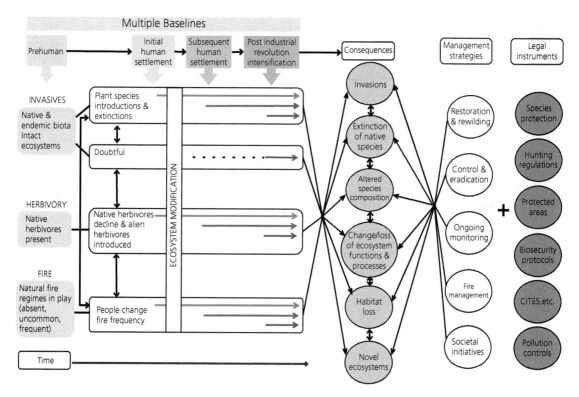

Figure 15.3 Meshing science with conservation strategies and legal instruments on islands. From the top left, reading down and across, a conceptual model of how knowledge of past ecological baselines, human agencies of change and their biodiversity impacts, may feed into considerations of potential conservation management strategies, which also need to be aligned with and supported by legal instruments (right) to provide a basis for conservation action. The selection of strategies and instruments is illustrative rather than comprehensive. Diagram based on Figure 2 from Nogué et al. (2017) with additional boxes added.

humans are embedded within nature and are concerned with attaining healthy ecosystems and ecological sustainability. In essence, this implies that compositionalism is about restoring the pre-human composition, which is often impractical in island ecosystems, while functionalism is about restoring healthy ecosystems, which may involve acceptance of ongoing changes and loss of some features and can be hard to capture in terms of conservation management plans as it lacks clear objectives as to what the landscape should look like. It is precisely this vision, however, that is encapsulated in many rewilding initiatives, a movement with gathering force in conservation in many parts of the world at the time of writing. What relevance do these ideas have specifically to islands?

First, it should be said that, in practice, conservation management plans often incorporate essentially 'compositionalist' targets for what the landscape should look like and what biodiversity features it should contain while fully recognizing that the landscape concerned has been shaped for hundreds or thousands of years by people. This is true also of many island conservation management initiatives. Hence, compositionalism doesn't have to exclude people or seek a pristine pre-human contact baseline, although some conservation initiatives may do. Nogué et al. (2017) develop this argument by suggesting that rather than seeking a single 'pre-contact' baseline for island systems, palaeoecologists could aim to identify multiple 'baselines'; that is, descriptions of different phases in the human transformation of island systems, reflecting the cultural interactions between different historical phases of island societies and their island natures (Fig. 15.3). Conservation organizations in many places (insular and continental) already follow such a practice and it is, for example, an approach deeply embedded in much of the work of the National Trust, a major natural and cultural heritage charity in the UK.

Second, rewilding, in many mainland (or large land-bridge island) settings, often involves initiatives to return free-ranging populations of large mammalian herbivores and/or predators to restore top-down regulation of ecosystem functions with minimal human control thereafter. The best examples of this approach in an oceanic island

setting involve the re-introduction of giant tortoises (herbivores), into islands that have lost their native tortoise species (Hansen et al. 2010; and see below). Alongside such gentle giants, island conservationists may also be keen to reintroduce or restore populations of avian predators and scavengers (e.g. the effective reversal of the ecological extinction of Mauritian kestrel, a conservation success story; or potentially the reintroduction to the Canaries of the red kite, extirpated c. 1971). However, in most remote island systems, the goal of restoring more natural ecosystem function requires the removal of problematic vertebrates that 'don't belong'. As Russell and Kaiser-Bunbury (2019) put it: 'freed from biotic constraints in their native range, species introduced to islands no longer experience top-down limitation, instead becoming limited by and disrupting bottom-up processes that dominate on resource-limited islands. ... Whereas on continents the focus of conservation is on restoring native apex species and top-down limitation, on islands the focus must instead be on removing introduced animal and plant species to restore bottom-up limitation'.

Hence, we see the compositionalist/functionalist dichotomy as over-simplifying a complex reality. In practice, conservation management of island landscapes, including adjacent seascapes, involves compromise, the finding of negotiated solutions that recognize pre-existing stakeholders and that are capable of gaining societal support. This will typically involve the deployment of a range of conservation measures and philosophies. In a later section, we return to Canarian conservation practices to illustrate and amplify these points.

15.8 Key instruments in science and policy implementation

Conservation action requires a solid scientific base, societal and political support at all levels, and it requires funding. It follows that conservationists and environmentalists need to engage in educational, outreach, and public engagement programmes to ensure that the public and polity are well informed and to build and maintain support from relevant stakeholder communities. Conservation action, whether focused on biosecurity, establishment and management of protected areas, eradication of non-natives, or prevention of hunting/trapping of endemic species, must also be based upon legal instruments, embedded within international and national binding agreements and legislation. None of this is unique to islands, but all of it is relevant to islands.

The large catalogue of extinctions of native and endemic island species is bound to increase. Yet, it is not a hopeless task. Species can be saved by measures such as habitat protection and restoration, pest or predator control, and translocation of endangered island species (Franklin and Steadman 1991; Marrero-Gómez et al. 2003). Although solutions need to be tailored to the particular circumstances of each island system, field managers and conservationists often face the same problems: feral animals, habitat loss and burning, growth of tourism pressures, etc. (Figures 13.4b, 15.1). Since the start of the 21st century, improved computer power, internet connectivity, and online access to scientific data and literature has facilitated a phase shift not only in information exchange but also in the power and sophistication of analyses of extinction drivers on islands, which now can consider large numbers of islands and species in a single analysis (e.g. Wetzel et al. 2013; Courchamp et al. 2014; McCreless et al. 2016; Spatz et al. 2017). These developments have gone hand in hand with improved connectivity among island researchers and conservation managers, facilitated by developments such as the foundation (c. 2014–2019) of the Society for Island Biology (Fernández-Palacios et al. 2021a). In the following sections we briefly review some of the key conservation policy options.

15.9 Tackling invasive species

The increased connectivity of remote islands has and will continue to bring disruption through introductions of non-native species (Lenzner et al. 2020; Spatz et al. 2022). It is a challenging and imprecise science to predict which non-native species will become invasive and problematic on particular islands. However, Adsersen (1995) notes the prevalence of the same invasive plants in archipelagos as diverse and remote from one another as the Mascarenes, the Galápagos, the Canaries, and the

Bahamas (and see Table 14.1)—the same can be said of animals and especially of introductions of vertebrates. A recent survey of expert opinion highlighted that the key tools for mitigating these problems include systematic and well-resourced biosecurity measures, effective communication strategies, continued scientific research and monitoring, and increased pro-active management (Lenzner et al. 2020). We also concur with Russell et al. (2016) that there are circumstances in which lethal control, carried out within stringent ethical guidelines, is necessary.

Generally, islands are beset by multiple non-native species, which can interact, often synergistically (Bellard et al. 2017). Rodents and cats in combination are a problem on many islands, including many where they have been present for a long time. For example, within the Azores, nest predation by rats and cats was found to have reduced breeding success of the endemic Azores woodpigeon (*Columba palumbus azorica*) to as low as 9% (Terceira Island) and 19% (Pico Island) in camera trap studies of 56 nests (Lamelas-López et al. 2020).

The extent of the problem is shown by Spatz et al. (2017), who searched the literature for 1288 islands that have highly threatened (Red-listed) vertebrate taxa, for 1030 of which data were also available for invasive non-native vertebrates. Of these 1030 islands, 76% possess one or more invasive species, most commonly rats. They estimate that adopting control measures could benefit 39% of the world's most threatened vertebrate species and they stressed that biosecurity measures are crucial to preventing further introductions. Utilizing the same database, expanded by inclusion of recently extinct species, McCreless et al. (2016) analyse extirpation probabilities in relation to the threatening invasive species and to island properties such as size, precipitation, and human presence (Fig. 15.4). They found that removal of the invasive species could be central to preventing 40–75% of likely future island extinctions, with the greatest benefits arising from interventions on small, dry islands: actions on larger, wet islands, occupied by humans, are unlikely alone to bring the same benefits.

Invasive mammal eradication programmes on islands have benefitted greatly from systematic analyses reviewing the efficacy of mitigation

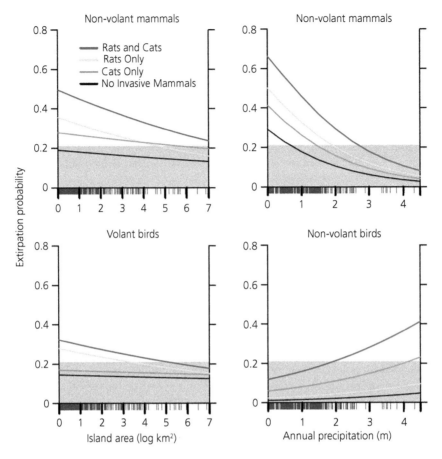

Figure 15.4 Modelled extinction probabilities for mammals and birds on uninhabited islands in relation to island area and precipitation given the presence of rats and/or cats, or no invasive mammals. Grey shading represents the overall predicted persistence level; rug plots on the x axes provide the distribution of values.

From McCreless et al. (2016).

measures (e.g. Veitch et al. 2019; Spatz et al. 2022). The problem of cats and rodents is a case in point. Mesopredator suppression theory (the corollary of mesopredator release: see Section 12.4) suggests that numbers of a small non-native predator (rodents) are likely to be suppressed by a larger one (cats), thus reducing extinction risk of native species where both are present. However, the analyses by McCreless et al. (2016) suggest that while this effect occurs, it is probably uncommon. Multispecies interactions involving several non-native mammals and island prey species can be complex and management interventions need to be carefully researched and coordinated (Fig. 15.5; Russell and Kaiser-Bunbury 2019). If there is a generalization

to be drawn it is that feral cats and non-native rodents, in tandem or isolation, are bad news and that efforts to remove only one species, when two or more are involved in the food chain, may fail to produce the desired benefits.

A classic example is provided by sub-Antarctic Marion Island, to which mice were inadvertently introduced in the early 19th century, followed in 1949 by cats, introduced to control mice at a weather station. The cats subsequently became feral, leading to local species extinctions and to decreased populations of burrowing petrels. As a result, a cat-eradication programme was undertaken, which concluded successfully by 1993. Since then, evidence of mouse predation of chicks of

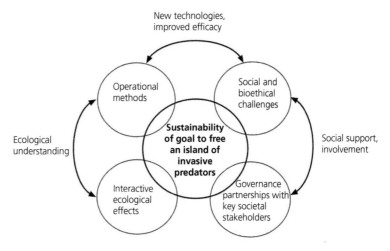

Figure 15.5 Attainment of long-term goals to eradicate multiple non-native vertebrate predators from islands and to maintain that status requires a concerted, long-lasting endeavour, combining scientific, social, and broad-based political support. Schematic modified from that developed in relation to the goal of a predator-free New Zealand. Based on Fig. 1 from Peltzer et al. (2019). Copyright © The Royal Society of New Zealand, reprinted by permission of Taylor & Francis Ltd, http://www.tandfonline.com on behalf of The Royal Society of New Zealand.

Tristan albatrosses and burrowing petrels has increased, with recent records of attacks also on adult birds on Marion and on Gough Islands (Jones and Ryan 2010; Jones et al. 2019). This highlights that systematic monitoring programmes are crucial to ensure that management interventions are adaptive and ultimately successful. However, the evidence from systematic efforts to review the effects of goat and rabbit eradication schemes is that only a small fraction of published studies to date have included reports on vegetation monitoring following eradication. Such data may be key to identifying unforeseen negative outcomes that can follow management interventions (Schweizer et al. 2016).

Notwithstanding these concerns, it is important to emphasize that eradication schemes can be highly effective (Spatz et al. 2022). For example, Benkwitt et al. (2021) demonstrate how rat removal from islands, particularly when accompanied by active management, can lead to the relatively rapid recovery of seabird biomass and restoration of nutrient subsidies between marine and terrestrial systems, with cascading ecological benefits for island and adjacent coral reef systems within two decades. The most ambitious island eradication scheme announced to date has to be the goal of a predator-free New Zealand by 2050, announced by the New Zealand government in July 2016 and requiring the eradication of rats, possums, and stoats on a scale never before attempted (Murphy et al. 2019; Peltzer et al. 2019). Achieving this goal will require not only improved methods for eradication at scale,

but successful engagement with the social and bioethical challenges that will unfold and thus interaction with a wide range of stakeholders from different communities (Fig. 15.5).

Biological control—a dangerous but necessary weapon

Biological control involves the introduction of a biological agent that targets a problematic non-native organism without seriously affecting non-target species. Although in many cases biological control is the only practical approach, there are associated risks (Veitch et al. 2019). Classic examples include introductions of carnivorous snails to the Society Islands between 1974 and the 1990s. This resulted in the extinction of around half of the 55 species of partulid snails (Haponski et al. 2019). Another classic example is the past use of the small Indian mongoose as a pest-control agent, again on Pacific islands. The mongoose has shown limited success in controlling rats but has been effective in devastating populations of many native island bird and reptile species, particularly ground-foraging skinks and snakes (Hays and Conant 2007).

These notorious failures show how crucial rigorous screening and testing is to ensure that the control agent's effects are highly specific to the target species. For example, field and laboratory studies of the parasitoid wasp *Conura annulifera* indicate sufficient specificity of feeding niche to be a potential solution to the damaging impact of a parasitic fly, *Philornis downsi*, introduced to the Galápagos in 1964. The larvae of the fly feed on nestling birds and are a leading cause of recent declines in species of Darwin's finches and other endemic land birds (Boulton et al. 2019). A successful illustration of classical biological control in the Galápagos comes from a post-release evaluation of the ladybird, *Rodolia cardinalis*, introduced to control the cottony cushion scale *Icerya purchasi*, a problematic non-native plant pest. The ladybird was found to reduce infestations without predating native invertebrates, although some outbreaks continued to occur, especially where the cushion scale is tended by non-native ants, which protect the colonies as a source of honeydew (Hoddle et al. 2013). The continued success of the programme relies upon regular (five-yearly) monitoring to ensure that the control agent populations remain viable.

In their review of eradication methods applied to islands, Simberloff et al. (2018) point to increasing success rates as lessons have been learned from earlier failures, and this is especially true of biological control methods. Where thoroughly researched, systematically carried out and adequately resourced, non-target downsides can be avoided, while suppression or eradication is achieved in a high proportion of cases. Successful approaches often involve multipronged initiatives. An example is provided by a project on Christmas Island targeting the yellow crazy ant (*Anoplolepis gracilipes*), which has had devastating impact on a keystone species, the endemic land crab (*Gecarcoidea natalis*). The project involves the introduction of a parasitoid to control another non-native, a scale insect, which the ants tend for honeydew: the idea being that by reducing this important source of energy for the ant colonies, it will provide a form of indirect biological control on the ants (Ong et al. 2019). Previous control efforts, involving spraying of fipronil from helicopters, have the disadvantages of not being fully effective and having non-target impacts.

For further illustrations and review of the application of biological control of invasive animals and plants on islands alongside other eradication methods, such as toxic baiting/poisoning and trapping, see: Smith-Ramírez et al. (2017); Simberloff et al.

(2018); Murphy et al. (2019); Veitch et al. (2019); and Spatz et al. (2022).

15.10 Translocation, reintroductions, and rewilding

Translocation and repatriation from *ex situ* breeding programmes have both been employed on islands. Translocation may be used to remove a species from an overwhelming local threat or to re-establish a population on an offshore island free from exotic predators. This approach has been used extensively for the conservation of Australian mammals endangered by cat and fox predation (Legge et al. 2018) and to create safe havens for many endemic bird species in New Zealand, 80% of which are classed as at risk or threatened (Murphy et al. 2019). In general, translocated wild animals establish more successfully than captive-bred ones. Griffith et al. (1989) surveyed 198 bird or mammal translocation programmes undertaken in Australia, Canada, Hawaii, mainland USA, and New Zealand over the period 1973–1986. Only about 46% of release programmes of threatened/endangered species were successful, a far lower percentage than for translocations of native game species, for which the success rate was 86%. The most important determinant of success was the number of animals released. They also found that herbivores were significantly more likely to be successfully translocated than carnivores or omnivores, that it was better to put an animal back into the centre of its historical range than the periphery, and that wild-caught animals fared better than captive-reared animals.

In New Zealand, a number of threatened bird species have been translocated into small offshore islands that either lacked or have been cleared of introduced predators, such as the rats, mustelids, and possums, which are prevalent on the mainland (Murphy et al. 2019). By 2015, one third of New Zealand's offshore islands had been cleared of mammalian predators, but between 2000 and 2017, there were around 390 pest-incursions (mostly of rodents) into predator-free islands, some of which proved costly in both monetary terms and population impacts (Murphy et al. 2019). Nonetheless, translocation programmes have been crucial in avoiding the final extinction of species such as the

South Island saddleback (*Philesturnus carunculatus*) (Garcia-R and Di Marco 2020).

The kakapo (*Strigops habroptilus*), a flightless, nocturnal parrot, was once considered to be effectively extinct, until the discovery of a small population on Stewart Island (Clout and Craig 1995). They continued to decline following rediscovery, to fewer than 50 individuals, mostly due to predation by feral cats. All known kakapo were therefore transferred to three predator-free island refuges, where, supported by supplementary feeding, they appeared to have better prospects. However, recovery was initially disappointing due to the bias of survivors to older birds, a delay in resuming breeding after translocation, and periodic invasions of stoats on one of the islands. The overall status of kakapo is currently described as deteriorating (Garcia-R and Di Marco 2020). It is not alone, as some 23 species (64%) of New Zealand's bird species have declined in their Red-list status since 1985, with flightless, ground-nesting, and larger birds of longer incubation periods generally faring worst. Nonetheless, thanks to predator eradication, translocation, and other conservation measures, the last decade has seen signs of a turning of the tide.

To improve prospects for threatened species it is recommended that: (i) action is taken before populations reach critically low levels; (ii) several separate populations are maintained; (iii) any translocations are into good quality habitat for the species; and (iv) threats in the target area are minimal or can be mitigated. A useful case study is provided by Bakker et al. (2020), who appraise the potential benefits of a combination of translocation to a second island and vaccination against West Nile virus as alternative or combined means of improving chances of persistence of the scrub-jay (*Aphelocoma insularis*), a single-island endemic species from Santa Cruz Island, California.

While some translocation programmes use small islands as 'safe havens', in other cases the reintroduction of a species to an island amounts to the reversal of its extirpation from that island and may form part of an ecological restoration or rewilding agenda. In such cases, an important question is from where do you source the animals (or plants) that you are reintroducing? Conservation scientists typically aim to use local genetic stock, but this is not always possible, and individuals from a different population, or different ESU (evolutionarily significant unit) or variety, or even a close relative or functional analogue of the original species may be considered (e.g. Franklin and Steadman 1991; Falcón and Hansen 2018). For example, Franklin and Steadman (1991) advocated the translocation of the parrots *Vini australis* and *Phigys solitarius* into the island of 'Eua from elsewhere in Tonga, on the grounds that the fossil record demonstrates their former presence. Similarly, when the white-tailed laurel pigeon (*Columba junoniae*) was reintroduced into Gran Canaria, the translocated individuals were sourced from the island of La Palma, where the species is abundant, rather than from the nearest island, Tenerife, where it is comparatively rare.

The prime example of insular translocation as 'rewilding' is the introduction of Aldabra giant tortoise, *Aldabrachelys gigantea*, into other Indian Ocean islands, as 'functional analogues' of extinct species (Hansen et al. 2010; Falcón and Hansen 2018). Until the Late Pleistocene, giant tortoises were prevalent on many oceanic islands, including in the Caribbean, the Atlantic (Canary Islands and Cabo Verde), Pacific Ocean (Galápagos), and Indian Ocean. Although some went extinct prior to human arrival, the major cause of extinction since the Late Pleistocene has been human arrival. Giant tortoises were the largest herbivores on a number of remote islands prior to human arrival and, as can be seen today on the Galápagos, they can have key ecosystem engineer roles through browsing, nutrient cycling/distribution, and seed dispersal. By good fortune, the Aldabra tortoise persisted in the wild and within Aldabra the culling of goats has enabled the tortoise population to persist in reasonable numbers.

Although rewilding has lately gained attention as a revolutionary conservation approach, it has been around for quite a while, even on islands. Between 1978 and 1982, 250 Aldabra tortoises were translocated to Curieuse in the granitic Seychelles (Hambler 1994). Mortality was quite high in that project, with only 117 animals surviving to 1990. Losses were attributed to poaching and possibly resource deficiencies. To ensure success, measures may be taken to ensure that non-native mammalian predators are controlled, and that the tortoises have adequate shade and water resources as they become accustomed to their new environments.

Introductions of Aldabran giant tortoise to Ile aux Aigrettes in 2000 and to Round Island in 2007 (in the latter case alongside an unrelated Madagascan tortoise species) have produce encouraging results, with evidence that they are selectively grazing non-native species, including an invasive legume (*Leucaena leucocephala*) on Ile aux Aigrettes, where the tortoises began breeding soon after their introduction (Hansen et al. 2010). The transformative effect on the vegetation cover noted in this study has been observed elsewhere. Giant tortoise translocation and rewilding projects involving endemic Galápagos tortoise (genus *Chelonoidis*) have begun within the Galápagos and have also been proposed as options for use in the Caribbean, where many islands possessed related species of giant tortoise prior to human colonization (Falcón and Hansen 2018). The introduction of giant tortoises is a low-risk option to replace lost ecosystem functions given that they are not difficult to control should unforeseen negative consequences arise. They thus provide 'flagship' examples of island rewilding, although typically, additional conservation measures and interventions are required to ensure that the desired outcomes are attained and sustainable. This is generally true for *ex situ* breeding and translocation programmes, as stressed above.

Programmes based on *ex situ* breeding and translocation back into the wild thus have their place in the conservation of island endemics. However, they have to be undertaken as part of a highly organized, managed programme. Christian (1993) noted that as of the early 1990s, several parrots have been taken out of St Vincent and the Grenadines for captive breeding, but there was no evidence of any captive-bred parrots being returned to St Vincent for re-release. According to a 2016 Birdlife International fact sheet, that remains the case and, indeed, this is a frequent criticism of *ex situ* breeding programmes. As an insurance against final extinction of a species, such programmes provide a vital safety net, but ultimately, it is crucial to mitigate the *in situ* threats to enable successful reintroduction. For examples, see Sanz and Grajal (2008) on the reintroduction of an endangered parrot to Margarita Island, Venezuela, and Daltry et al. (2017) on the reintroduction of an endangered snake (the Antiguan racer *Alsophis antiguae*) to small islands off Antigua following predator eradication initiatives.

Endangered plants may also be the subject of translocation. Although vegetative propagation is sometimes used, translocation efforts often involve the collection of seeds from a remnant population or seedbank store and experiments to determine optimal germination, growing on and establishment prior to planting out. Such work typically involves botanic gardens and other institutions with appropriate greenhouse facilities and combines *ex situ* and *in situ* action. An exemplary case is provided by efforts to conserve a rare endemic shrub/small tree, *Bencomia exstipulata*, which is found only within two high-elevation National Parks on La Palma and Tenerife (Canary Islands). Conservation actions over a 15-year period have involved population supplementation and detailed genetic and field monitoring work to assess threats and to assess population viability through analysis of pollination and dispersal of the species (Marrero-Gómez et al. 2019). These efforts have resulted in improvements in the conservation status of the species and a clear set of recommendations for further action. Fortunately, this species does not appear to be particularly impacted by rabbit, mouflon, or Barbary sheep, which threaten survival of several other endangered endemic species in these environments (e.g. Martín-Esquivel et al. 2020).

It is encouraging to see more such initiatives coming on stream, as exemplified by Fenu et al. (2019), who describe the early stages of the Care-Mediflora project, a collaborative effort focused on several dozen plant species on six Mediterranean islands, involving translocation either to reinforce existing populations or to form new ones. As the authors stress, such initiatives benefit from knowledge sharing, building on experience gained in previous studies. They also depend, of course, on appropriate financial resources and on successful engagement with key stakeholders to ensure continuing support for conservation goals and measures.

Restoration of native vegetation on islands (as elsewhere) often involves active efforts to replant with native species, but in places can follow land-abandonment through spontaneous invasion by native species. This can be observed, for example, in some previously cultivated areas within the mid-elevations on Tenerife, where pioneers of the laurel forest can be seen encroaching on abandoned terraces in the Anaga and Teno peninsulas. Whether it is through active or passive management, the restoration of connected suites of natural habitats that provide for elevational range adjustments is an important conservation goal in the context of global climate change (e.g. Pouteau and Birnbaum 2016; Falcón and Hansen 2018).

15.11 Protected area and species protection systems: the Canarian example

One of the most important conservation tools on islands, as elsewhere, is the establishment of protected areas. The International Union for the Conservation of Nature (IUCN) formally defines protected areas (PAs) as: 'A clearly defined geographical space, recognised, dedicated and managed, through legal or other effective means, to achieve the long-term conservation of nature with associated ecosystem services and cultural values'

(Dudley 2013). As illustrated below, this definition encompasses a range of values and management goals. For practical purposes, the IUCN currently recognizes six distinct management categories (arguably seven, as one type is further subdivided). They range from strict nature reserves in which human visitation and use are strictly controlled to areas in which there are numerous settlements and substantial levels of economic activity. Given which, and also reflecting varying resourcing and enforcement, the degree of protection afforded by PA status is highly variable. Nonetheless, it is encouraging that island PA coverage has increased markedly since the adoption of the Convention for Biological Diversity in Rio in 1992. Mouillot et al. (2020) have collated data for c. 2500 inhabited islands across the world, reporting that on average 22% of the land area and 13% of marine areas of these islands were designated as PAs: both figures being higher than the respective Aichi Target 11 figures of 17% and 10% by 2020. However, as indicated in Figure 15.6, the coverage is very uneven.

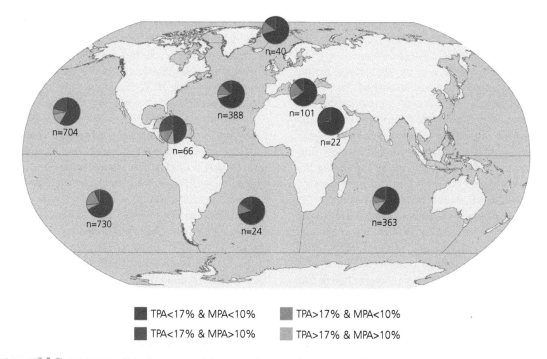

Figure 15.6 The proportions of islands meeting (or failing to meet) the 17% land area and 10% marine area Aichi targets for Protected Areas to be established by 2020, across the world's oceans, based on data for 2438 inhabited islands. TPA = terrestrial protected areas; MPA = marine protected areas.
Figure 18 from Mouillot et al. (2020).

Indeed, 57% of the islands in the survey have no terrestrial PAs. Given the prevailing threats to island biodiversity, there is a demonstrable need to increase the protection of island ecosystems and, indeed, Fernández-Palacios et al. (2021a) call for an increase to 30% (by area) of each major habitat type to be protected within each island group, in line with the recent call by the UN Convention on Biological Diversity (CBD) Secretariat for this to be set as the global target.

The Canaries provide an interesting example of a well-developed insular protected area system. These islands form an autonomous region of Spain and conservation actions within the archipelago are thus subject to a range of insular, archipelagic, national (Spain), and supra-national (EU, international) legislation. The combination of high biodiversity value and severe pressure on natural environments and biota within the archipelago has been recognized by governmental and NGOs at all levels. For example, Birdlife International's Endemic Bird Areas scheme rates the islands a 'high priority', and they are included within an expanded Mediterranean hotspot in the Conservation International 'biodiversity hotspots' scheme (see Box 4.1).

Teide National Park was the first Canarian PA, designated in January 1954. A few small reserves were added in 1975 and by 1985 three further National Parks had been established. In 1987, a new indigenous conservation network, the Red Canaria de Espacios Naturales Protegidos (ENP network), was put into law, instituting a representative network of PAs across the archipelago. Subsequent laws were passed, slightly expanding this network and providing the basis for their management designation, so that by the turn of the century, management plans were being drafted and implemented for PAs covering some 40% of the land area (Machado and Martín 1993; Castro-Torres 2019). These plans were guided by the IUCN protected areas model (see Dudley 2013). The network thus includes both strictly protected sites (e.g. 'integral natural reserves' and 'natural monuments'), and 'rural parks' (Table 15.3). In the latter, the emphasis is on sustaining the rural populations and their culture, while improving livelihoods and standards of living. Thus, the goals of species and habitat conservation within these parks are to be achieved within 'cultural landscapes' of traditional resources use (Martín-Esquivel et al. 1995; Castro-Torres 2019). It is worthy of note that throughout this period, the development of the PA network in the Canaries was at the forefront of PA planning and expansion within Spain as a whole, showing that sometimes island societies are able to take bold initiatives.

The establishment of the EU's Natura 2000 network was used by the regional administration to underscore the existing terrestrial network, to complement it with a few carefully selected additional habitats, increasing the land area covered from 40% to 44.5%, and to incorporate the first marine PAs. Further additions have followed (Table 15.3, final row). The development of these networks has involved important efforts of stakeholder engagement and political support. The resulting governance arrangements are complex and multi-level and generate numerous challenges to ensure effective management for conservation and development (Fernandes et al. 2015; Castro-Torres 2019), but what has been achieved is in many ways remarkable. Moreover, there are several other categories of PA, often overlapping to provide multiple layers of designation and protection for high-value areas. Under UNESCO designation systems, the Canaries contain seven Biosphere Reserves, two natural World Heritage sites, one Ramsar Convention wetland (Table 15.3), and three (possibly soon four) marine areas incorporating no-take zones. There are four National Parks, one quarter of the Spanish network, although the archipelago constitutes just 1.5% of Spain's land area. Teide National Park, which received 4 million visitors in 2016, accounts for 30% of all visitors to the Spanish National Park network.

In addition to these 'area-based' schemes, there are two catalogues of protected species, one designated at national (Spanish state) level, seeking to protect 173 Canarian species and the other at the Canarian (autonomous region) level, seeking to protect 450 endemic species or Canarian populations of non-endemic species with a high emblematic value for global conservation. A review of the second of these catalogues in 2004 concluded that the application of crude thresholds of geographical range derived from continental contexts had led to the inclusion of some 200 species that were not

Table 15.3 The multiple layers of Protected Area (PA) designation in the Canaries: the networks they belong to, their IUCN (International Union for the Conservation of Nature) PA management designation, their essential characteristics, the number of PAs per category, and their total area. The first eight categories form the Canarian ENP network (see text). Compiled partly from Castro-Torres (2019) and partly from official sources.

Protected Area type (and network)	IUCN CATEGORY	CHARACTERISTICS	No.	Area (ha)
National Park (Spanish and Canarian networks)	IUCN II	Large areas relatively untransformed by human activity, with high importance due to the singularity of their biota, geology, or geomorphology and representing the main natural Spanish ecosystems.	4	27 352
Natural Park (Canarian network)	IUCN II	Large areas with similar characteristics to the National Parks, representative of the Canarian natural heritage.	11	111 022
Rural Park (Canarian network)	IUCN V–VI	Large areas where agricultural and livestock activities coexist with zones of great natural and ecological interest.	7	80 401
Integral Natural Reserve (Canarian network)	IUCN Ia	Small natural areas protecting populations, communities, ecosystems or geological elements of special value for their rareness or fragility. Only scientific activities allowed.	11	7 492
Special Natural Reserve (Canarian network)	IUCN Ib	Similar to 'integral' but allowing educational and recreational activities together with scientific ones.	15	14 944
Site of Scientific Interest (Canarian network)	IUCN IV	Small, isolated sites comprising populations of threatened species. [Not precise IUCN equivalence; inspired by UK Sites of Special Scientific Interest]	19	1402
Natural Monument (Canarian network)	IUCN III	Small areas characterized by geological or palaeontological elements of special singularity.	52	28 972
Protected Landscape (Canarian network)	IUCN V	Areas with outstanding aesthetic or cultural values.	27	39 333
Special Area of Conservation (EU Natura 2000 network)	–	Areas which contribute to maintaining or restoring natural habitat types or species in a favourable conservation status. The 151 terrestrial SACs largely overlap with the ENP (Canarian) network, adding 300 km^2 to it; the 23 marine SACs contributed 1800 km^2.	174	462 425
Special Protected Area (EU Natura 2000 network)	–	Areas which contribute to preserve, maintain or restore the diversity and extension of the proper habitats for the 44 Canarian bird species included in the European directives.	27	277 310
Biosphere Reserves (UNESCO network)	–	Areas protecting spaces where human activity constitutes an integral component of the territory, and where the management should focus on the sustainable development of the resources.	7	820 848
World Heritage Site (UNESCO network)	–	Outstanding natural areas with a singularity at a world scale.	2	17 555
Geoparks (UNESCO network)	–	Clearly defined geographical areas of high geological significance, managed through a holistic approach combining protection and conservation, education, and sustainable development through a bottom-up approach.	2	2783
Ramsar Convention Protected Wetlands (UNESCO network)	[cf. IUCN IV]	Wetlands offering important ecological services, such as the regulation of water regimes as well as important sources of biodiversity.	1	127
Marine Reserves of Fishery Interest (Spanish Fisheries Ministry Network)	–	Reserves managed by the Canarian Government or Spanish Fisheries Ministry, where fishing is banned within a 'no take area', where only scuba diving is allowed.	3	74 644

particularly endangered, while a further 40 species were removed from the list based on reassessment of their taxonomic or distributional status (Martín-Esquivel et al. 2005). This example highlights that Red-list criteria require some adjustment for the generally smaller ranges and often uncertain taxonomic status of insular species for application to island settings. See proposals to deal with this challenge provided by Martín (2009) and Gray et al (2019).

Being able to reduce the designated threat level for island species, whether as the result of conservation action or improved data, is always welcome news. However, political imperatives often remain focused on short-term economic considerations and a model based around increased tourism, development, and urbanization. This economic model casts a dense shadow of uncertainty over the future of the natural resource base and biodiversity of the Canaries, particularly of the warm, dry climate belt so popular with European tourists. These concerns are multiplied by the uncertain ramifications of global climate change, with evidence of sensitivity of these systems to interactions between non-native species and changing insular climate (Martín-Esquivel et al. 2020). Overall, the pressures on the natural resources of the Canaries remain intense (García-Falcón and Medina-Muñoz 1999; Fernández-Palacios and de Nascimento 2011; Fernandes et al. 2015) and political battles often need to be refought. These pressures include efforts to reduce the protection afforded to particular PAs that have potential commercial value for development: thus far resisted. Nonetheless, the various networks of protected area status and the other conservation measures discussed in this section show how it is possible to tailor protected area and Red-listing models to an insular context and for them to gain a remarkable degree of acceptance.

15.12 Recap and synthesis: island conservation depends on societal support

At the outset of this chapter, we suggested that remote islands have distinctive conservation problems, arising ultimately from their small size, isolation, distinctive biology (high proportion of endemics, disharmonic, etc.), and environments (climatically distinctive, peculiar geologies, limited habitat diversity, etc.). Insular peoples have commonly evolved distinctive cultures and retain a strong allegiance to both home and culture (Beller et al. 1990). While there are many points of difference across the world's remote islands, their societies often operate within a common set of constraints, facing similar challenges (Table 15.1).

The remedies identified in the literature are often the same: (i) control harmful non-native species, (ii) halt or reverse habitat loss, (iii) prevent over-hunting, (iii) assist breeding of endangered species and by a range of *ex situ* and *in situ* measures, improve their immediate prospects, (iv) enlist political and legislative support, and (v) educate and enthuse the people to improve compliance and cooperation in meeting management goals (Box 15.3). This is, of course, quite an agenda! Knowledge sharing and pooling of expertise across comparable island nations and groupings is thus a critical step to successful conservation management (Caujapé-Castells et al. 2010; Fernández-Palacios et al. 2021a).

Box 15.3 Ten suggested conservation priorities for remote island systems

Adapted from a fuller list of priorities identified by Fernández-Palacios et al. (2021a), their box 1.

(1) Fund and co-ordinate scientific surveys to complete the **biotic inventory of less well-surveyed islands** (e.g. Guinea Gulf, Red Sea, Andaman, Nicobar, Micronesia, Desventuradas).

(2) Establish a minimum threshold of **30% of territory with some level of protection** in islands worldwide.

(3) Prepare **red** (threatened species) and **black** (invasive non-native species) **lists** for all island territories worldwide.

(4) Implement **biosecurity barriers** on islands to reduce the risk of introduction of non-native species.

(5) Develop and implement targeted non-native species **control/eradication programmes** (e.g. New Zealand's predator-free 2050 target) alongside long-term monitoring efforts, and including use of translocation of populations of threatened endemic species into predator-free islands or areas.

(6) Promote **habitat restoration** to: (a) avoid paying extinction debts on islands and (b) to restore ecosystem services.

(7) Create **elevational corridors** that facilitate species movements in response to climate change.

(8) Develop captive **breeding programmes and germplasm banks** for all critically endangered island species.

(9) **Reintroduce extirpated species** where threats have been mitigated, and perform controlled rewilding experiments for restoration of ecosystem functions using appropriate taxon substitutes where keystone species have been extinguished.

(10) **Build local conservation capacity** and engage with island communities to encourage their stewardship of island biodiversity and ecosystem services.

It is important, finally, to consider whether there are any special considerations on the societal, economic, and political side of the equation that must be included in conservation thinking. For this, we return to the basket of island traits set out in Table 15.1. While it is undeniable that prehistoric societies transformed island environments and generated large numbers of insular extinctions and in some cases failed, the abandonment of a marginal habitation in favour of better prospects elsewhere is not especially remarkable: some small Scottish islands were abandoned as recently as the 20th century. Rather, than insular maladaptation, it is fragility to outside contact that is the more dominant theme of our review, continuing to the present day. At the time of writing, over the course of the Covid-19 pandemic (*c.* 2020–2022), we have seen that many island nations and societies have been caught in a dilemma, able for a while at least to prevent the arrival of an unmanageable disease by cutting links to the outside world, but unable to sustain this

for any length of time because of their interdependency on global economic links (e.g. Davila et al. 2020; Kumar and Patel 2021). Indeed, the dependence today of many islands on tourism threatens another classic 'bust' phase, highlighted by the way the travel industry globally reeled from an enforced collapse in demand. In the context of reduced fisheries and other resource limitations, climate change, existential threats from rising sea level, and the natural disasters that have, for example, afflicted many Caribbean islands in recent years, the challenges for many island governments are enormous.

The outside world presents both opportunities and risks for small island states, as witnessed during the major wars of the 20th century when many small Pacific islands were the location of brutal conflict, and as evident again today in jostling for support from the major powers, China and the USA, for strategic influence in parts of the Pacific and South China Sea (e.g. see media reports by Doherty 2020; Lubold 2020). Given such environmental, geopolitical, and economic realities, the importance of concerted international actions through respected well-supported international institutions and mechanisms, such as the CBD, IUCN, IPCC, etc., are essential to providing the framework for sustainable development (Table 15.4; Andersson et al. 2019) and successful biodiversity conservation action in small-island contexts.

Within islands themselves, it is essential that conservationists engage with local communities

Table 15.4 Sustainable development options for small islands should aim at increasing self-reliance and can be categorized under six basic headings according to Hess (1990), from which this is modified. Some options (e.g. tax havens) may come with significant downsides and risks. All rely upon effective governance regimes.

Categories	Examples
Resource preservation	Conservation zones, multiple-use options, control of hunting, introduction of 'no-take' zones, mangrove protection schemes, deployment of coastline protection schemes to mitigate sea-level increase
Resource restoration	Replanting, re-introduction, alien herbivore and predator removal, improve quality of water discharge
Resource enhancement	Freshwater re-use, seawater desalination
Sustainable resource development	Small-scale diversified, closely managed forms of resource-based enterprises (agriculture, fisheries, eco-tourism)
Provision of human services	Alternative energy generation sources and distribution systems, waste disposal
Non-resource-dependent development options	Financial services (tax havens), light industries processing imported materials, rental of fishing rights, encouragement of the internet economy

and stakeholders to build and maintain support for conservation action, including through outreach and educational activities. We mentioned earlier how the first rabbits were introduced to the Madeiran island of Porto Santo by Christopher Columbus' father-in-law in the 1420s, with seriously damaging effects on the native vegetation and ecology of the island, including the extinction of species. The lessons learnt from this oft-repeated experiment may have been well learnt by the scientific community, but it cannot be assumed that these insights are shared. In recent years, the rabbit population on Porto Santo crashed owing to a combination of the rabbit haemorrhagic disease and the *Myxoma* virus. It appears to us deeply ironic that rather than seizing this as a conservation opportunity, the regional government drew up plans to boost the rabbit population by translocation from Madeira, to align with the hunting lobby, for which rabbits represent the main game species (Rocha et al. 2017).

The task facing island conservationists and environmental managers is not a simple one. In many ways the ecological management goals are the easiest part of the equation to specify (e.g. Box 15.3), while managing the societal, economic, and political issues can appear intractable (Beller et al. 1990; González et al. 2008; Connell 2018; Andersson et al. 2019). To provide a different illustration, concepts of ownership among island nations can be both extremely varied and complex. In Papua New Guinea, the Solomon Islands, and Vanuatu, ownership is defined by oral traditions, land is typically owned by family groups rather than individuals, and there may be secondary and tertiary levels of 'ownership' providing rights of use but not implying actual ownership of the land (Keast and Miller 1996). While this is a very specific point, the need to gain and keep societal support is of universal importance. Success depends on how well insular resource planning and management can provide solutions tailored to local circumstances and cultures and whether they carry with them the support of the island communities (Nunn 2004; Murphy et al. 2019; Wald et al. 2019). The involvement of major resource users, farmers, fishermen, hunters, charter operators, etc., is one element in this. The mobilization of NGOs in island problems is another element. Perhaps above all, it

is necessary to foster a general public awareness of the environmental and ecological constraints and problems, and the significance to the island societies themselves of the ecological 'goods' at stake (Christian 1993; González et al. 2008; Fernandes et al. 2015). There are remedies, but the solutions need to be tailored to the islands and not simply exported from the continents.

15.13 Summary

This final chapter addresses contemporary problems and threats to island environments, ecosystems, species, and societies, and reviews some important elements of the conservation remedies that have been developed. At the outset, we identify a set of distinctive 'island societal syndromes' that characterize the interactions between remote islands and the outside world and which island societies have to respond to and plan for.

Among key challenges, we review the implications of 21st-century climate change, with the Maldives an example of the existential threat provided to many small, low-lying islands by combinations of sea-level rise, ocean warming and acidification, regional climate change, and non-native species. The loss of even tiny islands can be of massive economic and geopolitical importance where they lead to a substantial reduction in a nation's EEZ and rights over undersea resources.

In an increasingly globalized world, the effective isolation of many oceanic islands has been broken and increasingly so. We review the devastation of Nauru through mining of phosphate-rich guano deposits over the last century in relation to the geopolitical context of colonial-era and post-colonial resource extraction and power: an extreme example of the vulnerability of island socio-economic and ecological systems to the outside world. A very different example of the imposition of continental ecological footprints into islands is provided by mass tourism on the Canary Islands, which has fuelled huge increases in population size and infrastructure and placed ever-increasing demands on the natural environment. The growth of tourism, in forms ranging from inexpensive mass tourism to higher-end eco-tourism and luxury cruises, has increasingly impinged on many islands across the globe and is a key engine of population expansion on the Galápagos, where conflicts between agriculture, fishing, and conservation have manifested in recent decades, with important implications for efforts to control and eliminate non-native species.

We move on to ask what is it we wish to prioritize in island conservation? One key goal is to avoid species extinctions, but choices often need to be made as to which species to prioritize for action. Other choices include whether to manage towards past baselines that value not only the 'pristine' but also historical cultural landscapes, or whether to work towards a more dynamic, less regulated future, prioritizing 'healthy' functional ecosystems over specific compositional states. Key conservation instruments include biosecurity measures, control of non-native species, regulation of hunting, and establishment of protected areas.

In the last quarter-century, methods for controlling invasive species have advanced considerably, aided by systematic efforts to trial alternative treatments, to monitor outcomes and to share best practice. Interactive (often synergistic) effects between combinations of non-native species require careful analysis to avoid unwanted outcomes from management interventions. Islands provide some classic examples of ill-considered biological control, but increasingly numerous instances of effective treatments and in combination with other measures, biological control can be seen as a key tool.

Similarly, coordinated *ex situ* and *in situ* measures are needed for successful translocation and reintroduction programmes. In cases, as exemplified by introductions of the giant tortoise within Indian Ocean islands, these projects can be termed 'rewilding', although in general the emphasis on remote islands is on removing large herbivores and predators rather than reintroducing them.

The establishment of Protected Area systems is another key conservation measure, and one that has now been adopted within numerous archipelagos. We illustrate, through the example of the multi-layered Canarian PA system, how it is possible—and arguably crucial—to adapt the practical and legal instruments of protected area planning and species protection, for application in insular contexts.

We close the book by reflecting on the special context of biodiversity conservation on remote islands. Here, again, improved science and knowledge exchange from multiple island systems provides a range of options that can be tailored to meet the context. To be successful, conservation actions must ultimately be part of broader management for the sustainable development of island ecosystems and this in turn requires political and legislative support (from local level to international treaties), and investment in educational programmes to engender the interest and support of island peoples. Sustained success requires societal support and that can never be taken for granted but needs continual renewal. We should take encouragement from the many individual success stories and the improved conservation science underpinning island conservation and redouble our efforts.

References

Abdelkrim, J., Pascal, M., and Samadi, S. (2005). Island colonization and founder effects: the invasion of the Guadeloupe Islands by ship rats (*Rattus rattus*). *Molecular Ecology*, **14**, 2923–2931.

Abe, T. (2006). Threatened pollination systems in native flora of the Ogasawara (Bonin) Islands. *Annals of Botany*, **98**, 317–334.

Abbott, I. and Grant, P. R. (1976). Nonequilibrial bird faunas on islands. *American Naturalist*, **110**, 507–528.

Adler, G. H. (1992). Endemism in birds of tropical Pacific islands. *Evolutionary Ecology*, **6**, 296–306.

Adler, G. H. (1994). Avifaunal diversity and endemism on tropical Indian Ocean islands. *Journal of Biogeography*, **21**, 85–95.

Adler, G. H., Austin, C. A., and Dudley, R. (1995). Dispersal and speciation of skinks among archipelagos in the tropical Pacific Ocean. *Evolutionary Ecology*, **9**, 529–541.

Adler, G. H. and Dudley, R. (1994). Butterfly biogeography and endemism on tropical Pacific Islands. *Biological Journal of the Linnean Society*, **52**, 151–162.

Adler, G. H. and Levins, R. (1994). The island syndrome in rodent populations. *Quarterly Review of Biology*, **69**, 473–490.

Adsersen, H. (1995). Research on islands: classic, recent, and prospective approaches. In *Islands: Biological Diversity and Ecosystem Function* (ed. P. M. Vitousek, L. L. Loope, and H. Adsersen), Ecological Studies **115**, pp. 7–21. Springer-Verlag, Berlin.

Aguilée, R., Claessen, D., and Lambert, A. (2013). Adaptive radiation driven by the interplay of eco-evolutionary and landscape dynamics. *Evolution*, **67**, 1291–1306.

Aguilée, R., Pellerin, F., Soubeyrand, M., Choin, J., and Thébaud, C. (2021). Biogeographic drivers of community assembly on oceanic islands: the importance of archipelago structure and history. *Journal of Biogeography*, **48**, 2616–2628.

Aguilera, F., Brito, A., Castilla, C., et al. (1994). *Canarias: economía ecología y medio ambiente*. Francisco Lemus Editor, La Laguna.

Akbar, P. G., Nugroho, T. W., Suranto, M. et al. (2021). No longer an enigma: rediscovery of Black-browed Babbler *Malacocincla perspicillata* in Kalimantan, Indonesia. *Journal of Asian Ornithology*, **37**, 1–5

Alarcón, M., Aldasoro, J. J., Roquet, C. and Olesen, J. M. (2014). The evolution and pollination of oceanic bellflowers (Campanulaceae). In *Evolutionary Biology: Genome Evolution, Speciation, Coevolution and Origin of Life* (ed. P. Pontarotti), pp. 301–332. Springer, Switzerland.

Alava, J. J. and Paladines, F. (2017). Illegal fishing on the Galápagos high seas. *Science*, **357**, 1362.

Alcover, J. A., Pieper, H., Pereira, F., and Rando, J. C. (2015). Five new extinct species of rails (Aves: Gruiformes: Rallidae) from the Macaronesian Islands (North Atlantic Ocean). *Zootaxa*, **4057**, 151–190.

Ali, J. R. (2012). Colonizing the Caribbean: is the GAARlandia land-bridge hypothesis gaining a foothold? *Journal of Biogeography*, **39**, 431–433.

Ali, J. R. (2017). Islands as biological substrates: classification of the biological assemblage components and the physical island types. *Journal of Biogeography*, **44**, 984–994.

Ali, J. R. and Aitchison, J. C. (2014). Exploring the combined role of eustasy and oceanic island thermal subsidence in shaping biodiversity on the Galápagos. *Journal of Biogeography*, **41**, 1227–1241.

Ali, J. R., Aitchison, J. C., and Meiri, S. (2020). Redrawing Wallace's Line based on the fauna of Christmas Island, eastern Indian Ocean. *Biological Journal of the Linnean Society*, **130**, 225–237.

Ali, J. R. and Heaney, L. R. (2021). Wallace's line, Wallacea, and associated divides and areas: history of a tortuous tangle of ideas and labels. *Biological Reviews*, **96**, 922–942.

Ali, J. R. and Hedges, S. B. (2021). Colonizing the Caribbean: new geological data and an updated land-vertebrate colonization record challenge the GAARlandia land-bridge hypothesis. *Journal of Biogeography*, **48**, 2699–2707.

Ali, J. R. and Hedges, S. B. (2022). A review of geological evidence bearing on proposed Cenozoic land connections between Madagascar and Africa and its relevance to biogeography. *Earth-Science Reviews*, **232**, 104103.

Almeida-Gomes, M., Gotelli, N. J., Rocha, C. F. D., Vieira, M. V., and Prevedello, J. Y. (2022). Random placement models explain species richness and dissimilarity of frog assemblages within Atlantic Forest fragments. *Journal of Animal Ecology*, **91**, 618–629.

Almeida-Neto, M., Guimarães, P. R., Jr, and Lewinsohn, T. M. (2007). On nestedness analyses: rethinking matrix temperature and anti-nestedness. *Oikos*, **116**, 716–722.

Alonso, J. C., Alonso, J. A., and Muñoz-Pulido, R. (1994). Mitigation of bird collisions with transmission lines through groundwire marking. *Biological Conservation*, **67**, 129–134.

Alsos, I. G., Ehrich, D., Eidesen, P. B., et al. (2015). Long-distance plant dispersal to North Atlantic islands: colonization routes and founder effect. *AoB PLANTS*, **7**, plv036.

Amborella Genome Project (2013). The *Amborella* genome and the evolution of flowering plants. *Science*, **342**, 1241089.

Ancochea, E., Hernán, F., Cendrero, A., et al. (1994). Constructive and destructive episodes in the building of a young oceanic island, La Palma, Canary Islands, and the genesis of the Caldera de Taburiente. *Journal of Volcanology and Geothermal Research*, **60**, 243–262.

Anderson, C. L., Channing, A., and Zamuner, A. B. (2009). Life, death and fossilization on Gran Canaria – implications for Macaronesian biogeography and molecular dating. *Journal of Biogeography*, **36**, 2189–2201.

Anderson, G. J. and Bernardello, G. (2018). Reproductive biology. In *Plants of Oceanic Islands: Evolution, Biogeography, and Conservation of the Flora of the Juan Fernández (Robinson Crusoe Islands)* (ed. T. F. Stuessy, D. J. Crawford, P. López-Sepúlveda, C. M. Baeza, and E. A. Ruiz), pp. 193–206. Cambridge University Press, Cambridge.

Anderson, W. B. and Wait, D. A. (2001). Subsidized island biogeography hypothesis: another new twist on an old theory. *Ecology Letters*, **4**, 289–291.

Andersson, A. J., Venn, A. A., Pendleton, L., et al. (2019). Ecological and socioeconomic strategies to sustain Caribbean coral reefs in a high-CO_2 world. *Regional Studies in Marine Science*, **29**, 100677.

Anguita, F., Márquez, A., Castiñeiras, P., and Hernán, F. (2002). *Los volcanes de Canarias. Guía geológica e itinerarios*. Ed. Rueda, Madrid.

Anton, A., Geraldi, N. R., Ricciardi, A. and Dick, J. T. A. (2020). Global determinants of prey naiveté to exotic predators. *Proceeding of the Royal Society* B, **287**, 20192978.

Arendt, W. J., Gibbons, D. W., and Gray, G. (1999). Status of the volcanically threatened Montserrat Oriole *Icterus oberi* and other forest birds in Montserrat, West Indies. *Bird Conservation International*, **9**, 351–372.

Arnedo, M. A., Oromí, P., Martín de Abréu, S. and Ribera, C. (2008). Biogeographical and evolutionary patterns in the Macaronesian shield-backed katydid genus *Calliphona* Krauss, 1892 (Orthoptera: Tettigoniidae) and allies as inferred from phylogenetic analyses of multiple mitochondrial genes. *Systematic Entomology*, **33**, 145–158.

Arrhenius, O. (1920). Yta och arter. I. *Svensk Botanisk Tidsskrift*, **14**, 327–329.

Atkinson, C. T., Saili, K. S., Utzurrum, R. B., and Jarvi, S. I. (2013). Experimental evidence for evolved tolerance to avian malaria in a wild population of low elevation Hawai'i 'amakihi (*Hemignathus virens*). *EcoHealth*, **10**, 366–375.

Ávila, S. P., Melo, C., Berning, B., et al. (2019). Towards a 'Sea-Level Sensitive' dynamic model: impact of island ontogeny and glacio-eustasy on global patterns of marine island biogeography. *Biological Reviews*, **94**, 1116–1142.

Bahn, P. and Flenley, J. R. (1992). *Easter Island, Earth Island*. Thames & Hudson, London.

Baier, F. and Hoekstra, H. E. (2019). The genetics of morphological and behavioural island traits in deer mice. *Proceedings of the Royal Society B*, **286**, 20191697.

Bailey G., Momber, G., Bell, M., et al. (2020). Great Britain: the intertidal and underwater archaeology of Britain's submerged landscapes. In *The Archaeology of Europe's Drowned Landscapes* (ed. G. Bailey, N. Galanidou, H. Peeters, H. Jöns, and M. Mennenga), pp. 189–219. Coastal Research Library, vol. 35.

Bairlein, F. (2016). Migratory birds under threat. *Science*, **354**, 547–548.

Baiser, B. and Li, D. (2018). Comparing species–area relationships of native and exotic species. *Biological Invasions*, **20**, 3647–3658.

Baker, H. G. (1955). Self-compatibility and establishment after long-distance dispersal. *Evolution*, **9**, 347–349.

Bakker, V. J., Sillett, T. S., Boyce, W. M., et al. (2020). Translocation with targeted vaccination is the most effective strategy to protect an island endemic bird threatened by West Nile virus. *Diversity and Distributions*, **26**, 1104–1115.

Baldursson, S. and Ingadóttir, Á. (eds) (2007). *Nomination of Surtsey for the UNESCO World Heritage List*. Icelandic Institute of Natural History, Reykjavík.

Ballmer, M.D., van Hunen, J., Ito, G., Bianco, T. A., and Tackley, P. J. (2009). Intraplate volcanism with complex age-distance patterns: a case for small-scale sublithospheric convection. *Geochemistry, Geophysics, Geosystems*, **10**, Q06015.

Bambrick, H. (2017). Resource extractivism, health and climate change in small islands. *International Journal of Climate Change Strategies and Management*, **10**, 272–288.

Barajas-Barbosa, M. P., Weigelt, P., Borregaard, M. K., Keppel, G., and Kreft, H. (2020). Environmental heterogeneity dynamics drive plant diversity on oceanic islands. *Journal of Biogeography*, **47**, 2248–2260.

Barreto, E., Rangel, T. F., Pellissier, L., and Graham, C. H. (2021). Area, isolation and climate explain the diversity of mammals on islands worldwide. *Proceedings of the Royal Society B*, **288**, 20211879.

Barrett, S. C. H. (1996). The reproductive biology and genetics of island plants. *Philosophical Transactions of the Royal Society of London*, B **351**, 725–733.

Barrett, S. C. H. and Hough, J. (2013). Sexual dimorphism in flowering plants. *Journal of Experimental Botany*, **64**, 67–82.

Baselga, A. (2010). Partitioning the turnover and nestedness components of beta diversity. *Global Ecology and Biogeography*, **19**, 134–143.

Baselga, A. (2012). The relationship between species replacement, dissimilarity derived from nestedness, and nestedness. *Global Ecology and Biogeography*, **21**, 1223–1232.

Beauchamp, G. (2021). Do avian species survive better on islands? *Biology Letters*, **16**, 20200643.

Bell, R. C., Drewes, R. C., Channing, A., et al. (2015). Overseas dispersal of *Hyperolius* reed frogs from Central Africa to the oceanic islands of São Tomé and Príncipe. *Journal of Biogeography*, **42**, 65–75.

Bellard, C., Rysman, J.-F., Leroy, B., Claud, C., and Mace, G. M. (2017). A global picture of biological invasion threat on islands. *Nature Ecology and Evolution*, **1**, 1862–1869.

Beller, W., d'Ayala, P., and Hein, P. (ed.) (1990). *Sustainable Development and Environmental Management of Small Islands*, Vol. 5, Man and the Biosphere Series. UNESCO/Parthenon Publishing, Paris.

Bellwood, P. (2017). *First Islanders: Prehistory and Human Migration in Island Southeast Asia*. Wiley, Oxford.

Benchimol, M. and Peres, C. A. (2013). Anthropogenic modulators of species–area relationships in Neotropical primates: a continental-scale analysis of fragmented forest landscapes. *Diversity and Distributions*, **19**, 1339–1352.

Benchimol, M. and Peres, C. A. (2015). Predicting local extinctions of Amazonian vertebrates in forest islands created by a mega dam. *Biological Conservation*, **187**, 61–72.

Benchimol, M. and Peres, C. A. (2021). Determinants of population persistence and abundance of terrestrial and arboreal vertebrates stranded in tropical forest land-bridge islands. *Conservation Biology*, **35**, 870–883.

Benítez-López, A., Alkemade, R., and Verweij, P. A. (2010). The impacts of roads and other infrastructure on mammal and bird populations: a meta-analysis. *Biological Conservation*, **143**, 1307–1316.

Benítez-López, A., Santini, L., Gallego-Zamorano, J., et al. (2021). The island rule explains consistent patterns of body size evolution in terrestrial vertebrates. *Nature Ecology and Evolution*, **5**, 768–786.

Benkwitt, C. E., Gunn, R. L., Le Corre, M., Carr, P., and Graham, N. J. (2021). Rat eradication restores nutrient subsidies from seabirds across terrestrial and marine ecosystems. *Current Biology*, **31**, 2704–2711.

Benn-Torres, J., Bonilla, C., Robbins, C. M., et al. (2008) Admixture and population stratification in African Caribbean populations. *Annals of Human Genetics*, **72**, 90–98.

Benton, T. and Spencer, T. (ed.) (1995). *The Pitcairn Islands: Biogeography, Ecology and Prehistory*. Academic Press, London.

Bernardello, G., Anderson, G. J., Stuessy, T. F., and Crawford, D. J. (2001). A survey of floral traits, breeding systems, floral visitors, and pollination systems of the angiosperms of the Juan Fernández Islands (Chile). *Botanical Review*, **67**, 255–308.

Bess E. C., Catanach, T. A., and Johnson, K. P. (2013). The importance of molecular dating analyses for inferring Hawaiian biogeographical history: a case study with bark lice (Psocidae: *Ptycta*). *Journal of Biogeography* **41**, 158–167.

Beyhl, F. E. (1995). The dendroid spurges of Macaronesia. Some functional and biogeographical aspects. *Boletim do Museu Municipal do Funchal*, **4**, 95–100.

Biddick, M. and Burns, K. C. (2019). Reply to Brian and Walker-Hale: support for the island rule does not hide morphological disparity in insular plants. *Proceedings of the National Academy of Sciences, USA*, **116**, 29431–29432.

Biddick, M., Hendriks, A. and Burns, K. C. (2019). Plants obey (and disobey) the island rule. *Proceedings of the National Academy of Sciences, USA*, **116**, 17632–17634.

Biedermann, R. (2003). Body size and area–incidence relationships: is there a general pattern? *Global Ecology and Biogeography*, **12**, 381–387.

Billerman, S. M., Keeney, B. K., Rodewald, P. G., and Schulenberg, T. S. (2022). Birds of the world. https://birdsoftheworld.org/bow/home.

BirdLife Australia (2022). *BirdLife Kangaroo Island: endemics*. Available at: https://birdlife.org.au/locations/birdlife-kangaroo-island/endemics-ki

Blackburn, T. M., Cassey, P., and Pyšek, P. (2021). Species–area relationships in alien species: pattern and process. In *The Species–Area Relationship: Theory and Application* (ed. T. J. Matthews, K. A. Triantis, and R. J. Whittaker), pp. 133–154. Cambridge University Press, Cambridge.

Bliard, L., Paquet, M., Robert, A., et al. (2020). Examining the link between relaxed predation and bird coloration on islands. *Biology Letters*, **16**, 20200002.

Blonder, B. (2017). Hypervolume concepts in niche- and trait-based ecology. *Ecography*, **41**, 1441–1455.

Blonder, B., Baldwin, B. G., Enquist, B. J. and Robichaux, R. H. (2016). Variation and macroevolution in leaf functional traits in the Hawaiian silversword alliance (Asteraceae). *Journal of Ecology*, **104**, 219–228.

Böhle, U.-R., Hilger, H. H., and Martin, W. F. (1996). Island colonization and evolution of the insular woody habit in *Echium* L. (Boraginaceae). *Proceedings of the National Academy of Sciences, USA*, **93**, 11740–11745.

Borges, P. A. V., Cardoso, P., Fattorini, S., et al. (2018a). Community structure of woody plants on islands along a bioclimatic gradient. *Frontiers of Biogeography*, **10**, e40295.

Borges, P. A. V., Cardoso, P., Kreft, H., et al. (2018b). Global Island Monitoring Scheme (GIMS): a proposal for the long-term coordinated survey and monitoring of native island forest biota. *Biodiversity and Conservation*, **27**, 2567–2586.

Borges, P. A. V. and Hortal, J. (2009). Time, area and isolation: factors driving the diversification of Azorean arthropods. *Journal of Biogeography*, **36**, 178–191.

Borregaard, M. K., Amorim, I. R., Borges, P. A. V., et al. (2017). Oceanic island biogeography through the lens of the general dynamic model: assessment and prospect. *Biological Reviews*, **92**, 830–853.

Borregaard, M. K., Matthews, T. J., and Whittaker, R. J. (2016). The general dynamic model: towards a unified theory of island biogeography? *Global Ecology and Biogeography*, **25**, 805–816.

Borrero, J. C., Solihuddin, T., Fritz, H. M., et al. (2020). Field survey and numerical modelling of the December 22, 2018 Anak Krakatau tsunami. *Pure and Applied Geophysics*, **177**, 2457–2475.

Borroto-Páez, R. and Mancina, C. A. (2017). Biodiversity and conservation of Cuban mammals: past, present, and invasive species. *Journal of Mammalogy*, **98**, 964–985.

Boulton, R. A., Bulgarella, M., Ramirez, I. E., Causton, C. E., and Heimpel, G. E. (2019). Management of an invasive avian parasitic fly in the Galapagos Islands: is biological control a viable option? In *Island Invasives: Scaling Up to Meet the Challenge* (ed. C. R. Veitch, M. N. Clout, A. R. Martin, J. C. Russell, and C. J. West), pp. 360–363. Occasional Paper SSC no. 62. IUCN, Gland, Switzerland.

Bouzat, J. L. (2010). Conservation genetics of population bottlenecks: the role of chance, selection, and history. *Conservation Genetics*, **11**, 463–478.

Bowen, L. and van Vuren, D. (1997). Insular endemic plants lack defenses against herbivores. *Conservation Biology*, **11**, 1249–1254.

Boyer, A. G. (2008). Extinction patterns in the avifauna of the Hawaiian Islands. *Diversity and Distributions*, **14**, 509–517.

Boyer, A. G. and Jetz, W. (2014). Extinctions and the loss of ecological function in island bird communities. *Global Ecology and Biogeography*, **23**, 679–688.

Braby, M. F., Eastwood, R., and Murray, N. (2012). The subspecies concept in butterflies: has its application in taxonomy and conservation biology outlived its usefulness? *Biological Journal of the Linnean Society*, **106**, 699–716.

Braje T. D., Leppard T. P., Fitzpatrick S. M., and Erlandson J. M. (2017). Archaeology, historical ecology and anthropogenic island ecosystems. *Environmental Conservation*, **44**, 286–297.

Brian, J. I. and Walker-Hale, N. (2019). Focus on an island rule may hide morphological disparity in insular plants. *Proceedings of the National Academy of Sciences, USA*, **116**, 24919–24930.

Brock, K. M., Bednekoff, P. A., Pafilis, P., and Foufopoulos, J. (2015). Evolution of antipredator behavior in an island lizard species, *Podarcis erhardii* (Reptilia: Lacertidae): The sum of all fears? *Evolution*, **69**, 216–231.

Brooke, M. de L., Bonnaud, E., Dilley, E. J., et al. (2018). Seabird population changes following mammal eradication on islands. *Animal Conservation*, **21**, 3–12.

Brown, B. L., Swan, C. M., Auerbach, D. A., et al. (2011). Metacommunity theory as a multispecies, multiscale framework for studying the influence of river network structure on riverine communities and ecosystems. *Journal of the North American Benthological Society*, **30**, 310–327.

Brown, J. H. (1971). Mammals on mountaintops: nonequilibrium insular biogeography. *American Naturalist*, **105**, 467–478.

Brown, J. H. (1981). Two decades of homage to Santa Rosalia: toward a general theory of diversity. *American Zoologist*, **21**, 877–888.

Brown, J. H. (1995). *Macroecology*. Chicago University Press, Chicago.

Brown, J. H. (1999). Macroecology: progress and prospect. *Oikos*, **87**, 3–13.

Brown, J. H. and Kodric-Brown, A. (1977). Turnover rates in insular biogeography: effect of immigration on extinction. *Ecology*, **58**, 445–449.

Brown, S., Wadey, M. P., Nicholls, R. J., et al. (2020). Land raising as a solution to sea-level rise: an analysis of coastal flooding on an artificial island in the Maldives. *Journal of Flood Risk Management*, **13**, e12567.

Brown, W. L., Jr and Wilson, E. O. (1956). Character displacement. *Systematic Zoology*, **7**, 49–64.

Buckley, L. B. and Jetz, W. (2007). Insularity and the determinants of lizard population density. *Ecology Letters*, **10**, 481–489.

Buckley, R. C. and Knedlhans, S. B. (1986). Beachcomber biogeography: interception of dispersing propagules by islands. *Journal of Biogeography*, **13**, 68–70.

Bulman, C. R., Wilson, R. J., Holt, A. R., et al. (2007). Minimum viable metapopulation size, extinction debt, and the conservation of a declining species. *Ecological Applications*, **17**, 1460–1473.

Bunnefeld, N. and Phillimore, A. B. (2012). Island, archipelago and taxon effects: mixed models as a means of dealing with the imperfect design of nature's experiments. *Ecography*, **35**, 15–22.

Burjachs, F., Pérez-Obiol, R., Picornell-Gelabert, L., et al. (2017). Overview of environmental changes and human colonization in the Balearic Islands (Western Mediterranean) and their impacts on vegetation composition during the Holocene. *Journal of Archaeological Science: Reports*, **12**, 845–859.

Burkey, T. V. (1993). Edge effects in seed and egg predation at two Neotropical rainforest sites. *Biological Conservation*, **66**, 139–143.

Burns, F. A., Bonadonna, C., Podi, L., Cole, P. D., and Stinton, A. (2017). Ash aggregation during the 11 February 2010 partial dome collapse of the Soufrière Hills Volcano, Montserrat. *Journal of Volcanology and Geothermal Research*, **335**, 92–112.

Burns, K. C. (2016). Size changes in island plants: independent trait evolution in *Alyxia ruscifolia* (Apocynaceae) on Lord Howe Island. *Biological Journal of the Linnean Society*, **119**, 847–855.

Burns, K. C. (2019). *Evolution in Isolation: The Search for an Island Syndrome in Plants*. Cambridge University Press, Cambridge.

Burns, K. C. (2022). The paradox of island evolution. *Journal of Biogeography*, **49**, 248–253.

Burns, K. C., Herold, N. and Wallace, B. (2012). Evolutionary size changes in plants of the south-west Pacific. *Global Ecology and Biogeography*, **21**, 819–828.

Bush, M. B., Rozas-Davila, A., Raczka, M., et al. (2022). A palaeoecological perspective on the transformation of the tropical Andes by early human activity. *Philosophical Transactions of the Royal Society B*, **377**, 20200497.

Bush, M. B. and Whittaker, R. J. (1991). Krakatau: colonization patterns and hierarchies. *Journal of Biogeography*, **18**, 341–356.

Bush, M. B. and Whittaker, R. J. (1993). Non-equilibration in island theory of Krakatau. *Journal of Biogeography*, **20**, 453–458.

Butler, R. W. (1980). The concept of a tourism area cycle of evolution: implications for the management of resources. *Canadian Geographer*, 24, 5–12.

Cabral, J. S., Whittaker, R. J., Wiegand, K., and Kreft, H. (2019b). Assessing predicted isolation effects from the general dynamic model of island biogeography with an eco-evolutionary model for plants. *Journal of Biogeography*, **46**, 1569–1581.

Cabral, J. S., Wiegand, K., and Kreft, H. (2019a). Interactions between ecological, evolutionary and environmental processes unveil complex dynamics of insular plant diversity. *Journal of Biogeography*, **46**, 1582–1597.

Callicott, J.B., Crowder, L.B., and Mumford, K. (1999). Normative concepts in conservation. *Conservation Biology*, **13**, 22–35.

Camacho, J. P. (2012). *The Mysterious People of the Canary Islands: Guanches Legend and Reality*. 3rd edn. Weston.

Cameron, R. A. D., Cook, L. M., and Hallows, J. D. (1996). Land snails on Porto Santo: adaptive and non-adaptive radiation. *Philosophical Transactions of the Royal Society of London*, B **351**, 309–337.

Cameron, R. A. D., Triantis, K. A., Parent, C. E., et al. (2013). Snails on oceanic islands: testing the general dynamic model of oceanic island biogeography using linear mixed effect models. *Journal of Biogeography*, **40**, 117–130.

Campbell, K. and Donlan, C. J. (2005). Feral goat eradications on islands. *Conservation Biology*, **19**, 1362–1374.

Canals, M., Urgelés, R., Masson, D. G., and Casamor, J. L. (2000). Los deslizamientos submarinos de las Islas Canarias. *Makaronesia*, **2**, 57–69.

Cardoso, P., Arnedo, M., Triantis, K. A., and Borges, P. A. V. (2010). Drivers of diversity in Macaronesian spiders and the role of species extinctions. *Journal of Biogeography*, **37**, 1034–1046.

Cardoso, P., Rigal, F., Carvalho, J. C., et al. (2014). Partitioning taxon, phylogenetic and functional beta diversity into replacement and richness difference components. *Journal of Biogeography*, **41**, 749–761.

Carey, M., Boland, J., Weigelt, P., and Keppel, G. (2020). Towards an extended framework for the general dynamic theory of biogeography. *Journal of Biogeography*, **47**, 2554–2566.

Carine, M. (2005). Relationships of the Macaronesian flora: a relictual series or window of opportunity? *Taxon*, **54**, 895–903.

Carine, M. A., Russell, S. J., Santos-Guerra, A., and Francisco-Ortega, J. (2004). Relationships of the Macaronesian and Mediterranean floras: molecular evidence for multiple colonizations into Macaronesia and back-colonization of the continent in *Convolvulus* (Convolvulaceae). *American Journal of Botany*, **91**, 1070–1085.

Carine, M. A., Santos-Guerra, A., Guma, I. R., and Reyes-Betancourt, J. A. (2010). Endemism and evolution of the Macaronesian flora. In *Beyond Cladistics: The Branching of a Paradigm* (ed. D. M. Williams and S. Knapp), pp. 101–124. University of California Press, Oakland.

Carlquist, S. (1965). *Island Life: A Natural History of the Islands of the World*. Natural History Press, New York.

Carlquist, S. (1974). *Island Biology*. Columbia University Press, New York.

Carlquist, S. (1995). Introduction. In *Hawaiian Biogeography: Evolution on a Hot Spot Archipelago* (ed. W. L. Wagner and V. A. Funk), pp. 1–13. Smithsonian Institution Press, Washington, DC.

Carlquist, S. (2009). Darwin on island plants. *Botanical Journal of the Linnean Society*, **161**, 20–25.

Carpenter, J. K., Wilmshurst, J. M., McConkey, K. R., et al. (2020). The forgotten fauna: native vertebrate seed predators on islands. *Functional Ecology*, **34**, 1802–1813.

Carr, G. D., Robichaux, R. H, Witter, M. S., and Kyhos D. W. (1989). Adaptive radiation of the Hawaiian silversword alliance (Compositae—Madiinae): a comparsion with Hawaiian picture-winged *Drosophila*. In *Genetics, Speciation and the Founder Principle* (ed. L. Y. Giddings, K. Y. Kaneshiro, and W. W. Anderson), pp. 79–95. Oxford University Press, New York.

Carracedo, J. C., Day, S. J., Guillou, H., and Torrado, F. J. P. (1999). Giant Quaternary landslides in the evolution of La Palma and El Hierro, Canary Islands. *Journal of Volcanology and Geothermal Research*, **94**, 169–190.

Carracedo, J. C. and Tilling, R. I. (2003). *Geología y volcanología de islas volcánicas oceánicas. Canarias–Hawaii*. CajaCanarias–Gobierno de Canarias.

Carracedo, J. C. and Troll, V. R. (2021). North-East Atlantic Islands: The Macaronesian Archipelagos. In *Encyclopedia of Geology*, 2nd edn (ed. D. Alterton and S. A. Elias), pp. 674–699. Elsevier.

Carranza, S., Harris, D. J., Arnold, E. N., Batista, V., and Gonzalez de la Vega, J. P. (2006). Phylogeography of the lacertid lizard *Psammodromos algirus*, in Iberia and across the Strait of Gibraltar. *Journal of Biogeography*, **33**, 1279–1288.

Carrera-Martínez, R., Ruiz-Arocho, J., Aponte-Díaz, L., Jenkins, D. A., and O'Brien, J. J. (2020). Effects of fire on native columnar and globular cacti of Puerto Rico: a case study of El Faro, Cabo Rojo. *Fire Ecology*, **16**, article 14.

Carrion, V., Donlan, C. J., Campbell, K. J., Lavoie, C., and Cruz, F. (2011). Archipelago-wide island restoration in the Galápagos Islands: reducing costs of invasive mammal eradication programs and reinvasion risk. *PLOS ONE*, **6**, e18835.

Carson, H. L. (1992). Genetic change after colonization. *Geojournal*, **28**, 297–302.

Carson, H. L., Lockwood, J. P., and Craddock, E. M. (1990). Extinction and recolonization of local populations on a growing shield volcano. *Proceedings of the National Academy of Sciences, USA*, **87**, 7055–7057.

Carstensen, D. W., Dalsgaard, B., Svenning, J-C., et al. (2012). Biogeographical modules and island roles: a comparison of Wallacea and the West Indies. *Journal of Biogeography*, **39**, 739–749.

Carter, Z. T., Perry, G. L. W., and Russell, J. C. (2020). Determining the underlying structure of insular isolation measures. *Journal of Biogeography*, **47**, 955–967.

Carvajal, A. and Adler, G. H. (2005). Biogeography of mammals on tropical Pacific islands. *Journal of Biogeography*, **32**, 1561–1569.

Carvajal-Endara, S., Hendry, A. P., Emery, N. C., and Davies, T. J. (2017). Habitat filtering not dispersal limitation shapes oceanic island floras: species assembly of the Galápagos archipelago. *Ecology Letters*, **20**, 495–504.

Carvalho, J. C., Cardoso, P., and Gomes, P. (2012). Determining the relative roles of species replacement and species richness differences in generating beta-diversity patterns. *Global Ecology and Biogeography*, **21**, 760–771.

Carvalho, J. C., Malumbres-Olarte, J., Arnedo, M. A., Crespo, L. C., Domenech, M., and Cardoso, P. (2020). Taxonomic divergence and functional convergence in Iberian spider forest communities: insights from beta diversity partitioning. *Journal of Biogeography*, **47**, 288–300.

Castilla-Beltrán, A., de Nascimento, L., Fernández-Palacios, J. M., et al. (2021). Anthropogenic transitions from forested to human-dominated landscapes in southern Macaronesia. *Proceedings of the National Academy of Sciences, USA*, **118**, e2022215118.

Castilla-Beltrán, A., Duarte, I., de Nascimento, L., et al. (2020). Using multiple palaeoecological indicators to guide biodiversity conservation in tropical dry islands: the case of São Nicolau, Cabo Verde. *Biological Conservation*, **242**, 108397.

Castro-Torres, S. (2019). *An evaluation of the governance arrangement of the Rural Parks in the Canary Islands: an institutional diagnostic analysis*. MSc thesis. University of Oxford.

Cauclí, J. P., Moncada, S., Bambrick, H., and Correa-Velez, I. (2020). Coping with environmental hazards and shocks in

Kiribati: experiences of climate change by atoll communities in the Equatorial Pacific. *Environmental Development*, **37**, e100549.

Caujapé-Castells, J. (2011). Jesters, red queens, boomerangs and surfers: a molecular outlook on the diversity of the Canarian endemic flora. In *The Biology of Island Floras* (ed. D. Bramwell and J. Caujapé-Castells), pp. 284–324. Cambridge University Press, Cambridge.

Caujapé-Castells, J., García-Verdugo, C., Marrero-Rodríguez, Á., Fernández-Palacios, J. M., Crawford, D. J., and Mort, M. E. (2017). Island ontogenies, syngameons, and the origins and evolution of genetic diversity in the Canarian endemic flora. *Perspectives in Plant Ecology, Evolution and Systematics*, **27**, 9–22.

Caujapé-Castells, J., García-Verdugo, C., Sanmartín, I., et al. (2022). The late Pleistocene endemicity increase hypothesis and the origins of diversity in the Canary Islands Flora. *Journal of Biogeography*, **49**, 1469–1480.

Caujapé-Castells, J., Tye, A., Crawford, D. J. et al. (2010). Conservation of oceanic island floras: present and future global challenges. *Perspectives in Plant Ecology, Evolution and Systematics*, **12**, 107–129.

Caves, E. M., Jennings, S. B., HilleRisLambers, J., Tewksbury, J. J., and Rogers, H. S. (2013). Natural experiment demonstrates that bird loss leads to cessation of dispersal of native seeds from intact to degraded forests. *PLOS ONE*, **8**, e65618.

Čeřňanský, A., Klembara, J., and Smith, K. T. (2016). Fossil lizard from central Europe resolves the origin of large body size and herbivory in giant Canary Island lacertids. *Zoological Journal of the Linnean Society*, **176**, 861–877.

Chadwick, O. A., Derry, L., Vitousek, P. M., Huebert, B. J., and Hedin, L. O. (1999). Changing sources of nutrients during four million years of ecosystem development. *Nature*, **397**, 491–497.

Chamberland, L., McHugh, A., Kechejian, S., et al. (2018). From Gondwana to GAARlandia: evolutionary history and biogeography of ogre-faced spiders (*Deinopis*). *Journal of Biogeography*, **45**, 2442–2457.

Chamorro, S., Heleno, R., Olesen, J., McMullen, C. K., and Traveset, A. (2012). Pollination patterns and plant breeding systems in the Galápagos: a review. *Annals of Botany*, **110**, 1489–1501.

Chang, M. (2000). Exclusive economic zones. In http://geography.about.com/library/misc/uceez.htm (visited March 2006).

Chase, J. M., Gooriah, L., May, F., et al. (2019). A framework for disentangling ecological mechanisms underlying the island species–area relationship. *Frontiers of Biogeography*, **11**, e40844.

Chatterjee, S., Goswami, A. and Scotese, C. R. (2013). The longest voyage: tectonic, magmatic, and paleoclimate evolution of the Indian plate during its northward flight from Gondwana to Asia. *Gondwana Research*, **23**, 238–267.

Chen, C., Xu, A., and Wang, Y. (2021). Area threshold and trait–environment associations of butterfly assemblages in the Zhoushan Archipelago, China. *Journal of Biogeography*, **48**, 785–797.

Chen, C., Yang, X, Tan, X., and Wang, Y. (2020). The role of habitat diversity in generating the small-island effect. *Ecography*, **43**, 1241–1249.

Cherry, J. F. and Leppard, T. P. (2018). Patterning and its causation in the pre-Neolithic colonization of the Mediterranean islands (late Pleistocene to early Holocene). *The Journal of Island and Coastal Archaeology*, **13**, 191–205.

Chetcuti, J., Kunin, W. E., and Bullock, J. M. (2022). Species' movement influence responses to habitat fragmentation. *Diversity and Distributions*, **28**, 2215–2228.

Chiba, S. (2004). Ecological and morphological patterns in communities of land snails of the genus *Mandarina* from the Bonin Islands. *Journal of Evolutionary Biology*, **17**, 131–143.

Chisholm, R. A., Fung, T., Chimalakonda, D., and O'Dwyer, J. P. (2016). Maintenance of biodiversity on islands. *Proceedings of the Royal Society B*, **283**, 20160102.

Christenhusz, M. J. M. and Byng, J. W. (2020). The number of known plant species in the world and its annual increase. *Phytotaxa*, **261**, 201–217.

Christian, C. S. (1993). The challenge of parrot conservation in St Vincent and the Grenadines. *Journal of Biogeography*, **20**, 463–469.

Chua, M. A. H. and Aziz, S. A. (2019). Into the light: atypical diurnal foraging activity of Blyth's horseshoe bat, *Rhinolophus lepidus* (Chiroptera: Rhinolophidae) on Tioman Island, Malaysia. *Mammalia*, **83**, 78–83.

Clague, D. A. (1996). The growth and subsidence of the Hawaiian-Emperor volcanic chain. In *The Origin and Evolution of Pacific Island Biotas, New Guinea to Eastern Polynesia: Patterns and Processes* (ed. A. Keast, and S. E. Miller), pp. 35–50. SPB Academic Publishing, Amsterdam.

Clarke, B. and Grant, P. R. (ed.) (1996). Evolution on islands. *Philosophical Transactions of the Royal Society of London*, B, **351**, 723–854.

Clegg, S. M., Degnan, S. M., Kikkawa, J., Moritz, C., Estoup, A., and Owens, I. P. F. (2002). Genetic consequences of sequential founder events by an island-colonizing bird. *Proceedings of the National Academy of Sciences, USA*, **99**, 8127–8132.

Clifford, M. J., Ali, S. H., and Matsubae, K. (2019). Mining, land restoration and sustainable development in isolated islands: an industrial ecology perspective on extractive transitions on Nauru. *Ambio*, **48**, 397–408.

Climent, J., Tapias, R., Pardos, J. A., and Gil, L. (2004). Fire adaptations in the Canary Islands pine (*Pinus canariensis*). *Plant Ecology*, **171**, 185–196.

Clout, M. N. and Craig, J. L. (1995). The conservation of critically endangered flightless birds in New Zealand. *Ibis*, **137**, S181–190.

Cody, M. L. and Overton, J. McC. (1996). Short-term evolution of reduced dispersal in island plant populations. *Journal of Ecology*, **84**, 53–61.

Colinvaux, P. A. (1972). Climate and the Galápagos Islands. *Nature*, **219**, 590–594.

Collyns, D. (2020). Seascape: the state of our oceans Galápagos Islands. *The Guardian*, 6 August 2020. www.theguardian.com/environment/2020/aug/06/chinese-fleet-fishing-galapagos-islands-environment

Compton, S. G., Ross, S. J., and Thornton, I. W. B. (1994). Pollinator limitation of fig tree reproduction on the island of Anak Krakatau (Indonesia). *Biotropica*, **26**, 180–186.

Condit, R., Aguilar, S., Hernandez, A., et al. (2004). Tropical forest dynamics across a rainfall gradient and the impact of an El Niño dry season. *Journal of Tropical Ecology*, **20**, 51–72.

Connell, J. (2018). Islands: balancing development and sustainability. *Environmental Conservation*, **45**, 111–124.

Connor, E. F. and McCoy, E. D. (1979). The statistics and biology of the species–area relationship. *American Naturalist*, **113**, 791–833.

Connor, E. F. and Simberloff, D. (1979). The assembly of species communities: chance or competition? *Ecology*, **60**, 1132–1140.

Connor, S. E., van Leeuwen, J. F. N., Rittentour, T. M., van der Knapp, W. O., Ammann, B., and Björck, S. (2012). The ecological impact of oceanic island colonization – a palaeoecological perspective from the Azores. *Journal of Biogeography*, **39**, 1007–1023.

Conroy, J. L., Overpeck, J. T., Cole, J. E., Shanahan, T. M., and Steinitz-Kannan, M. (2008). Holocene changes in eastern tropical Pacific climate inferred from a Galápagos lake sediment record. *Quaternary Science Reviews*, **27**, 1166–1180.

Cook, L. G. and Crisp, M. D. (2005). Directional asymmetry of long-distance dispersal and colonization could mislead reconstructions of biogeography. *Journal of Biogeography*, **32**, 741–754.

Cooper, W. E. Jr, Pyron, A., and Garland, T. Jr (2014). Island tameness: living on islands reduces flight initiation distance. *Proceedings of the Royal Society B*, **281**, 20133019.

Cordeiro, N. J., Borghesio, L., Joho, M. P., Monoski, T. J., Mkongewa, V. J., and Dampf, C. J. (2015). Forest fragmentation in an African biodiversity hotspot impacts mixed-species bird flocks. *Biological Conservation*, **188**, 61–71.

Cosson, J.-F., Pons, J.-M., and Masson, D. (1999a). Effects of forest fragmentation on frugivorous and nectarivorous bats in French Guiana. *Journal of Tropical Ecology*, **15**, 515–534.

Cosson, J.-F., Ringuet, S., Claeseens, O., et al. (1999b). Ecological changes in recent land-bridge islands in French Guiana, with emphasis on vertebrate communities. *Biological Conservation*, **91**, 213–222.

Costa, A. C. G., Hildenbrand, A., Marques, F. O., Sibrant, A. L. R., and Santos de Campos, A. (2015). Catastrophic flank collapses and slumping in Pico Island during the last 130 kyr (Pico-Faial ridge, Azores Triple Junction). *Journal of Volcanology and Geothermal Research*, **302**, 33–46.

Cotoras, D. D., Bi, K., Brewer, M. S., Lindberg, D. R., Prost, S., and Gillespie, R. G. (2018). Co-occurrence of ecologically similar species of Hawaiian spiders reveals critical early phase of adaptive radiation. *BMC Evolutionary Biology*, **18**, 100.

Cotoras, D. D., Suenaga, M., and Mikheyev, A. S. (2021). Intraspecific niche partition without speciation: individual level web polymorphism within a single island spider population. *Proceedings of the Royal Society B*, **287**, 20203138.

Cottee-Jones, H. E. W., Bajpai, O., Chaudhary, L. B., and Whittaker, R. J. (2016a). The importance of *Ficus* (Moraceae) trees for tropical forest restoration. *Biotropica*, **48**, 413–419.

Cottee-Jones, H. E. W., Matthews, T. J., and Whittaker, R. J. (2016b). The movement shortfall in bird conservation: accounting for nomadic, dispersive and irruptive species. *Animal Conservation*, **19**, 227–234.

Cottee-Jones, H. E. W., Matthews, T. J., Bregman, T. P., Barua, M. Tamuly, J., and Whittaker, R. J. (2015). Are protected areas required to maintain functional diversity in human-modified landscapes? *PLOS ONE*, **10**, e0123952.

Courchamp, F., Hoffmann, B. D., Russell, J. C. Leclerc, C., and Bellard, C. (2014). Climate change, sea-level rise, and conservation: keeping island biodiversity afloat. *Trends in Ecology & Evolution*, **29**, 127–130.

Covas, R. (2012). Evolution of reproductive life histories in island birds worldwide. *Proceedings of the Royal Society B*, **279**, 1531–1537.

Cox, C., Carranza, S., and Brown, R. P. (2010). Divergence times and colonization of the Canary Islands by *Gallotia* lizards. *Molecular Phylogenetics and Evolution*, **56**, 747–757.

Cox, C. B. and Moore, P. D. (1993). *Biogeography: An Ecological and Evolutionary Approach* (5th edn). Blackwell Scientific Publications, Oxford.

Cox, G. W. (1999). *Alien Species in North America and Hawaii: Impacts on Natural Ecosystems*. Island Press, Washington, DC.

Cox, M. P., Nelson, M. G., Tumonggor, M. K., Ricaut, F.-X., and Sudoyo, H. (2012). A small cohort of island Southeast Asian women founded Madagascar. *Proceedings of the Royal Society B*, **279**, 2761–2768.

Cox, P. A. and Elmqvist, T. (2000). Pollinator extinction in the Pacific Islands. *Conservation Biology*, **14**, 1237–1239.

Craig, D. A. (2003). Geomorphology, development of running water habitats, and evolution of black flies on Polynesian Islands. *Bioscience*, **53**, 1079–1093.

Cramp, S. and Perrins, C. M. (eds) (1994). *Handbook of the Birds of Europe, the Middle East and North Africa. The Birds of the Western Palearctic. Vol. VIII. Crows to Finches*. Oxford University Press, Oxford.

Craven, D., Knight, T. M., Barton, K. E., Bialic-Murphy, L., and Chase, J. M. (2019). Dissecting macroecological and macroevolutionary patterns of forest biodiversity across the Hawaiian archipelago. *Proceedings of the National Academy of Sciences, USA*, **116**, 16436–16441.

Craven, D., Weigelt, P., Wolkis, D., and Kreft, H. (2021). Niche properties constrain occupancy but not abundance patterns of native and alien woody species across Hawaiian forests. *Journal of Vegetation Science*, **32**, e13025.

Crawford, D. J., Anderson, G. J., and Bernardello, G. (2011). The reproductive biology of island plants. In *The Biology of Island Floras* (ed. D. Bramwell and J. Caujapé-Castells), pp. 11–36. Cambridge University Press, Cambridge.

Cronk, Q. C. B. (1992). Relict floras of Atlantic islands: patterns assessed. *Biological Journal of the Linnean Society*, **46**, 91–103.

Cronk, Q. C. B. and Fuller, J. L. (1995). *Plant Invaders*. Chapman & Hall, London.

Crooks, K. R. and Soulé, M. E. (1999). Mesopredator release and avifaunal extinctions in a fragmented system. *Nature*, **400**, 563–566.

Crosby, A. W. (2004). *Ecological Imperialism: The Biological Expansion of Europe, 900–1900*, 2nd edn. Cambridge University Press, New York.

Crowell, K. L. (1962). Reduced interspecific competition among the birds of Bermuda. *Ecology*, **43**, 75–88.

Crowley, B. E., Yanes, Y., Mosher, S. G., and Rando, J. C. (2019). Revisiting the foraging ecology and extinction history of two endemic vertebrates from Tenerife, Canary Islands. *Quaternary*, **2**, quat2010010.

Cruz, F., Donlan, C. J., Campbell, K., and Carrión, V. (2005). Conservation action in the Galápagos: feral pig (*Sus scrofa*) eradication from Santiago Island. *Biological Conservation*, **121**, 473–478.

Cruzado-Caballero, P., Ruiz, G. C., Bolet, A., et al. (2019). First nearly complete skull of *Gallotia auaritae* (lower-middle Pleistocene, Squamata, Gallotiinae) and a morphological phylogenetic analysis of the genus *Gallotia*. *Scientific Reports*, **9**, 16629.

Cubas, J., Martín-Esquivel, J. L., Nogales, M., et al. (2018). Contrasting effects of invasive rabbits on endemic plants driving vegetation change in a subtropical alpine insular environment. *Biological Invasions*, **20**, 793–807.

Cubas, J., Irl, S. D. H., Villafuerte, R., et al. (2019). Endemic plant species are more palatable to introduced herbivores than non-endemics. *Proceedings of the Royal Society B*, **286**, 20190136.

Cunliffe, B. (2012). *Britain Begins*. Oxford University Press, Oxford.

Czekanski-Moir, J. E. and Rundell, R. J. (2019). The ecology of nonecological speciation and nonadaptive radiations. *Trends in Ecology & Evolution*, **34**, 400–415.

D'Antonio, C. M., Ostertag, R., Cordell, S., and Yelenik, S. (2017). Interactions among invasive plants: lessons from Hawai'i. *Annual Review of Ecology, Evolution and Systematics*, **48**, 521–541.

Da Re, D., Tordoni, E., De Pascalis, F., et al. (2020). Invasive fountain grass (*Pennisetum setaceum* (Forssk.) Chiov.) increases its potential area of distribution in Tenerife Island under future climatic scenarios. *Plant Ecology*, **221**, 867–882.

Daily, G. C., Ceballos, G., Pacheco, J., Suzán, G., and Sánchez-Azofeifa, A. (2003). Countryside biogeography of Neotropical mammals: conservation opportunities in agricultural landscapes of Costa Rica. *Conservation Biology*, **17**, 1814–1826.

Dalsgaard, B., Carstensen, D. W., Fjeldså, J., et al. (2014). Determinants of bird species richness, endemism, and island network roles in Wallacea and the West Indies: is geography sufficient or does current and historical climate matter? *Ecology and Evolution*, **20**, 4019–4031.

Daltry, J. C., Lindsay, K., Lawrence, S. N., Morton, M. N., Otto, A., and Thibou, A. (2017). Successful reintroduction of the Critically Endangered Antiguan racer *Alsophis antiguae* to offshore islands in Antigua, West Indies. *International Zoo Yearbook*, **51**, 91–106.

Damuth, J. (1993). Cope's rule, the island rule and the scaling of mammalian population density. *Nature*, **365**, 748–750.

Darlington, P. J. (1957). *Zoogeography: The Geographical Distribution of Animals*. Wiley, New York.

Darwell, C. T., Fischer, G., Sarnat, E. M., et al. (2020). Genomic and phenomic analysis of island ant community assembly. *Molecuar Ecology*, **29**, 1611–1627.

Darwin C. (1839). *Journal of Researches into the Geology and Natural History of the Various Countries Visited by H.M.S. Beagle, under the Command of Captain Fitzroy, R. N. from 1832–1836*. Henry Colburn, London.

Darwin, C. (1842). *The Structure and Distribution of Coral Reefs. Being the First Part of the Geology of the Voyage of the Beagle, under the Command of Capt. Fitzroy, R. N., during the Years 1832–36*. Smith, Elder, & Company, London.

Darwin, C. (1859). *On the Origin of Species by Means of Natural Selection*. J. Murray, London. (Page numbers cited here are from an edition published by Avenel, New York, in 1979 under the title *The Origin of Species*.)

Davila, F., Crimp, S., and Wilkes, B. (2020). A systematic assessment of COVID-19 impacts on Pacific Islands' food systems. *Human Ecology Review*, **26**, 5–17.

Davis, S. D., Heywood, V. H., and Hamilton, A. C. (ed.) (1995). *Centres of Plant Diversity: A Guide and Strategy for their Conservation. Volume 2: Asia, Australasia and the Pacific*. WWF and IUCN, Cambridge.

Dawson, D. (1994). Are habitat corridors conduits for animals and plants in a fragmented landscape? A review of the scientific evidence. *English Nature Research Reports*, No. 94.

Dawson, E. Y. (1966). Cacti on the Galapagos Islands, with special reference to their relations with tortoises. In *The Galapagos* (ed. R. I. Bowman), pp. 209–214. University of California Press, Berkeley.

Dawson, H. (2014). *Mediterranean Voyages: The Archaeology of Island Colonisation and Abandonment*. Publications of the Institute of Archaeology, University College London. West Coast Press Inc, Walnut Creek, CA.

Dawson, M. N. (2016). Island and island-like marine environments. *Global Ecology and Biogeography*, **25**, 831–846.

de Nascimento, L., Nogué, S., Naranjo-Cigala, A., et al. (2020). Human impact and ecological changes during prehistoric settlement on the Canary Islands. *Quaternary Science Reviews*, **239**, 106332.

de Nascimento, L., Willis, K. J., Fernández-Palacios, J. M., Criado, C., and Whittaker, R. J. (2009). The long-term ecology of the lost forest of La Laguna, Tenerife (Canary Islands). *Journal of Biogeography*, **36**, 499–514.

de Nicolás, J. P., Fernández-Palacios, J. M., Ferrer, F. J., and Nieto, E. (1989). Inter-island floristic similarities in the Macaronesian Region. *Vegetatio*, **84**, 117–125.

Decker, R. W. and Decker, B. B. (1991). *Mountains of Fire: The Nature of Volcanoes*. Cambridge University Press, Cambridge.

del Arco, M. J. and Rodríguez-Delgado, O. (2018). *The Vegetation of the Canary Islands*. Springer, New York.

del Castillo, R. F. (2015). A conceptual framework to describe the ecology of fragmented landscapes and implications for conservation and management. *Ecological Applications*, **25**, 1447–1455.

Delavaux, C. S., Weigelt, P., Dawson, W., et al. (2021). Mycorrhizal types influence island biogeography of plants. *Communications Biology*, **4**, article 1128.

Delcourt, H. R. and Delcourt, P. A. (1991). *Quaternary Ecology: A Paleoecological Perspective*. Chapman & Hall, London.

Delgado, J. D., Arévalo, J. R. and Fernández-Palacios, J. M. (2007). Road edge effect on the abundance of the lizard *Gallotia galloti* (Sauria: Lacertidae) in two Canary Islands forests. *Biodiversity and Conservation*, **16**, 2949–2963.

Delgado, J. D., Riera, R., Rodríguez, R. A., González-Moreno, P., and Fernández-Palacios, J. M. (2017). A reappraisal of the role of humans in the biotic disturbance of islands. *Environmental Conservation*, **44**, 371–380.

Dellinger, A. S., Pérez-Barrales, R., Michelangeli, F. A., Penneys, D. S., Fernández-Fernández, D. M., and Schönenberger, J. (2021). Low bee visitation rates explain pollinator shifts to vertebrates in tropical mountains. *New Phytologist*, **231**, 864–877.

den Ouden, N., Reumer, J. W. F., and van den Hoek Ostende, L. W. (2012). Did mammoth end up a lilliput? Temporal body size trends in Late Pleistocene mammoths, *Mammuthus primigenius* (Blumenbach, 1799) inferred from dental data. *Quaternary International*, **255**, 53–58.

Dengler, J. (2010). Robust methods for detecting a small island effect. *Diversity and Distributions*, **16**, 256–266.

Désamoré, A., Laenen, B., González-Mancebo, J. M., et al. (2012). Inverted patterns of genetic diversity in continental and island populations of the heather *Erica scoparia* s.l. *Journal of Biogeography*, **39**, 574–584.

Develey, P. F. and Phalan, B. T. (2021). Bird extinctions in Brazil's Atlantic Forest and how they can be prevented. *Frontiers in Ecology and Evolution*, **9**, 624587.

Devenish-Nelson, E. S., Weidemann, D., Townsend, J., and Nelson, H. P. (2019). Patterns in island endemic forest-dependent birds research: the Caribbean as a case-study. *Biodiversity and Conservation*, **28**, 1885–1904.

Diamond, J. M. (1972). Biogeographical kinetics: estimation of relaxation times for avifaunas of southwest Pacific islands. *Proceedings of the National Academy of Sciences, USA*, **69**, 3199–3203.

Diamond, J. M. (1974). Colonization of exploded volcanic islands by birds: the supertramp strategy. *Science*, **184**, 803–806.

Diamond, J. M. (1975). Assembly of species communities. In *Ecology and Evolution of Communities* (ed. M. L. Cody and J. M. Diamond), pp. 342–444. Harvard University Press, Cambridge, MA.

Diamond, J. M. (2005). *Collapse: How Societies Choose to Fail or Survive.* Allen Lane/Penguin, London.

Diamond, J. M. and Gilpin, M. E. (1983). Biogeographical umbilici and the origin of the Philippine avifauna. *Oikos*, **41**, 307–321.

Diamond, J. M. and May, R. M. (1977). Species turnover rates on islands: dependence on census interval. *Science*, **197**, 266–270.

Diamond, J. M. and May, R. M. (1981). Island biogeography and the design of nature reserves. In *Theoretical Ecology* (2nd edn) (ed. R. M. May), pp. 228–252. Blackwell, Oxford.

Diamond, J. M., Pimm, S. L., Gilpin, M. E., and LeCroy, M. (1989). Rapid evolution of character displacement in Myzomelid Honeyeaters. *American Naturalist*, **134**, 675–708.

Diamond, J. M., Pimm, S. L., and Sanderson, J. G. (2015). The checkered history of checkerboard distributions: comment. *Ecology*, **96**, 3386–3388.

Dias, R. A., Bastazini, V. A. G., de Castro Knopp, B., Bonow, F. C., Gonçalves, M. S. S., and Guanuca, A. T. (2020). Species richness and patterns of overdispersion, clustering and randomness shape phylogenetic and functional diversity–area relationships in habitat islands. *Journal of Biogeography*, **47**, 1638–1648.

Dickinson, W. R. (2009). Pacific atoll living: how long already and until when? *GSA Today*, **19**, 4–10.

Diehl, S. and Bush, G. L. (1989). The role of habitat preference in adaptation and speciation. In *Speciation and its Consequences* (ed. D. Otte and J. A. Endler), pp. 345–365. Sinauer, Sunderland, MA.

Diniz-Filho, J. A. F., Meiri, S., Hortal, J., Santos, A. M. C., and Raia, P. (2021). Too simple models may predict the island rule for the wrong reasons. *Ecology Letters*, **24**, 2521–2523.

Doherty, B. (2020). Photo of Chinese ambassador to Kiribati walking across backs of locals 'misinterpreted'. *The Guardian*, 18 August 2020. www.theguardian.com/world/2020/aug/18/photo-of-chinese-ambassador-to-kiribati-walking-across-backs-of-locals-misinterpreted

Donner, S. D. and Potere, D. (2007). The inequity of the global threat to coral reefs. *BioScience*, **57**, 214–215.

Dória, L. C., Podadera, D. S., del Arco, M., et al. (2018). Insular woody daisies (*Argyranthemum*, Asteraceae) are more resistant to drought-induced hydraulic failure than their herbaceous relatives. *Functional Ecology*, **32**, 1467–1478.

dos Reis, M., Donoghue, P. C. J., and Yang, Z. (2016). Bayesian molecular clock dating of species divergences in the genomics era. *Nature Reviews Genetics*, **17**, 71–80.

Doutrelant, C., Paquet, M., Renoult, J. P., Grégoire, A., Crochet, P.-A., and Covas, R. (2016). Worldwide patterns of bird colouration on islands. *Ecology Letters*, **19**, 537–545.

Douville, E., Paterne, M., Cabioch, G., et al. (2010). Abrupt sea surface pH change at the end of the Younger Dryas in the central sub-equatorial Pacific inferred from boron isotope abundance in corals (*Porites*). *Biogeosciences*, **7**, 2445–2459.

Doxa, A., Devictor, V., Baumel, A., Pavon, D., Médail, F., and Leriche, A. (2020). Beyond taxonomic diversity: revealing spatial mismatches in phylogenetic and functional diversity facets in Mediterranean tree communities in southern France. *Forest Ecology and Management*, **474**, 118318.

Draper, J. T., Haigh, T., Atakan, O., et al. (2021). Extreme host range in an insular bee supports the super-generalist hypothesis with implications for both weed invasion and crop pollination. *Arthropod-Plant Interactions*, **15**, 13–22.

Driscoll, D. A. and Lindenmayer, D. B. (2012). Framework to improve the application of theory in ecology and conservation. *Ecological Monographs*, **82**, 129–147.

Dudley, N. (ed.) (2013). *Guidelines for Applying Protected Area Management Categories.* (Revised edition). IUCN, Gland, Switzerland.

Dufour, C. M. S., Herrel, A., and Losos, J. B. (2018). Ecological character displacement between a native and an introduced species: the invasion of *Anolis cristatellus* in Dominica. *Biological Journal of the Linnean Society*, **123**, 43–54.

Dulin, M. W. and Kirchoff, B. K. (2010). Paedomorphosis, secondary woodiness and insular woodiness in plants. *Botanical Review*, **76**, 405–490.

Duncan, R. P., Blackburn, T. M., and Worthy, T. H. (2002). Prehistoric bird extinctions and human hunting. *Proceedings of the Royal Society B*, **269**, 517–521.

Duncan, R. P., Boyer, A. G., and Blackburn, T. M. (2013). Magnitude and variation of prehistoric bird extinctions in the Pacific. *Proceedings of the National Academy of Sciences, USA*, **110**, 6436–6441.

Dutson, G., Magsalay, P., and Timmins, R. (1993). The rediscovery of the Cebu flowerpecker *Dicaeum quadricolor*, with notes on other forest birds on Cebu, Philippines. *Bird Conservation International*, **3**, 235–243.

Ecker, A. (1976). Groundwater behaviour in Tenerife, volcanic island (Canary Islands, Spain). *Journal of Hydrology*, **28**, 73–86.

Economo, E. P. and Sarnat, E. M. (2012). Revisiting the ants of Melanesia and the taxon cycle: historical and human-mediated invasions of a tropical archipelago. *The American Naturalist*, **180**, E1–E16.

Economo, E. P., Sarnat, E. M., Janda, M., et al. (2015). Breaking out of biogeographical modules: range expansion and taxon cycles in the hyperdiverse ant genus *Pheidole*. *Journal of Biogeography*, **42**, 2289–2301.

Eddy, T. D., Friedlander, A. M., and Salinas de León, P. (2019). Ecosystem effects of fishing & El Niño at the Galápagos Marine Reserve. *PeerJ*, **7**, e6878.

Edgar, G. J., Banks, S. A., Brandt, M., et al. (2010). El Niño, grazers and fisheries interact to greatly elevate extinction risk for Galapagos marine species. *Global Change Biology*, **16**, 2876–2890.

Ehrlich, P. R. and Hanski, I. (eds) (2004). *On the Wings of Checkerspots: A Model System for Population Biology*. Oxford University Press, Oxford.

Elias, R. B., Connor, S. E., Góis-Marques, C. A., et al. (2022). Is there solid evidence of widespread landscape disturbance in the Azores before the arrival of the Portuguese? *Proceedings of the National Academy of Sciences, USA*, **119**, e2119218119.

Elmqvist, T., Rainey, W. E., Pierson, E. D., and Cox, P. A. (1994). Effects of tropical cyclones Ofa and Val on the structure of a Samoan lowland rain forest. *Biotropica*, **26**, 384–391.

Elton, C. S. (1958). *The Ecology of Invasions by Animals and Plants*. Methuen, London.

Emerson, B. C. and Kolm, N. (2005a). Species diversity can drive speciation. *Nature*, **434**, 1015–1017.

Emerson, B. C. and Kolm, N. (2005b). Is speciation driven by species diversity? (Reply). *Nature*, **438**, doi: 10.1038/nature04309.

Emerson, B. C. and Patiño, J. (2018). Anagenesis, cladogenesis, and speciation on islands. *Trends in Ecology & Evolution*, **33**, 488–491.

Engeman, R. M., Shiels, A. B., and Clark, C. S. (2018). Objectives and integrated approaches for the control of brown tree snakes: an updated review. *Journal of Environmental Management*, **219**, 115–124.

Eriksson, A., Elías-Wolff, F., Mehlig, B., and Manica, A. (2014). The emergence of the rescue effect from explicit within- and between-patch dynamics in a metapopulation. *Proceedings of the Royal Society B*, **281**, 20133127.

Escoriza, D. (2020). Organization of Squamata (Reptilia) assemblages in Mediterranean archipelagos. *Ecology and Evolution*, **10**, 1592–1601.

Everson, K. M., Soarimalala, V., Goodman, S. V., and Olson, L. E. (2016). Multiple loci and complete taxonomic sampling resolve the phylogeny and biogeographic history of Tenrecs (Mammalia: Tenrecidae) and reveal higher speciation rates in Madagascar's humid forests. *Systematic Biology*, **65**, 890–909.

Ewers, R. M. and Didham, R. K. (2006). Confounding factors in the detection of species responses to habitat fragmentation. *Biological Reviews*, **81**, 117–142.

Faeth, S. H. and Kane, T. C. (1978). Urban biogeography: city parks as islands for Diptera and Coleoptera. *Oecologia*, **32**, 127–133.

Fahrig, L. (2013). Rethinking patch size and isolation effects: the habitat amount hypothesis. *Journal of Biogeography*, **40**, 1649–1663.

Fahrig, L. (2017). Ecological responses to habitat fragmentation per se. *Annual Review of Ecology, Evolution and Systematics*, **48**, 1–23.

Faith, D. P. (1992). Conservation evaluation and phylogenetic diversity. *Biological Conservation*, **61**, 1–10.

Falcón, W. and Hansen, D. M. (2018). Island rewilding with giant tortoises in an era of climate change. *Philosophical Transactions of the Royal Society B*, **373**, article 20170442.

Falcón, W., Moll, D., and Hansen, D. M. (2020). Frugivory and seed dispersal by chelonians: a review and synthesis. *Biological Reviews*, **95**, 142–166.

Farkas, T. E., Hendry, A. P., Nosil, P., and Beckerman, A. P. (2015). How maladaptation can structure biodiversity: eco-evolutionary island biogeography. *Trends in Ecology & Evolution*, **30**, 154–160.

Farrington, H. L., Lawson, L. P., Clark, C. M., and Petren, K. (2014). The evolutionary history of Darwin's finches: speciation, gene flow, and introgression in a fragmented landscape. *Evolution*, **68**, 2932–2944.

Fattorini, S. (2009). On the general dynamic model of oceanic island biogeography. *Journal of Biogeography*, **36**, 1100–1110.

Fattorini, S. (2014). Island biogeography of urban insects: Tenebrionid beetles from Rome tell a different story. *Journal of Insect Conservation*, **18**, 729–735.

Fattorini, S., Mantoni, C., De Simoni, L., and Galassi, D. (2018). Island biogeography of insect conservation in urban green spaces. *Environmental Conservation*, **45**, 1–10.

Fattorini, S., Rigal, F., Cardoso, P., and Borges, P. A. V. (2016). Using species abundance distribution models and diversity indices for biogeographical analyses. *Acta Oecologica*, **70**, 21–28.

Faurby, S. and Svenning, J.-C. (2016). Resurrection of the island rule: human-driven extinctions have obscured a basic evolutionary pattern. *The American Naturalist*, **187**, 812–820.

Fensome, A. G. and Matthews, F. (2016). Roads and bats: a meta-analysis and review of the evidence on vehicle collisions and barrier effects. *Mammal Review*, **46**, 311–323.

Fenu, G., Baccheta, G., Charalambos, S. C., et al. (2019). An early evaluation of translocation actions for endangered plant species on Mediterranean islands. *Plant Diversity*, **41**, 94–104.

Fernandes, J. P., Guiomar, N., and Gil, A. (2015). Strategies for conservation planning and management of terrestrial ecosystems in small islands (exemplified for the Macaronesian islands). *Environmental Science and Policy*, **51**, 1–22.

Fernández-Mazuecos, M. and Vargas, P. (2011). Genetically depauperate in the continent but rich in Oceanic Islands: *Cistus monspeliensis* (Cistaceae) in the Canary Islands. *PLOS ONE*, **6**, e17172.

Fernández-Palacios, J. M. (2010). Why islands? In *Islands and Evolution* (ed. V. Pérez Mellado, V. & C. Ramón), pp. 85–109. Institut Menorquí d'Estudis, Mahón, Menorca.

Fernández-Palacios, J. M., Arévalo, J. R., Balguerías, E., et al. (2019). *The Laurisilva. Canaries, Madeira and Azores*. Editorial Macaronesia, Santa Cruz de Tenerife, 417 pp.

Fernández-Palacios, J. M., Arévalo J. R., Delgado, J. D., and Otto, R. (2004). *Canarias: Ecología, medio ambiente y desarrollo*. Centro de la Cultura Popular de Canarias, La Laguna.

Fernández-Palacios, J. M. and de Nascimento, L. (2011). Political erosion dismantles the conservation network existing in the Canary Islands. *Frontiers of Biogeography*, **3**, 106–110.

Fernández-Palacios, J. M., de Nascimento, L., Otto, R., et al. (2011). A reconstruction of Palaeo-Macaronesia, with particular reference to the long-term biogeography of the Atlantic Island laurel forests. *Journal of Biogeography*, **38**, 226–246.

Fernández-Palacios, J. M. and de Nicolás, J. P. (1995). Altitudinal pattern of vegetation variation on Tenerife. *Journal of Vegetation Science*, **6**, 183–190.

Fernández-Palacios, J. M., Kreft, H., Irl, S. D. H., et al. (2021a). Scientists' warning – the outstanding biodiversity of islands is in peril. *Global Ecology and Conservation*, **31**, e01847.

Fernández-Palacios, J. M., Nogué, S., Criado, C., et al. (2016b). Climate change and human impact in Macaronesia. *Past Global Changes Magazine*, **24**, 68–69.

Fernández-Palacios, J. M., Otto, R., Borregaard, M. K., et al. (2021b). Evolutionary winners are ecological losers among oceanic island plants. *Journal of Biogeography*, **48**, 2186–2198.

Fernández-Palacios, J. M., Otto, R., Thebaud, C., and Price, J. P. (2014). Overview of habitat history in subtropical oceanic island summit ecosystems. *Arctic, Antarctic, and Alpine Research*, **46**, 801–809.

Fernández-Palacios, J. M., Rijsdijk, K. F., Norder, S. J., et al. (2016a). Towards a glacial-sensitive model of island biogeography. *Global Ecology and Biogeography*, **25**, 817–830.

Fernández-Palacios, J. M., Schrader, J., de Nascimento, L., Irl, S. D. H., Sánchez-Pinto, L., and Otto, R. (2023). Are plant communities on the Canary Islands resistant to plant invasion? *Diversity and Distributions*, **29**, 51–60.

Fernández-Palacios, J. M. and Whittaker, R. J. (2020). Early recognition by Ball and Hooker in 1878 of plant back-colonization (boomerang) events from Macaronesia to Africa. *Frontiers of Biogeography*, **12**, e45375.

Figueiredo, L., Krauss, J., Steffan-Dewenter, I., and Cabral, J. S. (2019). Understanding extinction debts: spatio–temporal scales, mechanisms and a roadmap for future research. *Ecography*, **42**, 1973–1990.

Finkelstein, M. E., Wolf, S., Goldman, M., et al. (2010). The anatomy of a (potential) disaster: volcanoes, behavior, and population viability of the short-tailed albatross (*Phoebastria albatrus*). *Biological Conservation*, **143**, 321–331.

Fischer, J., Abson, D. J., Butsic, V., et al. (2014). Land sharing versus land sharing: moving forward. *Conservation Letters*, **7**, 149–157.

Fisher, R. A., Corbet, A. S., and Williams, C. B. (1943). The relation between the number of species and the number of individuals in a random sample of an animal population. *Journal of Animal Ecology*, **12**, 42–58.

Flantua, S. G. A., Payne, D., Borregaard, M. K., et al. (2020). Snapshot isolation and isolation history challenge the analogy between mountains and islands used to understand endemism. *Global Ecology and Biogeography*, **29**, 1651–1673.

Fleming, T. H., Geiselman, C., and Kress, W. J. (2009). The evolution of bat pollination: a phylogenetic perspective. *Annals of Botany*, **104**, 1017–1043.

Fleming, T. H. and Racey, P. A. (eds) (2009). *Island Bats: Evolution, Ecology, & Conservation*. Chicago University Press, Chicago.

Fletcher, R. J. Jr, Didham, R. K., Banks-Leite, C., et al. (2018). Is habitat fragmentation good for biodiversity? *Biological Conservation*, **226**, 9–15.

Florencio, M., Patiño, J., Nogué, S., et al. (2021). Macaronesia as a fruitful arena for ecology, evolution, and conservation biology. *Frontiers in Ecology and Evolution*, **9**, 718169.

Flores, M. A., Schlatter, R. P., and Hucke-Gaete, R. (2014). Seabirds of Easter Island, Salas y Gómez Island and Desventuradas Islands, southeastern Pacific Ocean. *Latin American Journal of Aquatic Research*, **42**, 752–759.

Forsman, A. (1991). Adaptive variation in head size in *Vipera berus* L. populations. *Biological Journal of the Linnean Society*, **43**, 281–296.

Forster, G. (1777). *A Voyage Round the World, Vol. I* (ed. N. Thomas and O. Berghof; 2000). University of Hawai'i Press, Honolulu.

Forster, J. R. (1778). *Observations Made during a Voyage Round the World, on Physical Geography, Natural History and Ethic Philosophy*. G. Robinson, London.

Foster, J. B. (1964). Evolution of mammals on islands *Nature*, **202**, 234–245.

Francisco-Ortega, J., Santos-Guerra, A., Kim, S. C., and Crawford, D. J. (2000). Plant genetic diversity in the Canary Islands: a conservation perspective. *American Journal of Botany*, **87**, 909–919.

Frankham, R., Bradshaw, C. J. A., and Brook, B. W. (2014). Genetics in conservation management: revised recommendations for the 50/500 rules, Red List criteria and population viability analyses. *Biological Conservation*, **170**, 56–63.

Franklin, J. and Steadman, D. W. (1991). The potential for conservation of Polynesian birds through habitat mapping and species translocation. *Conservation Biology*, **5**, 506–521.

Franklin, J. and Steadman, D. W. (2008). Prehistoric species richness of birds on oceanic islands. *Oikos*, **117**, 1885–1891.

Fraser, C. I., Dutoit, L., Morrison, A. K., et al. (2022). Southern Hemisphere coasts are biologically connected by frequent, long-distance rafting events. *Current Biology*, **32**, 3154–3160.

Fresnillo, B. and Ehlers, B. K. (2008). Variation in dispersability among mainland and island populations of three wind dispersed plant species. *Plant Systematics and Evolution*, **270**, 243–255.

Frishkoff, L. O., Ke, A., Martins, I. S., Olimpi, E. M., and Karp, D. S. (2019). Countryside biogeography: the controls of species distributions in human-dominated landscapes. *Current Landscape Ecology Reports*, **4**, 15–30.

Fritts, T. H. and Leasman-Tanner, D. (2001). The brown treesnake on Guam: how the arrival of one invasive species damaged the ecology, commerce, electrical systems, and human health on Guam: a comprehensive information source. Available online: www.fort.usgs.gov/resources/education/bts/bts_home.asp (visited June 2005).

Fromm, A. and Meiri, S. (2021). Big, flightless, insular and dead: characterising the extinct birds of the Quaternary. *Journal of Biogeography*, **48**, 2350–2359.

Funk, V. A. and Wagner, W. L. (1995). Biogeographic patterns in the Hawaiian Islands. In *Hawaiian Biogeography. Evolution on a Hot Spot Archipelago* (ed. W. L. Wagner and V. A. Funk). pp. 379–419. Smithsonian, Washington, DC.

Furrer, R. D. and Pasinelli, G. (2016). Empirical evidence for source–sink populations: a review on occurrence, assessments and implications. *Biological Reviews*, **91**, 782–795.

Fuster, F., Kaiser-Bunbury, C., Olesen, J. M., and Traveset, A. (2019). Global patterns of the double mutualism phenomenon. *Ecography*, **42**, 826–835.

Futuyma, D. (1986). *Evolutionary Biology*. Sinauer Associates, Sunderland, MA.

Gale, S. J. (2019). Lies and misdemeanours: Nauru, phosphate and global geopolitics. *The Extractive Industries and Society*, **6**, 737–746.

Gallardo, A., Fernández-Palacios, J. M., Bermúdez, A., et al. (2020). The pedogenic Walker and Syers model under high atmospheric P deposition rates. *Biogeochemistry*, **148**, 237–253.

Galley, C. and Linder, H. P. (2006). Geographical affinities of the Cape flora, South Africa. *Journal of Biogeography*, **33**, 236–250.

Gao, D., Cao, Z., Xu, P., and Perry, G. (2019). On piecewise models and species–area patterns. *Ecology and Evolution*, **9**, 8351–8361.

García-Carrasco, J.-M., Tapia, W., and Muñoz, A. R. (2020). Roadkill of birds in Galapagos Islands: a growing need for solutions. *Avian Conservation and Ecology*, **15**, article 19.

García-Falcón, J. M. and Medina-Muñoz, D. (1999). Sustainable tourism development in islands: a case study of Gran Canaria. *Business Strategy and the Environment*, **8**, 336–357.

García-Maroto, F., Mañas-Fernández, A., Garrido-Cárdenas, J. A., et al. (2009). Δ^6-Desaturase sequence evidence for explosive Pliocene radiations within the adaptive radiation of Macaronesian *Echium* (Boraginaceae). *Molecular Phylogenetics and Evolution*, **52**, 563–574.

García-Olivares, V., López, H., Patiño, J., et al. (2017). Evidence for mega-landslides as drivers of island colonization. *Journal of Biogeography*, **44**, 1053–1064.

Garcia-R, J. C. and Di Marco, M. (2020). Drivers and trends in the extinction risk of New Zealand's endemic birds. *Biological Conservation*, **249**, 108730.

García-Verdugo, C., Caujapé-Castells, J., Mairal, M., and Monroy, P. (2019b). How repeatable is microevolution on islands? Patterns of dispersal and colonization-related plant traits in a phylogeographical context. *Annals of Botany*, **123**, 557–568.

García-Verdugo, C., Caujapé-Castells, J., and Sanmartín, I. (2019a). Colonization time on island settings: lessons from the Hawaiian and Canary Island floras. *Botanical Journal of the Linnean Society*, **191**, 155–163.

García-Verdugo, C., Mairal, M., Monroy, P., Sajeva, M., and Caujapé-Castells, J. (2017). The loss of dispersal on islands hypothesis revisited: implementing phylogeography to investigate evolution of dispersal traits in *Periploca* (Apocynaceae). *Journal of Biogeography*, **44**, 2595–2606.

Garfield, R. (2015). Three islands of the Portuguese Atlantic: their economic rise, fall and (sometimes) rerise. *Shima*, **9**, 47–59.

Gascuel, F., Laroche, F., Bonnet-Lebrun, A.-S., and Rodrigues, A. S. L. (2016). The effects of archipelago spatial structure on island diversity and endemism: predictions from a spatially-structured neutral model. *Evolution*, **70**, 2657–2666.

Gavrilets, S. (2014). Models of speciation: where are we now? *Journal of Heredity*, **105**, 743–755.

Geffen, E. and Yom-Tov, Y. (2019). Pacific island invasions: how do settlement time, latitude, island area and number of competitors affect body size of the kiore (Polynesian rat) across the Pacific? *Biological Journal of the Linnean Society*, **126**, 462–470.

Gerlach, J. (2008). Preliminary conservation status and needs of an oceanic island fauna: the case of Seychelles insects. *Journal of Insect Conservation*, **12**, 293–305.

Gerstner, K., Dormann, C. F., Václavík, T., Kreft, H., and Seppelt, R. (2014). Accounting for geographical variation in species–area relationships improves the prediction of plant species richness at the global scale. *Journal of Biogeography*, **41**, 261–273.

Gibbs, G. (2016). *Ghosts of Gondwana. The History of Life in New Zealand*. Potton & Burton, Nelson.

Gibbs, H. L. and Grant, P. R. (1987). Ecological consequences of an exceptionally strong El Niño event on Darwin's finches. *Ecology*, **68**, 1735–1746.

Gibbs, J. P., Sterling, E. J., and Zabala, F. J. (2010). Giant tortoises as ecological engineers: a long term quasi-experiment in the Galápagos Islands. *Biotropica*, **42**, 208–214.

Gibson, L., Lynam, A. J., Bradshaw, C. J. A., et al. (2013). Near-complete extinction of native small mammal fauna 25 years after forest fragmentation. *Science*, **341**, 1508–1510.

Gibson, L. A., Cowan, M. A., Lyons, M. N., Palmer, R., Pearson, D. J., and Doughty, P. (2017). Island refuges: conservation significance of the biodiversity patterns resulting from 'natural' fragmentation. *Biological Conservation*, **212**, 349–356.

Gilbert, F. S. (1980). The equilibrium theory of island biogeography, fact or fiction? *Journal of Biogeography*, **7**, 209–235.

Gilbert-Norton, L., Wilson, R., Stevens, J. R., and Beard, K. H. (2010). A meta-analytic review of corridor effectiveness. *Conservation Biology*, **24**, 660–668.

Gill, F., Donsker, D., and Rasmussen, P. C. (2022). *IOC World bird list (v12.1)*. doi:10.14344/IOC.ML.12.1. Available at www.worldbirdnames.org/new/

Gillespie, R. (2004). Community assembly through adaptive radiation in Hawaiian spiders. *Science* 303, 356–359.

Gillespie, R. (2016). Island time and the interplay between ecology and evolution in species diversification. *Evolutionary Applications*, 9, 53–73.

Gillespie, R. G., Bennett, G. M., De Meester, L., et al. (2020). Comparing adaptive radiations across space, time, and taxa. *Journal of Heredity*, 111, 1–20.

Gillespie, R. G. and Clague, D. A. (eds) (2009). *Encyclopedia of Islands*. University of California Press, Berkeley.

Gillespie, R. G. and Roderick, G. K. (2002). Arthropods on islands: colonization, speciation, and conservation. *Annual Review of Entomology*, 47, 595–632.

Gillson, L. and Willis, K. J. (2004). 'As Earth's testimonies tell': wilderness conservation in a changing world. *Ecology Letters*, 7, 990–998.

Gilpin, M. E. and Diamond, J. M. (1982). Factors contributing to non-randomness in species co-occurences on islands. *Oecologia*, 52, 75–84.

Gittenberger, E. (1991). What about non-adaptive radiation? *Biological Journal of the Linnean Society*, 43, 263–272.

Givnish, T. J. (1998). Adaptive plant evolution on islands. In *Evolution on Islands* (ed. P. R. Grant) pp. 281–304. Oxford University Press, Oxford.

Givnish, T. J. (2010). Ecology of plant speciation. *Taxon*, 59, 1326–1366.

Givnish, T. J., Millam, K. C., Mast, A. R., et al. (2009). Origin, adaptive radiation and diversification of the Hawaiian lobeliads (Asterales: Campanulaceae). *Proceedings of the Royal Society B*, 276, 407–416.

Gleditsch, J., Behm, J. E., Ellers, J., Jesse, W., and Helmus, M. (2023). Contemporizing island biogeography theory with anthropogenic drivers of species richness. *Global Ecology and Biogeography*, 33, 233–249.

Glor, R. E. (2011). Remarkable new evidence for island radiation in birds. *Molecular Ecology*, 20, 4823–4826.

Gohli, J., Leder, E. H., Garcia-del-Rey, E., et al. (2015). The evolutionary history of Afrocanarian blue tits inferred from genomewide SNPs. *Molecular Ecology*, 24, 180–191.

Goldberg, J., Trewick, S.A., and Paterson, A.M. (2008). Evolution of New Zealand's terrestrial fauna: a review of molecular evidence. *Philosophical Transactions of the Royal Society B*, 363, 3319–3334.

Golinski, M. and Boecklen, W. J. (2006). A model-independent test for the presence of regulatory equilibrium and non-random structure in island species trajectories. *Journal of Biogeography*, 33, 1566–1570.

González, J., Delgado Castro, G., Garcia-del-Rey, E., Berger, C., and Wink, M. (2009). Use of mitochondrial and nuclear genes to infer the origin of two endemic pigeons from the Canary Islands. *Journal of Ornithology*, 150, 357–367.

González, J. A., Montes, C., Rodríguez, J., and Tapia, W. (2008) Rethinking the Galapagos Islands as a complex social-ecological system: implications for conservation and management. *Ecology and Society*, 13, article 13.

González-Castro, A., Calviño-Cancela, M., and Nogales, M. (2015). Comparing seed dispersal effectiveness by frugivores at the community level. *Ecology*, 96, 808–818.

González-Pérez, M. A., Caujapé-Castells, J., and Sosa, P. A. (2004). Molecular evidence of hybridisation between the endemic *Phoenix canariensis* and the widespread *Phoenix dactylifera* with Random Amplified Polymorphic DNA (RAPD). *Plant Systematics and Evolution*, 247, 165–175.

Gooriah, L., Blowes, S. A., Sagouis, A., et al. (2021). Synthesis reveals that island species–area relationships emerge from processes beyond passive sampling. *Global Ecology and Biogeography*, 30, 2119–2131.

Gotelli, N. J. and McCabe, D. J. (2002). Species co-occurrence: a meta-analysis of J. M. Diamond's assembly rules model. *Ecology*, 83, 2091–2096.

Gouhier, M. and Paris, R. (2019). SO_2 and tephra emissions during the December 22, 2018 Anak Krakatau flank-collapse eruption. *Volcanica*, 2, 91–103.

Graham, N. A. J., Wilson, S. K., Carr, P., Hoey, A. S., Jennings, S., and MacNeil, M. A. (2018). Seabirds enhance coral reef productivity and functioning in the absence of invasive rats. *Nature*, 559, 250–253.

Graham, R. E., Reyes-Betancort, A., Chapman, M. A., and Carine, M. A. (2021). Inter-island differentiation and contrasting patterns of diversity in the iconic Canary Island sub-alpine endemic *Echium wildpretii* (Boraginaceae). *Systematics and Biodiversity*, 19, 507–525.

Grant, B. R. and Grant, P. R. (1996a). High survival of Darwin's finch hybrids: effects of beak morphology and diets. *Ecology*, 77, 500–509.

Grant, P. R. (1994). Population variation and hybridization: comparison of finches from two archipelagos. *Evolutionary Ecology*, 8, 598–617.

Grant, P. R. and Grant, B. R. (1989). Sympatric speciation and Darwin's finches. In *Speciation and its Consequences* (ed. D. Otte and J. A. Endler), pp. 433–457. Sinauer Associates, Sunderland, MA.

Grant, P. R. and Grant, B. R. (1994). Phenotypic and genetic effects of hybridization in Darwin's finches. *Evolution*, 48, 297–316.

Grant, P. R. and Grant, B. R. (1996b). Speciation and hybridization in island birds. *Philosophical Transactions of the Royal Society of London*, B 351, 765–772.

Grant, P. R. and Grant, B. R. (2005). Darwin's finches. *Current Biology*, 15, R614–R615.

Grant, P. R. and Grant, B. R. (2006). Evolution of character displacement in Darwin's finches. *Science*, 313, 224–226.

Grant, P. R. and Grant, B.R. (2008). *How and Why Species Multiply. The Radiation of Darwin's Finches*. Princeton University Press, Princeton.

Gravel, D., Massol, F., Canard, E., Mouillot, D., and Mouquet, N. (2011). Trophic theory of island biogeography. *Ecology Letters*, 14, 1010–1016.

Gray, A. and Cavers, S. (2014). Island biogeography, the effects of taxonomic effort and the importance of island niche diversity to single-island endemic species. *Systematic Biology*, 63, 55–65.

Gray, A., Wilkins, V., Price, D., et al (2019). The status of the invertebrate fauna on the South Atlantic island of St Helena:

problems, analysis, and recommendations. *Biodiversity and Conservation*, **28**, 275–296.

Green, S. J., Akins, J. L., Malijković, A., and Côté, I. M. (2012). Invasive lionfish drive Atlantic coral reef fish declines. *PLOS ONE*, **7**, e32596.

Green, R. E., Cornell, S. J., Scharlemann, J. P. W., and Balmford, A. (2005). Farming and the fate of wild nature. *Science*, **307**, 550–555.

Grehan, J. R. (2017). Biogeographic relationships between Macaronesia and the Americas. *Australian Systematic Botany*, **29**, 447–472.

Grelle, C. E. V., Alves, M. A. S., Bergallo, H. G., et al. (2005). Prediction of threatened tetrapods based on the species–area relationship in Atlantic forest, Brazil. *Journal of the Zoological Society of London*, **265**, 359–364.

Griffith, B., Scott, J. M., Carpenter, J. W., and Reed, C. (1989). Translocation as a species conservation tool: status and strategy. *Science*, **245**, 477–480.

Griffiths, O., Cook, A., and Wells, S. M. (1993). The diet of the introduced carnivorous snail *Euglandina rosea* in Mauritius and its implications for threatened island gastropod faunas. *Journal of Zoology (London)*, **229**, 78–89.

Groombridge, B. (ed.) (1992). *Global Biodiversity: Status of the Earth's Living Resources*. (A report compiled by the World Conservation Monitoring Centre.) Chapman & Hall, London.

Groombridge, B. and Jenkins, M. D. (2002). *World Atlas of Biodiversity. Earth's Living Resources in the 21st Century*. United Nations Environmental Programme–World Conservation Monitoring Center, University of California Press, Berkeley.

Grossenbacher, D. L., Brandvain, Y., Auld, J. R., et al. (2017). Self-compatibility is over-represented on islands. *New Phytologist*, **215**, 469–478.

Gruner, D. S. (2007). Geological age, ecosystem development, and local resource constraints on arthropod community structure in the Hawaiian Islands. *Biological Journal of the Linnean Society*, **90**, 551–570.

Guevara, S. and Laborde, J. (1993). Monitoring seed dispersal at isolated standing trees in tropical pastures: consequences for local species availability. *Plant Ecology*, **107/108**, 319–338.

Gupta, S., Collier, J. S., Garcia-Moreno, D., et al. (2017). Two-stage opening of the Dover Strait and the origin of island Britain. *Nature Communications*, **8**, 15101.

Haddad, N. M. (2008). Finding the corridor more travelled. *Proceedings of the National Academy of Sciences, USA*, **105**, 19569–19570.

Haddad, N. M., Brudvig, L. A., Clobert, J., et al. (2015). Habitat fragmentation and its lasting impact on Earth's ecosystems. *Scientific Advances*, **1**, e1500052.

Haddad, N. M., Gonzalez, A., Brudvig, L. A., Burt, M. A., Levey, D. J., and Damschen, E. I. (2017). Experimental evidence does not support the Habitat Amount Hypothesis. *Ecography*, **40**, 48–55.

Haila, Y. (1990). Towards an ecological definition of an island: a northwest European perspective. *Journal of Biogeography*, **17**, 561–568.

Haila, Y. (2002). A conceptual genealogy of fragmentation research: from island biogeography to landscape ecology. *Ecological Applications*, **12**, 321–334.

Halley, J. M., Sgardeli, V., and Triantis, K. A. (2014). Extinction debt and the species–area relationship: a neutral perspective. *Global Ecology and Biogeography*, **23**, 113–123.

Halligan, J. J., Waters, M. R., Perrotti, A., et al. (2016). Pre-Clovis occupation 14,550 years ago at the Page-Ladson site, Florida, and the peopling of the Americas. *Science Advances*, **2**, e1600375.

Hamann, O. (2011). Ecology, demography and conservation in the Galapagos flora. In *The Biology of Island Floras* (ed. D. Bramwell and J. Caujapé-Castells), pp. 385–424. Cambridge University Press, Cambridge

Hambler, C. (1994). Giant tortoise *Geochelone gigantea* translocation to Curieuse Island (Seychelles): Success or failure? *Biological Conservation*, **69**, 293–299.

Hammond, J. O. S., Kendall, J.-M., Collier, J. S., and Rümpker, G. (2013). The extent of continental crust beneath the Seychelles. *Earth and Planetary Science Letters*, **381**, 166–176.

Hammoud, C., Kougioumoutzis, K., Rijsdijk, K. F., et al. (2021). Past connections with the mainland structure patterns of insular species richness in a continental-shelf archipelago (Aegean Sea, Greece). *Ecology and Evolution*, **11**, 5441–5458.

Hance, J. (2017). 'Guardian of the forest' routinely culled in Madeira. *The Guardian*, 11 July 2017.

Hanley, K. A., Bolger, D. T., and Case, T. J. (1994). Comparative ecology of sexual and asexual gecko species (*Lepidodactylus*) in French Polynesia. *Evolutionary Ecology*, **8**, 438–454.

Hanna, E. and Cardillo, M. (2014). Island mammal extinctions are determined by interactive effects of life history, island biogeography and mesopredator suppression. *Global Ecology and Biogeography*, **23**, 395–404.

Hansen, D. M., Donlan, C. J., Griffiths, C. J., and Campbell, K. J. (2010). Ecological history and latent conservation potential: large and giant tortoises as a model for taxon substitutions. *Ecography* **33**, 272–284.

Hansen, D. M. and Müller, C. B. (2009). Reproductive ecology of the endangered enigmatic Mauritian endemic *Roussea simplex* (Rousseaceae). *International Journal of Plant Sciences*, **170**, 42–52.

Hansford, J. P., Wright, P. C., Pérez, V. R., Muldoon, K. N., Turvey, S. T., and Godfrey, L. R. (2020). Evidence for early human arrival in Madagascar is robust: a response to Mitchell. *The Journal of Island and Coastal Archaeology*, **15**, 596–602.

Hanski, I. (1986). Population dynamics of shrews on small islands accord with the equilibrium model. *Biological Journal of the Linnean Society*, **28**, 23–36.

Hanski, I. (2010). The theories of island biogeography and metapopulation dynamics. In *The Theory of Island Biogeography Revisited* (ed. J. B. Losos and R. E. Ricklefs), pp. 186–213. Princeton University Press, Princeton.

Hanski, I., Moilanen, A., and Gyllenberg, M. (1996). Minimum viable metapopulation size. *American Naturalist*, **147**, 527–541.

Haponski, A. E., Lee, T., and Ó Foighil, D. (2019). Deconstructing an infamous extinction crisis: survival of *Partula* species on Moorea and Tahiti. *Evolutionary Applications*, **12**, eva.12778.

Harris, L. D. (1984). *The Fragmented Forest: Island Biogeography Theory and the Preservation of Biotic Diversity*. University of Chicago Press, Chicago.

Harrison, S., Murphy, D. D., and Ehrlich, P. R. (1988). Distribution of the Bay Checkerspot Butterfly, *Euphydryas editha bayensis*: evidence for a metapopulation model. *American Naturalist*, **132**, 360–382.

Harrison, S. and Noss, R. (2017). Endemism hotspots are linked to stable climatic refugia. *Annals of Botany*, **119**, 207–214.

Hatfield, J. H., Harrison, M. L. K., and Banks-Leite, C. (2018). Functional diversity metrics: how they are affected by landscape change and how they represent ecosystem functioning in the tropics. *Current Landscape Ecology Reports*, **3**, 35–42.

Hayes, S. E., Tuiwawa, M., Stevens, M. I., and Schwarz, M. P. (2019). A recipe for weed disaster in islands: a supergeneralist native pollinator aided by a 'Parlourmaid' plant welcome new arrivals in Fiji. *Biological Invasions*, **21**, 1643–1655.

Hays, W. S. T. and Conant, S. (2007). Biology and impacts of Pacific island invasive species. 1. A worldwide review of effects of the Small Indian Mongoose, *Herpestes javanicus* (Carnivora: Herpestidae). *Pacific Science*, **6**, 3–16.

He, F. and Hubbell, S. P. (2011). Species–area relationships always overestimate extinction rates from habitat loss. *Nature*, **473**, 368–371.

Heads, M. (2011). Old taxa on young islands: a critique of the use of island age to date island-endemic clades and calibrate phylogenies. *Systematic Biology*, **60**, 204–218.

Heads, M. (2017). Metapopulation vicariance explains old endemics on young volcanic islands. *Cladistics*, **2017**, 1–20.

Heaney, L. R. (2000). Dynamic disequilibrium: a long-term, large-scale perspective on the equilibrium model of island biogeography. *Global Ecology and Biogeography*, **34**, 59–74.

Heaney, L. R., Balete, D. S., Duya, M. R. M., et al. (2016). Doubling diversity: a cautionary tale of previously unsuspected mammalian diversity on a tropical oceanic island. *Frontiers of Biogeography*, **8**, e29667.

Heaney, L. R., Balete, D. S., and Rickart, E. A. (2013). Models of oceanic island biogeography: changing perspectives on biodiversity dynamics in archipelagoes. *Frontiers of Biogeography*, **5**, 249–257.

Heaney, L. R., Balete, D. S., and Rickart, E. A. (2016). *The Mammals of Luzon Island: Biogeography and Natural History of a Philippine Fauna*. Johns Hopkins University Press, Baltimore.

Heaney, L. R., Walsh, J. S., Jr, and Peterson, A. T. (2005). The roles of geological history and colonization abilities in genetic differentiation between mammalian populations in the Philippine archipelago. *Journal of Biogeography*, **32**, 229–247.

Hébert, K., Millien, V., and Lessard, J.-P. (2021). Source pool diversity and proximity shape the compositional uniqueness of insular mammal assemblages worldwide. *Journal of Biogeography*, **48**, 2337–2349.

Heinen, J. H., van Loon, E. E., Hansen, D. M., and Kissling, W. D. (2018). Extinction-driven changes in frugivore communities on oceanic islands. *Ecography*, **41**, 1245–1255.

Heleno, R. and Vargas, P. (2015). How do islands become green? *Global Ecology and Biogeography*, **24**, 518–526.

Helmus, M. R., Mahler, D. L., and Losos, J. B. (2014). Island biogeography of the Anthropocene. *Nature*, **513**, 543–547.

Hembry, D. H. and Balukjian, B. (2016). Molecular phylogeography of the Society Islands (Tahiti; South Pacific) reveals departures from hotspot archipelago models. *Journal of Biogeography*, **43**, 1372–1387.

Hembry, D. H., Bennett, G., Bess, E., et al. (2021). Insect radiations on islands: biogeographic pattern and evolutionary process in Hawaiian insects. *The Quarterly Review of Biology*, **96**, 247–296.

Henderson, S., Dawson, T. P., and Whittaker, R. J. (2006). Progress in invasive plants research. *Progress in Physical Geography*, **30**, 25–46.

Hernández, E., Nogales, M., and Martín, A. (2000). Discovery of a new lizard in the Canary Islands, with a multivariate analysis of *Gallotia*. *Herpetologica*, **56**, 63–76.

Herrmann, N. C., Stroud, J. T., and Losos, J. B. (2021). The evolution of 'Ecological Release' into the 21st Century. *Trends in Ecology & Evolution*, **36**, 206–215.

Hervías-Parejo, S., Nogales, M., Guzmán, B., et al. (2020). Potential role of lava lizards as pollinators across the Galápagos Islands. *Integrative Zoology*, **15**, 144–148.

Hess, A. L. (1990). Overview: sustainable development and environmental management of small islands. In *Sustainable Development and Environmental Management of Small Islands*. Vol. **5**, Man and the Biosphere Series (ed. W. Beller, P. d'Ayala, and P. Hein), pp. 3–14. UNESCO/Parthenon Publishing, Paris.

Higuchi, H. (1976). Comparative study on the breeding of mainland and island subspecies of the varied tit (*Parus varius*). *Tori*, **25**, 11–20.

Hildenbrand, A., Weis, D., Madureira, P., and Marques, F. O. (2014). Recent plate re-organization at the Azores Triple Junction: Evidence from combined geochemical and geochronological data on Faial, S. Jorge and Terceira volcanic islands. *Lithos*, **210–211**, 27–39.

Hilton, G. M., Atkinson, P. W., Gray, G. A. L., Arendt, W. J., and Gibbons, D. W. (2003). Rapid decline of the volcanically threatened Montserrat oriole. *Biological Conservation*, **111**, 79–89.

Hinsley, S. A., Bellamy, P. E., Newton, I., and Sparks, T. H. (1994). Factors influencing the presence of individual breeding bird species in woodland fragments. *English Nature Research Reports*, No. 99.

Hipsley, C. A. and Müller, J. (2014). Beyond fossil calibrations: realities of molecular clock practices in evolutionary biology. *Frontiers in Genetics*, **5**, 138.

Hiraiwa, M. K. and Ushimaru, A. (2017). Low functional diversity promotes niche changes in natural island pollinator communities. *Proceedings of the Royal Society B*, **284**, 20162218.

Hnatiuk, S. H. (1978). Plant dispersal by the Aldabran Giant tortoise *Geochelone gigantea* (Schweigger). *Oecologia*, **36**, 345–350.

Ho, S. Y. W., Tong, K. J., Foster, C. S. P., Ritchie, A. M., Lo, N., and Crisp, M. D. (2015). Biogeographic calibrations for the molecular clock. *Biology Letters*, **11**, 20150194.

Hoddle, M. S., Ramírez, C. C., Hoddle, C. D., et al. (2013). Post release evaluation of *Rodolia cardinalis* (Coleoptera: Coccinellidae) for control of *Icerya purchasi* (Hemiptera: Monophlebidae) in the Galápagos Islands. *Biological Control*, **67**, 262–274.

Holloway, J. D. (1996). The Lepidoptera of Norfolk Island, actual and potential, their origins and dynamics. In *The Origin and Evolution of Pacific Island Biotas, New Guinea to Eastern Polynesia: Patterns and Processes* (ed. A. Keast, and S. E. Miller), pp. 123–151. SPB Academic Publishing, Amsterdam.

Holt, B. G., Lessard, J.-P., Borregaard, M. K., et al. (2013). An update of Wallace's zoogeographic regions of the world. *Science*, **339**, 74–78.

Holt, R. D. (2010). Toward a trophic island biogeography: reflections on the interface of island biogeography and food web ecology. In *The Theory of Island Biogeography Revisited* (ed. J. B. Losos and R. E. Ricklefs), pp. 143–158. Princeton University Press, Princeton.

Holt, R. D., Gravel, D., Stier, A., and Rosindell, J. (2021). On the interface of food webs and spatial ecology: the trophic dimension of species–area relationships. In *The Species–Area Relationship: Theory and Application* (ed. T. J. Matthews, K. A. Triantis and R. J. Whittaker), pp. 289–318. Cambridge University Press, Cambridge.

Honnay, O., Hermy, M., and Coppin, P. (1999). Nested plant communities in deciduous forest fragments: species relaxation or nested habitats? *Oikos*, **84**, 119–129.

Hooft van Huysduynen, A., Janseens, S., Merckx, V., et al. (2021). Temporal and palaeoclimatic context of the evolution of insular woodiness in the Canary Islands. *Ecology and Evolution*, **11**, 12220–12231.

Hooker, J. D. (1866). *Insular Floras*. Lecture delivered at the British Association for the Advancement of Science meeting at Nottingham, on 27 August 1866. Subsequently reprinted in the *Gardeners' Chronicle* in Jan. 1867.

Horváth, Z., Ptacnik, R., Vad, C. F., and Chase, J. M. (2019). Habitat loss over six decades accelerates regional and local biodiversity loss via changing landscape connectance. *Ecology Letters*, **22**, 1019–1027.

Hosken, D. J. and House, C. M. (2011). Sexual selection. *Current Biology*, **21**, R62–R65.

Hubbell, S. P. (2001). *The Unified Neutral Theory of Biodiversity and Biogeography*. Princeton University Press, Princeton, NJ.

Hubbell, S. P. (2010). Neutral theory and the theory of island biogeography. In *The Theory of Island Biogeography Revisited* (ed. J. B. Losos and R. E. Ricklefs), pp. 264–292. Princeton University Press, Princeton.

Hughes, C. and Eastwood, R. (2006). Island radiation on a continental scale: exceptional rates of plant diversification after uplift of the Andes. *Proceedings of the National Academy of Sciences, USA*, **103**, 10334–10339.

Hughes, C. E. and Atchison, G. W. (2015). The ubiquity of alpine plant radiations: from the Andes to the Hengduan Mountains. *New Phytologist*, **207**, 275–282.

Hume, J. P. (2017). *Extinct Birds*. Bloomsbury, London.

Hunt, J. E., Cassidy, M., and Talling, P. J. (2018). Multi-stage volcanic island flank collapses with coeval explosive caldera-forming eruptions. *Scientific Reports*, **8**, 1146.

Hunt, J. E. and Jarvis, I. (2017). Prodigious submarine landslides during the inception and early growth of volcanic islands. *Nature Communications*, **8**, 2061.

Hunt, T. (2006). Rethinking the fall of Easter Island. *American Scientist*, **94**, 412–419.

Hürlimann A., Martí J., and Ledesma, A. (2004). Morphological and geological aspects related to large slope failures on oceanic islands — the huge La Orotava landslides on Tenerife, Canary Islands. *Geomorphology*, **62**, 143–158.

Hutchinson, G. E. (1957). Concluding remarks. *Cold Spring Harbor Symposia on Quantitative Biology*, **22**, 415–427.

Ibanez, T., Keppel, G., Baider, C., et al. (2018). Regional forcing explains local species diversity and turnover on tropical islands. *Global Ecology and Biogeography*, **27**, 474–486.

Ibanez, T., Keppel, G., Baider, C., et al. (2020). Tropical cyclones and island area shape species abundance distributions of local tree communities. *Oikos*, **129**, 1856–1866.

Ibáñez, J.-J. and Effland, W. R. (2011). Toward a theory of island pedogeography: testing the driving forces for pedological assemblages in archipelagos of different origins. *Geomorphology*, **135**, 215–223.

Igea, J., Bogarín, D., Papadopulos, A. S. T., and Savolainen, V. (2015). A comparative analysis of island floras challenges taxonomy-based biogeographical models of speciation. *Evolution*, **69**, 482–491.

Illera, J. C., Rando, J. C., Richardson, D. S., and Emerson, B. C. (2012). Age, origins and extinctions of the avifauna of Macaronesia: a synthesis of phylogenetic and fossil information. *Quaternary Science Reviews*, **50**, 14–22.

Illera, J. C., Spurgin, L. G., Rodriguez-Exposito, Nogales, M., and Rando, J. C. (2016). What are we learning about speciation and extinction from the Canary Islands? *Ardeola*, **63**, 15–33.

Inchausti-Sintes, F., Voltes-Dorta, A., and Suau-Sánchez, P. (2020). The income elasticity gap and its implications for economic growth and tourism development: the Balearic vs the Canary Islands. *Current Issues in Tourism*, **24**, 98–116.

Inoue, K. and Kawahara, T. (1996). Evolution of *Campanula* flowers in relation to insect pollinators on islands. In *Floral Biology: Studies on Floral Evolution in Animal-Pollinated Plants* (ed. D. G. Lloyd and S. C. H. Barrett), pp. 377–400. Chapman & Hall, New York.

Ioannidis, A. G., Blanco-Portillo, J., Sandoval, K., et al. (2020). Native American gene flow into Polynesia predating Easter Island settlement. *Nature*, **583**, 572–577.

Ioannidis, A. G., Blanco-Portillo, J., Sandoval, K., et al. (2021). Paths and timing of the peopling of Polynesia inferred from genomic networks. *Nature*, **597**, 522–526.

Itescu, Y. (2019). Are island-like systems biologically similar to islands? A review of the evidence. *Ecography*, **42**, 1298–1314.

Itescu, Y., Foufopoulos, J., Pafilis, P., and Meiri, S. (2019). The diverse nature of island isolation and its effect on land bridge insular faunas. *Global Ecology and Biogeography*, **29**, 262–280.

Iturralde-Vinent, M. A. (2006). Meso-Cenozoic Caribbean paleogeography: implications for the historical biogeography of the region. *International Geology Review*, **48**, 791–827.

IUCN/SSC Invasive Species Specialist Group (2004). www.issg.org (visited June 2005).

Iwamura, T., Lambin, E. F., Silvius, K. M., Luzar, J. B., and Fragoso, J. M. V. (2016). Socio-environmental sustainability of indigenous lands: simulating coupled human–natural systems in the Amazon. *Frontiers in Ecology and the Environment*, **14**, 77–83.

Izquierdo, I., Martín, J. L., Zurita, N., and Arechavaleta, M. (eds) (2004). *Lista de especies silvestres de Canarias (hongos, plantas y animales terrestres) 2004*. Consejería de Medio Ambiente y Ordenación Territorial, Gobierno de Canarias, Santa Cruz de Tenerife.

Jackson, E. D., Silver, E. A., and Dalrymple, G. B. (1972). Hawaiian-Emperor chain and its relation to Cenozoic circumpacific tectonics. *Geological Society of America, Bulletin*, **83**, 601–618.

Jackson, M. H. (1995). *Galápagos: A Natural History*. University of Calgary Press, Calgary.

Jacquet, C., Mouillot, D., Kulbicki, M., and Gravel, D. (2016). Extensions of Island Biogeography Theory predict the scaling of functional trait composition with habitat area and isolation. *Ecology Letters*, **20**, 135–146.

Jaleel, A. (2013). The status of the coral reefs and the management approaches: the case of the Maldives. *Ocean & Coastal Management*, **82**, 104–118.

James, J. E., Lanfear, R., and Eyre-Walker, A. (2016). Molecular evolutionary consequences of island colonization. *Genome Biology and Evolution*, **8**, 1876–1888.

Jamieson, I. G. and Allendorf, F. W. (2012). How does the 50/500 rule apply to MVPs? *Trends in Ecology & Evolution*, **27**, 578–584.

Jaramillo, M., Donaghy-Cannon, M., Vargas, F. H., and Parker, P. G. (2016). The diet of the Galapagos hawk (*Buteo galapagoensis*) before and after goat eradication. *Journal of Raptor Research*, **50**, 33–44.

Jarman, C. L., Larsen, T. L., Hunt, T., et al. (2016). Diet of the prehistoric population of Rapa Nui (Easter Island, Chile) shows environmental adaptation and resilience. *American Journal of Physical Anthropology*, **164**, 343–361.

Jenkyns, H. C. and Wilson, P. A. (1999). Stratigraphy, paleoceanography, and evolution of Cretaceous Pacific guyots: relics from a greenhouse earth. *American Journal of Science*, **299**, 341–392.

Jetz, W., Thomas, G. H., Joy, J. B., Hartmann, K., and Mooers, A. O. (2012). The global diversity of birds in space and time. *Nature*, **491**, 444–448.

Jõks, M., Kreft, H., Weigelt, P., and Pärtel, M. (2021). Legacy of archipelago history in modern island biodiversity – an agent-based simulation model. *Global Ecology and Biogeography*, **30**, 247–261.

Jones, C. W., Risi, M. M., Cleeland, J., and Ryan, P. G. (2019). First evidence of mouse attacks on adult albatrosses and petrels breeding on sub-Antarctic Marion and Gough Islands. *Polar Biology*, **42**, 619–623.

Jones, K. E., Reyes-Betancourt, J. A., Hiscock, S. J., and Carine, M. J. (2014). Allopatric diversification, multiple habitat shifts, and hybridization in the evolution of *Pericallis* (Asteraceae), a Macaronesian endemic genus. *American Journal of Botany*, **101**, 637–651.

Jones, M. G. W. and Ryan, P. G. (2010). Evidence of mouse attacks on albatross chicks on sub-Antarctic Marion Island. *Antarctic Science*, **22**, 39–42.

Jónsson, K. A., Borregaard, M. K., Carstensen, D. W., et al. (2017). Biogeography and biotic assembly of Indo-Pacific corvoid passerine birds. *Annual Review of Ecology, Evolution, and Systematics*, **48**, 231–253.

Jónsson, K. A., Fabre, P.-H., Fritz, S. A., et al. (2012). Ecological and evolutionary determinants for the adaptive radiation of the Madagascan vangas. *Proceedings of the National Academy of Sciences, USA*, **109**, 6620–6625.

Jónsson, K. A., Fabre. P.-H., Ricklefs, R. E., and Fjeldså, J. (2011). Major global radiation of corvoid birds originated in the proto-Papuan archipelago. *Proceedings of the National Academy of* Sciences, USA, **108**, 2328–2333.

Jónsson, K. A., Irestedt, M., Christidis, L., Clegg, S. M., Holt, B. G., and Fjeldså, J. (2014). Evidence of taxon cycles in an Indo-Pacific passerine bird radiation (Aves: *Pachycephala*). *Proceedings of the Royal Society B*, **281**, 20131727.

Jordan, S., Simon, C., Foote, D., and Englund, R. A. (2005). Phylogeographic patterns of Hawaiian *Megalagrion* damselflies (Odonata: Coenagrionidae) correlate with Pleistocene island boundaries. *Molecular Ecology*, **14**, 3457–3470.

Juette, T., Garnt, D., Jameson, J. W., and Réale, D. (2020). The island syndrome hypothesis is only partially validated in two rodent species in an inland–island system. *Oikos*, **129**, 1739–1751.

Kalmar, A. and Currie, D. J. (2006). A global model of island biogeography. *Global Ecology and Biogeography*, **15**, 72–81.

Kaneshiro, K. Y. (1995). Evolution, speciation, and the genetic structure of island populations. In *Islands: Biological Diversity and Ecosystem Function* (ed. P. M. Vitousek, L. L. Loope, and H. Adsersen), Ecological Studies **115**, pp. 22–23. Springer-Verlag, Berlin.

Katinas, L., Crisci, J. V., Hoch, P., Telleria, M. C., and Apodaca, M. J. (2013). Trans-oceanic dispersal and evolution of early composites. *Perspectives in Plant Ecology and Evolutionary Systematics*, **15**, 269–280.

Kearney, M., Fujita, M. K., and Ridenour, J. (2009). Lost sex in the reptiles: constraints and correlations. In *Lost Sex* (ed. I. Schön, K. Martens, and P. Dijk), pp. 447–474. Springer, Dordrecht.

Keast, A. and Miller, S. E. (ed.) (1996). *The Origin and Evolution of Pacific Island Biotas, New Guinea to Eastern Polynesia: Patterns and Processes*. SPB Academic Publishing, Amsterdam.

Keegan, W. F. and Hofman, C. L. (2017). *The Caribbean before Columbus*. Oxford University Press, New York.

Keeley, S. C. and Funk, V. A. (2011). Origin and evolution of Hawaiian endemics: new patterns revealed by molecular phylogenetic studies. In *The Biology of Island Floras* (ed. D. Bramwell and J. Caujapé-Castells), pp. 57–58. Cambridge University Press, Cambridge.

Keinath, D. A., Doak, D. F., Hodges, K. E., et al. (2017). A global analysis of traits predicting species sensitivity to habitat fragmentation. *Global Ecology and Biogeography*, **26**, 115–117.

Kellman, M. (1996). Redefining roles: plant community reorganization and species preservation in fragmented systems. *Global Ecology and Biogeography Letters*, **5**, 111–116.

Kennedy, J. D., Marki, P. Z., Reeve, A. H., et al. (2022). Diversification and community assembly of the world's largest tropical island. *Global Ecology and Biogeography*, **31**, 1078–1089.

Kennedy, S. R., Lin, J.Y., Adams, S. A., Krehenwinkel, H., and Gillespie, R. G. (2022). What is adaptive radiation? Many manifestations of the phenomenon in an iconic lineage of Hawaiian spiders. *Molecular Phylogenetics and Evolution*, **175**, 107564.

Keppel, G., Buckley, Y. M., and Possingham, H. P. (2010). Drivers of lowland rain forest community assembly, species diversity and forest structure on islands in the tropical South Pacific. *Journal of Ecology*, **98**, 87–95.

Kim, S.-C., Crawford, D. J., Francisco-Ortega, J., and Santos-Guerra, A. (1996). A common origin for woody *Sonchus* and five related genera in the Macaronesian islands: molecular evidence for extensive radiation. *Proceedings of the National Academy of Sciences, USA*, **93**, 7743–7748.

Kimball, B. A., Stelting, S. A., McAuliffe, T. W., Stahl, R. S., Garcia, R. A., and Pitt, W. C. (2016). Development of artificial bait for brown treesnake suppression. *Biological Invasions*, **18**, 359–369.

Kis, E. and Schweitzer, F. (2010). Dust accumulation and loess formation under the oceanic semiarid climate of Tenerife, Canary Islands. *Hungarian Geographical Bulletin*, **59**, 207–230.

Kisel, Y. and Barraclough, T. G. (2010). Speciation has a spatial scale that depends on levels of gene flow. *American Naturalist*, **175**, 316–334.

Kitayama, K. and Mueller-Dombois, D. (1995). Biological invasion on an oceanic island mountain: do alien plant species have wider ecological ranges than native species? *Journal of Vegetation Science*, **6**, 667–674.

Kitchener, D. J., Chapman, A., Muir, B. G., and Palmer, M. (1980). The conservation value for mammals of reserves in the Western Australian wheatbelt. *Biological Conservation*, **18**, 179–207.

Kitson, J. J. N., Warren, B. H., Thébaud, C., Strasberg, D., and Emerson, B. C. (2018). Community assembly and diversification in a species-rich radiation of island weevils (Coleoptera: Cratopini). *Journal of Biogeography*, **45**, 2016–2026.

Kleindorfer, S., Connor, J. A., Dudaniec, R. Y., Myers, S. A., Robertson, J., and Sulloway, F. J. (2014). Species collapse via hybridization in Darwin's tree finches. *The American Naturalist*, **183**, 325–341.

Kleinkopf, J. A., Roberts, W. R., Wagner, W. L., and Roalson, E. H. (2019). Diversification of Hawaiian *Cyrtandra* (Gesneriaceae) under the influence of incomplete lineage sorting and hybridization. *Journal of Systematics and Evolution*, **57**, 561–578.

Knope, M. L., Bellinger, M. R., Datlof, E. M., Gallaher, T. J., and Johnson, M. A. (2020a). Insights into the evolutionary history of the Hawaiian *Bidens* (Asteraceae) adaptive radiation revealed through phylogenomics. *Journal of Heredity*, **111**, 119–137.

Knope, M. L., Funk, V. A., Johnson, M. A., et al. (2020b). Dispersal and adaptive radiation of *Bidens* (Compositae) across the remote archipelagoes of Polynesia. *Journal of Systematics and Evolution*, **58**, 805–822.

Knope, M. L., Morden, C. W., Funk, V. A., and Fukami, T. (2012). Area and the rapid radiation of Hawaiian *Bidens* (Asteraceae). *Journal of Biogeography*, **39**, 1206–1216.

Knowlton, J. L., Flaspohler, D. J., Mcinerney, N. C. R., and Fleischer, R. C. (2014). First record of hybridization in the Hawaiian honeycreepers: 'I'iwi (*Vestiaria coccinea*) × 'Apapane (*Himatione sanguinea*). *The Wilson Journal of Ornithology*, **126**, 562–568.

Kny, L. (1867). Über die Flora oceanischer Inseln. *Zeitschrift der Gesselschaft für Erdkunde zu Berlin*, **2**, 208–226.

Kondraskov, P., Schütz, N, Schüßler, C., et al. (2015). Biogeography of Mediterranean hotspot biodiversity: re-evaluating the 'Tertiary relict' hypothesis of Macaronesian laurel forests. *PLOS ONE*, **10**, e0132091.

König, C., Weigelt, P., Taylor, A., et al. (2020). Source pools and disharmony of the world's island floras. *Ecography*, **44**, 44–55.

Koutroumpa, K., Warren, B. H., Theodoridis, S., et al. (2021). Geo-climatic changes and apomixis as major drivers of diversification in the Mediterranean sea lavenders (*Limonium* Mill.). *Frontiers in Plant Science*, **11**, 612258.

Kraemer, A. C., Roell, Y. E., Shoobs, N. F., and Parent, C. E. (2022). Does island ontogeny dictate the accumulation of both species richness and functional diversity? *Global Ecology and Biogeography*, **31**, 123–137.

Kreft, H., Jetz, W., Mutke, J., Kier, G., and Barthlott, W. (2008). Global diversity of island floras from a macroecological perspective. *Ecology Letters*, **11**, 116–127.

Kudoh, H., Takayama, K., and Kachi, N. (2013). Loss of seed buoyancy in *Hibiscus glaber* on the oceanic Bonin Islands. *Pacific Science*, **67**, 591–597.

Kumar, N. N. and Patel, A. (2021). Modelling the impact of COVID-19 in small Pacific island communities. *Current Issues in Tourism*, 25, 394–404.

Kurle, C. M., Zilliacus, K. M., Sparks, J., et al. (2021). Indirect effects of invasive rat removal result in recovery of island rocky intertidal community structure. *Scientific Reports*, **11**, e5395.

Kuussaari, M., Bommarco, R., Heikkinen, R. K., et al. (2009). Extinction debt: a challenge for biodiversity conservation. *Trends in Ecology & Evolution*, **24**, 564–571.

Kvist, L., Broggi, J., Illera, J. C., and Koivula, K. (2005). Colonisation and diversification of the blue tits (*Parus caeruleus teneriffae*-group) in the Canary Islands. *Molecular Phylogenetics and Evolution*, **34**, 501–511.

Lachlan, R. F., Verzijden, M. N., Bernard, C. S., et al. (2013). The progressive loss of syntactical structure in bird song along an island colonization chain. *Current Biology*, **23**, 1896–1901.

Lack, D. (1947a). *Darwin's Finches: An Essay on the General Biological Theory of Evolution*. Cambridge University Press, Cambridge

Lack, D. (1947b). The significance of clutch size. *Ibis*, **89**, 302–352.

Lack, D. (1969). The numbers of bird species on islands. *Bird Study*, **16**, 193–209.

Lack, D. (1970). The endemic ducks of remote islands. *Wildfowl*, **21**, 5–10.

Ladle, R. J. (2009). Forecasting extinctions: uncertainties and limitations. *Diversity*, **1**, 133–150.

Ladle, R. J. and Jepson, P. R. (2008). Toward a biocultural theory of avoided extinction. *Conservation Letters*, **1**, 111–118.

Ladle, R. J. and Whittaker, R. J. (eds) (2011). *Conservation Biogeography*. Wiley-Blackwell, Chichester.

Lambdon, P. and Cronk, Q. (2020). Extinction dynamics under extreme conservation threat: the flora of St Helena. *Frontiers in Ecology and Evolution*, **8**, article 41.

Lambeck, K. and Chappell, J. (2001). Sea level change through the last glacial cycle. *Science*, **292**, 679–686.

Lambeck, K., Rouby, H., Purcell, A., Sun, Y., and Sambridge, M. (2014). Sea level and global ice volumes from the Last Glacial Maximum to the Holocene. *Proceedings of the National Academy of Sciences, USA*, **111**, 15296–15303.

Lamelas-López, L., Fontaine, R., Borges, P. A. V., and Gonçalves, D. (2020). Impact of introduced nest predators on insular endemic birds: the case of the Azores Wood-pigeon (*Columba palumbus azorica*). *Biological Invasions*, **22**, 3593–3608.

Lamichhaney, S., Berglund, J., Almén, M. S., et al. (2015). Evolution of Darwin's finches and their beaks revealed by genome sequencing. *Nature*, **518**, 371–375.

Lamichhaney, S., Han, F., Berglund, J., et al. (2016). A beak size locus in Darwin's finches facilitated character displacement during a drought. *Science*, **352**, 470–474.

Lamichhaney, S., Han, F., Webster, M. T., Andersson, L., Grant, B. R., and Grant, P. R. (2018). Rapid hybrid speciation in Darwin's finches. *Science*, **359**, 224–228.

Lapiedra, O., Sayol, F., Garcia-Porta, J., and Sol, D. (2021). Niche shifts after island colonization spurred adaptive diversification and speciation in a cosmopolitan bird clade. *Proceedings of the Royal Society B*, **288**, 20211022.

Larrue, S., Butaud, J-F., Daehler, C. C., Ballet, S., Chadeyron, J., and Oyono, R. (2018). Persistence at the final stage of volcanic island ontogeny: abiotic predictors explain native plant species richness on 111 remote Pacific atolls. *Ecology and Evolution*, **8**, 12208–12220.

Larter, R. D. and Leat, P. T. (2003). Intra-oceanic subduction systems: an introduction. *Geological Society, London, Special Publications*, **219**, 1–17.

Laurance, W. F. (2002). Hyperdynamism in fragmented habitats. *Journal of Vegetation Science*, **13**, 595–602.

Laurance, W. F., Camargo, J. L. C., Fearnside, P. M., et al. (2017). An Amazonian rainforest and its fragments as a laboratory of global change. *Biological Reviews*, **93**, 223–247.

Lawlor, T. E., Hafner, D. J., Stapp, P. T., Riddle, B. R., and Alvarez-Castaneda, S. T. (2002). The mammals. In *A New Island Biogeography of the Sea of Cortés* (ed. T. J. Case, M. L. Cody, and E. Ezcurra), pp. 326–361. Oxford University Press, New York.

Le Friant, A., Harford, C. L., Deplus, C., et al. (2004). Geomorphological evolution of Montserrat (West Indies): importance of flank collapse and erosional processes. *Journal of the Geological Society*, **161**, 147–160.

Le Pepke, M., Irestedt, M., Fjeldså, J., Rahbek, C., and Jønsson, K. A. (2019). Reconciling supertramps, great speciators and relict species with the taxon cycle stages of a large island radiation (Aves: Campephagidae). *Journal of Biogeography*, **46**, 1214–1225.

Le Roux, J. J., Strasberg, D., Rouget, M., et al. (2014). Relatedness defies biogeography: the tale of two island endemics

(*Acacia heterophylla* and *A. koa*). *New Phytologist*, **204**, 230–242.

Leader-Williams, N. and Walton, D. (1989). The isle and the pussycat. *New Scientist*, **121** (1651), 11 February, 48–51.

Leberg, P. L. (1991). Influence of fragmentation and bottlenecks on genetic divergence of wild turkey populations. *Conservation Biology*, **5**, 522–530.

Leclerc, C., Courchamp, F., and Bellard, C. (2018). Insular threat associations within taxa worldwide. *Scientific Reports*, **8**, 6393.

Lee, M. S. Y. and Ho, S. Y. W. (2016). Molecular clocks. *Current Biology*, **26**, R399–R402.

Lees, A. C. and Gilroy, J. J. (2014). Vagrancy fails to predict colonization of oceanic islands. *Global Ecology and Biogeography*, **23**, 405–413.

Legge, S., Woinarski, J. C. Z., Burbidge, A. A., et al. (2018). Havens for threatened Australian mammals: the contributions of fenced areas and offshore islands to the protection of mammal species susceptible to introduced predators. *Wildlife Research*, **45**, 627–644.

Leibold, M. A., and Chase, J. M. (2018). *Metacommunity Ecology*. Princeton University Press, Princeton.

Leibold, M. A., Holyoak, M., Mouquet, N., et al. (2004). The metacommunity concept: a framework for multi-scale community ecology. *Ecology Letters*, **7**, 601–613.

Leihy, R. I. and Chown, S. L. (2020). Wind plays a major but not exclusive role in the prevalence of insect flight loss on remote islands. *Proceedings of the Royal Society B*, **287**, 20202121.

Leihy, R. I., Duffy, G. A., and Chown, S. L. (2018). Species richness and turnover among indigenous and introduced plants and insects of the Southern Ocean Islands. *Ecosphere*, **9**, e02358.

Lens, F., Davin, N., Smets, E., and del Arco, M. (2013). Insular woodiness on the Canary Islands: a remarkable case of convergent evolution. *International Journal of Plant Sciences*, **174**, 992–1013.

Lenzner, B., Latombe, G., Capinha, C., et al. (2020). What will the future bring for biological invasions on islands? An expert-based assessment. *Frontiers in Ecology and Evolution*, **8**, article 820.

Lenzner, B., Latombe, G., Schertler, A., et al. (2022). Naturalized alien floras still carry the legacy of European colonialism. *Nature Ecology and Evolution*, **6**, 1723–1732.

Lenzner, B., Weigelt, P., Kreft, H., Beierkuhnlein, C., and Steinbauer, M. J. (2017). The general dynamic model of island biogeography revisited at the level of major flowering plant families. *Journal of Biogeography*, **44**, 1029–1040.

Leppard, T. P. (2016). Between deterministic and random process in prehistoric Pacific island abandonment. *Journal of Pacific Archaeology*, **7**, 20–25.

Leppard, T. P. (2017). The biophysical effects of Neolithic island colonization: general dynamics and sociocultural implications. *Human Ecology*, **45**, 555–568.

Leppard, T. P., Cochrane, E. E., Gaffney, D., et al. (2022). Global patterns in island colonization during the Holocene. *Journal of World Prehistory*, **35**, 163–232.

Lerner, H. R. L., Meyer, M., James, H. F., Hofreiter, M., and Fleischer, R. C. (2011). Multilocus resolution of phylogeny and timescale in the extant adaptive radiation of Hawaiian honeycreepers. *Current Biology*, **21**, 1838–1844.

Leroy, T., Rousselle, M., Tilak, M.-K., et al. (2021). Island songbirds as windows into evolution in small populations. *Current Biology*, **31**, 1303–1310.e4.

Lesnek, A. J., Briner, J. P., Lindqvist, C., Baichtal, J. F., and Heaton, T. H. (2018). Deglaciation of the Pacific coastal corridor directly preceded the human colonization of the Americas. *Science Advances*, **4**, eaar5040.

Leuschner, C. (1996). Timberline and alpine vegetation on the tropical and warm-temperate oceanic islands of the world: elevation, structure, and floristics. *Vegetatio*, **123**, 193–206.

Levey, D. J., Bolker, B. M., Tewksbury, J. J., Sargent, S., and Haddad, N. M. (2005). Effects of landscape corridors on seed dispersal by birds. *Science*, **309**, 146–148.

Levin, I. I., Zwiers, P., Deem, S. L., et al. (2013). Multiple lineages of avian malaria parasites (*Plasmodium*) in the Galapagos Islands and evidence for arrival via migratory birds. *Conservation Biology*, **27**, 1366–1377.

Li, E., Bellard, C., Hu, F., and Li, H. (2020). Effect of distance, area, and climate on the frequency of introduction and extinction events on islands and archipelagos. *Ecosphere*, **11**, e03008.

Lichtenberg, E. M., Kennedy, C. M., Kremen, C., et al. (2017). A global synthesis of the effects of diversified farming systems on arthropod diversity within fields and across agricultural landscapes. *Global Change Biology*, **23**, 4946–4957.

Lifjeld, J. T., Anmarkrud, J. A., Calabuig, P., et al. (2016). Species-level divergences in multiple functional traits between the two endemic subspecies of Blue Chaffinches *Fringilla teydea* in Canary Islands. *BMC Zoology*, **1**, 1–19.

Lim, J. and Marshall, C. R. (2017). The true tempo of evolutionary radiation and decline revealed on the Hawaiian archipelago. *Nature*, **543**, 710–713.

Lindenmayer, D. B. and Fischer, J. (2007). Tackling the habitat fragmentation panchreston. *Trends in Ecology & Evolution*, **22**, 127–132.

Lindenmayer, D. B., Wood, J., McBurney, L., Blair, D., and Banks, S. C. (2015). Single large versus several small: the SLOSS debate in the context of bird responses to a variable retention logging experiment. *Forest Ecology and Management*, **339**, 1–10.

Linder, P. and Baker, N. P. (2014). Does polyploidy facilitate long-distance dispersal? *Annals of Botany*, **113**, 1175–1183.

Littleford-Colquhoun, B., Clemente, C., Whiting, M. J., Ortiz-Barrientos, D., and Frère, C. H. (2017). Archipelagos of the Anthropocene: rapid and extensive differentiation of native terrestrial vertebrates in a single metropolis. *Molecular Ecology*, **26**, 2466–2481.

Liu, C., Sarnat, E. M., Friedman, N. R., et al. (2020). Colonize, radiate, decline: unravelling the dynamics of island community assembly with Fijian trap-jaw ants. *Evolution*, **74**, 1082–1097.

Lokatis, S. and Jeschke, J. M. (2018). The island rule: an assessment of biases and research trends. *Journal of Biogeography*, **45**, 289–303.

Lomolino, M. V. (1984). Immigrant selection, predation, and the distribution of *Microtus pennsylvanicus* and *Blarina brevicauda* on islands. *American Naturalist*, **123**, 468–483.

Lomolino, M. V. (1985). Body size of mammals on islands: the island rule reexamined. *American Naturalist*, **125**, 310–316.

Lomolino, M. V. (1986). Mammalian community structure on islands: the importance of immigration, extinction and interactive effects. *Biological Journal of the Linnean Society*, **28**, 1–21.

Lomolino, M. V. (1990). The target area hypothesis: the influence of island area on immigration rates of non-volant mammals. *Oikos*, **57**, 297–300.

Lomolino, M. V. (2000a). A call for a new paradigm of island biogeography. *Global Ecology and Biogeography*, **9**, 1–6.

Lomolino, M. V. (2000b). Ecology's most general, yet protean pattern: the species–area relationship. *Journal of Biogeography*, **27**, 17–26.

Lomolino, M. V. (2002). '... there are areas too small, and areas too large, to show clear diversity patterns ... ' R. H. MacArthur (1972: 191). *Journal of Biogeography*, **29**, 555–557.

Lomolino, M. V. (2005). Body size evolution in insular vertebrates: generality of the island rule. *Journal of Biogeography*, **32**, 1683–1699.

Lomolino, M. V., Riddle, B. R., and Whittaker, R. J. (2017). *Biogeography*, 5th edn. Sinauer, Sunderland, MA.

Lomolino, M. V., Sax, D. F., and Brown, J. H. (eds) (2004). *Foundations of Biogeography: Classic Papers with Commentaries*. The University of Chicago Press, Chicago.

Lomolino, M. V., Sax, D. F., Palombo, M. R., and van der Geer, A. A. (2012). Of mice and mammoths: evaluations of causal explanations for body size evolution in insular mammals. *Journal of Biogeography*, **39**, 842–854.

Lomolino, M. V., Tomlinson, S., Wood, J., Wilmshurst, J., and Fordham, D. A. (2021). Geographic and ecological segregation in an extinct guild of flightless birds: New Zealand's moa. *Frontiers of Biogeography*, **13**, e53416.

Lomolino, M. V., van der Geer, A. A., Lyras, G. A., Palombo, M. R., Sax, D. F., and Rozzi, R. (2013). Of mice and mammoths: generality and antiquity of the island rule. *Journal of Biogeography*, **40**, 1427–1439.

Lomolino, M. V. and Weiser, M. D. (2001). Towards a more general species–area relationship: diversity on all islands, great and small. *Journal of Biogeography*, **28**, 431–445.

Long, A. J., Crosby, M. J., Stattersfield, A. J., and Wege, D. C. (1996). Towards a global map of biodiversity: patterns in the distribution of range-restricted birds. *Global Ecology and Biogeography Letters*, **5**, 281–304.

Longpré, M. A. and Felpeto, A. (2021). Historical volcanism in the Canary Islands; part 1: a review of precursory and eruptive activity, eruption parameter estimates, and implications for hazard assessment. *Journal of Volcanology and Geothermal Research*, **419**, 107363.

Lorenzo-Carballa, M. A., Hassall, C., Encalada, A. C., Sanmartín-Villar, I., Torres-Cambas, Y., and Cordero-Rivera, A. (2017). Parthenogenesis did not consistently evolve in insular populations of *Ischnura hastata* (Odonata, Coenagrionidae). *Ecological Entomology*, **42**, 67–76.

Losos, J. B. (2009). *Lizards in an Evolutionary Tree: Ecology and Adaptive Radiation of* Anoles. University of California Press, Berkeley.

Losos, J. B. and Mahler, D. L. (2010). Adaptive radiation: the interaction of ecological opportunity, adaptation, and speciation. In *Evolution since Darwin: The First 150 Years* (ed. M. Bell, D. Futuyma, W. Eanes, and J. Levinton), pp. 381–420. Sinauer, Sunderland.

Losos, J. B. and Ricklefs, R. E. (eds) (2010). *The Theory of Island Biogeography Revisited.* Princeton University Press, Princeton.

Losos, J. B. and Schluter, D. (2000). Analysis of an evolutionary species–area relationship. *Nature*, **408**, 847–850.

Louiseau, C., Melo, M., Lee, Y., et al. (2018). High endemism of mosquitoes on São Tomé and Príncipe Islands: evaluating the general dynamic model in a worldwide island comparison. *Insect Conservation and Diversity*, **12**, 69–79.

Lovejoy, T. E., Bierregaard, R. O., Rylands, A. B., et al. (1986). Edge and other effects of isolation on Amazon forest fragments. In *Conservation Biology: The Science of Scarcity and Diversity* (ed. M. Soulé), pp. 257–285. Sinauer Associates, Sunderland, MA.

Lu, M., Vasseur, D., and Jetz, W. (2019). Beta diversity patterns derived from island biogeography theory. *The American Naturalist*, **194**, e52–e65.

Lubold, G. (2020). U.S. military is offered new bases in the Pacific. *The Wall Street Journal*, 8 September 2020. www.wsj.com/articles/u-s-military-is-offered-new-bases-in-the-pacific-11599557401

Lugo, A. E. (1988). Ecological aspects of catastrophes in Caribbean islands. *Acta Científica*, **2**, 24–31.

Lynch, J. D. and Johnson, N. V. (1974). Turnover and equilibria in insular avifaunas, with special reference to the California Channel Islands. *Condor*, **76**, 370–384.

Lyons, S. K., Miller, J. H., Fraser, D., et al. (2016). The changing role of mammal life histories in Late Quaternary extinction vulnerability on continents and islands. *Biology Letters*, **12**, article 20160342.

Mabberley, D. J. (1979). Pachycaul plants and islands. In *Plants and Islands* (ed. D. Bramwell), pp. 259–277. Academic Press, London.

MacArthur, R. H. (1957). On the relative abundance of bird species. *Proceedings of the National Academy of Sciences, USA*, **43**, 293–295.

MacArthur, R. H., Diamond, J. M., and Karr, J. (1972). Density compensation in island faunas. *Ecology*, **53**, 330–342.

MacArthur, R. H. and Wilson, E. O. (1963). An equilibrium theory of insular zoogeography. *Evolution*, **17**, 373–387.

MacArthur, R. H. and Wilson, E. O. (1967). *The Theory of Island Biogeography*. Princeton University Press, Princeton.

Machado, A. (2022). *The Macaronesian* Laparocerus *(Coleoptera, Curculionidae, Entiminae). Taxonomy, Phylogeny, and Natural History*. Turquesa Ediciones, Santa Cruz de Tenerife.

Machado, A. and Martín, J. L. (1993). The 'Fenix project'. A protected areas network for the Canary Islands. Unpublished paper. www.antoniomachado.net/production/technical/#conservation_publications

Machado, A., Rodríguez-Expósito, E., López, M., and Hernández. M. (2017). Phylogenetic analysis of the genus Laparocerus, with comments on colonisation and diversification in Macaronesia (Coleoptera, Curculionidae, Entiminae). Zookeys, 651, 1–77.

Magnússon, B., Gudmundsson, G. A., Metúsalemsson, S., and Granquist, S. M. (2020). Seabirds and seals as drivers of plant succession on Surtsey. *Surtsey Research*, **14**, 115–130.

Magnússon, B., Magnússon, S. H., and Fridriksson, S. (2009). Developments in plant colonization and succession on Surtsey during 1999–2008. *Surtsey Research*, **12**, 57–76.

Magurran, A. E. (2004). *Measuring Biological Diversity*. Blackwell, Oxford.

Majeský, Ľ., Krahulec, F., and Vašut, R. J. (2017). How apomictic taxa are treated in current taxonomy: a review. *Taxon*, **66**, 1017–1040.

Mammola, S., Carmona, C. P., Guillerme, T., and Cardoso, P. (2021). Concepts and applications in functional diversity. *Functional Ecology*, **35**, 1869–1885.

Manne, L. L., Pimm, S. L., Diamond, J. M., and Reed, T. M. (1998). The form of the curves: a direct evaluation of MacArthur & Wilson's classic theory. *Journal of Animal Ecology*, **67**, 784–794.

Margalef, O., Cañellas-Boltà, N., Pla-Rabes, S., et al. (2013). A 70,000 year multiproxy record of climatic and environmental change from Rano Aroi peatland (Easter Island). *Global and Planetary Change*, **108**, 72–84.

Marino, C., Leclerc, C., and Bellard, C. (2022). Profiling insular vertebrates prone to biological invasions: what makes them vulnerable? *Global Change Biology*, **28**, 1077–1090.

Marques, D. A., Meier, J. I., and Seehausen, O. (2019). A combinatorial view on speciation and adaptive radiation. *Trends in Ecology & Evolution*, **34**, 531–544.

Marrero, M. V., Oostermeijer, G., Nogales, M., et al. (2019). Comprehensive population viability study of a rare endemic shrub from the high mountain zone of the Canary Islands and its conservation implications. *Journal for Nature Conservation*, **47**, 65–76.

Marrero-Gómez, M. V., Bañares-Baudet, A., and Carqué-Alamo, E. (2003). Plant resource conservation planning in protected natural areas: an example from the Canary Islands, Spain. *Biological Conservation*, **113**, 399–410.

Marshall, B. E. (2018). Guilty as charged: Nile perch was the cause of the haplochromine decline in Lake Victoria. *Canadian Journal of Fisheries and Aquatic Sciences*, **75**, 1542–1559.

Marshall, H. D., Baker, A. J., and Grant, A. R. (2013). Complete mitochondrial genomes from four subspecies of common chaffinch (*Fringilla coelebs*): new inferences about mitochondrial rate heterogeneity, neutral theory, and phylogenetic relationships within the order Passeriformes. *Gene*, **517**, 37–45.

Martín, A. and Lorenzo, J. A. (2001). *Aves del Archipiélago Canario*. Francisco Lemus Ed., La Laguna.

Martín, J. L. (2009). Are the IUCN standard home-range thresholds for species a good indicator to prioritise conservation urgency in small islands? A case study in the Canary Islands (Spain). *Journal for Nature Conservation*, **17**, 87–98.

Martín-Esquivel, J. L., Fajardo, S., Cabrera, M. A., et al. (2005). *Evaluación 2004 de especies amenazadas de Canarias. Especies en peligro de extinción, sensibles a la alteración de su hábitat y vulnerables*. Consejería de Medio ambiente y ordenación territorial, Gobierno de Canarias, Santa Cruz de Tenerife.

Martín-Esquivel, J. L., García, H., Redondo, C., García, I., and Carralero, I. (1995). *La red Canaria de espacios naturales protegidos*. Viceconsejería de Medio Ambiente, Gobierno de Canarias, Santa Cruz de Tenerife.

Martín-Esquivel, J. L., Marrero-Gómez, M., Cubas, J., González-Mancebo, J. M., Olano, J. M., and del Arco, M. (2020). Climate warming and introduced herbivores disrupt alpine plant community of an oceanic island (Tenerife, Canary Islands). *Plant Ecology*, **221**, 1117–1131.

Martínez-Garza, C. and Howe, H. F. (2003). Restoring tropical diversity: beating the time tax on species loss. *Journal of Applied Ecology*, **40**, 423–429.

Masson, D. G., Watts, A. B., Gee, M. J. R., et al. (2002). Slope failures in the flanks of the western Canary Islands. *Earth-Science Reviews*, **57**, 1–35.

Masters, J. C., Génin, F., Zhang, Y., et al. (2020). Biogeographic mechanisms involved in the colonization of Madagascar by African vertebrates: rifting, rafting and runways. *Journal of Biogeography*, **48**, 492–510.

Matos-Maraví, P., Matzke, N. J., Larabee, F. J., et al. (2018). Taxon cycle predictions supported by model-based inference in Indo-Pacific trap-jaw ants (Hymenoptera: Formicidae: Odontomachus). *Molecular Ecology*, **27**, 4090–4107.

Matsubayashi, K. W. and Yamaguchi, R. (2022). The speciation view: disentangling multiple causes of adaptive and non-adaptive radiation in terms of speciation. *Population Ecology*, **64**, 95–107.

Matthews, T. J. (2021). On the biogeography of habitat islands: the importance of matrix effects, noncore species, and source-sink dynamics. *The Quarterly Review of Biology*, **96**, 73–104.

Matthews, T. J., Aspin, T. W. H., Ulrich, W., et al. (2019b). Can additive beta-diversity be reliably partitioned into nestedness and turnover components? *Global Ecology and Biogeography*, **28**, 1146–1154.

Matthews, T. J., Borges, P. A. V., de Azevedo, E. B., and Whittaker, R. J. (2017). A biogeographical perspective on species abundance distributions: recent advances and opportunities for future research. *Journal of Biogeography*, **44**, 1705–1710.

Matthews, T. J., Cottee-Jones, H. E. W., and Whittaker, R. J. (2014a). Habitat fragmentation and the species–area relationship: a focus on total species richness obscures the impact of habitat loss on habitat specialists. *Diversity and Distributions*, **20**, 1136–1146.

Matthews, T. J., Cottee-Jones, H. E. W., and Whittaker, R. J. (2015a). Quantifying and interpreting nestedness in habitat islands: a synthetic analysis of multiple datasets. *Diversity and Distributions*, **21**, 392–404.

Matthews, T. J., Guilhaumon, F., Triantis, K. A., Borregaard, M. K., and Whittaker, R. J. (2016a). On the form of species–area relationships in habitat islands and true islands. *Global Ecology and Biogeography*, **25**, 847–858.

Matthews, T. J. and Rigal, F. (2021). Thresholds and the species–area relationship: a set of functions for fitting, evaluating and plotting a range of commonly used piecewise models in R. *Frontiers of Biogeography*, **13**, e49404.

Matthews, T. J., Rigal, F., Kougioumoutzis, K., Trigas, P., and Triantis, K. A. (2020). Unravelling the small-island effect

through phylogenetic community ecology. *Journal of Biogeography*, **47**, 2341–2352.

Matthews, T. J., Rigal, F., Proios, K., Triantis, K. A., and Whittaker, R. J. (2021b). Explaining variation in island species–area relationship (ISAR) model parameters between different island types: expanding a global model of ISARs. In *The Species–Area Relationship: Theory and Application* (ed. T. J. Matthews, K. A. Triantis and R. J. Whittaker), pp. 51–77. Cambridge University Press, Cambridge.

Matthews, T. J., Rigal, F., Triantis, K. A., and Whittaker, R. J. (2019a). A global model of island species–area relationships. *Proceedings of the National Academy of Sciences, USA*, **116**, 12337–12342.

Matthews, T. J., Sheard, C., Cottee-Jones, H. E. W., Bregman, T. P., Tobias, J. A., and Whittaker, R. J. (2015b). Ecological traits reveal functional nestedness of bird communities in habitat islands: a global survey. *Oikos*, **124**, 817–826.

Matthews, T. J., Steinbauer, M. J., Tzirkalli, E., Triantis, K. A., and Whittaker, R. J. (2014b). Thresholds and the species–area relationship: a synthetic analysis of habitat island datasets. *Journal of Biogeography*, **41**, 1018–1028.

Matthews, T. J. and Triantis, K. A. (2021). Island biogeography. *Current Biology*, **31**, R1201–R1207.

Matthews, T. J., Triantis, K. A., Rigal, F., Borregaard, M. K., Guilhaumon, F., and Whittaker, R. J. (2016b). Island species–area relationships and species accumulation curves are not equivalent: an analysis of habitat island datasets. *Global Ecology and Biogeography*, **25**, 607–618.

Matthews, T. J., Triantis, K. A., and Whittaker, R. J. (eds) (2021a). *The Species–Area Relationship: Theory and Application*. Cambridge University Press, Cambridge.

Matthews, T. J., Wayman, J. P., Cardoso, P., et al. (2022). Threatened and extinct island endemic birds of the world: distribution, threats and functional diversity. *Journal of Biogeography*, **49**, 1920–1940.

Matthews, T. J. and Whittaker, R. J. (2014). Fitting and comparing competing models of the species abundance distribution: assessment and prospect. *Frontiers of Biogeography*, **6**, 67–82.

Matthews, T. J., Whittaker, R. J., and Borges, P. A. V. (2014c). Multimodal species abundance distributions: a deconstruction approach reveals the processes behind the pattern. *Oikos*, **123**, 533–544.

Maturana, C. S., Segovia, N. I., González-Wevar, C. A., et al. (2020). Evidence of strong small-scale population structure in the Antarctic freshwater copepod *Boeckella poppei* in lakes on Signy Island, South Orkney Islands. *Limnology and Oceanography*, **65**, 2024–2040.

Mawdsley, N. A., Compton, S. G., and Whittaker, R. J. (1998). Population persistence, pollination mutualisms, and figs in fragmented tropical landscapes. *Conservation Biology*, **12**, 1416–1420.

Maxwell, S. L., Fuller, R. A., Brooks, T. M., and Watson, J. E. M. (2016). The ravages of guns, nets and bulldozers. *Nature*, **536**, 143–145.

May, F., Rosenbaum, B., Schurr, F. M., and Chase, J. E. (2019). The geometry of habitat fragmentation: effects of species distribution patterns on extinction risk due to habitat conversion. *Ecology and Evolution*, **9**, 2775–2790.

Mayfield, M. M. and Levine, J. M. (2010). Opposing effects of competitive exclusion on the phylogenetic structure of communities. *Ecology Letters*, **13**, 1085–1093.

Mayr, E. (1942). *Systematics and the Origin of Species*. Columbia University Press, New York.

Mayr, E. (1954). Change of genetic environment and evolution. In *Evolution as a Process* (ed. J. S. Huxley, A. C. Hardy, and E. B. Ford), pp. 156–180. Allen & Unwin, London.

Mayr, E. (1963). *Animal Species and Evolution*. Harvard University Press, Cambridge, MA.

McClure, K. M., Fleischer, R. C., and Kilpatrick, A. M. (2020). The role of native and introduced birds in transmission of avian malaria in Hawaii. *Ecology*, **101**, e03038.

McCollin, D. (2017). Turnover dynamics of breeding land birds on islands: is island biogeography theory 'true but trivial' over decadal time-scales? *Diversity*, **9**, doi:10.3390/d9010003.

McCreless, E. E., Huff, D. D., Croll, D. A., et al. (2016). Past and estimated future impact of invasive alien mammals on insular threatened vertebrate populations. *Nature Communications*, **7**, 12488.

McCulloch, G. A., Wallis, G. P., and Waters, J. M. (2016). A time-calibrated phylogeny of southern hemisphere stoneflies: testing for Gondwanan origins. *Molecular Phylogenetics and Evolution*, **96**, 150–160.

McFarlane, S. E. and Pemberton, J. M. (2019). Detecting the true extent of introgression during anthropogenic hybridization. *Trends in Ecology & Evolution*, **34**, 315–326.

McGee, M. D., Borstein, S. R., Meier, J. I., et al. (2020). The ecological and genomic basis of explosive adaptive radiation. *Nature*, **586**, 75–79.

McGeoch, M. A. and Gaston, K. J. (2002). Occupancy frequency distributions: patterns, artefacts and mechanisms. *Biological Reviews*, **77**, 311–331.

McGill, B. J., Etienne, R. S., Gray, J. S., et al. (2007). Species abundance distributions: moving beyond single prediction theories to integration within an ecological framework. *Ecology Letters*, **10**, 995–1015.

McGlone, M.S. (2002). The Late Quaternary peat, vegetation and climate history of the Southern Oceanic Islands of New Zealand. *Quaternary Science Reviews*, **21**, 683–707.

McGuinness, K. A. (1984). Equations and explanations in the study of species–area curves. *Biological Reviews*, **59**, 423–440.

McInerny, C. J., Musgrove, A. J., Stoddart, A., Harrop, A. H. J., Dudley, S. P., and The British Ornithologists' Union Records Committee (2018). The British List: a checklist of birds of Britain (9th edn). *Ibis*, **160**, 190–240.

McWerthy, D. B., Whitlock, C., Wilmshurst, J. M., et al. (2010). Rapid landscape transformation in South Island, New Zealand, following initial Polynesian settlement. *Proceedings of the National Academy of Sciences, USA*, **107**, 21343–21348.

Médail, F. (2022). Plant biogeography and vegetation patterns of the Mediterranean Islands. *The Botanical Review*, **88**, 63–129.

Medina, F. M., Bonnaud, E., Vidal, E., et al. (2011). A global review of the impacts of invasive cats on island endangered vertebrates. *Global Change Biology*, **17**, 3503–3510.

Meier, J. I., Marques, D. A., Mwaiko, S., Wagner, C. E., Excoffier, L., and Seehausen, O. (2017). Ancient hybridization fuels rapid cichlid fish adaptive radiations. *Nature Communications*, **8**, 14363.

Meier, J. I., Stelkens, R. B., Joyce, D. A., et al. (2019). The coincidence of ecological opportunity with hybridization explains rapid adaptive radiation in Lake Mweru cichlid fishes. *Nature Communications*, **10**, 5391.

Meiri, S., Dayan, T., and Simberloff, D. (2006). The generality of the island rule re-examined. *Journal of Biogeography*, **33**, 1571–1577.

Melchett, P. (2017). Pesticides – experts ignore the most serious threat to UK wildlife. *Biodiversity*, **18**, 60–63.

Menard, H. W. (1986). *Islands*. Scientific American Library, New York.

Menegotto, A., Rangel, T. R., Schrader, J., Weigelt, P., and Kreft, H. (2020). A global test of the subsidized island biogeography hypothesis. *Global Ecology and Biogeography*, **29**, 320–330.

Menezes de Sequeira, M., Jardim, R., Gouveia, M., Góis-Marques, C. A., and Eddie, W. M. M. (2021). Population decline in the Critically Endangered *Musschia isambertoi* (Campanulaceae) endemic to Desertas Islands (Madeira Archipelago) calls for urgent conservation management. *Journal of Natural Conservation*, **60**, 125955.

Meredith, F. L., Tindall, M. L., Hemmings, F. A., and Moles, A. T. (2019). Prickly pairs: the proportion of spinescent species does not differ between islands and mainlands. *Journal of Plant Ecology*, **12**, 941–948.

Meudt, H. M., Albach, D. C., Tanentzap, A. J., et al. (2021). Polyploidy on islands: its emergence and importance for diversification. *Frontiers in Plant Science*, **12**, 637214.

Meusel, H. (1952). Über Wuchsformen Verbreitung und Phylogenie einiger mediterran-mitteleuropäischer Angiospermen Gattungen. *Flora*, **139**, 333–393.

Meyer, J.-Y. and Florence, J. (1996). Tahiti's native flora endangered by the invasion of *Miconia calvescens* DC. (Melastomataceae). *Journal of Biogeography*, **23**, 775–781.

Michaux, B. (2010). Biogeology of Wallacea: geotectonic models, areas of endemism, and natural biogeographical units. *Biological Journal of the Linnean Society*, **101**, 193–212.

Mielke, H. W. (1989). *Patterns of Life: Biogeography of a Changing World*. Unwin Hyman, Boston, MA.

Milberg, P. and Tyrberg, T. (1993). Naive birds and noble savages—a review of man-caused prehistoric extinctions of island birds. *Ecography*, **16**, 229–250.

Miller, C. R. and Waits, L. P. (2003). The history of effective population size and genetic diversity in the Yellowstone grizzly (*Ursus arctos*): Implications for conservation. *Proceedings of the National Academy of Sciences, USA*, **100**, 4334–4339.

Miller, M. S., O'Driscoll, L. J., Butcher, A. J., and Thomas, C. (2015). Imaging Canary Island hotspot material beneath the lithosphere of Morocco and southern Spain. *Earth and Planetary Science Letters*, **431**, 186–194.

Millien, V. (2006). Morphological evolution is accelerated among island mammals. *PLOS Biology*, **4**, e321.

Miskelly, C. M. (1990). Effects of the 1982–83 El Niño event on two endemic landbirds on the Snares Islands, New Zealand. *Emu*, **90**, 24–27.

Mittermeier, R. A., Turner, W. R., Larsen, F. W., Brooks, T. M., and Gascon, C. (2011). Global biodiversity conservation: the critical role of hotspots. In *Biodiversity Hotspots* (ed. F. E. Zachos and J. C. Habel), pp. 3–22. Springer-Verlag, Berlin.

Monroy, P. and García-Verdugo, C. (2019). Testing the hypothesis of loss of defenses on islands across a wide latitudinal gradient of *Periploca laevigata* populations. *American Journal of Botany*, **106**, 303–312.

Moore, D. M. (1979). The origins of temperate island floras. In *Plants and Islands* (ed. D. Bramwell), pp. 69–86. Academic Press, London.

Moore, J. G., Normark, W. R., and Holcomb, R. T. (1994). Giant Hawaiian underwater landslides. *Science*, **264**, 46–47.

Morales, J., Rodríguez, A., Alberto, V., Machado, C., and Criado, C. (2009).The impact of human activities on the natural environment of the Canary Islands (Spain) during the pre-Hispanic stage (3rd–2nd Century BC to 15th Century AD): an overview. *Environmental Archaeology*, **14**, 27–36.

Moran, R. (1996). The flora of Guadalupe Island. *Memoirs of the Californian Academy of Science*, **19**, 1–190.

Morand, S. (2000). Geographic distance and the role of island area and habitat diversity in the species–area relationships of four Lesser Antillean faunal groups: a complementary note to Ricklefs and Lovette. *Journal of Animal Ecology*, **69**, 1117–1119.

Moreira, X. and Abdala-Roberts, L. (2022). A roadmap for future research on insularity effects on plant–herbivore interactions. *Global Ecology and Biogeography*, **31**, 602–610.

Moreira, X., Abdala-Roberts, L., Castagneyrol, B., et al. (2022). A phylogenetically controlled test does not support the prediction of lower putative anti-herbivore leaf traits for insular woody species. *Journal of Biogeography*, **49**, 274–285.

Moreira, X., Castagneyrol, B., García-Verdugo, C., and Abdala-Roberts, L. (2021). A meta-analysis of insularity effects on herbivory and plant defences. *Journal of Biogeography*, **48**, 386–393.

Moreira-Muñoz, A., Francioli, S. E., Hobohm, C., and Menezes de Sequeira, M. (2014). Endemism on islands - case studies. In *Endemism in Vascular Plants* (ed. C. Hobohm), pp. 165–204. Springer, Dordrecht.

Moreno, Á. C., Carrascal, L. M., Delgado, A., Suárez, V., and Seoane, J. (2018). Striking resilience of an island endemic bird to a severe perturbation: the case of the Gran Canaria blue chaffinch. *Animal Biodiversity and Conservation*, **41**, 131–140.

Morici, C. (2004). Palmeras e islas: la insularidad en una de las familias más diversas del reino vegetal. In *Ecología Insular/Island Ecology* (ed. J. M. Fernández-Palacios and C. Morici), pp. 81–122. Asociación Española de Ecología Terrestre-Cabildo Insular de La Palma, Santa Cruz de La Palma.

Morinay, J., Cardoso, G., Doutrelant, C., and Covas, R. (2013). The evolution of birdsong on islands. *Ecology and Evolution*, **3**, 5127–5140.

Morrison, L. W. (2002a). Island biogeography and metapopulation dynamics of Bahamian ants. *Journal of Biogeography*, **29**, 387–394.

Morrison, L. W. (2002b). Determinants of plant species richness on small Bahamian islands. *Journal of Biogeography*, **29**, 931–941.

Morrison, L. W. (2014). The small-island effect: empty islands, temporal variability and the importance of species composition. *Journal of Biogeography*, **41**, 1007–1017.

Mortelliti, A. and Lindenmayer, D. B. (2015). Effects of landscape transformation on bird colonization and extinction patterns in a large-scale, long-term natural experiment. *Conservation Biology*, **29**, 1314–1326.

Mortensen, J. L. and Reed, J. M. (2016). Population viability and vital rate sensitivity of an endangered avian cooperative breeder, the white-breasted thrasher (*Ramphocinclus brachyurus*). *PLOS ONE*, **11**, e0148928.

Mortimer, N., Campbell, H. J., Tulloch, A. J., et al. (2017). Zealandia: Earth's hidden continent. *GSA Today*, **27**, 27–35.

Moser, D., Lenzner, B., Weigelt, P., et al. (2018). Remoteness promotes biological invasions on islands worldwide. *Proceedings of the National Academy of Sciences, USA*, **115**, 9270–9275.

Mouillot, D., Velez, L., Maire, E., et al. (2020). Global correlates of terrestrial and marine coverage by protected areas on islands. *Nature Communications*, **11**, 4438.

Moya, O., Contreras-Díaz, H. G., Oromí, P., and Juan, C. (2004). Genetic structure, phylogeography and demography of two ground-beetle species endemic to the Tenerife laurel forest (Canary Islands). *Molecular Ecology*, **13**, 3153–3167.

Mueller-Dombois, D. and Fosberg, F. R. (1998). *Vegetation of the Tropical Pacific Islands*. Springer-Verlag, New York.

Muhs, D. R., Simmons, K. R., Groves, L. T., McGeehin, J. P., Schumann, R. R., and Agenbroad, L. D. (2015). Late Quaternary sea-level history and the antiquity of mammoths (*Mammuthus exilis* and *Mammuthus columbi*), Channel Islands National Park, California, USA. *Quaternary Research*, **83**, 502–521.

Müller, R. D., Sdrolias, M., Gaina, C., and Roest, W. R. (2008). Age, spreading rates, and spreading asymmetry of the world's ocean crust. *Geochemistry, Geophysics, Geosystems*, **9**, Q04006.

Münkemüller, T., Gallien, L., Pollock, L. J., et al. (2020). Dos and don'ts when inferring assembly rules from diversity patterns. *Global Ecology and Biogeography*, **29**, 1212–1229.

Murphy, E. C., Russell, J. C., Broome, K. G., Ryan, G. J., and Dowding, J. E. (2019). Conserving New Zealand's native fauna: a review of tools being developed for the Predator Free 2050 programme. *Journal of Ornithology*, **160**, 883–892.

Musgrave, R. J. (2013). Evidence for Late Eocene emplacement of the Malaita Terrane, Solomon Islands: implications for an even larger Ontong Java Nui oceanic plateau. *Journal of Geophysical Research, Solid Earth*, **118**, 2670–2686.

Myers, N., Mittermeier, R. A., Mittermeier, C. G., da Fonseca, G. A. B., and Kent, J. (2000). Biodiversity hotspots for conservation priorities. *Nature*, **403**, 853–859.

Napolitano, M. F., DiNapoli, R. J., Stone, J. H., et al. (2019). Reevaluating human colonization of the Caribbean using chronometric hygiene and Bayesian modelling. *Science Advances*, **5**, eaar7806.

Navarro-Pérez, M. L., López, J., Fernández-Mazuecos, M., Rodríguez-Riaño, T., Vargas, P., and Ortega-Olivencia, A. (2013). The role of birds and insects in pollination shifts of *Scrophularia* (Scrophulariaceae). *Molecular Phylogenetics and Evolution*, **69**, 239–254.

Newsome, D., Moore, S. A., and Dowling, R. K. (2013). *Natural Area Tourism: Ecology, Impacts and Management*, 2nd edn. Channel View Publications, Bristol.

Niering, W. A. (1956). Bioecology of Kapingamarangi Atoll, Caroline Islands: terrestrial aspects. *Atoll Research Bulletin*, **49**, 1–32.

Nietlisbach, P., Wandeler, P., Parker, P. G., et al. (2013). Hybrid ancestry of an island subspecies of Galápagos mockingbird explains discordant gene trees. *Molecular Phylogenetics and Evolution*, **69**, 581–592.

Nilsson, I. N. and Nilsson, S. G. (1985). Experimental estimates of census efficiency and pseudo-turnover on islands: error trend and between-observer variation when recording vascular plants. *Journal of Ecology*, **73**, 65–70.

Nogales, M., Heleno, R., Traveset, A., and Vargas, P. (2012). Evidence for overlooked mechanisms of long-distance seed dispersal to and between oceanic islands. *New Phytologist*, **194**, 313–317.

Nogales, M., Hernández, E. C., and Valdés, F. (1999). Seed dispersal by common ravens *Corvus coraxa* among island habitats (Canarian Archipelago). *Ecoscience*, **5**, 56–61.

Nogales, M., Nieves, C., Illera, J. C., Padilla, D. P., and Traveset, A. (2005). Effect of native and alien vertebrate frugivores on seed viability and germination patterns of *Rubia fruticosa* (Rubiaceae) in the eastern Canary Islands. *Functional Ecology*, **19**, 429–436.

Nogué, S., de Nascimento, L., Fernández-Palacios, J. M., Whittaker, R. J., and Willis, K. J. (2013). The ancient forests of La Gomera, Canary Islands, and their sensitivity to environmental change. *Journal of Ecology*, **101**, 368–377.

Nogué, S., de Nascimento, L., Froyd, C. A., et al. (2017). Island biodiversity conservation needs palaeoecology. *Nature Ecology and Evolution*, **1**, 0181.

Nogué, S., Santos, A. M. C., Birks, H. J. B., et al. (2021). The human dimension of biodiversity changes on islands. *Science*, **372**, 488–491.

Norder, S. N., Proios, K. V., Whittaker, R. J., et al. (2019). Beyond the Last Glacial Maximum: island endemism is best explained by long-lasting archipelago configurations. *Global Ecology and Biogeography*, **28**, 184–197.

Noss, R. F., Platt, W. J., Weakley, A. S., et al. (2015). How global biodiversity hotspots may go unrecognized: lessons from the North American Coastal Plain. *Diversity and Distributions*, **21**, 236–244.

Novosolov, M. and Meiri, S. (2013). The effect of island type on lizard reproductive traits. *Journal of Biogeography*, **40**, 2385–2395.

Novosolov, M., Raia, P., and Meiri, S. (2013). The island syndrome in lizards. *Global Ecology and Biogeography*, **22**, 184–191.

Novosolov, M., Rodda, G. H., Feldman, A., Kadison, A. E., Dor, R., and Meiri, S. (2016). Power in numbers. Drivers of high population density in insular lizards. *Global Ecology and Biogeography*, **25**, 87–95.

Nunn, P. D. (1994). *Oceanic Islands*. Blackwell, Oxford.

Nunn, P. D. (1997). Late Quaternary environmental changes on Pacific islands: controversy, certainty and conjecture. *Journal of Quaternary Science*, **12**, 443–450.

Nunn, P. D. (2000). Illuminating sea-level fall around AD 1220–1510 (730–440 cal yr BP) in the Pacific Islands: implications for environmental change and cultural transformation. *New Zealand Geographer*, **56**, 46–54.

Nunn, P. D. (2004). Through a mist on the ocean: human understanding of island environments. *Tijdschrift voor Economische en Sociale Geografie*, **95**, 311–325.

Nunn, P. D., Kumar, L., Eliod, I., and McLean, R. F. (2016). Classifying Pacific islands. *Geoscience Letters*, **3**, article 7.

Nürk, N. M., Atchison, G. W., and Hughes, C. E. (2019). Island woodiness underpins accelerated disparification in plant radiations. *New Phytologist*, **224**, 518–531.

O'Dowd, D. J., Green, P. T., and Lake, P. S. (2003). Invasional 'meltdown' on an oceanic island. *Ecology Letters*, **6**, 812–817.

O'Grady, P. and DeSalle, R. (2018). Hawaiian *Drosophila* as an evolutionary model clade: days of future past. *BioEssays*, **40**, 1700246.

Ó Margaich, F. (2022). *Dispersal and Speciation in the Avian Archipelago*. D.Phil. thesis. Trinity College, Dublin.

Ó Marcaigh, F., O'Connell, D. P., Analuddin, K., et al. (2022). Tramps in transition: genetic differentiation between populations of an iconic 'supertramp' taxon in the Central Indo-Pacific. *Frontiers of Biogeography*, **14.2**, e54512.

Obrist, D. S., Hanly, P. J., Kennedy, J. C., et al. (2020). Marine subsidies mediate patterns in avian island biogeography. *Proceedings of the Royal Society B*, **287**, 20200108.

Olesen, J. M. (1992). How do plants reproduce on their range margin? In *Plant–Animal Interactions in Mediterranean-Type Ecosystems*. (ed. C. A. Thanos), pp. 217–222. University of Athens, Athens.

Olesen, J. M., Bascompte, J., Dupont, Y. L., and Jordano, P. (2007). The modularity of pollination networks. *Proceedings of the National Academy of Sciences, USA*, **104**, 19891–19896.

Olesen, J. M., Damgaard, C. F., Fuster, F., et al. (2018). Disclosing the double mutualist role of birds on Galápagos. *Scientific Reports*, **8**, 57.

Olesen, J. M., Eskildsen, L. I., and Venkatasamy, S. (2002). Invasion of pollination networks on oceanic islands: importance of invader complexes and endemic super-generalists. *Diversity and Distributions*, **8**, 181–192.

Olesen, J. M. and Jordano, P. (2002). Geographical patterns in plant-pollinator mutualistic networks. *Ecology*, **83**, 2416–2424.

Olesen, J. M. and Valido, A. (2003). Lizards as pollinators and seed dispersers: an island phenomenon. *Trends in Ecology & Evolution*, **18**, 177–181.

Olesen, J. M. and Valido, A. (2004). Lizards and birds as generalized pollinators and seed dispersers of island plants. In *Ecología Insular/Island Ecology* (ed. J. M. Fernández-Palacios and C. Morici), pp. 229–249. Asociación Española de Ecología Terrestre–Cabildo Insular de La Palma, Santa Cruz de La Palma.

Oliver, P. M., Brown, R. M., Kraus, F., Rittmeyer, E., Travers, S. L., and Siler, C. D. (2018). Lizards of the lost arcs: mid-Cenozoic diversification, persistence and ecological marginalization in the West Pacific. *Proceedings of the Royal Society B*, **258**, 20171760.

Ollerton, J., Cranmer, L., Stelzer, R., Sullivan, S., and Chittka, L. (2008). Bird Pollination of Canary Island Endemic Plants. *Nature Precedings*, https://doi.org/10.1038/npre.2008.1977.1.

Ollerton, J., Winfree, R., and Tarrant, S. (2011). How many flowering plants are pollinated by animals? *Oikos*, **120**, 321–326.

Ollier, C. D. (1988). *Volcanoes*. Blackwell, Oxford.

Ong, S. P., O'Dowd, D. J. and Green, P. T. (2019). Production and export of the parasitoid *Tachardiaephagus somervilli* (Chalcidoidea: Encyrtidae), a biological control agent for the yellow lac scale, *Tachardina aurantiaca* (Hemiptera: Coccoidea: Kerriidae). *Journal of Asia-Pacific Entomology*, **22**, 543–548.

Oppel, S., Cassini, A., Fenton, C., Daley, J., and Gray, G. (2014). Population status and trend of the Critically Endangered Montserrat Oriole. *Bird Conservation International*, **24**, 252–261.

Ottaviani, G, Keppel, G., Götzenberger, L., et al. (2020). Linking plant functional ecology to island biogeography. *Trends in Plant Science*, **25**, 329–339.

Otte, D. (1989). Speciation in Hawaiian crickets. In *Speciation and its Consequences* (ed. D. Otte and J. A. Endler), pp. 482–526. Sinauer Associates, Sunderland, MA.

Otto, R., Fernández-Lugo, S., Blandino, C., Manganelli, G., Chiarucci, A., and Fernández-Palacios, J. M. (2020). Biotic homogenization of oceanic islands depends on taxon, spatial scale and the quantification approach. *Ecography*, **43**, 747–758.

Otto, R., Krüsi, B. O., Burga, C. A., and Fernández-Palacios, J. M. (2006). Old-field succession along a precipitation gradient in the semi-arid coastal region of Tenerife. *Journal of Arid Environments*, **65**, 156–178.

Otto, R., Whittaker, R. J., von Gaisberg, M., et al. (2016). Transferring and implementing the general dynamic model of oceanic island biogeography at the scale of island fragments: the roles of geological age and topography in plant diversification in the Canaries. *Journal of Biogeography*, **43**, 911–922.

Ovaskainen, O. and Saastamoinen, M. (2018). Frontiers in metapopulation biology: the legacy of Ilkka Hanski. *Annual Review of Ecology, Evolution, and Systematics*, **49**, 231–252.

Padilla, D. P., González-Castro, A., and Nogales, M. (2012). Significance and extent of secondary seed dispersal by predatory birds on oceanic islands: the case of the Canary archipelago. *Journal of Ecology*, **100**, 416–427.

Paine, R. T. (1985). Re-establishment of an insular winter wren population following a severe freeze. *Condor*, **87**, 558–559.

Pannell, J. R., Auld, J. R., Brandvain, Y., et al. (2015). The scope of Baker's Law. *New Phytologist*, **208**, 656–667.

Paravisini-Gebert, L. (2018). The parrots of the Caribbean. *ReVista (Cambridge)*, **18**, 45–47.

Parent, C. E., Caccone, A., and Petren, K. (2008). Colonization and diversification of Galápagos terrestrial fauna: a phylogenetic and biogeographical synthesis. *Philosophical Transactions of the Royal Society B*, **363**, 3347–3361.

Paris, R., Ramalho, R. S., Madeira, J., et al. (2018). Megatsunami conglomerates and flank collapses of ocean island volcanoes. *Marine Geology*, **395**, 168–187.

Parrish, T. (2002). *Krakatau: Genetic Consequences of Island Colonization*. University of Utrecht and Netherlands Institute of Ecology, Heteren.

Partomihardjo, T., Mirmanto, E., and Whittaker, R. J. (1992). Anak Krakatau's vegetation and flora circa 1991, with observations on a decade of development and change. *Geojournal*, **28**, 233–248.

Patiño, J., Bisang, I., Hedenäs, L., et al. (2013). Baker's law and the island syndromes in bryophytes. *Journal of Ecology*, **101**, 1245–1255.

Patiño, J., Carine, M., Mardulyn, P., et al. (2015). Approximate Bayesian computation reveals the crucial role of oceanic islands for the assembly of continental biodiversity. *Systematic Biology*, **64**, 579–589.

Patiño, J., Guilhaumon, F., Whittaker, R. J., et al. (2013). Accounting for data heterogeneity in patterns of biodiversity: an application of Linear Mixed Effect Models to the oceanic island biogeography of spore-producing plants. *Ecography*, **36**, 904–913.

Patrick, L. (2001). Introduced Species Summary Project: Brown tree snake (*Boiga irregularis*). www.columbia.edu/itc/cerc/danoff-burg/invasion_bio/inv_spp_summ/boiga_irregularis.html (last visited March 2006).

Patterson, B. D. and Atmar, W. (1986). Nested subsets and the structure of insular mammalian faunas and archipelagos. *Biological Journal of the Linnean Society*, **28**, 65–82.

Paulay, G. (1994). Biodiversity on oceanic islands: its origin and extinction. *American Zoologist*, **34**, 134–144.

Peck, S. B. (1990). Eyeless arthropods of the Galapagos Islands, Ecuador: composition and origin of the cryptozoic fauna of a young, tropical, oceanic archipelago. *Biotropica*, **22**, 366–381.

Pedersen, P. B. M., Olsen, J. B., Sandel, B. and Svenning J.-C. (2019). Wild steps in a semi-wild setting? Habitat selection and behavior of European bison reintroduced to an enclosure in an anthropogenic landscape. *PLOS ONE*, **14**, e0198308.

Pe'er, G., Tsianou, M. A., Franz, K. W., et al. (2014). Toward better application of minimum area requirements in conservation planning. *Biological Conservation*, **170**, 92–102.

Peltzer, D. A., Bellingham, P. J., Dickie, I. A., et al. (2019). Scale and complexity implications of making New Zealand predator-free by 2050. *Journal of the Royal Society of New Zealand*, **49**, 412–439.

Percy, D. M., Garver, A. M., Wagner, W. L., et al. (2008). Progressive island colonization and ancient origin of Hawaiian *Metrosideros* (Myrtaceae). *Proceedings of the Royal Society B*, **275**, 1479–1490.

Pereira, H. M., Borda-de-Água, L., and Martins, I. S. (2012). Geometry and scale in species–area relationships. *Nature*, **482**, E3–E4.

Pereira, H. M. and Navarro, L. M. (eds) (2015). *Rewilding European Landscapes*. Springer Open, Dordrecht.

Pregill, G. K. and Olson, S. L. (1981). Zoogeography of West Indian vertebrates in relation to Pleistocene climate cycles. *Annual Review of Ecology and Systematics*, **12**, 75–98.

Peres, C. A. (2001). Synergistic effects of subsistence hunting and habitat fragmentation on Amazonian forest vertebrates. *Conservation Biology*, **15**, 1490–1505.

Péréz-Mendez, N., Jordano, P., and Valido, A. (2015). Downsized mutualisms: consequences of seed dispersers' body-size reduction for early plant recruitment. *Perspectives in Plant Ecology, Evolution and Systematics*, **17**, 151–159.

Pérez-Torrado, F. J., Paris, R., Cabrera, M. C., et al. (2006). Tsunami deposits related to flank collapse in oceanic volcanoes: the Agaete Valley evidence, Gran Canaria, Canary Islands. *Marine Geology*, **227**, 135–149.

Petchey, F., Clark, G., Winter, O., O'Day, P., and Litster, M. (2017). Colonisation of Remote Oceania: new dates for the Bapot-1 site in the Mariana Islands. *Archaeology in Oceania*, **52**, 108–126.

Petchey, O. L. and Gaston, K. J. (2006). Functional diversity: back to basics and looking forward. *Ecology Letters*, **9**, 741–758.

Pfennig, D. W. and Pfennig, K. S. (2020). Character displacement. *Current Biology*, **30**, R1023–R1024.

Phillimore, A. B. (2010). Chapter 4: subspecies origination and extinction in birds. *Ornithological Monographs*, **67**, 42–53.

Phillimore, A. B. (2013). Geography, range evolution, and speciation. In *The Princeton Guide to Evolution* (ed. J. B. Losos, D. A. Baum, D. J. Futuyma, et al.), pp. 504–511. Princeton University Press, Princeton.

Pickett, S. T. A., Parker, V. T., and Fiedler, P. L. (1992). The new paradigm in ecology: implications for conservation biology above the species level. In *Conservation Biology: The Theory and Practice of Nature Conservation and Management* (ed. P. L. Fiedler and S. K. Jain), pp. 91–125. Chapman & Hall, New York.

Pickett, S. T. A. and White, P. S. (ed.) (1985). *The Ecology of Natural Disturbance and Patch Dynamics*. Academic Press, Orlando, FL.

Pierson, E. D., Elmqvist, T., Rainey, W. E., and Cox, P. A. (1996). Effects of tropical cyclonic storms on flying fox populations on the South Pacific islands of Samoa. *Conservation Biology*, **10**, 438–451.

Pigot, A. L., Sheard, C., Miller, E. T., et al. (2020). Macroevolutionary convergence connects morphological form to ecological function in birds. *Nature Ecology and Evolution*, **4**, 230–239.

Pimm, S. L., Jones, H. L., and Diamond, J. (1988). On the risk of extinction. *American Naturalist*, **132**, 757–785.

Pinheiro, H. T., Bernardi, G., Simon, T., et al. (2017). Island biogeography of marine organisms. *Nature*, **549**, 82–85.

Poe, S., Nieto-Montes de Oca, A., Torres-Carvajal, O., et al. (2018). Comparative evolution of an archetypal adaptive radiation: innovation and opportunity in *Anolis* lizards. *The American Naturalist*, **191**, E185–E194.

Pokorny, L., Riina, R., Mairal, M., et al. (2015). Living on the edge: timing of Rand flora disjunctions congruent with ongoing aridification in Africa. *Frontiers in Genetics*, **6**, Article 154.

Pouteau, R. and Birnbaum, P. (2016). Island biodiversity hotspots are getting hotter: vulnerability of tree species to climate change in New Caledonia. *Biological Conservation*, **201**, 111–119.

Porter, D. M. (1979). Endemism and evolution in Galapagos Islands vascular plants. In *Plants and Islands* (ed. D. Bramwell), pp. 225–256. Academic Press, London.

Post, D. M., Pace, M. L., and Hairston, N. G. Jr (2000). Ecosystem size determines food-chain length in lakes. *Nature*, **405**, 1047–1049.

Potter, B. A., Chatters, J. C., Prentiss, A. M., et al. (2022). Current understanding of the earliest human occupations in the Americas: evaluation of Becerra-Valdivia and Higham (2020). *PaleoAmerica*, **8**, 62–76.

Poulakakis, N., Glaberman, S., Russello, M., et al. (2008). Historical DNA analysis reveals living descendants of an extinct species of Galápagos tortoise. *Proceedings of the National Academy of Sciences, USA*, **105**, 15464–15469.

Power, D. M. (1972). Numbers of bird species on the California islands. *Evolution*, **26**, 451–463.

Pratt, H. D. (2005). *The Hawaiian Honeycreepers*. Oxford University Press, Oxford.

Pratt, H. D. (2014). A consensus taxonomy for the Hawaiian honeycreepers. *Occasional Papers of the Museum of Natural Science, Louisiana State University*, **1**, number 85, doi: 10.31390/opmns.085.

Prebble, M., Whitau, R., Meyer, J-Y., Sibley-Punnett, L., Fallon, S., and Porch, N. (2016). Abrupt late Pleistocene ecological and climate change on Tahiti (French Polynesia). *Journal of Biogeography*, **43**, 2438–2453.

Prebble, M. and Wilmshurst, J. M. (2009). Detecting the initial impact of humans and introduced species on island environments in Remote Oceania using palaeoecology. *Biological Invasions*, **11**, 1529–1556.

Pregill, G. K. and Steadman, D. W. (2009). The prehistory and biogeography of terrestrial vertebrates on Guam, Mariana Islands. *Diversity and Distributions*, **15**, 983–996.

Preston, F. W. (1948). The commonness, and rarity, of species. *Ecology*, **48**, 254–283.

Preston, F. W. (1962). The canonical distribution of commonness and rarity. *Ecology*, **43**, part I, pp. 185–215; part II, pp. 410–432.

Price, J. P. and Clague, D. A. (2002). How old is the Hawaiian biota? Geology and phylogeny suggest recent divergence. *Proceedings of the Royal Society B*, **269**, 2429–2435.

Price, J. P. and Elliott-Fisk, D. (2004). Topographic history of the Maui Nui complex, Hawai'i, and its implications for biogeography. *Pacific Science*, **58**, 27–45.

Price, J. P., Otto, R., Menezes de Sequeira, M., et al. (2018). Colonization and diversification shape species–area relationships in three Macaronesian archipelagos. *Journal of Biogeography*, **45**, 2027–2039.

Price, J. P. and Wagner, W. L. (2004). Speciation in Hawaiian angiosperm lineages: cause, consequence, and mode. *Evolution*, **58**, 2185–2200.

Proença, V. and Pereira, H. M. (2013). Species–area models to assess biodiversity change in multi-habitat landscapes: the importance of species habitat affinity. *Basic and Applied Ecology*, **14**, 102–114.

Proios, K. (2021). Global biogeographic patterns of insular land snails. PhD dissertation. National and Kapodistrian University of Athens, Athens, Greece.

Proios, K., Cameron, R. A., and Triantis, K. A. (2021). Land snails on islands: building a global inventory. *Frontiers of Biogeography*, **13**, e51126.

Prugh, L. R., Hodges, K. E., Sinclair, A. R. E., and Brashares, J. S. (2008). Effect of habitat area and isolation on fragmented animal populations. *Proceedings of the National Academy of Sciences, USA*, **105**, 20770–20775.

Puleston, C. O., Ladefoged, T. N., Haoa, S., Chadwick. O. A., Vitousek, P. M., and Stevenson, C. M. (2017). Rain, sun, soil, and sweat: a consideration of population limits on Rapa Nui (Easter Island) before European contact. *Frontiers in Ecology and Evolution*, **5**, a69.

Rabelo, R. M., Aragón, S., Bicca-Marques, J. C., and Nelson, B. W. (2019). Habitat amount hypothesis and passive sampling explain mammal species composition in Amazonian river islands. *Biotropica*, **51**, 84–92.

Racine, C. H. and Downhower, J. F. (1974). Vegetative and reproductive strategies of *Opuntia* (Cactaceae) in the Galapagos Islands. *Biotropica*, **6**, 175–186.

Ramalho, R. S., Helffrich, G., Madeira, J., et al. (2017). Emergence and evolution of Santa Maria Island (Azores)—the conundrum of uplifted islands revisited. *GSA Bulletin*, **129**, 372–391.

Ramírez-Delgado, J. P., Di Marco, M., Watson, J. E. M., et al. (2022). Matrix condition mediates the effects of habitat fragmentation on species extinction risk. *Nature Communications*, **13**, 595.

Ramstad, K. M., Colbourne, R. M., Robertson, H. A., Allendorf, F. W. and Daugherty, C. H. (2013). Genetic consequences of a century of protection: serial founder events and survival of the little spotted kiwi (*Apteryx owenii*). *Proceedings of the Royal Society B*, **280**, 20130576.

Rando, J. C., Alcover, J. A., Pieper, H., Olsson, S. L., Hernández, C. N., and López-Jurado, F. (2020). Unforeseen diversity of quails (Galliformes: Phasianidae: *Coturnix*) in oceanic islands provided by the fossil record of Macaronesia. *Zoological Journal of the Linnean Society*, **188**, 1296–1317.

Rando, J. C., López, M., and Seguí, B. (1999). A new species of extinct flightless Passerine (Emberizidae: *Emberiza*) from the Canary Islands. *The Condor*, **101**, 1–13.

Rando, J. C., Pieper, H., and Alcover, J. A. (2014). Radiocarbon evidence for the presence of mice on Madeira Island (North Atlantic) one millennium ago. *Proceedings of the Royal Society B*, **281**, article 20133126.

Raposeiro, P. M., Hernández, A., Pla-Rabes, S., et al. (2021). Climate change facilitated the early colonization of the Azores Archipelago during medieval times. *Proceedings of the National Academy of Sciences, USA*, **118**, e2108236118.

Ravazzi, C., Mariani, M., Criado, C., et al. (2021). The influence of natural fire and cultural practices on island ecosystems: insights from a 4800 year record from Gran Canaria, Canary Islands. *Journal of Biogeography*, **48**, 276–290.

Rawlinson, P. A., Zann, R. A., van Balen, S., and Thornton, I. W. B. (1992). Colonization of the Krakatau islands by vertebrates. *Geojournal*, **28**, 225–231.

Rayner, M. J., Hauber, M. E., Imber, M. J., Stamp, R. K., and Clout, M. N. (2007). Spatial heterogeneity of mesopredator release within an oceanic island system. *Proceedings of the National Academy of Sciences, USA*, **104**, 20862–20865.

Razanajatovo, M., van Kleunen, M., Kreft, H., et al. (2019). Autofertility and self-compatibility moderately benefit island colonization of plants. *Global Ecology and Biogeography*, **28**, 341–352.

Reaney, A. M., Bouchenak-Khelladi, Y., Tobias, J. A., and Abzhanov, A. (2020). Ecological and morphological determinants of evolutionary diversification in Darwin's finches and their relatives. *Ecology and Evolution*, **10**, 14020–14032.

Recuerda, M., Illera, J. C., Blanco, G., Zardoya, R., and Milá, B. (2021). Sequential colonization of oceanic archipelagos led to a species-level radiation in the common chaffinch complex (Aves: *Fringilla coelebs*). *Molecular Phylogenetics and Evolution*, **164**, 107291.

Reding, D. M., Foster, J. T., James, H. F., Pratt, H. D., and Fleischer, R. C. (2009). Convergent evolution of 'creepers' in the Hawaiian honeycreeper radiation. *Biology Letters*, **5**, 221–224.

Reed, D. H. (2007). Extinction of island endemics: it is not inbreeding depression. *Animal Conservation*, **10**, 145–146.

Reed, D. H., O'Grady, J. J., Brook, B. W., Ballou, J. D., and Frankham, R. (2003). Estimates of minimum viable population sizes for vertebrates and factors influencing those estimates. *Biological Conservation*, **113**, 23–34.

Reeder-Myers, L., Erlandson, J. M., Muhs, D. R., and Rick, T. C. (2015). Sea level, paleogeography, and archeology on California's Northern Channel Islands. *Quaternary Research*, **83**, 263–272.

Rees, D. J., Emerson, B. C., Oromí, P., and Hewitt, G. M. (2001). The diversification of the genus *Nesotes* (Coleoptera: Tenebrionidae) in the Canary Islands: evidence from mtDNA. *Molecular Phylogenetics and Evolution*, **21**, 321–326.

Regelous, M., Hofman, A. W., Abouchami, W., and Galer, S. J. G. (2003). Geochemistry of lavas from the Emperor Seamounts, and the geochemical evolution of Hawaiian magmatism from 85 to 42 Ma. *Journal of Petrology*, **44**, 113–140.

Régnier, C., Bouchet, P., Hayes, K. A., et al. (2015). Extinction in a hyperdiverse endemic Hawaiian land snail family and implications for the underestimation of invertebrate extinction. *Conservation Biology*, **29**, 1715–1723.

Reis, S. M., Marimon, B. S., Marimom, B. H. Jr, et al. (2018). Climate and fragmention affect forest structure at the southern border of Amazonia. *Plant Ecology and Diversity*, **11**, 13–25.

Renom, P., de-Dios, T., Civit, S., et al. (2021). Genetic data from the extinct giant rat from Tenerife (Canary Islands) points to a recent divergence from mainland relatives. *Biology Letters*, **17**, 20210533.

Renvoize, S. A. (1979). The origins of Indian Ocean island floras. In *Plants and Islands* (ed. D. Bramwell), pp. 107–129. Academic Press, London.

Resasco, J. (2019). Meta-analysis on a decade of testing corridor efficacy: what new have we learned? *Current Landscape Ecology Reports*, **4**, 61–69.

Reudink, M. W., Pageau, C., Fisher, M., et al. (2021). Evolution of song and color in island birds. *The Wilson Journal of Ornithology*, **133**, 1–10.

Rey, J. R. (1984). Experimental tests of island biogeographic theory. In *Ecological Communities: Conceptual Issues and the Evidence* (ed. D. R. Strong, Jr, D. Simberloff, L. G. Abele, and A. B. Thistle), pp. 101–112. Princeton University Press, Princeton.

Rey, J. R. (1985). Insular ecology of salt marsh arthropods: species level patterns. *Journal of Biogeography*, **12**, 97–107.

Rheindt, F. E., Prawiradilaga, D. M., Ashari, H., et al. (2020). A lost world in Wallacea: description of a montane archipelagic avifauna. *Science*, **367**, 167–170.

Rhodin, A. G. J., Thomson, S., Georgalis, G. L., et al. (2015). Turtles and tortoises of the world during the rise and global spread of humanity: first checklist and review of extinct Pleistocene and Holocene chelonians. *Chelonian Research Monographs*, **No 5**.

Ribon, R., Simon, J. E., and de Mattos, G. T. (2003). Bird extinctions in Atlantic forest fragments of the Viçosa region, southeastern Brazil. *Conservation Biology*, **17**, 1827–1839.

Rice, A., Šmarda, P., Novosolov, M., et al. (2019). The global biogeography of polyploid plants. *Nature Ecology and Evolution*, **3**, 265–273.

Richmond, J. Q., Ota, H., Grismer, L. L., and Fisher, R. N. (2021). Influence of niche breadth and position on the historical biogeography of seafaring scincid lizards. *Biological Journal of the Linnean Society*, **132**, 74–92.

Ricketts, T. H. (2001). The matrix matters: effective isolation in fragmented landscapes. *American Naturalist*, **158**, 87–99.

Ricklefs, R. E. (1980). Geographical variation in clutch size among passerine birds: Ashmole's hypothesis. *Auk*, **97**, 38–49.

Ricklefs, R. E. (1987). Community diversity: relative roles of local and regional processes. *Science*, **235**, 167–171.

Ricklefs, R. E. (2008). Disintegration of the ecological community. *The American Naturalist*, **172**, 741–750.

Ricklefs, R. E. (2017). Historical biogeography and extinction in the Hawaiian honeycreepers. *The American Naturalist*, **190**, E106–E111.

Ricklefs, R. E. and Bermingham, E. (2001). Nonequilibrium diversity dynamics of the Lesser Antillean avifauna. *Science*, **294**, 1522–1524.

Ricklefs, R. E. and Bermingham, E. (2002). The concept of the taxon cycle in biogeography. *Global Ecology and Biogeography*, **11**, 353–362.

Ricklefs, R. E. and Cox, G. W. (1972). Taxon cycles in the West Indian avifauna. *American Naturalist*, **106**, 195–219.

Ricklefs, R. E. and Cox, G. W. (1978). Stage of taxon cycle, habitat distribution, and population density in the avifauna of the West Indies. *American Naturalist*, **112**, 875–895.

Ricklefs, R. E., Soares, L., Ellis, V. A., and Latta, S. C. (2016). Haemosporidian parasites and avian host population abundance in the Lesser Antilles. *Journal of Biogeography*, **43**, 1277–1286.

Ridley, H. N. (1930). *The Dispersal of Plants throughout the World*. Reeve, Ashford England,

Ridley, M. (ed.) (1994). *A Darwin Selection*. Fontana Press, London.

Rieseberg, L. H., Archer, M. A., and Wayne, R. K. (1999). Transgressive segregation, adaptation and speciation. *Heredity*, **83**, 363–372.

Rijsdijk, K.F., Hengl, T., Norder, S., et al. (2014). Quantifying surface area changes of volcanic islands driven by Pleistocene sea level cycles: biogeographic implications for Macaronesian archipelagos, Atlantic Ocean. *Journal of Biogeography*, **41**, 1227–1439.

Riskin, D. K., Parsons, S., Schutt, W. A. Jr, Carter, G. G., and Hermanson, J. W. (2006). Terrestrial locomotion of the New Zealand short-tailed bat *Mystacina tuberculata* and the common vampire bat *Desmodus rotundus*. *Journal of Experimental Biology*, **209**, 1725–1736.

Ritchie, E. G. and Johnson, C. N. (2009). Predator interactions, mesopredator release and biodiversity conservation. *Ecology Letters*, **12**, 982–998.

Robinson, W. D. and Sherry, T. W. (2012). Mechanisms of avian population decline and species loss in tropical forest fragments. *Journal of Ornithology*, **153**, 141–152.

Rocha, R., Sequeira, M. M., Douglas, L. R., et al. (2017). Extinctions of introduced game species on oceanic islands: curse for hunters or conservation opportunities? *Biodiversity and Conservation*, **26**, 2517–2520.

Rodda, G. H., Fritts, T. H., McCoid, M. J., and Campbell, E. W. III (1999). An overview of the biology of the brown treesnake (*Boiga irregularis*), a costly introduced pest on Pacific Islands. In *Problem Snake Management: The Habu and the Brown Treesnake* (ed. G. H. Rodda, Y. Sawai, D. Chiszar, and H. Tanaka), pp. 44–80. Cornell University Press, Ithaca, NY.

Rodda, G. H., Perry, G., Rondeau, R. J., and Lazell, J. (2001). The densest terrestrial vertebrate. *Journal of Tropical Ecology*, **17**, 331–338.

Rodrigues, P., Lopes, R. J., Drovetski, S. V., Reis, S., Ramos, J. A., and da Cunha, R. T. (2013). Phylogeography and genetic diversity of the robin (*Erithacus rubecula*) in the Azores Islands: evidence of a recent colonisation. *Journal of Ornithology*, **154**, 889–900.

Rodríguez de la Fuente, F. (1980). *Enciclopedia Salvat de la Fauna*. Salvat Ediciones, Pamplona.

Rodríguez, F. and Moreno, A. (2004). Pinzón Azul de Gran Canaria. *Fringilla teydea polatzeki*. In *Libro Rojo de las Aves de España* (ed. A. Madroño, C. González, and J. C. Atienza), pp. 370–372. Dirección General para la Biodiversdad, SEO/Birdlife, Madrid.

Rodríguez, J. R. O., Parra-López, E., and Yanes-Esévez, V. (2008). The sustainability of island destinations: tourism area life cycle and teleological perspectives. The case of Tenerife. *Tourism Management*, **29**, 53–65.

Rodríguez-Rodríguez, M. C. (2015). Opportunistic pollination by birds and lizards in the Canary Islands. PhD. La Laguna University, La Laguna, Spain.

Roff, D. A. (1990). The evolution of flightlessness in insects. *Ecological Monographs*, **60**, 389–421.

Rohling, E. J., Foster, G. L., Grant, K. M., et al. (2014). Sea-level and deep-sea-temperature variability over the past 5.3 million years. *Nature*, **508**, 477–482.

Romano, A., Séchaud, R., and Roulin, A. (2021). Evolution of wing length and melanin-based coloration in insular populations of a cosmopolitan raptor. *Journal of Biogeography*, **48**, 961–973.

Romano, M., Manucci, F., and Palombo, M. P. (2021). The smallest of the largest: new volumetric body mass estimate and in-vivo restoration of the dwarf elephant *Palaeoloxodon* ex gr. *P. falconeri* from Spinagallo Cave (Sicily). *Historical Biology*, **33**, 340–353.

Roncal, J., Nieto-Blázquez, M. A., Cardona, A., and Bacon, C. D. (2020). Historical biogeography of Caribbean plants revises regional paleogeography. In *Neotropical Diversification: Patterns and Processes*. (ed. V. Rull and A. Carnaval), pp. 521–546. Springer, Cham.

Ronco, F., Matschiner, M., Böhne, A., et al. (2021). Drivers and dynamics of a massive adaptive radiation in cichlid fishes. *Nature*, **589**, 76–81.

Root, R. B. (1967). The niche exploitation pattern of the blue-gray gnatcatcher. *Ecological Monographs*, **37**, 317–350.

Rösch, V., Tscharntke, T., Scherber, C., and Batáry, P. (2015). Biodiversity conservation across taxa and landscapes requires many small as well as single large habitat fragments. *Oecologia*, **179**, 209–222.

Rosenzweig, M. L. (1995). *Species Diversity in Space and Time*. Cambridge University Press, Cambridge.

Rosenzweig, M. L. (2003). *Win-Win Ecology: How the Earth's Species Can Survive in the Midst of Human Enterprise*. Oxford University Press, New York.

Rosenzweig, M. L. and Ziv, Y. (1999). The echo pattern of species diversity: pattern and processes. *Ecography*, **22**, 614–628.

Rosindell, J. and Harmon, L. J. (2013). A unified model of species immigration, extinction and abundance on islands. *Journal of Biogeography*, **40**, 1107–1118.

Rosindell, J., Hubbell, S. P., and Etienne, R. S. (2011). The unified neutral theory of biodiversity and biogeography at age ten. *Trends in Ecology & Evolution*, **26**, 340–348.

Roslin, T., Várkonyi, G., Koponen, M., Vikberg, V., and Nieminen, M. (2014). Species–area relationships across four trophic levels–decreasing island size truncates food chains. *Ecography*, **37**, 443–453.

Ross, S. R. P.-J., Friedman, N. R., Janicki, J., and Economo, E. P. (2019). A test of trophic and functional island biogeography theory with the avifauna of a continental archipelago. *Journal of Animal Ecology*, **88**, 1392–1405.

Rosselló, J. A. and Castro, M. (2008). Karyological evolution of the angiosperm endemic flora of the Balearic Islands. *Taxon*, **57**, 259–273.

Roughgarden, J. (1989). The structure and assembly of communities. In *Perspectives in Ecological Theory* (ed. J. Roughgarden, R. M. May, and S. A. Levin), pp. 203–226. Princeton University Press, Princeton.

Roughgarden, J. and Pacala, S. (1989). Taxon cycle among *Anolis* lizard populations: review of evidence. In *Speciation and its Consequences* (ed. D. Otte and J. A. Endler), pp. 403–432. Sinauer Associates, Sunderland, MA.

Rowe, T. B., Stafford, T. W. Jr, Fisher, D. C., et al. (2022). Human occupation of the North American Colorado Plateau ~37,000 years ago. *Frontiers in Ecology and Evolution*, **10**, article 903795.

Rozzi, R., Varela, S., Bover, P., and Martin, J. M. (2020). Causal explanations for the evolution of 'low gear' locomotion in insular ruminants. *Journal of Biogeography*, **47**, 2274–2285.

Rubin, C.-J., Enbody, E. D., Dobreva, M. P., et al. (2022). Rapid adaptive radiation of Darwin's finches depends on ancestral genetic modules. *Science Advances*, **8**, eabm5982.

Rull, V. (2019). The deforestation of Easter Island. *Biological Reviews*, **95**, 124–141.

Rull, V., Cañellas-Boltà, N., Margalef, O., Pla-Rabes, S., Sáez, A., and Giralt, S. (2016). Three millennia of climatic, ecological, and cultural change on Easter Island: an integrative overview. *Frontiers in Ecology and Evolution*, **4**, article 29.

Rull, V. and Giralt, S. (eds) (2018). Paleoecology of Easter Island: natural and anthropogenic drivers of ecological change. *Frontiers in Ecology and Evolution*, **6**, ebook; doi: 10.3389/978-2-88945-562-1.

Rull, V. and Stevenson, C. (eds) (2022a). *Developments in Paleoenvironmental Research 22, The Prehistory of Rapa Nui (Easter Island)*. Springer, Cham.

Rull, V. and Stevenson, C. (2022b). Chapter 24: Towards a holistic approach to Easter Island's prehistory. In *Developments in Paleoenvironmental Research 22, The Prehistory of Rapa Nui (Easter Island)* (ed. V. Rull and C. Stevenson), pp. 611–623. Springer, Cham.

Rundell, R. J. and Price, T. D. (2009). Adaptive radiation, non-adaptive radiation, ecological speciation and nonecological speciation. *Trends in Ecology & Evolution*, **24**, 394–399.

Runemark, A., Brydegaard, M., and Svensson, E. I. (2014). Does relaxed predation drive phenotypic divergence among insular populations? *Journal of Evolutionary Biology*, **27**, 1676–1690.

Runge, C. A., Martin, T. G., Possingham, H. P., Willis, S. G., and Fuller, R. A. (2014). Conserving mobile species. *Frontiers in Ecology and Environment*, **12**, 395–402.

Russell, G. J., Diamond, J. M., Pimm, S. L., and Reed, T. M. (1995). A century of turnover: community dynamics at three timescales. *Journal of Animal Ecology*, **64**, 628–641.

Russell, J. C., Jones, H. P., and Armstrong, D. P. (2016). Importance of lethal control of invasive predators for island conservation. *Conservation Biology*, **30**, 670–672.

Russell, J. C. and Kaiser-Bunbury, C. N. (2019). Consequences of multispecies introductions on island ecosystems. *Annual Review of Ecology, Evolution, and Systematics*, **50**, 169–190.

Russell, J. C. and Kueffer, C. (2019). Island biodiversity in the Anthropocene. *Annual Review of Environment and Resources*, **44**, 31–60.

Russell, J. C., Ringler, D., Trombini, A., and Le Corre, M. (2011). The island syndrome and population dynamics of introduced rats. *Oecologia*, **167**, 667–676.

Russo, D., Maglio, G., Rainho, A., Meyer, C. F. J., and Palmeirim, J. M. (2011). Out of the dark: diurnal activity in the bat *Hipposideros ruber* on São Tomé Island (West Africa). *Mammalian Biology*, **76**, 701–708.

Rybicki, J. and Hanski, I. (2013). Species–area relationships and extinctions caused by habitat loss and fragmentation. *Ecology Letters*, **16**, 27–38.

Sadler, J. P. (1999). Biodiversity on oceanic islands: a palaeoecological assessment. *Journal of Biogeography*, **26**, 75–87.

Sáez, A., Valero-Garcés, B. L., Giralt, S., et al. (2009). Glacial to Holocene climate changes in the SE Pacific. The Raraku

Lake sedimentary record (Easter Island, 27°S). *Quaternary Science Reviews*, **28**, 2743–2759.

Sakai, A. K., Wagner, W. L., Ferguson, D. M., and Herbst, D. R. (1995). Biogeographical and ecological correlates of dioecy in the Hawaiian flora. *Ecology*, **76**, 2530–2543.

Sakamoto, M. and Venditti, C. (2018). Phylogenetic non-independence in rates of trait evolution. *Biology Letters*, **14**, 20180502.

Salces-Castellano, A., Patiño, J., Alvarez, N., et al. (2020). Climate drives community-wide divergence within species over a limited spatial scale: evidence from an oceanic island. *Ecology Letters*, **23**, 305–315.

Salinas-de-León, P., Arnés-Urgellés, C., Bermudez, J. R., et al. (2020). Evolution of the Galapagos in the Anthropocene. *Nature Climate Change*, **10**, 380–382.

Samonds, K. E., Godfrey, L. R., Ali, J. R., et al. (2013). Imperfect isolation: factors and filters shaping Madagascar's extant vertebrate fauna. *PLOS ONE*, **8**, e620866.

Samuelson, G. A. (2003). *Review of* Rhyncogonus *of the Hawaiian Islands (Coleoptera: Curculionidae)*. Bishop Museum, Honolulu.

Sanderson, J. G., Diamond, J., and Pimm, S. L. (2009). Pairwise co-existence of Bismarck and Solomon landbird species. *Evolutionary Ecology Research*, **11**, 771–786.

Sangster, G., Luksenburg, J. A., Päckert, M., Roselaar, C. S., Irestedt, M., and Pericson, P. G. P. (2022). Integrative taxonomy documents two additional cryptic *Erithacus* species on the Canary Islands (Aves). *Zoologica Scripta*, **6**, 629–642.

Sanmartín, I., Anderson, C. L., Alarcon, M., Ronquist, F., and Aldasoro, J. J. (2010). Bayesian island biogeography in a continental setting: the Rand Flora case. *Biology Letters*, **6**, 703–707.

Santos, A. M. C., Cianciaruso, M. V., and De Marco, P. Jr (2016). Global patterns of functional diversity and assemblage structure of island parasitoid faunas. *Global Ecology and Biogeography*, **25**, 869–879.

Santos, A. M. C., Whittaker, R. J., Triantis, K. A., et al. (2010). Are species–area relationships from entire archipelagos congruent with those of their constituent islands? *Global Ecology and Biogeography*, **19**, 527–540.

Santos-Guerra, A. (1999). Origen y evolución de la flora canaria. In *Ecología y cultura en Canarias* (ed. J. M. Fernández-Palacios, J. J. Bacallado, and J. A. Belmonte), pp. 107–129. Museo de las Ciencias y el Cosmos, Cabildo Insular de Tenerife, Santa Cruz de Tenerife.

Santos-Guerra, A. (2001). Flora vascular nativa. In *Naturaleza de las Islas Canarias. Ecología y Conservación*. (ed. J. M. Fernández-Palacios and J. L. Martín Esquivel), pp. 185–192. Turquesa Ediciones, Santa Cruz de Tenerife.

Sanz, C. and Grajal, A. (2008). Successful reintroduction of captive-raised yellow-shouldered Amazon parrots on Margarita Island, Venezuela. *Conservation Biology*, **12**, 430–441.

Satterfield, D. and Johnson, D. W. (2020). Local adaptation of antipredator behaviors in populations of a temperate reef fish. *Oecologia*, **194**, 571–584.

Saunders, D. A. and Hobbs, R. J. (1989). Corridors for conservation. *New Scientist*, **121** (1648), 28 January, 63–68.

Saunders, D. A., Hobbs, R. J., and Margules, C. R. (1991). Biological consequences of ecosystem fragmentation: a review. *Conservation Biology*, **5**, 18–32.

Saunders, S. P., Cuthbert, F. J., and Ziplin, E. F. (2018). Evaluating population viability and efficacy of conservation management using integrated population models. *Journal of Applied Ecology*, **55**, 1380–1392.

Saura, S. (2021). The habitat amount hypothesis predicts that fragmentation poses a threat to biodiversity: a reply to Fahrig. *Journal of Biogeography*, **48**, 1536–1540.

Sax, D. F., Brown, J. H., and Gaines, S. D. (2002). Species invasions exceed extinctions on islands world-wide: a comparative study of plants and birds. *American Naturalist*, **160**, 776–783.

Sayol, F., Cooke, R. S. C., Pigot, A. L., et al. (2021). Loss of functional diversity through anthropogenic extinctions of island birds is not offset by biotic invasions. *Science Advances*, **7**, eabj5790.

Sayol, F., Downing, P. A., Iwaniuk, A. N., Maspons, J., and Sol, D. (2018). Predictable evolution towards larger brains in birds colonizing oceanic islands. *Nature Communications*, **9**, 2820.

Sayol, F., Steinbauer, M. J., Blackburn, T. M., Antonelli, A., and Faurby, S. (2020). Anthropogenic extinctions conceal widespread evolution of flightlessness in birds. *Science Advances*, **6**, eabb6095.

Scatena, F. N. and Larsen, M. C. (1991). Physical aspects of Hurricane Hugo in Puerto Rico. *Biotropica*, **23**, 317–323.

Scheiner, S. M. (2003). Six types of species-area curves. *Global Ecology and Biogeography*, **12**, 441–447.

Schenk, J. J. (2021). The next generation of adaptive radiation studies in plants. *International Journal of Plant Science*, **182**, 245–262.

Schimper, A. F. W. (1907). Die Canarischen Federbuschgewächse. In *Beiträge zur Kenntnis der Vegetation der Canarischen Inseln. Mit Einfügung hinterlassener Schriften A. F. W. Schimpers* (ed. H. Schenck), pp. 47–63. Deutsche Tiefsee-Expedition: 1898-1899. Bd. II. 1. Teil. Gustav Fischer, Jena.

Schlesinger, W. H., Bruijnzeel, L. A., Bush, M. B., et al. (1998). The biogeochemistry of phosphorus after the first century of soil development on Rakata Island, Krakatau, Indonesia. *Biogeochemistry*, **40**, 37–55.

Schlessman, M. A., Vary, L. B., Munzinger, J., and Lowry II, P. P. (2014). Incidence, correlates, and origins of dioecy in the island Flora of New Caledonia. *International Journal of Plant Sciences*, **175**, 271–286.

Schluter, D. (1988). Character displacement and the adaptive divergence of finches on islands and continents. *American Naturalist*, **131**, 799–824.

Schluter, D. (2000). *The Ecology of Adaptive Radiation*. Oxford University Press, Oxford.

Schneider-Maunoury, L., Fefebvre, V., Ewers, R. M., et al. (2016). Abundance signals of amphibians and reptiles indicate strong edge effects in Neotropical fragmented forest landscapes. *Biological Conservation*, **200**, 207–215.

Schnell, J. K., Harris, G. M., Pimm, S. L., and Russell, F. J. (2013). Estimating extinction risk with metapopulation models of large scale fragmentation. *Conservation Biology*, **27**, 520–530.

Schoener, T. W. (1975). Presence and absence of habitat shift in some widespread lizard species. *Ecological Monographs*, **45**, 233–258.

Schoener, T. W. (1983). Rate of species turnover decreases from lower to higher organisms: a review of the data. *Oikos*, **41**, 372–377.

Schoener, T. W. and Spiller, D. A. (1987). High population persistence in a system with high turnover. *Nature*, **330**, 474–477.

Schrader, J., Craven, D., Sattler, C., Cámara-Leret, R., Moeljono, S., and Kreft, H. (2021). Life-history dimensions indicate non-random assembly processes in tropical island tree communities. *Ecography*, **44**, 469–480.

Schrader, J., König, C., Triantis, K. A., Trigas, P., Kreft, H., and Weigelt, P. (2020). Species–area relationships on small islands differ among plant growth forms. *Global Ecology and Biogeography*, **29**, 814–829.

Schrader, J., Westoby, M., Wright, I. J., and Kreft, H. (2021a). Disentangling direct and indirect effects of island area on plant functional trait distributions. *Journal of Biogeography*, **48**, 2098–2110.

Schrader, J., Wright, I. J., Kreft, H., and Westoby, M. (2021b). A roadmap to plant functional island biogeography. *Biological Reviews*, **96**, 2851–2870.

Schumm, S. A. (1991). *To Interpret the Earth: Ten Ways to Be Wrong*. Cambridge University Press, Cambridge.

Schwaner, T. D. and Sarre, S. D. (1988). Body size of Tiger Snakes in Southern Australia, with particular reference to *Notechis ater serventyi* (Elapidae) on Chappell Island. *Journal of Herpetology*, **22**, 24–33.

Schwarz, R., Itescu, Y., Antonopoulos, A., et al. (2020). Isolation and predation drive gecko life-history evolution on islands. *Biological Journal of the Linnean Society*, **129**, 618–629.

Schwarz, R. and Meiri, S. (2017). The fast-slow life-history continuum in insular lizards: a comparison between species with invariant and variable clutch sizes. *Journal of Biogeography*, **44**, 2808–2815.

Schweizer, D., Jones, H. P., and Holmes, N. D. (2016). Literature review and meta-analysis of vegetation responses to goat and European rabbit eradications on islands. *Pacific Science*, **70**, 55–71.

Scott, T. A. (1994). Irruptive dispersal of black-shouldered kites to a coastal island. *Condor*, **96**, 197–200.

Seebens, H., Blackburn, T. M., Dyer, E. E., et al. (2017). No saturation in the accumulation of alien species worldwide. *Nature Communications*, **8**, 14435.

Seehausen, O. (2004). Hybridization and adaptive radiation. *Trends in Ecology & Evolution*, **19**, 198–207.

Şekercioğlu, C. H., Loarie, S. R., Oviedo-Brenes, F., Mendenhall, C. D., Daily, G. C., and Ehrlich, P. R. (2015). Tropical countryside riparian corridors provide critical habitat and connectivity for seed-dispersing forest birds in a fragmented landscape. *Journal of Ornithology*, **156, Supplement 1**, 343–353.

Sendell-Price, A. T., Ruegg, K. C., Robertson, B. C., and Clegg, S. M. (2021). An island-hopping bird reveals how founder events shape genome-wide divergence. *Molecular Ecology*, **30**, 2495–2510.

Sequeira, A. S., Lanteri, A. A., Scataglini, M. A., Confalonieri, V. A., and Farrell, B. D. (2000). Are flightless *Galapaganus* weevils older than the Galápagos Islands they inhabit? *Heredity*, **85**, 20–29.

Sfenthourakis, S., Paflis, P., Parmakelis, A., Poulakakis, N., and Triantis, K. A. (2018). *Biogeography and Biodiversity of the Aegean*. Broken Hill Publishers, Nicosia, Cyprus.

Sfenthourakis, S. and Triantis, K. A. (2017). The Aegean archipelago: a natural laboratory of evolution, ecology and civilisations. *Journal of Biological Research-Thessaloniki*, **24**, article 4.

Sfenthourakis, S., Triantis, K. A., Proios, K., and Rigal, F. (2021). The role of ecological specialization in shaping patterns of insular communities. *Journal of Biogeography*, **48**, 243–252.

Shilton, L. A., Altringham, J. D., Compton, S. G., and Whittaker, R. J. (1999). Old World fruit bats can be long-distance seed dispersers through extended retention of viable seeds in the gut. *Proceedings of the Royal Society B*, **266**, 219–223.

Si, X., Baselga, A., and Ding, P. (2015). Revealing beta-diversity patterns of breeding bird and lizard communities on inundated land-bridge islands by separating the turnover and nestedness components. *PLOS ONE*, **10**, e0127692.

Si, X., Cadotte, M. W., Davies, T. J., et al. (2022). Phylogenetic and functional clustering illustrate the roles of adaptive radiation and dispersal filtering in jointly shaping late-Quaternary mammal assemblages on oceanic islands. *Ecology Letters*, **25**, 1250–1262.

Si, X., Cadotte, M. W., Zeng, D., et al. (2017). Functional and phylogenetic structure of island bird communities. *Journal of Animal Ecology*, **86**, 532–542.

Si, X., Pimm, S. L., Russell, G. J., and Ding, P. (2014). Turnover of breeding bird communities on islands in an inundated lake. *Journal of Biogeography*, **41**, 2283–2292.

Siegel, P. E., Jones, J. G., Pearsall, D. M. et al. (2015). Paleoenvironmental evidence for first human colonization of the eastern Caribbean. *Quaternary Science Reviews*, **129**, 275–295.

Siliceo, I. and Díaz, J. A. (2010). A comparative study of clutch size, range size, and the conservation status of island vs. mainland lacertid lizards. *Conservation Biology*, **143**, 2601–2608.

Simberloff, D. (1976). Species turnover and equilibrium island biogeography. *Science*, **194**, 572–578.

Simberloff, D. (1978). Using island biogeographic distributions to determine if colonization is stochastic. *American Naturalist*, **112**, 713–726.

Simberloff, D. (1992). Do species–area curves predict extinction in fragmented forests? In *Tropical Deforestation and Species Extinction* (ed. T. C. Whitmore and J. A. Sayer), pp. 119–142. Chapman & Hall, London.

Simberloff, D., Farr, J. A., Cox, J., and Mehlman, D. W. (1992). Movement corridors: conservation bargains or poor investments? *Conservation Biology*, **6**, 493–504.

Simberloff, D., Keitt, B., Will, D., Holmes, N., Pickett, E., and Genovesi, P. (2018). Yes we can! Exciting progress and prospects for controlling invasives on islands and beyond. *Western North American Naturalist*, **78**, 942–958.

Simberloff, D. and Levin, B. (1985). Predictable sequences of species loss with decreasing island area—land birds in two archipelagoes. *New Zealand Journal of Ecology*, **8**, 11–20.

Simberloff, D. and Wilson, E. O. (1969). Experimental zoogeography of islands. The colonisation of empty islands. *Ecology*, **50**, 278–296.

Simberloff, D. and Wilson, E. O. (1970). Experimental zoogeography of islands. A two year record of colonization. *Ecology*, **51**, 934–937.

Simpson, G. G. (1940). Mammals and land bridges. *Journal of the Washington Academy of Sciences*, **30**, 137–163.

Singh, B. N. (2012). Concepts of species and modes of speciation. *Current Science*, **103**, 784–790.

Sirabani, M. C., Di Marco, M., Rondinini, C., and Kark, S. (2019). Measuring the surrogacy potential of charismatic megafauna species across taxonomic, phylogenetic and functional diversity on a megadiverse island. *Journal of Applied Ecology*, **56**, 1220–1231.

Smith-Ramírez, C., Vargas, R., Castillo, J., Mora, J. P., and Arellano-Cataldo, G. (2017). Woody plant invasions and restoration in forests of island ecosystems: lessons from Robinson Crusoe Island, Chile. *Biodiversity and Conservation*, **26**, 1507–1524.

Soares, F. C., de Lima, R. F., Palmeirim, J. M., Cardoso, P. and Rodrigues, A. S. L. (2022). Combined effects of bird extinctions and introductions in oceanic islands: decreased functional diversity despite increased species richness. *Global Ecology and Biogeography*, **31**, 1172–1183.

Sondaar, P. Y. (1977). Insularity and its effect on mammal evolution. In *Major Patterns in Vertebrate Evolution* (ed. M. K. Hecht, P. C. Goody, and B. M. Hecht), pp. 671–707. Springer, Boston, MA.

Soulé, M. E., Bolger, D. T., Alberts, A. C., Wright, J., Sorice, M., and Hill, S. (1988). Reconstructed dynamics of rapid extinctions of Chaparral-requiring birds in urban habitat islands. *Conservation Biology*, **2**, 75–92.

Spatz, D. R., Holmes, N. D., Will, D. J., et al. (2022). The global contribution of invasive vertebrate eradication as a key island restoration tool. *Scientific Reports*, **12**, 13391.

Spatz, D. R., Zilliacus, K. M., Holmes, N. D., et al. (2017). Globally threatened vertebrates on islands with invasive species. *Science Advances*, **3**, e1603080.

Spencer-Smith, D., Ramos, S. J., McKenzie, F., Munroe, E., and Miller, L. D. (1988). Biogeographical affinities of the butterflies of a 'forgotten' island: Mona (Puerto Rico). *Bulletin of the Allyn Museum*, No. **121**, pp. 1–35.

Spurgin, L. G., Illera, J. C., Jorgensen, T. H., Dawson, D. A., and Richardson, D. S. (2014). Genetic and phenotypic divergence in an island bird: isolation by distance, by colonization or by adaptation? *Molecular Ecology*, **23**, 1028–1039.

Stace, C. A. (1989). Dispersal versus vicariance—no contest! *Journal of Biogeography*, **16**, 201–202.

Staddon, P., Lindo, Z., Crittenden, P. D., Gilbert, F., and Gonzalez, A. (2010). Connectivity, non-random extinction and ecosystem function in experimental metacommunities. *Ecology Letters*, **13**, 543552.

Stamps, J. A. and Buechner, M. (1985) The territorial defence hypothesis and the ecology of insular vertebrates. *Quarterly Review of Biology*, **60**, 155–181.

Steadman, D. W. (1997). Human-caused extinctions of birds. In *Biodiversity II: Understanding and Protecting our Biological Resources* (ed. M. L. Reaka-Kudla, W. E. Wilson, and W. O. Wilson), pp. 139–161. Joseph Henry Press, Washington, DC.

Steadman, D. W. (2006). *Extinction and Biogeography of Tropical Pacific Birds*. Chicago University Press, Chicago.

Steadman, D. W. and Franklin, J. (2020). Bird populations and species lost to Late Quaternary environmental change and human impact in the Bahamas. *Proceedings of the National Academy of Sciences, USA*, **117**, 26833–26841.

Steers, J. A. and Stoddart, D. R. (1977). The origin of fringing reefs, barrier reefs and atolls. In *Biology and Geology of Coral Reefs* (ed. O. A. Jones and R. Endean), pp. 21–57. Academic Press, New York.

Steffen, W., Broadgate, W., Deutsch, L., Gaffney, O., and Ludwig, C. (2015). The trajectory of the Anthropocene: the Great Acceleration. *The Anthropocene Review*, **2**, 81–98.

Steinbauer, M. J., Dolos, K., Field, R., Reineking, B., and Beierkuhnlein, C. (2013). Re-evaluating the general dynamic theory of oceanic island biogeography. *Frontiers of Biogeography*, **5**, 185–194.

Stervander, M., Illera, J. C., Kvist, L., et al. (2015). Disentangling the complex evolutionary history of the Western Palearctic blue tits (*Cyanistes* spp.) – phylogenomic analyses suggest radiation by multiple colonization events and subsequent isolation. *Molecular Ecology*, **24**, 2477–2494.

Stoddart, D. R. and Walsh, R. P. D. (1992). Environmental variability and environmental extremes as factors in the island ecosystem. *Atoll Research Bulletin*, No. 356.

Stone, L. and Roberts, A. (1992). Competitive exclusion, or species aggregation? *Oecologia*, **91**, 419–424.

Storlazzi, C. D., Gingerich, S. B., van Dongeren, A. P., et al. (2018). Most atolls will be uninhabitable by the mid-21st century because of sea-level rise exacerbating wave-driven flooding. *Science Advances*, **4**, eaap974.

Streelman, J. T. and Danley, P. D. (2003). The stages of vertebrate evolutionary radiation. *Trends in Ecology & Evolution*, **18**, 126–131.

Strobl, E. (2012). The economic growth impact of natural disasters in developing countries: evidence from hurricane strikes in the Central American and Caribbean regions. *Journal of Development Economics*, **97**, 130–141.

Strona, G., Ulrich, W., and Gotelli, N. J. (2017). Bi-dimensional null model analysis of presence-absence binary matrices. *Ecology*, **99**, 103–115.

Stroud, J. T. and Losos, J. B. (2016). Ecological opportunity and adaptive radiation. *Annual Review of Ecology, Evolution, and Systematics*, **47**, 507–532.

Stuart, Y. E. and Losos, J. B. (2013). Ecological character displacement: glass half full or half empty? *Trends in Ecology & Evolution*, **28**, 402–408.

Stuessy, T. F. (2007). Evolution of specific and genetic diversity during ontogeny of island floras: the importance of understanding process for interpreting island biogeographic patterns. In *Biogeography in a Changing World* (ed. M. C. Ebach and R. S. Tangney), pp. 117–133. CRC Press, Boca Raton, FL.

Stuessy, T. F., Crawford, D. J., and Marticorena, C. (1990). Patterns of phylogeny in the endemic vascular flora of the Juan Fernandez Islands, Chile. *Systematic Botany*, **15**, 338–346.

Stuessy, T. F., Crawford, D. J., Greimler, J., et al. (2022). Metamorphosis of flora and vegetation during ontogeny of the Juan Fernández (Robinson Crusoe) Islands. *Botanical Journal of the Linnean Society*, **199**, 609–645.

Stuessy, T. F., Crawford, D. J., Marticorena, C., and Rodríguez, R. (1998). Island biogeography of angiosperms of the Juan Fernandez archipelago. In *Evolution and Speciation of Island Plants* (ed. T. F. Stuessy and M. Ono), pp. 121–138. Cambridge University Press, Cambridge.

Stuessy, T. F., Jakubowsky, G., Salguero-Gómez, R., et al. (2006). Anagenetic evolution in island plants. *Journal of Biogeography*, **33**, 1259–1265.

Stuessy, T. F., Takayama, K., López-Sepúlveda, P., and Crawford, D. J. (2014). Interpretation of patterns of genetic variation in endemic plant species of oceanic islands. *Botanical Journal of the Linnean Society*, **174**, 276–288.

Sturt, F., Garrow, D., and Bradley, C. (2013). New models of North West European Holocene palaeogeography and inundation. *Journal of Archaeological Science*, **40**, 3963–3976.

Sun, Y., Li, Y., Vargas-Mendoza, C. F., Wang, F., and Zing, F. (2016). Colonization and diversification of the *Euphorbia* species (sect. *Aphyllis* subsect. Macaronesicae) on the Canary Islands. *Scientific Reports*, **6**, Article 34454.

Szabo, J. K., Khwaja, N., Garnett, S. T., and Butchart, S. H. M. (2012). Global patterns and drivers of avian extinctions at the species and subspecies level. *PLOS ONE*, **7**, e47080.

Tabak, M. A., Poncet, S., Passfield, K., and del Rio, C. M. (2014). Invasive species and land bird diversity on remote South Atlantic islands. *Biological Invasions*, **16**, 341–352.

Takayama, K., Crawford, D. J., López-Sepúlveda, P., Greimler, J., and Stuessy, T. F. (2018). Factors driving adaptive radiation in plants of oceanic islands: a case study from the Juan Fernández Archipelago. *Journal of Plant Research*, **131**, 469–485.

Takayama, K., López-Sepúlveda, P., Greimler, J., et al. (2015). Genetic consequences of cladogenetic vs. anagenetic speciation in endemic plants of oceanic islands, *AoB PLANTS*, **7**, plv102.

Talavera, M., Arista, M., and Ortiz, P. L. (2012). Evolution of dispersal traits in a biogeographical context: a study using the heterocarpic *Rumex bucephalophorus* as a model. *Journal of Ecology*, **100**, 1194–1203.

Terborgh, J. (1992). Maintenance of diversity in tropical forests. *Biotropica*, **24**, 283–292.

Terborgh, J. (2010). The trophic cascade on islands. In *The Theory of Island Biogeography Revisited* (ed. J. B. Losos and R. E. Ricklefs), pp. 116–142. Princeton University Press, Princeton.

Terborgh J., Lopez, L., Nuñez, P., et al. (2001). Ecological meltdown in predator-free forest fragments. *Science*, **294**, 1923–1926.

Terzopoulou, S., Rigal, F., Whittaker, R. J., Borges, P. A. V., and Triantis, K. A. (2015). Drivers of extinction: the case of Azorean beetles. *Biology Letters*, **11**, 20150273.

Theuerkauf, J., Chartendrault, V., Desmoulins, F., Barré, N., and Gula, R. (2017). Positive range–abundance relationships in Indo-Pacific bird communities. *Journal of Biogeography*, **44**, 2161–2163.

Thomas, C. D. (1994). Extinction, colonization, and metapopulations: environmental tracking by rare species. *Conservation Biology*, **8**, 373–378.

Thomas, C. D. and Harrison, S. (1992). Spatial dynamics of a patchily distributed butterfly species. *Journal of Animal Ecology*, **61**, 437–446.

Thornton, I. W. B. (1996). *Krakatau—The Destruction and Reassembly of an Island Ecosystem*. Harvard University Press, Cambridge, MA.

Thornton, I. W. B., Partomihardjo, T., and Yukawa, J. (1994). Observations on the effects, up to July 1993, of the current eruptive episode of Anak Krakatau. *Global Ecology and Biogeography Letters*, **4**, 88–94.

Thornton, I. W. B., Ward, S. A., Zann, R. A., and New, T. R. (1992). Anak Krakatau—a colonization model within a colonization model? *Geojournal*, **28**, 271–286.

Thornton, I. W. B., Zann, R. A., and van Balen, S. (1993). Colonization of Rakata (Krakatau Is.) by non-migrant land birds from 1883 to 1992 and implications for the value of island equilibrium theory. *Journal of Biogeography*, **20**, 441–452.

Tjørve, E. (2010). How to resolve the SLOSS debate: lessons from species-diversity models. *Journal of Theoretical Biology*, **264**, 604–612.

Tjørve, E., Matthews, T. J., and Whittaker, R. J. (2021). The history of the species–area relationship. In *The Species–Area Relationship: Theory and Application* (ed. T. J. Matthews, K. A. Triantis, and R. J. Whittaker), pp. 20–48. Cambridge University Press, Cambridge.

Tjørve, E. and Tjørve, K. M. C. (2021). Mathematical expressions for the species–area relationship and the assumptions behind the models. In *The Species–Area Relationship: Theory and Application* (ed. T. J. Matthews, K. A. Triantis, and R. J. Whittaker), pp. 157–184. Cambridge University Press, Cambridge.

Tobias, J. A., Sheard, C., Pigot, A. L., et al. (2022). AVONET: morphological, ecological and geographical data for all birds. *Ecology Letters*, **25**, 581–597.

Toft, C. A. and Schoener, T. W. (1983). Abundance and diversity of orb spiders on 106 Bahamian islands: biogeography at an intermediate trophic level. *Oikos*, **41**, 411–426.

Tokita, M., Yano, W., James, H. F., and Abzhanov, A. (2017). Cranial shape evolution in adaptive radiations of birds: comparative morphometrics of Darwin's finches and Hawaiian honeycreepers. *Philosophical Transactions of the Royal Society B*, **372**, 20150481.

Toomey, M. R., Ashton, A. D., Raymo, M. E., and Perron, J. T. (2016). Late Cenozoic sea level and the rise of modern rimmed atolls. *Palaeogeography, Palaeoclimatology, Palaeoecology*, **451**, 73–83.

Toral-Granda, M. V., Causton, C. E., Jäger, H., et al. (2017). Alien species pathways to the Galapagos Islands, Ecuador. *PLOS ONE*, **12**, e0184379.

Torre, G., Fernández Lugo, S., Guarino, R., and Fernández-Palacios, J. M. (2019). Network analysis by simulated annealing of taxa and islands of Macaronesia (North Atlantic Ocean). *Ecography*, **42**, 768–779.

Touchton, J. M. and Smith, J. N. M. (2011). Species loss, delayed numerical responses, and functional compensation in an antbird guild. *Ecology*, **92**, 1126–1136.

Traill, L. W., Bradshaw, C. J. A., and Brook, B. W. (2007). Minimum viable population size: a meta-analysis of 30 years of published estimates. *Biological Conservation*, **139**, 159–166.

Traveset, A. (2001). Ecología reproductiva de plantas en condiciones de insularidad: consecuencias ecológicas y evolutivas del aislamiento geográfico. In *Ecosistemas mediterráneos: análisis funcional* (ed. R. Zamora and F. Puignaire), pp. 269–289. Consejo Superior de Investigaciones Científicas–Asociación Española de Ecología Terrestre, Granada.

Traveset, A., Fernández-Palacios, J. M., Kueffer, C., Bellingham, P. J., Morden, C., and Drake, D. R. (2016). Introduction to the Special Issue: advances in island plant biology since Sherwin Carlquist's Island Biology. *AoB Plants*, **8**, plv148.

Traveset, A., Olesen, J. M., Nogales, M., et al. (2015). Bird–flower visitation networks in the Galápagos unveil a widespread interaction release. *Nature Communications*, **6**, 6376.

Traveset, A. and Santamaría, L. (2004). Alteración de mutualismos planta-animal debido a la introducción de especies exóticas en ecosistemas insulares. In *Ecología Insula /Island Ecology* (ed. J. M. Fernández-Palacios and C. Morici), pp. 251–276. Asociación Española de Ecología Terrestre–Cabildo Insular de La Palma, Santa Cruz de La Palma.

Traveset, A., Tur, C., Trøjelsgaard, K., Heleno, R., Castro-Urgal, R., and Olesen, J. M. (2016). Global patterns of mainland and insular pollination networks. *Global Ecology and Biogeography*, **25**, 880–890.

Triantis, K. A. and Bhagwat, S. (2011). Applied island biogeography. In *Conservation Biogeography* (ed. R. J. Ladle and R. J. Whittaker), pp. 190–223. Wiley-Blackwell, Chichester.

Triantis, K. A., Borges, P. A. V., Ladle, R. J., et al. (2010). Extinction debt on oceanic islands. *Ecography*, **33**, 285–294.

Triantis, K. A., Economo, E. P., Guilhaumon, F., and Ricklefs, R. E. (2015). Diversity regulation at macro-scales: species richness on oceanic archipelagos. *Global Ecology and Biogeography*, **24**, 594–605.

Triantis, K. A., Guilhaumon, F., and Whittaker, R. J. (2012). The island species–area relationship: biology and statistics. *Journal of Biogeography*, **39**, 215–231.

Triantis, K. A., Mylonas, M., Lika, K., and Vardinoyannis, K. (2003). A model for the species–area–habitat relationship. *Journal of Biogeography*, **30**, 19–27.

Triantis, K. A., Mylonas, M., and Whittaker, R. J. (2008). Evolutionary species–area curves as revealed by single-island endemics: insights for the inter-provincial species–area relationship. *Ecography*, **31**, 401–407.

Triantis, K. A., Rigal, F., Whittaker, R. J., et al. (2022). Deterministic assembly and anthropogenic extinctions drive convergence of island bird communities. *Global Ecology and Biogeography*, **31**, 1741–1755.

Triantis, K. A., Vardinoyannis, K., Tsolaki, E. P., Botsaris, I., Lika, K., and Mylonas, M. (2006). Re-approaching the small island effect. *Journal of Biogeography*, **33**, 914–923.

Trøjelsgaard, K., Báez, K. M., Espadaler, X., et al. (2013). Island biogeography of mutualistic interaction networks. *Journal of Biogeography*, **40**, 2020–2031.

Troll, V. R., Deegan, F. M., Burhardt, S., et al. (2015). Nanno-fossils: the smoking gun for the Canarian hotspot. *Geology Today*, **31**, 137–145.

Trusty, J. L., Olmstead, R. G., Santos-Guerra, A., Sá-Fontinha, S., and Francisco-Ortega, J. (2005). Molecular phylogenetics of the Macaronesian-endemic genus *Bystropogon* (Lamiaceae): palaeo-islands, ecological shifts and interisland colonizations. *Molecular Ecology*, **14**, 1177–1189.

Tucker, C. M., Cadotte, M. W., Carvalho, S. B., et al. (2017). A guide to phylogenetic metrics for conservation, community ecology and macroecology. *Biological Reviews*, **92**, 698–715.

Tudge, C. (1991). Time to save rhinoceroses. *New Scientist*, **131** (1788), 30–35.

Turner, G. F. (2007). Adaptive radiation of cichlid fish. *Current Biology*, **17**, R827–R831.

Turvey, S. T., Kennerley, R. J., Nuñez-Miño, J. M., and Young, R. P. (2017). The last survivors: current status and conservation of the non-volant land mammals of the insular Caribbean. *Journal of Mammalogy*, **98**, 918–936.

Ulrich, W. and Gotelli, N. J. (2007). Null model analysis of species nestedness patterns. *Ecology*, **88**, 1824–1831.

Ulrich, W. and Gotelli, N. J. (2012). A null model algorithm for presence–absence matrices based on proportional resampling. *Ecological Modelling*, **244**, 20–27.

Ulrich, W., Lens, L., Tobias, J. A., and Habel, J. C. (2016). Contrasting patterns of species richness and functional diversity in bird communities of East African cloud forest fragments. *PLOS ONE*, **11**, e0163338.

Upham, N. S. (2017). Past and present of insular Caribbean mammals: understanding Holocene extinctions to inform modern biodiversity conservation. *Journal of Mammalogy*, **98**, 913–917.

USFWS (2011). *Revised Recovery Plan for the Northern Spotted Owl (Strix occidentalis caurina)*. U.S. Fish and Wildlife Service, Portland, Oregon.

Uy, J. A. C. and Vargas-Castro, L. E. (2015). Island size predicts the frequency of melanic birds in the color-polymorphic flycatcher *Monarcha castaneiventris* of the Solomon Islands. *The Auk*, **132**, 787–794.

Vacchi, M., Ghilardi, M., Melis, R. T., et al. (2018). New relative sea-level insights into the isostatic history of the Western Mediterranean. *Quaternary Science Reviews*, **201**, 396–408.

Valente, L. M., Etienne, R. S., and Phillimore, A. B. (2014). The effects of island ontogeny on species diversity and phylogeny. *Proceedings of the Royal Society B*, **281**, 20133227.

Valente, L. M., Illera, J. C., Havenstein, K., Pallein, T., Etienne, R. S., and Tiedemann, R. (2017). Equilibrium bird species diversity in Atlantic Islands. *Current Biology*, **27**, 1660–1666.

Valente, L. M., Phillimore, A. B., and Etienne, R. S. (2015). Equilibrium and non-equilibrium dynamics simultaneously operate in the Galápagos Islands. *Ecology Letters*, **18**, 844–852.

Valente, L. M., Phillimore, A. B., and Etienne, R. S. (2018). Using molecular phylogenies in island biogeography: it's about time. *Ecography*, **41**, 1684–1686.

Valente, L., Phillimore, A. B., Melo, M., et al. (2020). A simple dynamic model explains the diversity of island birds worldwide. *Nature*, **579**, 92–96.

Valido, A., Dupont, Y. L., and Hansen, D. M. (2002). Native birds and insects, and introduced honey bees visiting *Echium wildpretii* (Boraginaceae) in the Canary Islands. *Acta Oecologica*, **23**, 413–419.

Valido, A. and Nogales, M. (1994). Frugivory and seed dispersal by the lizard *Gallotia galloti* (Lacertidae) in a xeric habitat of the Canary Islands. *Oikos*, **70**, 403–411.

Valido, A., Nogales, M., and Medina, F. M. (2003). Fleshy fruits in the diet of Canarian lizards *Gallotia galloti* (Lacertidae) in a xeric habitat of the island of Tenerife. *Journal of Herpetology*, **37**, 741–747.

Valido, A. and Olesen, J. M. (2019). Frugivory and seed dispersal by lizards: a global review. *Frontiers in Ecology and Evolution*, **7**, 49.

Valido, A., Rando, J. C., Nogales, M., and Martín, A. (2000). 'Fossil' lizard found alive in the Canary Islands. *Oryx*, **34**, 71–72.

Vamosi, J. C., Otto, S. P., and Barrett, S. C. H. (2003). Phylogenetic analysis of the ecological correlates of dioecy in angiosperms. *Journal of Evolutionary Biology*, **16**, 1006–1018.

van Damme, R. (1999). Evolution of herbivory in lacertid lizards: effects of insularity and body size. *Journal of Herpetology*, **33**, 663–674.

van den Bogaard, P. (2013). The origin of the Canary Island Seamount Province – new ages of old seamounts. *Scientific Reports* **3**, srep2107.

van der Geer, A. A., Lomolino, M. V., and Lyras, G. A. (2018a). 'On being the right size'. Do aliens follow the rules? *Journal of Biogeography*, **45**, 515–529.

van der Geer, A. A., Lyras, G. A., Lomolino, M. V., Palombo, M. R., and Sax, D. F. (2013). Body size evolution of palaeo-insular mammals: temporal variations and interspecific interactions. *Journal of Biogeography*, **40**, 1440–1450.

van der Geer, A. A., Lyras, G. A., Mitteroecker, P., and MacPhee, R. D. E. (2018b). From Jumbo to Dumbo: cranial shape changes in elephants and hippos during phyletic dwarfing. *Evolutionary Biology*, **45**, 303–317.

van der Geer, A. A., van den Bergh, G. D., Lyras, G. A., et al. (2016). The effect of area and isolation on insular dwarf proboscideans. *Journal of Biogeography*, **43**, 1656–1666.

Van Houtan, K. S., Pimm, S. L., Halley, J. M., Bierregaard, R. O. Jr, and Lovejoy, T. E. (2007). Dispersal of Amazonian birds in continuous and fragmented forest. *Ecology Letters*, **9**, 1–11.

Van Schmidt, N. D. and Beissinger, S. R. (2020). The rescue effect and inference from isolation–extinction relationships. *Ecology Letters*, **23**, 598–606.

Vargas, P., Arjona, Y., Nogales, M., and Heleno, R. H. (2015). Long-distance dispersal to oceanic islands: success of plants with multiple diaspore specializations. *AoB Plants*, **7**, plv073.

Vargas, P., Heleno, R., Traveset, A., and Nogales, M. (2012). Colonization of the Galápagos Islands by plants with no specific syndromes for long-distance dispersal: a new perspective. *Ecography*, **35**, 33–43.

Veitch, C. R., Clout, M. N., Martin, A. R., Russell, J. C., and West, C. J. (eds) (2019). Island invasives: scaling up to meet the challenge, Occasional Paper SSC no. 62. IUCN, Gland, Switzerland.

Vellend, M., Verheyen, K., Jacquemyn, H., et al. (2006). Extinction debt of forest plants persist for more than a century following habitat fragmentation. *Ecology*, **87**, 542–548.

Veron, S., Haevermans, T., Govaerts, R., Mouchet, M., and Pellens, R. (2019). Distribution and relative age of endemism across islands worldwide. *Scientific Reports*, **9**, 11693.

Viana, D. S., Gangoso, L., Bouten, W., and Figuerola, J. (2016). Overseas seed dispersal by migratory birds. *Proceedings of the Royal Society B*, **283**, 20152406.

Villéger, S., Mason, N. W. H., and Mouillot, D. (2008). New multidimensional functional diversity indices for a multifaceted framework in functional ecology. *Ecology*, **89**, 2290–2301.

Vitousek, P. M. (2004). *Nutrient Cycling and Limitation. Hawai'i as a Model System*. Princeton University Press, Princeton.

Vitousek, P. M., Loope, L. L., and Adsersen, H. (ed.) (1995). *Islands: Biological Diversity and Ecosystem Function*, Ecological Studies **115**. Springer-Verlag, Berlin.

Vitousek, P. M., Loope, L. L., and Stone, C. P. (1987a). Introduced species in Hawaii: biological effects and opportunities for ecological research. *Trends in Ecology & Evolution*, **2**, 224–227.

Vitousek, P. M. and Walker, L. R. (1989). Biological invasion by *Myrica faya* in Hawai'i: plant demongraphy, nitrogen fixation, ecosystem effects. *Ecological Monographs*, **59**, 247–265.

Vitousek, P. M., Walker, L. R., Whiteaker, L. D., Mueller-Dombois, D., and Matson, P. A. (1987b). Biological invasion by *Myrica faya* alters ecosystem development in Hawaii. *Science*, **238**, 802–804.

Vizentin-Bugoni, J., Tarwater, C. E., Foster, J. T., et al. (2019). Structure, spatial dynamics, and stability of novel seed dispersal mutualistic networks in Hawai'i. *Science*, **364**, 78–82.

Vögeli, M., Serraon, D., Pacios, F., and Tella, J. L. (2010). The relative importance of patch habitat quality and landscape attributes on a declining steppe-bird metapopulation. *Biological Conservation*, **143**, 1057–1067.

Volonec, Z. M. and Dobson, A. P. (2020). Conservation value of small reserves. *Conservation Biology*, **34**, 66–79.

von Buch, L. (1825). *Physicalische Beschreibung der Canarischen Inseln*. Druckerei der Königliche Akademie der Wissenschaften, Berlin.

Wada, S., Kawakami, K., and Chiba, S. (2012). Snails can survive passage through a bird's digestive system. *Journal of Biogeography*, **39**, 69–73.

Wagner, C. E., Harmon, L. H., and Seehausen, O. (2012). Ecological opportunity and sexual selection together predict adaptive radiation. *Nature*, **487**, 366–370.

Wagner, C. E., Harmon, L. J., and Seehausen, O. (2014). Cichlid species-area relationships are shaped by adaptive radiations that scale with area. *Ecology Letters*, **17**, 583–592.

Wagner, W. L., Weller, S. G., and Sakai, A. K. (2005). Monograph of *Schiedea* (Caryophyllaceae-Alsinoideae). *Systematic Botany Monographs*, **72**, 1–169.

Waide, R. B. (1991). Summary of the responses of animal populations to hurricanes in the Caribbean. *Biotropica*, **23**, 508–512.

Wald, D. M., Nelson, K. A., Gawel, A. M., and Rogers, H. S. (2019). The role of trust in public attitudes toward invasive species management on Guam: a case study. *Journal of Environmental Management*, **229**, 133–144.

Walker, L. R. and del Moral, R. (2003). *Primary Succession and Ecosystem Rehabilitation*. Cambridge University Press, Cambridge.

Walker, L. R., Lodge, D. J., Brokaw, N. V. L., and Waide, R. B. (1991). An introduction to hurricanes in the Caribbean. *Biotropica*, **23**, 313–316.

Walker, L. R., Silver, W. L., Willig, M. R., and Zimmerman, J. K. (ed.) (1996). Special issue: long term responses of Caribbean ecosystems to disturbance. *Biotropica*, **28**, 414–613.

Walker, T. W. and Syers, J. K. (1976). The fate of phosphorus during pedogenesis. *Geoderma*, **15**, 1–19.

Wallace, A. R. (1869). *The Malay Archipelago*. Macmillan, London.

Wallace, A. R. (1878). *Tropical Nature and Other Essays*. Macmillan, London.

Wallace, A. R. (1902). *Island Life* (3rd edn). Macmillan, London.

Walsh, J., Campagna, L., Feeney, W. E., King, J., and Webster, M.S. (2021). Patterns of genetic divergence and demographic history shed light on island-mainland population dynamics and melanic plumage evolution in the white-winged fairywren. *Evolution*, **75**, 1348–1360.

Wang, Y., Bao, Y., Yu, M., Xu, G., and Ding, P. (2010). Nestedness for different reasons: the distributions of birds, lizards and small mammals on islands of an inundated lake. *Diversity and Distributions*, **16**, 862–873.

Wang, Y., Chen, S., and Ding, P. (2011). Testing multiple assembly rule models in avian communities on islands of an inundated lake, Zhejiang Province, China. *Journal of Biogeography*, **38**, 1330–1344.

Wang, Y., Chen, C., and Millien, V. (2018). A global synthesis of the small-island effect in habitat islands. *Proceedings of the Royal Society B*, **285**, 20181868.

Wang, Y., Millien, V., and Ding, P. (2016). On empty islands and the small-island effect. *Global Ecology and Biogeography*, **25**, 1333–1345.

Wang, Z., Wen, M., Qian, X., Pei, N., and Zhang, D. (2020). Plants are visited by more pollinator species than pollination syndromes predicted in an oceanic island community. *Scientific Reports*, **10**, 13918.

Ward, S. N. and Day, S. (2001). Cumbre Vieja Volcano—potential collapse and tsunami at La Palma, Canary Islands. *Geophysical Research Letters*, **28**, 3397–3400.

Wardle, D. A., Zackrisson, O., Hörnberg, G., and Gallet, C. (1997). The influence of island area on ecosystem properties. *Science*, **277**, 1296–1299.

Warren, B. H., Simberloff, D., Ricklefs, R. E., et al. (2015). Islands as model systems in ecology and evolution: prospects fifty years after MacArthur–Wilson. *Ecology Letters*, **18**, 200–217.

Warren, B. H., Strasberg, D., Bruggemann, J. H., Prys-Jones, R. P., and Thébaud C. (2010). Why does the biota of the Madagascar region have such a strong Asiatic flavor? *Cladistics*, **26**, 526–538.

Waterloo, M. J., Schelleken, J., Bruijnzeel, L. A., Vugts, H. F., Assenberg, P. N., and Rawaqa, T. T. (1997). Chemistry of bulk precipitation in southwestern Viti Levu, Fiji. *Journal of Tropical Ecology*, **13**, 427–447.

Waters, J. M., Emerson, B. C., Arribas, P., and McCulloch G. A. (2020). Dispersal reduction: causes,

genomic mechanisms, and evolutionary consequences. *Trends in Ecology & Evolution*, **35**, 512–522.

Watson, D. M. (2003). Long-term consequences of habitat fragmentation – highland birds in Oaxaca, Mexico. *Biological Conservation*, **111**, 283–303.

Watson, J. E. M. (2004). Bird responses to habitat fragmentation: illustrations from Madagascan and Australian case studies. D.Phil. thesis. University of Oxford.

Watson, J. E. M., Venter, O., Lee, J., et al. (2018). Protect the last of the wild. *Nature*, **563**, 27–30.

Watson, J. E. M., Whittaker, R. J., and Dawson, T. P. (2004a). Habitat structure and proximity to forest edge affect the abundance and distribution of forest-dependent birds in tropical coastal forests of southeastern Madagascar. *Biological Conservation*, **120**, 311–327.

Watson, J. E. M., Whittaker, R. J., and Dawson, T. P. (2004b). Avifaunal responses to habitat fragmentation in the threatened littoral forests of south-eastern Madagascar. *Journal of Biogeography*, **31**, 1791–1807.

Watson, J. E. M., Whittaker, R. J., and Freudenberger, D. (2005). Bird community responses to habitat fragmentation: how consistent are they across landscapes? *Journal of Biogeography*, **32**, 1353–1370.

Watts, D. (1970). Persistence and change in the vegetation of oceanic islands: an example from Barbados, West Indies. *Canadian Geographer*, **14**, 91–109.

Watts, P. C., Buley, K. R., Sanderson, S., Boardman, W., Ciofi, C., and Gibson, R. (2006). Parthenogenesis in Komodo dragons. *Nature*, **444**, 1021–1022.

Wayman, J. P., Sadler, J. P., Pugh, T. A. M., Martin, T. E., Tobias, J. A., and Matthews, T. J. (2021). Identifying the drivers of spatial taxonomic and functional beta-diversity of British breeding birds. *Frontiers in Ecology and Evolution*, **9**, 620062.

Weaver, M. and Kellman, M. (1981). The effects of forest fragmentation on woodlot tree biotas in Southern Ontario. *Journal of Biogeography*, **8**, 199–210.

Webb, C. J., Lloyd, D. G., and Delph, L. F. (1999). Gender dimorphism in indigenous New Zealand seed plants. *New Zealand Journal of Botany*, **37**, 119–130.

Weigelt, P., Jetz, W., and Kreft, H. (2013). Bioclimatic and physical characterization of the world's islands. *Proceedings of the National Academy of Sciences, USA*, **110**, 15307–15312.

Weigelt, P., Kissling, W. D., Kisel, Y., et al. (2015). Global patterns and drivers of phylogenetic structure in island floras. *Scientific Reports*, **5**, srep12213.

Weigelt, P., König, C., and Kreft, H. (2020). GIFT – A Global Inventory of Floras and Traits for macroecology and biogeography. *Journal of Biogeography*, **47**, 16–43.

Weigelt, P. and Kreft, H. (2013). Quantifying island isolation – insights from global patterns of insular plant species richness. *Ecography*, **36**, 417–429.

Weigelt, P., Steinbauer, M. J., Cabral, J. S., and Kreft, H. (2016). Late Quaternary climate change shapes island biodiversity. *Nature*, **532**, 99–102.

Weiher, E. and Keddy, P. A. (1995). Assembly rules, null models, and trait dispersion: new questions from old patterns. *Oikos*, **74**, 159–164.

Weisler, M. I. (1995). Henderson Island prehistory: colonization and extinction on a remote Polynesian island. *Biological Journal of the Linnean Society*, **56**, 377–404.

Wenger, S. J., Leasure, D. R., and Dauwalter, D. C. (2017). Viability analysis for multiple populations. *Biological Conservation*, **216**, 67–77.

Werner, T. K. and Sherry, T. W. (1987). Behavioural feeding specialization in *Pinaroloxias inornata*, the 'Darwin's finch' of Cocos Island, Costa Rica. *Proceedings of the National Academy of Sciences, USA*, **84**, 5506–5510.

Wessel, P., Sandwell, D. T., and Kim, S.-S. (2010). The global seamount census. *Oceanography*, **23**, 24–33.

West, K., Collins, C., Kardailsky, O., et al. (2017). The Pacific rat race to Easter Island: tracking the prehistoric dispersal of *Rattus exulans* using ancient mitochondrial genomes. *Frontiers in Ecology and Evolution*, **5**, article 52.

Wetzel, F. T., Beissmann, H., Penn, D. J., and Jetz, W. (2013). Vulnerability of terrestrial island vertebrates to projected sea-level rise. *Global Change Biology*, **19**, 2058–2070.

Whelan, F. and Kelletat, D. (2003). Submarine slides on volcanic islands—a source for mega-tsunamis in the Quaternary. *Progress in Physical Geography*, **27**, 198–216.

White, G. B. (1981). Semispecies, sibling species and superspecies. In *The Evolving Biosphere (Chance, Chance and Challenge)* (ed. P. L. Forey), pp. 21–28. British Museum (Natural History) and Cambridge University Press, Cambridge.

White, O. W., Reyes-Betancort, J. A., Chapman, M. A., and Carine, M. A. (2018). Independent homoploid hybrid speciation events in the Macaronesian endemic genus *Argyranthemum*. *Molecular Ecology*, **27**, 4856–4874.

White, O. W., Reyes-Betancort, J. A. Chapman, M. A., and Carine, M. A. (2020). Geographical isolation, habitat shifts and hybridisation in the diversification of the Macaronesian endemic genus *Argyranthemum* (Asteraceae). *New Phytologist*, **228**, 1953–1971.

Whitehead, D. R. and Jones, C. E. (1969). Small islands and the equilibrium theory of insular biogeography. *Evolution*, **23**, 171–179.

Whittaker, R. H. (1977). Evolution of species diversity in land communities. In *Evolutionary Biology* (ed M. K. Hecht, W. C. Steere, and B. Wallace), pp 250–268. Plenum Press, New York.

Whittaker, R. J. (1998). *Island Biogeography: Ecology, Evolution, and Conservation*, 1st edn. Oxford University Press, Oxford.

Whittaker, R. J. (2004). The island biogeography of a long-running natural experiment: Krakatau, Indonesia. In *Ecología Insular/Island Ecology* (ed. J. M. Fernández-Palacios and C. Morici), pp. 57–79. Asociación Española de Ecología Terrestre (AEET)-Cabildo Insular de La Palma, Santa Cruz de La Palma, La Palma.

Whittaker, R. J., Araújo, M. B., Jepson, P., Ladle, R. J., Watson, J. E. M., and Willis, K. J. (2005). Conservation biogeography: assessment and prospect. *Diversity and Distributions*, **11**, 3–23.

Whittaker, R. J., Bush, M. B., and Richards, K. (1989). Plant recolonization and vegetation succession on the Krakatau Islands, Indonesia. *Ecological Monographs*, **59**, 59–123.

Whittaker, R. J. and Fernández-Palacios, J. M. (2007). *Island Biogeography: Ecology, Evolution, and Conservation*, 2nd edn. Oxford University Press, Oxford.

Whittaker, R. J., Fernández-Palacios, J. M., Matthews, T. J., Borregaard, M. K., and Triantis, K. A. (2017). Island Biogeography: taking the long view of nature's laboratories. *Science*, **357**, eaam8326.

Whittaker, R. J., Fernández-Palacios, J. M., Matthews, T. J., Rigal, F., and Triantis, K. A. (2018). Archipelagos and meta-archipelagos. *Frontiers of Biogeography*, **10**, e41470.

Whittaker, R. J., Field, R., and Partomihardjo, T. (2000). How to go extinct: lessons from the lost plants of Krakatau. *Journal of Biogeography*, **27**, 1049–1064.

Whittaker, R. J. and Jones, S. H. (1994a). Structure in rebuilding insular ecosystems: an empirically derived model. *Oikos*, **69**, 524–529.

Whittaker, R. J. and Jones, S. H. (1994b). The role of frugivorous bats and birds in the rebuilding of a tropical forest ecosystem, Krakatau, Indonesia. *Journal of Biogeography*, **21**, 689–702.

Whittaker, R. J., Jones, S. H., and Partomihardjo, T. (1997). The re-building of an isolated rain forest assemblage: how disharmonic is the flora of Krakatau? *Biodiversity and Conservation*, **6**, 1671–1696.

Whittaker, R. J., Ladle, R. J., Araújo, M. B., Fernández-Palacios, J. M., Delgado, J. D., and Arévalo, J. R. (2007). The island immaturity – speciation pulse model of island evolution: an alternative to the 'diversity begets diversity' model. *Ecography*, **30**, 321–327.

Whittaker, R. J., Rigal, F., Borges, P. A. V., et al. (2014). Functional biogeography of oceanic islands and the scaling of functional diversity in the Azores. *Proceedings of the National Academy of Sciences, USA*, **111**, 13709–13714.

Whittaker, R. J., Triantis, K. A., and Ladle, R. J. (2008). A general dynamic theory of oceanic island biogeography. *Journal of Biogeography*, **35**, 977–994.

Whittaker, R. J., Triantis, K. A. and Ladle, R. J. (2010). A general dynamic theory of oceanic island biogeography: extending the MacArthur–Wilson theory to accommodate the rise and fall of volcanic islands. In *The Theory of Island Biogeography Revisited* (ed. J. B. Losos and R. E. Ricklefs), pp. 88–115. Princeton University Press, Princeton NJ.

Whittaker, R. J., Willis, K. J., and Field, R. (2001). Scale and species richness: towards a general, hierarchical theory of species diversity. *Journal of Biogeography*, **28**, 453–470.

Wilcove, D. S., McLellan, C. H., and Dobson, A. P. (1986). Habitat fragmentation in the temperate zone. In *Conservation Biology: The Science of Scarcity and Diversity* (ed. M. Soulé), pp. 237–256. Sinauer Associates, Sunderland, MA.

Wilk, R. J., Lesmeister, D. B., and Forsman, E. D. (2018). Nest trees of northern spotted owls (*Strix occidentalis caurina*) in Washington and Oregon, USA. *PLOS ONE*, **13**, e0197887.

Williams, B. R. M., Schaefer, H., Menezes de Sequeira, M., Reyes-Betancort, J. A., Patiño, J., and Carine, M. A. (2015). Are there any widespread endemic flowering plant species in Macaronesia? Phylogeography of *Ranunculus cortusifolius*. *American Journal of Botany*, **102**, 1736–1746.

Williams, E. E. (1972). The origin of faunas. Evolution of lizard congeners in a complex island fauna: A trial analysis. *Evolutionary Biology*, **6**, 47–88.

Williamson, M. H. (1981). *Island Populations*. Oxford University Press, Oxford.

Williamson, M. H. (1989a). The MacArthur and Wilson theory today: true but trivial. *Journal of Biogeography*, **16**, 3–4.

Williamson, M. H. (1989b). Natural extinction on islands. *Philosophical Transactions of the Royal Society of London*, B, **325**, 457–468.

Willis, E. O. (1974). Populations and local extinctions of birds on Barro Colorado Island, Panama. *Ecological Monographs*, **44**, 153–169.

Willis, K. J., Gillson, L., and Brncic, T. M. (2004). How 'virgin' is virgin rainforest? *Science*, **304**, 402–403.

Wilmshurst, J. M., Hunt, T. L., Lipo, C. P., and Anderson, A. J. (2011). High-precision radiocarbon dating shows recent and rapid initial human colonization of East Polynesia. *Proceedings of the National Academy of Sciences, USA*, **108**, 1815–1820.

Wilson, D. E. and Graham, G. L. (ed.) (1992). *Pacific island flying foxes: proceedings of an international conservation conference*. US Fish and Wildlife Service, Biological Report, **90** (23), Washington, DC.

Wilson, E. O. (1959). Adaptive shift and dispersal in a tropical ant fauna. *Evolution*, **13**, 122–144.

Wilson, E. O. (1961). The nature of the taxon cycle in the Melanesian ant fauna. *American Naturalist*, **95**, 169–193.

Wilson, E. O. (1969). The species equilibrium. *Brookhaven Symposia in Biology*, **22**, 38–47.

Wilson, E. O. and Bossert, W. H. (1971). *A Primer of Population Biology*. Sinauer Associates, Stamford, CT.

Wilson, J. T. (1963). A possible origin of the Hawaiian Islands. *Canadian Journal of Physics*, **41**, 863–870.

Wintle, B. A., Kujala, H., Whitehead, et al. (2019). Global synthesis of conservation studies reveals the importance of small habitat patches for biodiversity. *Proceedings of the National Academy of Sciences, USA*, **116**, 909–914.

Wolanski, E., Choukroun, S., and Nhan, N. H. (2020). Island building and overfishing in the Spratly Islands archipelago are predicted to decrease larval flow and impact the whole system. *Estuarine, Coastal and Shelf Science*, **233**, 106545.

Wood, J. R., Alcover, J. A., Blackburn, T. M., et al. (2017). Island extinctions: processes, patterns, and potential for ecosystem restoration. *Environmental Conservation*, 44, 348–358.

Woodford, D. J. and McIntosh, A. R. (2010). Evidence of source–sink metapopulations in a vulnerable native galaxiid fish driven by introduced trout. *Ecological Applications*, **20**, 967–977.

Worthen, W. B. (1996). Community composition and nested-subset analyses: basic descriptors for community ecology. *Oikos*, **76**, 417–426.

Wragg, G. M. (1995). The fossil birds of Henderson Island, Pitcairn Group: natural turnover and human impact, a synopsis. *Biological Journal of the Linnean Society*, **56**, 405–414.

Wright, D. H. (1983). Species–energy theory: an extension of species–area theory. *Oikos*, **41**, 496–506.

Wright, D. H., Patterson, B. D., Mikkelson, G. M., Cutler, A., and Atmar, W. (1998). A comparative analysis of nested subset patterns of species composition. *Oecologia*, **113**, 1–20.

Wright, N. A., Steadman, D. W., and Witt, C. C. (2016). Predictable evolution toward flightlessness in volant island birds. *Proceedings of the National Academy of Sciences, USA*, **113**, 4765–4770.

Wright, S. (1932). The roles of mutation, inbreeding, crossbreeding, and selection in evolution. *Proceedings of the XI International Congress of Genetics*, **1**, 356–366.

Wright, S. J. (1980). Density compensation in island avifaunas. *Oecologia*, **45**, 385–389.

Wu, L., Si, X., Didham, R. K., Ge, D., and Ding, P. (2017). Dispersal modality determines the relative partitioning of beta diversity in spider assemblages on subtropical land-bridge islands. *Journal of Biogeography*, **44**, 2121–2131.

Wylie, J. L. and Currie, D. J. (1993). Species–energy theory and patterns of species richness: I. Patterns of bird, angiosperm, and mammal richness on islands. *Biological Conservation*, **63**, 137–144.

Xu, A., Han, X., Zhang, X., Millien, V., and Wang, Y. (2017). Nestedness of butterfly assemblages in the Zhoushan Archipelago, China: area effects, life-history traits and conservation implications. *Biodiversity and Conservation*, **26**, 1375–1392.

Young, T. P. (1994). Natural die-offs of large mammals: implications for conservation. *Conservation Biology*, **8**, 410–418.

Zanne, A. E., Tank, D. C., Cornwell, W. K., et al. (2014). Three keys to the radiation of angiosperms into freezing environments. *Nature*, **506**, 89–92.

Zavaleta, E. S., Hobbs, R. J., and Money, H. A. (2001). Viewing invasive species removal in a whole-ecosystem context. *Trends in Ecology & Evolution*, **16**, 454–459.

Zhao, Y., Dunn, R. R., Zhou, H., Si, Z., and Ding, P. (2020). Island area, not isolation, drives taxonomic, phylogenetic and functional diversity of ants on land-bridge islands. *Journal of Biogeography*, **47**, 1627–1637.

Zhao, Y., Sanders, N. J., Liu, J., et al. (2021). β diversity among ant communities on fragmented habitat islands: the roles of species trait, phylogeny and abundance. *Ecography*, **44**, 1568–1578.

Zimmerman, B. L. and Bierregaard, R. O. Jr (1986). Relevance of the equilibrium theory of island biogeography and species–area relations to conservation with a case from Amazonia. *Journal of Biogeography*, **13**, 133–143.

Zizka, A., Onstein, R. E., Rozzi, R., et al. (2022). The evolution of insular woodiness. *Proceedings of the National Academy of Sciences, USA*, **119**, e2208629119.

Zuccon, D., Prŷs-Jones, R., Rasmussen, P. C., and Ericson, P. G. P. (2012). The phylogenetic relationships and generic limits of finches (Fringillidae). *Molecular Phylogenetics and Evolution*, **62**, 581–596.

Index

Note: Tables, figures, and boxes are indicated by an italic *t*, *f*, and *b* following the page number.